W9-ACQ-533

DYNAMICS OF STRUCTURES

PRENTICE-HALL INTERNATIONAL SERIES IN CIVIL ENGINEERING AND ENGINEERING MECHANICS
William J. Hall, Editor

Au and Christiano, *Fundamentals of Structural Analysis*
Au and Christiano, *Structural Analysis*
Barson and Rolfe, *Fracture and Fatigue Control in Structures, 2/e*
Bathe, *Finite Element Procedures in Engineering Analysis*
Berg, *Elements of Structural Dynamics*
Biggs, *Introduction to Structural Engineering*
Chajes, *Structural Analysis, 2/e*
Chopra, *Dynamics of Structures: Theory and Applications to Earthquake Engineering*
Collins and Mitchell, *Prestressed Concrete Structures*
Cooper and Chen, *Designing Steel Structures*
Cording et al., *The Art and Science of Geotechnical Engineering*
Gallagher, *Finite Element Analysis*
Hendrickson and Au, *Project Management for Construction*
Higdon et al., *Engineering Mechanics, 2nd Vector Edition*
Hultz and Kovacs, *Introduction in Geotechnical Engineering*
Humar, *Dynamics of Structures*
Johnston, Lin, and Galambos, *Basic Steel Design, 3/e*
Kelkar and Sewell, *Fundamentals of the Analysis and Design of Shell Structures*
MacGregor, *Reinforced Concrete: Mechanics and Design, 2/e*
Mehta and Monteiro, *Concrete: Structure, Properties and Materials, 2/e*
Melosh, *Structural Engineering Analysis by Finite Elements*
Meredith et al., *Design and Planning of Engineering Systems, 2/e*
Mindess and Young, *Concrete*
Nawy, *Prestressed Concrete*
Nawy, *Reinforced Concrete: A Fundamental Approach, 2/e*
Pfeffer, *Solid Waste Management*
Popov, *Engineering Mechanics of Solids*
Popov, *Introduction to the Mechanics of Solids*
Popov, *Mechanics of Materials, 2/e*
Schneider and Dickey, *Reinforced Masonry Design, 2/e*
Wang and Salmon, *Introductory Structural Analysis*
Weaver and Johnson, *Structural Dynamics by Finite Elements*
Wolf, *Dynamic Soil-Structure Interaction*
Wray, *Measuring Engineering Properties of Soils*
Yang, *Finite Element Structural Analysis*

DYNAMICS OF STRUCTURES

Theory and Applications to Earthquake Engineering

Anil K. Chopra

University of California at Berkeley

PRENTICE HALL

Englewood Cliffs, New Jersey 07632

Library of Congress Cataloging-in-Publication Data

Chopra, Anil K.
Dynamics of structures : theory and applications to earthquake engineering / Anil K. Chopra.
p. cm.
Includes index.
ISBN 0-13-855214-2
1. Earthquake engineering. 2. Structural dynamics. I. Title.
TA654.6.C466 1995
624.1′ 762—dc20 94-46527
CIP

Acquisitions Editor: Bill Stenquist
Production Editor: Kurt Scherwatzky
Production Coordinator: Bayani Mendoza de Leon
Editorial-Production Service: Electronic Publishing Services, Inc.
Buyer: Bill Scazzero
Cover Designer: Douglas DeLuca
Cover Photo: Transamerica Building, San Francisco, California. The motions shown are accelerations recorded during the Loma Prieta earthquake of October 17, 1989 at basement, twenty-ninth floor, and forty-ninth floor. (Courtesy of Transamerica Corporation.)

Printed in the United States of America

10 9 8 7 6 5 4 3 2 1

ISBN 0-13-855214-2

Prentice-Hall International (UK) Limited, *London*
Prentice-Hall of Australia Pty. Limited, *Sydney*
Prentice-Hall Canada Inc., *Toronto*
Prentice-Hall Hispanoamericana, S.A., *Mexico*
Prentice-Hall of India Private Limited, *New Delhi*
Prentice-Hall of Japan, Inc., *Tokyo*
Simon & Schuster Asia Pte. Ltd., *Singapore*
Editora Prentice Hall do Brasil, Ltda., *Rio de Janeiro*

Dedicated to Hamida and Nasreen with gratitude for suggesting the idea of working on a book and with appreciation for patiently enduring and sharing these years of preparation with me, especially the past year and a half. Their presence and encouragement made this idea a reality.

Overview

PART I SINGLE-DEGREE-OF-FREEDOM SYSTEMS **1**

1 *Equations of Motion, Problem Statement, and Solution Methods* *3*

2 *Free Vibration* *35*

3 *Response to Harmonic and Periodic Excitations* *61*

4 *Response to Arbitrary, Step, and Pulse Excitations* *119*

5 *Numerical Evaluation of Dynamic Response* *155*

6 *Earthquake Response of Linear Systems* *187*

7 *Earthquake Response of Inelastic Systems* *241*

8 *Generalized Single-Degree-of-Freedom Systems* *277*

PART II MULTI-DEGREE-OF-FREEDOM SYSTEMS **311**

9 Equations of Motion, Problem Statement, and Solution Methods *313*

10 Free Vibration *365*

11 Damping in Structures *409*

12 Dynamic Analysis and Response of Linear Systems *429*

13 Earthquake Analysis of Linear Systems *467*

14 Reduction of Degrees of Freedom *549*

15 Numerical Evaluation of Dynamic Response *565*

16 Systems with Distributed Mass and Elasticity *585*

17 Introduction to the Finite Element Method *613*

PART III EARTHQUAKE RESPONSE AND DESIGN OF MULTISTORY BUILDINGS **639**

18 Earthquake Response of Linearly Elastic Buildings *641*

19 Earthquake Response of Inelastic Buildings *659*

20 Earthquake Dynamics of Base-Isolated Buildings *683*

21 Structural Dynamics in Building Codes *703*

Contents

Foreword xix

Preface xxi

Acknowledgments xxvii

PART I SINGLE-DEGREE-OF-FREEDOM SYSTEMS 1

1 Equations of Motion, Problem Statement, and Solution Methods 3

1.1 Simple Structures 3

1.2 Single-Degree-of-Freedom System 7

1.3 Force–Displacement Relation 8

1.4 Damping Force 13

1.5 Equation of Motion: External Force 14

1.6 Mass–Spring–Damper System 18

1.7 Equation of Motion: Earthquake Excitation 20

1.8 Problem Statement and Element Forces 23

1.9 Combining Static and Dynamic Responses 25

1.10 Methods of Solution of the Differential Equation 25

1.11 Study of SDF Systems: Organization 29

Appendix 1: Stiffness Coefficients for a Flexural Element 30

2 Free Vibration **35**

2.1 Undamped Free Vibration 35

2.2 Viscously Damped Free Vibration 44

2.3 Energy in Free Vibration 52

2.4 Coulomb-Damped Free Vibration 53

3 Response to Harmonic and Periodic Excitations **61**

Part A: Viscously Damped Systems: Basic Results 62

3.1 Harmonic Vibration of Undamped Systems 62

3.2 Harmonic Vibration with Viscous Damping 68

Part B: Viscously Damped Systems: Applications 80

3.3 Response to Vibration Generator 80

3.4 Natural Frequency and Damping from Harmonic Tests 83

3.5 Force Transmission and Vibration Isolation 85

3.6 Response to Ground Motion and Vibration Isolation 87

3.7 Vibration-Measuring Instruments 91

3.8 Energy Dissipated in Viscous Damping 94

3.9 Equivalent Viscous Damping 98

Part C: Systems with Nonviscous Damping 100

3.10 Harmonic Vibration with Rate-Independent Damping 100

3.11 Harmonic Vibration with Coulomb Friction 104

Part D: Response to Periodic Excitation 108

3.12 Fourier Series Representation 109

3.13 Response to Periodic Force 109

Appendix 3: Four-Way Logarithmic Graph Paper 113

4 *Response to Arbitrary, Step, and Pulse Excitations* *119*

Part A: Response to Arbitrarily Time-Varying Forces 119

4.1 Response to Unit Impulse 120

4.2 Response to Arbitrary Force 121

Part B: Response to Step and Ramp Forces 123

4.3 Step Force 123

4.4 Ramp or Linearly Increasing Force 125

4.5 Step Force with Finite Rise Time 126

Part C: Response to Pulse Excitations 129

4.6 Solution Methods 129

4.7 Rectangular Pulse Force 131

4.8 Half-Cycle Sine Pulse Force 137

4.9 Symmetrical Triangular Pulse Force 142

4.10 Effects of Pulse Shape and Approximate Analysis for Short Pulses 144

4.11 Effects of Viscous Damping 147

4.12 Response to Ground Motion 149

5 *Numerical Evaluation of Dynamic Response* *155*

5.1 Time-Stepping Methods 155

5.2 Methods Based on Interpolation of Excitation 157

5.3 Central Difference Method 161

5.4 Newmark's Method 164

5.5 Stability and Computational Error 170

5.6 Analysis of Nonlinear Response: Central Difference Method 174

5.7 Analysis of Nonlinear Response: Newmark's Method 174

6 Earthquake Response of Linear Systems 187

6.1 Earthquake Excitation 187

6.2 Equation of Motion 193

6.3 Response Quantities 194

6.4 Response History 195

6.5 Response Spectrum Concept 197

6.6 Deformation, Pseudo-velocity, and Pseudo-acceleration Response Spectra 198

6.7 Peak Structural Response from the Response Spectrum 206

6.8 Response Spectrum Characteristics 211

6.9 Elastic Design Spectrum 217

6.10 Comparison of Design and Response Spectra 225

6.11 Distinction between Design and Response Spectra 227

6.12 Velocity and Acceleration Response Spectra 228

Appendix 6: El Centro, 1940 Ground Motion 232

7 Earthquake Response of Inelastic Systems 241

7.1 Force–Deformation Relations 242

7.2 Normalized Yield Strength, Yield Reduction Factor, and Ductility Factor 248

7.3 Equation of Motion and Controlling Parameters 249

7.4 Effects of Yielding 250

7.5 Response Spectrum for Yield Deformation and Yield Strength 257

7.6 Design Strength and Deformation from the Response Spectrum 261

7.7 Design Yield Strength 261

7.8 Relative Effects of Yielding and Damping 263

7.9 Dissipated Energy 264

7.10 Inelastic Design Spectrum 269

7.11 Comparison of Design and Response Spectra 274

8 *Generalized Single-Degree-of-Freedom Systems* 277

8.1 Generalized SDF Systems 277

8.2 Rigid-Body Assemblages 279

8.3 Systems with Distributed Mass and Elasticity 281

8.4 Lumped-Mass System: Shear Building 292

8.5 Natural Vibration Frequency by Rayleigh's Method 298

8.6 Selection of Shape Function 302

Appendix 8: Inertia Forces for Rigid Bodies 306

PART II MULTI-DEGREE-OF-FREEDOM SYSTEMS 311

9 *Equations of Motion, Problem Statement, and Solution Methods* 313

9.1 Simple System: Two-Story Shear Building 313

9.2 General Approach for Linear Systems 318

9.3 Static Condensation 334

9.4 Planar or Symmetric-Plan Systems: Ground Motion 337

9.5 Unsymmetric-Plan Buildings: Ground Motion 342

9.6 Symmetric-Plan Buildings: Torsional Excitation 350

9.7 Multiple Support Excitation 351

9.8 Inelastic Systems 355

9.9 Problem Statement 356

9.10 Element Forces 356

9.11 Methods for Solving the Equations of Motion: Overview 357

10 *Free Vibration* 365

Part A: Natural Vibration Frequencies and Modes 366

10.1 Systems without Damping 366

10.2 Natural Vibration Frequencies and Modes 368

10.3 Modal and Spectral Matrices 370

10.4 Orthogonality of Modes 371

10.5 Interpretation of Modal Orthogonality 372

10.6 Normalization of Modes 372

10.7 Modal Expansion of Displacements 382

Part B: Free Vibration Response 383

10.8 Solution of Free Vibration Equations: Undamped Systems 383

10.9 Free Vibration of Systems with Damping 386

10.10 Solution of Free Vibration Equations: Classically Damped Systems 390

Part C: Computation of Vibration Properties 392

10.11 Solution Methods for the Eigenvalue Problem 392

10.12 Rayleigh's Quotient 394

10.13 Inverse Vector Iteration Method 394

10.14 Vector Iteration with Shifts: Preferred Procedure 399

10.15 Transformation of $\mathbf{k}\phi = \omega^2\mathbf{m}\phi$ to the Standard Form 404

11 Damping in Structures ***409***

Part A: Experimental Data and Recommended Modal Damping Ratios 409

11.1 Vibration Properties of Millikan Library Building 409

11.2 Estimating Modal Damping Ratios 414

Part B: Construction of Damping Matrix 416

11.3 Damping Matrix 416

11.4 Classical Damping Matrix 417

11.5 Nonclassical Damping Matrix 425

12 *Dynamic Analysis and Response of Linear Systems* 429

Part A: Two-Degree-of-Freedom Systems 429

12.1 Analysis of Two-DOF Systems without Damping 429

12.2 Vibration Absorber or Tuned Mass Damper 432

Part B: Modal Analysis 434

12.3 Modal Equations for Undamped Systems 434

12.4 Modal Equations for Damped Systems 436

12.5 Displacement Response 438

12.6 Element Forces 438

12.7 Modal Analysis: Summary 439

Part C: Modal Response Contributions 444

12.8 Modal Expansion of Excitation Vector $\mathbf{p}(t) = \mathbf{s}p(t)$ 444

12.9 Modal Analysis for $\mathbf{p}(t) = \mathbf{s}p(t)$ 447

12.10 Modal Contribution Factors 448

12.11 Modal Contributions to Response 449

Part D: Special Analysis Procedures 455

12.12 Static Correction Method 455

12.13 Mode Acceleration Superposition Method 458

12.14 Analysis of Nonclassically Damped Systems 459

13 *Earthquake Analysis of Linear Systems* 467

Part A: Response History Analysis 468

13.1 Modal Analysis 468

13.2 Multistory Buildings with Symmetric Plan 474

13.3 Multistory Buildings with Unsymmetric Plan 492

13.4 Torsional Response of Symmetric-Plan Buildings 503

13.5 Response Analysis for Multiple Support Excitation 508

13.6 Structural Idealization and Earthquake Response 513

Part B: Response Spectrum Analysis 514

13.7 Peak Response from Earthquake Response Spectrum 514

13.8 Multistory Buildings with Symmetric Plan 519

13.9 Multistory Buildings with Unsymmetric Plan 532

14 Reduction of Degrees of Freedom 549

14.1 Kinematic Constraints 550

14.2 Static Condensation 551

14.3 Rayleigh–Ritz Method 551

14.4 Selection of Ritz Vectors 554

14.5 Dynamic Analysis Using Ritz Vectors 560

15 Numerical Evaluation of Dynamic Response 565

15.1 Time-Stepping Methods 565

15.2 Analysis of Linear Systems with Nonclassical Damping 567

15.3 Analysis of Nonlinear Systems 574

16 Systems with Distributed Mass and Elasticity 585

16.1 Equation of Undamped Motion: Applied Forces 586

16.2 Equation of Undamped Motion: Support Excitation 587

16.3 Natural Vibration Frequencies and Modes 588

16.4 Modal Orthogonality 595

16.5 Modal Analysis of Forced Dynamic Response 596

16.6 Earthquake Response History Analysis 600

16.7 Earthquake Response Spectrum Analysis 604

16.8 Difficulty in Analyzing Practical Systems 607

17 Introduction to the Finite Element Method 613

Part A: Rayleigh–Ritz Method 613

17.1 Formulation Using Conservation of Energy 613

17.2 Formulation Using Virtual Work 617

17.3 Disadvantages of Rayleigh–Ritz Method 618

Part B: Finite Element Method 619

17.4 Finite Element Approximation 619

17.5 Analysis Procedure 621

17.6 Element Degrees of Freedom and Interpolation Functions 622

17.7 Element Stiffness Matrix 624

17.8 Element Mass Matrix 625

17.9 Element (Applied) Force Vector 626

17.10 Comparison of Finite Element and Exact Solutions 630

17.11 Dynamic Analysis of Structural Continua 632

PART III EARTHQUAKE RESPONSE AND DESIGN OF MULTISTORY BUILDINGS 639

18 Earthquake Response of Linearly Elastic Buildings 641

18.1 Systems Analyzed, Design Spectrum, and Response Quantities 641

18.2 Influence of T_1 and ρ on Response 646

18.3 Modal Contribution Factors 647

18.4 Influence of T_1 on Higher-Mode Response 649

18.5 Influence of ρ on Higher-Mode Response 652

18.6 Heightwise Variation of Higher-Mode Response 653

18.7 How Many Modes to Include 655

19 Earthquake Response of Inelastic Buildings 659

19.1 Allowable Ductility and Ductility Demand 660

19.2 Buildings with “Weak” or “Soft” First Story 665

19.3 Buildings Designed for Code Force Distribution 670

19.4 Limited Scope 680

20 Earthquake Dynamics of Base-Isolated Buildings ***683***

20.1 Isolation Systems 683

20.2 Base-Isolated One-Story Buildings 686

20.3 Effectiveness of Base Isolation 691

20.4 Base-Isolated Multistory Buildings 695

20.5 Applications of Base Isolation 701

21 Structural Dynamics in Building Codes ***703***

Part A: Building Codes and Structural Dynamics 704

21.1 *Uniform Building Code* (United States), 1994 704

21.2 *National Building Code of Canada*, 1995 707

21.3 *Mexico Federal District Code*, 1987 711

21.4 Structural Dynamics in Building Codes 713

Part B: Evaluation of Building Codes 720

21.5 Base Shear 720

21.6 Story Shears and Equivalent Static Forces 723

21.7 Overturning Moments 726

21.8 Concluding Remarks 728

A Notation ***A.1***

B Answers to Selected Problems ***B.1***

Index ***I.1***

Foreword

The need for a textbook on earthquake engineering was first pointed out by the eminent consulting engineer, John R. Freeman (1855–1932). Following the destructive Santa Barbara, California earthquake of 1925, he became interested in the subject and searched the Boston Public Library for relevant books. He found that not only was there no textbook on earthquake engineering, but the subject itself was not mentioned in any of the books on structural engineering. Looking back, we can see that in 1925 engineering education was in an undeveloped state with computing done by slide rule and curricula that did not prepare the student for understanding structural dynamics. In fact, no instruments had been developed for recording strong ground motions, and society appeared to be unconcerned about earthquake hazards.

In recent years books on earthquake engineering and structural dynamics have been published, but the present book by Professor Anil K. Chopra fills a niche that exists between more elementary books and books for advanced graduate studies. The author is a well-known expert in earthquake engineering and structural dynamics, and his book will be valuable to students not only in earthquake-prone regions but also in other parts of the world, for a knowledge of structural dynamics is essential for modern engineering. The book presents material on vibrations and the dynamics of structures and demonstrates the application to structural motions caused by earthquake ground shaking. The material in the book is presented very clearly with numerous worked-out illustrative examples so that even a student at a university where such a course is not given should be able to study the book on his or her own time. Readers who are now practicing engineering should have no difficulty in studying the subject by means of this book. An especially interesting feature of the book is the application of structural dynamics theory to important issues in the seismic response and design of multistory buildings.

The information presented in this book will be of special value to those engineers who are engaged in actual seismic design and want to improve their understanding of the subject.

Although the material in the book leads to earthquake engineering, the information presented is also relevant to wind-induced vibrations of structures, as well as man-made motions such as those produced by drophammers or by heavy vehicular traffic. As a textbook on vibrations and structural dynamics, this book has no competitors and can be recommended to the serious student. I believe that this is the book for which John R. Freeman was searching.

George W. Housner
California Institute of Technology

PHILOSOPHY AND OBJECTIVES

This book on dynamics of structures is conceived as a textbook for courses in civil engineering. It includes many topics in the theory of structural dynamics, and applications of this theory to earthquake analysis, response, and design of structures. No prior knowledge of structural dynamics is assumed in order to make this book suitable for the reader learning the subject for the first time. The presentation is sufficiently detailed and carefully integrated by cross-referencing to make the book suitable for self-study. This feature of the book, combined with a practically motivated selection of topics, should interest professional engineers, especially those concerned with analysis and design of structures in earthquake country.

In developing this book, much emphasis has been placed on making structural dynamics easily accessible to students and professional engineers because many find this subject to be difficult. In order to achieve this goal, the presentation is characterized by several features: The mathematics is kept as simple as each topic would permit. Analytical procedures are summarized to emphasize the key steps and to facilitate their implementation by the reader. These procedures are illustrated by over 100 worked-out examples, including many comprehensive and realistic examples where the physical interpretation of results is stressed. Some 400 figures have been carefully designed and executed to be pedagogically effective; many of them involve extensive computer simulations of dynamic response of structures. Photographs of structures and structural motions recorded during earthquakes are included to relate the presentation to the real world.

The preparation of this book has been inspired by several objectives:

- Relate the structural idealizations studied to the properties of real structures.
- Present the theory of dynamic response of structures in a manner that emphasizes physical insight into the analytical procedures.
- Illustrate applications of the theory to solutions of problems motivated by practical applications.
- Interpret the theoretical results to understand the response of structures to various dynamic excitations, with emphasis on earthquake excitation.
- Apply structural dynamics theory to conduct parametric studies that bring out several fundamental issues in the earthquake response and design of multistory buildings.

This mode of presentation should help the reader to achieve a deeper understanding of the subject and to apply with confidence structural dynamics theory in tackling practical problems, especially in earthquake analysis and design of structures, thus narrowing the gap between theory and practice.

SUBJECTS COVERED

This book is organized into three parts: I. Single-Degree-of-Freedom Systems; II. Multi-Degree-of-Freedom Systems; and III. Earthquake Response and Design of Multistory Buildings.

Part I includes eight chapters. In the opening chapter the structural dynamics problem is formulated for simple elastic and inelastic structures, which can be idealized as single-degree-of-freedom (SDF) systems, and four methods for solving the differential equation governing the motion of the structure are reviewed briefly. We then study the dynamic response of linearly elastic systems (1) in free vibration (Chapter 2), (2) to harmonic and periodic excitations (Chapter 3), and (3) to step and pulse excitations (Chapter 4). Included in Chapters 2 and 3 is the dynamics of SDF systems with Coulomb damping, a topic that is normally not included in civil engineering texts, but one that has become relevant to earthquake engineering, because energy-dissipating devices based on friction are being used in earthquake-resistant construction. After presenting numerical time-stepping methods for calculating the dynamic response of systems (Chapter 5), the earthquake response of linearly elastic systems and of inelastic systems is studied in Chapters 6 and 7, respectively. Coverage of these topics is more comprehensive than in texts presently available; included are details on the construction of response and design spectra, effects of damping and yielding, and the distinction between response and design spectra. The analysis of complex systems treated as generalized SDF systems is the subject of Chapter 8.

Part II includes Chapters 9 through 17 on the dynamic analysis of multi-degree-of-freedom (MDF) systems. In the opening chapter of Part II the structural dynamics problem is formulated for structures idealized as systems with a finite number of degrees of freedom and illustrated by numerous examples; also included is an overview of methods for solving the differential equations governing the motion of the structure. Chapter 10 is concerned with free vibration of systems with classical damping and with the numerical calculation of natural vibration frequencies and modes of the structure. Also included are differences in free vibration of systems with classical damping and of systems with nonclassical damping, a topic not normally discussed in textbooks. Chapter 11 addresses several issues that arise in defining the damping properties of structures, including experimental data—from forced vibration tests on structures and recorded motions of structures during earthquakes—that provide a basis for estimating modal damping ratios, and analytical procedures to construct the damping matrix, if necessary. Chapter 12 is concerned with the dynamics of linear systems, where the classical modal analysis procedure is emphasized. Part C of this chapter represents a "new" way of looking at modal analysis that facilitates understanding of how modal response contributions are influenced by the spatial distribution and the time variation of applied forces, leading to practical criteria on the number of modes to include in response calculation. In Chapter 13, modal analysis procedures for earthquake analysis of structures are developed; both response history analysis and response spectrum analysis procedures are presented in a form that provides physical interpretation. The presentation and application of modal combination rules to estimate the peak response of MDF systems directly from the earthquake response or design spectrum is more comprehensive than in textbooks presently available. The procedures are illustrated by numerous examples, including coupled lateral-torsional response of unsymmetric-plan buildings and torsional response of nominally symmetric buildings.

Chapter 14 is devoted to the practical computational issue of reducing the number of degrees of freedom in the structural idealization required for static analysis in order to recognize that the dynamic response of many structures can be well represented by their first few natural modes. In Chapter 15 numerical time-stepping methods are presented for MDF systems not amenable to classical modal analysis: systems with nonclassical damping or systems responding into the range of nonlinear behavior. Chapter 16 is concerned with classical problems in the dynamics of distributed-mass systems; only one-dimensional systems are included. In Chapter 17 two methods are presented for discretizing one-dimensional distributed-mass systems: the Rayleigh–Ritz method and the finite element method. The consistent mass matrix concept is introduced, and the accuracy and convergence of the approximate natural frequencies of a cantilever beam, determined by the finite element method, are demonstrated.

Part III of the book contains four chapters concerned with earthquake response and design of multistory buildings, a subject not normally included in structural dynamics texts. Several important and practical issues are addressed using analytical procedures developed in the preceding chapters. In Chapter 18 the earthquake response of linearly elastic multistory buildings is presented for a wide range of two key parameters: fundamental natural vibration period and beam-to-column stiffness ratio. Based on these results, we develop an understanding of how these parameters affect the earthquake response of buildings and, in particular, the relative response contributions of the various natural modes, leading to practical information on the number of higher modes to include in earthquake response calculations. Chapter 19 is concerned with the important subject of earthquake response of multistory buildings deforming into their inelastic range. It includes discussion of the heightwise variation of story ductility demands, large ductility demand in the first story if it is "weak" or "soft" relative to the upper stories, ductility demands for buildings designed according to the lateral force distribution of the 1994 *Uniform Building Code,* and how these demands compare with allowable ductility. The currently active and expanding subject of base isolation is the subject of Chapter 20. Our goal is to study the dynamic behavior of buildings supported on base isolation systems with the limited objective of understanding why and under what conditions isolation is effective in reducing the earthquake-induced forces in a structure. In Chapter 21 we present the seismic force provisions in three building codes—*Uniform Building Code* (United States), *National Building Code of Canada*, and *Mexico Federal District Code*—together with their relationship to the theory of structural dynamics developed in Chapters 6, 7, 8, and 13. Subsequently, the code provisions are evaluated in light of the results of dynamic analysis of buildings presented in Chapters 18 and 19.

A NOTE FOR INSTRUCTORS

This book is suitable for courses at the graduate level and at the senior undergraduate level. No previous knowledge of structural dynamics is assumed. The necessary background is available through the usual courses required of civil engineering undergraduates. These include:

- Static analysis of structures, including statically indeterminate structures and matrix formulation of analysis procedures (background needed primarily for Part II)
- Structural design
- Rigid-body dynamics
- Mathematics: ordinary differential equations (for Part I), linear algebra (for Part II), and partial differential equations (for Chapter 16 only)

By providing an elementary but thorough treatment of a large number of topics, this book permits unusual flexibility in selection of the course content at the discretion of the instructor. Several courses can be developed based on the material in this book. Here are a few examples.

Almost the entire book can be covered in a one-year course:

- *Title:* Dynamics of Structures I (1 semester)

  Syllabus: Chapters 1 and 2; Parts A and B of Chapter 3; Chapter 4; selected topics from Chapter 5; Sections 1 to 7 of Chapter 6; Sections 1 to 7 of Chapter 7; selected topics from Chapter 8; Sections 1 to 4 and 8 to 11 of Chapter 9; Parts A and B of Chapter 10; Parts A and B of Chapter 12; Sections 1, 2, 7, and 8 of Chapter 13; and selected topics from Part A of Chapter 21

- *Title:* Dynamics of Structures II (1 semester)

  Syllabus: Chapter 6 (including review of Sections 1 to 7); Chapter 7 (including review of Sections 1 to 7); Sections 5 to 7 of Chapter 9; Part C of Chapter 10; Chapter 11; Parts C and D of Chapter 12; Sections 3 to 9 of Chapter 13; and Chapters 14 to 21

The selection of topics for the first course has been dictated in part by the need to provide comprehensive coverage, including dynamic and earthquake analysis of MDF systems, for students taking only one course.

An abbreviated version of the outline above covering two quarters can be organized as follows:

- *Title:* Dynamics of Structures I (1 quarter)

  Syllabus: Chapter 1; Sections 1 and 2 of Chapter 2; Sections 1 to 4 of Chapter 3; Sections 1 to 5 of Chapter 4; selected topics from Chapter 5; Sections 1 to 7 of Chapter 6; Sections 1 to 7 of Chapter 7; selected topics from Chapter 8; Sections 1 to 4 and 8 to 11 of Chapter 9; Parts A and B of Chapter 10; Parts A and B of Chapter 12; Sections 1, 2, 7, and 8 (excluding the CQC method) of Chapter 13; and selected topics from Part A of Chapter 21

- *Title:* Dynamics of Structures II (1 quarter)

  Syllabus: Chapter 6 (including review of Sections 1 to 7); Chapter 7 (including review of Sections 1 to 7); Sections 5 to 7 of Chapter 9; Chapter 11; Parts C and D of Chapter 12; Sections 3 to 9 of Chapter 13; and Chapters 18 to 21

A one-semester course emphasizing earthquake engineering can be organized as follows:

- *Title:* Structural Dynamics and Earthquake Engineering

  Syllabus: Chapter 1; Sections 1 and 2 of Chapter 2; Sections 1 to 4 of Chapter 3; Sections 1 to 5 of Chapter 4; Chapters 6 and 7; selected topics from Chapter 8; Sections 1 to 4 and 8 to 11 of Chapter 9; Parts A and B of Chapter 10; Part A of Chapter 11; Parts A and B of Chapter 12; Sections 1, 2, 7, and 8 of Chapter 13; and Part A of Chapter 21

As every instructor knows, solving problems is essential for students who are learning structural dynamics. For this purpose the first 17 chapters include 233 problems. Chapters 18 through 21 do not include problems, for two reasons: (1) no new dynamic analysis procedures are introduced in these chapters; (2) this material does not lend itself to short, meaningful problems. However, the reader will find it instructive to work through the examples presented in Chapters 18 to 21 and to reproduce some of the results. Most of the problems can be solved with an electronic hand calculator and a sufficient quantity of patience and perseverance; a computer is most helpful, of course. The computer is essential for solving some of the problems, and these have been identified. In solving these problems, it is assumed that the student will have access to computer programs such as MATLAB or CAL. A solutions manual is available.

In my lectures at Berkeley I use transparencies of figures in this book. Instructors wishing to utilize these visual aids can make transparencies from the enlarged versions of the figures available from the publisher.

A NOTE FOR PROFESSIONAL ENGINEERS

Many professional engineers have encouraged me to prepare a book more comprehensive than *Dynamics of Structures, A Primer,* a monograph published in 1981 by the Earthquake Engineering Research Institute. This need, I hope, is filled by the present book. Having been conceived as a textbook, it includes the formalism and detail necessary for students, but these features should not deter the professional from using the book because, to the extent possible, its philosophy and style are akin to those of the monograph.

For professional engineers interested in earthquake analysis, response, and design of structures, I suggest the following reading path through the book: Chapters 1 and 2; Parts A and B of Chapter 3; Chapters 6 to 9; Parts A and B of Chapter 10; Part A of Chapter 11; and Chapters 13, 18, 19, 20, and 21.

REFERENCES

In this introductory text it is impractical to acknowledge sources for the information presented. References have been omitted to avoid distracting the reader. However, I have included occasional comments to add historical perspective and, at the end of almost every chapter, a brief list of publications suitable for further reading.

YOUR COMMENTS ARE INVITED

Since this is a new book, I request that instructors, students, and professional engineers write to me if they have questions, suggestions for improvements or clarifications, or if they identify errors. I thank you in advance for taking the time and interest to do so.

Anil K. Chopra

Acknowledgments

I am grateful to the many people who helped in the preparation of this book.

- Dr. Rakesh K. Goel, a partner from beginning to end, assisted in numerous ways and played an important role. His most significant contribution was to develop and execute the computer software necessary to generate the numerical results and create the over 450 figures.
- Professor Gregory L. Fenves read the first draft, discussed it with me weekly, and provided substantive suggestions for improvement.
- Six reviewers—Professors Luis Esteva, William J. Hall, George W. Housner, Donald E. Hudson, Rafael Riddell, and C. C. Tung—examined a final draft. They provided encouragement as well as perceptive suggestions for improvement.
- Professor W. K. Tso reviewed Chapter 21 and advised on interpretation of the *National Building Code of Canada.*
- The late Professor Emilio Rosenblueth provided much valuable advice and background on the building code for the Mexico Federal District.
- Several students, present and former, assisted in preparing solutions for the worked-out examples and end-of-chapter problems: Juan Chavez, Juan Carlos De la Llera, Rakesh K. Goel, and Tsung-Li Tai. Han-Chen Tan did the word processing and graphics for the Solutions Manual.
- Julie Reynolds and Eric Eisman did the word processing of the text in TeX. Mr. Eisman also assisted in the preliminary editing of some of the material.
- Ms. Katherine Frohmberg helped in selecting and collecting several photographs.

- Professor Joseph Penzien assumed my duties as Associate Editor of *Earthquake Engineering and Structural Dynamics* from June 1993 until August 1994 while I was working on the book.

I also wish to express my deep appreciation to Professors Ray W. Clough, Jr., Joseph Penzien, Emilio Rosenblueth, and A. S. Veletsos for the influence they have had on my professional growth. In the early 1960s, Professors Clough, Penzien, and Rosenblueth exposed me to their enlightened views and their superbly organized courses on structural dynamics and earthquake engineering. Subsequently, Professor Veletsos, through his research, writing, and lectures, influenced my teaching and research philosophy. His work, in collaboration with the late Professor Nathan M. Newmark, defined the approach adopted for parts of Chapters 6 and 7.

This book has been influenced by my own research experience in collaboration with my students. Since 1969, several organizations have supported my research in earthquake engineering, including the National Science Foundation, U.S. Army Corps of Engineers, and California Strong Motion Instrumentation Program. I am especially grateful to the National Science Foundation, in particular Dr. S. C. Liu, for sustained support.

This book was prepared during a year of sabbatical leave, a privilege for which I am grateful to the University of California at Berkeley.

Anil K. Chopra

DYNAMICS OF STRUCTURES

PART I

Single-Degree-of-Freedom Systems

Equations of Motion, Problem Statement, and Solution Methods

PREVIEW

In this opening chapter, the structural dynamics problem is formulated for simple structures that can be idealized as a system with a lumped mass and a massless supporting structure. Linearly elastic structures as well as inelastic structures subjected to applied dynamic force or earthquake-induced ground motion are considered. Then four methods for solving the differential equation governing the motion of the structure are reviewed briefly. The chapter ends with an overview of how our study of the dynamic response of single-degree-of-freedom systems is organized in the chapters to follow.

1.1 SIMPLE STRUCTURES

We begin our study of structural dynamics with *simple* structures, such as the pergola shown in Fig. 1.1.1 and the elevated water tank of Fig. 1.1.2. We are interested in understanding the vibration of these structures when subjected to a lateral (or horizontal) force at the top or horizontal ground motion due to an earthquake.

We call these structures *simple* because they can be idealized as a concentrated or lumped mass m supported by a massless structure with stiffness k in the lateral direction. Such an idealization is appropriate for this pergola with a heavy concrete roof supported by light-steel-pipe columns, which can be assumed as massless. The concrete roof is very stiff and the flexibility of the structure in lateral (or horizontal) motion is provided entirely by the columns. The idealized system is shown in Fig. 1.1.3a with a pair of columns supporting the tributary length of the concrete roof. This system has a lumped

Figure 1.1.1 This pergola at the Macuto-Sheraton Hotel near Caracas, Venezuela was damaged by the earthquake of July 29, 1967. The Magnitude 6.5 event, which was centered about 15 miles from the hotel, overstrained the steel pipe columns. (Courtesy of G. W. Housner.)

mass m equal to the mass of the roof shown and its lateral stiffness k is equal to the sum of the stiffnesses of individual pipe columns. A similar idealization, shown in Fig. 1.1.3b, is appropriate for the tank when it is full of water. With sloshing of water not possible in a full tank, it is a lumped mass m supported by a relatively light tower that can be assumed as massless. The cantilever tower supporting the water tank provides lateral stiffness k to the structure. For the moment we will assume that the lateral motion of these structures is small in the sense that the supporting structures deform within their linear elastic limit.

We shall see later in this chapter that the differential equation governing the lateral displacement $u(t)$ of these idealized structures without any external excitation—applied force or ground motion—is

$$m\ddot{u} + ku = 0 \tag{1.1.1}$$

where an overdot denotes differentiation with respect to time; thus $\dot{u}$ denotes the velocity of the mass and $\ddot{u}$ its acceleration. The solution of this equation, presented in Chapter 2, will show that if the mass of the idealized systems of Fig. 1.1.3 is displaced through some initial displacement $u(0)$, then released and permitted to vibrate freely, the structure will oscillate or vibrate back and forth about its initial equilibrium position. As shown in Fig. 1.1.3c, the same maximum displacement occurs oscil-

Figure 1.1.2 This reinforced-concrete tank on a 40-ft.-tall single concrete column, located near the Valdivia Airport, was undamaged by the Chilean earthquakes of May 1960. When the tank is full of water, the structure can be analyzed as a single-degree-of-freedom system. (From K. V. Steinbrugge Collection, courtesy of the Earthquake Engineering Research Center, University of California at Berkeley.)

lation after oscillation; these oscillations continue forever and these idealized systems would never come to rest. This is unrealistic, of course. Intuition suggests that if the roof of the pergola or the top of the water tank were pulled laterally by a rope and the rope were suddenly cut, the structure would oscillate with ever-decreasing amplitude

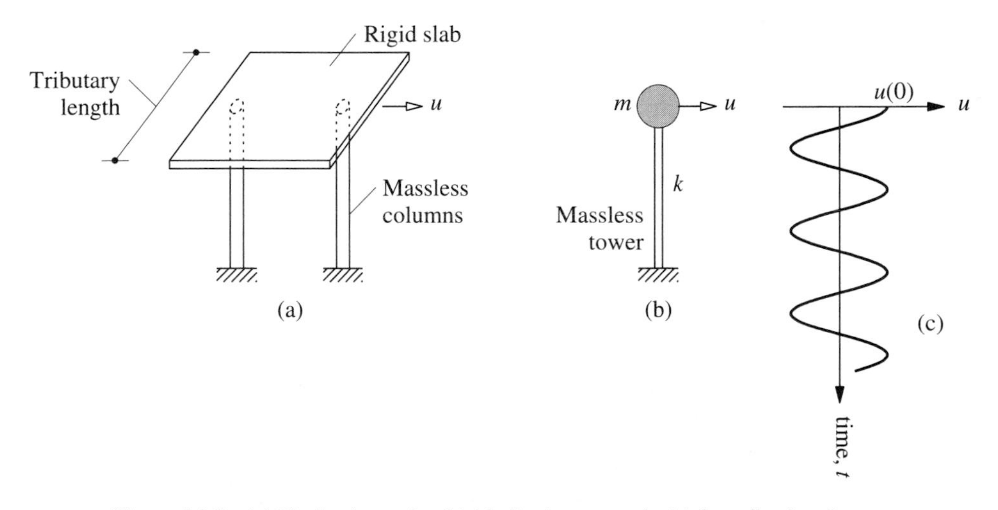

Figure 1.1.3 (a) Idealized pergola; (b) idealized water tank; (c) free vibration due to initial displacement.

displacement—for dynamic analysis if it is idealized with mass concentrated at one location, typically the roof level. Thus we call this a *single-degree-of-freedom* (SDF) *system.*

Two types of dynamic excitation will be considered: (1) external force $p(t)$ in the lateral direction (Fig. 1.2.1a), and (2) earthquake-induced ground motion $u_g(t)$ (Fig. 1.2.1b). In both cases u denotes the relative displacement between the mass and the base of the structure.

1.3 FORCE–DISPLACEMENT RELATION

Consider the system shown in Fig. 1.3.1a with no dynamic excitation subjected to an externally applied static force f_S along the DOF u as shown. The internal force resisting the displacement u is equal and opposite to the external force f_S (Fig. 1.3.1b). It is desired to determine the relationship between the force f_S and the relative displacement u associated with deformations in the structure. This force–displacement relation would be linear at small deformations but would become nonlinear at larger deformations (Fig. 1.3.1c); both nonlinear and linear relations are considered (Fig. 1.3.1c and d).

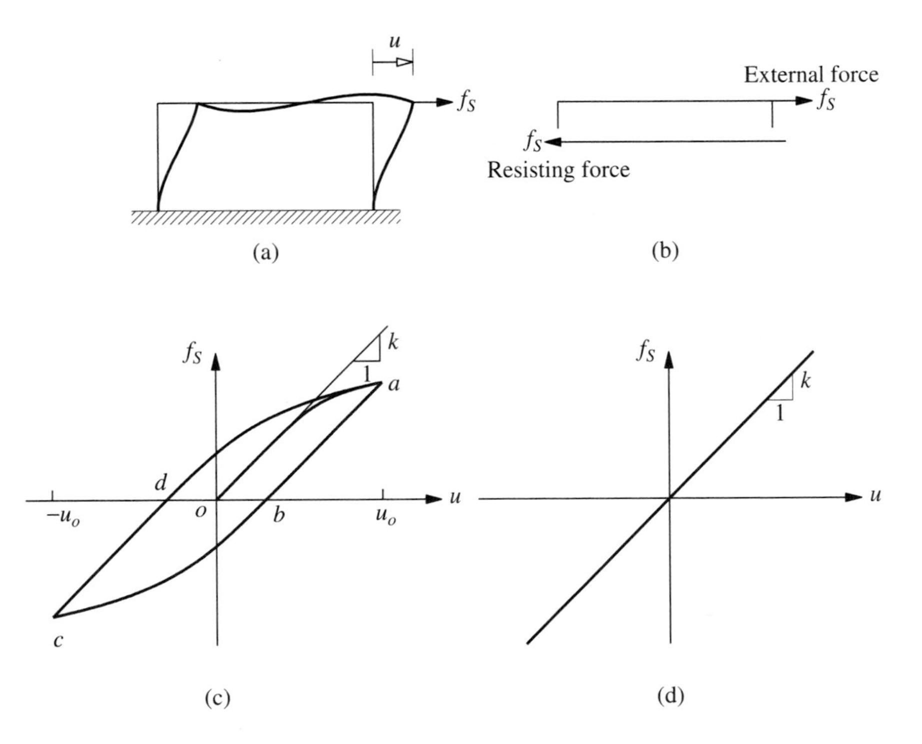

Figure 1.3.1

To determine the relationship between f_S and u is a standard problem in static structural analysis, and we assume that the reader is familiar with such analyses. Thus the presentation here is brief and limited to those aspects that are essential.

1.3.1 Linearly Elastic Systems

For a *linear system* the relationship between the lateral force f_S and resulting deformation u is linear, that is,

$$f_S = ku \tag{1.3.1}$$

where k is the lateral stiffness of the system; its units are force/length. Implicit in Eq. (1.3.1) is the assumption that the linear f_S–u relationship determined for small deformations of the structure is also valid for larger deformations. Because the resisting force is a single-valued function of u, the system is *elastic*; hence we use the term *linearly elastic system.*

Consider the frame of Fig. 1.3.2a with bay width L, height h, elastic modulus E, and second moment of the cross-sectional area (or moment of inertia)† about the axis of bending $= I_b$ and I_c for the beam and columns, respectively; the columns are clamped (or fixed) at the base. The *lateral stiffness* of the frame can readily be determined for the two extreme cases: If the beam is rigid [i.e., flexural rigidity $EI_b = \infty$ (Fig. 1.3.2b)],

$$k = \sum_{\text{columns}} \frac{12EI_c}{h^3} = 24\frac{EI_c}{h^3} \tag{1.3.2}$$

On the other hand, for a beam with no stiffness [i.e., $EI_b = 0$ (Fig. 1.3.2c)],

$$k = \sum_{\text{columns}} \frac{3EI_c}{h^3} = 6\frac{EI_c}{h^3} \tag{1.3.3}$$

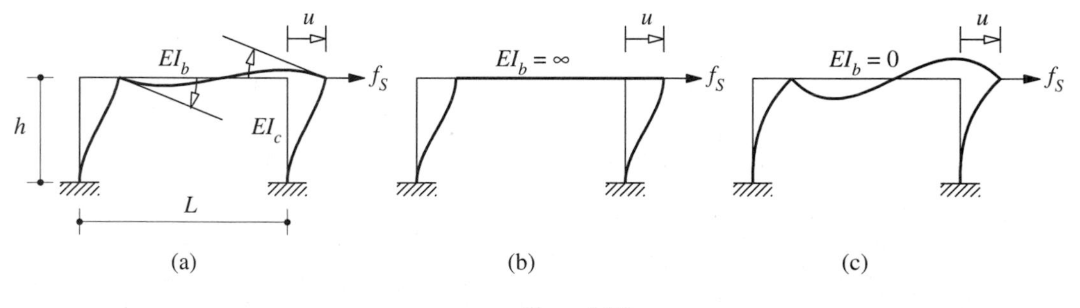

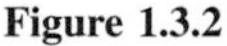
Figure 1.3.2

†In this book the preferred term for I is *second moment of area* instead of the commonly used *moment of inertia*; the latter will be reserved for defining inertial effects associated with rotational motion of rigid bodies.

Observe that for the two extreme values of beam stiffness, the lateral stiffness of the frame is independent of L, the beam length or bay width.

The lateral stiffness of the frame with an intermediate, realistic stiffness of the beam can be calculated by standard procedures of static structural analysis. The stiffness matrix of the frame is formulated with respect to three DOFs: the lateral displacement u and the rotations of the two beam–column joints (Fig. 1.3.2a). By static condensation or elimination of the rotational DOFs, the lateral force–displacement relation of Eq. (1.3.1) is determined. Applying this procedure to a frame with $L = 2h$ and $EI_b = EI_c$, its lateral stiffness is obtained (see Example 1.1):

$$k = \frac{96}{7}\frac{EI_c}{h^3} \tag{1.3.4}$$

The lateral stiffness of the frame can be computed similarly for any values of I_b and I_c using the frame stiffness coefficients developed in Appendix 1. If shear deformations in elements are neglected, the result can be written in the form

$$k = \frac{24EI_c}{h^3}\frac{12\rho + 1}{12\rho + 4} \tag{1.3.5}$$

where $\rho = I_b/4I_c$ is the *beam-to-column stiffness ratio* [see Eq. (18.1.1)]. For $\rho = 0, \infty$, and $\frac{1}{4}$, Eq. (1.3.5) reduces to the results of Eqs. (1.3.3), (1.3.2), and (1.3.4), respectively. The lateral stiffness is plotted as a function of ρ in Fig. 1.3.3; it increases by a factor of 4 as ρ increases from zero to infinity.

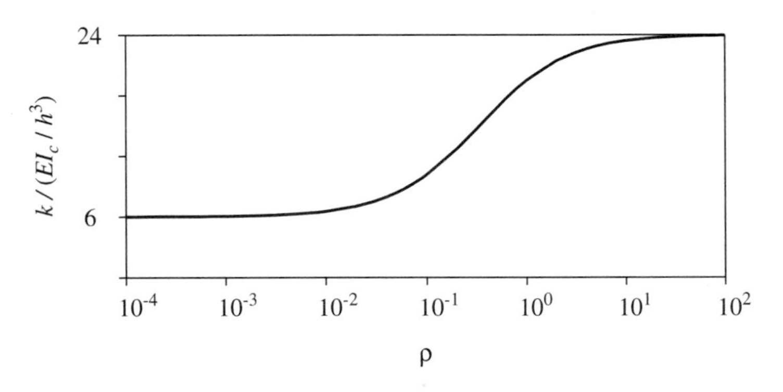

Figure 1.3.3 Variation of lateral stiffness, k, with beam-to-column stiffness ratio, ρ.

Example 1.1

Calculate the lateral stiffness for the frame shown in Fig. E1.1a, assuming the elements to be axially rigid.

Solution This structure can be analyzed by any of the standard methods, including moment distribution. Here we use the definition of stiffness influence coefficients to solve the problem.

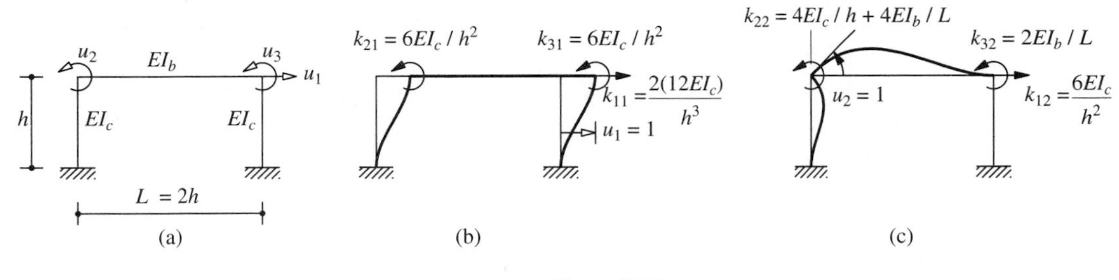

Figure E1.1

The system has the three DOFs shown in Fig. E1.1a. To obtain the first column of the 3×3 stiffness matrix, we impose unit displacement in DOF u_1, with $u_2 = u_3 = 0$. The forces k_{i1} required to maintain this deflected shape are shown in Fig. E1.1b. These are determined using the stiffness coefficients for a uniform flexural element presented in Appendix 1. The elements k_{i2} in the second column of the stiffness matrix are determined by imposing $u_2 = 1$ with $u_1 = u_3 = 0$; see Fig. E1.1c. Similarly, the elements k_{i3} in the third column of the stiffness matrix can be determined by imposing displacements $u_3 = 1$ with $u_1 = u_2 = 0$. Thus the 3×3 stiffness matrix of the structure is known and the equilibrium equations can be written. For a frame with $I_b = I_c$ subjected to lateral force f_S, they are

$$\frac{EI_c}{h^3}\begin{bmatrix} 24 & 6h & 6h \\ 6h & 6h^2 & h^2 \\ 6h & h^2 & 6h^2 \end{bmatrix}\begin{Bmatrix} u_1 \\ u_2 \\ u_3 \end{Bmatrix} = \begin{Bmatrix} f_S \\ 0 \\ 0 \end{Bmatrix} \tag{a}$$

From the second and third equations, the joint rotations can be expressed in terms of lateral displacement as follows:

$$\begin{Bmatrix} u_2 \\ u_3 \end{Bmatrix} = -\begin{bmatrix} 6h^2 & h^2 \\ h^2 & 6h^2 \end{bmatrix}^{-1}\begin{bmatrix} 6h \\ 6h \end{bmatrix} u_1 = -\frac{6}{7h}\begin{bmatrix} 1 \\ 1 \end{bmatrix} u_1 \tag{b}$$

Substituting Eq. (b) into the first of three equations in Eq. (a) gives

$$f_S = \left(\frac{24EI_c}{h^3} - \frac{EI_c}{h^3}\frac{6}{7h}\langle 6h \quad 6h\rangle\begin{bmatrix} 1 \\ 1 \end{bmatrix}\right) u_1 = \frac{96}{7}\left(\frac{EI_c}{h^3}\right) u_1 \tag{c}$$

Thus the lateral stiffness of the frame is

$$k = \frac{96}{7}\frac{EI_c}{h^3} \tag{d}$$

This procedure to eliminate joint rotations, known as the *static condensation method*, is presented in textbooks on static analysis of structures. We return to this topic in Chapter 9.

1.3.2 Inelastic Systems

Force–deformation relations for typical structures undergoing cyclic deformations are shown in Fig. 1.3.4. The initial loading curve is nonlinear at the larger amplitudes of deformation, and the unloading and reloading curves differ from the initial loading

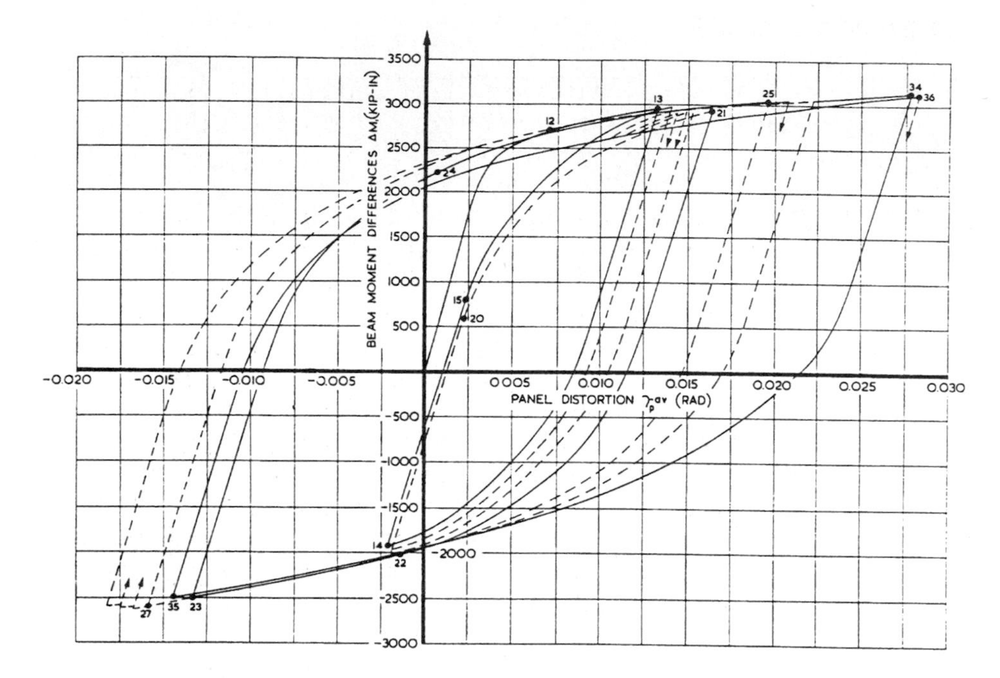

Figure 1.3.4 Force–deformation relation for a structural steel component. (From H. Krawinkler, V. V. Bertero, and E. P. Popov, "Inelastic Behavior of Steel Beam-to-Column Subassemblages," *Report No. EERC 71-7*, University of California, Berkeley, Calif., 1971.)

branch. This implies that the force f_S corresponding to deformation u is not single valued and depends on the history of the deformations and on whether the deformation is increasing (positive velocity) or decreasing (negative velocity). Thus the resisting force can be expressed as

$$f_S = f_S(u, \dot{u}) \tag{1.3.6}$$

The force–deformation relation for the idealized one-story frame (Fig. 1.3.1a) deforming into the inelastic range can be determined in one of two ways. One approach is to use methods of nonlinear static structural analysis. For example, in analyzing a steel structure with an assumed stress–strain law, the analysis keeps track of the initiation and spreading of yielding at critical locations and formation of plastic hinges to obtain the initial loading curve (*o–a*) shown in Fig. 1.3.1c. The unloading (*a–c*) and reloading (*c–a*) curves can be computed similarly or can be defined from the initial loading curve using existing hypotheses. Another approach is to define the inelastic force–deformation relation as an idealized version of the experimental data, such as in Fig. 1.3.4.

We are interested in studying the dynamic response of inelastic systems because many structures are designed with the expectation that they will undergo some cracking, yielding, and damage during intense ground shaking caused by earthquakes.

1.4 DAMPING FORCE

As mentioned earlier, the process by which free vibration steadily diminishes in amplitude is called *damping*. In damping, the energy of the vibrating system is dissipated by various mechanisms, and often more than one mechanism may be present at the same time. In simple "clean" systems such as the laboratory models of Fig. 1.1.4, most of the energy dissipation presumably arises from the thermal effect of repeated elastic straining of the material and from the internal friction when a solid is deformed. In actual structures, however, many other mechanisms also contribute to the energy dissipation. In a vibrating building these include friction at steel connections, opening and closing of microcracks in concrete, friction between the structure itself and nonstructural elements such as partition walls. It seems impossible to identify or describe mathematically each of these energy-dissipating mechanisms in an actual building.

As a result, the damping in actual structures is usually represented in a highly idealized manner. For many purposes the actual damping in a SDF structure can be idealized satisfactorily by a linear viscous damper or dashpot. The damping coefficient is selected so that the vibrational energy it dissipates is equivalent to the energy dissipated in all the damping mechanisms, combined, present in the actual structure. This idealization is therefore called *equivalent viscous damping*, a concept developed further in Chapter 3.

Figure 1.4.1a shows a linear viscous damper subjected to a force f_D along the DOF u. The internal force in the damper is equal and opposite to the external force f_D (Fig. 1.4.1b). As shown in Fig. 1.4.1c, the damping force f_D is related to the velocity $\dot{u}$ across the linear viscous damper by

$$f_D = c\dot{u} \tag{1.4.1}$$

where the constant c is the *viscous damping coefficient*; it has units of force × time/length.

Unlike the stiffness of a structure, the damping coefficient cannot be calculated from the dimensions of the structure and the sizes of the structural elements. This should not be surprising because, as we noted earlier, it is not feasible to identify all the mechanisms that dissipate vibrational energy of actual structures. Thus vibration experiments on actual structures provide the data for evaluating the damping coefficient. These may be free

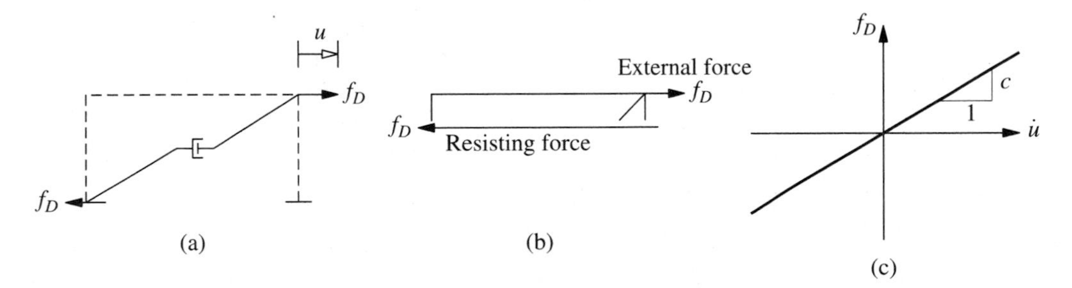

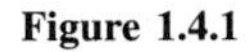
Figure 1.4.1

vibration experiments that lead to data such as those shown in Fig. 1.1.4; the measured rate at which motion decays in free vibration will provide a basis for evaluating the damping coefficient, as we shall see in Chapter 2. The damping property may also be determined from forced vibration experiments, a topic that we study in Chapter 3.

The equivalent viscous damper is intended to model the energy dissipation at deformation amplitudes within the linear elastic limit of the overall structure. Over this range of deformations, the damping coefficient c determined from experiments may vary with the deformation amplitude. This nonlinearity of the damping property is usually not considered explicitly in dynamic analyses. It may be handled indirectly by selecting a value for the damping coefficient that is appropriate for the expected deformation amplitude, usually taken as the deformation associated with the linearly elastic limit of the structure.

Additional energy is dissipated due to inelastic behavior of the structure at larger deformations. Under cyclic forces or deformations, this behavior implies formation of a force–deformation hysteresis loop (Fig. 1.3.1c). The damping energy dissipated during one deformation cycle between deformation limits $\pm u_o$ is given by the area within the hysteresis loop *abcda* (Fig. 1.3.1c). This energy dissipation is usually not modeled by a viscous damper, especially if the excitation is earthquake ground motion, for reasons we note in Chapter 7. Instead, the most common and direct approach to account for the energy dissipation through inelastic behavior is to recognize the inelastic relationship between resisting force and deformation, such as shown in Figs. 1.3.1c and 1.3.4, in solving the equation of motion (Chapter 5). Such force–deformation relationships are obtained from experiments on structures or structural components at slow rates of deformation, thus excluding any energy dissipation arising from rate-dependent effects. The usual approach is to model this damping in the inelastic range of deformations by the same viscous damper that was defined earlier for smaller deformations within the linearly elastic range.

1.5 EQUATION OF MOTION: EXTERNAL FORCE

Figure 1.5.1a shows the idealized one-story frame introduced earlier subjected to an externally applied dynamic force $p(t)$ in the direction of the DOF u. This notation indicates that the force p varies with time t. The resulting displacement of the mass

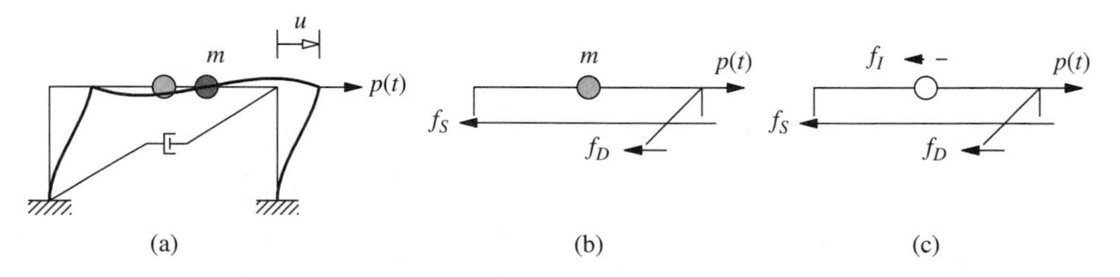

Figure 1.5.1

also varies with time; it is denoted by $u(t)$. In Sections 1.5.1 and 1.5.2 we derive the differential equation governing the displacement $u(t)$ by two methods using (1) Newton's second law of motion, and (2) dynamic equilibrium. An alternative point of view for the derivation is presented in Section 1.5.3.

1.5.1 Using Newton's Second Law of Motion

The forces acting on the mass at some instant of time are shown in Fig. 1.5.1b. These include the external force $p(t)$, the elastic (or inelastic) resisting force f_S, and the damping force f_D. The external force is taken to be positive in the direction of the x-axis, and the displacement $u(t)$, velocity $\dot{u}(t)$, and acceleration $\ddot{u}(t)$ are also positive in the direction of the x-axis. The elastic and damping forces are shown acting in the opposite direction because they are internal forces that resist the deformation and velocity, respectively.

The resultant force along the x-axis is $p - f_S - f_D$ and Newton's second law of motion gives

$$p - f_S - f_D = m\ddot{u} \quad \text{or} \quad m\ddot{u} + f_D + f_S = p(t) \tag{1.5.1}$$

This equation after substituting Eqs. (1.3.1) and (1.4.1) becomes

$$m\ddot{u} + c\dot{u} + ku = p(t) \tag{1.5.2}$$

This is the equation of motion governing the deformation or displacement $u(t)$ of the idealized structure of Fig. 1.5.1a, assumed to be linearly elastic, subjected to an external dynamic force $p(t)$. The units of mass are force/acceleration.

This derivation can readily be extended to inelastic systems. Equation (1.5.1) is still valid and all that needs to be done is to replace Eq. (1.3.1), restricted to linear systems, by Eq. (1.3.6), valid for inelastic systems. For such systems, therefore, the equation of motion is

$$m\ddot{u} + c\dot{u} + f_S(u, \dot{u}) = p(t) \tag{1.5.3}$$

1.5.2 Dynamic Equilibrium

Having been trained to think in terms of equilibrium of forces, structural engineers may find D'Alembert's principle of *dynamic equilibrium* particularly appealing. This principle is based on the notion of a fictitious *inertia force*, a force equal to the product of mass times its acceleration and acting in a direction opposite to the acceleration. It states that with inertia forces included, a system is in equilibrium at each time instant. Thus a free-body diagram of a moving mass can be drawn, and principles of statics can be used to develop the equation of motion.

Figure 1.5.1c is the free-body diagram at time t with the mass replaced by its inertia force, which is shown by a dashed line to distinguish this "fictitious" force from the real forces. Setting the sum of all the forces equal to zero gives Eq. (1.5.1b),[†] which was derived earlier by using Newton's second law of motion.

1.5.3 Stiffness, Damping, and Mass Components

In this section the governing equation for the idealized one-story frame is formulated based on an alternative viewpoint. Under the action of external force $p(t)$ the state of the system is described by displacement $u(t)$, velocity $\dot{u}(t)$, and acceleration $\ddot{u}(t)$; see Fig. 1.5.2a. Now visualize the system as the combination of three pure components: (1) the stiffness component: the frame without damping or mass (Fig. 1.5.2b); (2) the damping component: the frame with its damping property but no stiffness or mass (Fig. 1.5.2c); and (3) the mass component: the roof mass without the stiffness or damping of the frame (Fig. 1.5.2d). The external force f_S on the stiffness component is related to the displacement u by Eq. (1.3.1) if the system is linearly elastic, the external force f_D on the damping component is related to the velocity $\dot{u}$ by Eq. (1.4.1), and the external force f_I on the mass component is related to the acceleration by $f_I = m\ddot{u}$. The external force $p(t)$ applied to the complete system may therefore be visualized as distributed among the three components of the structure, and $f_S + f_D + f_I$ must equal the applied force $p(t)$ leading to Eq. (1.5.1b). Although this alternative viewpoint may seem unnecessary for the simple system of Fig. 1.5.2a, it is useful for complex systems (Chapter 9).

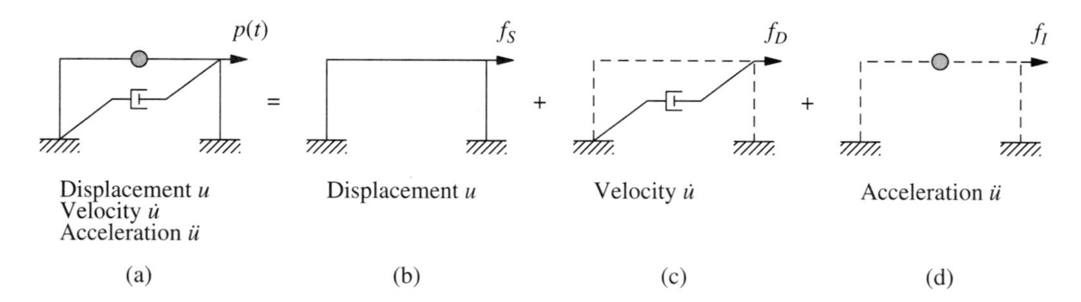

Figure 1.5.2 (a) System; (b) stiffness component; (c) damping component; (d) mass component.

Example 1.2

A small one-story industrial building, 20 by 30 ft in plan, is shown in Fig. E1.2 with moment frames in the north–south direction and braced frames in the east–west direction. The weight of the structure can be idealized as 30 lb/ft^2 lumped at the roof level. The horizontal cross bracing is at the bottom chord of the roof trusses. All columns are W8 $\times$ 24 sections; their second moments of cross-sectional area about the x and y axes are I_x = 82.8 in^4 and I_y = 18.3 in^4, respectively; for steel, E = 29,000 ksi. The vertical cross-bracings are made of 1-in.-diameter rods. Formulate the equation governing free vibration in (**a**) the north–south direction and (**b**) the east–west direction.

[†]Two or more equations in the same line with the same equation number will be referred to as equations a, b, c, etc., from left to right.

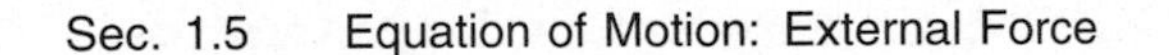
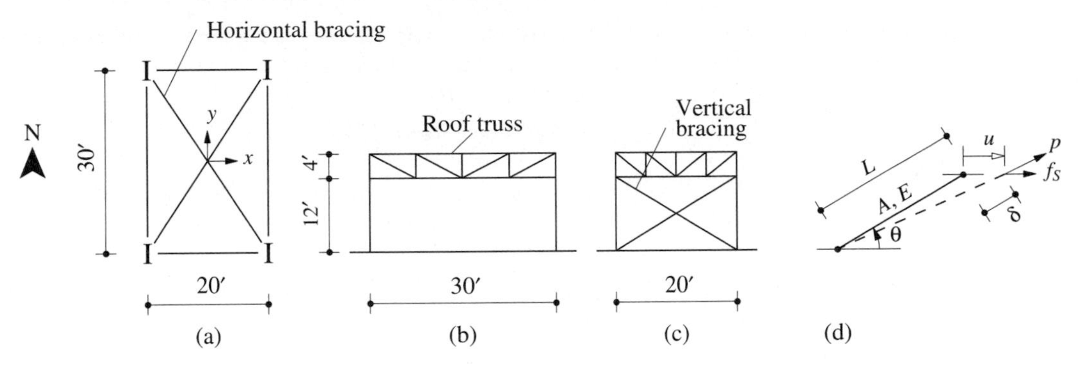

Figure E1.2 (a) Plan; (b) east and west elevations; (c) north and south elevations; (d) cross brace.

Solution The mass lumped at the roof is

$$m = \frac{w}{g} = \frac{30 \times 30 \times 20}{386} = 46.63 \text{ lb-sec}^2/\text{in.} = 0.04663 \text{ kip-sec}^2/\text{in.}$$

Because of the horizontal cross-bracing, the roof can be treated as a rigid diaphragm.

(**a**) *North–south direction.* The lateral stiffness of the two moment frames (Fig. E1.2b) is

$$k_{\text{N-S}} = 4\left(\frac{12EI_x}{h^3}\right) = 4\frac{12(29 \times 10^3)(82.8)}{(12 \times 12)^3} = 38.58 \text{ kips/in.}$$

and the equation of motion is

$$m\ddot{u} + (k_{\text{N-S}})\,u = 0 \tag{a}$$

(**b**) *East–west direction.* Braced frames, such as those shown in Fig. E1.2c, are usually designed as two superimposed systems: an ordinary rigid frame that supports vertical (dead and live) loads, plus a vertical bracing system, generally regarded as a pin-connected truss that resists lateral forces. Thus the lateral stiffness of a braced frame can be estimated as the sum of the lateral stiffnesses of individual braces. The stiffness of a brace (Fig. E1.2d) is $k_{\text{brace}} = (AE/L)\cos^2\theta$. This can be derived as follows.

We start with the axial force–deformation relation for a brace:

$$p = \frac{AE}{L}\delta \tag{b}$$

By statics $f_S = p\cos\theta$ and by kinematics $u = \delta/\cos\theta$. Substituting $p = f_S/\cos\theta$ and $\delta = u\cos\theta$ in Eq. (b) gives

$$f_S = k_{\text{brace}}u \qquad k_{\text{brace}} = \frac{AE}{L}\cos^2\theta \tag{c}$$

For the brace in Fig. E1.2c, $\cos\theta = 20/\sqrt{12^2 + 20^2} = 0.8575$, $A = 0.785$ in.^{2}, $L = 23.3$ ft, and

$$k_{\text{brace}} = \frac{0.785(29 \times 10^3)}{23.3 \times 12}(0.8575)^2 = 59.8 \text{ kips/in.}$$

Although each frame has two cross-braces, only the one in tension will provide lateral resistance; the one in compression will buckle at small axial force and will contribute little to the lateral stiffness. Considering the two frames

$$k_{\text{E-W}} = 2 \times 59.8 = 119.6 \text{ kips/in.}$$

and the equation of motion is

$$m\ddot{u} + (k_{\text{E-W}})\, u = 0 \tag{d}$$

Observe that the error in neglecting the stiffness of columns is small: $k_{\text{col}} = 12EI_y/h^3 = 2.13$ kips/in. versus $k_{\text{brace}} = 59.8$ kips/in.

1.6 MASS–SPRING–DAMPER SYSTEM

We have introduced the SDF system by idealizing a one-story structure (Fig. 1.5.1a), an approach that should appeal to structural engineering students. However, the classic SDF system is the mass–spring–damper system of Fig. 1.6.1a. The dynamics of this system is developed in textbooks on mechanical vibration and elementary physics. If we consider the spring and damper to be massless, the mass to be rigid, and all motion to be in the direction of the x-axis, we have an SDF system. Figure 1.6.1b shows the forces acting on the mass; these include the elastic resisting force, $f_S = ku$, exerted by a linear spring of stiffness k, and the damping resisting force, $f_D = c\dot{u}$, due to a linear viscous damper. Newton's second law of motion then gives Eq. (1.5.1b). Alternatively, the same equation is obtained using D'Alembert's principle and writing an equilibrium equation for forces in the free-body diagram, including the inertia force (Fig. 1.6.1c). It is clear that the equation of motion derived earlier for the idealized one-story frame of Fig. 1.5.1a is also valid for the mass–spring–damper system of Fig. 1.6.1a.

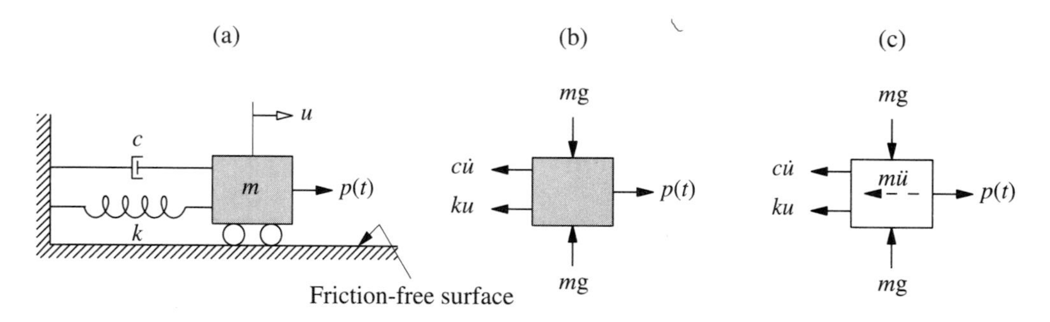

Figure 1.6.1 Mass–spring–damper system.

Example 1.3

Derive the equation of motion of the weight w suspended from a spring at the free end of a cantilever steel beam shown in Fig. E1.3a. For steel, $E = 29{,}000$ ksi. Neglect the mass of the beam and spring.

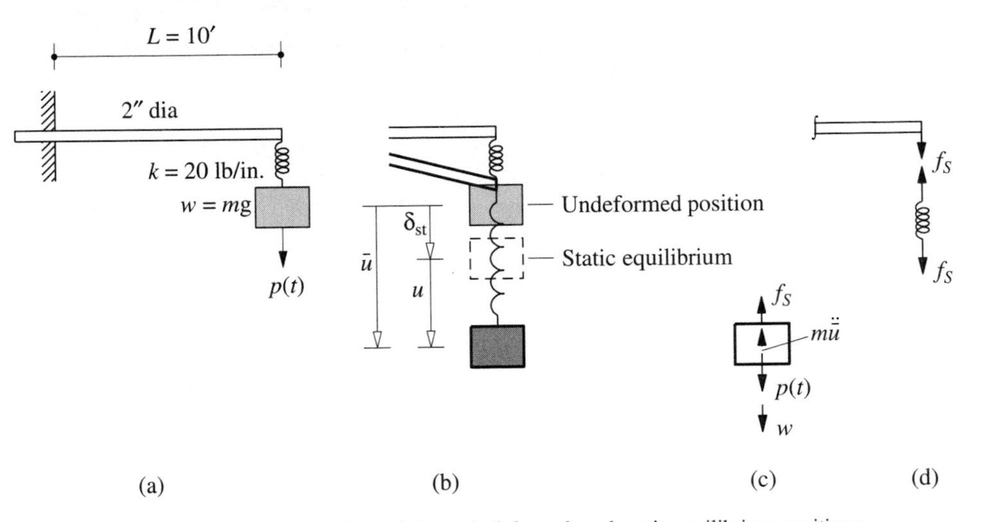

Figure E1.3 (a) System; (b) undeformed, deformed, and static equilibrium positions; (c) free-body diagram; (d) spring and beam forces.

Solution Figure E1.3b shows the deformed position of the free end of the beam, spring, and mass. The displacement of the mass $\bar{u}$ is measured from its initial position with the beam and spring in their original undeformed configuration. Equilibrium of the forces of Fig. E1.3c gives

$$m\ddot{\bar{u}} + f_S = w + p(t) \tag{a}$$

where

$$f_S = k_e \bar{u} \tag{b}$$

and the effective stiffness k_e of the system remains to be determined. The equation of motion is

$$m\ddot{\bar{u}} + k_e \bar{u} = w + p(t) \tag{c}$$

The displacement $\bar{u}$ can be expressed as

$$\bar{u} = \delta_{st} + u \tag{d}$$

where δ_{st} is the static displacement due to weight w and u is measured from the position of static equilibrium. Substituting Eq. (d) in Eq. (a) and noting that (1) $\ddot{\bar{u}} = \ddot{u}$ because δ_{st} does not vary with time, and (2) $k_e \delta_{st} = w$ gives

$$m\ddot{u} + k_e u = p(t) \tag{e}$$

Observe that this is the same as Eq. (1.5.2) with $c = 0$ for a spring–mass system oriented in the horizontal direction (Fig. 1.6.1). Also note that the equation of motion (e) governing u, measured from the static equilibrium position, is unaffected by gravity forces.

For this reason we usually formulate a dynamic analysis problem with the static equilibrium position as the reference position. The displacement $u(t)$ and associated internal forces in the system will represent the dynamic response of the system. If the system is linear, the total displacements and forces are obtained by adding the corresponding static quantities to the dynamic response.

The effective stiffness k_e remains to be determined. It relates the static force f_S to the resulting displacement $\overline{u}$ by

$$f_S = k_e \overline{u} \tag{f}$$

where

$$\overline{u} = \delta_{\text{spring}} + \delta_{\text{beam}} \tag{g}$$

where δ_{beam} is the deflection of the right end of the beam and δ_{spring} is the deformation in the spring. With reference to Fig. E1.3d,

$$f_S = k\delta_{\text{spring}} = k_{\text{beam}}\delta_{\text{beam}} \tag{h}$$

In Eq. (g), substitute for $\overline{u}$ from Eq. (f) and the δ's from Eq. (h) to obtain

$$\frac{f_S}{k_e} = \frac{f_S}{k} + \frac{f_S}{k_{\text{beam}}} \qquad \text{or} \qquad k_e = \frac{k k_{\text{beam}}}{k + k_{\text{beam}}} \tag{i}$$

Now $k = 20$ lb/in. and

$$k_{\text{beam}} = \frac{3EI}{L^3} = \frac{3(29 \times 10^6)[\pi(1)^4/4]}{(10 \times 12)^3} = 39.54 \text{ lb/in.}$$

Substituting for k and k_{beam} in Eq. (i) gives

$$k_e = 13.39 \text{ lb/in.}$$

1.7 EQUATION OF MOTION: EARTHQUAKE EXCITATION

In earthquake-prone regions, the principal problem of structural dynamics that concern structural engineers is the behavior of structures subjected to earthquake-induced motion of the base of the structure. The displacement of the ground is denoted by u_g, the total (or absolute) displacement of the mass by u^t, and the relative displacement between the mass and ground by u (Fig. 1.7.1). At each instant of time these displacements are related by

$$u^t(t) = u(t) + u_g(t) \tag{1.7.1}$$

Both u^t and u_g refer to the same inertial frame of reference and their positive directions coincide.

The equation of motion for the idealized one-story system of Fig. 1.7.1a subjected to earthquake excitation can be derived by any one of the approaches introduced in

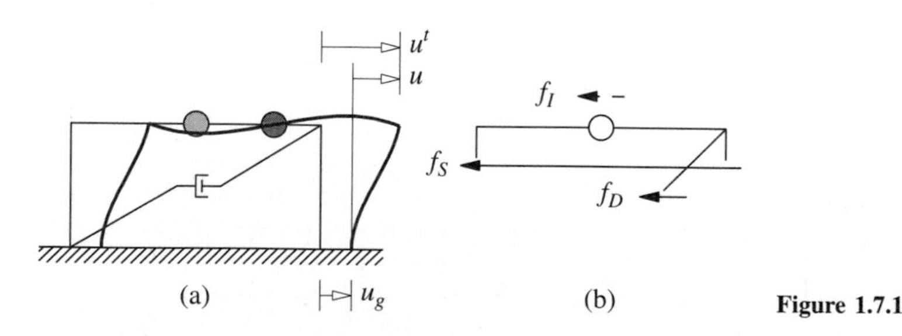

Figure 1.7.1

Section 1.5. Here we choose to use the concept of dynamic equilibrium. From the free-body diagram including the inertia force f_I, shown in Fig. 1.7.1b, the equation of dynamic equilibrium is

$$f_I + f_D + f_S = 0 \tag{1.7.2}$$

Only the relative motion u between the mass and the base due to structural deformation produces elastic and damping forces (i.e., the rigid-body component of the displacement of the structure produces no internal forces). Thus for a linear system Eqs. (1.3.1) and (1.4.1) are still valid. The inertia force f_I is related to the acceleration $\ddot{u}^t$ of the mass by

$$f_I = m\ddot{u}^t \tag{1.7.3}$$

Substituting Eqs. (1.3.1), (1.4.1), and (1.7.3) in Eq. (1.7.2) and using Eq. (1.7.1) gives

$$m\ddot{u} + c\dot{u} + ku = -m\ddot{u}_g(t) \tag{1.7.4}$$

This is the equation of motion governing the relative displacement or deformation $u(t)$ of the linear structure of Fig. 1.7.1a subjected to ground acceleration $\ddot{u}_g(t)$.

For inelastic systems, Eq. (1.7.2) is valid but Eq. (1.3.1) should be replaced by Eq. (1.3.6). The resulting equation of motion is

$$m\ddot{u} + c\dot{u} + f_S(u, \dot{u}) = -m\ddot{u}_g(t) \tag{1.7.5}$$

Comparison of Eqs. (1.5.2) and (1.7.4), or of Eqs. (1.5.3) and (1.7.5), shows that the equations of motion for the structure subjected to two separate excitations—ground acceleration $\ddot{u}_g(t)$ and external force $= -m\ddot{u}_g(t)$—are one and the same. Thus the relative displacement or deformation $u(t)$ of the structure due to ground acceleration $\ddot{u}_g(t)$ will be identical to the displacement $u(t)$ of the structure if its base were stationary and if it were subjected to an external force $= -m\ddot{u}_g(t)$. As shown in Fig. 1.7.2, the ground motion can therefore be replaced by the *effective earthquake force* (indicated by the subscript "eff"):

$$p_{\text{eff}}(t) = -m\ddot{u}_g(t) \tag{1.7.6}$$

This force is equal to mass times the ground acceleration, acting opposite to the acceleration. It is important to recognize that the effective earthquake force is proportional to the mass of the structure. Thus the structural designer increases the effective earthquake force if the structural mass is increased.

Although the rotational components of ground motion are not measured during earthquakes, they can be estimated from the measured translational components and it is

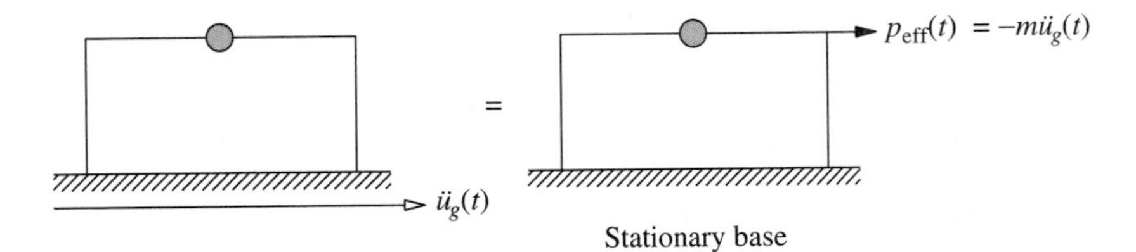

Figure 1.7.2 Effective earthquake force: horizontal ground motion.

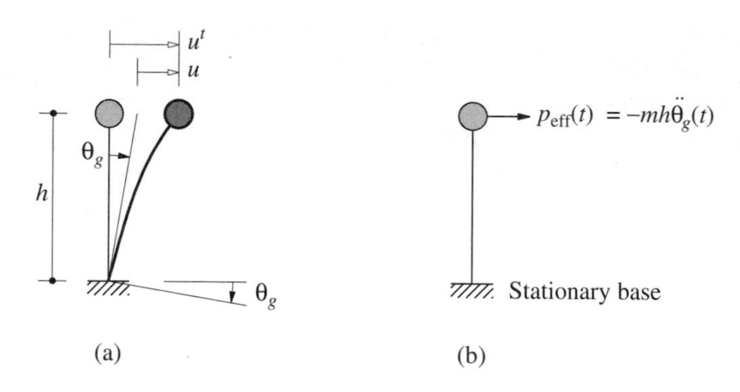

Figure 1.7.3 Effective earthquake force: rotational ground motion.

of interest to apply the preceding concepts to this excitation. For this purpose, consider the cantilever tower of Fig. 1.7.3a, which may be considered as an idealization of the water tank of Fig. 1.1.2, subjected to base rotation θ_g. The total displacement u^t of the mass is made up of two parts: u associated with structural deformation and a rigid-body component $h\theta_g$, where h is the height of the mass above the base. At each instant of time these displacements are related by

$$u^t(t) = u(t) + h\theta_g(t) \tag{1.7.7}$$

Equations (1.7.2) and (1.7.3) are still valid but the total acceleration $\ddot{u}^t(t)$ must now be determined from Eq. (1.7.7). Putting all these equations together leads to

$$m\ddot{u} + c\dot{u} + ku = -mh\ddot{\theta}_g(t) \tag{1.7.8}$$

The effective earthquake force associated with ground rotation is

$$p_{\text{eff}}(t) = -mh\ddot{\theta}_g(t) \tag{1.7.9}$$

Example 1.4

A uniform rigid slab of total mass m is supported on four columns of height h rigidly connected to the top slab and to the foundation slab. Each column has a rectangular cross section with second moments of area I_x and I_y for bending about the x and y axes, respectively. Determine the equation of motion for this system subjected to rotation $u_{g\theta}$ of the foundation about a vertical axis. Neglect the mass of the columns.

Solution The elastic resisting torque or torsional moment f_S acting on the mass is shown in Fig. E1.4b, and Newton's second law gives

$$-f_S = I_O\ddot{u}_\theta^t \tag{a}$$

where

$$u_\theta^t(t) = u_\theta(t) + u_{g\theta}(t) \tag{b}$$

Here u_θ is the rotation of the roof slab relative to the ground and $I_O = m(b^2 + d^2)/12$ is the moment of inertia of the roof slab about the axis normal to the slab passing through its center of mass O. The units of moment of inertia are force $\times$ (length)^2/acceleration. The torque f_S and relative rotation u_θ are related by

$$f_S = k_\theta u_\theta \tag{c}$$

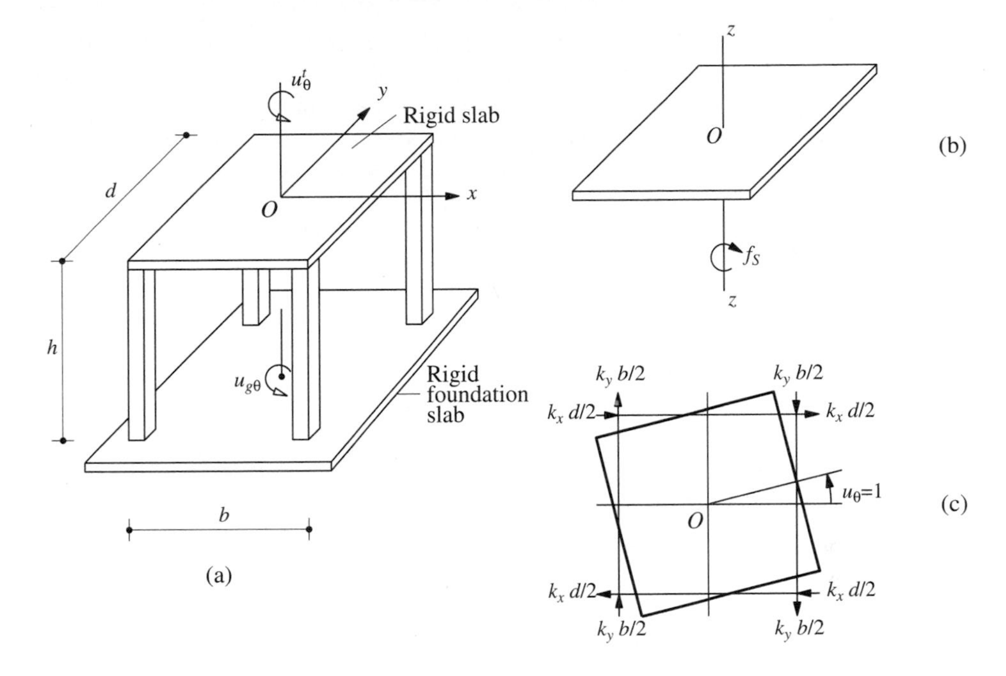

Figure E1.4

where k_θ is the torsional stiffness. To determine k_θ, we introduce a unit rotation, $u_\theta = 1$, and identify the resisting forces in each column (Fig. E1.4c). For a column with both ends clamped, $k_x = 12EI_y/h^3$ and $k_y = 12EI_x/h^3$. The torque required to equilibrate these resisting forces is

$$k_\theta = 4\left(k_x \frac{d}{2}\frac{d}{2}\right) + 4\left(k_y \frac{b}{2}\frac{b}{2}\right) = k_x d^2 + k_y b^2 \tag{d}$$

Substituting Eqs. (c), (d), and (b) in (a) gives

$$I_O \ddot{u}_\theta + (k_x d^2 + k_y b^2) u_\theta = -I_O \ddot{u}_{g\theta} \tag{e}$$

This is the equation governing the relative rotation u_θ of the roof slab due to rotational acceleration $\ddot{u}_{g\theta}$ of the foundation slab.

1.8 PROBLEM STATEMENT AND ELEMENT FORCES

1.8.1 Problem Statement

Given the mass m, the stiffness k of a linearly elastic system, or the force–deformation relation $f_S(u, \dot{u})$ for an inelastic system, the damping coefficient c, and the dynamic excitation—which may be an external force $p(t)$ or ground acceleration $\ddot{u}_g(t)$—a fundamental problem in structural dynamics is to determine the response of an SDF system: the idealized one-story system or the mass–spring–damper system. The term *response* is used in a general sense to include any response quantity, such as displacement, velocity, or acceleration of the mass; also, an internal force or internal stress in the structure. When the excitation is an external force, the response quantities of interest are the displacement or deformation $u(t)$, velocity $\dot{u}(t)$, and acceleration $\ddot{u}(t)$ of the mass. For earthquake excitation, both the total (or absolute) and the relative values of these quantities may be needed. The relative displacement $u(t)$ associated with deformations of the structure is the most important since the internal forces in the structure are directly related to $u(t)$.

1.8.2 Element Forces

Once the deformation response history $u(t)$ has been evaluated by dynamic analysis of the structure, the element forces and stresses needed for structural design can be determined by static analysis of the structure at each instant in time (i.e., no additional dynamic analysis is necessary). This static analysis of a one-story frame can be visualized in two ways:

1. At each instant, the lateral displacement u is known to which joint rotations are related and hence they can be determined; see Eq. (b) of Example 1.1. From the known displacement and rotation of each end of a structural element (beam and column) the element forces (bending moments and shears) can be determined through the element stiffness properties (Appendix 1); and stresses can be obtained from element forces.

2. The second approach is to introduce the *equivalent static force*, a central concept in earthquake response of structures, as we shall see in Chapter 6. At any instant of time t this force f_S is the external force that will produce the deformation u at the same t in the stiffness component of the structure [i.e., the system without mass or damping (Fig. 1.5.2b)]. Thus

$$f_S(t) = ku(t) \tag{1.8.1}$$

where k is the lateral stiffness of the structure. Element forces or stresses can be determined at each time instant by static analysis of the structure subjected to the force f_S determined from Eq. (1.8.1). It is unnecessary to introduce the equivalent static force concept for the mass–spring–damper system because the spring force, also given by Eq. (1.8.1), can readily be visualized.

For inelastic systems the element forces can be determined by appropriate modifications of these procedures to recognize that such systems are typically analyzed by incremental time-stepping procedures (Chapter 5). The change Δu_i in displacement u over a time step t_i to $t_i + \Delta t$ is determined by dynamic analysis. The element forces associated with displacement Δu_i are computed from the linear force–deformation relation valid over the time step by implementing the first of the two approaches mentioned in the preceding paragraph. The displacements and forces at time t_i are combined with their increments over the time step to determine their values at $t_i + \Delta t$.

Why is the lateral force not defined as $f_S(t) + f_D(t) = ku(t) + c\dot{u}(t)$? It is inappropriate to include the velocity-dependent damping force because for structural design the computed element stresses are to be compared with allowable stresses that are specified based on static tests on materials (i.e., tests conducted at slow loading rates).

1.9 COMBINING STATIC AND DYNAMIC RESPONSES

In practical application we need to determine the total forces in a structure, including those existing before dynamic excitation of the structure and those resulting from the dynamic excitation. For a linear system the total forces can be determined by combining the results of two separate analyses: (1) static analysis of the structure due to dead and live loads, temperature changes, and so on; and (2) analysis of dynamic response due to the time-varying excitation. This direct superposition of the results of two analyses is valid only for linear systems.

The analysis of nonlinear systems cannot, however, be separated into two parts. The dynamic analysis of such a system must recognize the forces and deformations already existing in the structure before the onset of dynamic excitation. This is necessary to establish the initial stiffness property of the structure required to start the dynamic analysis.

1.10 METHODS OF SOLUTION OF THE DIFFERENTIAL EQUATION

The equation of motion for a linear SDF system subjected to external force is the second-order differential equation derived earlier:

$$m\ddot{u} + c\dot{u} + ku = p(t) \tag{1.10.1}$$

The initial displacement $u(0)$ and initial velocity $\dot{u}(0)$ at time zero must be specified to define the problem completely. Typically, the structure is at rest before the onset of dynamic excitation, so that the initial velocity and displacement are zero. A brief review of four methods of solution is given in the following sections.

1.10.1 Classical Solution

Complete solution of the linear differential equation of motion consists of the sum of the complementary solution $u_c(t)$ and the particular solution $u_p(t)$, that is, $u(t) = u_c(t) + u_p(t)$. Since the differential equation is of second order, two constants of integration are involved. They appear in the complementary function and are evaluated from a knowledge of the initial conditions.

Example 1.5

Consider a step force: $p(t) = p_o$, $t \geq 0$. In this case, the differential equation of motion for a system without damping (i.e., $c = 0$) is

$$m\ddot{u} + ku = p_o \tag{a}$$

The particular solution for Eq. (a) is

$$u_p(t) = \frac{p_o}{k} \tag{b}$$

and the complementary solution is

$$u_c(t) = A \cos \omega_n t + B \sin \omega_n t \tag{c}$$

where A and B are constants of integration and $\omega_n = \sqrt{k/m}$.

The complete solution is given by the sum of Eqs. (b) and (c):

$$u(t) = A \cos \omega_n t + B \sin \omega_n t + \frac{p_o}{k} \tag{d}$$

If the system is initially at rest, $u(0) = 0$ and $\dot{u}(0) = 0$ at $t = 0$. For these initial conditions the constants A and B can be determined:

$$A = -\frac{p_o}{k} \qquad B = 0 \tag{e}$$

Substituting Eq. (e) in Eq. (d) gives

$$u(t) = \frac{p_o}{k}(1 - \cos \omega_n t) \tag{f}$$

The classical solution will be the principal method we will use in solving the differential equation for free vibration and for excitations, which can be described analytically, such as harmonic, step, and pulse forces.

1.10.2 Duhamel's Integral

Another well-known approach to the solution of linear differential equations, such as the equation of motion of an SDF system, is based on representing the applied force as a sequence of infinitesimally short impulses. The response of the system to an applied force, $p(t)$, at time t is obtained by adding the responses to all impulses up to that time. We develop this method in Chapter 4, leading to the following result for an undamped SDF system:

$$u(t) = \frac{1}{m\omega_n} \int_0^t p(\tau) \sin[\omega_n (t - \tau)]\, d\tau \tag{1.10.2}$$

where $\omega_n = \sqrt{k/m}$. Implicit in this result are "at rest" initial conditions. Equation (1.10.2), known as *Duhamel's integral*, is a special form of the convolution integral found in textbooks on differential equations.

Example 1.6

Using Duhamel's integral, we determine the response of an SDF system, assumed to be initially at rest, to a step force, $p(t) = p_o$, $t \geq 0$. For this applied force, Eq. (1.10.2) specializes to

$$u(t) = \frac{p_o}{m\omega_n} \int_0^t \sin[\omega_n (t - \tau)]\, d\tau = \frac{p_o}{m\omega_n} \left[\frac{\cos \omega_n (t - \tau)}{\omega_n} \right]_{\tau=0}^{\tau=t} = \frac{p_o}{k}(1 - \cos \omega_n t)$$

This result is the same as that obtained in Section 1.10.1 by the classical solution of the differential equation.

Duhamel's integral provides an alternative method to the classical solution if the applied force $p(t)$ is defined analytically by a simple function that permits analytical evaluation of the integral. For complex excitations that are defined only by numerical values of $p(t)$ at discrete time instants, Duhamel's integral can be evaluated by numerical methods. Such methods are not included in this book, however, because more efficient numerical procedures are available to determine dynamic response; some of these are presented in Chapter 5.

1.10.3 Transform Methods

The Laplace and Fourier transforms provide powerful tools for the solution of linear differential equations, in particular the equation of motion for a linear SDF system. Because the two transform methods are similar in concept, here we mention only the Fourier transform method, which leads to the *frequency-domain method* of dynamic analysis.

The Fourier transform $\hat{p}(i\omega)$ of a known excitation function $p(t)$ is defined by

$$\hat{p}(i\omega) = \mathcal{F}[p(t)] = \int_{-\infty}^{\infty} e^{-i\omega t} p(t)\, dt \tag{1.10.3}$$

In solving the equation of motion by Fourier transformation, the first step is to transform the differential equation in variable t into an algebraic equation in the imaginary-valued variable $i\omega$. Then the algebraic equation is readily solved for $\hat{u}(i\omega)$, the transform of $u(t)$. Finally, the solution $u(t)$ of the differential equation is determined by an inverse transformation of $\hat{u}(i\omega)$. The process of inverse transformation is symbolized by

$$u(t) = \frac{1}{2\pi} \int_{-\infty}^{\infty} H(i\omega)\hat{p}(i\omega)e^{i\omega t}\, d\omega \tag{1.10.4}$$

where the complex frequency-response function $H(i\omega)$ describes the response of the system to harmonic excitation. For SDF systems the integral of Eq. (1.10.4) is evaluated by contour integration using the residue theorem of complex analysis. Closed-form results can be obtained if $p(t)$ is a simple function, and application of the Fourier transform method was restricted to such $p(t)$ until high-speed computers became available.

The Fourier transform method is now feasible for the dynamic analysis of linear systems to complicated excitations $p(t)$ or $\ddot{u}_g(t)$ that are described numerically. In such situations, the integrals of both Eqs. (1.10.3) and (1.10.4) are evaluated numerically by the discrete fast Fourier transform (DFFT) computational algorithm developed in the mid-1960s.

The frequency-domain method of dynamic analysis is symbolized by Eqs. (1.10.3) and (1.10.4). The first gives the amplitudes $\hat{p}(i\omega)$ of all the harmonic components that make up the excitation $p(t)$. The second equation can be interpreted as evaluating the response of the system to each component of the excitation and then superposing the harmonic responses to obtain the response $u(t)$. The frequency-domain method, which is an alternative to the *time-domain method* symbolized by Duhamel's integral, is especially useful and powerful for dynamic analysis of structures interacting with

Figure 1.10.1 These two reinforced-concrete dome-shaped containment structures house the nuclear reactors of the San Onofre power plant in California. For design purposes, their fundamental natural vibration period was computed to be 0.15 sec assuming the base as fixed, and 0.50 sec considering soil flexibility. This large difference in the period indicates the important effect of soil–structure interaction for these structures. (Courtesy of D. K. Ostrom.)

unbounded media. Examples are (1) the earthquake response analysis of a structure where the effects of interaction between the structure and the unbounded underlying soil are significant (Fig. 1.10.1), and (2) the earthquake response analysis of concrete dams interacting with the water impounded in the reservoir that extends to great distances in the upstream direction (Fig. 1.10.2). Because earthquake analysis of such complex structure–soil and structure–fluid systems is beyond the scope of this book, the frequency-domain method of dynamic analysis is not included.

1.10.4 Numerical Methods

The preceding three dynamic analysis methods are restricted to linear systems and cannot consider the inelastic behavior of structures anticipated during earthquakes if the ground shaking is intense. The only practical approach for such systems involves numerical

Figure 1.10.2 Morrow Point Dam, a 465-ft-high arch dam, on the Gunnison River, Colorado. Determined by forced vibrations tests, the fundamental natural vibration period of the dam in antisymmetric vibration is 0.268 sec with the reservoir partially full and 0.303 sec with a full reservoir. (Courtesy of U.S Bureau of Reclamation.)

time-stepping methods, which are presented in Chapter 5. These methods are also useful for evaluating the response of linear systems to excitation—applied force $p(t)$ or ground motion $\ddot{u}_g(t)$—which is too complicated to be defined analytically and is described only numerically.

1.11 STUDY OF SDF SYSTEMS: ORGANIZATION

We will study the dynamic response of linearly elastic SDF systems in free vibration (Chapter 2), to harmonic and periodic excitations (Chapter 3), to step and pulse excitations (Chapter 4), and to earthquake ground motion (Chapter 6). Because most structures are designed with the expectation that they will deform beyond the linearly elastic limit during major, infrequent earthquakes, the inelastic response of SDF systems is studied in Chapter 7. The time variation of response $r(t)$ to these various excitations will be of interest. For structural design purposes, the maximum value (over time) of response r

contains the crucial information, for it is related to the maximum forces and deformations that a structure must be able to withstand. We will be especially interested in the peak value of response, or for brevity, *peak response*, defined as the maximum of the absolute value of the response quantity:

$$r_o \equiv \max_t |r(t)| \tag{1.11.1}$$

By definition the peak response is positive; the algebraic sign is dropped because it is usually irrelevant for design. Note that the subscript o attached to a response quantity denotes its peak value.

FURTHER READING

Clough, R. W., and Penzien, J., *Dynamics of Structures*, McGraw-Hill, New York, 1993, Sections 4-3, 6-2, 6-3, and 12-6.

Humar, J. L., *Dynamics of Structures*, Prentice Hall, Englewood Cliffs, N.J., 1990, Chapter 9 and Section 13.5.

APPENDIX 1: STIFFNESS COEFFICIENTS FOR A FLEXURAL ELEMENT

To compute bending moments and shears in a structural element—beam or column—the stiffness coefficients for the element are required. These are presented in Fig. A1.1 for a uniform element of length L, second moment of area I, and elastic modulus E. The stiffness coefficients for joint rotation are shown in part (a) and those for joint translation in part (b) of the figure.

Now consider the uniform element shown in Fig. A1.1c with its two nodes identified as a and b that is assumed to be axially inextensible. Its four degrees of freedom are the nodal displacements u_a and u_b and nodal rotations θ_a and θ_b. The bending moments at the two nodes are

$$M_a = \frac{4EI}{L}\theta_a + \frac{2EI}{L}\theta_b + \frac{6EI}{L^2}u_a - \frac{6EI}{L^2}u_b \tag{A1.1}$$

$$M_b = \frac{2EI}{L}\theta_a + \frac{4EI}{L}\theta_b + \frac{6EI}{L^2}u_a - \frac{6EI}{L^2}u_b \tag{A1.2}$$

The shearing forces at the two nodes are

$$V_a = \frac{12EI}{L^3}u_a - \frac{12EI}{L^3}u_b + \frac{6EI}{L^2}\theta_a + \frac{6EI}{L^2}\theta_b \tag{A1.3}$$

$$V_b = -\frac{12EI}{L^3}u_a + \frac{12EI}{L^3}u_b - \frac{6EI}{L^2}\theta_a - \frac{6EI}{L^2}\theta_b \tag{A1.4}$$

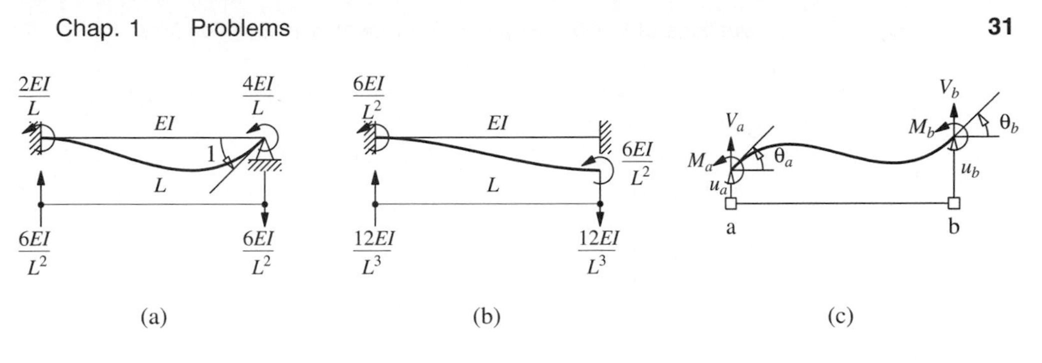

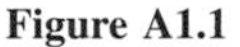

Figure A1.1

At each instant of time, the nodal forces M_a, M_b, V_a, and V_b are calculated from u_a, u_b, θ_a, and θ_b. The bending moment and shear at any other location along the element are determined by statics applied to the element of Fig. A1.1c.

PROBLEMS

1.1– 1.3 Starting from the basic definition of stiffness, determine the effective stiffness of the combined spring and write the equation of motion for the spring–mass systems shown in Figs. P1.1 to P1.3.

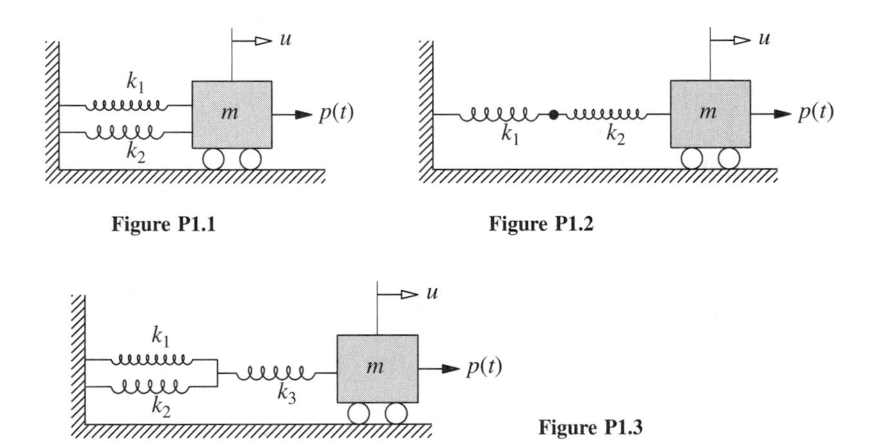

Figure P1.1

Figure P1.2

Figure P1.3

1.4 Develop the equation governing the longitudinal motion of the system of Fig. P1.4. The rod is made of an elastic material with elastic modulus E; its cross-sectional area is A and its length is L. Ignore the mass of the rod and measure u from the static equilibrium position.

1.5 A rigid disk of mass m is mounted at the end of a flexible shaft (Fig. P1.5). Neglecting the weight of the shaft and neglecting damping, derive the equation of free torsional vibration of the disk. The shear modulus (of rigidity) of the shaft is G.

1.6– 1.8 Write the equation governing the free vibration of the systems shown in Figs. P1.6 to P1.8. Assuming the beam to be massless, each system has a single DOF defined as the vertical

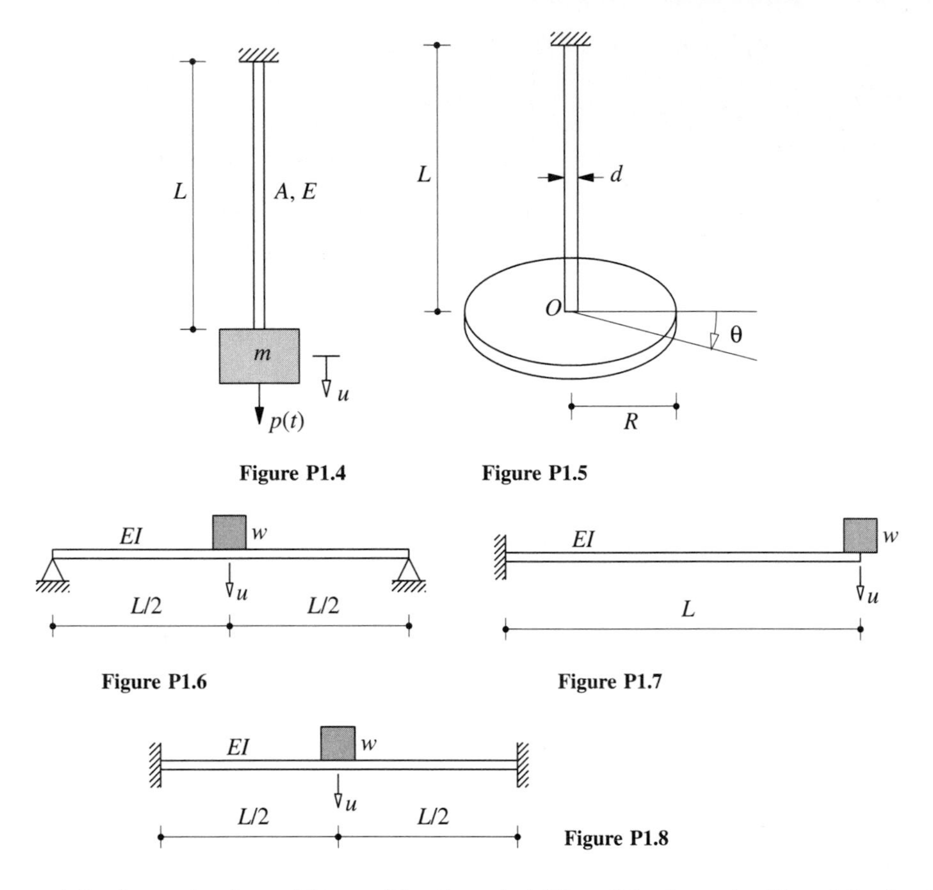

Figure P1.4

Figure P1.5

Figure P1.6

Figure P1.7

Figure P1.8

deflection under the weight w. The flexural rigidity of the beam is EI and the length is L.

1.9– 1.10 Write the equation of motion of the one-story, one-bay frame shown in Figs. P1.9 and P1.10. The flexural rigidity of the beam and columns is as noted. The mass lumped at the beam is m; otherwise, assume the frame to be massless and neglect damping.

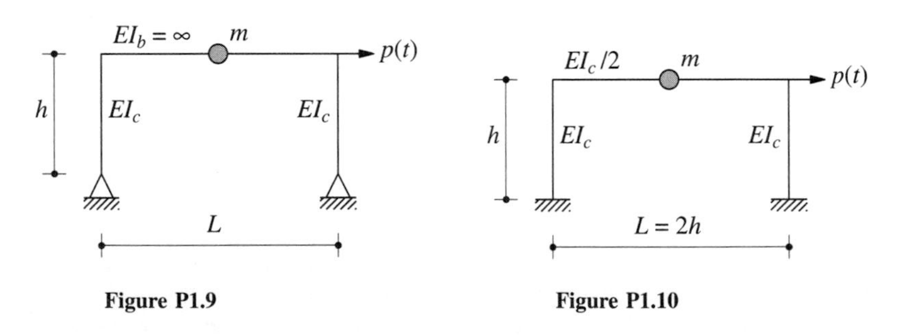

Figure P1.9

Figure P1.10

1.11 A heavy rigid platform of weight w is supported by four columns, hinged at the top and the bottom, and braced laterally in each side panel by two diagonal steel wires as shown in the Fig. P1.11. Each diagonal wire is pretensioned to a high stress; its cross-sectional area is A and elastic modulus is E. Neglecting the mass of the columns and wires, derive the equation of motion governing free vibration in (**a**) the x-direction, and (**b**) the y-direction.

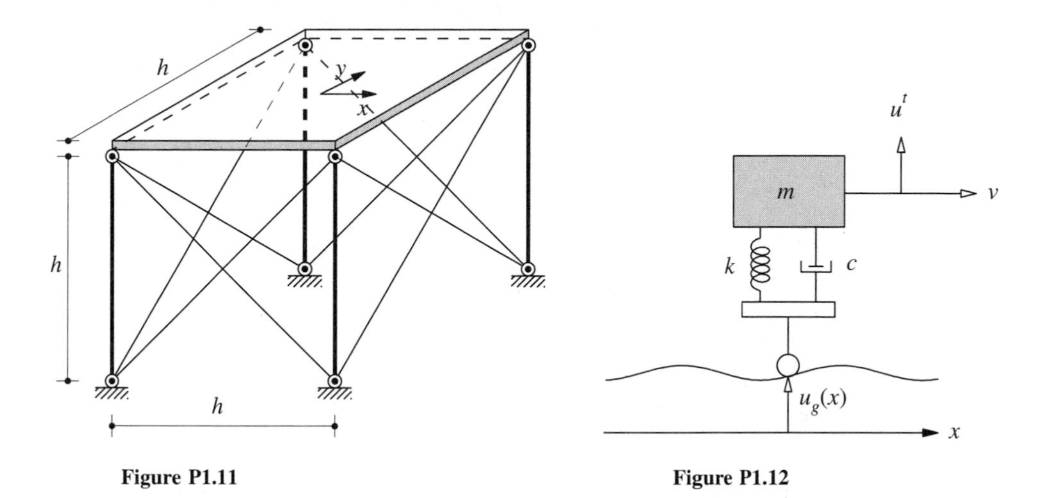

Figure P1.11 **Figure P1.12**

1.12 An automobile is crudely idealized as a lumped mass m supported on a spring–damper system as shown in Fig. P1.12. The automobile travels at constant speed v over a road whose roughness is known as a function of position along the road. Derive the equation of motion.

1.13 Derive the equation of motion governing the torsional vibration of the system of Fig. P1.11 about the vertical axis passing through the center of the platform.

Free Vibration

PREVIEW

A structure is said to be undergoing *free vibration* when it is disturbed from its static equilibrium position and then allowed to vibrate without any external dynamic excitation. In this chapter we study free vibration leading to the notions of the natural vibration frequency and damping ratio for an SDF system. We will see that the rate at which the motion decays in free vibration is controlled by the damping ratio. Thus the analytical results describing free vibration provide a basis to determine the natural frequency and damping ratio of a structure from experimental data of the type shown in Figs. 1.1.4.

Although damping in actual structures is due to several energy-dissipating mechanisms acting simultaneously, a mathematically convenient approach is to idealize them by equivalent viscous damping. Consequently, this chapter deals primarily with viscously damped systems. However, free vibration of systems in the presence of Coulomb friction forces is analyzed toward the end of the chapter.

2.1 UNDAMPED FREE VIBRATION

The motion of linear SDF systems, visualized as an idealized one-story frame or a mass–spring–damper system, subjected to external force $p(t)$ is governed by Eq. (1.5.2). Setting $p(t) = 0$ gives the differential equation governing free vibration of the system, which for systems without damping ($c = 0$) specializes to

$$m\ddot{u} + ku = 0 \qquad (2.1.1)$$

Free vibration is initiated by disturbing the system from its static equilibrium position by imparting the mass some displacement $u(0)$ and velocity $\dot{u}(0)$ at time zero, defined as the instant the motion is initiated:

$$u = u(0) \qquad \dot{u} = \dot{u}(0) \tag{2.1.2}$$

Subject to these initial conditions, the solution to the homogeneous differential equation is obtained by standard methods (see Derivation 2.1):

$$u(t) = u(0)\cos\omega_n t + \frac{\dot{u}(0)}{\omega_n}\sin\omega_n t \tag{2.1.3}$$

where

$$\omega_n = \sqrt{\frac{k}{m}} \tag{2.1.4}$$

Equation (2.1.3) is plotted in Fig. 2.1.1. It shows that the system undergoes vibratory (or oscillatory) motion about its static equilibrium (or undeformed, $u = 0$) position; and that this motion repeats itself after every $2\pi/\omega_n$ seconds. In particular, the state (displacement and velocity) of the mass at two time instants, t_1 and $t_1 + 2\pi/\omega_n$, is identical: $u(t_1) = u(t_1 + 2\pi/\omega_n)$ and $\dot{u}(t_1) = \dot{u}(t_1 + 2\pi/\omega_n)$. These equalities can easily be proved, starting with Eq. (2.1.3). The motion described by Eq. (2.1.3) and shown in Fig. 2.1.1 is known as *simple harmonic motion.*

The portion *a–b–c–d–e* of the displacement–time curve describes one cycle of free vibration of the system. From its static equilibrium (or undeformed) position at *a*, the mass moves to the right, reaching its maximum positive displacement u_o at *b*, at which time the velocity is zero and the displacement begins to decrease and the mass returns

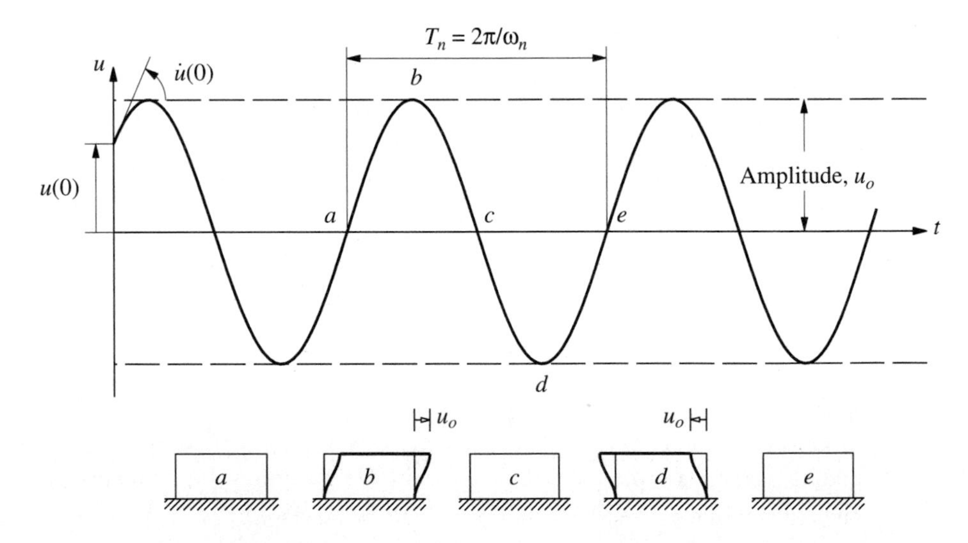

Figure 2.1.1 Free vibration of a system without damping.

back to its equilibrium position c, continues moving to the left, reaching its minimum displacement $-u_o$ at d, and then the displacement decreases again with the mass returning to its equilibrium position at e. At time instant e, $2\pi/\omega_n$ seconds after time instant a, the state (displacement and velocity) of the mass is the same as it was at time instant a, and the mass is ready to begin another cycle of vibration.

The time required for the undamped system to complete one cycle of free vibration is the *natural period of vibration* of the system, which we denote as T_n, in units of seconds. It is related to the *natural circular frequency of vibration*, ω_n, in units of radians per second:

$$T_n = \frac{2\pi}{\omega_n} \tag{2.1.5}$$

A system executes $1/T_n$ cycles in 1 sec. This *natural cyclic frequency of vibration* is denoted by

$$f_n = \frac{1}{T_n} \tag{2.1.6}$$

The units of f_n are hertz (Hz) [cycles per second (cps)]; f_n is obviously related to ω_n through

$$f_n = \frac{\omega_n}{2\pi} \tag{2.1.7}$$

The term *natural frequency of vibration* applies to both ω_n and f_n.

The natural vibration properties ω_n, T_n, and f_n depend only on the mass and stiffness of the structure; see Eqs. (2.1.4) to (2.1.6). The stiffer of two SDF systems having the same mass will have the higher natural frequency and the shorter natural period. Similarly, the heavier (more mass) of two structures having the same stiffness will have the lower natural frequency and the longer natural period. The qualifier *natural* is used in defining T_n, ω_n, and f_n to emphasize the fact that these are natural properties of the system when it is allowed to vibrate freely without any external excitation. Because the system is linear, these vibration properties are independent of the initial displacement and velocity. The natural frequency and period of the various structures of interest to us vary over a wide range, as shown in Figs. 1.10.1, 1.10.2, and 2.1.2a–f.

The natural circular frequency ω_n, natural cyclic frequency f_n, and natural period T_n defined by Eqs. (2.1.4) to (2.1.6) can be expressed in the alternative form

$$\omega_n = \sqrt{\frac{g}{\delta_{st}}} \qquad f_n = \frac{1}{2\pi}\sqrt{\frac{g}{\delta_{st}}} \qquad T_n = 2\pi\sqrt{\frac{\delta_{st}}{g}} \tag{2.1.8}$$

where $\delta_{st} = mg/k$, and where g is the acceleration due to gravity. This is the static deflection of the mass m suspended from a spring of stiffness k; it can be visualized as the system of Fig. 1.6.1 oriented in the vertical direction. In the context of the one-story frame of Fig. 1.2.1, δ_{st} is the lateral displacement of the mass due to lateral force mg.

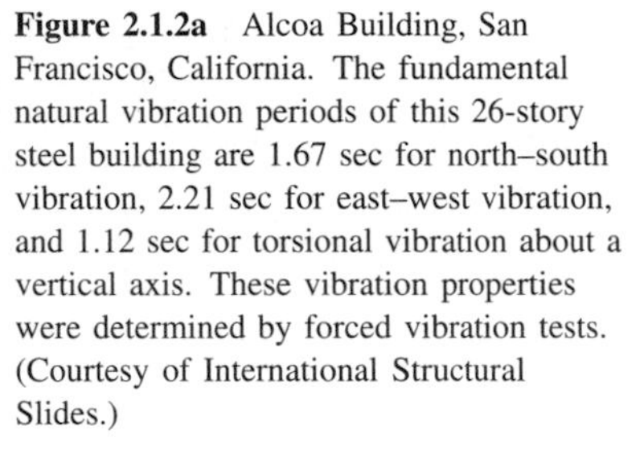

Figure 2.1.2a Alcoa Building, San Francisco, California. The fundamental natural vibration periods of this 26-story steel building are 1.67 sec for north–south vibration, 2.21 sec for east–west vibration, and 1.12 sec for torsional vibration about a vertical axis. These vibration properties were determined by forced vibration tests. (Courtesy of International Structural Slides.)

Figure 2.1.2b Transamerica Building, San Francisco, California. The fundamental natural vibration periods of this 60-story steel building, tapered in elevation, are 2.90 sec for north–south vibration and also for east–west vibration. These vibration properties were determined by forced vibration tests. (Courtesy of International Structural Slides.)

Figure 2.1.2c Medical Center Building, Richmond, California. The fundamental natural vibration periods of this three-story steel frame building are 0.63 sec for vibration in the long direction, 0.74 sec in the short direction, and 0.46 sec for torsional vibration about a vertical axis. These vibration properties were determined from motions of the building recorded during the 1989 Loma Prieta earthquake. (Courtesy of California Strong Motion Instrumentation Program.)

Figure 2.1.2d Pine Flat Dam on the Kings River, near Fresno, California. The fundamental natural vibration period of this 400-ft-high concrete gravity dam was measured by forced vibration tests to be 0.288 sec and 0.306 sec with the reservoir depth at 310 ft and 345 ft, respectively. (Courtesy of D. Rea.)

Figure 2.1.2e Golden Gate Bridge, San Francisco, California. The fundamental natural vibration periods of this suspension bridge with the main span of 4200 ft are 18.2 sec for transverse vibration, 10.9 sec for vertical vibration, 3.81 sec for longitudinal vibration, and 4.43 sec for torsional vibration. These vibration properties were determined from recorded motions of the bridge under ambient (wind, traffic, etc.) conditions. (Courtesy of International Structural Slides.)

Figure 2.1.2f Reinforced-concrete chimney, located in Aramon, France. The fundamental natural vibration period of this 250-m-high chimney is 3.57 sec; it was determined from records of wind-induced vibration. (Courtesy of Chimney Consultants, Inc. and SITES S.A.)

The undamped system oscillates back and forth between the maximum displacement u_o and minimum displacement $-u_o$. The magnitude u_o of these two displacement values is the same; it is called the *amplitude of motion* and given by

$$u_o = \sqrt{[u(0)]^2 + \left[\frac{\dot{u}(0)}{\omega_n}\right]^2} \tag{2.1.9}$$

The amplitude u_o depends on the initial displacement and velocity. Cycle after cycle it remains the same; that is, the motion does not decay. We had mentioned in Section 1.1 this unrealistic behavior of a system if a damping mechanism to represent dissipation of energy is not included.

The natural frequency of the one-story frame of Fig. 1.3.2a with lumped mass m and columns clamped at the base is

$$\omega_n = \sqrt{\frac{k}{m}} \qquad k = \frac{24EI_c}{h^3}\frac{12\rho + 1}{12\rho + 4} \tag{2.1.10}$$

where the lateral stiffness comes from Eq. (1.3.5) and $\rho = I_b/4I_c$. For the extreme cases of a rigid beam, $\rho = \infty$, and a beam with no stiffness, $\rho = 0$, the lateral stiffnesses are given by Eqs. (1.3.2) and (1.3.3) and the natural frequencies are

$$(\omega_n)_{\rho=\infty} = \sqrt{\frac{24EI_c}{mh^3}} \qquad (\omega_n)_{\rho=0} = \sqrt{\frac{6EI_c}{mh^3}} \tag{2.1.11}$$

The natural frequency is doubled as the beam-to-column stiffness ratio, ρ, increases from 0 to ∞; its variation with ρ is shown in Fig. 2.1.3.

The natural frequency is similarly affected by the boundary conditions at the base of the columns. If the columns are hinged at the base rather than clamped and the beam is rigid, $\omega_n = \sqrt{6EI_c/mh^3}$, which is one-half of the natural frequency of the frame with clamped-base columns.

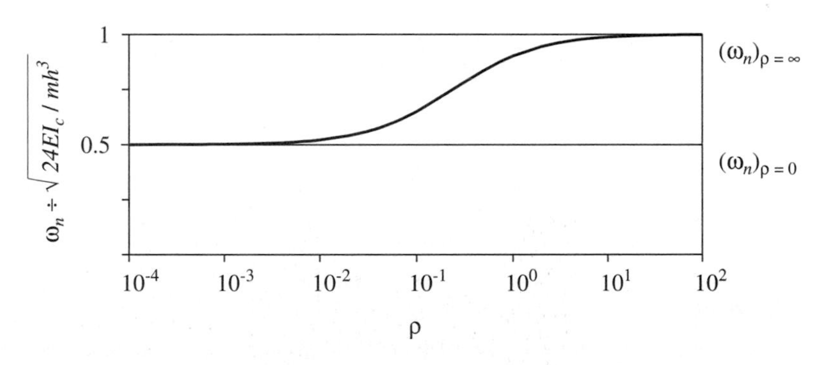

Figure 2.1.3 Variation of natural frequency, ω_n, with beam-to-column stiffness ratio, ρ.

Derivation 2.1

The solution of Eq. (2.1.1), a linear, homogeneous, second-order differential equation with constant coefficients, has the form

$$u = e^{st} \tag{a}$$

where the constant s is unknown. Substitution into Eq. (2.1.1) gives

$$(ms^2 + k)e^{st} = 0$$

The exponential term is never zero, so the characteristic equation is

$$(ms^2 + k) = 0 \qquad s_{1,2} = \pm i\,\omega_n \tag{b}$$

where $i = \sqrt{-1}$. The general solution of Eq. (2.1.1) is

$$u(t) = A_1 e^{s_1 t} + A_2 e^{s_2 t}$$

which after substituting Eq. (b) becomes

$$u(t) = A_1 e^{i\omega_n t} + A_2 e^{-i\omega_n t} \tag{c}$$

where A_1 and A_2 are constants yet undetermined. By using de Moivre's theorem,

$$\cos x = \frac{e^{ix} + e^{-ix}}{2} \qquad \sin x = \frac{e^{ix} - e^{-ix}}{2i}$$

Equation (c) can be rewritten as

$$u(t) = A\cos\omega_n t + B\sin\omega_n t \tag{d}$$

where A and B are constants yet undetermined. Equation (d) is differentiated to obtain

$$\dot{u}(t) = -\omega_n A\sin\omega_n t + \omega_n B\cos\omega_n t \tag{e}$$

Evaluating Eqs. (d) and (e) at time zero gives the constants A and B in terms of the initial displacement $u(0)$ and initial velocity $\dot{u}(0)$:

$$u(0) = A \qquad \dot{u}(0) = \omega_n B \tag{f}$$

Substituting for A and B from Eq. (f) into Eq. (d) leads to the solution given in Eq. (2.1.3).

Example 2.1

For the one-story industrial building of Example 1.2, determine the natural circular frequency, natural cyclic frequency, and natural period of vibration in **(a)** the north–south direction and **(b)** the east–west direction.

Solution (a) *North–south direction:*

$$(\omega_n)_{\text{N-S}} = \sqrt{\frac{38.58}{0.04663}} = 28.73 \text{ rad/sec}$$

$$(T_n)_{\text{N-S}} = \frac{2\pi}{28.73} = 0.219 \text{ sec}$$

$$(f_n)_{\text{N-S}} = \frac{1}{0.219} = 4.57 \text{ Hz}$$

(b) *East–west direction:*

$$(\omega_n)_{\text{E-W}} = \sqrt{\frac{119.6}{0.04663}} = 50.64 \text{ rad/sec}$$

$$(T_n)_{\text{E-W}} = \frac{2\pi}{50.64} = 0.124 \text{ sec}$$

$$(f_n)_{\text{E-W}} = \frac{1}{0.124} = 8.06 \text{ Hz}$$

Observe that the natural frequency is much higher (and the natural period much shorter) in the east–west direction because the vertical bracing makes the system much stiffer, although the columns of the frame are bending about their weak axis; the vibrating mass is the same in both directions.

Example 2.2

Determine the natural cyclic frequency and the natural period of vibration of a weight of 20 lb suspended as described in Example 1.3.

Solution

$$f_n = \frac{1}{2\pi}\sqrt{\frac{g}{\delta_{\text{st}}}} \qquad \delta_{\text{st}} = \frac{w}{k_e} = \frac{20}{13.39} = 1.494 \text{ in.}$$

$$f_n = \frac{1}{2\pi}\sqrt{\frac{386}{1.494}} = 2.56 \text{ Hz}$$

$$T_n = \frac{1}{f_n} = 0.391 \text{ sec}$$

Example 2.3

Consider the system described in Example 1.4 with $b = 30$ ft, $d = 20$ ft, $h = 12$ ft, slab weight $= 0.1$ kip/ft^2, and the lateral stiffnesses of each column in the x and y directions is $k_x = 1.5$ and $k_y = 1.0$, both in kips/in. Determine the natural frequency and period of torsional motion about the vertical axis.

Solution From Example 1.4, the torsional stiffness k_θ and the moment of inertia I_O are

$$k_\theta = k_x d^2 + k_y b^2 = 1.5(12)(20)^2 + 1.0(12)(30)^2 = 18,000 \text{ kip-ft/rad}$$

$$I_O = m\frac{b^2 + d^2}{12} = \frac{0.1(30 \times 20)}{(32.2)}\left[\frac{(30)^2 + (20)^2}{12}\right] = 201.86 \text{ kip-sec}^2\text{-ft}$$

$$\omega_n = \sqrt{\frac{k_\theta}{I_O}} = 9.44 \text{ rad/sec} \qquad f_n = 1.49 \text{ Hz} \qquad T_n = 0.67 \text{ sec}$$

2.2 VISCOUSLY DAMPED FREE VIBRATION

Setting $p(t) = 0$ in Eq. (1.5.2) gives the differential equation governing free vibration of SDF systems with damping:

$$m\ddot{u} + c\dot{u} + ku = 0 \tag{2.2.1a}$$

Dividing by m gives

$$\ddot{u} + 2\zeta\omega_n\dot{u} + \omega_n^2 u = 0 \tag{2.2.1b}$$

where $\omega_n = \sqrt{k/m}$ as defined earlier and

$$\zeta = \frac{c}{2m\omega_n} = \frac{c}{c_{\text{cr}}} \tag{2.2.2}$$

We will refer to

$$c_{\text{cr}} = 2m\omega_n = 2\sqrt{km} = \frac{2k}{\omega_n} \tag{2.2.3}$$

as the *critical damping coefficient* for reasons that will appear shortly; and ζ is the *damping ratio* or *fraction of critical damping*. The damping constant c is a measure of the energy dissipated in a cycle of free vibration or in a cycle of forced harmonic vibration (Section 3.8). However, the damping ratio—a dimensionless measure of damping—is a property of the system that also depends on its mass and stiffness. The differential equation (2.2.1) can be solved by standard methods (similar to Derivation 2.1) for given initial displacement $u(0)$ and velocity $\dot{u}(0)$. Before writing any formal solution, however, we examine the solution qualitatively.

2.2.1 Types of Motion

Figure 2.2.1 shows a plot of the motion $u(t)$ due to initial displacement $u(0)$ for three values of ζ. If $c = c_{\text{cr}}$ or $\zeta = 1$, the system returns to its equilibrium position without oscillating. If $c > c_{\text{cr}}$ $\zeta > 1$, again the system does not oscillate and returns to its equilibrium position, as in the $\zeta = 1$ case, but at a slower rate. If $c < c_{\text{cr}}$ or $\zeta < 1$, the system oscillates about its equilibrium position with a progressively decreasing amplitude.

The damping coefficient c_{cr} is called the *critical damping coefficient* because it is the smallest value of c that inhibits oscillation completely. It represents the dividing line between oscillatory and nonoscillatory motion.

The rest of this presentation is restricted to *underdamped systems* ($c < c_{\text{cr}}$) because structures of interest—buildings, bridges, dams, nuclear power plants, offshore structures,

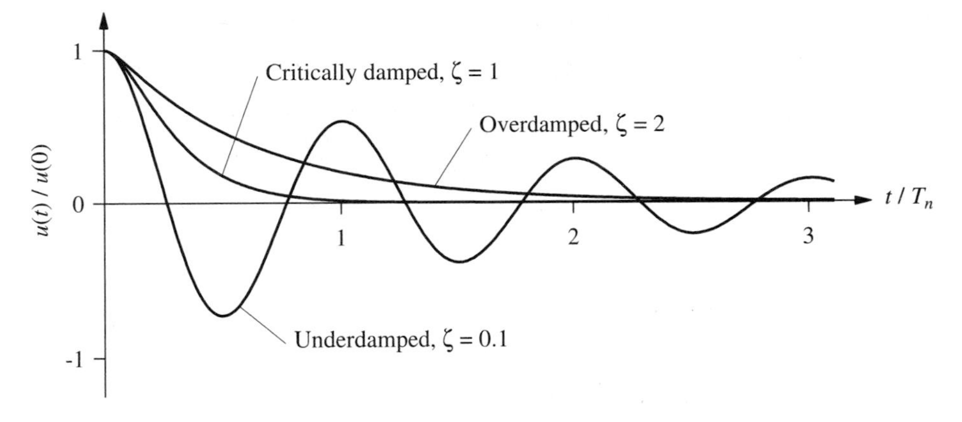

Figure 2.2.1 Free vibration of underdamped, critically damped, and overdamped systems.

etc.—all fall into this category because typically their damping ratio is less than 0.10. Therefore, we have little reason to study the dynamics of *critically damped systems* ($c = c_{cr}$) or *overdamped systems* ($c > c_{cr}$). Such systems do exist, however; for example, recoil mechanisms, such as the common automatic door closer, are overdamped; and instruments used to measure steady-state values, such as a scale measuring dead weight, are usually critically damped. Even for automobile shock absorber systems, however, damping is usually less than half of critical, $\zeta < 0.5$.

2.2.2 Underdamped Systems

The solution to Eq. (2.2.1) subject to the initial conditions of Eq. (2.1.2) for systems with $c < c_{cr}$ or $\zeta < 1$ is (see Derivation 2.2)

$$u(t) = e^{-\zeta\omega_n t}\left[u(0)\cos\omega_D t + \left(\frac{\dot{u}(0) + \zeta\omega_n u(0)}{\omega_D}\right)\sin\omega_D t\right] \tag{2.2.4}$$

where

$$\omega_D = \omega_n\sqrt{1-\zeta^2} \tag{2.2.5}$$

Observe that Eq. (2.2.4) specialized for undamped systems ($\zeta = 0$) reduces to Eq. (2.1.3).

Equation (2.2.4) is plotted in Fig. 2.2.2, which shows the free vibration response of an SDF system with damping ratio $\zeta = 0.05$, or 5%. Included for comparison is the free vibration response of the same system but without damping, presented earlier in Fig. 2.1.1. Free vibration of both systems is initiated by the same initial displacement $u(0)$ and velocity $\dot{u}(0)$, and hence both displacement–time plots start at $t = 0$ with the same ordinate and slope. Equation (2.2.4) and Fig. 2.2.2 indicate that the *natural frequency of damped vibration* is ω_D, and it is related by Eq. (2.2.5) to the natural frequency ω_n of the system without damping. The natural period of damped vibration, $T_D = 2\pi/\omega_D$, is related to the natural period T_n without damping by

$$T_D = \frac{T_n}{\sqrt{1-\zeta^2}} \tag{2.2.6}$$

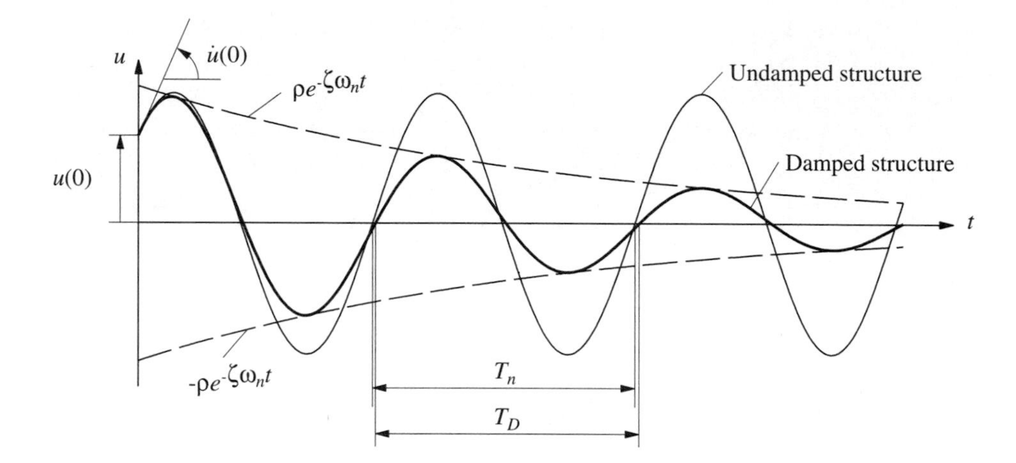

Figure 2.2.2 Effects of damping on free vibration.

The displacement amplitude of the undamped system is the same in all vibration cycles, but the damped system oscillates with amplitude, decreasing with every cycle of vibration. Equation (2.2.4) indicates that the displacement amplitude decays exponentially with time, as shown in Fig. 2.2.2. The envelope curves $\pm\rho e^{-\zeta\omega_n t}$, where

$$\rho = \sqrt{[u(0)]^2 + \left[\frac{\dot{u}(0) + \zeta\omega_n u(0)}{\omega_D}\right]^2} \tag{2.2.7}$$

touch the displacement–time curve at points slightly to the right of its peak values.

Damping has the effect of lowering the natural frequency from ω_n to ω_D and lengthening the natural period from T_n to T_D. These effects are negligible for damping ratios below 20%, a range that includes most structures, as shown in Fig. 2.2.3, where

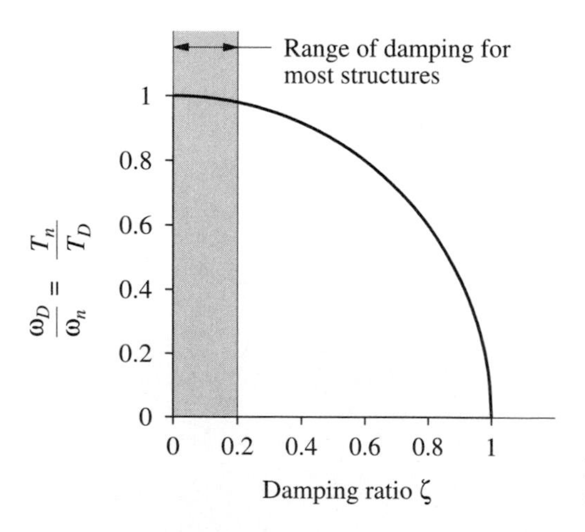

Figure 2.2.3 Effects of damping on the natural vibration frequency.

the ratio $\omega_D/\omega_n = T_n/T_D$ is plotted against ζ. For most structures the damped properties ω_D and T_D are approximately equal to the undamped properties ω_n and T_n, respectively. For systems with critical damping, $\omega_D = 0$ and $T_D = \infty$. This is another way of saying that the system does not oscillate, as shown in Fig. 2.2.1.

The more important effect of damping is on the rate at which free vibration decays. This is displayed in Fig. 2.2.4, where the free vibration due to initial displacement $u(0)$ is plotted for four systems having the same natural period T_n but differing damping ratios: $\zeta = 2$, 5, 10, and 20%.

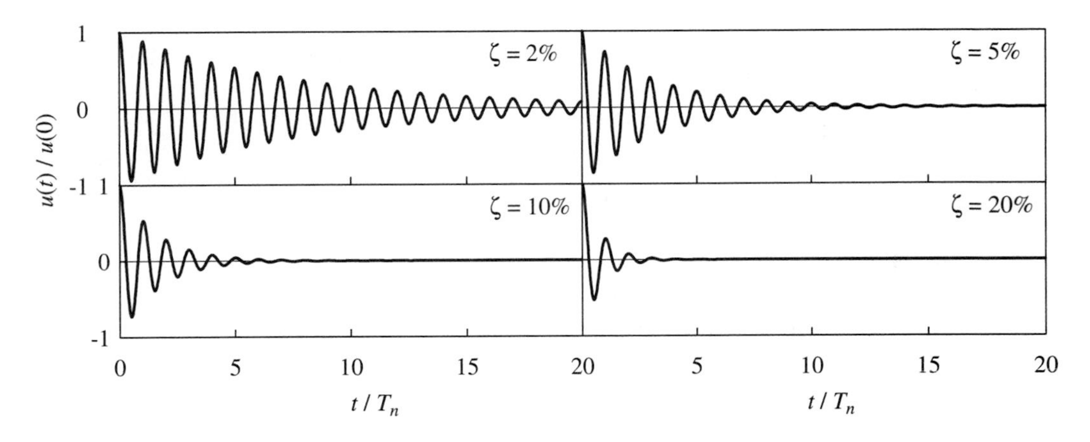

Figure 2.2.4 Free vibration of systems with four levels of damping: $\zeta = 2$, 5, 10, and 20%.

Derivation 2.2

The solution of the differential equation (2.2.1b) has the form

$$u = e^{st} \tag{a}$$

Substitution into Eq. (2.2.1b) gives

$$(s^2 + 2\zeta\omega_n s + \omega_n^2)e^{st} = 0$$

which is satisfied for all values of t if

$$s^2 + 2\zeta\omega_n s + \omega_n^2 = 0 \tag{b}$$

Equation (b), which is known as the *characteristic equation*, has two roots:

$$s_{1,2} = \omega_n\left(-\zeta \pm i\sqrt{1-\zeta^2}\right) \tag{c}$$

Hence the general solution is

$$u(t) = A_1 e^{s_1 t} + A_2 e^{s_2 t}$$

which after substituting Eq. (c) becomes

$$u(t) = e^{-\zeta\omega_n t}\left(A_1 e^{i\omega_D t} + A_2 e^{-i\omega_D t}\right) \tag{d}$$

where A_1 and A_2 are constants as yet undetermined and

$$\omega_D = \omega_n \sqrt{1 - \zeta^2} \tag{e}$$

As in Derivation 2.1, the term in parentheses in Eq. (d) can be rewritten in terms of trigonometric functions to obtain

$$u(t) = e^{-\zeta\omega_n t} \left(A \cos \omega_D t + B \sin \omega_D t\right) \tag{f}$$

where A and B are constants yet undetermined. These can be expressed in terms of the initial conditions by proceeding along the lines of Derivation 2.1:

$$A = u(0) \qquad B = \frac{\dot{u}(0) + \zeta\omega_n u(0)}{\omega_D} \tag{g}$$

Substituting for A and B in Eq. (f) leads to the solution given in Eq. (2.2.4).

2.2.3 Decay of Motion

In this section a relation between the ratio of two successive peaks of damped free vibration and the damping ratio is presented. The ratio of the displacement at time t to its value a full vibration period T_D later is independent of t. Derived from Eq. (2.2.4), this ratio is given by the first equality in

$$\frac{u(t)}{u(t + T_D)} = \exp(\zeta\omega_n T_D) = \exp\left(\frac{2\pi\zeta}{\sqrt{1 - \zeta^2}}\right) \tag{2.2.8}$$

and the second equality is obtained by utilizing Eqs. (2.2.6) and (2.1.5). This result also gives the ratio u_i/u_{i+1} of successive peaks (maxima) shown in Fig. 2.2.5, because these peaks are separated by period T_D:

$$\frac{u_i}{u_{i+1}} = \exp\left(\frac{2\pi\zeta}{\sqrt{1 - \zeta^2}}\right) \tag{2.2.9}$$

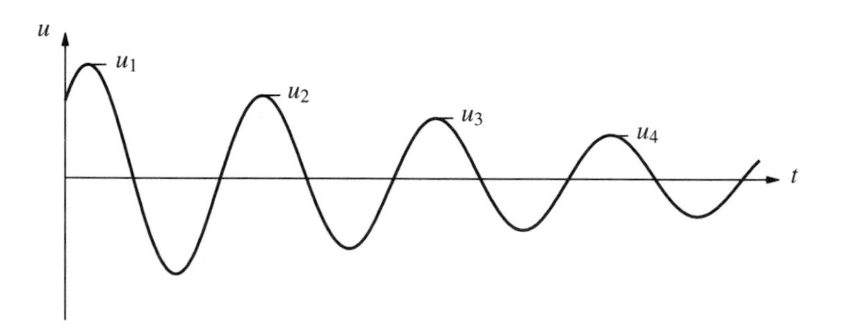

Figure 2.2.5

The natural logarithm of this ratio, called the *logarithmic decrement*, we denote by δ:

$$\delta = \ln \frac{u_i}{u_{i+1}} = \frac{2\pi\zeta}{\sqrt{1-\zeta^2}} \tag{2.2.10}$$

If ζ is small, $\sqrt{1-\zeta^2} \simeq 1$ and this gives an approximate equation

$$\delta \simeq 2\pi\zeta \tag{2.2.11}$$

Figure 2.2.6 shows a plot of the exact and approximate relations between δ and ζ. It is clear that Eq. (2.2.11) is valid for $\zeta < 0.2$, which covers most practical structures.

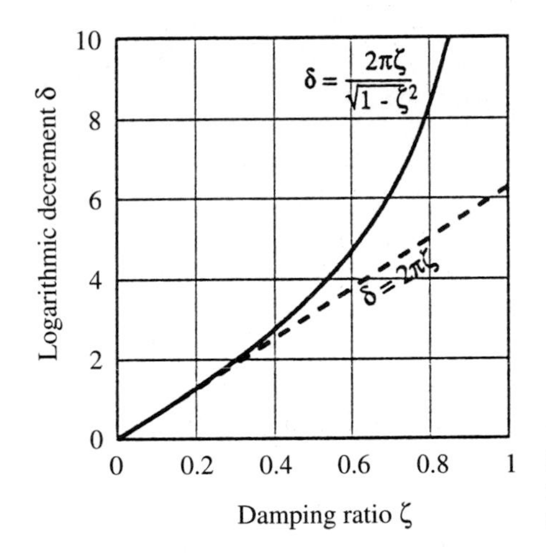

Figure 2.2.6 Exact and approximate relations between logarithmic decrement and damping ratio.

If the decay of motion is slow, as is the case for lightly damped systems such as the aluminum model in Fig. 1.1.4, it is desirable to relate the ratio of two amplitudes several cycles apart, instead of successive amplitudes, to the damping ratio. Over j cycles the motion decreases from u_1 to u_{j+1}. This ratio is given by

$$\frac{u_1}{u_{j+1}} = \frac{u_1}{u_2}\frac{u_2}{u_3}\frac{u_3}{u_4}\cdots\frac{u_j}{u_{j+1}} = e^{j\delta}$$

Therefore,

$$\delta = \frac{1}{j}\ln\frac{u_1}{u_{j+1}} \simeq 2\pi\zeta \tag{2.2.12}$$

To determine the number of cycles elapsed for a 50% reduction in displacement amplitude, we obtain the following relation from Eq. (2.2.12):

$$j_{50\%} \simeq \frac{0.11}{\zeta} \tag{2.2.13}$$

This equation is plotted in Fig. 2.2.7.

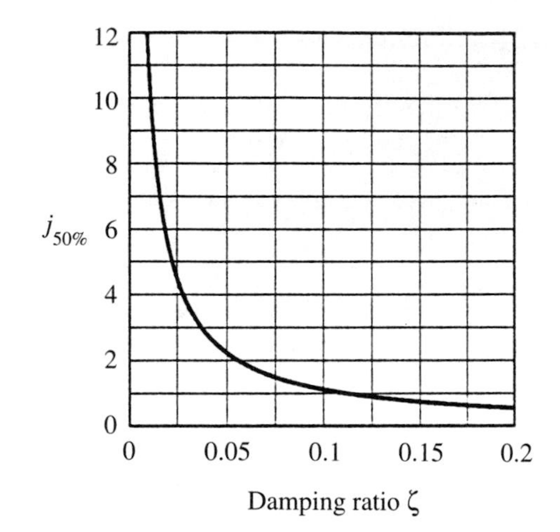

Figure 2.2.7 Number of cycles required to reduce the free vibration amplitude by 50%.

2.2.4 Free Vibration Tests

Because it is not possible to determine analytically the damping ratio ζ for practical structures, this elusive property should be determined experimentally. Free vibration experiments provide one means of determining the damping. Such experiments on two one-story models led to the free vibration records presented in Fig. 1.1.4; a part of such a record is shown in Fig. 2.2.8. For lightly damped systems the damping ratio can be determined from

$$\zeta = \frac{1}{2\pi j} \ln \frac{u_i}{u_{i+j}} \qquad \text{or} \qquad \zeta = \frac{1}{2\pi j} \ln \frac{\ddot{u}_i}{\ddot{u}_{i+j}} \tag{2.2.14}$$

The first of these equations is equivalent to Eq. (2.2.12), which was derived from the equation for $u(t)$. The second is a similar equation in terms of accelerations, which are easier to measure than displacements. It can be shown to be valid for lightly damped systems.

The natural period T_D of the system can also be determined from the free vibration record by measuring the time required to complete one cycle of vibration. Comparing

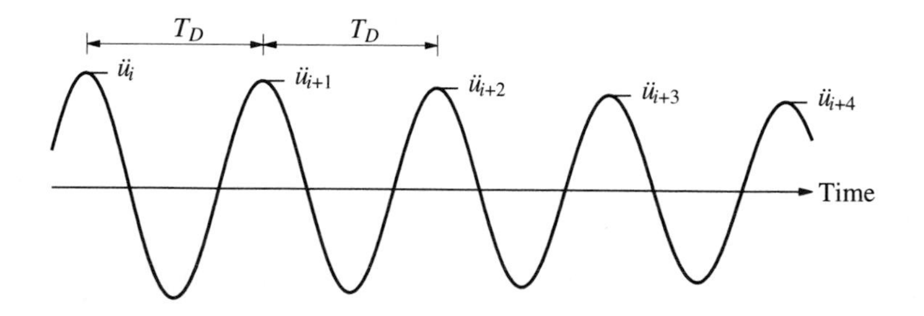

Figure 2.2.8 Acceleration record of a freely vibrating system.

this with the natural period obtained from the calculated stiffness and mass of an idealized system tells us how accurately these properties were calculated and how well the idealization represents the actual structure.

Example 2.4

Determine the natural vibration period and damping ratio of the plexiglass frame model (Fig. 1.1.4a) from the acceleration record of its free vibration shown in Fig. 1.1.4c.

Solution The peak values of acceleration and the time instants they occur can be read from the free vibration record or obtained from the corresponding data stored in a computer during the experiment. The latter provides the following data:

Peak	Time, t_i (sec)	Peak, $\ddot{u}_i$ (g)
1	1.110	0.915
11	3.844	0.076

$$T_n = \frac{3.844 - 1.110}{10} = 0.273 \text{ sec} \qquad \zeta = \frac{1}{2\pi(10)} \ln \frac{0.915\text{g}}{0.076\text{g}} = 0.0396 \text{ or } 3.96\%$$

Example 2.5

A free vibration test is conducted on an empty elevated water tank such as the one in Fig. 1.1.2. A cable attached to the tank applies a lateral (horizontal) force of 16.4 kips and pulls the tank horizontally by 2 in. The cable is suddenly cut and the resulting free vibration is recorded. At the end of four complete cycles, the time is 2.0 sec and the amplitude is 1 in. From these data compute the following: **(a)** damping ratio; **(b)** natural period of undamped vibration; **(c)** effective stiffness; **(d)** effective weight; **(e)** damping coefficient; and **(f)** number of cycles required for the displacement amplitude to decrease to 0.2 in.

Solution **(a)** Assuming small damping:

$$j_{50\%} \simeq \frac{0.11}{\zeta} \qquad \zeta = \frac{0.11}{4} = 0.0275 = 2.75\%$$

Assumption of small damping implicit in Eq. (2.2.13) is valid.

(b) $T_D = \dfrac{2.0}{4} = 0.5 \text{ sec}, \quad T_n \simeq T_D = 0.5 \text{ sec.}$

(c) $k = \dfrac{16.4}{2} = 8.2 \text{ kips/in.}$

(d) $\omega_n = \dfrac{2\pi}{T_n} = \dfrac{2\pi}{0.5} = 12.57,$

$$m = \frac{k}{\omega_n^2} = \frac{8.2}{(12.57)^2} = 0.0519 \text{ kip-sec}^2\text{/in.};$$

$$w = (0.0519)386 = 20.03 \text{ kips.}$$

(e) $c = \zeta(2\sqrt{km}) = 0.0275\left[2\sqrt{8.2(0.0519)}\right] = 0.0359 \text{ kip-sec/in.}$

(f) $\zeta \simeq \dfrac{1}{2\pi j} \ln \dfrac{u_1}{u_{1+j}}, \quad j \simeq \dfrac{1}{2\pi(0.0275)} \ln \dfrac{2}{0.2} = 13.32 \text{ cycles} \sim 13 \text{ cycles.}$

Example 2.6

The weight of water required to fill the tank of Example 2.5 is 80 kips. Determine the natural vibration period and damping ratio of the structure with the tank full.

Solution

$$w = 20.03 + 80 = 100.03 \text{ kips}$$

$$m = \frac{100.03}{386} = 0.2591 \text{ kip-sec}^2\text{/in.}$$

$$T_n = 2\pi\sqrt{\frac{m}{k}} = 2\pi\sqrt{\frac{0.2591}{8.2}} = 1.12 \text{ sec}$$

$$\zeta = \frac{c}{2\sqrt{km}} = \frac{0.0359}{2\sqrt{8.2(0.2591)}} = 0.0123 = 1.23\%$$

2.3 ENERGY IN FREE VIBRATION

The energy input to an SDF system by imparting to it the initial displacement $u(0)$ and initial velocity $\dot{u}(0)$ is

$$E_I = \frac{1}{2}k[u(0)]^2 + \frac{1}{2}m[\dot{u}(0)]^2 \tag{2.3.1}$$

At any instant of time the total energy in a freely vibrating system is made up of two parts, kinetic energy E_K of the mass and potential energy equal to the strain energy E_S of deformation in the spring:

$$E_K = \frac{1}{2}m[\dot{u}(t)]^2 \qquad E_S = \frac{1}{2}k[u(t)]^2 \tag{2.3.2}$$

Substituting $u(t)$ from Eq. (2.1.3) for an undamped system leads to

$$E_K(t) = \frac{1}{2}m\omega_n^2\left[-u(0)\sin\omega_n t + \frac{\dot{u}(0)}{\omega_n}\cos\omega_n t\right]^2 \tag{2.3.3}$$

$$E_S(t) = \frac{1}{2}k\left[u(0)\cos\omega_n t + \frac{\dot{u}(0)}{\omega_n}\sin\omega_n t\right]^2 \tag{2.3.4}$$

The total energy is

$$E_K(t) + E_S(t) = \frac{1}{2}k[u(0)]^2 + \frac{1}{2}m[\dot{u}(0)]^2 \tag{2.3.5}$$

wherein Eq. (2.1.4) has been utilized together with a well-known trigonometric identity. Thus, the total energy is independent of time and equal to the input energy of Eq. (2.3.1), implying conservation of energy during free vibration of a system without damping.

For systems with viscous damping the kinetic energy and potential energy could be determined by substituting $u(t)$ from Eq. (2.2.4) and its derivative $\dot{u}(t)$ into Eq. (2.3.2). The total energy will now be a decreasing function of time because of energy dissipated in viscous damping, which over the time duration 0 to t_1 is

$$E_D = \int_o^u f_D(t)\, du = \int_o^u c\dot{u}\, du = \int_o^{t_1} c\dot{u}^2\, dt \tag{2.3.6}$$

All the input energy will eventually get dissipated in viscous damping; as t_1 goes to ∞, the dissipated energy, Eq. (2.3.6), tends to the input energy, Eq. (2.3.1).

2.4 COULOMB-DAMPED FREE VIBRATION

In Section 1.4 we mentioned that damping in actual structures is due to several energy-dissipating mechanisms acting simultaneously, and a mathematically convenient approach is to idealize them by equivalent viscous damping. Although this approach is sufficiently accurate for practical analysis of most structures, it may not be appropriate when special friction devices have been introduced in a building to reduce its vibrations during earthquakes. Currently, there is much interest in such application and we return to them in Chapter 7. In this section the free vibration of systems under the presence of Coulomb friction forces is analyzed.

Coulomb damping results from friction against sliding of two dry surfaces. The friction force $F = \mu N$, where μ denotes the coefficients of static and kinetic friction, taken to be equal, and N the normal force between the sliding surfaces. The friction force is assumed to be independent of the velocity once the motion is initiated. The direction of the friction force opposes motion, and the sign of the friction force will change when the direction of motion changes. This necessitates formulation and solution of two differential equations, one valid for motion in one direction and the other valid when motion is reversed.

Figure 2.4.1 shows a mass–spring system with the mass sliding against a dry surface, and the free-body diagrams for the mass, including the inertia force, for two directions of motion. The equation governing the motion of the mass from right to left is

$$m\ddot{u} + ku = F \tag{2.4.1}$$

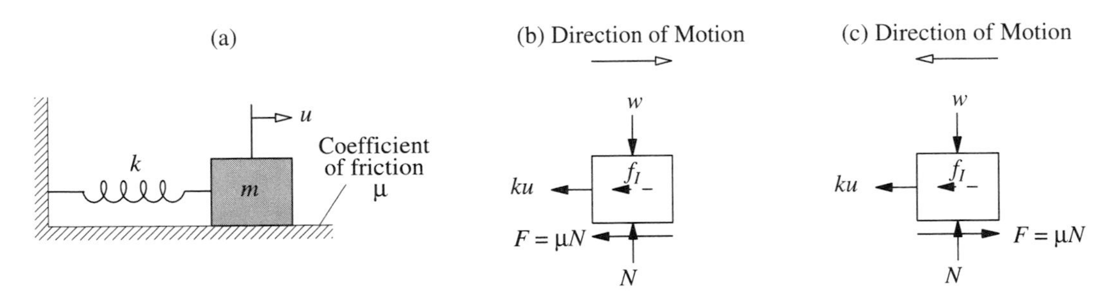

Figure 2.4.1

The lighter the damping, the larger is the number of cycles required to reach a certain percentage of u_o, the steady-state amplitude. For example, the number of cycles required to reach 95% of u_o is 48 for $\zeta = 0.01$, 24 for $\zeta = 0.02$, 10 for $\zeta = 0.05$, 5 for $\zeta = 0.10$, and 2 for $\zeta = 0.20$.

3.2.3 Maximum Deformation and Phase Lag

The steady-state deformation of the system due to harmonic force, described by Eqs. (3.2.3) and (3.2.4), can be rewritten as

$$u(t) = u_o \sin(\omega t - \phi) = \frac{p_o}{k} R_d \sin(\omega t - \phi) \tag{3.2.10}$$

where $u_o = \sqrt{C^2 + D^2}$ and $\phi = \tan^{-1}(-D/C)$. Substituting for C and D gives

$$R_d = \frac{u_o}{(u_{\text{st}})_o} = \frac{1}{\sqrt{[1 - (\omega/\omega_n)^2]^2 + [2\zeta(\omega/\omega_n)]^2}} \tag{3.2.11}$$

$$\phi = \tan^{-1} \frac{2\zeta(\omega/\omega_n)}{1 - (\omega/\omega_n)^2} \tag{3.2.12}$$

Equation (3.2.10) is plotted in Fig. 3.2.5 for three values of ω/ω_n and a fixed value of $\zeta = 0.20$. The values of R_d and ϕ computed from Eqs. (3.2.11) and (3.2.12) are identified. Also shown by dashed lines is the static deformation [Eq. (3.1.8)] due to $p(t)$, which varies with time just as does the applied force, except for the constant k. The steady-state motion is seen to occur at the forcing period $T = 2\pi/\omega$, but with a time lag = $\phi/2\pi$; ϕ is called the *phase angle* or *phase lag*.

A plot of the amplitude of a response quantity against the excitation frequency is called a *frequency-response curve*. Such a plot for deformation u is given by Fig. 3.2.6, wherein R_d [from Eq. (3.2.11)] is plotted as a function of ω/ω_n for a few values of ζ; all the curves are below the $\zeta = 0$ curve in Fig. 3.1.3. Damping reduces R_d and hence the deformation amplitude at all excitation frequencies. The magnitude of this reduction is strongly dependent on the excitation frequency and is examined next for three regions of the excitation-frequency scale:

1. If the frequency ratio $\omega/\omega_n \ll 1$ (i.e., the force is "slowly varying"), R_d is only slightly larger than 1 and is essentially independent of damping. Thus

$$u_o \simeq (u_{\text{st}})_o = \frac{p_o}{k} \tag{3.2.13}$$

This result implies that the dynamic response is essentially the same as the static deformation and is controlled by the stiffness of the system.

2. If $\omega/\omega_n \gg 1$ (i.e., the force is "rapidly varying"), R_d tends to zero as ω/ω_n increases and is essentially unaffected by damping. For large values of ω/ω_n, the $(\omega/\omega_n)^4$

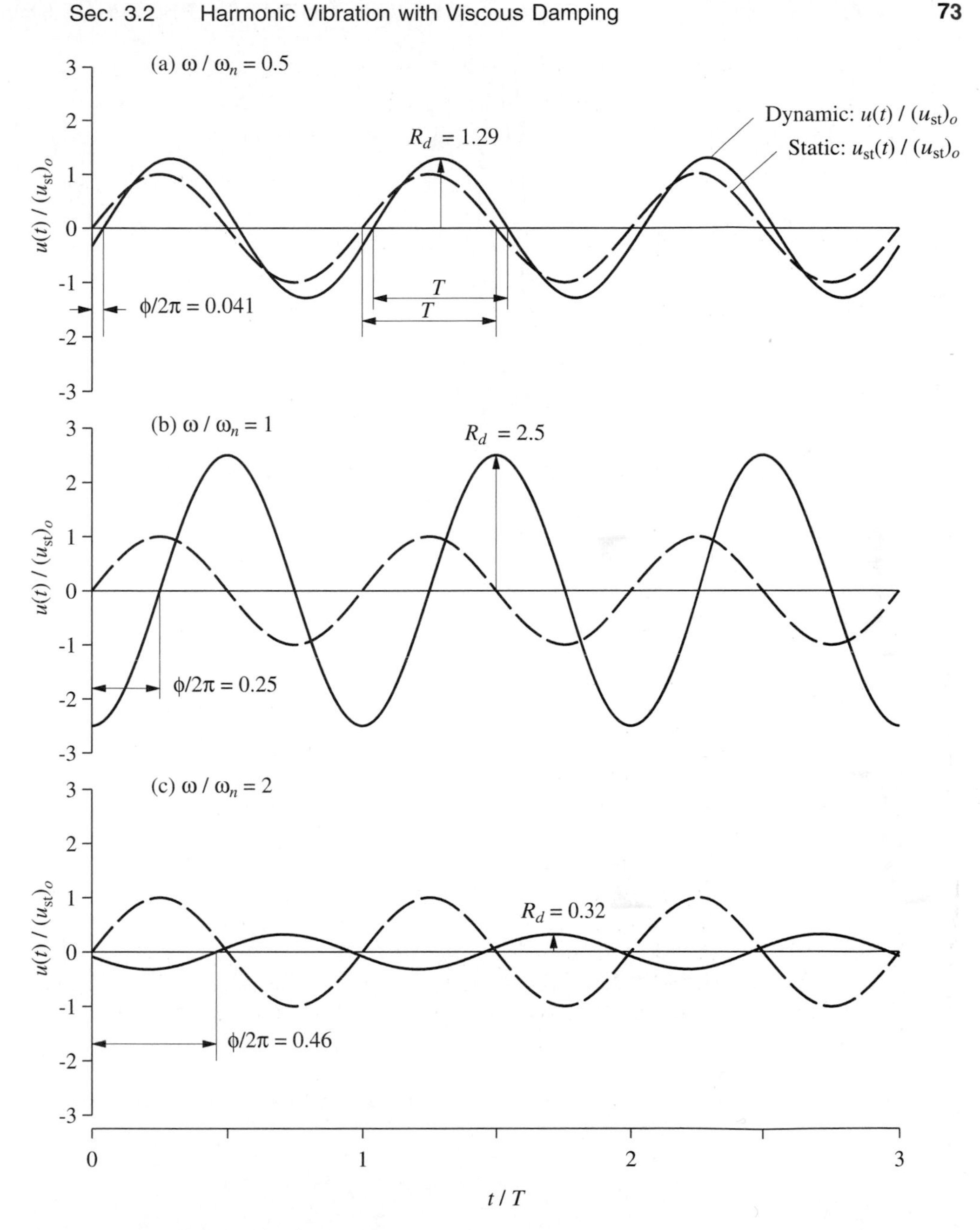

Figure 3.2.5 Steady-state response of damped systems ($\zeta = 0.2$) to sinusoidal force for three values of the frequency ratio: (a) $\omega/\omega_n = 0.5$, (b) $\omega/\omega_n = 1$, (c) $\omega/\omega_n = 2$.

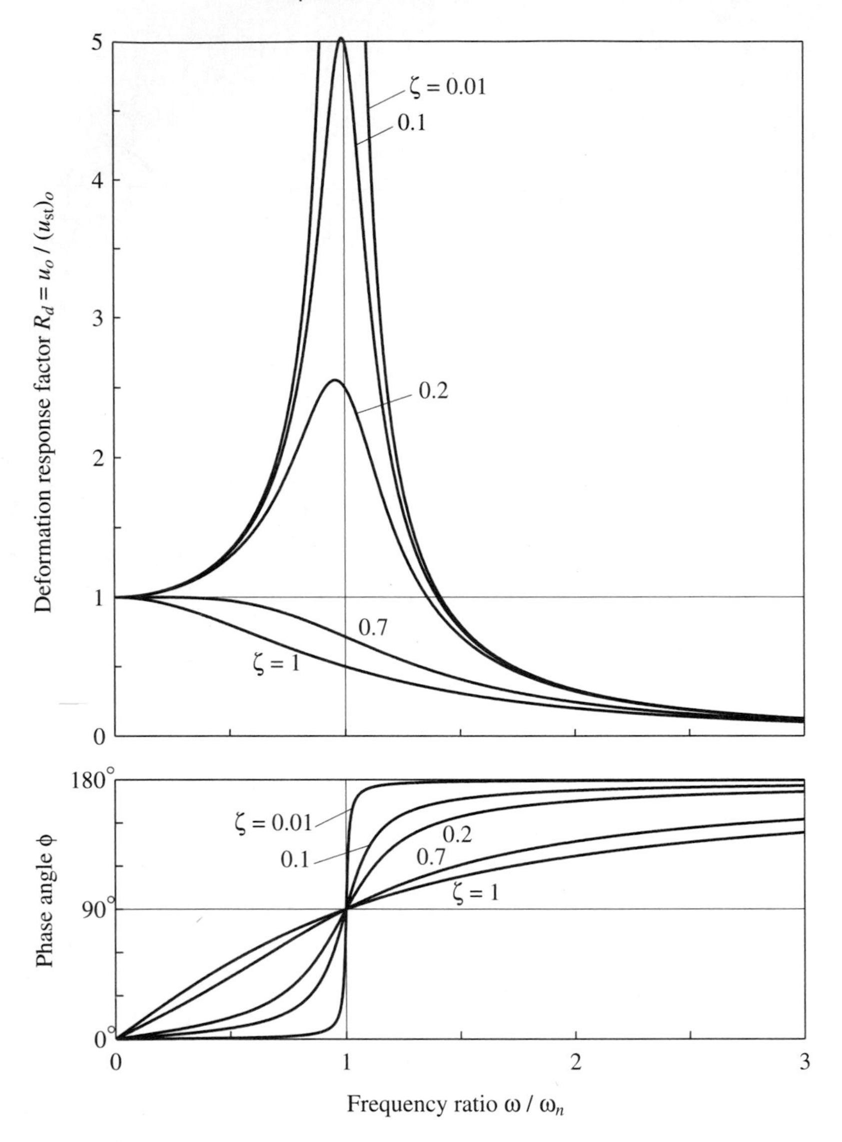

Figure 3.2.6 Deformation response factor and phase angle for a damped system excited by harmonic force.

term is dominant in Eq. (3.2.11), which can be approximated by

$$u_o \simeq (u_{st})_o \frac{\omega_n^2}{\omega^2} = \frac{p_o}{m\omega^2} \tag{3.2.14}$$

This result implies that the response is controlled by the mass of the system.

3. If $\omega/\omega_n \simeq 1$ (i.e., the forcing frequency is close to the natural frequency of the system), R_d is very sensitive to damping and, for the smaller damping values, R_d can be several times larger than 1, implying that the dynamic deformation can be much larger than the static deformation. If $\omega = \omega_n$, Eq. (3.2.11) gives

$$u_o = \frac{(u_{st})_o}{2\zeta} = \frac{p_o}{c\omega_n} \tag{3.2.15}$$

This result implies that the response is controlled by the damping of the system.

The phase angle ϕ, which defines the time by which the response lags behind the force, varies with ω/ω_n as shown in Fig. 3.2.6. It is examined next for the same three regions of the excitation-frequency scale:

1. If $\omega/\omega_n \ll 1$ (i.e., the force is "slowly varying"), ϕ is close to 0° and the displacement is essentially in phase with the applied force, as in Fig. 3.2.5a. When the force in Fig. 1.2.1a acts to the right, the system would also be displaced to the right.

2. If $\omega/\omega_n \gg 1$ (i.e., the force is "rapidly varying"), ϕ is close to 180° and the displacement is essentially out of phase relative to the applied force, as in Fig. 3.2.5c. When the force acts to the right, the system would be displaced to the left.

3. If $\omega/\omega_n = 1$ (i.e., the forcing frequency is equal to the natural frequency), $\phi = 90°$ for all values of ζ, and the displacement attains its peaks when the force passes through zeros, as in Fig. 3.2.5b.

Example 3.1

The displacement amplitude u_o of an SDF system due to harmonic force is known for two excitation frequencies. At $\omega = \omega_n$, $u_o = 5$ in.; at $\omega = 5\omega_n$, $u_o = 0.02$ in. Estimate the damping ratio of the system.

Solution At $\omega = \omega_n$, from Eq. (3.2.15),

$$u_o = (u_{st})_o \frac{1}{2\zeta} = 5 \tag{a}$$

At $\omega = 5\omega_n$, from Eq. (3.2.14),

$$u_o \simeq (u_{st})_o \frac{1}{(\omega/\omega_n)^2} = \frac{(u_{st})_o}{25} = 0.02 \tag{b}$$

From Eq. (b), $(u_{st})_o = 0.5$ in. Substituting in Eq. (a) gives $\zeta = 0.05$.

3.2.4 Dynamic Response Factors

In this section we introduce deformation (or displacement), velocity, and acceleration response factors that are dimensionless and define the amplitude of these three response quantities. The steady-state displacement of Eq. (3.2.10) is repeated for convenience:

$$\frac{u(t)}{p_o/k} = R_d \sin(\omega t - \phi) \tag{3.2.16}$$

where, as defined earlier, the *deformation response factor* R_d is the ratio of the amplitude u_o of the vibratory displacement to the static deformation $(u_{st})_o$.

Differentiating Eq. (3.2.16) gives an equation for the velocity response:

$$\frac{\dot{u}(t)}{p_o/\sqrt{km}} = R_v \cos(\omega t - \phi) \tag{3.2.17}$$

where the *velocity response factor* R_v is related to R_d by

$$R_v = \frac{\omega}{\omega_n} R_d \tag{3.2.18}$$

Differentiating Eq. (3.2.17) gives an equation for the acceleration response:

$$\frac{\ddot{u}(t)}{p_o/m} = -R_a \sin(\omega t - \phi) \tag{3.2.19}$$

where the *acceleration response factor* R_a is related to R_d by

$$R_a = \left(\frac{\omega}{\omega_n}\right)^2 R_d \tag{3.2.20}$$

Observe from Eq. (3.2.19) that R_a is the ratio of the amplitude of the vibratory acceleration to the acceleration due to force p_o acting on the mass.

The dynamic response factors R_d, R_v, and R_a are plotted as functions of ω/ω_n in Fig. 3.2.7. The plots of R_v and R_a are new, but the one for R_d is the same as that in Fig. 3.2.6. As mentioned earlier, the deformation response factor R_d is unity at $\omega/\omega_n = 0$, peaks at $\omega/\omega_n < 1$, and approaches zero as $\omega/\omega_n \to \infty$. The velocity response factor R_v is zero at $\omega/\omega_n = 0$, peaks at $\omega/\omega_n = 1$, and approaches zero as $\omega/\omega_n \to \infty$. The acceleration response factor R_a is zero at $\omega/\omega_n = 0$, peaks at $\omega/\omega_n > 1$, and approaches unity as $\omega/\omega_n \to \infty$. For $\zeta > 1/\sqrt{2}$ no peak occurs for R_d and R_a.

The simple relations among the dynamic response factors

$$\frac{R_a}{\omega/\omega_n} = R_v = \frac{\omega}{\omega_n} R_d \tag{3.2.21}$$

make it possible to present all three factors in a single graph. The R_v–ω/ω_n data in the linear plot of Fig. 3.2.7b is replotted as shown in Fig. 3.2.8 on four-way logarithmic graph paper. The R_d and R_a values can be read from the diagonally oriented logarithmic scales that are different from the vertical scale for R_v. This compact presentation makes it possible to replace the three linear plots of Fig. 3.2.7 by a single plot. The concepts underlying construction of this four-way logarithmic graph paper are presented in Appendix 3.

3.2.5 Resonant Frequencies and Resonant Responses

A *resonant frequency* is defined as the forcing frequency at which the largest response amplitude occurs. Figure 3.2.7 shows that the peaks in the frequency-response curves for displacement, velocity, and acceleration occur at slightly different frequencies. These resonant frequencies can be determined by setting to zero the first derivative of R_d, R_v,

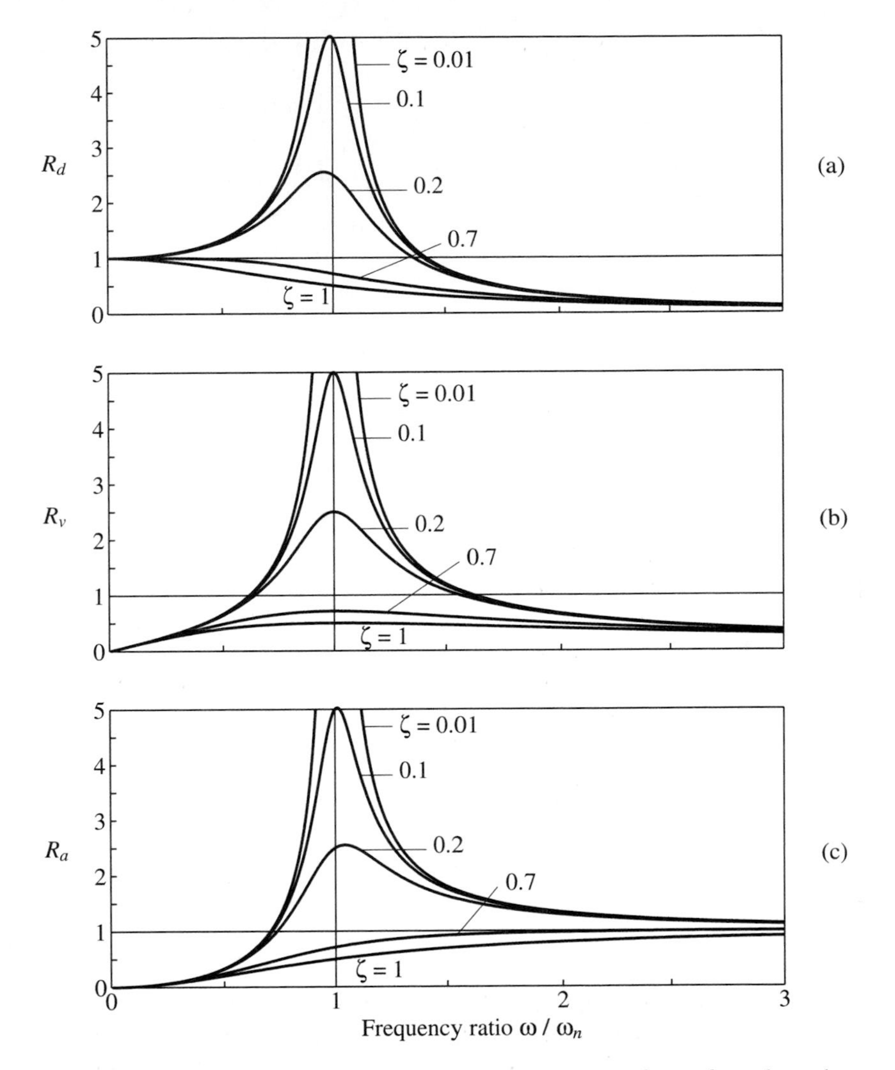

Figure 3.2.7 Deformation, velocity, and acceleration response factors for a damped system excited by harmonic force.

and R_a with respect to ω/ω_n; for $\zeta < 1/\sqrt{2}$ they are:

Displacement resonant frequency:	$\omega_n\sqrt{1-2\zeta^2}$
Velocity resonant frequency:	ω_n
Acceleration resonant frequency:	$\omega_n \div \sqrt{1-2\zeta^2}$

For an undamped system the three resonant frequencies are identical and equal to the natural frequency ω_n of the system. Intuition might suggest that the resonant

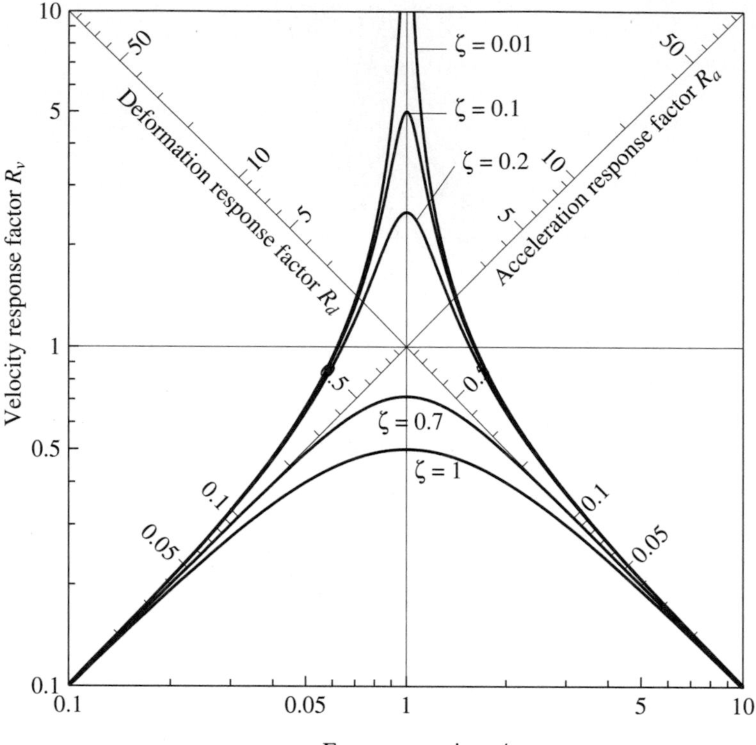

Figure 3.2.8 Four-way logarithmic plot of deformation, velocity, and acceleration response factors for a damped system excited by harmonic force.

frequencies for a damped system should be at its natural frequency $\omega_D = \omega_n\sqrt{1-\zeta^2}$, but this does not happen. The difference is small, however; for the degree of damping usually embodied in structures, typically well below 20%, the differences among the three resonant frequencies and the natural frequency are negligible.

The three dynamic response factors at their respective resonant frequencies are

$$R_d = \frac{1}{2\zeta\sqrt{1-\zeta^2}} \qquad R_v = \frac{1}{2\zeta} \qquad R_a = \frac{1}{2\zeta\sqrt{1-\zeta^2}} \tag{3.2.22}$$

3.2.6 Half-Power Bandwidth

An important property of the frequency response curve for R_d is shown in Fig. 3.2.9, where the *half-power bandwidth* is defined. If ω_a and ω_b are the forcing frequencies on either side of the resonant frequency at which the amplitude u_o is $1/\sqrt{2}$ times the resonant amplitude, then for small ζ

$$\frac{\omega_b - \omega_a}{\omega_n} = 2\zeta \tag{3.2.23}$$

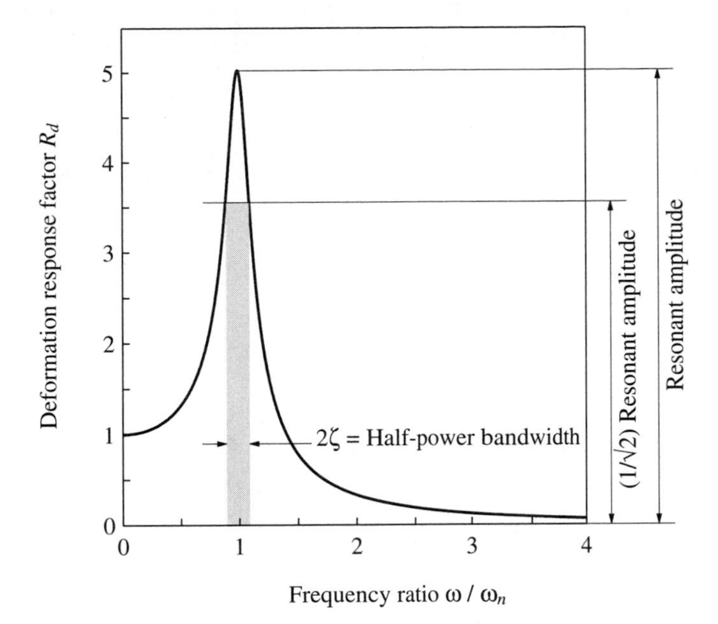

Figure 3.2.9 Definition of half-power bandwidth.

This result, derived in Derivation 3.4, can be rewritten as

$$\zeta = \frac{\omega_b - \omega_a}{2\omega_n} \qquad \text{or} \qquad \zeta = \frac{f_b - f_a}{2f_n} \tag{3.2.24}$$

where $f = \omega/2\pi$ is the cyclic frequency. This important result enables evaluation of damping from forced vibration tests without knowing the applied force (Section 3.4).

Derivation 3.4

Equating R_d from Eq. (3.2.11) and $1/\sqrt{2}$ times the resonant amplitude of R_d given by Eq. (3.2.22), by definition, the forcing frequencies ω_a and ω_b satisfy the condition

$$\frac{1}{\sqrt{\left[1-(\omega/\omega_n)^2\right]^2 + [2\zeta(\omega/\omega_n)]^2}} = \frac{1}{\sqrt{2}}\frac{1}{2\zeta\sqrt{1-\zeta^2}} \tag{a}$$

Inverting both sides, squaring them, and rearranging terms gives

$$\left(\frac{\omega}{\omega_n}\right)^4 - 2(1-2\zeta^2)\left(\frac{\omega}{\omega_n}\right)^2 + 1 - 8\zeta^2(1-\zeta^2) = 0 \tag{b}$$

Equation (b) is a quadratic equation in $(\omega/\omega_n)^2$, the roots of which are

$$\left(\frac{\omega}{\omega_n}\right)^2 = (1-2\zeta^2) \pm 2\zeta\sqrt{1-\zeta^2} \tag{c}$$

where the positive sign gives the larger root ω_b and the negative sign corresponds to the smaller root ω_a.

For the small damping ratios representative of practical structures, the two terms containing ζ^2 can be dropped and

$$\frac{\omega}{\omega_n} \simeq (1 \pm 2\zeta)^{1/2} \tag{d}$$

Taking only the first term in the Taylor series expansion of the right side gives

$$\frac{\omega}{\omega_n} \simeq 1 \pm \zeta \tag{e}$$

Subtracting the smaller root from the larger one gives

$$\frac{\omega_b - \omega_a}{\omega_n} \simeq 2\zeta \tag{f}$$

3.2.7 Steady-State Response to Cosine Force

The differential equation to be solved is

$$m\ddot{u} + c\dot{u} + ku = p_o \cos \omega t \tag{3.2.25}$$

The particular solution given by Eq. (3.2.3) still applies, but in this case the constants C and D are

$$\begin{aligned} C &= \frac{p_o}{k} \frac{2\zeta(\omega/\omega_n)}{\left[1 - (\omega/\omega_n)^2\right]^2 + [2\zeta(\omega/\omega_n)]^2} \\ D &= \frac{p_o}{k} \frac{1 - (\omega/\omega_n)^2}{\left[1 - (\omega/\omega_n)^2\right]^2 + [2\zeta(\omega/\omega_n)]^2} \end{aligned} \tag{3.2.26}$$

These are determined by the procedure of Derivation 3.3. The steady-state response given by Eqs. (3.2.3) and (3.2.26) can be expressed as

$$u(t) = u_o \cos(\omega t - \phi) = (u_{\text{st}})_o R_d \cos(\omega t - \phi) \tag{3.2.27}$$

where the amplitude u_o, the deformation response factor R_d, and the phase angle ϕ are the same as those derived in Section 3.2.3 for a sinusoidal force. This similarity in the steady-state responses to the two harmonic forces is not surprising since the two excitations are the same except for a time shift.

PART B: VISCOUSLY DAMPED SYSTEMS: APPLICATIONS

3.3 RESPONSE TO VIBRATION GENERATOR

Vibration generators (or shaking machines) were developed to provide a source of harmonic excitation appropriate for testing full-scale structures. In this section theoretical results for the steady-state response of an SDF system to a harmonic force caused by a vibration generator are presented. These results provide a basis for evaluating the natural frequency and damping of a structure from experimental data (Section 3.4).

3.3.1 Vibration Generator

Figure 3.3.1 shows a vibration generator having the form of two flat baskets rotating in opposite directions about a vertical axis. By placing various numbers of lead weights in the baskets, the magnitudes of the rotating weights can be altered. The two counterrotating masses, $m_e/2$, are shown schematically in Fig. 3.3.2 as lumped masses with eccentricity = e; their locations at $t = 0$ are shown in (a) and at some time t in (b). The x-components of the inertia forces of the rotating masses cancel out, and the y-components combine to produce a force

$$p(t) = (m_e e \omega^2) \sin \omega t \tag{3.3.1}$$

By bolting the vibration generator to the structure to be excited, this force can be transmitted to the structure. The amplitude of this harmonic force is proportional to the square of the excitation frequency ω. Therefore, it is difficult to generate force at low frequencies and impractical to obtain the static response of a structure.

Figure 3.3.1 Counterrotating eccentric weight vibration generator. (Courtesy of D. E. Hudson.)

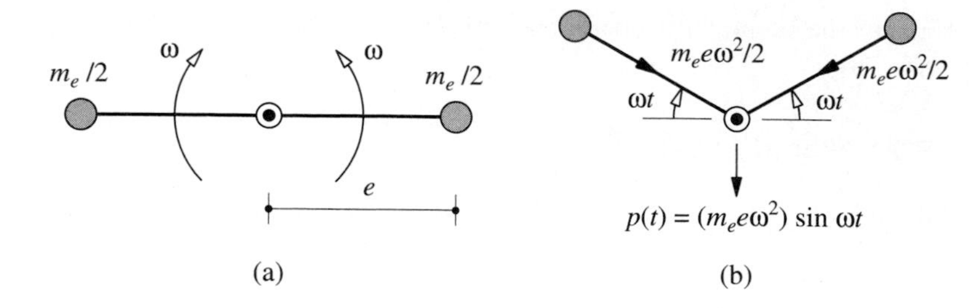

Figure 3.3.2 Vibration generator: (a) initial position; (b) position and forces at time t.

3.3.2 Structural Response

Assuming that the eccentric mass m_e is small compared to the mass m of the structure, the equation governing the motion of an SDF system excited by a vibration generator is

$$m\ddot{u} + c\dot{u} + ku = \left(m_e e\omega^2\right)\sin\omega t \tag{3.3.2}$$

The amplitudes of steady-state displacement and of steady-state acceleration of the SDF system are given by the maximum values of Eqs. (3.2.16) and (3.2.19) with $p_o = m_e e\omega^2$. Thus

$$u_o = \frac{m_e e}{k}\omega^2 R_d = \frac{m_e e}{m}\left(\frac{\omega}{\omega_n}\right)^2 R_d \tag{3.3.3}$$

$$\ddot{u}_o = \frac{m_e e}{m}\omega^2 R_a = \frac{m_e e\omega_n^2}{m}\left(\frac{\omega}{\omega_n}\right)^2 R_a \tag{3.3.4}$$

The acceleration amplitude of Eq. (3.3.4) is plotted as a function of the frequency ratio ω/ω_n in Fig. 3.3.3. For forcing frequencies ω greater than the natural frequency ω_n of the system, the acceleration increases rapidly with increasing ω because the amplitude of the exciting force, Eq. (3.3.1), is proportional to ω^2.

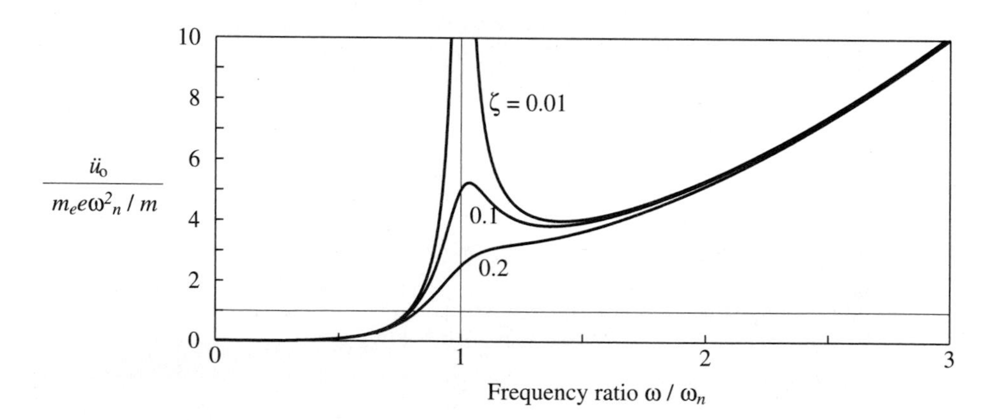

Figure 3.3.3

3.4 NATURAL FREQUENCY AND DAMPING FROM HARMONIC TESTS

The theory of forced harmonic vibration, presented in the preceding sections of this chapter, provides a basis to determine the natural frequency and damping of a structure from its measured response to a vibration generator. The measured damping provides data for an important structural property that cannot be computed from the design of the structure. The measured value of the natural frequency is the "actual" property of a structure against which values computed from the stiffness and mass properties of structural idealizations can be compared. Such research investigations have led to better procedures for developing structural idealizations that are representative of actual structures.

3.4.1 Resonance Testing

The concept of resonance testing is based on the result of Eq. (3.2.15), rewritten as

$$\zeta = \frac{1}{2}\frac{(u_{\mathrm{st}})_o}{(u_o)_{\omega=\omega_n}} \qquad (3.4.1)$$

The damping ratio ζ is calculated from experimentally determined values of $(u_{\mathrm{st}})_o$ and of u_o at forcing frequency equal to the natural frequency of the system.† Usually, the acceleration amplitude is measured and $u_o = \ddot{u}_o/\omega^2$. This seems straightforward except that the true value ω_n of the natural frequency is unknown. The natural frequency is detected experimentally by utilizing the earlier result that the phase angle is 90° if $\omega = \omega_n$. Thus the structure is excited at forcing frequency ω, the phase angle is measured, and the exciting frequency is progressively adjusted until the phase angle is 90°.

If the displacement due to static force p_o—the amplitude of the harmonic force—can be obtained, Eq. (3.4.1) provides the damping ratio. As mentioned earlier, it is difficult for a vibration generator to produce a force at low frequencies and impractical to obtain a significant static force. An alternative is to measure the static response by some other means, such as by pulling on the structure. In this case, Eq. (3.4.1) should be modified to recognize any differences in the force applied in the static test relative to the amplitude of the harmonic force.

3.4.2 Frequency-Response Curve

Because of the difficulty in obtaining the static structural response using a vibration generator, the natural frequency and damping ratio of a structure are usually determined by obtaining the frequency-response curve experimentally. The vibration generator is operated at a selected frequency, the structural response is observed until the transient part damps out, and the amplitude of the steady-state acceleration is measured. The frequency of the vibration generator is adjusted to a new value and the measurements are

†Strictly speaking, this is not the resonant frequency.

repeated. The forcing frequency is varied over a range that includes the natural frequency of the system. A frequency-response curve in the form of acceleration amplitude versus frequency may be plotted directly from the measured data. This curve is for a force with amplitude proportional to ω^2 and would resemble the frequency-response curve of Fig. 3.3.3. If each measured acceleration amplitude is divided by ω^2, we obtain the frequency–acceleration curve for a constant-amplitude force. This curve from measured data would resemble a curve in Fig. 3.2.7c. If the measured accelerations are divided by ω^4, the resulting frequency–displacement curve for a constant-amplitude force would be an experimental version of the curve in Fig. 3.2.7a.

The natural frequency and damping ratio can be determined from any one of the experimentally obtained versions of the frequency-response curves of Figs. 3.3.3, 3.2.7c, and 3.2.7a. For the practical range of damping the natural frequency f_n is essentially equal to the forcing frequency at resonance. The damping ratio is calculated by Eq. (3.2.24) using the frequencies f_a and f_b, determined, as illustrated in Fig. 3.4.1, from the experimental curve shown schematically. Although this equation was derived from the frequency–displacement curve for a constant-amplitude harmonic force, it is approximately valid for the other response curves mentioned earlier as long as the structure is lightly damped.

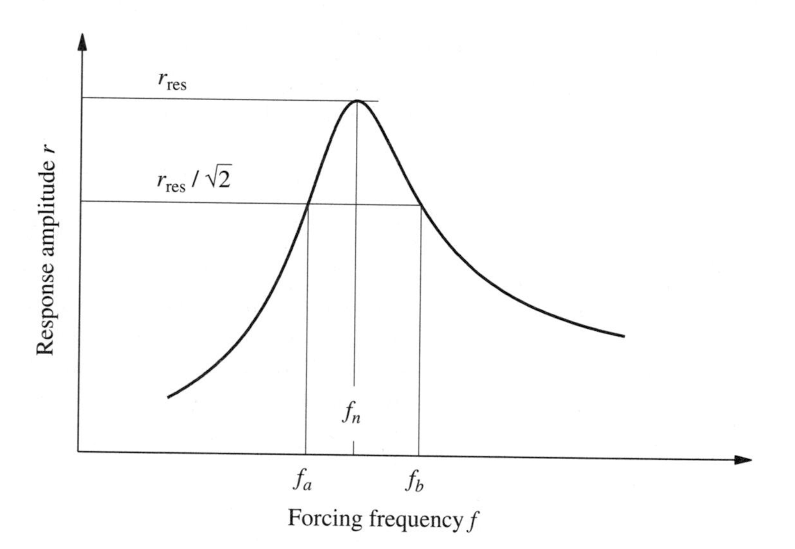

Figure 3.4.1 Evaluating damping from frequency-response curve.

Example 3.2

The plexiglass frame model of Fig. 1.1.4 is mounted on a shaking table which can apply harmonic base motions of specified frequencies and amplitudes. At each excitation frequency ω, acceleration amplitudes $\ddot{u}_{go}$ and $\ddot{u}_o^t$ of the table and the top of the frame, respectively, are recorded. The transmissibility $\text{TR} = \ddot{u}_o^t / \ddot{u}_{go}$ is compiled and the data are plotted in Fig. E3.2. Determine the natural frequency and damping ratio of the plexiglass frame from these data.

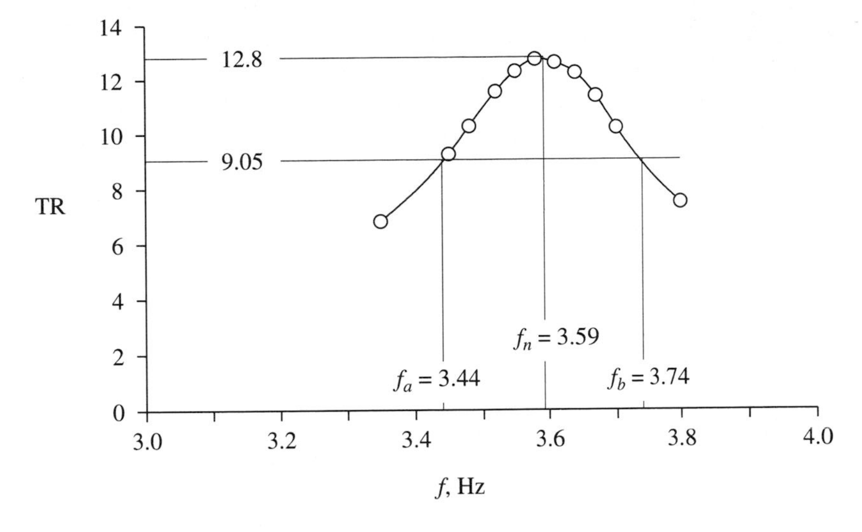

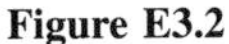

Figure E3.2

Solution The peak of the frequency-response curve occurs at 3.59 Hz. Assuming that the damping is small, the natural frequency $f_n = 3.59$ Hz.

The peak value of the transmissibility curve is 12.8. Now draw a horizontal line at $12.8/\sqrt{2} = 9.05$ as shown. This line intersects the frequency-response curve at $f_b = 3.74$ Hz and $f_a = 3.44$ Hz. Therefore, from Eq. (3.2.24),

$$\zeta = \frac{3.74 - 3.44}{2\,(3.59)} = 0.042 = 4.2\%$$

This damping value is slightly higher than the 3.96% determined from a free vibration test on the model (Example 2.4).

Note that we have used Eq. (3.2.24) to determine the damping ratio of the system from its transmissibility (TR) curve, whereas this equation had been derived from the frequency–displacement curve. This approximation is appropriate because at excitation frequencies in the range f_a to f_b, the numerical values of TR and R_d are close; this is left for the reader to verify after an equation for TR is presented in Section 3.6.

3.5 FORCE TRANSMISSION AND VIBRATION ISOLATION

Consider the mass–spring–damper system shown in the left inset in Fig. 3.5.1 subjected to a harmonic force. The force transmitted to the base is

$$f_T = f_S + f_D = ku(t) + c\dot{u}(t) \tag{3.5.1}$$

Substituting Eq. (3.2.10) for $u(t)$ and Eq. (3.2.17) for $\dot{u}(t)$ and using Eq. (3.2.18) gives

$$f_T(t) = (u_{\text{st}})_o R_d\,[k\sin(\omega t - \phi) + c\omega\cos(\omega t - \phi)] \tag{3.5.2}$$

The maximum value of $f_T(t)$ over t is

$$(f_T)_o = (u_{\text{st}})_o R_d \sqrt{k^2 + c^2\omega^2}$$

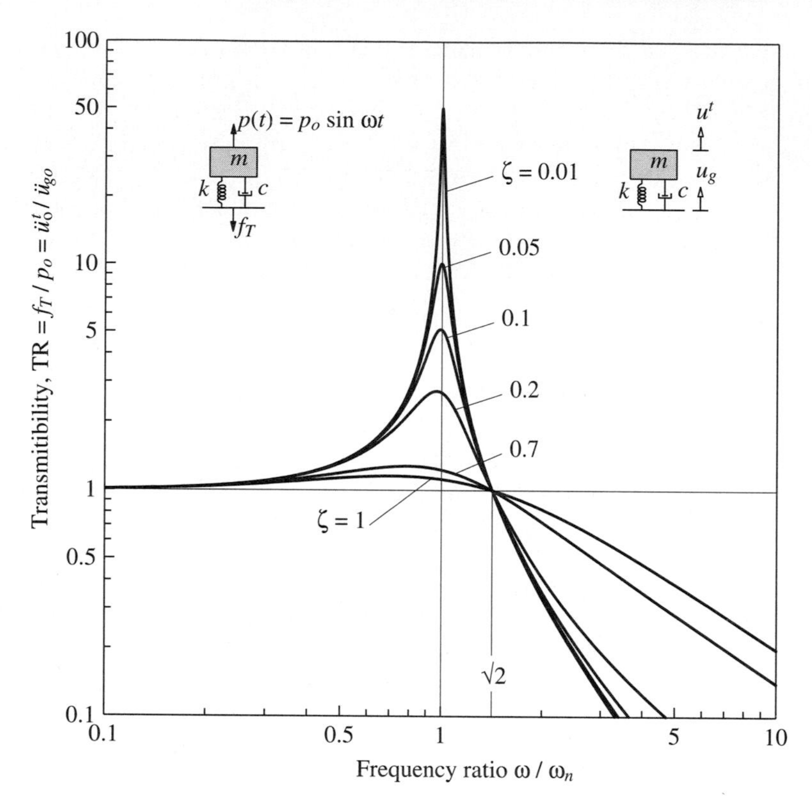

Figure 3.5.1 Transmissibility for harmonic excitation. Force transmissibility and ground motion transmissibility are identical.

which, after using $(u_{\text{st}})_o = p_o/k$ and $\zeta = c/2m\omega_n$, can be expressed as

$$\frac{(f_T)_o}{p_o} = R_d\sqrt{1 + \left(2\zeta\frac{\omega}{\omega_n}\right)^2}$$

Substituting Eq. (3.2.11) for R_d gives an equation for the ratio of the maximum transmitted force to the amplitude p_o of the applied force, known as the *transmissibility* (TR) of the system:

$$\text{TR} = \left\{\frac{1 + [2\zeta(\omega/\omega_n)]^2}{[1 - (\omega/\omega_n)^2]^2 + [2\zeta(\omega/\omega_n)]^2}\right\}^{1/2} \tag{3.5.3}$$

The transmissibility is plotted in Fig. 3.5.1 as a function of the frequency ratio ω/ω_n for several values of the damping ratio ζ. Logarithmic scales have been chosen to highlight the curves for large ω/ω_n, the region of interest. While damping decreases the amplitude of motion at all excitation frequencies (Fig. 3.2.6), damping decreases the transmitted force only if $\omega/\omega_n < \sqrt{2}$. For the transmitted force to be less than the applied force, the stiffness of the support system and hence the natural frequency should be small enough so that $\omega/\omega_n > \sqrt{2}$. No damping is desired in the support system because, in this frequency range, damping increases the transmitted force. This implies a trade-off between a soft spring to reduce the transmitted force and an acceptable static displacement.

If the applied force arises from a rotating machine, its frequency will vary as it starts to rotate and increases its speed to reach the operating frequency. In this case the choice of a flexible support system to minimize the transmitted force must be a compromise. It must have sufficient damping to limit the force transmitted while passing through resonance, but not enough to add significantly to the force transmitted at operating speeds. Luckily, natural rubber is a very satisfactory material and is often used for the isolation of vibration.

3.6 RESPONSE TO GROUND MOTION AND VIBRATION ISOLATION

In this section we determine the response of an SDF system (see the right inset in Fig. 3.5.1) to harmonic ground motion:

$$\ddot{u}_g(t) = \ddot{u}_{go} \sin \omega t \tag{3.6.1}$$

For this excitation the governing equation is Eq. (1.7.4), where the forcing function is $p_{\text{eff}}(t) = -m\ddot{u}_g(t) = -m\ddot{u}_{go} \sin \omega t$, the same as Eq. (3.2.1) for an applied harmonic force with p_o replaced by $-m\ddot{u}_{go}$. Making this substitution in Eq. (3.2.10) gives

$$u(t) = \frac{-m\ddot{u}_{go}}{k} R_d \sin(\omega t - \phi) \tag{3.6.2}$$

The acceleration of the mass is

$$\ddot{u}^t(t) = \ddot{u}_g(t) + \ddot{u}(t) \tag{3.6.3}$$

Substituting Eq. (3.6.1) and the second derivative of Eq. (3.6.2) gives an equation for $\ddot{u}^t(t)$ from which the amplitude or maximum value $\ddot{u}^t_o$ can be determined (see Derivation 3.5):

$$\text{TR} = \frac{\ddot{u}^t_o}{\ddot{u}_{go}} = \left\{ \frac{1 + [2\zeta(\omega/\omega_n)]^2}{[1 - (\omega/\omega_n)^2]^2 + [2\zeta(\omega/\omega_n)]^2} \right\}^{1/2} \tag{3.6.4}$$

The ratio of acceleration $\ddot{u}^t_o$ transmitted to the mass and amplitude $\ddot{u}_{go}$ of ground acceleration is also known as the *transmissibility* (TR) of the system. From Eqs. (3.6.4) and (3.5.3) it is clear that the transmissibility for the ground excitation problem is the same as for the applied force problem.

Therefore, Fig. 3.5.1 also gives the ratio $\ddot{u}^t_o/\ddot{u}_{go}$ as a function of the frequency ratio ω/ω_n. If the excitation frequency ω is much smaller than the natural frequency ω_n of the system, $\ddot{u}^t_o \simeq \ddot{u}_{go}$ (i.e., the mass moves rigidly with the ground, both undergoing the same acceleration). If the excitation frequency ω is much higher than the natural frequency ω_n of the system, $\ddot{u}^t_o \simeq 0$ (i.e., the mass stays still while the ground beneath it moves). This is the basic concept underlying isolation of a mass from a moving base by using a very flexible support system. For example, buildings have been mounted on natural rubber bearings to isolate them from ground-borne vertical vibration—typically with frequencies that range from 25 to 50 Hz—due to rail traffic.

Before closing this section, we mention without derivation the results of a related problem. If the ground motion is defined as $u_g(t) = u_{go} \sin \omega t$, it can be shown that the amplitude u_o^t of the total displacement $u^t(t)$ of the mass is given by

$$\mathrm{TR} = \frac{u_o^t}{u_{go}} = \left\{ \frac{1 + [2\zeta(\omega/\omega_n)]^2}{[1-(\omega/\omega_n)^2]^2 + [2\zeta(\omega/\omega_n)]^2} \right\}^{1/2} \tag{3.6.5}$$

Comparing this with Eq. (3.6.4) indicates that the transmissibility for displacements and accelerations is identical.

Example 3.3

A sensitive instrument with weight 100 lb is to be installed at a location where the vertical acceleration is 0.1g at a frequency of 10 Hz. This instrument is mounted on a rubber pad of stiffness 80 lb/in. and damping such that the damping ratio for the system is 10%. **(a)** What acceleration is transmitted to the instrument? **(b)** If the instrument can tolerate only an acceleration of 0.005g, suggest a solution assuming that the same rubber pad is to be used. Provide numerical results.

Solution **(a)** *Determine TR.*

$$\omega_n = \sqrt{\frac{80}{100/386}} = 17.58 \text{ rad/sec}$$

$$\frac{\omega}{\omega_n} = \frac{2\pi(10)}{17.58} = 3.575$$

Substituting these in Eq. (3.6.4) gives

$$\mathrm{TR} = \frac{\ddot{u}_o^t}{\ddot{u}_{go}} = \sqrt{\frac{1 + [2(0.1)(3.575)]^2}{[1-(3.575)^2]^2 + [2(0.1)(3.575)]^2}} = 0.104$$

Therefore, $\ddot{u}_o^t = (0.104)\ddot{u}_{go} = (0.104)0.1\text{g} = 0.01\text{g}$.

(b) *Determine the added mass to reduce acceleration.* The transmitted acceleration can be reduced by increasing ω/ω_n, which requires reducing ω_n by mounting the instrument on mass m_b. Suppose that we add a mass $m_b = 150$ lb/g; the total mass $= 250$ lb/g, and

$$\omega_n' = \sqrt{\frac{80}{250/386}} = 11.11 \text{ rad/sec} \qquad \frac{\omega}{\omega_n'} = 5.655$$

To determine the damping ratio for the system with added mass, we need the damping coefficient for the rubber pad:

$$c = \zeta(2m\omega_n) = 0.1(2)\left(\frac{100}{386}\right)17.58 = 0.911 \text{ lb-sec/in.}$$

Then

$$\zeta' = \frac{c}{2(m+m_b)\omega_n'} = \frac{0.911}{2(250/386)11.11} = 0.063$$

Substituting for ω/ω_n' and ζ' in Eq. (3.6.4) gives $\ddot{u}_o^t/\ddot{u}_{go} = 0.04$; $\ddot{u}_o^t = 0.004\text{g}$, which is satisfactory because it is less than 0.005g.

Instead of selecting an added mass by judgment, it is possible to set up a quadratic equation for the unknown mass which will give $\ddot{u}_o^t/\ddot{u}_{go} = 0.005\text{g}$.

Example 3.4

An automobile is traveling along a multispan elevated roadway supported every 100 ft. Long-term creep has resulted in a 6-in. deflection at the middle of each span (Fig. E3.4a). The roadway profile can be approximated as sinusoidal with an amplitude of 3 in. and period of 100 ft. The SDF system shown is a simple idealization of an automobile, appropriate for a "first approximation" study of the ride quality of the vehicle. When fully loaded, the weight of the automobile is 4 kips. The effective stiffness of the automobile suspension system is 800 lb/in., and its viscous damping coefficient is such that the damping ratio of the system is 40%. Determine **(a)** the amplitude u_o^t of vertical motion $u^t(t)$ when the automobile is traveling at 40 mph, and **(b)** the speed of the vehicle that would produce a resonant condition.

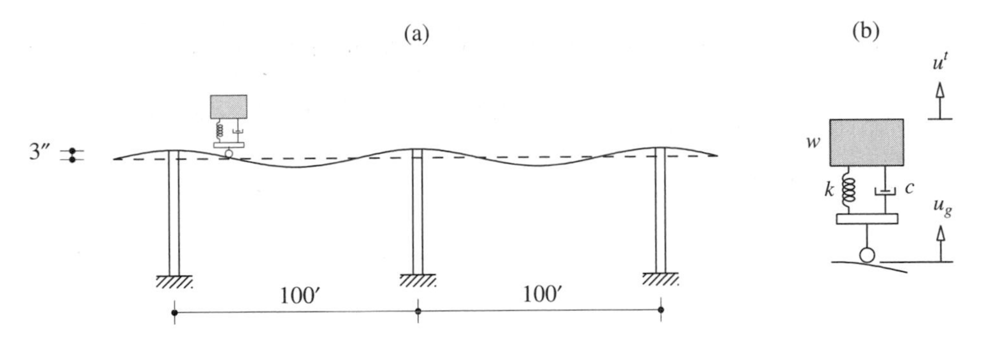

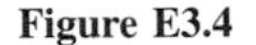
Figure E3.4

Solution Assuming that the tires are infinitely stiff and they remain in contact with the road, the problem can be idealized as shown in Fig. E3.4b. The vertical displacement of the tires is $u_g(t) = u_{go} \sin \omega t$, where $u_{go} = 3$ in. The forcing frequency $\omega = 2\pi/T$, where the forcing period $T = L/v$, the time taken by the automobile to cross the span; therefore, $\omega = 2\pi v/L$.

(a) *Determine* u_o^t.

$$v = 40 \text{ mph} = 58.67 \text{ ft/sec} \qquad \omega = 3.686 \text{ rad/sec}$$

$$\omega_n = \sqrt{\frac{k}{m}} = \sqrt{\frac{800}{4000/386}} = 8.786 \text{ rad/sec} \qquad \frac{\omega}{\omega_n} = 0.420$$

Substituting these data in Eq. (3.6.5) gives

$$\frac{u_o^t}{u_{go}} = \left\{ \frac{1 + [2(0.4)(0.420)]^2}{[1 - (0.420)^2]^2 + [2(0.4)(0.420)]^2} \right\}^{1/2} = 1.186$$

$$u_o^t = 1.186 u_{go} = 1.186(3) = 3.56 \text{ in.}$$

(b) *Determine the speed at resonance.* If ζ were small, resonance would occur approximately at $\omega/\omega_n = 1$. However, automobile suspensions have heavy damping, to reduce vibration. In this case, $\zeta = 0.4$, and for such large damping the resonant frequency is significantly different from ω_n. By definition, resonance occurs when TR (or TR^2) is maximum over all ω. Substituting $\zeta = 0.4$ in Eq. (3.6.5) and introducing $\beta = \omega/\omega_n$ gives

$$\mathrm{TR}^2 = \frac{1 + 0.64\beta^2}{(1 - 2\beta^2 + \beta^4) + 0.64\beta^2} = \frac{1 + 0.64\beta^2}{\beta^4 - 1.36\beta^2 + 1}$$

$$\frac{d(\mathrm{TR})^2}{d\beta} = 0 \Rightarrow \beta = 0.893 \Rightarrow \omega = 0.893\omega_n = 0.893(8.786) = 7.84 \text{ rad/sec}$$

Resonance occurs at this forcing frequency, which implies a speed of

$$v = \frac{\omega L}{2\pi} = \frac{(7.84)100}{2\pi} = 124.8 \text{ ft/sec} = 85 \text{ mph}$$

Example 3.5

Repeat part (a) of Example 3.4 if the vehicle is empty (driver only) with a total weight of 3 kips.

Solution Since the damping coefficient c does not change but the mass m does, we need to recompute the damping ratio for an empty vehicle from

$$c = 2\zeta_f\sqrt{km_f} = 2\zeta_e\sqrt{km_e}$$

where the subscripts f and e denote full and empty conditions, respectively. Thus

$$\zeta_e = \zeta_f\left(\frac{m_f}{m_e}\right)^{1/2} = 0.4\left(\frac{4}{3}\right)^{1/2} = 0.462$$

For an empty vehicle

$$\omega_n = \sqrt{\frac{k}{m}} = \sqrt{\frac{800}{3000/386}} = 10.15 \text{ rad/sec}$$

$$\frac{\omega}{\omega_n} = \frac{3.686}{10.15} = 0.363$$

Substituting for ω/ω_n and ζ in Eq. (3.6.5) gives

$$\frac{u_o^t}{u_{go}} = \left\{\frac{1 + [2(0.462)(0.363)]^2}{[1 - (0.363)^2]^2 + [2(0.462)(0.363)]^2}\right\}^{1/2} = 1.133$$

$$u_o^t = 1.133u_{go} = 1.133(3) = 3.40 \text{ in.}$$

Derivation 3.5

Equation (3.6.2) is first rewritten as a linear combination of sine and cosine functions. This can be accomplished by substituting Eqs. (3.2.11) and (3.2.12) for the R_d and ϕ, respectively, or by replacing p_o in Eq. (3.2.4) with $-m\ddot{u}_{go}$ and substituting in Eq. (3.2.3). Either way the relative displacement is

$$u(t) = \frac{-m\ddot{u}_{go}}{k}\left\{\frac{[1 - (\omega/\omega_n)^2]\sin\omega t - [2\zeta(\omega/\omega_n)]\cos\omega t}{[1 - (\omega/\omega_n)^2]^2 + [2\zeta(\omega/\omega_n)]^2}\right\} \quad \text{(a)}$$

Differentiating this twice and substituting it in Eq. (3.6.3) together with Eq. (3.6.1) gives

$$\ddot{u}^t(t) = \ddot{u}_{go}(C_1 \sin\omega t + D_1 \cos\omega t) \quad \text{(b)}$$

where

$$C_1 = \frac{1 - (\omega/\omega_n)^2 + 4\zeta^2(\omega/\omega_n)^2}{[1 - (\omega/\omega_n)^2]^2 + [2\zeta(\omega/\omega_n)]^2} \qquad D_1 = \frac{-2\zeta(\omega/\omega_n)^3}{[1 - (\omega/\omega_n)^2]^2 + [2\zeta(\omega/\omega_n)]^2} \tag{c}$$

The acceleration amplitude is

$$\ddot{u}_o^t = \ddot{u}_{go}\sqrt{C_1^2 + D_1^2} \tag{d}$$

This result, after substituting for C_1 and D_1 from Eq. (c) and some simplification, leads to Eq. (3.6.4).

3.7 VIBRATION-MEASURING INSTRUMENTS

Measurement of vibration is of great interest in many aspects of structural engineering. For example, measurement of ground shaking during an earthquake provides basic data for earthquake engineering, and records of the resulting motions of a structure provide insight into how structures respond during earthquakes. Although measuring instruments are highly developed and intricate, the basic element of these instruments is some form of a transducer. In its simplest form a transducer is a mass–spring–damper system mounted inside a rigid frame that is attached to the surface whose motion is to be measured. Figure 3.7.1 shows a schematic drawing of such an instrument to record the horizontal motion of a support point; three separate transducers are required to measure the three components of motion. When subjected to motion of the support point, the transducer mass moves relative to the frame, and this relative displacement is recorded after suitable magnification. It is the objective of this brief presentation to discuss the principle underlying the design of vibration-measuring instruments so that the measured relative displacement provides the desired support motion—acceleration or displacement.

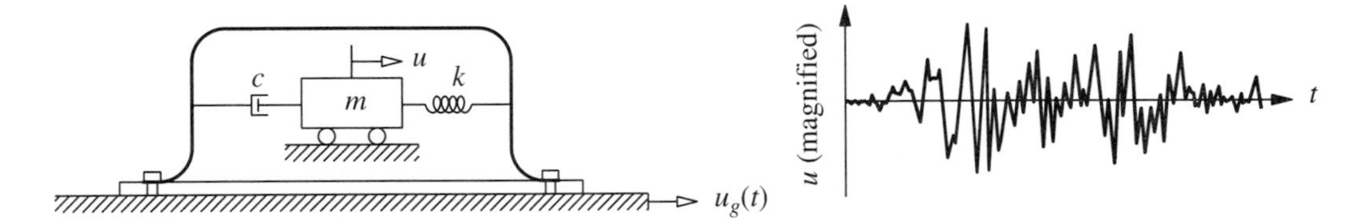

Figure 3.7.1 Schematic drawing of a vibration-measuring instrument and recorded motion.

3.7.1 Measurement of Acceleration

The motion to be measured generally varies arbitrarily with time and may include many harmonic components covering a wide range of frequencies. It is instructive, however, to consider first the measurement of simple harmonic motion described by Eq. (3.6.1). The displacement of the instrument mass relative to the moving frame is given by Eq. (3.6.2), which can be rewritten as

$$u(t) = -\left[\left(\frac{1}{\omega_n^2}\right) R_d\right] \ddot{u}_g\left(t - \frac{\phi}{\omega}\right) \tag{3.7.1}$$

The recorded $u(t)$ is the base acceleration modified by a factor $-R_d/\omega_n^2$ and recorded with a time lag ϕ/ω. As shown in Fig. 3.2.6, R_d and ϕ vary with the forcing frequency ω, but ω_n^2 is an instrument constant independent of the support motion.

The object of the instrument design is to make R_d and ϕ/ω as independent of excitation frequency as possible because then each harmonic component of acceleration will be recorded with the same proportionality factor and the same time lag. Then, even if the motion to be recorded consists of many harmonic components, the recorded $u(t)$ will have the same shape as the support motion with a constant shift of time. This constant time shift simply moves the time scale a little, which is usually not important. According to Fig. 3.7.2 (which is a magnified plot of Fig. 3.2.6 with additional damping values), if $\zeta = 0.7$, then over the frequency range $0 \leq \omega/\omega_n \leq 0.50$, R_d is close to 1 (less than 2.5% error) and the variation of ϕ with ω is close to linear, implying that ϕ/ω is essentially constant. Thus an instrument with a natural frequency of 50 Hz and damping ratio of 0.7 has a useful frequency range from 0 to 25 Hz with negligible error. These are the properties of modern, commercially available instruments designed to measure earthquake-induced ground acceleration. Because the measured amplitude of

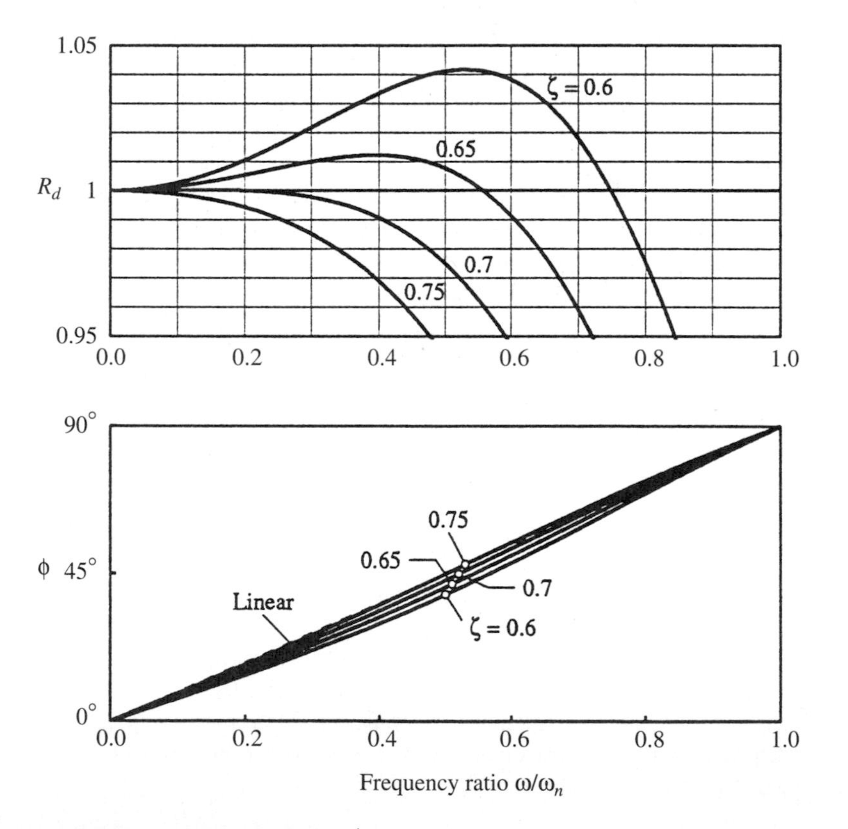

Figure 3.7.2 Variation of R_d and ϕ with frequency ratio ω/ω_n for $\zeta = 0.6$, 0.65, 0.7, and 0.75.

$u(t)$ is proportional to R_d/ω_n^2, a high-frequency instrument will result in a very small displacement that is substantially magnified in these instruments for proper measurement.

Figure 3.7.3 shows a comparison of the actual ground acceleration $\ddot{u}_g(t) = 0.1\text{g}\sin(2\pi ft)$ and the measured relative displacement of $R_d\ddot{u}_g(t - \phi/\omega)$, except for

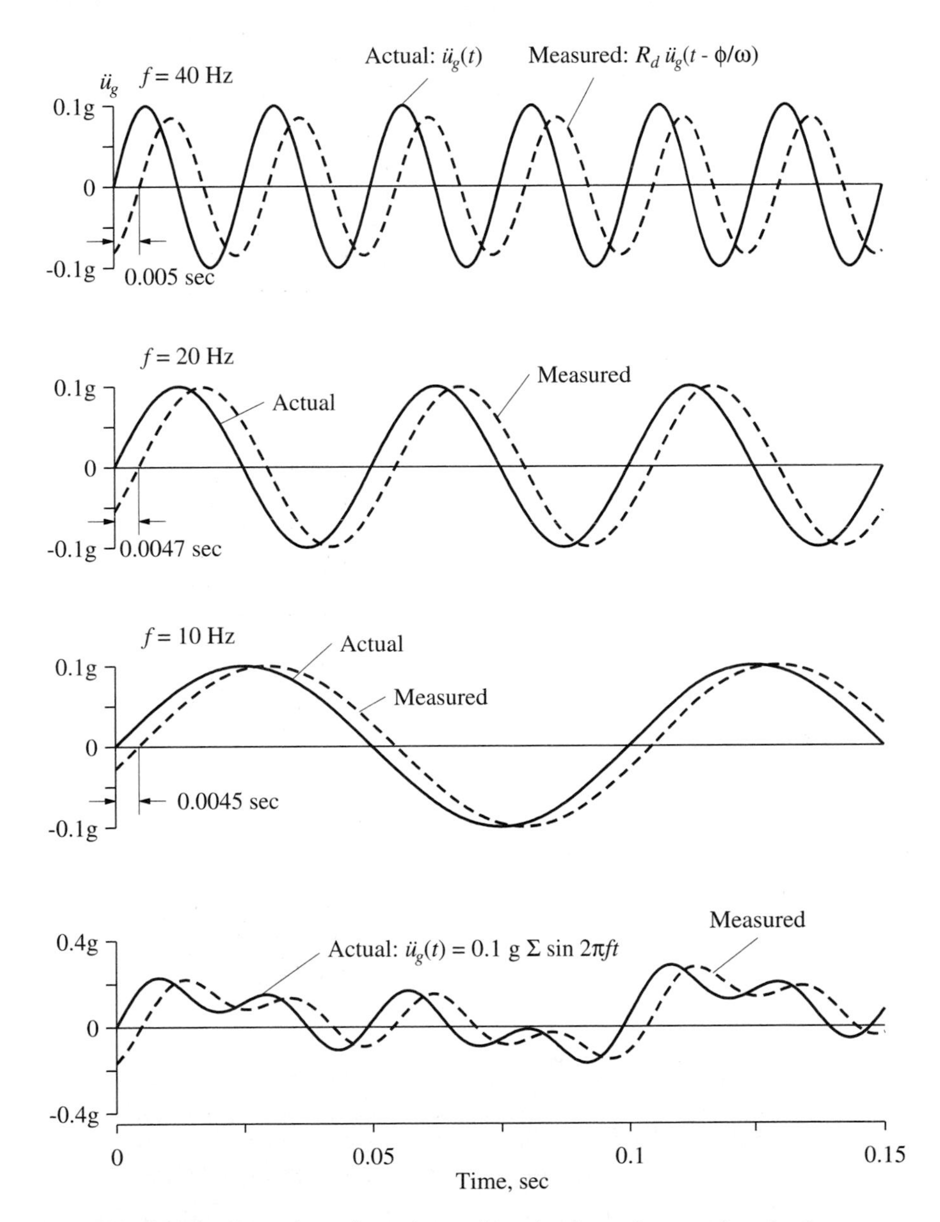

Figure 3.7.3 Comparison of actual ground acceleration and measured motion by an instrument with $f_n = 50$ Hz and $\zeta = 0.7$.

the instrument constant $-1/\omega_n^2$ in Eq. (3.7.1). For excitation frequencies $f = 20$ and 10 Hz, the measured motion has accurate amplitude, but the error at $f = 40$ Hz is noticeable; and the phase shift, although not identical for the three frequencies, is similar. If the ground acceleration is the sum of the three harmonic components, this figure shows that the recorded motion matches the ground acceleration satisfactorily in amplitude and shape. The accuracy of the recorded motion $u(t)$ can be improved, especially at higher frequencies, by calculating $\ddot{u}_g(t-\phi/\omega)$ from the measured $u(t)$ using Eq. (3.7.1) with R_d determined from Eq. (3.2.11) and known instrument properties ω_n and ζ. Such corrections are repeated for each harmonic component in $u(t)$, and the corrected components are then synthesized to obtain $\ddot{u}_g(t)$. These computations can be carried out by discrete fast Fourier transform procedures, which are beyond the scope of this book.

3.7.2 Measurement of Displacement

It is desired to design the transducer so that the relative displacement $u(t)$ measures the support displacement $u_g(t)$. This is achieved by making the transducer spring so flexible or the transducer mass so large, or both, that the mass stays still while the support beneath it moves. Such an instrument is unwieldy because of the heavy mass and soft spring, and because it must accommodate the anticipated support displacement, which may be as large as 12 to 36 in. during earthquakes.

To examine the basic concept further, consider harmonic ground displacement

$$u_g(t) = u_{go} \sin \omega t \tag{3.7.2}$$

With the forcing function $p_{\text{eff}}(t) = -m\ddot{u}_g(t) = m\omega^2 u_{go} \sin \omega t$, Eq. (1.7.4) governs the relative displacement of the mass; this governing equation is the same as Eq. (3.2.1) for applied harmonic force with p_o replaced by $m\omega^2 u_{go}$. Making this substitution in Eq. (3.2.10) and using Eq. (3.2.20) gives

$$u(t) = R_a u_{go} \sin(\omega t - \phi) \tag{3.7.3}$$

For excitation frequencies ω much higher than the natural frequency ω_n, R_a is close to unity (Fig. 3.2.7c) and ϕ is close to 180°, and Eq. (3.7.3) becomes

$$u(t) = -u_{go} \sin \omega t$$

The recorded displacement is the same as the support displacement except for the negative sign, which is usually inconsequential. Damping of the instrument is not a critical parameter because it has little effect on the recorded motion if ω/ω_n is very large.

3.8 ENERGY DISSIPATED IN VISCOUS DAMPING

Consider the steady-state motion of an SDF system due to $p(t) = p_o \sin \omega t$. The energy dissipated by viscous damping in one cycle of harmonic vibration is

$$E_D = \int f_D\, du = \int_0^{2\pi/\omega} (c\dot{u})\dot{u}\, dt = \int_0^{2\pi/\omega} c\dot{u}^2\, dt$$

$$= c\int_0^{2\pi/\omega}[\omega u_o\cos(\omega t-\phi)]^2\,dt = \pi c\omega u_o^2 = 2\pi\zeta\frac{\omega}{\omega_n}ku_o^2 \tag{3.8.1}$$

The energy dissipated is proportional to the square of the amplitude of motion, as shown in Fig. 3.8.1. It is not a constant value for any given amount of damping and amplitude since the energy dissipated increases linearly with excitation frequency.

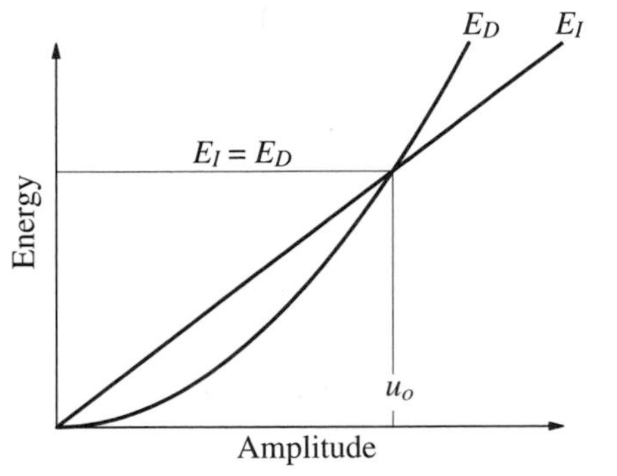

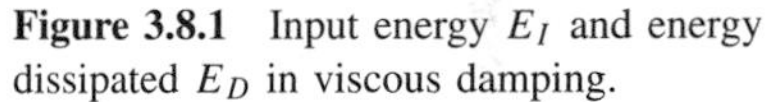
Figure 3.8.1 Input energy E_I and energy dissipated E_D in viscous damping.

The external force $p(t)$ inputs energy to the system, which for each cycle of vibration is

$$E_I = \int p(t)\,du = \int_0^{2\pi/\omega} p(t)\dot{u}\,dt$$

$$= \int_0^{2\pi/\omega}[p_o\sin\omega t][\omega u_o\cos(\omega t-\phi)]\,dt = \pi p_o u_o\sin\phi \tag{3.8.2}$$

Thus the input energy is proportional to the displacement amplitude, as shown in Fig. 3.8.1. Utilizing Eq. (3.2.12) for phase angle, this equation can be rewritten as (see Derivation 3.6)

$$E_I = 2\pi\zeta\frac{\omega}{\omega_n}ku_o^2 \tag{3.8.3}$$

Equations (3.8.1) and (3.8.3) indicate that in steady-state vibration, the energy input to the system due to the applied force is dissipated in viscous damping.

What about the potential energy and kinetic energy? Over each cycle of harmonic vibration the changes in potential energy (equal to the strain energy of the spring) and kinetic energy are zero. This can be confirmed as follows:

$$E_S = \int f_S\,du = \int_0^{2\pi/\omega}(ku)\dot{u}\,dt$$

$$= \int_0^{2\pi/\omega}k[u_o\sin(\omega t-\phi)][\omega u_o\cos(\omega t-\phi)]\,dt = 0 \tag{3.8.4}$$

$$E_K = \int f_I \, du = \int_0^{2\pi/\omega} (m\ddot{u})\dot{u} \, dt$$

$$= \int_0^{2\pi/\omega} m[-\omega^2 u_o \sin(\omega t - \phi)][\omega u_o \cos(\omega t - \phi)] \, dt = 0 \tag{3.8.5}$$

The displacement amplitude caused by harmonic force with $\omega = \omega_n$ grows until steady state is attained (Fig. 3.2.2), and this can be explained using energy concepts. As mentioned earlier and shown in Fig. 3.8.1, the input and dissipated energies vary linearly and quadratically, respectively, with the displacement amplitude. Before steady state is reached, the input energy per cycle exceeds the energy dissipated during the cycle by damping, leading to a larger amplitude of displacement in the next cycle. With growing displacement amplitude, the dissipated energy increases more rapidly than does the input energy. Eventually, the input and dissipated energies will match at the steady-state displacement amplitude u_o (Fig. 3.8.1), which will be bounded no matter how small the damping. This energy balance provides an alternative means of finding u_o; equating Eqs. (3.8.1) and (3.8.2) gives $u_o = p_o \sin\phi / c\omega$, and using Eq. (3.2.12) for ϕ gives

$$u_o = \frac{p_o}{k} R_d \tag{3.8.6}$$

This result agrees with Eq. (3.2.10), obtained by solving the equation of motion.

We will now present a graphical interpretation for the energy dissipated in viscous damping. For this purpose we first derive an equation relating the damping force f_D to the displacement u:

$$f_D = c\dot{u}(t) = c\omega u_o \cos(\omega t - \phi)$$

$$= c\omega\sqrt{u_o^2 - u_o^2 \sin^2(\omega t - \phi)}$$

$$= c\omega\sqrt{u_o^2 - [u(t)]^2}$$

This can be rewritten as

$$\left(\frac{u}{u_o}\right)^2 + \left(\frac{f_D}{c\omega u_o}\right)^2 = 1 \tag{3.8.7}$$

which is the equation of the ellipse shown in Fig. 3.8.2a. Observe that the f_D–u curve is not a single-valued function but a loop known as a *hysteresis loop*. The area enclosed by the ellipse is $\pi(u_o)(c\omega u_o) = \pi c\omega u_o^2$, which is the same as Eq. (3.8.1). Thus the area within the hysteresis loop gives the dissipated energy.

It is of interest to examine the total (elastic plus damping) resisting force because this is the force that is measured in an experiment:

$$f_S + f_D = ku(t) + c\dot{u}(t)$$

$$= ku + c\omega\sqrt{u_o^2 - u^2} \tag{3.8.8}$$

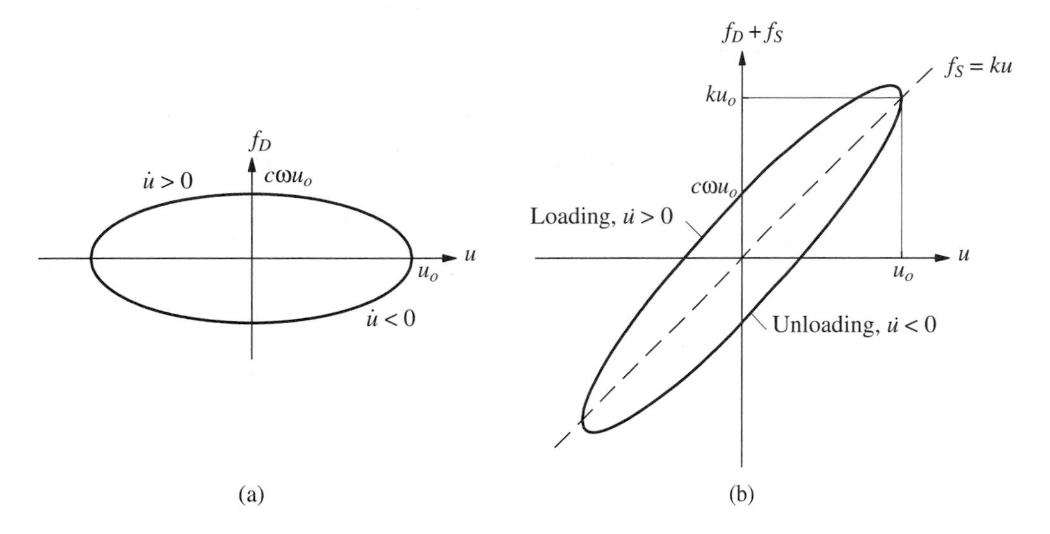

Figure 3.8.2 Hysterisis loops for (a) viscous damper; (b) spring and viscous damper in parallel.

A plot of $f_S + f_D$ against u is the ellipse of Fig. 3.8.2a rotated as shown in Fig. 3.8.2b because of the ku term in Eq. (3.8.8). The energy dissipated by damping is still the area enclosed by the ellipse because the area enclosed by the single-valued elastic force, $f_S = ku$, is zero.

The hysteresis loop associated with viscous damping is the result of *dynamic hysteresis* since it is related to the dynamic nature of the loading. The loop area is proportional to excitation frequency; this implies that the force–deformation curve becomes a single-valued curve (no hysteretic loop) if the cyclic load is applied slowly enough ($\omega = 0$). A distinguishing characteristic of dynamic hysteresis is that the hysteresis loops tend to be elliptical in shape rather than pointed, as in Fig. 1.3.1c, if they are associated with plastic deformations. In the latter case, the hysteresis loops develop even under static cyclic loads; this phenomenon is therefore known as *static hysteresis* because the force–deformation curve is insensitive to deformation rate.

In passing, we mention two measures of damping: *specific damping capacity* and the *specific damping factor*. The specific damping capacity, E_D/E_{So}, is that fractional part of the strain energy, $E_{So} = ku_o^2/2$, which is dissipated during each cycle of motion; both E_D and E_{So} are shown in Fig. 3.8.3. The specific damping factor, also known as the *loss factor,* is defined as

$$\xi = \frac{1}{2\pi} \frac{E_D}{E_{So}} \tag{3.8.9}$$

If the energy could be removed at a uniform rate during a cycle of simple harmonic motion (such a mechanism is not realistic), $E_D/2\pi$ could be interpreted as the energy loss per radian divided by the total energy in the system. These two measures of damping are not often used in structural vibration since they are most useful for very light damping (e.g., they are useful in comparing the damping capacity of materials).

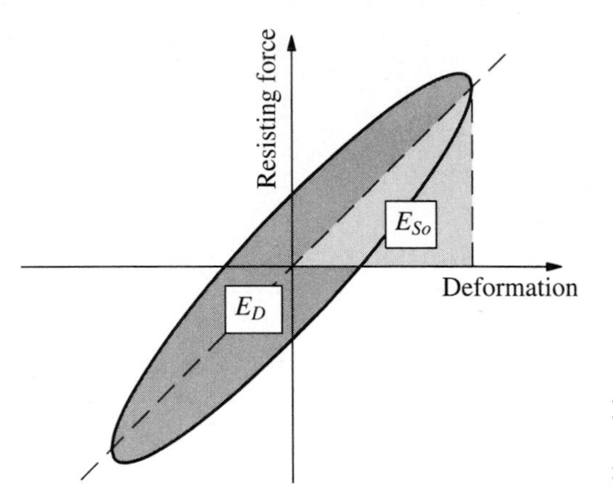

Figure 3.8.3 Definition of energy loss E_D in a cycle of harmonic vibration and maximum strain energy E_{So}.

Derivation 3.6

Equation (3.8.2) gives the input energy per cycle where the phase angle, defined by Eq. (3.2.12), can be expressed as

$$\sin\phi = \left(2\zeta\frac{\omega}{\omega_n}\right)R_d = \left(2\zeta\frac{\omega}{\omega_n}\right)\frac{u_o}{p_o/k}$$

Substituting this in Eq. (3.8.2) gives Eq. (3.8.3).

3.9 EQUIVALENT VISCOUS DAMPING

As introduced in Section 1.4, damping in actual structures is usually represented by equivalent viscous damping. It is the simplest form of damping to use since the governing differential equation of motion is linear and hence amenable to analytical solution, as seen in earlier sections of this chapter and in Chapter 2. The advantage of using a linear equation of motion usually outweighs whatever compromises are necessary in the viscous damping approximation. In this section we determine the damping coefficient for viscous damping so that it is equivalent in some sense to the combined effect of all damping mechanisms present in the actual structure; these were mentioned in Section 1.4.

The simplest definition of equivalent viscous damping is based on the measured response of a system to harmonic force at exciting frequency ω equal to the natural frequency ω_n of the system. The damping ratio ζ_{eq} is calculated from Eq. (3.4.1) using measured values of u_o and $(u_{\text{st}})_o$. This is the equivalent viscous damping since it accounts for all the energy-dissipating mechanisms that existed in the experiments.

Another definition of equivalent viscous damping is that it is the amount of damping that provides the same bandwidth in the frequency-response curve as obtained experimentally for an actual system. The damping ratio ζ_{eq} is calculated from Eq. (3.2.24) using the excitation frequencies f_a, f_b, and f_n (Fig. 3.4.1) obtained from an experimentally determined frequency-response curve.

The most common method for defining equivalent viscous damping is to equate the energy dissipated in a vibration cycle of the actual structure and an equivalent vis-

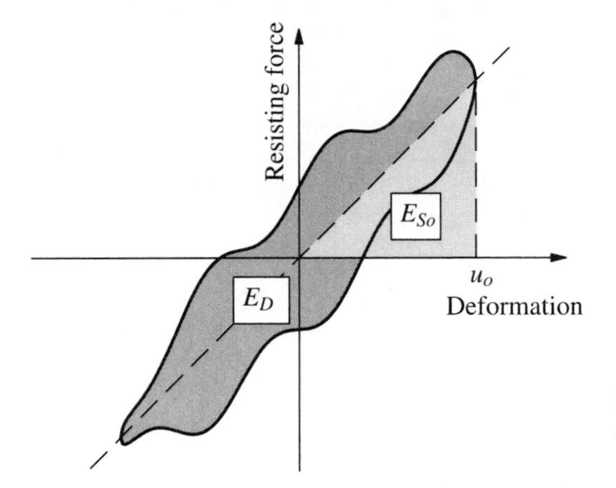

Figure 3.9.1 Energy dissipated E_D in a cycle of harmonic vibration determined from experiment.

cous system. For an actual structure the force-displacement relation obtained from an experiment under cyclic loading with displacement amplitude u_o is determined; such a relation of arbitrary shape is shown schematically in Fig. 3.9.1. The energy dissipated in the actual structure is given by the area E_D enclosed by the hysteresis loop. Equating this to the energy dissipated in viscous damping given by Eq. (3.8.1) leads to

$$4\pi\zeta_{\text{eq}}\frac{\omega}{\omega_n}E_{So} = E_D \quad \text{or} \quad \zeta_{\text{eq}} = \frac{1}{4\pi}\frac{1}{\omega/\omega_n}\frac{E_D}{E_{So}} \tag{3.9.1}$$

where the strain energy, $E_{So} = ku_o^2/2$, is calculated from the stiffness k determined by experiment.

The experiment leading to the force–deformation curve of Fig. 3.9.1 and hence E_D should be conducted at $\omega = \omega_n$, where the response of the system is most sensitive to damping. Thus Eq. (3.9.1) specializes to

$$\zeta_{\text{eq}} = \frac{1}{4\pi}\frac{E_D}{E_{So}} \tag{3.9.2}$$

The damping ratio ζ_{eq} determined from a test at $\omega = \omega_n$ would not be correct at any other exciting frequency, but it would be a satisfactory approximation (Section 3.10.2).

It is widely accepted that this procedure can be extended to model the damping in systems with many degrees of freedom. An equivalent viscous damping ratio is assigned to each natural vibration mode of the system (defined in Chapter 10) in such a way that the energy dissipated in viscous damping matches the actual energy dissipated in the system when the system vibrates in that mode at its natural frequency.

In this book the concept of equivalent viscous damping is restricted to systems vibrating at amplitudes within the linearly elastic limit of the overall structure. The energy dissipated in inelastic deformations of the structure have also been modeled as equivalent viscous damping in some research studies. This idealization is generally not satisfactory, however, for the large inelastic deformations of structures expected during strong earthquakes. We shall account for these inelastic deformations and the associated energy dissipation by nonlinear force–deformation relations, such as those shown in Fig. 1.3.4 (see Chapters 5 and 7).

Example 3.6

A body moving through a fluid experiences a resisting force that is proportional to the square of the speed, $f_D = \pm a\dot{u}^2$, where the positive sign applies to positive $\dot{u}$ and the negative sign to negative $\dot{u}$. Determine the equivalent viscous damping coefficient c_{eq} for such forces acting on an oscillatory system undergoing harmonic motion of amplitude u_o and frequency ω. Also find its displacement amplitude at $\omega = \omega_n$.

Solution If time is measured from the position of largest negative displacement, the harmonic motion is

$$u(t) = -u_o \cos \omega t$$

The energy dissipated in one cycle of motion is

$$E_D = \int f_D \, du = \int_0^{2\pi/\omega} f_D \dot{u} \, dt = 2\int_0^{\pi/\omega} f_D \dot{u} \, dt$$

$$= 2\int_0^{\pi/\omega} (a\dot{u}^2)\dot{u} \, dt = 2a\omega^3 u_o^3 \int_0^{\pi/\omega} \sin^3 \omega t \, dt = \tfrac{8}{3} a\omega^2 u_o^3$$

Equating this to the energy dissipated in viscous damping [Eq. (3.8.1)] gives

$$\pi c_{eq} \omega u_o^2 = \frac{8}{3} a\omega^2 u_o^3 \quad \text{or} \quad c_{eq} = \frac{8}{3\pi} a\omega u_o \tag{a}$$

Substituting $\omega = \omega_n$ in Eq. (a) and the c_{eq} for c in Eq. (3.2.15) gives

$$u_o = \left(\frac{3\pi}{8a} \frac{p_o}{\omega_n^2} \right)^{1/2} \tag{b}$$

PART C: SYSTEMS WITH NONVISCOUS DAMPING

3.10 HARMONIC VIBRATION WITH RATE-INDEPENDENT DAMPING

3.10.1 Rate-Independent Damping

Experiments on structural metals indicate that the energy dissipated internally in cyclic straining of the material is essentially independent of the cyclic frequency. Similarly, forced vibration tests on structures indicate that the equivalent viscous damping ratio is roughly the same for all natural modes and frequencies. Thus we refer to this type of damping as *rate-independent linear damping.* Other terms used for this mechanism of internal damping are *structural damping, solid damping,* and *hysteretic damping.* We prefer not to use these terms because the first two are not especially meaningful and the third is ambiguous because hysteresis is a characteristic of all materials or structural systems that dissipate energy.

Rate-independent damping is associated with static hysteresis due to plastic strain, localized plastic deformation, crystal plasticity, and plastic flow in a range of stresses within the apparent elastic limit. On the microscopic scale the inhomogeneity of stress distribution within crystals and stress concentration at crystal boundary intersections produce local stress high enough to cause local plastic strain even though the average "global" stress may be well below the elastic limit. This damping mechanism does not include the energy dissipation in global plastic deformations, which as mentioned earlier, is handled by a nonlinear relationship between force f_S and deformation u.

The simplest device that can be used to represent rate-independent linear damping is to assume that the damping force is proportional to velocity and inversely proportional to frequency:

$$f_D = \frac{\eta k}{\omega} \dot{u} \tag{3.10.1}$$

where k is the stiffness of the structure and η is a damping coefficient. The energy dissipated by this type of damping in a cycle of vibration at frequency ω is independent of ω (Fig. 3.10.1). It is given by Eq. (3.8.1) with c replaced by $\eta k/\omega$:

$$E_D = \pi \eta k u_o^2 = 2\pi \eta E_{So} \tag{3.10.2}$$

In contrast, the energy dissipated in viscous damping increases linearly with the forcing frequency as shown in Fig. 3.10.1.

Rate-independent damping is easily described if the excitation is harmonic and we are interested only in the steady-state response of this system. Difficulties arise in translating this damping mechanism back to the time domain. Thus it is most useful in the frequency-domain method of analysis (Section 1.10.3). The complex term $k(1 + i\eta)u$ represents both the elastic and damping forces at the same time; $k(1 + i\eta)$ is the *complex stiffness* of the system. There are advantages in this terminology if the reader is familiar with complex-variable notation. The complex stiffness has no physical meaning, however, in the same engineering sense as the elastic stiffness.

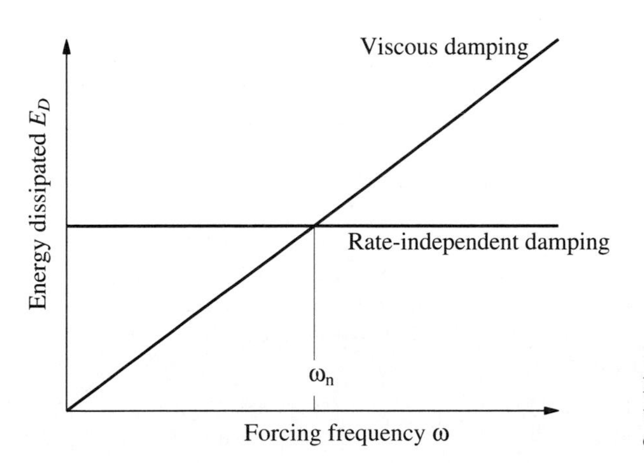

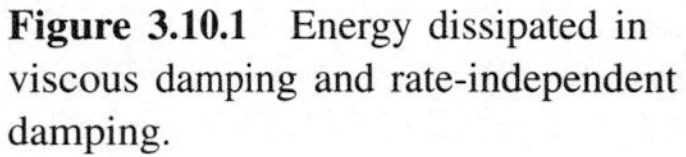
Figure 3.10.1 Energy dissipated in viscous damping and rate-independent damping.

3.10.2 Steady-State Response to Harmonic Force

The equation for an SDF system with rate-independent linear damping, denoted by a crossed box in Fig. 3.10.2, is Eq. (3.2.1) with the damping term replaced by Eq. (3.10.1):

$$m\ddot{u} + \frac{\eta k}{\omega}\dot{u} + ku = p(t) \tag{3.10.3}$$

The mathematical solution of this equation is quite complex for arbitrary $p(t)$. Here we consider only the steady-state motion due to a sinusoidal forcing function, $p(t) = p_o \sin \omega t$, which is described by

$$u(t) = u_o \sin(\omega t - \phi) \tag{3.10.4}$$

The amplitude u_o and phase angle ϕ are

$$u_o = (u_{\text{st}})_o \frac{1}{\sqrt{\left[1 - (\omega/\omega_n)^2\right]^2 + \eta^2}} \tag{3.10.5}$$

$$\phi = \tan^{-1} \frac{\eta}{1 - (\omega/\omega_n)^2} \tag{3.10.6}$$

These results are obtained by modifying the viscous damping ratio in Eqs. (3.2.11) and (3.2.12) to reflect the damping force associated with rate-independent damping, Eq. (3.10.1). In particular, ζ was replaced by

$$\zeta = \frac{c}{c_c} = \frac{\eta k/\omega}{2m\omega_n} = \frac{\eta}{2(\omega/\omega_n)} \tag{3.10.7}$$

Shown in Fig. 3.10.3 by solid lines are plots of $u_o/(u_{\text{st}})_o$ and ϕ as a function of the frequency ratio ω/ω_n for damping coefficient $\eta = 0$, 0.2, and 0.4; the dashed lines are described in the next section. Comparing these results with those in Fig. 3.2.6 for viscous damping, two differences are apparent: First, resonance (maximum amplitude) occurs at $\omega = \omega_n$, not at $\omega < \omega_n$. Second, the phase angle for $\omega = 0$ is $\phi = \tan^{-1} \eta$ instead of zero for viscous damping; this implies that motion with rate-independent damping can never be in phase with the forcing function.

These differences between forced vibration with rate-independent damping and forced vibration with viscous damping are not significant, but they are the source of some difficulty in reconciling physical data. In most damped vibration, damping is not viscous, and to assume that it is without knowing its real physical characteristics is an assumption of some error. In the next section this error is shown to be small when the real damping is rate independent.

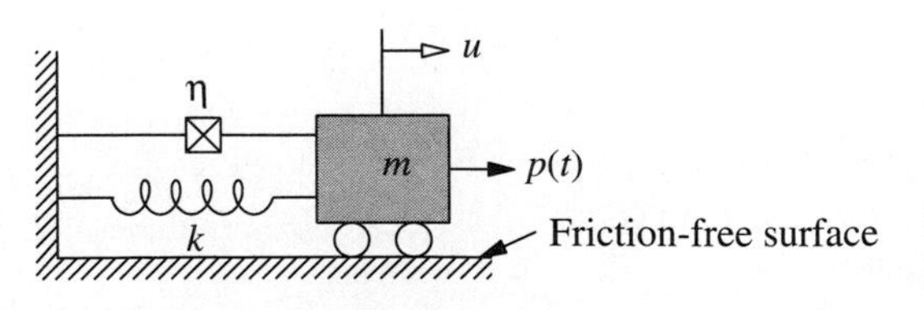

Figure 3.10.2 SDF system with rate-independent linear damping.

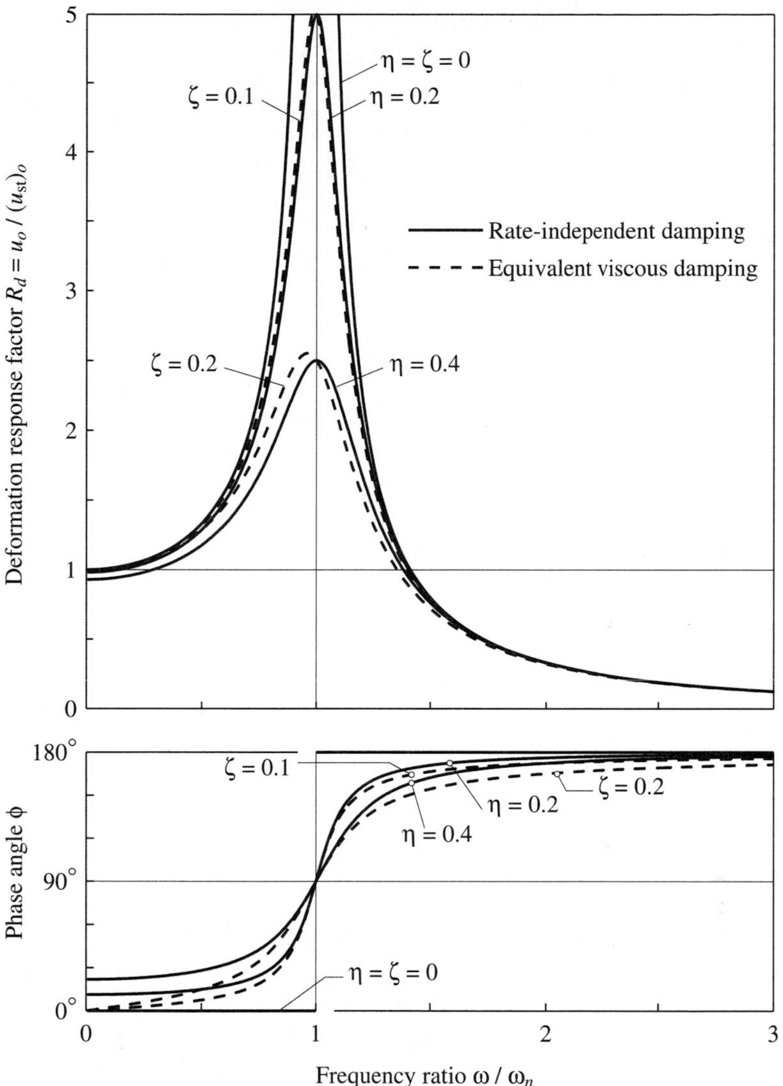

Figure 3.10.3 Response of system with rate-independent damping: exact solution and approximate solution using equivalent viscous damping.

3.10.3 Solution Using Equivalent Viscous Damping

In this section an approximate solution for the steady-state harmonic response of a system with rate-independent damping is obtained by modeling this damping mechanism as equivalent viscous damping.

Matching dissipated energies at $\omega = \omega_n$ (Fig. 3.10.1) led to Eq. (3.9.2), where E_D is given by Eq. (3.10.2), leading to the equivalent viscous damping ratio:

$$\zeta_{\text{eq}} = \frac{\eta}{2} \tag{3.10.8}$$

Substituting this ζ_{eq} for ζ in Eqs. (3.2.10) to (3.2.12) gives the system response. The resulting amplitude u_o and phase angle ϕ are shown by the dashed lines in Fig. 3.10.3. This approximate solution matches the exact result at $\omega = \omega_n$ because that was the criterion used in selecting ζ_{eq}. Over a wide range of excitation frequencies the approximate solution is seen to be accurate enough for many engineering applications. Thus Eq. (3.10.3)—which is difficult to solve for arbitrary force $p(t)$ that contains many harmonic components of different frequencies ω—can be replaced by the simpler Eq. (3.2.1) for a system with equivalent viscous damping defined by Eq. (3.10.8). This is the basic advantage of equivalent viscous damping.

3.11 HARMONIC VIBRATION WITH COULOMB FRICTION

3.11.1 Equation of Motion

Shown in Fig. 3.11.1 is a mass–spring system with Coulomb friction force $F = \mu N$ that opposes sliding of the mass. As defined in Section 2.4, the coefficients of static and kinetic friction are assumed to be equal to μ, and N is the normal force across the sliding surfaces. The equation of motion is obtained by including the exciting force in Eqs. (2.4.1) and (2.4.2) governing the free vibration of the system:

$$m\ddot{u} + ku \pm F = p(t) \tag{3.11.1}$$

The sign of the friction force changes with the direction of motion; the positive sign applies if the motion is from left to right ($\dot{u} > 0$) and the negative sign is for motion from right to left ($\dot{u} < 0$). Each of the two differential equations is linear, but the overall problem is nonlinear because the governing equation changes every half-cycle of motion. Therefore, exact analytical solutions would not be possible except in special cases.

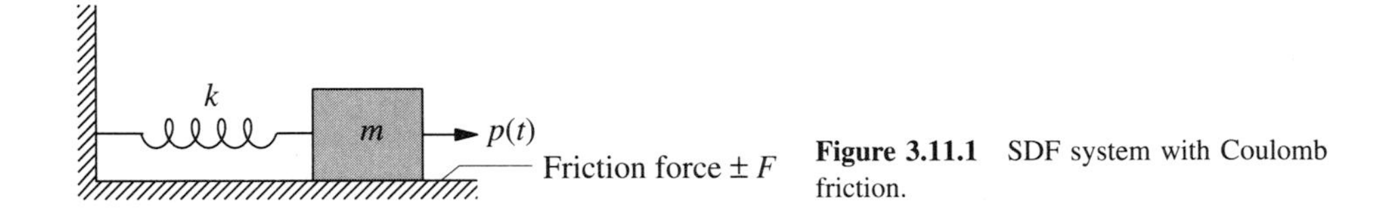

Figure 3.11.1 SDF system with Coulomb friction.

3.11.2 Steady-State Response to Harmonic Force

An exact analytical solution for the steady-state response of the system of Fig. 3.11.1 subjected to harmonic force was developed by J. P. Den Hartog in 1933. The analysis is not included here, but his results are shown by solid lines in Fig. 3.11.2; the

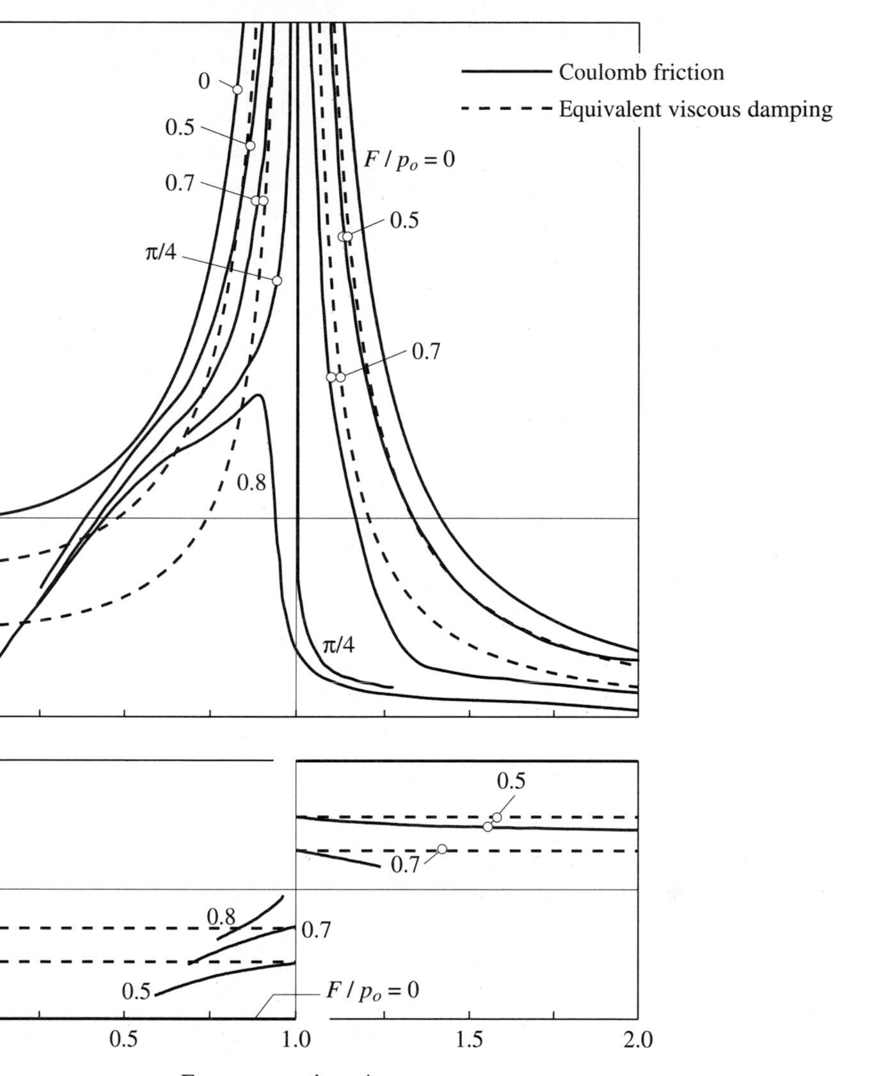

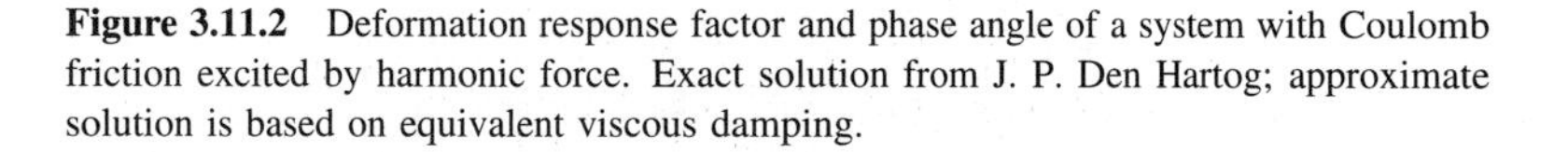

Figure 3.11.2 Deformation response factor and phase angle of a system with Coulomb friction excited by harmonic force. Exact solution from J. P. Den Hartog; approximate solution is based on equivalent viscous damping.

dashed lines are described in the next section. The displacement amplitude u_o, normalized relative to $(u_{st})_o = p_o/k$, and the phase angle ϕ are plotted as a function of the frequency ratio ω/ω_n for three values of F/p_o. If there is no friction, $F = 0$ and $u_o/(u_{st})_o = (R_d)_{\zeta=0}$, the same as in Eq. (3.1.11) for an undamped system. The friction force reduces the displacement amplitude u_o with the reduction depending on the frequency ratio ω/ω_n.

At $\omega = \omega_n$ the amplitude of motion is not limited by Coulomb friction if

$$\frac{F}{p_o} < \frac{\pi}{4} \tag{3.11.2}$$

which is surprising since $F = (\pi/4)\,p_o$ represents a large friction force, but can be explained by comparing the energy E_F dissipated in friction against the input energy E_I. The energy dissipated by Coulomb friction in one cycle of vibration with displacement amplitude u_o is the area of the hysteresis loop enclosed by the friction force–displacement diagram (Fig. 3.11.3):

$$E_F = 4Fu_o \tag{3.11.3}$$

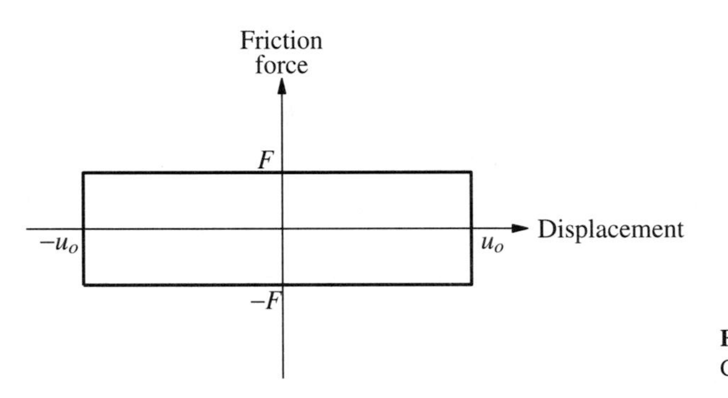

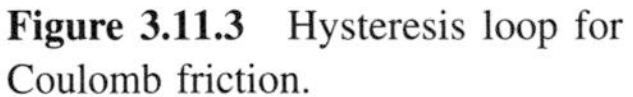
Figure 3.11.3 Hysteresis loop for Coulomb friction.

Observe that the dissipated energy in a vibration cycle is proportional to the amplitude of the cycle. The energy E_I input by the harmonic force applied at $\omega = \omega_n$ is also proportional to the displacement amplitude. If Eq. (3.11.2) is satisfied, it can be shown that

$$E_F < E_I$$

that is, the energy dissipated in friction per cycle is less than the input energy (Fig. 3.11.4). Therefore, the displacement amplitude would increase cycle after cycle and grow without bound. This behavior is quite different from that of systems with viscous damping or

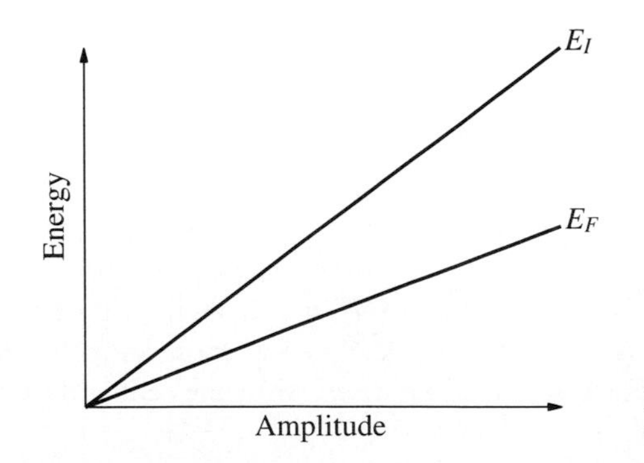

Figure 3.11.4 Input energy E_I and energy dissipated E_F by Coulomb friction.

rate-independent damping. For these forms of damping, as shown in Section 3.8, the dissipated energy increases quadratically with displacement amplitude, and the displacement amplitude is bounded no matter how small the damping. In connection with the fact that infinite amplitudes occur at $\omega = \omega_n$ if Eq. (3.11.2) is satisfied, the phase angle shows a discontinuous jump at $\omega = \omega_n$ (Fig. 3.11.2).

3.11.3 Solution Using Equivalent Viscous Damping

In this section an approximate solution for the steady-state harmonic response of a system with Coulomb friction is obtained by modeling this damping mechanism by equivalent viscous damping. Substituting E_F, the energy dissipated by Coulomb friction given by Eq. (3.11.3), for E_D in Eq. (3.9.1) provides the equivalent viscous damping ratio

$$\zeta_{\text{eq}} = \frac{2}{\pi}\frac{1}{\omega/\omega_n}\frac{u_F}{u_o} \tag{3.11.4}$$

where $u_F = F/k$. The approximate solution for the displacement amplitude u_o is obtained by substituting ζ_{eq} for ζ in Eq. (3.2.11):

$$\frac{u_o}{p_o/k} = \frac{1}{\left\{\left[1-(\omega/\omega_n)^2\right]^2 + [(4/\pi)(u_F/u_o)]^2\right\}^{1/2}}$$

This contains u_o on the right side also. Squaring and solving algebraically, the normalized displacement amplitude is

$$\frac{u_o}{(u_{\text{st}})_o} = \frac{\left\{1-[(4/\pi)(F/p_o)]^2\right\}^{1/2}}{1-(\omega/\omega_n)^2} \tag{3.11.5}$$

This approximate result is valid provided that $F/p_o < \pi/4$. The approximate solution cannot be used if $F/p_o > \pi/4$ because then the quantity under the radical is negative and the numerator is imaginary.

These approximate and exact solutions are compared in Fig. 3.11.2. If the friction force is small enough to permit continuous motion, this motion is practically sinusoidal and the approximate solution is close to the exact solution. If the friction force is large, discontinuous motion with stops and starts results, which is much distorted relative to a sinusoid, and the approximate solution is poor.

The approximate solution for the phase angle is obtained by substituting ζ_{eq} for ζ in Eq. (3.2.12):

$$\tan\phi = \frac{(4/\pi)(u_F/u_o)}{1-(\omega/\omega_n)^2}$$

Substituting for u_o from Eq. (3.11.5) gives

$$\tan\phi = \pm\frac{(4/\pi)(F/p_o)}{\left\{1-[(4/\pi)(F/p_o)]^2\right\}^{1/2}} \tag{3.11.6}$$

For a given value of F/p_o the $\tan\phi$ is constant but with a positive value if $\omega/\omega_n < 1$ and a negative value if $\omega/\omega_n > 1$. This is shown in Fig. 3.11.2, where it is seen that the phase angle is discontinuous at $\omega = \omega_n$ for Coulomb friction.

Example 3.7

The structure of Example 2.7 with friction devices deflects 2 in. under a lateral force of $p = 500$ kips. What would be the approximate amplitude of motion if the lateral force is replaced by the harmonic force $p(t) = 500 \sin \omega t$, where the forcing period $T = 1$ sec?

Solution The data (given and from Example 2.7) are

$$(u_{\text{st}})_o = \frac{p_o}{k} = 2 \text{ in.} \qquad u_F = 0.15 \text{ in.}$$

$$\frac{\omega}{\omega_n} = \frac{T_n}{T} = \frac{0.5}{1} = 0.5$$

Calculate u_o from Eq. (3.11.5).

$$\frac{F}{p_o} = \frac{F/k}{p_o/k} = \frac{u_F}{(u_{\text{st}})_o} = \frac{0.15}{2} = 0.075$$

Substituting for F/p_o in Eq. (3.11.5) gives

$$\frac{u_o}{(u_{\text{st}})_o} = \frac{\left\{1 - [(4/\pi)0.075]^2\right\}^{1/2}}{1 - (0.5)^2} = 1.327$$

$$u_o = 1.327(2) = 2.654 \text{ in.}$$

PART D: RESPONSE TO PERIODIC EXCITATION

A periodic function is one in which the portion defined over T_0 repeats itself indefinitely (Fig. 3.12.1). Many forces are periodic or nearly periodic. Under certain conditions, propeller forces on a ship, wave loading on an offshore platform, and wind forces induced by vortex shedding on tall, slender structures are nearly periodic. Earthquake ground

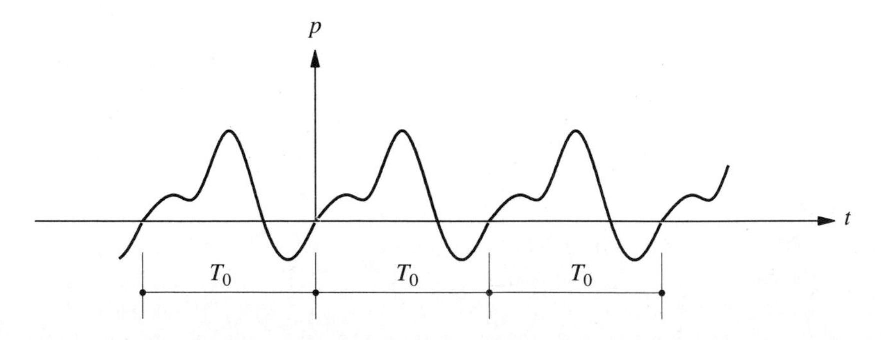

Figure 3.12.1 Periodic excitation.

motion has no resemblance to a periodic function. However, the base excitation arising from an automobile traveling on an elevated freeway that has settled because of long-term creep may be nearly periodic.

We are interested in analyzing the response to periodic excitation for yet another reason. The analysis can be extended to arbitrary excitations utilizing discrete fast Fourier transform techniques. These are not included in this book, however.

3.12 FOURIER SERIES REPRESENTATION

A function $p(t)$ is said to be periodic with period T_0 if it satisfies the following relationship:

$$p(t + jT_0) = p(t) \qquad j = -\infty, \ldots, -3, -2, -1, 0, 1, 2, 3, \ldots, \infty$$

A periodic function can be separated into its harmonic components using the *Fourier series:*

$$p(t) = a_0 + \sum_{j=1}^{\infty} a_j \cos(j\omega_0 t) + \sum_{j=1}^{\infty} b_j \sin(j\omega_0 t) \tag{3.12.1}$$

where the fundamental harmonic in the excitation has the frequency

$$\omega_0 = \frac{2\pi}{T_0} \tag{3.12.2}$$

The coefficients in the Fourier series can be expressed in terms of $p(t)$ because the sine and cosine functions are orthogonal:

$$a_0 = \frac{1}{T_0} \int_0^{T_0} p(t)\, dt \tag{3.12.3}$$

$$a_j = \frac{2}{T_0} \int_0^{T_0} p(t) \cos(j\omega_0 t)\, dt \qquad j = 1, 2, 3, \ldots \tag{3.12.4}$$

$$b_j = \frac{2}{T_0} \int_0^{T_0} p(t) \sin(j\omega_0 t)\, dt \qquad j = 1, 2, 3, \ldots \tag{3.12.5}$$

The coefficient a_0 is the average value of $p(t)$; coefficients a_j and b_j are the amplitudes of the jth harmonics of frequency $j\omega_0$.

Theoretically, an infinite number of terms is required for the Fourier series to converge to $p(t)$. In practice, however, a few terms are sufficient for good convergence. At a discontinuity, the Fourier series converges to a value that is the average of the values immediately to the left and to the right of the discontinuity.

3.13 RESPONSE TO PERIODIC FORCE

A periodic excitation implies that the excitation has been in existence for a long time, by which time the transient response associated with the initial displacement and velocity has decayed. Thus, just as for harmonic excitation, we are interested in finding the steady-state response. The response of a linear system to a periodic force can be determined by combining the responses to individual excitation terms in the Fourier series.

The response of an undamped system to constant force $p(t) = a_0$ is given by Eq. (f) of Example 1.5, in which the $\cos \omega t$ term will decay because of damping (see Section 4.3), leaving the steady-state solution.†

$$u_0(t) = \frac{a_0}{k} \tag{3.13.1}$$

The steady-state response of a viscously damped SDF system to harmonic cosine force $p(t) = a_j \cos(j\omega_0 t)$ is given by Eqs. (3.2.3) and (3.2.26) with ω replaced by $j\omega_0$:

$$u_j^c(t) = \frac{a_j}{k} \frac{2\zeta\beta_j \sin(j\omega_0 t) + (1 - \beta_j^2)\cos(j\omega_0 t)}{(1 - \beta_j^2)^2 + (2\zeta\beta_j)^2} \tag{3.13.2}$$

where

$$\beta_j = \frac{j\omega_0}{\omega_n} \tag{3.13.3}$$

Similarly, the steady-state response of the system to sinusoidal force $p(t) = b_j \sin(j\omega_0 t)$ is given by Eqs. (3.2.3) and (3.2.4) with ω replaced by $j\omega_0$:

$$u_j^s(t) = \frac{b_j}{k} \frac{(1 - \beta_j^2)\sin(j\omega_0 t) - 2\zeta\beta_j \cos(j\omega_0 t)}{(1 - \beta_j^2)^2 + (2\zeta\beta_j)^2} \tag{3.13.4}$$

If $\zeta = 0$ and one of $\beta_j = 1$, the steady-state response is unbounded and not meaningful because the transient response never decays (see Section 3.1); in the following it is assumed that $\zeta \neq 0$ and $\beta_j \neq 1$.

The steady-state response of a system with damping to periodic excitation $p(t)$ is the combination of responses to individual terms in the Fourier series:

$$u(t) = u_0(t) + \sum_{j=1}^{\infty} u_j^c(t) + \sum_{j=1}^{\infty} u_j^s(t) \tag{3.13.5}$$

Substituting Eqs. (3.13.1), (3.13.2), and (3.13.4) into (3.13.5) gives

$$u(t) = \frac{a_0}{k} + \sum_{j=1}^{\infty} \frac{1}{k} \frac{1}{(1 - \beta_j^2) + (2\zeta\beta_j)^2} \left\{ \left[a_j (2\zeta\beta_j) + b_j (1 - \beta_j^2) \right] \sin(j\omega_0 t) \right.$$
$$\left. + \left[a_j (1 - \beta_j^2) - b_j (2\zeta\beta_j) \right] \cos(j\omega_0 t) \right\} \tag{3.13.6}$$

The response $u(t)$ is a periodic function with period T_0.

The relative contributions of the various harmonic terms in Eq. (3.13.6) depend on two factors: (1) the amplitudes a_j and b_j of the harmonic components of the forcing function $p(t)$, and (2) the frequency ratio β_j. The response will be dominated by those

†The notation u_0 used here includes the subscript zero consistent with a_0; this should not be confused with u_o with the subscript "oh" used earlier to denote the maximum value of $u(t)$.

harmonic components for which β_j is close to unity [i.e., the forcing frequency $j\omega_0$ is close to the natural frequency (see Fig. 3.2.6)].

Example 3.8

The periodic force shown in Fig. E3.8a is defined by

$$p(t) = \begin{cases} p_o & 0 \le t \le T_0/2 \\ -p_o & T_0/2 \le t \le T_0 \end{cases} \tag{a}$$

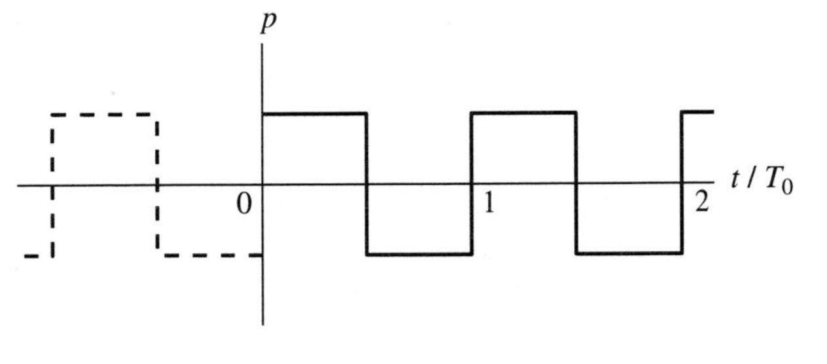

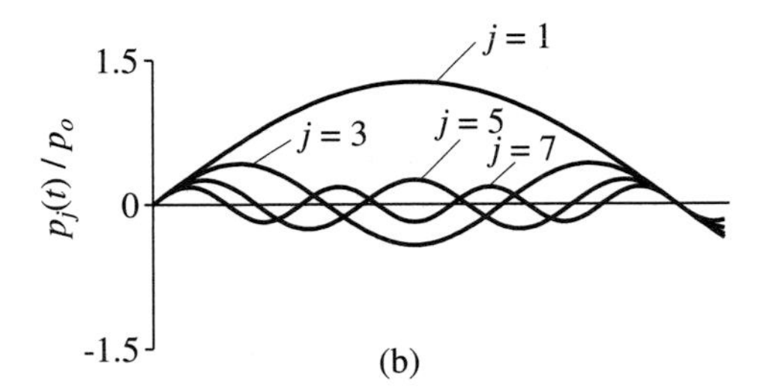

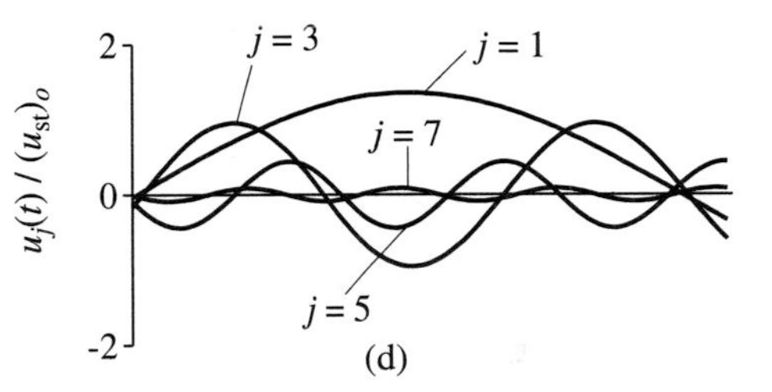

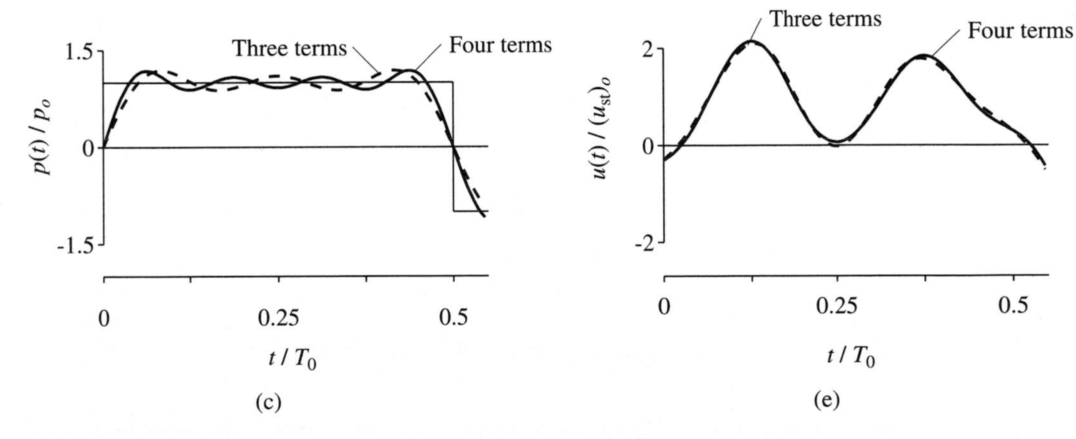

Figure E3.8

Substituting this in Eqs. (3.12.3) to (3.12.5) gives the Fourier series coefficients:

$$a_0 = \frac{1}{T_0}\int_0^{T_0} p(t)\, dt = 0 \tag{b}$$

$$a_j = \frac{2}{T_0}\int_0^{T_0} p(t)\cos(j\omega_0 t)\, dt$$

$$= \frac{2}{T_0}\left[p_0\int_0^{T_0/2} \cos(j\omega_0 t)\, dt + (-p_0)\int_{T_0/2}^{T_0} \cos(j\omega_0 t)\, dt\right] = 0 \tag{c}$$

$$b_j = \frac{2}{T_0}\int_0^{T_0} p(t)\sin(j\omega_0 t)\, dt$$

$$= \frac{2}{T_0}\left[p_0\int_0^{T_0/2} \sin(j\omega_0 t)\, dt + (-p_0)\int_{T_0/2}^{T_0} \sin(j\omega_0 t)\, dt\right]$$

$$= \begin{cases} 0 & j \text{ even} \\ 4p_0/j\pi & j \text{ odd} \end{cases} \tag{d}$$

Thus the Fourier series representation of $p(t)$ is

$$p(t) = \sum p_j(t) = \frac{4p_0}{\pi}\sum_{j=1,3,5}^{\infty} \frac{1}{j}\sin(j\omega_0 t) \tag{e}$$

The first four terms of this series are shown in Fig. E3.8b, where the frequencies and relative amplitudes—1, $\frac{1}{3}$, $\frac{1}{5}$, and $\frac{1}{7}$—of the four harmonics are apparent. The cumulative sum of the Fourier terms is shown in Fig. E3.8c, where four terms provide a reasonable representation of the forcing function. At $t = T_0/2$, where $p(t)$ is discontinuous, the Fourier series converges to zero, the average value of $p(T_0/2)$.

The response of an SDF system to the forcing function of Eq. (e) is obtained by substituting Eqs. (b), (c), and (d) in Eq. (3.13.6) to obtain

$$u(t) = (u_{st})_o \frac{4}{\pi}\sum_{j=1,3,5}^{\infty} \frac{1}{j}\frac{(1-\beta_j^2)^2\sin(j\omega_0 t) - 2\zeta\beta_j\cos(j\omega_0 t)}{(1-\beta_j^2)^2 + (2\zeta\beta_j)^2} \tag{f}$$

Shown in Fig. E3.8d are the responses of an SDF system with natural period $T_n = T_0/4$ to the first four loading terms in the Fourier series of Eq. (e). These are plots of individual terms in Eq. (f) with $\beta_j = j\omega_0/\omega_n = jT_n/T_0 = j/4$. The relative amplitudes of these terms are apparent. None of them is especially large because none of the β_j values is especially close to unity; note that $\beta_j = \frac{1}{4}, \frac{3}{4}, \frac{5}{4}, \frac{7}{4}$, and so on. The cumulative sum of the individual response terms of Eq. (f) is shown in Fig. E3.8e, where the contribution of the fourth term is seen to be small. The higher terms would be even smaller because the amplitudes of the harmonic components of $p(t)$ decrease with j and β_j would be even farther from unity.

FURTHER READING

Blake, R. E., "Basic Vibration Theory," Chapter 2 in *Shock and Vibration Handbook,* 3rd ed. (ed. C. M. Harris), McGraw-Hill, New York, 1988.

Hudson, D. E., *Reading and Interpreting Strong Motion Accelerograms,* Earthquake Engineering Research Institute, Berkeley, Calif., 1979.

Jacobsen, L. S., and Ayre, R. S., *Engineering Vibrations,* McGraw-Hill, New York, 1958, Section 5.8.

APPENDIX 3: FOUR-WAY LOGARITHMIC GRAPH PAPER

R_v is plotted as a function of ω/ω_n on log-log graph paper [i.e., $\log R_v$ is the ordinate and $\log(\omega/\omega_n)$ the abscissa]. Equation (3.2.21) gives

$$\log R_v = \log \frac{\omega}{\omega_n} + \log R_d \tag{A3.1}$$

If R_d is a constant, Eq. (A3.1) represents a straight line with slope of $+1$. Grid lines showing constant R_d would therefore be straight lines of slope $+1$, and the R_d-axis would be perpendicular to them (Fig. A3.1). Equation (3.2.21) also gives

$$\log R_v = -\log \frac{\omega}{\omega_n} + \log R_a \tag{A3.2}$$

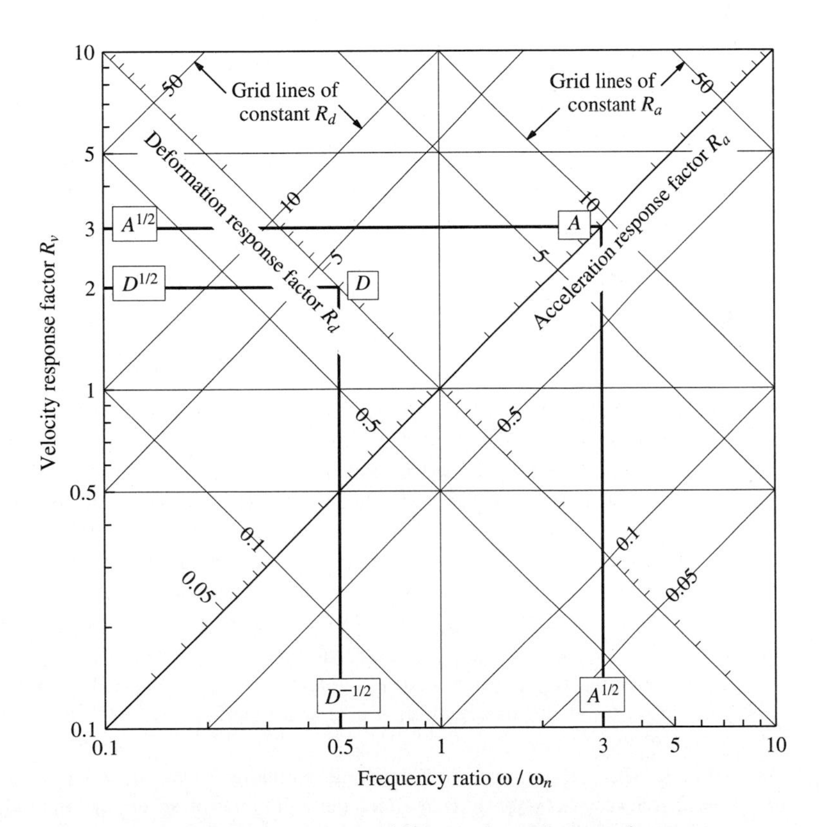

Figure A3.1 Construction of four-way logarithmic graph paper.

If R_a is a constant, Eq. (A3.2) represents a straight line with slope of -1. Grid lines showing constant R_a would be straight lines of slope -1, and the R_a-axis would be perpendicular to them (Fig. A3.1).

With reference to Fig. A3.1, the scales are established as follows:

1. With the point $(R_v = 1, \omega/\omega_n = 1)$ as the origin, draw a vertical R_v-axis and a horizontal ω/ω_n-axis with *equal* logarithmic scales.
2. The mark A on the R_a-axis would be located at the point $(R_v = A^{1/2}, \omega/\omega_n = A^{1/2})$ in order to satisfy

$$R_a = \frac{\omega}{\omega_n} R_v \tag{A3.3}$$

   R_v and ω/ω_n are taken to be equal because the R_a-axis has a slope of $+1$. This procedure is shown for $A = 9$, leading to the scale marks 3 on the R_v and ω/ω_n axes.
3. The mark D on the R_d-axis would be located at the point $(R_v = D^{1/2}, \omega/\omega_n = D^{-1/2})$ in order to satisfy

$$R_d = R_v \div \frac{\omega}{\omega_n} \tag{A3.4}$$

   and the condition that the R_d-axis has a slope of -1. This procedure is shown for $D = 4$, leading to the scale mark 2 on the R_v-axis and to the scale mark $\frac{1}{2}$ on the ω/ω_n-axis.

The logarithmic scales along the R_d and R_a axes are equal but not the same as the R_v and ω/ω_n scales.

PROBLEMS

Part A

3.1 The mass m, stiffness k, and natural frequency ω_n of an undamped SDF system are unknown. These properties are to be determined by harmonic excitation tests. At an excitation frequency of 4 Hz, the response tends to increase without bound (i.e., a resonant condition). Next, a weight $\Delta w = 5$ lb is attached to the mass m and the resonance test is repeated. This time resonance occurs at $f = 3$ Hz. Determine the mass and the stiffness of the system.

3.2 An SDF system is excited by a sinusoidal force. At resonance the amplitude of displacement was measured to be 2 in. At an exciting frequency of one-tenth the natural frequency of the system, the displacement amplitude was measured to be 0.2 in. Estimate the damping ratio of the system.

3.3 In a forced vibration test under harmonic excitation it was noted that the amplitude of motion at resonance was exactly four times the amplitude at an excitation frequency 20% higher than the resonant frequency. Determine the damping ratio of the system.

3.4 A machine is supported on four steel springs for which damping can be neglected. The natural frequency of vertical vibration of the machine–spring system is 200 cycles per minute. The machine generates a vertical force $p(t) = p_0 \sin \omega t$. The amplitude of the resulting steady-state vertical displacement of the machine is $u_o = 0.2$ in. when the machine is running at 20 revolutions per minute (rpm), 1.042 in. at 180 rpm, and 0.0248 in. at 600 rpm. Calculate the amplitude of vertical motion of the machine if the steel springs are replaced by four rubber isolators which provide the same stiffness, but introduce damping equivalent to $\zeta = 25\%$ for the system. Comment on the effectiveness of the isolators at various machine speeds.

3.5 An air-conditioning unit weighing 1200 lb is bolted at the middle of two parallel simply supported steel beams (Fig. P3.5). The clear span of the beams is 8 ft. The second moment of cross-sectional area of each beam is 10 in^4. The motor in the unit runs at 300 rpm and produces an unbalanced force of 60 lb at this speed. Neglect the weight of the beams and assume 1% viscous damping in the system; for steel $E = 30,000$ ksi. Determine the amplitudes of steady-state deflection and steady-state acceleration (in g's) of the beams at their midpoints which result from the unbalanced force.

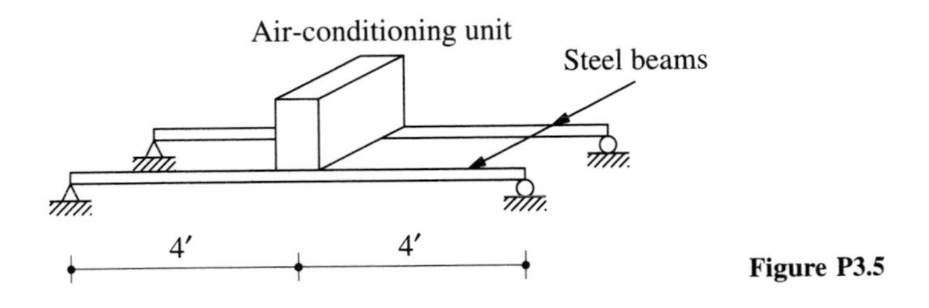

Figure P3.5

3.6 **(a)** Show that the steady-state response of an SDF system to a cosine force, $p(t) = p_o \cos \omega t$, is given by

$$u(t) = \frac{p_o}{k} \frac{\left[1 - (\omega/\omega_n)^2\right] \cos \omega t + \left[2\zeta(\omega/\omega_n)\right] \sin \omega t}{\left[1 - (\omega/\omega_n)^2\right]^2 + \left[2\zeta(\omega/\omega_n)\right]^2}$$

(b) Show that the maximum deformation due to cosine force is the same as that due to sinusoidal force.

3.7 **(a)** Show that $\omega_r = \omega_n(1 - 2\zeta^2)^{1/2}$ is the resonant frequency for displacement amplitude of an SDF system.

(b) Determine the displacement amplitude at resonance.

3.8 **(a)** Show that $\omega_r = \omega_n(1 - 2\zeta^2)^{-1/2}$ is the resonant frequency for acceleration amplitude of an SDF system.

(b) Determine the acceleration amplitude at resonance.

Part B

3.9 A one-story reinforced concrete building has a roof mass of 500 kips/g, and its natural frequency is 4 Hz. This building is excited by a vibration generator with two weights, each 50 lb, rotating about a vertical axis at an eccentricity of 12 in. When the vibration generator runs at the natural frequency of the building, the amplitude of roof acceleration is measured to be 0.02g. Determine the damping of the structure.

3.10 The steady-state acceleration amplitude of a structure caused by an eccentric-mass vibration generator was measured for several excitation frequencies. These data are as follows:

Frequency (Hz)	Acceleration (10^{-3}g)	Frequency (Hz)	Acceleration (10^{-3}g)
1.337	0.68	1.500	7.10
1.378	0.90	1.513	5.40
1.400	1.15	1.520	4.70
1.417	1.50	1.530	3.80
1.438	2.20	1.540	3.40
1.453	3.05	1.550	3.10
1.462	4.00	1.567	2.60
1.477	7.00	1.605	1.95
1.487	8.60	1.628	1.70
1.493	8.15	1.658	1.50
1.497	7.60		

Determine the natural frequency and damping ratio of the structure.

3.11 Consider an industrial machine of mass m supported on spring-type isolators of total stiffness k. The machine operates at a frequency of f hertz with a force unbalance p_o.
(a) Determine an expression giving the fraction of force transmitted to the foundation as a function of the forcing frequency f and the static deflection $\delta_{st} = mg/k$.
(b) Determine the static deflection δ_{st} for the force transmitted to be 10% of p_o if $f = 20$ Hz.

3.12 For the automobile in Example 3.4, determine the amplitude of the force developed in the spring of the suspension system.

3.13 A vibration isolation block is to be installed in a laboratory so that the vibration from adjacent factory operations will not disturb certain experiments (Fig. P3.13). If the isolation block weighs 2000 lb and the surrounding floor and foundation vibrate at 1500 cycles per minute, determine the stiffness of the isolation system such that the motion of the isolation block is limited to 10% of the floor vibration; neglect damping.

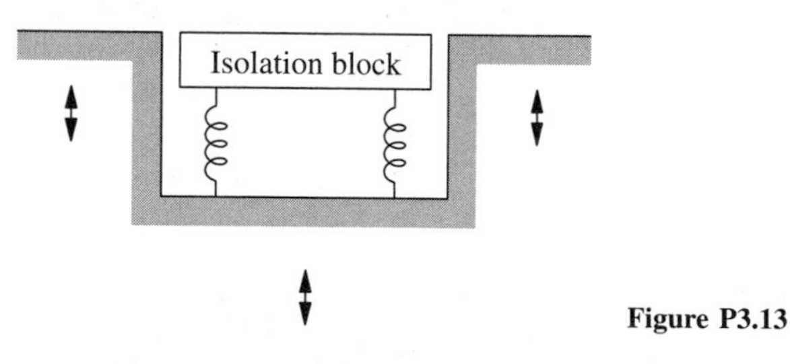

Figure P3.13

3.14 An SDF system is subjected to support displacement $u_g(t) = u_{go} \sin \omega t$. Show that the amplitude u_o^t of the total displacement of the mass is given by Eq. (3.6.5).

3.15 The natural frequency of an accelerometer is 50 Hz, and its damping ratio is 70%. Compute the recorded acceleration as a function of time if the input acceleration is $\ddot{u}_g(t) = 0.1\text{g} \sin(2\pi f t)$ for $f = 10$, 20, and 40 Hz. A comparison of the input and recorded accelerations was presented in Fig. 3.7.3. The accelerometer is calibrated to read the input acceleration correctly at very low values of the excitation frequency. What would be the error in the measured amplitude at each of the given excitation frequencies?

3.16 An accelerometer has the natural frequency $f_n = 25$ Hz and damping ratio $\zeta = 60\%$. Write an equation for the response $u(t)$ of the instrument as a function of time if the input acceleration is $\ddot{u}_g(t) = \ddot{u}_{go} \sin(2\pi f t)$. Sketch the ratio $\omega_n^2 u_o / \ddot{u}_{go}$ as a function of f/f_n. The accelerometer is calibrated to read the input acceleration correctly at very low values of the excitation frequency. Determine the range of frequencies for which the acceleration amplitude can be measured with an accuracy of $\pm 1\%$. Identify this frequency range on the above-mentioned plot.

3.17 The natural frequency of an accelerometer is $f_n = 50$ Hz, and its damping ratio is $\zeta = 70\%$. Solve Problem 3.16 for this accelerometer.

3.18 If a displacement-measuring instrument is used to determine amplitudes of vibration at frequencies very much higher than its own natural frequency, what would be the optimum instrument damping for maximum accuracy?

3.19 A displacement meter has a natural frequency $f_n = 0.5$ Hz and a damping ratio $\zeta = 0.6$. Determine the range of frequencies for which the displacement amplitude can be measured with an accuracy of $\pm 1\%$.

3.20 Repeat Problem 3.19 for $\zeta = 0.7$.

3.21 Show that the energy dissipated per cycle for viscous damping can be expressed by

$$E_D = \frac{\pi p_o^2}{k} \frac{2\zeta(\omega/\omega_n)}{\left[1 - (\omega/\omega_n)^2\right]^2 + [2\zeta(\omega/\omega_n)]^2}$$

3.22 Show that for viscous damping the loss factor ξ is independent of the amplitude and proportional to the frequency.

Part C

3.23 The properties of the SDF system of Fig. P2.16 are as follows: $w = 500$ kips, $F = 50$ kips, and $T_n = 0.25$ sec. Determine an approximate value for the displacement amplitude due to harmonic force with amplitude 100 kips and period 0.30 sec.

Part D

3.24 An SDF system with natural period T_n and damping ratio ζ is subjected to the periodic force shown in Fig. P3.24 with an amplitude p_o and period T_0.

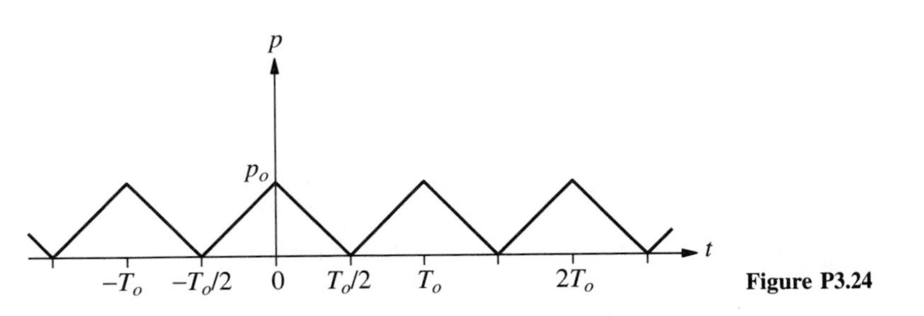

Figure P3.24

(a) Expand the forcing function in its Fourier series.
(b) Determine the steady-state response of an undamped system. For what values of T_0 is the solution indeterminate?
(c) For $T_0/T_n = 2$ determine and plot the response to individual terms in the Fourier series. How many terms are necessary to obtain reasonable convergence of the series solution?

Response to Arbitrary, Step, and Pulse Excitations

PREVIEW

In many practical situations the dynamic excitation is neither harmonic nor periodic. Thus we are interested in studying the dynamic response of SDF systems to excitations varying arbitrarily with time. A general result for linear systems, Duhamel's integral, is derived in Part A of this chapter. This result is used in Part B to study the response of systems to step force, linearly increasing force, and step force with finite rise time. These results demonstrate how the dynamic response of the system is affected by the rise time.

An important class of excitations that consist of essentially a single pulse is considered in Part C. The time variation of the response to three different force pulses is studied, and the concept of shock spectrum is introduced to present graphically the maximum response as a function of t_d/T_n, the ratio of pulse duration to the natural vibration period. It is then demonstrated that the response to short pulses is essentially independent of the pulse shape and that the response can be determined using only the pulse area. Most of the analyses and results presented are for systems without damping because the effect of damping on the response to a single pulse excitation is usually not important; this is demonstrated toward the end of the chapter.

PART A: RESPONSE TO ARBITRARILY TIME-VARYING FORCES

In this part a general procedure is developed to analyze the response of an SDF system subjected to force $p(t)$ varying arbitrarily with time. This result will enable analytical evaluation of response to forces described by simple functions of time.

We seek the solution of the differential equation of motion

$$m\ddot{u} + c\dot{u} + ku = p(t)$$

subject to the initial conditions

$$u(0) = 0 \qquad \dot{u}(0) = 0$$

In developing the general solution, $p(t)$ is interpreted as a sequence of impulses of infinitesimal duration, and the response of the system to $p(t)$ is the sum of the responses to individual impulses. These individual responses can conveniently be written in terms of the response of the system to a unit impulse.

4.1 RESPONSE TO UNIT IMPULSE

A very large force that acts for a very short time but with a time integral that is finite is called an *impulsive* force. Shown in Fig. 4.1.1 is the force $p(t) = 1/\varepsilon$, with time duration ε starting at the time instant $t = \tau$. As ε approaches zero the force becomes infinite; however, the *magnitude of the impulse*, defined by the time integral of $p(t)$, remains equal to unity. Such a force in the limiting case $\varepsilon \to 0$ is called the *unit impulse*. The *Dirac delta function* $\delta(t - \tau)$ mathematically defines a unit impulse centered at $t = \tau$.

According to Newton's second law of motion, if a force p acts on a body of mass m, the rate of change of momentum of the body is equal to the applied force, that is,

$$\frac{d}{dt}(m\dot{u}) = p \tag{4.1.1}$$

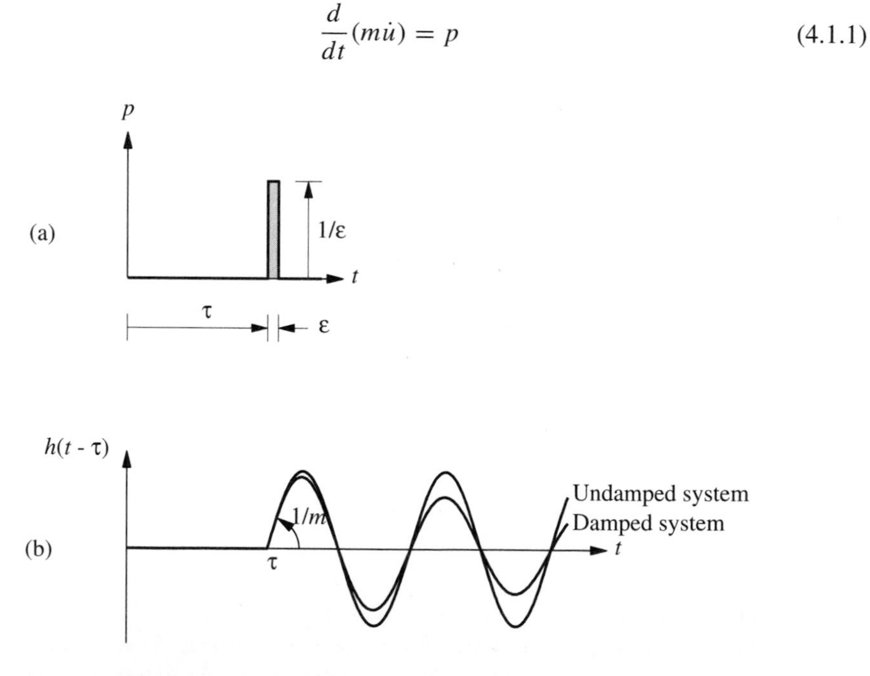

Figure 4.1.1 (a) Unit impulse; (b) response to unit impulse.

For constant mass, this equation becomes

$$p = m\ddot{u} \tag{4.1.2}$$

Integrating both sides with respect to t gives

$$\int_{t_1}^{t_2} p\,dt = m(\dot{u}_2 - \dot{u}_1) = m\,\Delta\dot{u} \tag{4.1.3}$$

The integral on the left side of this equation is the magnitude of the *impulse*. The product of mass and velocity is the *momentum*. Thus Eq. (4.1.3) states that the magnitude of the impulse is equal to the change in momentum.

This result is also applicable to an SDF mass–spring–damper system if the spring or damper has no effect. Such is the case if the force acts for an infinitesimally short duration so that the spring and damper have no time to respond. Thus a unit impulse at $t = \tau$ imparts to the mass, m, the velocity [from Eq. (4.1.3)]

$$\dot{u}(\tau) = \frac{1}{m} \tag{4.1.4}$$

but the displacement is zero prior to and up to the impulse:

$$u(\tau) = 0 \tag{4.1.5}$$

A unit impulse causes free vibration of the SDF system due to the initial displacement and velocity given by Eqs. (4.1.4) and (4.1.5). Substituting these in Eq. (2.1.3) gives the response of undamped systems:

$$h(t - \tau) \equiv u(t) = \frac{1}{m\omega_n}\sin[\omega_n(t - \tau)] \qquad t \geq \tau \tag{4.1.6}$$

Similarly, Eq. (2.2.4) provides the result for viscously damped systems:

$$h(t - \tau) \equiv u(t) = \frac{1}{m\omega_D}e^{-\zeta\omega_n(t-\tau)}\sin[\omega_D(t - \tau)] \qquad t \geq \tau \tag{4.1.7}$$

These *unit impulse response functions*, denoted by $h(t - \tau)$, are shown in Fig. 4.1.1b.

4.2 RESPONSE TO ARBITRARY FORCE

A force $p(t)$ varying arbitrarily with time can be represented as a sequence of infinitesimally short impulses (Fig. 4.2.1). The response of a linear dynamic system to one of these impulses, the one at time τ of magnitude $p(\tau)\,d\tau$, is this magnitude times the unit impulse response function:

$$du(t) = [p(\tau)\,d\tau]h(t - \tau) \qquad t > \tau \tag{4.2.1}$$

The response of the system at time t is the sum of the responses to all impulses up to that time. Thus

$$u(t) = \int_0^t p(\tau)h(t - \tau)\,d\tau \tag{4.2.2}$$

This is known as the *convolution integral*, a general result that applies to any linear dynamic system.

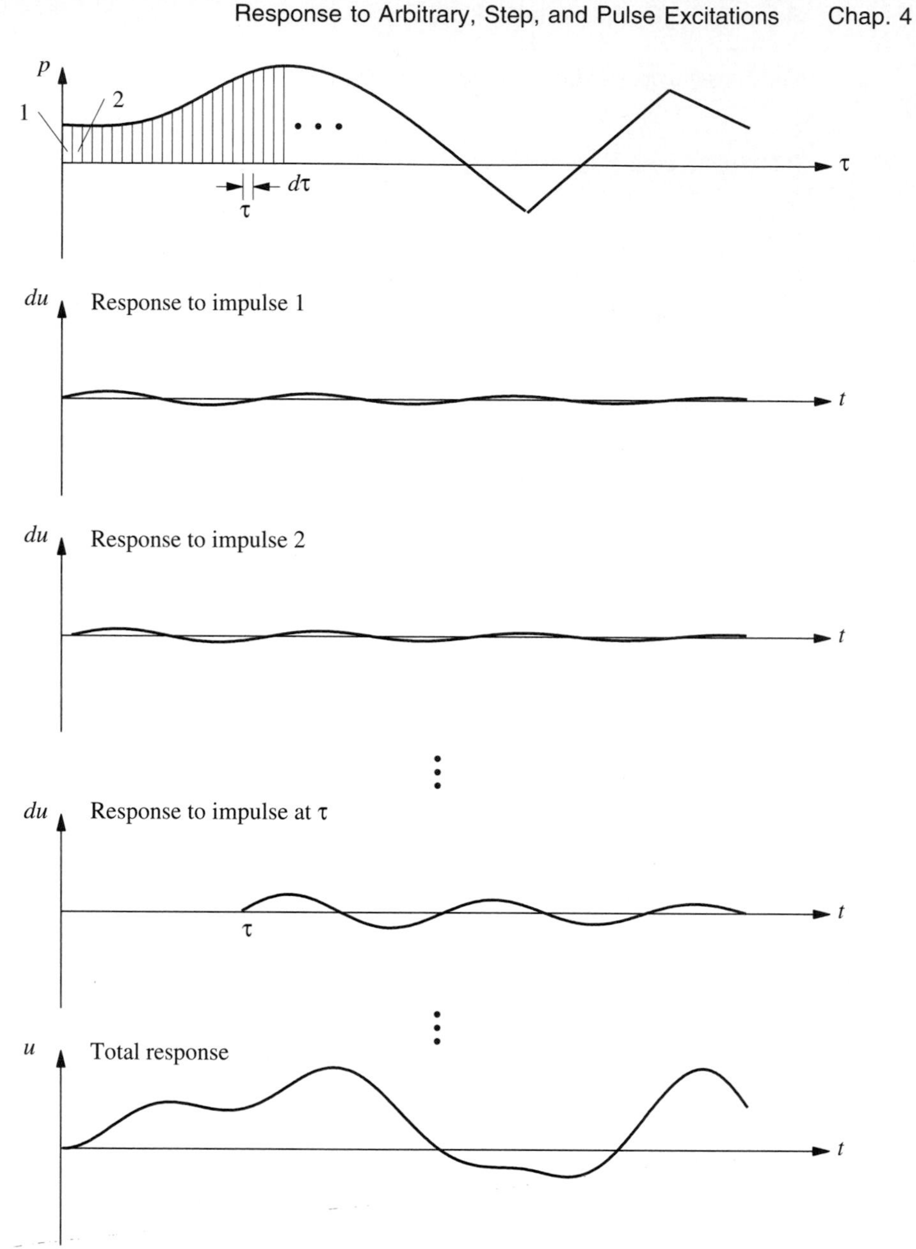

Figure 4.2.1 Schematic explanation of convolution integral.

Specializing Eq. (4.2.2) for the SDF system by substituting Eq. (4.1.7) for the unit impulse response function gives *Duhamel's integral*:

$$u(t) = \frac{1}{m\omega_D} \int_0^t p(\tau) e^{-\zeta\omega_n(t-\tau)} \sin[\omega_D(t-\tau)]\, d\tau \tag{4.2.3}$$

For an undamped system this result simplifies to

$$u(t) = \frac{1}{m\omega_n} \int_0^t p(\tau) \sin [\omega_n (t - \tau)] \, d\tau \tag{4.2.4}$$

Implicit in this result are "at rest" initial conditions, $u(0) = 0$ and $\dot{u}(0) = 0$. If the initial displacement and velocity are $u(0)$ and $\dot{u}(0)$, the resulting free vibration response given by Eqs. (2.2.4) and (2.1.3) should be added to Eqs. (4.2.3) and (4.2.4), respectively. Recall that we had used Eq. (4.2.4) in Section 1.10.2, where four methods for solving the equation of motion were introduced.

Duhamel's integral provides a general result for evaluating the response of a linear SDF system to arbitrary force. This result is restricted to linear systems because it is based on the principle of superposition. Thus it does not apply to structures deforming beyond their linearly elastic limit. If $p(\tau)$ is a simple function, closed-form evaluation of the integral is possible. Then the Duhamel's integral method is an alternative to the classical method for solving differential equations (Section 1.10.1). If $p(\tau)$ is a complicated function that is described numerically, evaluation of the integral requires numerical methods. These will not be presented in this book, however, because they are not particularly efficient. More effective methods for numerical solution of the equation of motion are presented in Chapter 5.

PART B: RESPONSE TO STEP AND RAMP FORCES

4.3 STEP FORCE

A *step force* jumps suddenly from zero to p_o and stays constant at that value (Fig. 4.3.1b). It is desired to determine the response of an undamped SDF system (Fig. 4.3.1a) starting

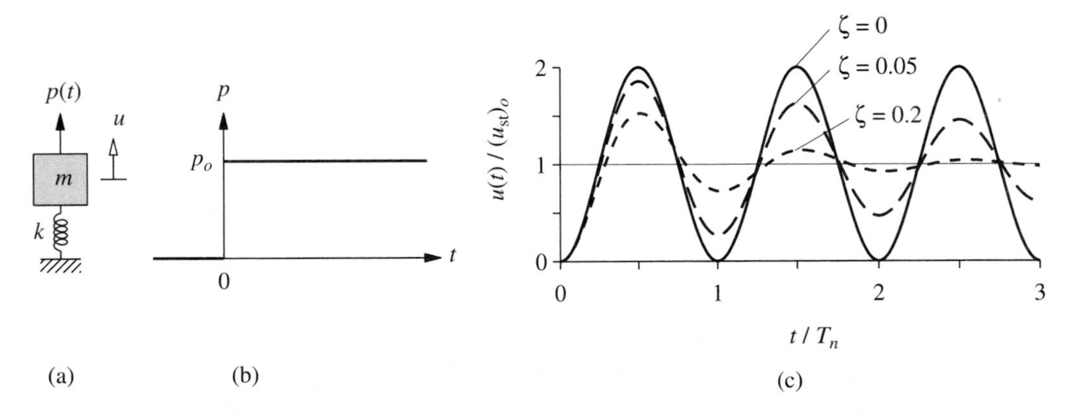

Figure 4.3.1 (a) SDF system; (b) step force; (c) dynamic response.

at rest to the step force:

$$p(t) = p_o \tag{4.3.1}$$

The equation of motion has been solved (Section 1.10.2) using Duhamel's integral to obtain

$$u(t) = (u_{\text{st}})_o(1 - \cos \omega_n t) = (u_{\text{st}})_o \left(1 - \cos \frac{2\pi t}{T_n}\right) \tag{4.3.2}$$

where $(u_{\text{st}})_o = p_o/k$, the static deformation due to force p_o.

The normalized deformation or displacement, $u(t)/(u_{\text{st}})_o$, is plotted against normalized time, t/T_n, in Fig. 4.3.1c. It is seen that the system oscillates at its natural period about a new equilibrium position, which is displaced through $(u_{\text{st}})_o$ from the original equilibrium position of $u = 0$. The maximum displacement can be determined by differentiating Eq. (4.3.2) and setting $\dot{u}(t)$ to zero, which gives $\omega_n \sin \omega_n t = 0$. The values t_o of t that satisfy this condition are

$$\omega_n t_o = j\pi \qquad \text{or} \qquad t_o = \frac{j}{2} T_n \tag{4.3.3}$$

where j is an odd integer; even integers correspond to minimum values of $u(t)$. The maximum value u_o of $u(t)$ is given by Eq. (4.3.2) evaluated at $t = t_o$; these maxima are all the same:

$$u_o = 2(u_{\text{st}})_o \tag{4.3.4}$$

Thus a suddenly applied force produces twice the deformation it would have caused as a slowly applied force.

The response of a system with damping can be determined by substituting Eq. (4.3.1) in Eq. (4.2.3) and evaluating Duhamel's integral to obtain

$$u(t) = (u_{\text{st}})_o \left[1 - e^{-\zeta\omega_n t}\left(\cos \omega_D t + \frac{\zeta}{\sqrt{1-\zeta^2}} \sin \omega_D t\right)\right] \tag{4.3.5}$$

For analysis of damped systems the classical method (Section 1.10.1) may be easier, however, than evaluating Duhamel's integral. The differential equation to be solved is

$$m\ddot{u} + c\dot{u} + ku = p_o \tag{4.3.6}$$

Its complementary solution is given by Eq. (f) of Derivation 2.2, the particular solution is $u_p = p_o/k$, and the complete solution is

$$u(t) = e^{-\zeta\omega_n t}(A \cos \omega_D t + B \sin \omega_D t) + \frac{p_o}{k} \tag{4.3.7}$$

where the constants A and B are to be determined from initial conditions. For a system starting from rest, $u(0) = \dot{u}(0) = 0$ and

$$A = -\frac{p_o}{k} \qquad B = -\frac{p_o}{k}\frac{\zeta}{\sqrt{1-\zeta^2}}$$

Substituting these constants in Eq. (4.3.7) gives the same result as Eq. (4.3.5). When specialized for undamped systems this result reduces to Eq. (4.3.2), already presented in Fig. 4.3.1c.

Equation (4.3.5) is plotted in Fig. 4.3.1c for two additional values of the damping ratio. With damping the overshoot beyond the static equilibrium position is smaller, and the oscillations about this position decay with time. The damping ratio determines the amount of overshoot and the rate at which the oscillations decay. Eventually, the system settles down to the static deformation. This is the steady-state deformation.

4.4 RAMP OR LINEARLY INCREASING FORCE

In Fig. 4.4.1b, the applied force $p(t)$ increases linearly with time. Naturally, it cannot increase indefinitely, but our interest is confined to the time duration where $p(t)$ is still small enough that the resulting spring force is within the linearly elastic limit of the spring.

While the equation of motion can be solved by any one of several methods, we illustrate use of Duhamel's integral to obtain the solution. The applied force

$$p(t) = p_o \frac{t}{t_r} \tag{4.4.1}$$

is substituted in Eq. (4.2.4) to obtain

$$u(t) = \frac{1}{m\omega_n} \int_0^t \frac{p_o}{t_r} \tau \sin \omega_n (t - \tau)\, d\tau$$

This integral is evaluated and simplified to obtain

$$u(t) = (u_{\text{st}})_o \left(\frac{t}{t_r} - \frac{\sin \omega_n t}{\omega_n t_r} \right) \tag{4.4.2}$$

where $(u_{\text{st}})_o = p_o/k$, the static deformation due to force p_o.

Equation (4.4.2) is plotted in Fig. 4.4.1c for $t_r/T_n = 2.5$, wherein the static deformation at each time instant,

$$u_{\text{st}}(t) = \frac{p(t)}{k} \tag{4.4.3}$$

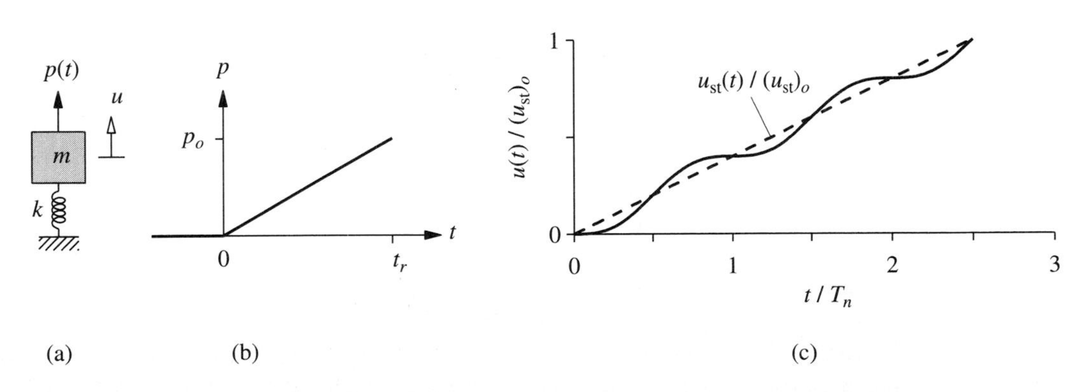

Figure 4.4.1 (a) SDF system; (b) ramp force; (c) dynamic and static responses.

is also shown; $u_{st}(t)$ varies with time in the same manner as $p(t)$ and the two differ by the scale factor $1/k$. It is seen that the system oscillates at its natural period T_n about the static solution.

4.5 STEP FORCE WITH FINITE RISE TIME

Since in reality a force can never be applied suddenly, it is of interest to consider a dynamic force that has a finite rise time, t_r, but remains constant thereafter, as shown in Fig. 4.5.1b:

$$p(t) = \begin{cases} p_o(t/t_r) & t \le t_r \\ p_o & t \ge t_r \end{cases} \tag{4.5.1}$$

The excitation has two phases: ramp or rise phase and constant phase.

For a system without damping starting from rest, the response during the ramp phase is given by Eq. (4.4.2), repeated here for convenience:

$$u(t) = (u_{st})_o \left(\frac{t}{t_r} - \frac{\sin \omega_n t}{\omega_n t_r} \right) \qquad t \le t_r \tag{4.5.2}$$

The response during the constant phase can be determined by evaluating Duhamel's integral after substituting Eq. (4.5.1) in Eq. (4.2.4). Alternatively, existing solutions for free vibration and step force could be utilized to express this response as

$$u(t) = u(t_r) \cos \omega_n (t - t_r) + \frac{\dot{u}(t_r)}{\omega_n} \sin \omega_n (t - t_r) + (u_{st})_o [1 - \cos \omega_n (t - t_r)] \tag{4.5.3}$$

The third term is the solution for a system at rest subjected to a step force starting at $t = t_r$; it is obtained from Eq. (4.3.2). The first two terms in Eq. (4.5.3) account for free vibration of the system resulting from its displacement $u(t_r)$ and velocity $\dot{u}(t_r)$ at the end of the ramp phase. Determined from Eq. (4.5.2), $u(t_r)$ and $\dot{u}(t_r)$ are substituted in Eq. (4.5.3) to obtain

$$u(t) = (u_{st})_o \left\{ 1 + \frac{1}{\omega_n t_r} \left[(1 - \cos \omega_n t_r) \sin \omega_n (t - t_r) - \sin \omega_n t_r \cos \omega_n (t - t_r) \right] \right\} \qquad t \ge t_r \tag{4.5.4a}$$

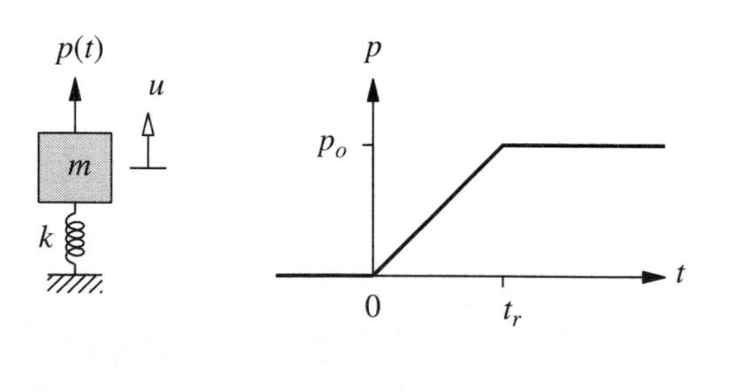

Figure 4.5.1 (a) SDF system; (b) step force with finite rise time.

This equation can be simplified, using a trigonometric identity, to

$$u(t) = (u_{st})_o \left\{ 1 - \frac{1}{\omega_n t_r} \left[\sin \omega_n t - \sin \omega_n (t - t_r) \right] \right\} \qquad t \geq t_r \qquad (4.5.4b)$$

The normalized deformation, $u(t)/(u_{st})_o$, is a function of the normalized time, t/T_n, because $\omega_n t = 2\pi(t/T_n)$. This function depends only on the ratio t_r/T_n because $\omega_n t_r = 2\pi(t_r/T_n)$, not separately on t_r and T_n. Figure 4.5.2 shows $u(t)/(u_{st})_o$ plotted against t/T_n for several values of t_r/T_n, the ratio of the rise time to the natural period. Each plot is valid for all combinations of t_r and T_n with the same ratio t_r/T_n. Also

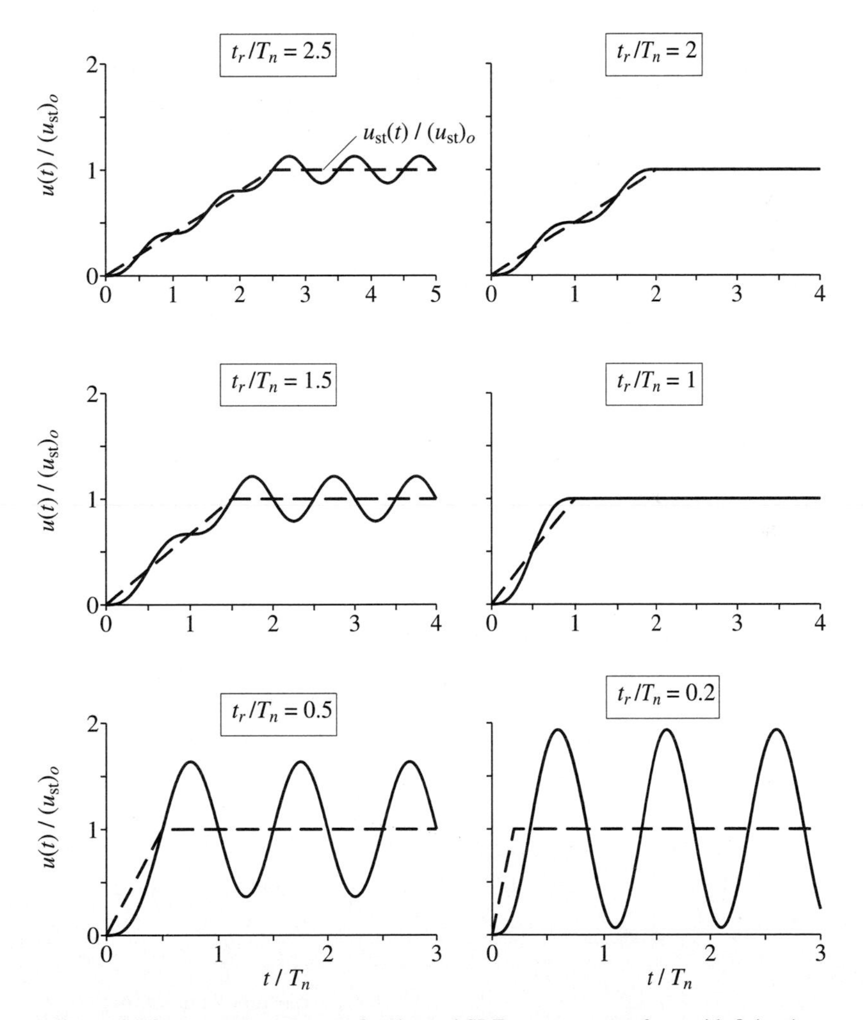

Figure 4.5.2 Dynamic response of undamped SDF system to step force with finite rise time; static solution is shown by dashed lines.

plotted is $u_{st}(t)$, the static deformation at each time instant [Eq. (4.4.3)]. These results permit several observations:

1. During the force-rise phase the system oscillates at the natural period T_n about the static solution.
2. During the constant-force phase the system oscillates also at the natural period T_n about the static solution.
3. If the velocity $\dot{u}(t_r)$ is zero at the end of the ramp, the system does not vibrate during the constant-force phase.
4. For smaller values of t_r/T_n (i.e., short rise time), the response is similar to that due to a sudden step force; see Fig. 4.3.1c.
5. For larger values of t_r/T_n, the dynamic displacement oscillates close to the static solution, implying that the dynamic effects are small (i.e., a force increasing slowly—relative to T_n—from 0 to p_o affects the system like a static force).

The deformation attains its maximum value during the constant-force portion of the response. From Eq. (4.5.4a) the maximum value of $u(t)$ is

$$u_o = (u_{st})_o \left\{ 1 + \frac{1}{\omega_n t_r} \sqrt{(1 - \cos \omega_n t_r)^2 + (\sin \omega_n t_r)^2} \right\} \tag{4.5.5}$$

Using trigonometric identities and $T_n = 2\pi/\omega_n$, Eq. (4.5.5) can be simplified to

$$R_d \equiv \frac{u_o}{(u_{st})_o} = 1 + \frac{|\sin(\pi t_r/T_n)|}{\pi t_r/T_n} \tag{4.5.6}$$

The deformation response factor R_d depends only on t_r/T_n, the ratio of the rise time to the natural period. A graphical presentation of this relationship, as in Fig. 4.5.3, is called the *response spectrum* for the step force with finite rise time.

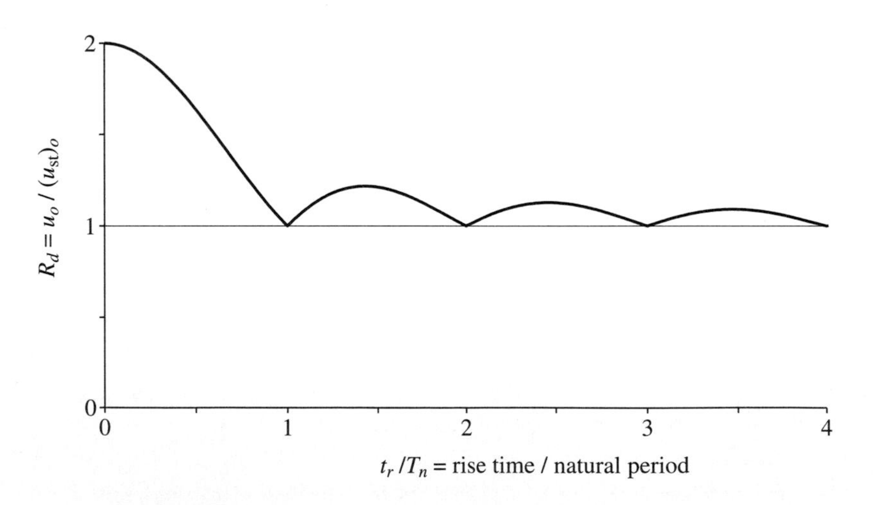

Figure 4.5.3 Response spectrum for step force with finite rise time.

This response spectrum characterizes the problem completely. In this case it contains information on the normalized maximum response, $u_o/(u_{st})_o$, of all SDF systems (without damping) due to any step force p_o with any rise time t_r. The response spectrum permits several observations:

1. If $t_r < T_n/4$ (i.e., a relatively short rise time), $u_o \simeq 2(u_{st})_o$, implying that the structure "sees" this excitation like a suddenly applied force.
2. If $t_r > 3T_n$ (i.e., a relatively long rise time), $u_o \simeq (u_{st})_o$, implying that this excitation affects the structure like a static force.
3. If $t_r/T_n = 1, 2, 3, \ldots, \dot{u}(t_r) = 0$ at the end of the force-rise phase, and the system does not oscillate during the constant-force phase; therefore, $u_o = (u_{st})_o$.

PART C: RESPONSE TO PULSE EXCITATIONS

We next consider an important class of excitations that consist of essentially a single pulse, such as shown in Fig. 4.6.1. Air pressures generated on a structure due to aboveground blasts or explosions are essentially a single pulse and can usually be idealized by simple shapes such as those shown in the left part of Fig. 4.6.2. The dynamics of structures subjected to such excitations was the subject of much work during the 1950s and 1960s.

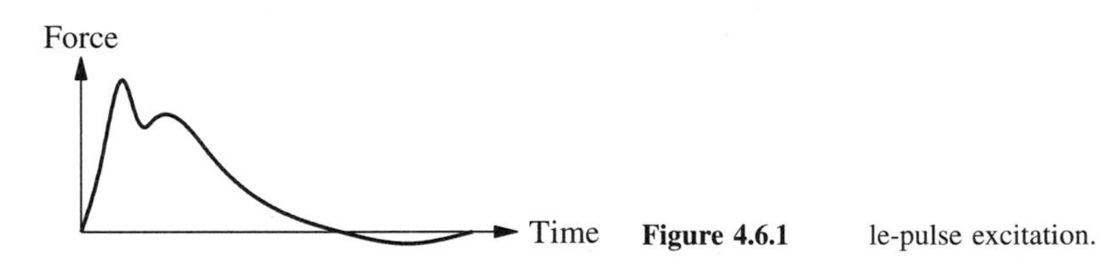

Figure 4.6.1 le-pulse excitation.

4.6 SOLUTION METHODS

The response of the system to such pulse excitations does not reach a steady-state condition; the effects of the initial conditions must be considered; but as will be shown later, damping can be ignored because it usually has little influence on the response. The response of the system to such pulse excitations can be determined by one of several analytical methods: (1) the classical method for solving differential equations, (2) evaluating Duhamel's integral, and (3) expressing the pulse as the superposition of two or more simpler functions for which response solutions are already available or easier to determine.

The last of these approaches is illustrated in Fig. 4.6.2 for three pulse forces. For example, the rectangular pulse is the step function $p_1(t)$ plus the step function $p_2(t)$ of equal amplitude, but after a time interval t_d has passed. The desired response is the sum of the responses to each of these step functions, and these responses can be determined

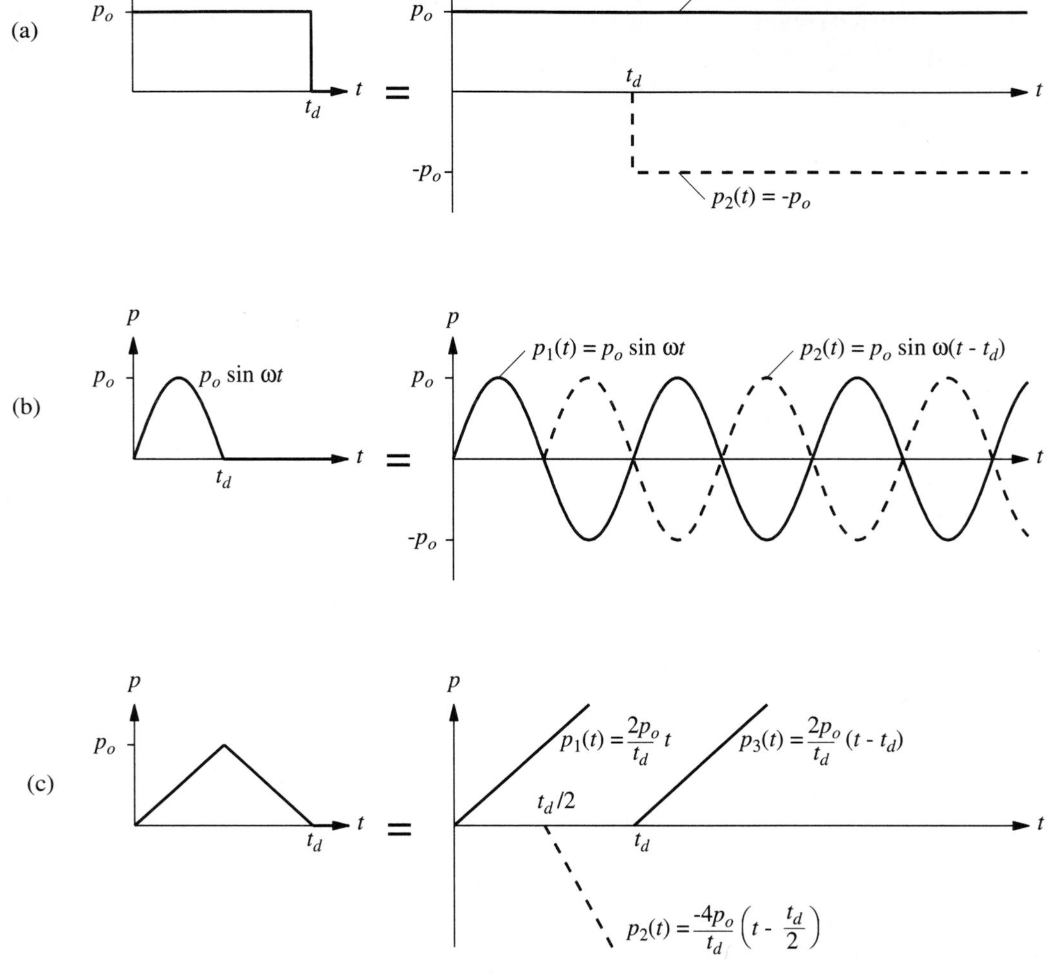

Figure 4.6.2 Expressing pulse force as superposition of simple functions: (a) rectangular pulse; (b) half-cycle sine pulse; (c) triangular pulse.

readily from the results of Section 4.3. A half-cycle sine pulse is the result of subtracting a sine function of amplitude p_o starting at $t = t_d$ from another sine function of the same frequency and amplitude starting at $t = 0$. The desired response is the difference of the transient responses to each sinusoidal force, obtained using the results of Section 3.1. Similarly, the response to the symmetrical triangular pulse is the sum of the responses to the three ramp functions in Fig. 4.6.2c; the individual responses come from Section 4.4. Thus the third method involves adapting existing results and manipulating them to obtain the desired response.

We prefer to use the classical method in evaluating the response of SDF systems to pulse forces because it is closely tied to the dynamics of the system. Using the classical method the response to pulse forces will be determined in two phases. The first is the forced vibration phase that covers the duration of the excitation. The second is the free vibration phase, which follows the end of the pulse force. Much of the presentation concerns systems without damping because, as will be shown in Section 4.11, damping has little influence on response to pulse excitations.

4.7 RECTANGULAR PULSE FORCE

We start with the simplest type of pulse, the rectangular pulse shown in Fig. 4.7.1. The equation to be solved is

$$m\ddot{u} + ku = p(t) = \begin{cases} p_o & t \le t_d \\ 0 & t \ge t_d \end{cases} \tag{4.7.1}$$

with at-rest initial conditions: $u(0) = \dot{u}(0) = 0$. The analysis is organized in two phases.

1. *Forced vibration phase.* During this phase, the system is subjected to a step force. The response of the system is given by Eq. (4.3.2), repeated for convenience:

$$\frac{u(t)}{(u_{\text{st}})_o} = 1 - \cos\omega_n t = 1 - \cos\frac{2\pi t}{T_n} \qquad t \le t_d \tag{4.7.2}$$

2. *Free vibration phase.* After the force ends at t_d, the system undergoes free vibration, defined by modifying Eq. (2.1.3) appropriately:

$$u(t) = u(t_d)\cos\omega_n(t - t_d) + \frac{\dot{u}(t_d)}{\omega_n}\sin\omega_n(t - t_d) \tag{4.7.3}$$

This free vibration is initiated by the displacement and velocity of the mass at $t = t_d$, determined from Eq. (4.7.2):

$$u(t_d) = (u_{\text{st}})_o[1 - \cos\omega_n t_d] \qquad \dot{u}(t_d) = (u_{\text{st}})_o\omega_n \sin\omega_n t_d \tag{4.7.4}$$

Substituting these in Eq. (4.7.3) gives

$$\frac{u(t)}{(u_{\text{st}})_o} = (1 - \cos\omega_n t_d)\cos\omega_n(t - t_d) + \sin\omega_n t_d \sin\omega_n(t - t_d) \qquad t \ge t_d$$

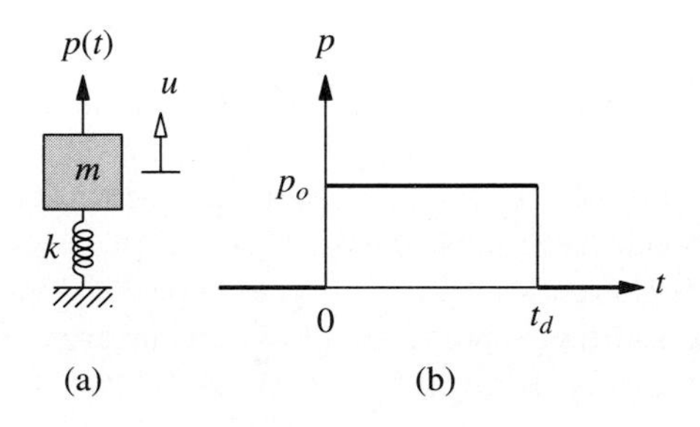

Figure 4.7.1 (a) SDF system; (b) rectangular pulse force.

which can be simplified, using a trigonometric identity, to

$$\frac{u(t)}{(u_{\text{st}})_o} = \cos\omega_n(t - t_d) - \cos\omega_n t \qquad t \geq t_d$$

Expressing $\omega_n = 2\pi/T_n$ and using trigonometric identities enables us to rewrite these equations as

$$\frac{u(t)}{(u_{\text{st}})_o} = \left(2\sin\frac{\pi t_d}{T_n}\right)\sin\left[2\pi\left(\frac{t}{T_n} - \frac{1}{2}\frac{t_d}{T_n}\right)\right] \qquad t \geq t_d \tag{4.7.5}$$

Response history. The normalized deformation $u(t)/(u_{\text{st}})_o$ given by Eqs. (4.7.2) and (4.7.5) is a function of t/T_n. It depends only on t_d/T_n, the ratio of the pulse duration to the natural vibration period of the system, not separately on t_d or T_n, and has been plotted in Fig. 4.7.2 for several values of t_d/T_n. Also shown in dashed lines is Eq. (4.4.3) for the static solution at each time instant due to $p(t)$. The nature of the response is seen to vary greatly by changing just the duration t_d of the pulse. However, no matter how long the duration, the dynamic response is not close to the static solution, because the force is suddenly applied.

While the force is applied to the structure, the system oscillates about the shifted position, $(u_{\text{st}})_o = p_o/k$, at its own natural period T_n. After the pulse has ended, the system oscillates freely about the original equilibrium position at its natural period T_n, with no decay of motion because the system is undamped. If $t_d/T_n = 1, 2, 3, \ldots$ the system stays still in its original undeformed configuration during the free vibration phase, because the displacement and velocity of the mass are zero when the force ends.

Each response result of Fig. 4.7.2 is applicable to all combinations of systems and forces with fixed t_d/T_n. Implicit in this figure, however, is the presumption that the natural period T_n of the system is constant and the pulse duration t_d varies. By modifying the time scale, the results can be presented for a fixed value of t_d and varying values of T_n.

Maximum response. Over each of the two phases, forced vibration and free vibration, separately, the maximum value of response is determined next. The larger of the two maxima is the overall maximum response.

The number of local maxima or peaks that develop in the forced vibration phase depends on t_d/T_n (Fig. 4.7.2); the longer the pulse duration, more such peaks occur. The first peak occurs at $t_o = T_n/2$ with the deformation

$$u_o = 2(u_{\text{st}})_o \tag{4.7.6}$$

consistent with the results derived in Section 4.3. Thus t_d must be longer than $T_n/2$ for at least one peak to develop during the forced vibration phase. If more than one peak develops during this phase, they all have this same value and occur at $t_o = 3T_n/2$, $5T_n/2$, and so on, again consistent with the results of Section 4.3.

As a corollary, if t_d is shorter than $T_n/2$ no peak will develop during the forced vibration phase (Fig. 4.7.2), and the response simply builds up from zero to $u(t_d)$. The displacement at the end of the pulse is given by Eq. (4.7.4a), rewritten to emphasize the

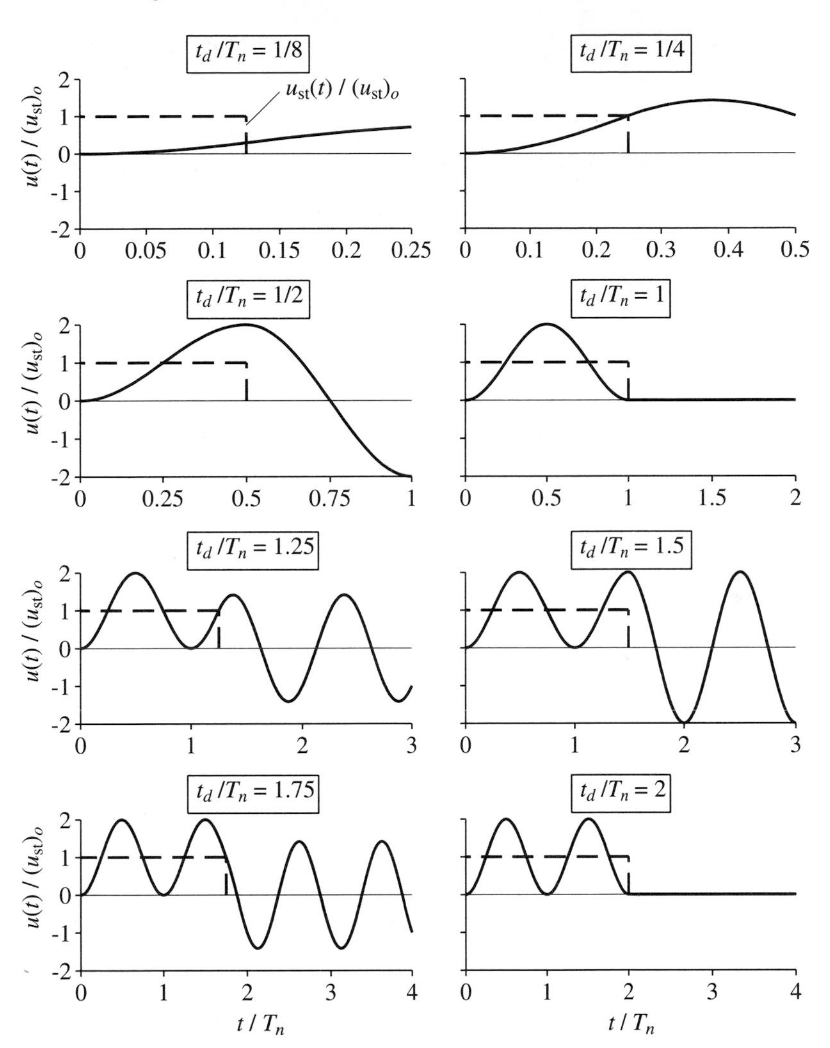

Figure 4.7.2 Dynamic response of undamped SDF system to rectangular pulse force; static solution is shown by dashed lines.

parameter t_d/T_n:

$$u(t_d) = (u_{st})_o \left(1 - \cos \frac{2\pi t_d}{T_n}\right) \tag{4.7.7}$$

The maximum deformation during the forced vibration phase, Eqs. (4.7.6) and (4.7.7), can be expressed in terms of the *deformation response factor*:

$$R_d = \frac{u_o}{(u_{st})_o} = \begin{cases} 1 - \cos(2\pi t_d/T_n) & t_d/T_n \leq \frac{1}{2} \\ 2 & t_d/T_n \geq \frac{1}{2} \end{cases} \tag{4.7.8}$$

This relationship is shown as "forced response" in Fig. 4.7.3a.

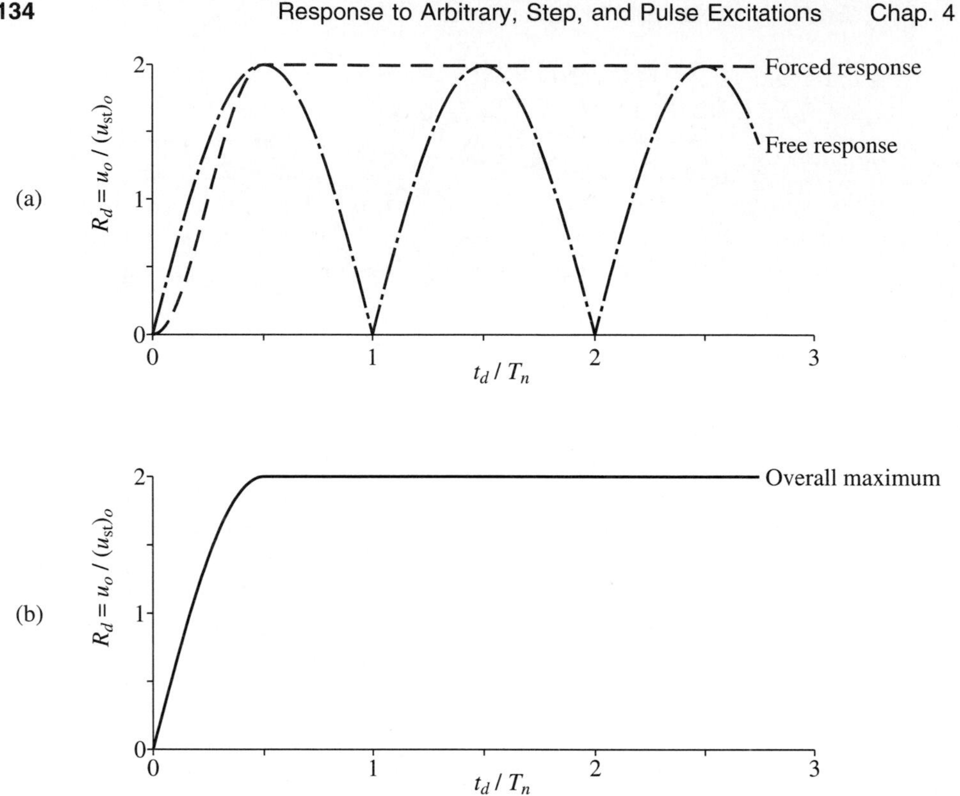

Figure 4.7.3 Response to rectangular pulse force: (a) maximum response during each of forced vibration and free vibration phases; (b) shock spectrum.

In the free vibration phase the system oscillates in simple harmonic motion, given by Eq. (4.7.3), with an amplitude

$$u_o = \sqrt{[u(t_d)]^2 + \left[\frac{\dot{u}(t_d)}{\omega_n}\right]^2} \tag{4.7.9}$$

which after substituting Eq. (4.7.4) and some manipulation becomes

$$u_o = 2(u_{\text{st}})_o \left|\sin \frac{\pi t_d}{T_n}\right| \tag{4.7.10}$$

The corresponding deformation response factor:

$$R_d \equiv \frac{u_o}{(u_{\text{st}})_o} = 2\left|\sin \frac{\pi t_d}{T_n}\right| \tag{4.7.11}$$

depends only on t_d/T_n and is shown as "free response" in Fig. 4.7.3a.

Having determined the maximum response during each of the forced and free vibration phases, we now determine the overall maximum. Figure 4.7.3a shows that if $t_d/T_n > \frac{1}{2}$, the overall maximum is the peak (or peaks because all are equal) in $u(t)$

that develops during the forced vibration phase because it will not be exceeded in free vibration. This observation can also be deduced from the mathematical results: the R_d of Eq. (4.7.11) for the free vibration phase can never exceed the $R_d = 2$, Eq. (4.7.8b), for the forced vibration phase.

If $t_d/T_n < \frac{1}{2}$, Fig. 4.7.3a shows that the overall maximum is the peak (or peaks because all are equal) in $u(t)$ that develops during the free vibration phase. In this case the response during the forced vibration phase has built up from zero at $t = 0$ to $u(t_d)$ at the end of the pulse, Eq. (4.7.7), and $\dot{u}(t_d)$ given by Eq. (4.7.4b) is positive; see Fig. 4.7.2 for $t_d/T_n = \frac{1}{8}$ or $\frac{1}{4}$. As a result, the first peak in free vibration is larger than $u(t_d)$.

Finally, if $t_d/T_n = \frac{1}{2}$, Fig. 4.7.3a shows that the overall maximum is given by either the forced-response maximum or the free-response maximum because the two are equal. The first peak occurs exactly at the end of the forced vibration phase (Fig. 4.7.2), the velocity $\dot{u}(t_d) = 0$, and the peaks in free vibration are the same as $u(t_d)$. This observation is consistent with Eqs. (4.7.8) and (4.7.11) because all of them give $R_d = 2$ for $t_d/T_n = \frac{1}{2}$.

In summary, the deformation response factor that defines the overall maximum response is

$$R_d = \frac{u_o}{(u_{\text{st}})_o} = \begin{cases} 2 \sin \pi t_d/T_n & t_d/T_n \le \frac{1}{2} \\ 2 & t_d/T_n \ge \frac{1}{2} \end{cases} \tag{4.7.12}$$

Clearly, R_d depends only on t_d/T_n, the ratio of the pulse duration to the natural period of the system. This relationship is shown in Fig. 4.7.3b.

Such a plot, which shows the maximum deformation of an SDF system as a function of the natural period T_n of the system (or a related parameter), is called a *response spectrum*. When the excitation is a single pulse, the terminology *shock spectrum* is also used for the response spectrum. Figure 4.7.3b then is the shock spectrum for a rectangular pulse force. The shock spectrum characterizes the problem completely; in this case it contains information on the maximum response of all SDF systems (without damping) to any rectangular pulse force p_o with any duration t_d.

The maximum deformation of an undamped SDF system having a natural period T_n to a rectangular pulse force of amplitude p_o and duration t_d can readily be determined if the shock spectrum for this excitation is available. Corresponding to the ratio t_d/T_n, the deformation response factor R_d is read from the spectrum, and the maximum deformation is computed from

$$u_o = (u_{\text{st}})_o R_d = \frac{p_o}{k} R_d \tag{4.7.13}$$

The maximum value of the equivalent static force (Section 1.8.2) is

$$f_{So} = k u_o = p_o R_d \tag{4.7.14}$$

that is, the applied force p_o multiplied by the deformation response factor. As mentioned in Section 1.8.2, static analysis of the structure subjected to f_{So} gives the internal forces and stresses.

Example 4.1

A one-story building, idealized as a 12-ft-high frame with two columns hinged at the base and a rigid beam, has a natural period of 0.5 sec. Each column is an American standard wide-flange steel section W8 × 18. Its properties for bending about its major axis are $I_x = 61.9$ in^4, $S = I_x/c = 15.2$ in^3; $E = 30{,}000$ ksi. Neglecting damping, determine the maximum response of this frame due to a rectangular pulse force of amplitude 4 kips and duration $t_d = 0.2$ sec. The response quantities of interest are displacement at the top of the frame and maximum bending stress in the columns.

Solution

1. *Determine R_d.*

$$\frac{t_d}{T_n} = \frac{0.2}{0.5} = 0.4$$

$$R_d = \frac{u_o}{(u_{st})_o} = 2\sin\frac{\pi t_d}{T_n} = 2\sin(0.4\pi) = 1.902$$

2. *Determine the lateral stiffness of the frame.*

$$k_{col} = \frac{3EI}{L^3} = \frac{3(30{,}000)61.9}{(12 \times 12)^3} = 1.865 \text{ kips/in.}$$

$$k = 2 \times 1.865 = 3.73 \text{ kips/in.}$$

3. *Determine $(u_{st})_o$.*

$$(u_{st})_o = \frac{p_o}{k} = \frac{4}{3.73} = 1.07 \text{ in.}$$

4. *Determine the maximum dynamic deformation.*

$$u_o = (u_{st})_o R_d = (1.07)(1.902) = 2.04 \text{ in.}$$

5. *Determine the bending stress.* The resulting bending moments in each column are shown in Fig. E4.1c. At the top of the column the bending moment is largest and is given by

$$M = \frac{3EI}{L^2}u_o = \left[\frac{3(30{,}000)61.9}{(12 \times 12)^2}\right]2.04 = 547.8 \text{ kip-in.}$$

Alternatively, we can find the bending moment from the equivalent static force:

$$f_{So} = p_o R_d = 4(1.902) = 7.61 \text{ kips}$$

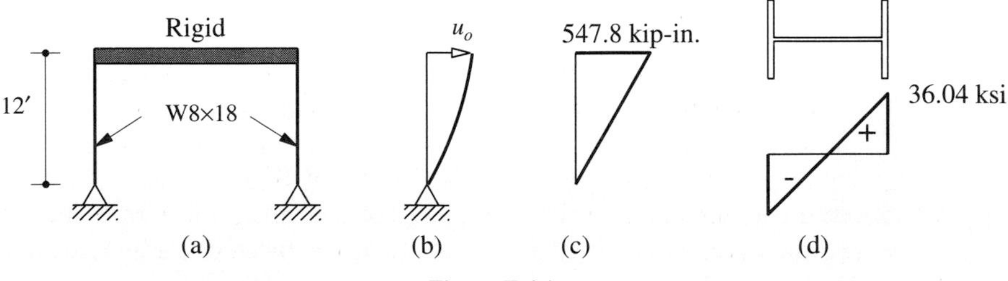

Figure E.4.1

Because both columns are identical in cross section and length, the force f_{So} will be shared equally. The bending moment at the top of the column is

$$M = \frac{f_{So}}{2} h = \left(\frac{7.61}{2} \right) 12 \times 12 = 547.8 \text{ kip-in.}$$

The bending stress is largest at the outside of the flanges at the top of the columns:

$$\sigma = \frac{M}{S} = \frac{547.8}{15.2} = 36.04 \text{ ksi}$$

The stress distribution is shown in Fig. E4.1d.

4.8 HALF-CYCLE SINE PULSE FORCE

The next pulse we consider is a half-cycle of sinusoidal force (Fig. 4.8.1b). The response analysis procedure for this pulse is the same as developed in Section 4.7 for a rectangular pulse, but the mathematical details become more complicated. The solution of the governing equation

$$m\ddot{u} + ku = p(t) = \begin{cases} p_o \sin(\pi t / t_d) & t \leq t_d \\ 0 & t \geq t_d \end{cases} \tag{4.8.1}$$

with at-rest initial conditions is presented separately for (1) $\omega \neq \omega_n$ or $t_d/T_n \neq \frac{1}{2}$ and (2) $\omega = \omega_n$ or $t_d/T_n = \frac{1}{2}$. For each case the analysis is organized in two phases: forced vibration and free vibration.

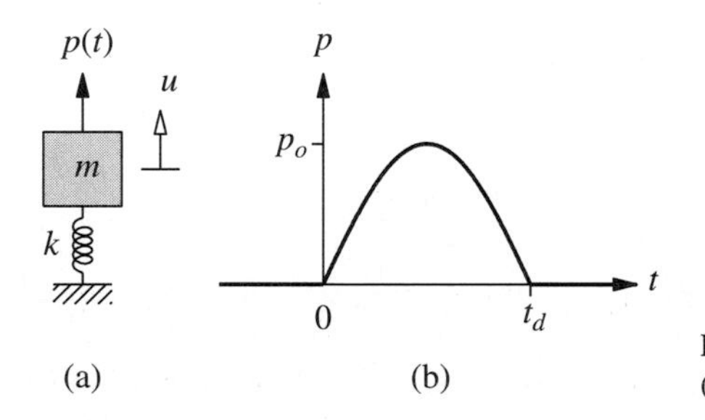

Figure 4.8.1 (a) SDF system; (b) half-cycle sine pulse force.

Case 1: $t_d/T_n \neq \frac{1}{2}$

Forced Vibration Phase. The force is the same as the harmonic force $p(t) = p_o \sin \omega t$ considered earlier with frequency $\omega = \pi / t_d$. The response of an undamped SDF system to such a force is given by Eq. (3.1.6b) in terms of ω and ω_n, the excitation and natural frequencies. The excitation frequency ω is not the most meaningful way of characterizing the pulse because, unlike a harmonic force, it is not a periodic function. A better characterization is the pulse duration t_d, which will be emphasized here. Using the relations $\omega = \pi / t_d$ and $\omega_n = 2\pi / T_n$, and defining $(u_{\text{st}})_o = p_o / k$, as before, Eq. (3.1.6b) becomes

$$\frac{u(t)}{(u_{st})_o} = \frac{1}{1-(T_n/2t_d)^2}\left[\sin\left(\pi\frac{t}{t_d}\right) - \frac{T_n}{2t_d}\sin\left(2\pi\frac{t}{T_n}\right)\right] \qquad t \le t_d \tag{4.8.2}$$

Free Vibration Phase. After the force pulse ends, the system vibrates freely with its motion described by Eq. (4.7.3). The displacement $u(t_d)$ and velocity $\dot{u}(t_d)$ at the end of the pulse are determined from Eq. (4.8.2). Substituting these in Eq. (4.7.3), using trigonometric identities, and manipulating the mathematical quantities, we obtain

$$\frac{u(t)}{(u_{st})_o} = \frac{(T_n/t_d)\cos(\pi t_d/T_n)}{(T_n/2t_d)^2 - 1}\sin\left[2\pi\left(\frac{t}{T_n} - \frac{1}{2}\frac{t_d}{T_n}\right)\right] \qquad t \ge t_d \tag{4.8.3}$$

Case 2: $t_d/T_n = \frac{1}{2}$

Forced Vibration Phase. The forced response is now given by Eq. (3.1.13b), repeated here for convenience:

$$\frac{u(t)}{(u_{st})_o} = \frac{1}{2}\left(\sin\frac{2\pi t}{T_n} - \frac{2\pi t}{T_n}\cos\frac{2\pi t}{T_n}\right) \qquad t \le t_d \tag{4.8.4}$$

Free Vibration Phase. After the force pulse ends at $t = t_d$, free vibration of the system is initiated by the displacement $u(t_d)$ and velocity $\dot{u}(t_d)$ at the end of the force pulse. Determined from Eq. (4.8.4), these are

$$\frac{u(t_d)}{(u_{st})_o} = \frac{\pi}{2} \qquad \dot{u}(t_d) = 0 \tag{4.8.5}$$

The second equation implies that the displacement in the forced vibration phase reaches its maximum at the end of this phase. Substituting Eq. (4.8.5) in Eq. (4.7.3) gives the response of the system after the pulse has ended:

$$\frac{u(t)}{(u_{st})_o} = \frac{\pi}{2}\cos 2\pi\left(\frac{t}{T_n} - \frac{1}{2}\right) \qquad t \ge t_d \tag{4.8.6}$$

Response history. The time variation of the normalized deformation, $u(t)/(u_{st})_o$, given by Eqs. (4.8.2) and (4.8.3) is plotted in Fig. 4.8.2 for several values of t_d/T_n. For the special case of $t_d/T_n = \frac{1}{2}$, Eqs. (4.8.4) and (4.8.6) describe the response of the system and these are also plotted in Fig. 4.8.2. The nature of the response is seen to vary greatly by changing just the duration t_d of the pulse. Also plotted in Fig. 4.8.2 is Eq. (4.4.3), the static solution. The difference between the two curves is an indication of the dynamic effects, which are seen to be small for $t_d = 3T_n$ because this implies that the force is varying slowly relative to the natural period T_n of the system.

The response during the force pulse contains both frequencies ω and ω_n and it is positive throughout. After the force pulse has ended, the system oscillates freely about its undeformed configuration with constant amplitude for lack of damping. If $t_d/T_n = 1.5, 2.5, \ldots$ the mass stays still after the force pulse ends because both the displacement and velocity of the mass are zero when the force pulse ends.

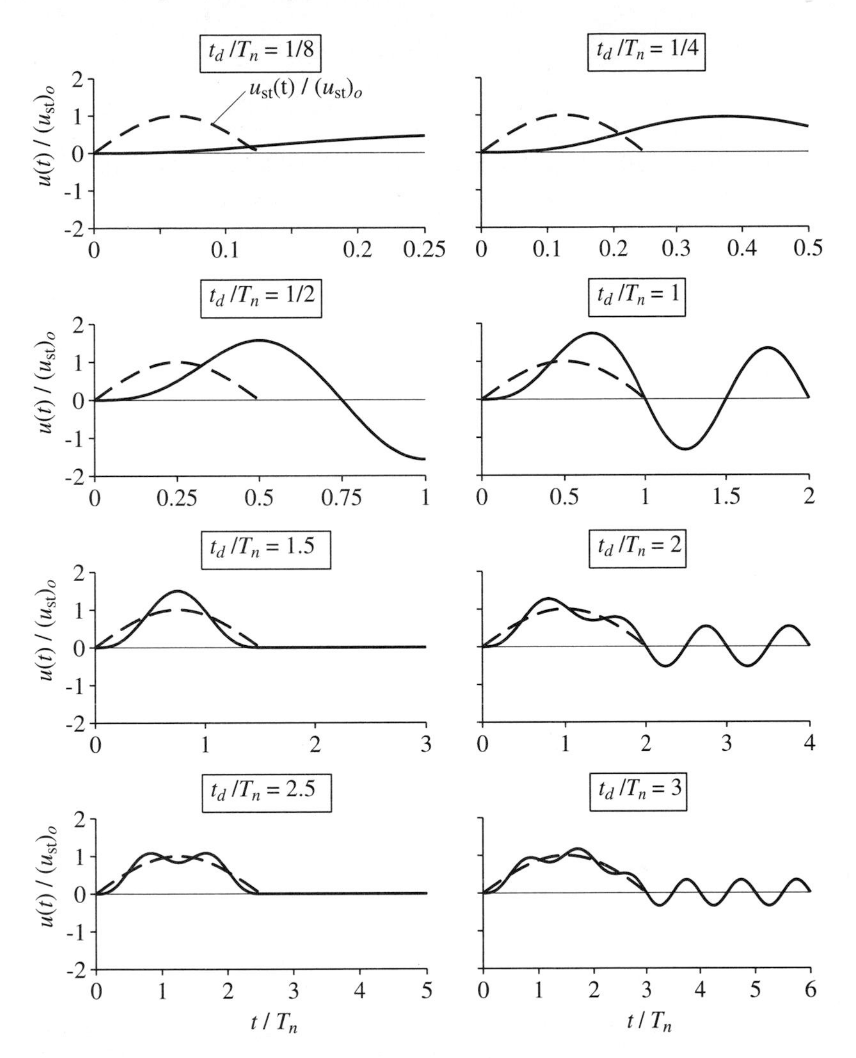

Figure 4.8.2 Dynamic response of undamped SDF system to half-cycle sine pulse force; static solution is shown by dashed lines.

Maximum response. As in the preceding section, the maximum values of response over each of the two phases, forced vibration and free vibration, separately, are determined. The larger of the two maxima is the overall maximum response.

During the forced vibration phase, the number of local maxima or peaks that develop depends on t_d/T_n (Fig. 4.8.2); the longer the pulse duration, more such peaks occur. The time instants t_o when the peaks occur are determined by setting to zero the velocity associated with $u(t)$ of Eq. (4.8.2), leading to

$$\cos \frac{\pi t_o}{t_d} = \cos \frac{2\pi t_o}{T_n}$$

This transcendental equation is satisfied by

$$(t_o)_l = \frac{2l}{1 \mp 2(t_d/T_n)} t_d \qquad l = 1, 2, 3, \ldots \tag{4.8.7}$$

where the negative sign is associated with local minima and the positive sign with local maxima. Thus the local maxima occur at time instants

$$(t_o)_l = \frac{2l}{1 + 2(t_d/T_n)} t_d \qquad l = 1, 2, 3, \ldots \tag{4.8.8}$$

While this gives an infinite number of $(t_o)_l$ values, only those that do not exceed t_d are relevant. For $t_d/T_n = 3$, Eq. (4.8.8) gives three relevant time instants $t_o = \frac{2}{7}t_d$, $\frac{4}{7}t_d$, and $\frac{6}{7}t_d$; $l = 4$ gives $t_o = \frac{8}{7}t_d$, which is not valid because it exceeds t_d. Substituting in Eq. (4.8.2) the $(t_o)_l$ values of Eq. (4.8.8) gives the local maxima u_o, which can be expressed in terms of the deformation response factor:

$$R_d = \frac{u_o}{(u_{\text{st}})_o} = \frac{1}{1 - (T_n/2t_d)^2} \left(\sin \frac{2\pi l}{1 + 2t_d/T_n} - \frac{T_n}{2t_d} \sin \frac{2\pi l}{1 + T_n/2t_d} \right) \tag{4.8.9}$$

Figure 4.8.3a shows these peak values plotted as a function of t_d/T_n. For each t_d/T_n value the above-described computations were implemented and then repeated for many t_d/T_n values. If $0.5 \le t_d/T_n \le 1.5$, only one peak, $l = 1$, occurs during the force pulse. A second peak develops if $t_d/T_n > 1.5$, but it is smaller than the first peak if $1.5 < t_d/T_n < 2.5$. A third peak develops if $t_d/T_n > 2.5$. The second peak is larger than the first and third peaks if $2.5 < t_d/T_n < 4.5$. Usually, we will be concerned only with the largest peak because that controls the design of the system. The shock spectrum for the largest peak of the forced response is shown in Fig. 4.8.3b. While the number of peaks increases as t_d becomes longer, for $t_d/T_n > 3$ all the peaks are only slightly larger than the static solution $(u_{\text{st}})_o$.

If $t_d/T_n < \frac{1}{2}$, no peak occurs during the forced vibration phase (Fig. 4.8.2). This becomes clear by examining the time of the first peak, Eq. (4.8.8), with $l = 1$:

$$t_o = \frac{2}{1 + 2t_d/T_n} t_d$$

If this t_o exceeds t_d, and it does for all $t_d < T_n/2$, no peak develops during the force pulse; the response builds up from zero to $u(t_d)$, obtained by evaluating Eq. (4.8.2) at $t = t_d$:

$$\frac{u(t_d)}{(u_{\text{st}})_o} = \frac{T_n/2t_d}{(T_n/2t_d)^2 - 1} \sin\left(2\pi \frac{t_d}{T_n}\right) \tag{4.8.10}$$

This is the maximum response during the forced vibration phase and it defines the deformation response factor over the range $0 \le t_d/T_n < \frac{1}{2}$ in Fig. 4.8.3a.

In the free vibration phase the response of a system is given by the sinusoidal function of Eq. (4.8.3), and its amplitude is

$$R_d = \frac{u_o}{(u_{\text{st}})_o} = \frac{(T_n/t_d)\cos(\pi t_d/T_n)}{(T_n/2t_d)^2 - 1} \tag{4.8.11}$$

This equation describes the maximum response during the free vibration phase and is plotted in Fig. 4.8.3b.

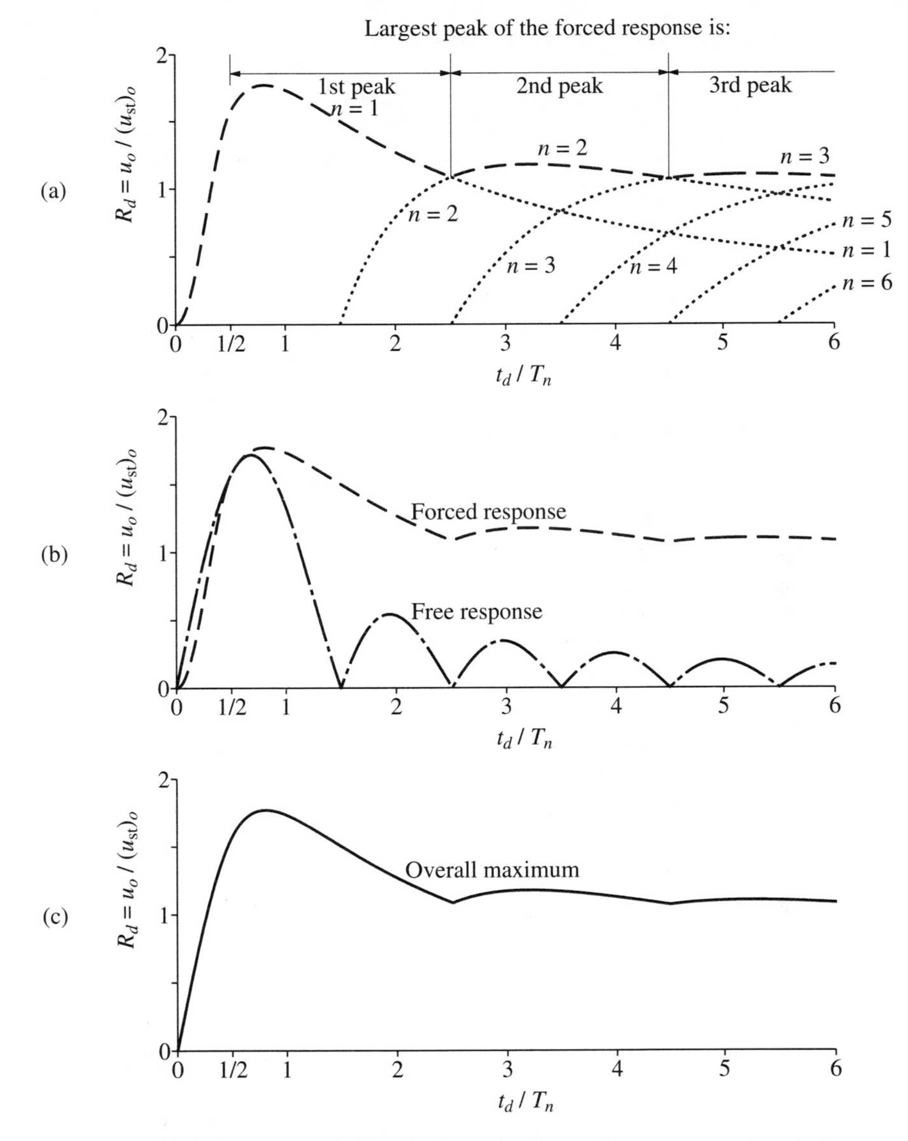

Figure 4.8.3 Response to half-cycle sine pulse force: (a) response maxima during forced vibration phase; (b) maximum responses during each of forced vibration and free vibration phases; (c) shock spectrum.

For the special case of $t_d/T_n = \frac{1}{2}$, the maximum response during each of the forced and free vibration phases can be determined from Eqs. (4.8.4) and (4.8.6), respectively; the two maxima are the same:

$$R_d = \frac{u_o}{(u_{st})_o} = \frac{\pi}{2} \tag{4.8.12}$$

The overall maximum response is the larger of the two maxima determined separately for the forced and free vibration phases. Figure 4.8.3b shows that if $t_d > T_n/2$, the overall maximum is the largest peak that develops during the force pulse. On the other hand, if $t_d < T_n/2$, the overall maximum is given by the peak response during the free vibration phase. For the special case of $t_d = T_n/2$, as mentioned earlier, the two individual maxima are equal. The overall maximum response is plotted against t_d/T_n in Fig. 4.8.3c; for each t_d/T_n it is the larger of the two plots of Fig. 4.8.3b. This is the shock spectrum for the half-cycle sine pulse force. If it is available the maximum deformation and elastic resisting force can readily be determined using Eqs. (4.7.13) and (4.7.14).

4.9 SYMMETRICAL TRIANGULAR PULSE FORCE

Consider next an SDF system initially at rest and subjected to the symmetrical triangular pulse shown in Fig. 4.9.1. The response of an undamped SDF system to this pulse could be determined by any of the methods mentioned in Section 4.6. For example, the classical method could be implemented in three separate phases: $0 \le t \le t_d/2$, $t_d/2 \le t \le t_d$, and $t \ge t_d$. The classical method was preferred in Sections 4.8 and 4.9 because it is closely tied to the dynamics of the system, but is abandoned here for expedience. Perhaps the easiest way to solve the problem is to express the triangular pulse as the superposition of three ramp functions shown in Fig. 4.6.2c. The response of the system to each of these ramp functions can readily be determined by appropriately adapting Eq. (4.4.2) to recognize the slope and starting time of each of the three ramp functions. These three individual responses are added to obtain the response to the symmetrical triangular pulse. The final result is

$$\frac{u(t)}{(u_{\text{st}})_o} = \begin{cases} 2\left(\frac{t}{t_d} - \frac{T_n}{2\pi t_d}\sin 2\pi \frac{t}{T_n}\right) & 0 \le t \le \frac{t_d}{2} \quad (4.9.1a) \\ 2\left\{1 - \frac{t}{t_d} + \frac{T_n}{2\pi t_d}\left[2\sin\frac{2\pi}{T_n}\left(t - \frac{1}{2}t_d\right) - \sin 2\pi\frac{t}{T_n}\right]\right\} & \frac{t_d}{2} \le t \le t_d \quad (4.9.1b) \\ 2\left\{\frac{T_n}{2\pi t_d}\left[2\sin\frac{2\pi}{T_n}\left(t - \frac{1}{2}t_d\right) - \sin\frac{2\pi}{T_n}(t - t_d) - \sin 2\pi\frac{t}{T_n}\right]\right\} & t \ge t_d \quad (4.9.1c) \end{cases}$$

The variation of the normalized dynamic deformation $u(t)/(u_{\text{st}})_o$ and of the static solution $u_{\text{st}}(t)/(u_{\text{st}})_o$ with time is shown in Fig. 4.9.2 for several values of t_d/T_n. The dynamic

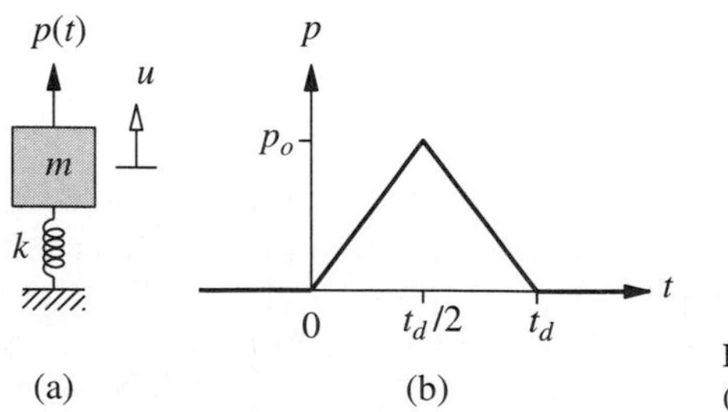

Figure 4.9.1 (a) SDF system; (b) triangular pulse force.

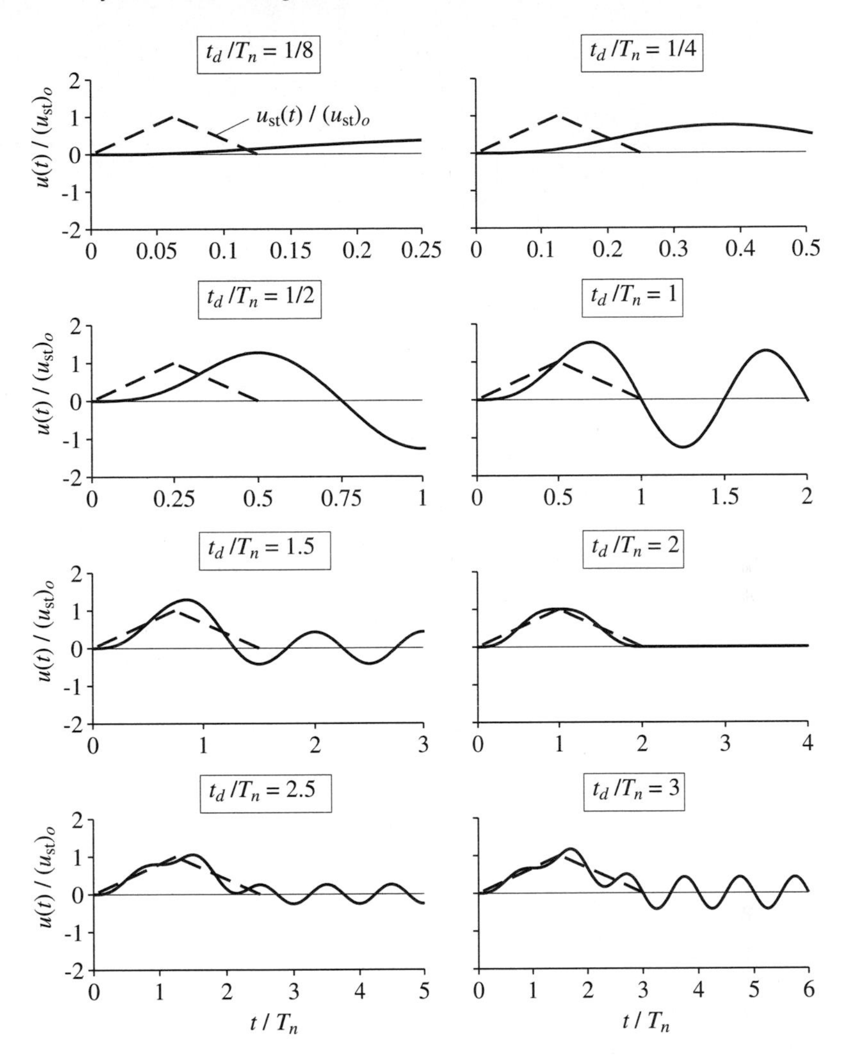

Figure 4.9.2 Dynamic response of undamped SDF system to triangular pulse force; static solution is shown by dashed lines.

effects are seen to decrease as the pulse duration t_d increases beyond $2T_n$. The first peak develops right at the end of the pulse if $t_d = T_n/2$, during the pulse if $t_d > T_n/2$, and after the pulse if $t_d < T_n/2$. The maximum response during free vibration (Fig. 4.9.3a) was obtained by finding the maximum value of Eq. (4.9.1c). The corresponding plot for maximum response during the forced vibration phase (Fig. 4.9.3a) was obtained by finding the largest of the local maxima of Eq. (4.9.1b), which is always larger than the maximum value of Eq. (4.9.1a).

The overall maximum response is the larger of the two maxima determined separately for the forced and free vibration phases. Figure 4.9.3a shows that if $t_d > T_n/2$, the

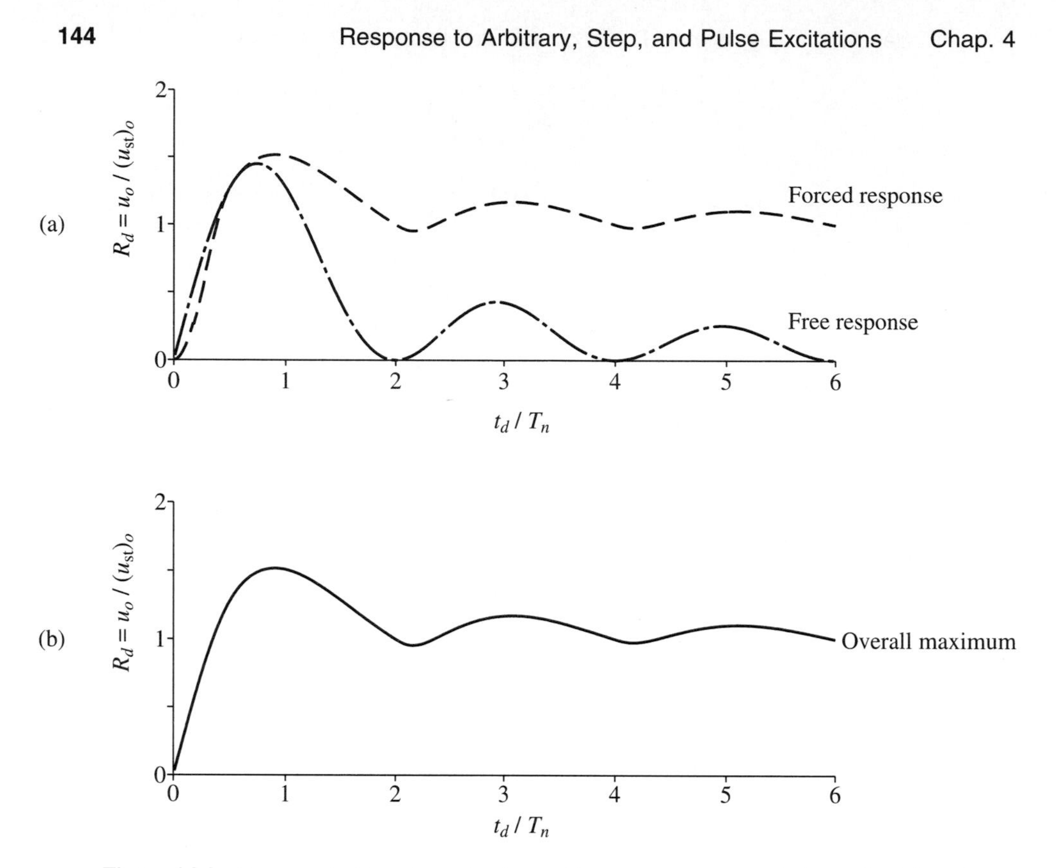

Figure 4.9.3 Response to triangular pulse force: (a) maximum response during each of forced vibration and free vibration phases; (b) shock spectrum.

overall maximum is the largest peak that develops during the force pulse. On the other hand, if $t_d < T_n/2$, the overall maximum is the peak response during the free vibration phase, and if $t_d = T_n/2$, the forced and free response maxima are equal. The overall maximum response is plotted against t_d/T_n in Fig. 4.9.3b. This is the shock spectrum for the symmetrical triangular pulse force.

4.10 EFFECTS OF PULSE SHAPE AND APPROXIMATE ANALYSIS FOR SHORT PULSES

The shock spectra for the three pulses of rectangular, half-cycle sine, and triangular shapes, each with the same value of maximum force p_o, are presented together in Fig. 4.10.1. As shown in the preceding sections, if the pulse duration t_d is longer than $T_n/2$, the overall maximum deformation occurs during the pulse and the pulse shape is of great significance. For the larger values of t_d/T_n, this overall maximum is influenced by the rapidity of the loading. The rectangular pulse in which the force increases suddenly from zero to p_o produces the largest deformation. The triangular pulse in which the

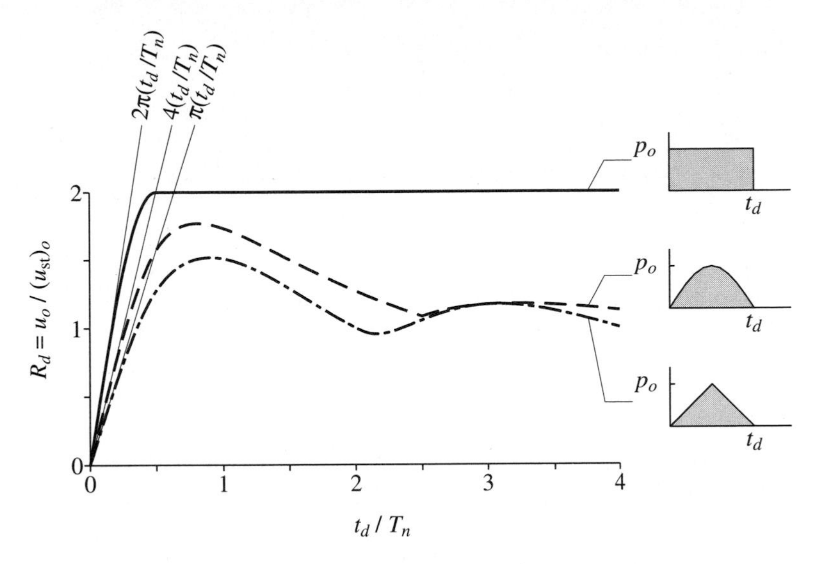

Figure 4.10.1 Shock spectra for three force pulses of equal amplitude.

increase in force is initially slowest among the three pulses produces the smallest deformation. The half-cycle sine pulse in which the force initially increases at an intermediate rate causes deformation that for many values of t_d/T_n is larger than the response to the triangular pulse.

If the pulse duration t_d is shorter than $T_n/2$, the overall maximum response of the system occurs during its free vibration phase and is controlled by the time integral of the pulse. This can be demonstrated by considering the limiting case as t_d/T_n approaches zero. As the pulse duration becomes extremely short compared to the natural period of the system, it becomes a pure impulse of magnitude

$$\mathcal{I} = \int_0^{t_d} p(t)\, dt \tag{4.10.1}$$

The response of the system to this impulsive force is the unit impulse response of Eq. (4.1.6) times $\mathcal{I}$:

$$u(t) = \mathcal{I}\left(\frac{1}{m\omega_n} \sin \omega_n t\right) \tag{4.10.2}$$

The maximum deformation,

$$u_o = \frac{\mathcal{I}}{m\omega_n} = \frac{\mathcal{I}}{k}\frac{2\pi}{T_n} \tag{4.10.3}$$

is proportional to the magnitude of the impulse.

Thus the maximum deformation due to the rectangular impulse of magnitude $\mathcal{I} = p_o t_d$ is

$$\frac{u_o}{(u_{st})_o} = 2\pi \frac{t_d}{T_n} \tag{4.10.4}$$

that due to the half-cycle sine pulse with $\mathcal{I} = (2/\pi)p_o t_d$ is

$$\frac{u_o}{(u_{\text{st}})_o} = 4\frac{t_d}{T_n} \tag{4.10.5}$$

and that due to the triangular pulse of magnitude $\mathcal{I} = p_o t_d/2$ is

$$\frac{u_o}{(u_{\text{st}})_o} = \pi\frac{t_d}{T_n} \tag{4.10.6}$$

These pure impulse solutions, which vary linearly with t_d/T_n (Fig. 4.10.1), are exact if $t_d/T_n = 0$; for all other values of t_d, they provide an upper bound to the true maximum deformation since the effect of the pulse has been overestimated by assuming it to be concentrated at $t = 0$ instead of being spread out over 0 to t_d. Over the range of $t_d/T_n < \frac{1}{4}$ the pure impulse solution is close to the exact response. The two solutions differ increasingly as t_d/T_n increases up to $\frac{1}{2}$. For larger values of t_d/T_n, the deformation attains its maximum during the pulse and the pure impulse solution is meaningless because it assumes that the maximum occurs in free vibration.

The preceding observations suggest that if the pulse duration is much shorter than the natural period, say $t_d < T_n/4$, the maximum deformation should be essentially controlled by the pulse area, independent of its shape. This expectation is confirmed by considering the rectangular pulse of amplitude $p_o/2$, the triangular pulse of amplitude p_o, and the half-cycle sine pulse of amplitude $(\pi/4)p_o$; these three pulses have the same area: $\frac{1}{2}p_o t_d$. For these three pulses, the shock spectra, determined by appropriately scaling the plots of Fig. 4.10.1, are presented in Fig. 4.10.2; observe that the quantity plotted now is $u_o \div p_o/k$, where p_o is the amplitude of the triangular pulse but not of the other two. Equation (4.10.2) with $\mathcal{I} = \frac{1}{2}p_o t_d$ gives the approximate result

$$\frac{u_o}{p_o/k} = \pi\frac{t_d}{T_n} \tag{4.10.7}$$

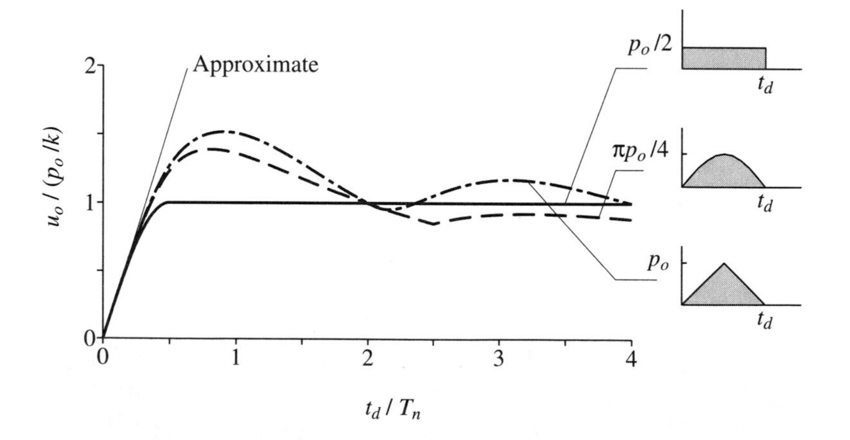

Figure 4.10.2 Shock spectra for three force pulses of equal area.

which is also shown in Fig. 4.10.2. It is clear that for $t_d < T_n/4$, the shape of a symmetrical pulse has little influence on the response and the response can be determined using only the pulse area.

Example 4.2

The 80-ft-high full water tank of Example 2.6 is subjected to the force $p(t)$ shown in Fig. E4.2, caused by an aboveground explosion. Determine the maximum base shear and bending moment at the base of the tower supporting the tank.

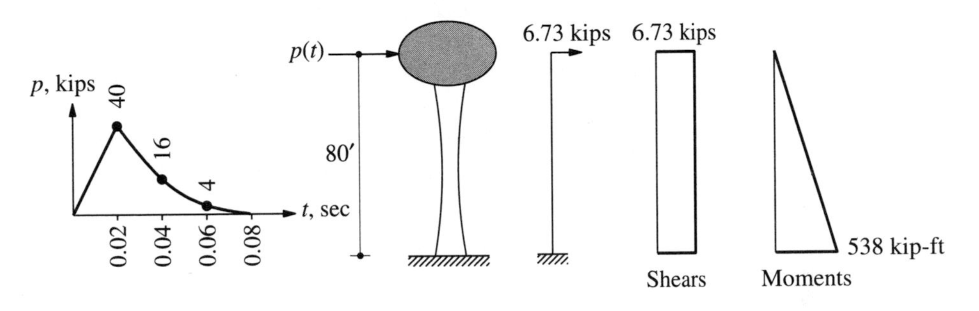

Figure E4.2

Solution For this water tank, weight $w = 100.03$ kips, $k = 8.2$ kips/in. $T_n = 1.12$ sec, and $\zeta = 1.23\%$. The ratio $t_d/T_n = 0.08/1.12 = 0.071$. Because $t_d/T_n < 0.25$, the forcing function may be treated as a pure impulse of magnitude

$$\mathcal{I} = \int_0^{0.08} p(t)\,dt = \frac{0.02}{2}\,[0 + 2(40) + 2(16) + 2(4) + 0] = 1.2 \text{ kip-sec}$$

where the integral is calculated by the trapezoidal rule. Neglecting the effect of damping, the maximum displacement is

$$u_o = \frac{\mathcal{I}}{k}\frac{2\pi}{T_n} = \frac{(1.2)2\pi}{(8.2)(1.12)} = 0.821 \text{ in.}$$

The equivalent static force f_{So} associated with this displacement is [from Eq. (1.8.1)]

$$f_{So} = ku_o = (8.2)0.821 = 6.73 \text{ kips}$$

The resulting shearing forces and bending moments over the height of the tower are shown in Fig. E4.2. The base shear and moment are

$$V_b = 6.73 \text{ kips} \qquad M_b = 538 \text{ kip-ft}$$

4.11 EFFECTS OF VISCOUS DAMPING

If the excitation is a single pulse, the effect of damping on the maximum response is usually not important unless the system is highly damped. This is in contrast to the results of Chapter 3, where damping was seen to have an important influence on the maximum steady-state response of systems to harmonic excitation at or near resonance.

For example, if the excitation frequency of harmonic excitation is equal to the natural frequency of the system, a tenfold increase in the damping ratio ζ, from 1% to 10%, results in a tenfold decrease in the deformation response factor R_d, from 50 to 5. Damping is so influential because of the cumulative energy dissipated in the many (the number depends on ζ) vibration cycles prior to attainment of steady state; see Figs. 3.2.2, 3.2.3, and 3.2.4.

In contrast, the energy dissipated by damping is small in systems subjected to pulse-type excitations. Consider a viscously damped system subjected to a half-cycle sine pulse with $t_d/T_n = \frac{1}{2}$ (which implies that $\omega = \omega_n$) and $\zeta = 0.1$. The variation of deformation with time (Fig. 4.11.1a) indicates that the maximum deformation (point b) is attained at the end of the pulse before completion of a single vibration cycle. The total (elastic plus damping) force–deformation diagram of Fig. 4.11.1b indicates that before the maximum response is reached, the energy dissipated in viscous damping is only the small shaded area multiplied by p_o^2/k. Thus the influence of damping on maximum response is expected to be small.

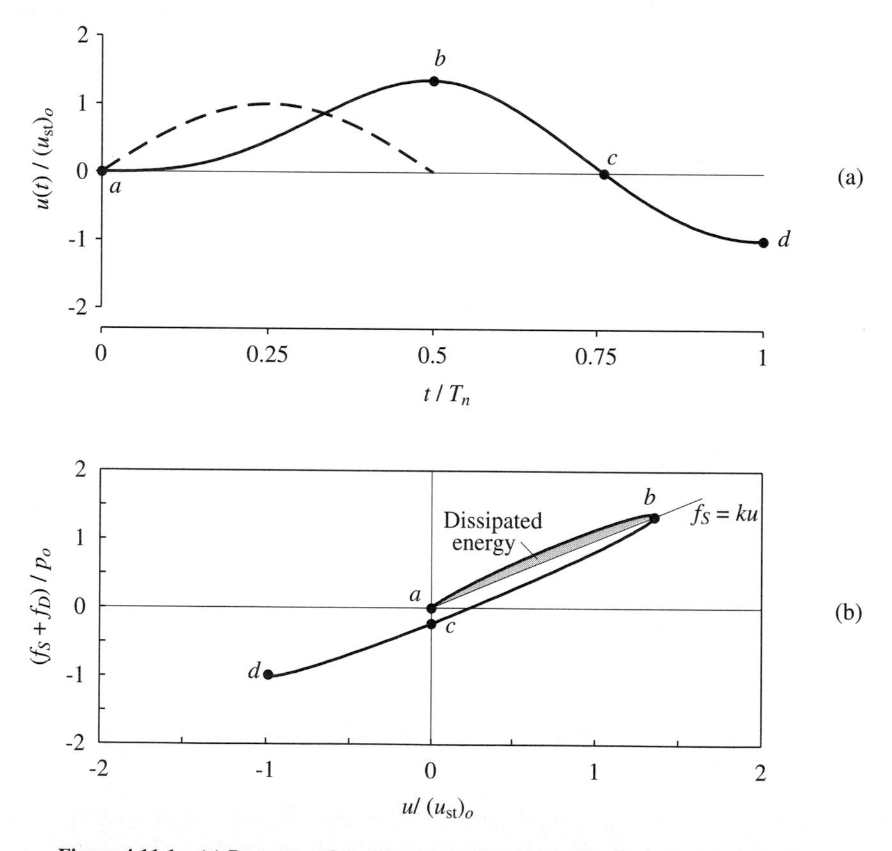

Figure 4.11.1 (a) Response of damped system ($\zeta = 0.1$) to a half-cycle sine pulse force with $t_d/T_n = \frac{1}{2}$; (b) force–deformation diagram showing energy dissipated in viscous damping.

This prediction is confirmed by the shock spectrum for a half-cycle sine pulse presented in Fig. 4.11.2. For $\zeta = 0$, this spectrum is the same as the spectrum of Fig. 4.8.3c for undamped systems. For $\zeta \neq 0$ and for each value of t_d/T_n the dynamic response of the damped system was computed by a numerical time-stepping procedure (Chapter 5) and the maximum deformation was determined. In the case of the system acted upon by a half-cycle sine pulse of duration $t_d = T_n/2$, increase in the damping ratio from 1% to 10% reduces the maximum deformation by only 12%. Thus a conservative but not overly conservative estimate of the response of many practical structures with damping to pulse-type excitations may be obtained by neglecting damping and using the earlier results for undamped systems.

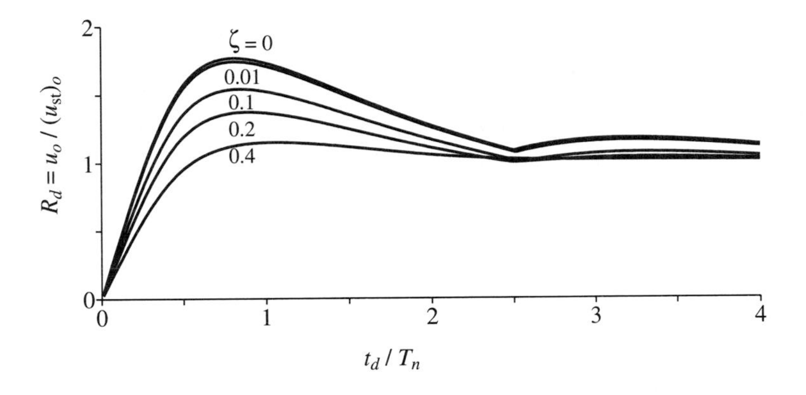

Figure 4.11.2 Shock spectra for a half-cycle sine pulse force for five damping values.

4.12 RESPONSE TO GROUND MOTION

The response spectrum characterizing the maximum response of SDF systems to ground motion $\ddot{u}_g(t)$ can be determined from the response spectrum for the applied force $p(t)$ with the same time variation as $\ddot{u}_g(t)$. This is possible because as shown in Eq. (1.7.6), the ground acceleration can be replaced by the effective force, $p_{\text{eff}}(t) = -m\ddot{u}_g(t)$.

The response spectrum for applied force $p(t)$ is a plot of $R_d = u_o/(u_{\text{st}})_o$, where $(u_{\text{st}})_o = p_o/k$, versus the appropriate system and excitation parameters: ω/ω_n for harmonic excitation and t_d/T_n for pulse-type excitation. Replacing p_o by $(p_{\text{eff}})_o$ gives

$$(u_{\text{st}})_o = \frac{(p_{\text{eff}})_o}{k} = \frac{m\ddot{u}_{go}}{k} = \frac{\ddot{u}_{go}}{\omega_n^2} \tag{4.12.1}$$

where $\ddot{u}_{go}$ is the maximum value of $\ddot{u}_g(t)$ and the negative sign in $p_{\text{eff}}(t)$ has been dropped. Thus

$$R_d = \frac{u_o}{(u_{\text{st}})_o} = \frac{\omega_n^2 u_o}{\ddot{u}_{go}} \tag{4.12.2}$$

Therefore, the response spectra presented in Chapters 3 and 4 showing the response $u_o/(u_{\text{st}})_o$ due to applied force also give the response $\omega_n^2 u_o/\ddot{u}_{go}$ to ground motion.

For undamped systems subjected to ground motion, Eqs. (1.7.4) and (1.7.3) indicate that the total acceleration of the mass is related to the deformation through $\ddot{u}^t(t) = -\omega_n^2 u(t)$. Thus the maximum values of the two responses are related by $\ddot{u}_o^t = \omega_n^2 u_o$. Substituting in Eq. (4.12.2) gives

$$R_d = \frac{\ddot{u}_o^t}{\ddot{u}_{go}} \tag{4.12.3}$$

Thus the earlier response spectra showing the response $u_o/(u_{st})_o$ of undamped systems subjected to applied force also display the response $\ddot{u}_o^t/\ddot{u}_{go}$ to ground motion.

As an example, the response spectrum of Fig. 4.8.3c for a half-cycle sine pulse force also gives the maximum values of responses $\omega_n^2 u_o/\ddot{u}_{go}$ and $\ddot{u}_o^t/\ddot{u}_{go}$ due to ground acceleration described by a half-cycle sine pulse.

Example 4.3

Consider the SDF model of an automobile described in Example 3.4 running over the speed bump shown in Fig. E4.3 at velocity v. Determine the maximum force developed in the suspension spring and the maximum vertical acceleration of the mass if **(a)** $v = 5$ mph, and **(b)** $v = 10$ mph.

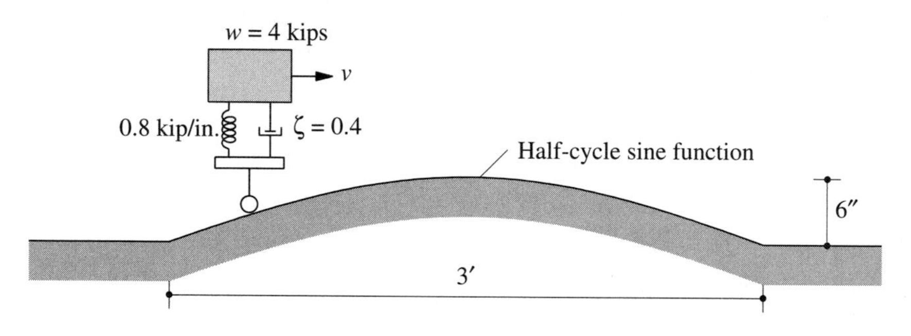

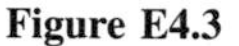

Figure E4.3

Solution

1. *Determine the system and excitation parameters.*

$$m = \frac{4000}{386} = 10.363 \text{ lb-sec}^2\text{/in.}$$

$$k = 800 \text{ lb/in.}$$

$$\omega_n = 8.786 \text{ rad/sec} \qquad T_n = 0.715 \text{ sec}$$

$$v = 5 \text{ mph} = 7.333 \text{ ft/sec} \qquad t_d = \frac{3}{7.333} = 0.4091 \text{ sec} \qquad \frac{t_d}{T_n} = 0.572$$

$$v = 10 \text{ mph} = 14.666 \text{ ft/sec} \qquad t_d = \frac{3}{14.666} = 0.2046 \text{ sec} \qquad \frac{t_d}{T_n} = 0.286$$

$$u_g(t) = 6 \sin \frac{\pi t}{t_d} \qquad \ddot{u}_{go} = \frac{6\pi^2}{t_d^2}$$

2. *Determine R_d for the t_d/T_n values above from Fig. 4.11.2.*

$$R_d = \begin{cases} 1.015 & v = 5 \text{ mph} \\ 0.639 & v = 10 \text{ mph} \end{cases}$$

Obviously, R_d cannot be read accurately to three or four significant digits; these values are from the numerical data used in plotting Fig. 4.11.2.

3. *Determine the maximum force, f_{So}.*

$$u_o = \frac{\ddot{u}_{go}}{\omega_n^2} R_d = 1.5 \left(\frac{T_n}{t_d}\right)^2 R_d$$

$$u_o = \begin{cases} 1.5\left(\dfrac{1}{0.572}\right)^2 1.015 = 4.65 \text{ in.} & v = 5 \text{ mph} \\ 1.5\left(\dfrac{1}{0.286}\right)^2 0.639 = 11.7 \text{ in.} & v = 10 \text{ mph} \end{cases}$$

$$f_{So} = ku_o = 0.8u_o = \begin{cases} 3.72 \text{ kips} & v = 5 \text{ mph} \\ 9.37 \text{ kips} & v = 10 \text{ mph} \end{cases}$$

Observe that the force in the suspension is much larger at the higher speed. The large deformation of the suspension suggests that it may deform beyond its linearly elastic limit.

4. *Determine the maximum acceleration, $\ddot{u}_o^t$.* Equation (4.12.3) provides a relation between $\ddot{u}_o^t$ and R_d that is exact for systems without damping but is approximate for damped systems. These approximate results can readily be obtained for this problem:

$$\ddot{u}_o^t = \ddot{u}_{go} R_d = \frac{6\pi^2}{t_d^2} R_d$$

$$\ddot{u}_o^t = \begin{cases} \left[\dfrac{6\pi^2}{(0.4091)^2}\right] 1.015 = 359.1 \text{ in./sec}^2 & v = 5 \text{ mph} \\ \left[\dfrac{6\pi^2}{(0.2046)^2}\right] 0.639 = 903.7 \text{ in./sec}^2 & v = 10 \text{ mph} \end{cases}$$

Observe that the acceleration of the mass is much larger at the higher speed; in fact, it exceeds 1 g, indicating that the SDF model would lift off from the road.

To evaluate the error in the approximate solution for $\ddot{u}_o^t$, a numerical solution of the equation of motion was carried out, leading to the "exact" value of $\ddot{u}_o^t = 422.7$ in./sec^2 for $v = 5$ mph.

FURTHER READING

Ayre, R. S., "Transient Response to Step and Pulse Functions," Chapter 8 in *Shock and Vibration Handbook,* 3rd ed. (ed. C. M. Harris), McGraw-Hill, New York, 1988.

Jacobsen, L. S., and Ayre, R. S., *Engineering Vibrations*, McGraw-Hill, New York, 1958, Chapters 3 and 4.

PROBLEMS

Part A

4.1 Show that the maximum deformation u_0 of an SDF system due to a unit impulse force, $p(t) = \delta(t)$, is

$$u_0 = \frac{1}{m\omega_n} \exp\left(-\frac{\zeta}{\sqrt{1-\zeta^2}} \tan^{-1} \frac{\sqrt{1-\zeta^2}}{\zeta}\right)$$

Plot this result as a function of ζ. Comment on the influence of damping on the maximum response.

4.2 Consider the deformation response $g(t)$ of an SDF system to a unit step function $p(t) = 1$, $t \geq 0$, and $h(t)$ due to a unit impulse $p(t) = \delta(t)$. Show that $h(t) = \dot{g}(t)$.

Part B

4.3 Using the classical method for solving differential equations, derive Eq. (4.4.2), which describes the response of an undamped SDF system to a linearly increasing force; the initial conditions are $u(0) = \dot{u}(0) = 0$.

4.4 An elevator is idealized as a weight of mass m supported by a spring of stiffness k. If the upper end of the spring begins to move with a steady velocity v, show that the distance u^t that the mass has risen in time t is governed by the equation

$$m\ddot{u}^t + ku^t = kvt$$

If the elevator starts from rest, show that the motion is

$$u^t(t) = vt - \frac{v}{\omega_n} \sin \omega t$$

Plot this result.

4.5 The deformation response of an undamped SDF system to a step force having finite rise time is given by Eqs. (4.5.2) and (4.5.4). Derive these results using Duhamel's integral.

4.6 Derive Eqs. (4.5.2) and (4.5.4) by considering the excitation as the sum of two ramp functions (Fig. P4.6). For $t \leq t_r$, $u(t)$ is the solution of the equation of motion for the first ramp function. For $t \geq t_r$, $u(t)$ is the sum of the responses to the two ramp functions.

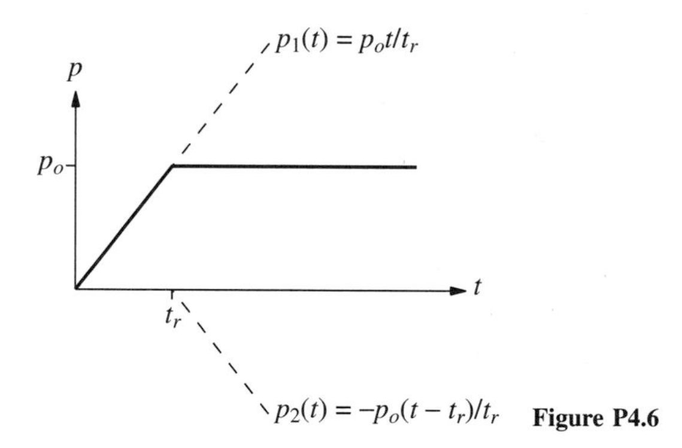

Figure P4.6

4.7 The elevated water tank of Fig. P4.7 weighs 100.03 kips when full with water. The tower has a lateral stiffness of 8.2 kips/in. Treating the water tower as an SDF system, estimate the maximum lateral displacement due to each of the two dynamic forces shown without any "exact" dynamic analysis. Instead, use your understanding of how the maximum response depends on the ratio of the rise time of the applied force to the natural vibration period of the system; neglect damping.

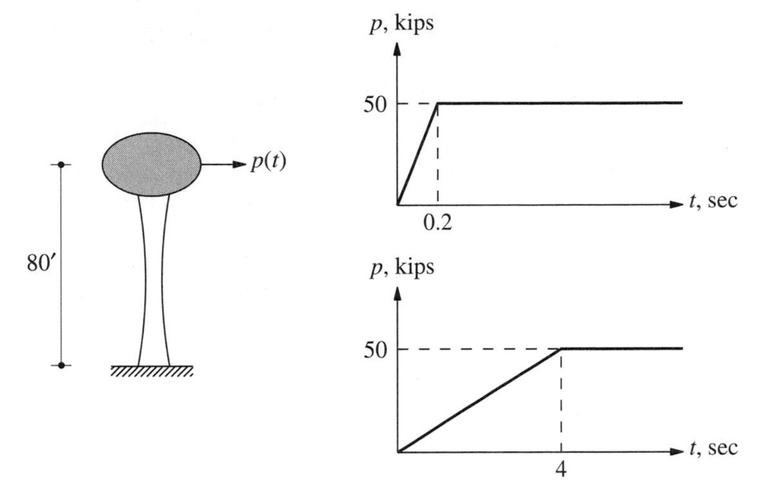

Figure P4.7

Part C

4.8 Determine the response of an undamped system to a rectangular pulse force of amplitude p_o and duration t_d by considering the pulse as the superposition of two step excitations (Fig. 4.6.2).

4.9 Using Duhamel's integral, determine the response of an undamped system to a rectangular pulse force of amplitude p_o and duration t_d.

4.10 Determine the response of an undamped system to a half-cycle sine pulse force of amplitude p_o and duration t_d by considering the pulse as the superposition of two sinusoidal excitations (Fig. 4.6.2); $t_d/T_n \neq \frac{1}{2}$.

4.11 The one-story building of Example 4.1 is modified so that the columns are clamped at the base instead of hinged. For the same excitation determine the maximum displacement at the top of the frame and maximum bending stress in the columns. Comment on the effect of base fixity.

4.12 Determine the maximum response of the frame of Example 4.1 to a half-cycle sine pulse force of amplitude $p_o = 5$ kips and duration $t_d = 0.25$ sec. The response quantities of interest are: displacement at the top of the frame and maximum bending stress in columns.

4.13 Derive equations (4.9.1) for the displacement response of an undamped SDF system to a symmetrical triangular pulse by considering the pulse as the superposition of three ramp functions.

4.14 Derive equations for the deformation $u(t)$ of an undamped SDF system due to the force $p(t)$ shown in Fig. P4.14 for each of the following time ranges: $t \leq t_1$, $t_1 \leq t \leq 2t_1$, $2t_1 \leq t \leq 3t_1$, and $t \geq 3t_1$.

4.15 An SDF system is subjected to the force shown in Fig. P4.14. Determine the maximum response during free vibration of the system and the time instant the first peak occurs.

4.16 To determine the maximum response of an undamped SDF system to the force of Fig. P4.14 for a particular value of t_d/T_n, where $t_d = 3t_1$, you would need to identify the time range

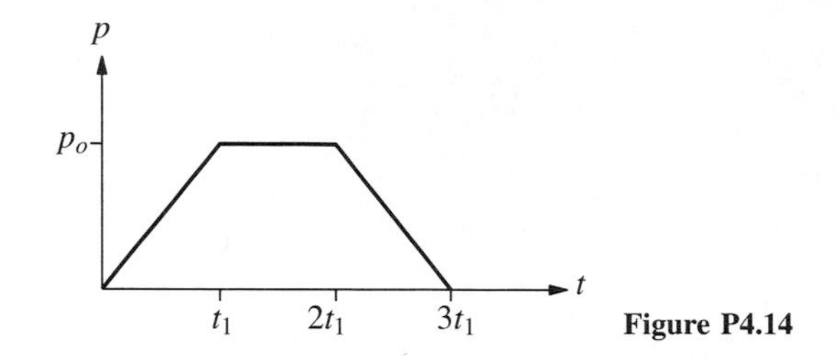

Figure P4.14

among the four mentioned in Problem 4.14 during which the overall maximum response would occur, and then find the value of that maximum. Such analyses would need to be repeated for many values of t_d/T_n in order to determine the complete shock spectrum. Obviously, this is time consuming but necessary if one wishes to determine the complete shock spectrum. However, the spectrum for small values of t_d/T_n can be determined by treating the force as an impulse. Determine the shock spectrum by this approach and plot it. What is the error in this approximate result for $t_d/T_n = \frac{1}{4}$?

4.17 **(a)** Determine the response of an undamped SDF system to the force shown in Fig. P4.17 for each of the following time intervals: (i) $0 \le t \le t_d/2$, (ii) $t_d/2 \le t \le t_d$, and (iii) $t \ge t_d$. Assume that $u(0) = \dot{u}(0) = 0$.

(b) Determine the maximum response u_o during free vibration of the system. Plot the deformation response factor $R_d = u_o/(u_{st})_o$ as a function of t_d/T_n over the range $0 \le t_d/T_n \le 4$.

(c) If $t_d \ll T_n$, can the maximum response be determined by treating the applied force as a pure impulse? State reasons for your answer.

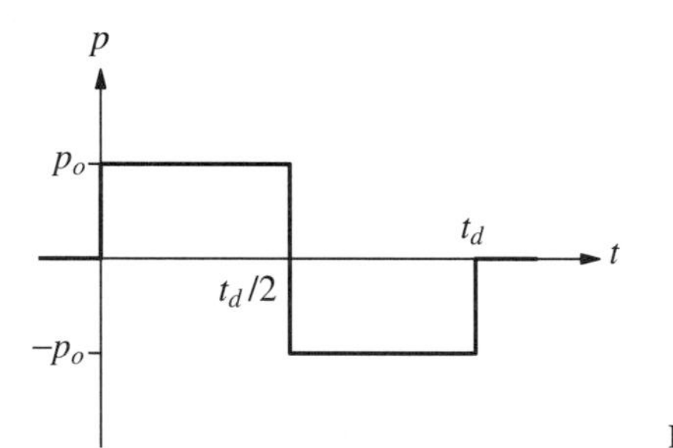

Figure P4.17

4.18 The 80-ft-high water tank of Examples 2.5 and 2.6 is subjected to the force $p(t)$ shown in Fig. E4.2a. The maximum response of the structure with the tank full was determined in Example 4.2.

(a) If the tank is empty, calculate the maximum base shear and bending moment at the base of the tower supporting the tank.

(b) By comparing these results with those for the full tank (Example 4.2), comment on the effect of mass on the response to impulsive forces. Explain the reason.

5

Numerical Evaluation of Dynamic Response

PREVIEW

Analytical solution of the equation of motion for a single-degree-of-freedom system is usually not possible if the excitation—applied force $p(t)$ or ground acceleration $\ddot{u}_g(t)$—varies arbitrarily with time or if the system is nonlinear. Such problems can be tackled by numerical time-stepping methods for integration of differential equations. A vast body of literature, including major chapters of several textbooks, exists about these methods for solving various types of differential equations that arise in the broad subject area of applied mechanics. The literature includes the mathematical development of these methods; their accuracy, convergence, and stability properties; and computer implementation.

Only a brief presentation of a very few methods that are especially useful in dynamic response analysis of SDF systems is included here, however. This presentation is intended to provide only the basic concepts underlying these methods and to provide a few computational algorithms. While these would suffice for many practical problems and research applications, the reader should recognize that a wealth of knowledge exists on this subject.

5.1 TIME-STEPPING METHODS

For an inelastic system the equation of motion to be solved numerically is

$$m\ddot{u} + c\dot{u} + f_S(u, \dot{u}) = p(t) \tag{5.1.1}$$

subject to the initial conditions

$$u = u(0) \qquad \dot{u} = \dot{u}(0)$$

The system is assumed to have linear viscous damping, but other forms of damping, including nonlinear damping, could be considered, as it would become obvious later. This is rarely done for lack of information on damping, especially at large amplitudes of motion. The applied force $p(t)$ is given by a set of discrete values $p_i = p(t_i)$, $i = 0$ to N (Fig. 5.1.1). The time interval

$$\Delta t_i = t_{i+1} - t_i \tag{5.1.2}$$

is usually taken to be constant, although this is not necessary. The response is determined at the discrete time instants t_i, denoted as time i; the displacement, velocity, and acceleration of the SDF system are u_i, $\dot{u}_i$, and $\ddot{u}_i$, respectively. These values, assumed to be known, satisfy Eq. (5.1.1) at time i:

$$m\ddot{u}_i + c\dot{u}_i + (f_S)_i = p_i \tag{5.1.3}$$

where $(f_S)_i$ is the resisting force at time i; $(f_S)_i = ku_i$ for a linearly elastic system but would depend on the prior history of displacement and velocity if the system were inelastic. The numerical procedures to be presented will enable us to determine the response quantities u_{i+1}, $\dot{u}_{i+1}$, and $\ddot{u}_{i+1}$ at the time instant $i+1$ that satisfy Eq. (5.1.1) at time $i+1$:

$$m\ddot{u}_{i+1} + c\dot{u}_{i+1} + (f_S)_{i+1} = p_{i+1} \tag{5.1.4}$$

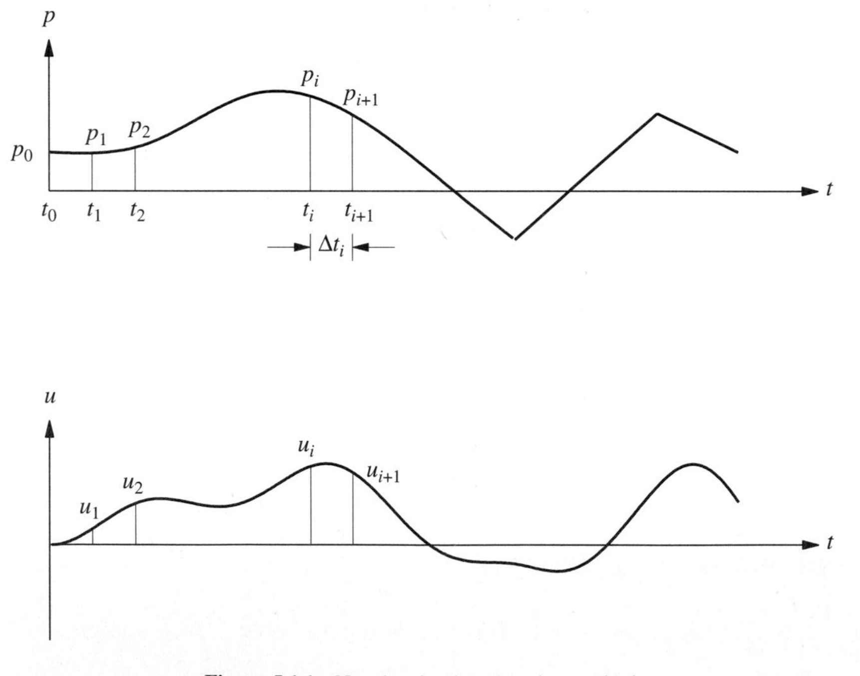

Figure 5.1.1 Notation for time-stepping methods.

When applied successively with $i = 0, 1, 2, 3, \ldots$, the time-stepping procedure gives the desired response at all time instants $i = 1, 2, 3, \ldots$. The known initial conditions provide the information necessary to start the procedure.

Stepping from time i to $i+1$ is usually not an exact procedure. Many approximate procedures are possible that are implemented numerically. The three important requirements for a numerical procedure are (1) convergence—as the time step decreases, the numerical solution should approach the exact solution, (2) stability—the numerical solution should be stable in the presence of numerical round-off errors, and (3) accuracy—the numerical procedure should provide results that are close enough to the exact solution. Although these are important issues, they are mentioned only briefly in this book.

Three types of time-stepping procedures are presented in this chapter: (1) methods based on interpolation of the excitation function, (2) methods based on finite difference expressions of velocity and acceleration, and (3) methods based on assumed variation of acceleration. Only one method is presented in each of the first two categories and two from the third group.

5.2 METHODS BASED ON INTERPOLATION OF EXCITATION

A highly efficient numerical procedure can be developed for linear systems by interpolating the excitation over each time interval and developing an exact solution for a linear system using the methods of Chapter 4. If the time intervals are short, linear interpolation is satisfactory. Figure 5.2.1 shows that over the time interval $t_i \le t \le t_{i+1}$, the excitation function is given by

$$p(\tau) = p_i + \frac{\Delta p_i}{\Delta t_i}\tau \tag{5.2.1a}$$

where

$$\Delta p_i = p_{i+1} - p_i \tag{5.2.1b}$$

and the time variable τ varies from 0 to Δt_i. For algebraic simplicity, we first consider systems without damping; later, the procedure will be extended to include damping. The

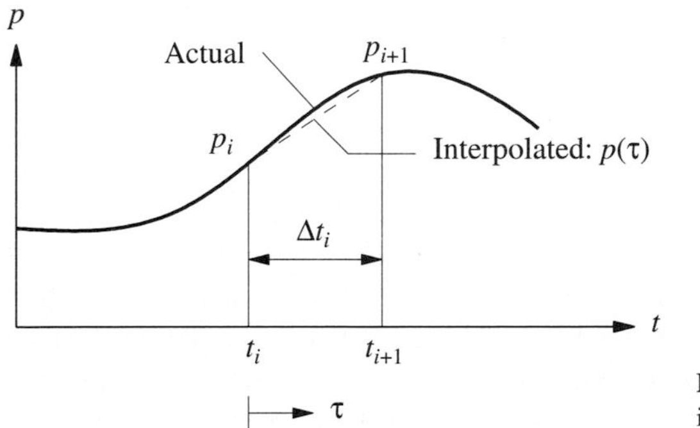

Figure 5.2.1 Notation for linearly interpolated excitation.

equation to be solved is

$$m\ddot{u} + ku = p_i + \frac{\Delta p_i}{\Delta t_i}\tau \tag{5.2.2}$$

The response $u(\tau)$ over the time interval $0 \leq \tau \leq \Delta t_i$ is the sum of three parts: (1) free vibration due to initial displacement u_i and velocity $\dot{u}_i$ at $\tau = 0$, (2) response to step force p_i with zero initial conditions, and (3) response to ramp force $(\Delta p_i/\Delta t_i)\tau$ with zero initial conditions. Adapting the available solutions for these three cases from Sections 2.1, 4.3, and 4.4, respectively, gives

$$u(\tau) = u_i \cos\omega_n\tau + \frac{\dot{u}_i}{\omega_n}\sin\omega_n\tau + \frac{p_i}{k}(1 - \cos\omega_n\tau) + \frac{\Delta p_i}{k}\left(\frac{\tau}{\Delta t_i} - \frac{\sin\omega_n\tau}{\omega_n\,\Delta t_i}\right) \tag{5.2.3a}$$

and

$$\frac{\dot{u}(\tau)}{\omega_n} = -u_i \sin\omega_n\tau + \frac{\dot{u}_i}{\omega_n}\cos\omega_n\tau + \frac{p_i}{k}\sin\omega_n\tau + \frac{\Delta p_i}{k}\frac{1}{\omega_n\,\Delta t_i}(1 - \cos\omega_n\tau) \tag{5.2.3b}$$

Evaluating these equations at $\tau = \Delta t_i$ gives the displacement u_{i+1} and velocity $\dot{u}_{i+1}$ at time $i + 1$:

$$\begin{aligned} u_{i+1} &= u_i \cos(\omega_n\,\Delta t_i) + \frac{\dot{u}_i}{\omega_n}\sin(\omega_n\,\Delta t_i) \\ &\quad + \frac{p_i}{k}[1 - \cos(\omega_n\,\Delta t_i)] + \frac{\Delta p_i}{k}\frac{1}{\omega_n\,\Delta t_i}[\omega_n\,\Delta t_i - \sin(\omega_n\,\Delta t_i)] \end{aligned} \tag{5.2.4a}$$

$$\begin{aligned} \frac{\dot{u}_{i+1}}{\omega_n} &= -u_i \sin(\omega_n\,\Delta t_i) + \frac{\dot{u}_i}{\omega_n}\cos(\omega_n\,\Delta t_i) \\ &\quad + \frac{p_i}{k}\sin(\omega_n\,\Delta t_i) + \frac{\Delta p_i}{k}\frac{1}{\omega_n\,\Delta t_i}[1 - \cos(\omega_n\,\Delta t_i)] \end{aligned} \tag{5.2.4b}$$

These equations can be rewritten after substituting Eq. (5.2.1b) as recurrence formulas:

$$u_{i+1} = Au_i + B\dot{u}_i + Cp_i + Dp_{i+1} \tag{5.2.5a}$$

$$\dot{u}_{i+1} = A'u_i + B'\dot{u}_i + C'p_i + D'p_{i+1} \tag{5.2.5b}$$

These formulas also apply to damped systems with the expressions for the coefficients $A, B, \ldots, D'$ given in Table 5.2.1 for under-critically damped systems (i.e., $\zeta < 1$). They depend on the system parameters ω_n, k, and ζ, and on the time interval $\Delta t \equiv \Delta t_i$.

Since the recurrence formulas are derived from exact solution of the equation of motion, the only restriction on the size of the time step Δt is that it permit a close approximation to the excitation function and that it provide response results at closely spaced time intervals so that the response peaks are not missed. This numerical procedure is especially useful when the excitation is defined at closely spaced time intervals—as for earthquake ground acceleration—so that the linear interpolation is essentially perfect. If the time step Δt is constant, the coefficients $A, B, \ldots, D'$ need to be computed only once.

TABLE 5.2.1 COEFFICIENTS IN RECURRENCE FORMULAS ($\zeta < 1$)

$$A = e^{-\zeta\omega_n \Delta t}\left(\frac{\zeta}{\sqrt{1-\zeta^2}}\sin\omega_D \Delta t + \cos\omega_D \Delta t\right)$$

$$B = e^{-\zeta\omega_n \Delta t}\left(\frac{1}{\omega_D}\sin\omega_D \Delta t\right)$$

$$C = \frac{1}{k}\left\{\frac{2\zeta}{\omega_n \Delta t} + e^{-\zeta\omega_n \Delta t}\left[\left(\frac{1-2\zeta^2}{\omega_D \Delta t} - \frac{\zeta}{\sqrt{1-\zeta^2}}\right)\sin\omega_D \Delta t - \left(1 + \frac{2\zeta}{\omega_n \Delta t}\right)\cos\omega_D \Delta t\right]\right\}$$

$$D = \frac{1}{k}\left[1 - \frac{2\zeta}{\omega_n \Delta t} + e^{-\zeta\omega_n \Delta t}\left(\frac{2\zeta^2 - 1}{\omega_D \Delta t}\sin\omega_D \Delta t + \frac{2\zeta}{\omega_n \Delta t}\cos\omega_D \Delta t\right)\right]$$

$$A' = -e^{-\zeta\omega_n \Delta t}\left(\frac{\omega_n}{\sqrt{1-\zeta^2}}\sin\omega_D \Delta t\right)$$

$$B' = e^{-\zeta\omega_n \Delta t}\left(\cos\omega_D \Delta t - \frac{\zeta}{\sqrt{1-\zeta^2}}\sin\omega_D \Delta t\right)$$

$$C' = \frac{1}{k}\left\{-\frac{1}{\Delta t} + e^{-\zeta\omega_n \Delta t}\left[\left(\frac{\omega_n}{\sqrt{1-\zeta^2}} + \frac{\zeta}{\Delta t\sqrt{1-\zeta^2}}\right)\sin\omega_D \Delta t + \frac{1}{\Delta t}\cos\omega_D \Delta t\right]\right\}$$

$$D' = \frac{1}{k\,\Delta t}\left[1 - e^{-\zeta\omega_n \Delta t}\left(\frac{\zeta}{\sqrt{1-\zeta^2}}\sin\omega_D \Delta t + \cos\omega_D \Delta t\right)\right]$$

The exact solution of the equations of motion required in this numerical procedure is feasible only for linear systems. It is conveniently developed for SDF systems, as shown above, but would be cumbersome for MDF systems unless their response is obtained as superposition of modal responses (Chapters 12 and 13).

Example 5.1

An SDF system has the following properties: $m = 0.2533$ kip-sec^2/in., $k = 10$ kips/in., $T_n = 1$ sec ($\omega_n = 6.283$ rad/sec), and $\zeta = 0.05$. Determine the response $u(t)$ of this system to $p(t)$ defined by the half-cycle sine pulse force shown in Fig. E5.1 by (**a**) using piecewise-linear interpolation of $p(t)$ with $\Delta t = 0.1$ sec, and (**b**) evaluating the theoretical solution.

Solution

1. *Initial calculations*

$$e^{-\zeta\omega_n \Delta t} = 0.9691 \qquad \omega_D = \omega_n\sqrt{1-\zeta^2} = 6.275$$

$$\sin\omega_D \Delta t = 0.5871 \qquad \cos\omega_D \Delta t = 0.8095$$

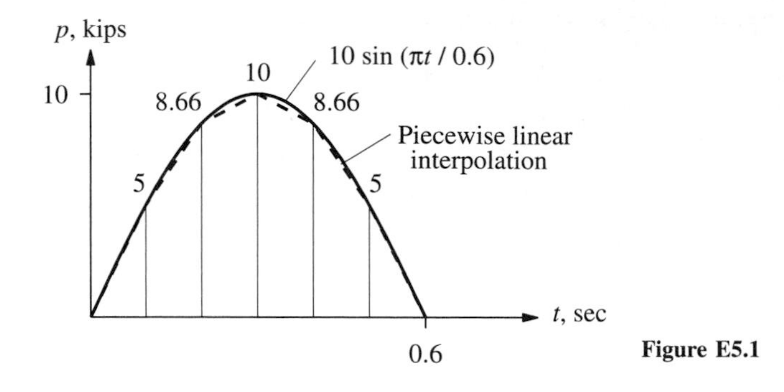

Figure E5.1

Substituting these in Table 5.2.1 gives

$$A = 0.8129 \quad B = 0.09067 \quad C = 0.01236 \quad D = 0.006352$$

$$A' = -3.5795 \quad B' = 0.7559 \quad C' = 0.1709 \quad D' = 0.1871$$

2. *Apply the recurrence equations (5.2.5).* The resulting computations are summarized in Tables E5.1a and E5.1b.

TABLE E5.1a NUMERICAL SOLUTION USING LINEAR INTERPOLATION OF EXCITATION

t_i	p_i	Cp_i	Dp_{i+1}	$B\dot{u}_i$	$\dot{u}_i$	Au_i	u_i	Theoretical u_i
0.0	0.0000	0.0000	0.0318	0.0000	0.0000	0.0000	0.0000	0.0000
0.1	5.0000	0.0618	0.0550	0.0848	0.9354	0.0258	0.0318	0.0328
0.2	8.6602	0.1070	0.0635	0.2782	3.0679	0.1849	0.2274	0.2332
0.3	10.0000	0.1236	0.0550	0.4403	4.8558	0.5150	0.6336	0.6487
0.4	8.6603	0.1070	0.0318	0.4290	4.7318	0.9218	1.1339	1.1605
0.5	5.0000	0.0618	0.0000	0.1753	1.9336	1.2109	1.4896	1.5241
0.6	0.0000	0.0000	0.0000	−0.2735	−3.0159	1.1771	1.4480	1.4814
0.7	0.0000	0.0000	0.0000	−0.6767	−7.4631	0.7346	0.9037	0.9245
0.8	0.0000	0.0000	0.0000	−0.8048	−8.8765	0.0471	0.0579	0.0593
0.9	0.0000	0.0000	0.0000	−0.6272	−6.9177	−0.6160	−0.7577	−0.7751
1.0	0.0000				−2.5171		−1.2432	−1.2718

3. *Compute the theoretical response.* Equation (3.2.5)—valid for $t \leq 0.6$ sec, Eq. (2.2.4) modified appropriately—valid for $t \geq 0.6$ sec, and the derivatives of these two equations are evaluated for each t_i; the results are given in Tables E5.1a and E5.1b.

4. *Check the accuracy of the numerical results.* The numerical solution based on piecewise linear interpolation of the excitation agrees reasonably well with the theoretical solution. The discrepancy arises because the half-cycle sine curve has been replaced by the series of straight lines shown in Fig. E5.1. With a smaller Δt the piecewise linear approximation would be closer to the half-cycle sine curve, and the numerical solution would be more accurate.

TABLE E5.1b NUMERICAL SOLUTION USING LINEAR INTERPOLATION OF EXCITATION

t_i	p_i	$C'p_i$	$D'p_{i+1}$	$A'u_i$	u_i	$B'\dot{u}_i$	$\dot{u}_i$	Theoretical $\dot{u}_i$
0.0	0.0000	0.0000	0.9354	0.0000	0.0000	0.0000	0.0000	0.0000
0.1	5.0000	0.8544	1.6201	−0.1137	0.0318	0.7071	0.9354	0.9567
0.2	8.6602	1.4799	1.8707	−0.8140	0.2274	2.3192	3.0679	3.1383
0.3	10.0000	1.7088	1.6201	−2.2679	0.6336	3.6708	4.8558	4.9674
0.4	8.6603	1.4799	0.9354	−4.0588	1.1339	3.5771	4.7318	4.8408
0.5	5.0000	0.8544	0.0000	−5.3320	1.4896	1.4617	1.9336	1.9783
0.6	0.0000	0.0000	0.0000	−5.1832	1.4480	−2.2799	−3.0159	−3.0848
0.7	0.0000	0.0000	0.0000	−3.2347	0.9037	−5.6418	−7.4631	−7.6346
0.8	0.0000	0.0000	0.0000	−0.2074	0.0579	−6.7103	−8.8765	−9.0808
0.9	0.0000	0.0000	0.0000	2.7124	−0.7577	−5.2295	−6.9177	−7.0771
1.0	0.0000				−1.2432		−2.5171	−2.5754

5.3 CENTRAL DIFFERENCE METHOD

This method is based on a finite difference approximation of the time derivatives of displacement (i.e., velocity and acceleration). Taking constant time steps, $\Delta t_i = \Delta t$, the central difference expressions for velocity and acceleration at time i are

$$\dot{u}_i = \frac{u_{i+1} - u_{i-1}}{2\Delta t} \qquad \ddot{u}_i = \frac{u_{i+1} - 2u_i + u_{i-1}}{(\Delta t)^2} \tag{5.3.1}$$

Substituting these approximate expressions for velocity and acceleration into Eq. (5.1.3), specialized for linearly elastic systems, gives

$$m\frac{u_{i+1} - 2u_i + u_{i-1}}{(\Delta t)^2} + c\frac{u_{i+1} - u_{i-1}}{2\Delta t} + ku_i = p_i \tag{5.3.2}$$

In this equation u_i and u_{i-1} are assumed known (from implementation of the procedure for the preceding time steps). Transferring these known quantities to the right side leads to

$$\left[\frac{m}{(\Delta t)^2} + \frac{c}{2\Delta t}\right]u_{i+1} = p_i - \left[\frac{m}{(\Delta t)^2} - \frac{c}{2\Delta t}\right]u_{i-1} - \left[k - \frac{2m}{(\Delta t)^2}\right]u_i \tag{5.3.3}$$

or

$$\hat{k}u_{i+1} = \hat{p}_i \tag{5.3.4}$$

where

$$\hat{k} = \frac{m}{(\Delta t)^2} + \frac{c}{2\Delta t} \tag{5.3.5}$$

and

$$\hat{p}_i = p_i - \left[\frac{m}{(\Delta t)^2} - \frac{c}{2\Delta t}\right]u_{i-1} - \left[k - \frac{2m}{(\Delta t)^2}\right]u_i \tag{5.3.6}$$

The unknown u_{i+1} is then given by

$$u_{i+1} = \frac{\hat{p}_i}{\hat{k}} \tag{5.3.7}$$

The solution u_{i+1} at time $i+1$ is determined from the equation of motion at time i without using Eq. (5.1.4), the equation of motion at time $i+1$. In particular, the elastic and damping forces can be computed explicitly using known displacements u_i and u_{i-1} and velocities $\dot{u}_i$ and $\dot{u}_{i-1}$. Such methods are called *explicit methods.*

Thus u_0 and u_{-1} are required to determine u_1; the specified initial displacement u_0 is known. To determine u_{-1}, we specialize Eq. (5.3.1) for $i = 0$ to obtain

$$\dot{u}_0 = \frac{u_1 - u_{-1}}{2\Delta t} \qquad \ddot{u}_0 = \frac{u_1 - 2u_0 + u_{-1}}{(\Delta t)^2} \tag{5.3.8}$$

Solving for u_1 from the first equation and substituting in the second gives

$$u_{-1} = u_0 - \Delta t(\dot{u}_0) + \frac{(\Delta t)^2}{2}\ddot{u}_0 \tag{5.3.9}$$

The initial displacement u_0 and initial velocity $\dot{u}_0$ are given, and the equation of motion at time 0 ($t_0 = 0$),

$$m\ddot{u}_0 + c\dot{u}_0 + ku_0 = p_0$$

provides the acceleration at time 0:

$$\ddot{u}_0 = \frac{p_0 - c\dot{u}_0 - ku_0}{m} \tag{5.3.10}$$

Table 5.3.1 summarizes the above-described procedure as it might be implemented on the computer.

The central difference method will "blow up," giving meaningless results, in the presence of numerical round-off if the time step chosen is not short enough. The specific requirement for stability is

$$\frac{\Delta t}{T_n} < \frac{1}{\pi}$$

This is never a constraint for SDF systems because a much smaller time step should be chosen to obtain results that are accurate. Typically, $\Delta t/T_n \le 0.1$ to define the response adequately, and in most earthquake response analyses even a shorter time step, typically $\Delta t = 0.01$ to 0.02 sec, is chosen to define the ground acceleration $\ddot{u}_g(t)$ accurately.

TABLE 5.3.1 CENTRAL DIFFERENCE METHOD

1.0 *Initial calculations*

1.1 $\ddot{u}_0 = \dfrac{p_0 - c\dot{u}_0 - ku_0}{m}$.

1.2 $u_{-1} = u_0 - \Delta t \dot{u}_0 + \dfrac{(\Delta t)^2}{2}\ddot{u}_0$.

1.3 $\hat{k} = \dfrac{m}{(\Delta t)^2} + \dfrac{c}{2\Delta t}$.

1.4 $a = \dfrac{m}{(\Delta t)^2} - \dfrac{c}{2\Delta t}$.

1.5 $b = k - \dfrac{2m}{(\Delta t)^2}$.

2.0 *Calculations for time step i*

2.1 $\hat{p}_i = p_i - au_{i-1} - bu_i$.

2.2 $u_{i+1} = \dfrac{\hat{p}_i}{\hat{k}}$.

2.3 If required: $\dot{u}_i = \dfrac{u_{i+1} - u_{i-1}}{2\Delta t}$, $\ddot{u}_i = \dfrac{u_{i+1} - 2u_i + u_{i-1}}{(\Delta t)^2}$.

3.0 *Repetition for the next time step*

Replace i by $i + 1$ and repeat steps 2.1, 2.2, and 2.3 for the next time step.

Example 5.2

Solve Example 5.1 by the central difference method using $\Delta t = 0.1$ sec.

Solution

1.0 *Initial calculations*

$$m = 0.2533 \qquad k = 10 \qquad c = 0.1592$$

$$u_0 = 0 \qquad \dot{u}_0 = 0$$

1.1 $\ddot{u}_0 = \dfrac{p_0 - c\dot{u}_0 - ku_0}{m} = 0$.

1.2 $u_{-1} = u_0 - (\Delta t)\dot{u}_0 + \dfrac{(\Delta t)^2}{2}\ddot{u}_0 = 0$.

1.3 $\hat{k} = \dfrac{m}{(\Delta t)^2} + \dfrac{c}{2\Delta t} = 26.13$.

1.4 $a = \dfrac{m}{(\Delta t)^2} - \dfrac{c}{2\Delta t} = 24.53$.

1.5 $b = k - \dfrac{2m}{(\Delta t)^2} = -40.66$.

2.0 *Calculations for each time step*

2.1 $\hat{p}_i = p_i - au_{i-1} - bu_i = p_i - 24.53u_{i-1} + 40.66u_i$.

2.2 $u_{i+1} = \dfrac{\hat{p}_i}{\hat{k}} = \dfrac{\hat{p}_i}{26.13}$.

3.0 Computational steps 2.1 and 2.2 are repeated for $i = 0, 1, 2, 3, \ldots$ leading to Table E5.2, wherein the theoretical result (from Table E5.1a) is also included.

TABLE E5.2 NUMERICAL SOLUTION BY CENTRAL DIFFERENCE METHOD

t_i	p_i	u_{i-1}	u_i	$\hat{p}_i$ [Eq. (2.1)]	u_{i+1} [Eq. (2.2)]	Theoretical u_{i+1}
0.0	0.0000	0.0000	0.0000	0.0000	0.0000	0.0328
0.1	5.0000	0.0000	0.0000	5.0000	0.1914	0.2332
0.2	8.6602	0.0000	0.1914	16.4419	0.6293	0.6487
0.3	10.0000	0.1914	0.6293	30.8934	1.1825	1.1605
0.4	8.6603	0.6293	1.1825	41.3001	1.5808	1.5241
0.5	5.0000	1.1825	1.5808	40.2649	1.5412	1.4814
0.6	0.0000	1.5808	1.5412	23.8809	0.9141	0.9245
0.7	0.0000	1.5412	0.9141	−0.6456	−0.0247	0.0593
0.8	0.0000	0.9141	−0.0247	−23.4309	−0.8968	−0.7751
0.9	0.0000	−0.0247	−0.8968	−35.8598	−1.3726	−1.2718
1.0	0.0000	−0.8968	−1.3726	−33.8058	−1.2940	−1.2674

5.4 NEWMARK'S METHOD

5.4.1 Basic Procedure

In 1959, N. M. Newmark developed a family of time-stepping methods based on the following equations:

$$\dot{u}_{i+1} = \dot{u}_i + [(1-\gamma)\,\Delta t]\,\ddot{u}_i + (\gamma\,\Delta t)\ddot{u}_{i+1} \tag{5.4.1a}$$

$$u_{i+1} = u_i + (\Delta t)\dot{u}_i + \left[(0.5-\beta)(\Delta t)^2\right]\ddot{u}_i + \left[\beta(\Delta t)^2\right]\ddot{u}_{i+1} \tag{5.4.1b}$$

The parameters β and γ define the variation of acceleration over a time step and determine the stability and accuracy characteristics of the method. Typical selection for γ is $\frac{1}{2}$ and $\frac{1}{6} \le \beta \le \frac{1}{4}$ is satisfactory from all points of view, including that of accuracy. These two equations, combined with the equilibrium equation (5.1.4) at the end of the time step, provide the basis for computing u_{i+1}, $\dot{u}_{i+1}$, and $\ddot{u}_{i+1}$ at time $i+1$ from the known u_i, $\dot{u}_i$, and $\ddot{u}_i$ at time i. Iteration is required to implement these computations because the unknown $\ddot{u}_{i+1}$ appears in the right side of Eq. (5.4.1).

For linear systems it is possible to modify Newmark's original formulation, however, to permit solution of Eqs. (5.4.1) and (5.1.4) without iteration. Before describing this modification, we demonstrate that two special cases of Newmark's method are the well-known average acceleration and linear acceleration methods.

5.4.2 Special Cases

For these two methods, Table 5.4.1 summarizes the development of the relationship between responses u_{i+1}, $\dot{u}_{i+1}$, and $\ddot{u}_{i+1}$ at time $i+1$ to the corresponding quantities at time i. Equation (5.4.2) describes the assumptions that the variation of acceleration over

TABLE 5.4.1 AVERAGE ACCELERATION AND LINEAR ACCELERATION METHODS

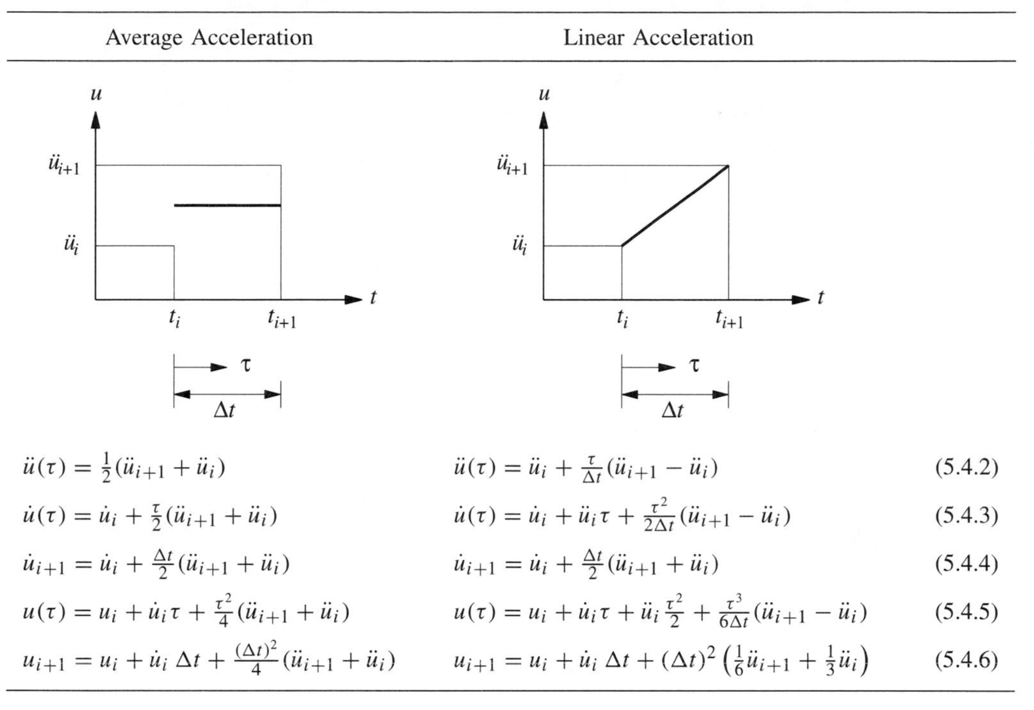

Average Acceleration	Linear Acceleration	
$\ddot{u}(\tau) = \frac{1}{2}(\ddot{u}_{i+1} + \ddot{u}_i)$	$\ddot{u}(\tau) = \ddot{u}_i + \frac{\tau}{\Delta t}(\ddot{u}_{i+1} - \ddot{u}_i)$	(5.4.2)
$\dot{u}(\tau) = \dot{u}_i + \frac{\tau}{2}(\ddot{u}_{i+1} + \ddot{u}_i)$	$\dot{u}(\tau) = \dot{u}_i + \ddot{u}_i\tau + \frac{\tau^2}{2\Delta t}(\ddot{u}_{i+1} - \ddot{u}_i)$	(5.4.3)
$\dot{u}_{i+1} = \dot{u}_i + \frac{\Delta t}{2}(\ddot{u}_{i+1} + \ddot{u}_i)$	$\dot{u}_{i+1} = \dot{u}_i + \frac{\Delta t}{2}(\ddot{u}_{i+1} + \ddot{u}_i)$	(5.4.4)
$u(\tau) = u_i + \dot{u}_i\tau + \frac{\tau^2}{4}(\ddot{u}_{i+1} + \ddot{u}_i)$	$u(\tau) = u_i + \dot{u}_i\tau + \ddot{u}_i\frac{\tau^2}{2} + \frac{\tau^3}{6\Delta t}(\ddot{u}_{i+1} - \ddot{u}_i)$	(5.4.5)
$u_{i+1} = u_i + \dot{u}_i\,\Delta t + \frac{(\Delta t)^2}{4}(\ddot{u}_{i+1} + \ddot{u}_i)$	$u_{i+1} = u_i + \dot{u}_i\,\Delta t + (\Delta t)^2\left(\frac{1}{6}\ddot{u}_{i+1} + \frac{1}{3}\ddot{u}_i\right)$	(5.4.6)

a time step is constant, equal to the average acceleration, or linear. Integration of $\ddot{u}(\tau)$ gives Eq. (5.4.3) for the variation $\dot{u}(\tau)$ of velocity over the time step in which $\tau = \Delta t$ is substituted to obtain Eq. (5.4.4) for the velocity $\dot{u}_{i+1}$ at time $i+1$. Integration of $\dot{u}(\tau)$ gives Eq. (5.4.5) for the variation $u(\tau)$ of displacement over the time step in which $\tau = \Delta t$ is substituted to obtain Eq. (5.4.6) for the displacement u_{i+1} at time $i+1$. Comparing Eqs. (5.4.4) and (5.4.6) with Eq. (5.4.1) demonstrates that Newmark's equations with $\gamma = \frac{1}{2}$ and $\beta = \frac{1}{4}$ are the same as those derived assuming constant average acceleration, and those with $\gamma = \frac{1}{2}$ and $\beta = \frac{1}{6}$ correspond to the assumption of linear variation of acceleration.

5.4.3 Noniterative Formulation

We now return to Eq. (5.4.1) and reformulate it to avoid iteration and to use incremental quantities:

$$\Delta u_i \equiv u_{i+1} - u_i \qquad \Delta\dot{u}_i \equiv \dot{u}_{i+1} - \dot{u}_i \qquad \Delta\ddot{u}_i \equiv \ddot{u}_{i+1} - \ddot{u}_i \tag{5.4.7}$$

$$\Delta p_i = p_{i+1} - p_i \tag{5.4.8}$$

While the incremental form is not necessary for analysis of linear systems, it is introduced because it provides a convenient extension to nonlinear systems. Equation (5.4.1) can be rewritten as

$$\Delta\dot{u}_i = (\Delta t)\,\ddot{u}_i + (\gamma\,\Delta t)\,\Delta\ddot{u}_i \qquad \Delta u_i = (\Delta t)\dot{u}_i + \frac{(\Delta t)^2}{2}\ddot{u}_i + \beta(\Delta t)^2\,\Delta\ddot{u}_i \tag{5.4.9}$$

The second of these equations can be solved for

$$\Delta\ddot{u}_i = \frac{1}{\beta(\Delta t)^2}\Delta u_i - \frac{1}{\beta\,\Delta t}\dot{u}_i - \frac{1}{2\beta}\ddot{u}_i \tag{5.4.10}$$

Substituting Eq. (5.4.10) into Eq. (5.4.9a) gives

$$\Delta\dot{u}_i = \frac{\gamma}{\beta\,\Delta t}\Delta u_i - \frac{\gamma}{\beta}\dot{u}_i + \Delta t\left(1 - \frac{\gamma}{2\beta}\right)\ddot{u}_i \tag{5.4.11}$$

Next, Eqs. (5.4.10) and (5.4.11) are substituted into the incremental equation of motion:

$$m\,\Delta\ddot{u}_i + c\,\Delta\dot{u}_i + k\,\Delta u_i = \Delta p_i \tag{5.4.12}$$

obtained by subtracting Eq. (5.1.3) from Eq. (5.1.4), both specialized to linear systems with $(f_S)_i = ku_i$ and $(f_S)_{i+1} = ku_{i+1}$. This substitution gives

$$\hat{k}\,\Delta u_i = \Delta\hat{p}_i \tag{5.4.13}$$

where

$$\hat{k} = k + \frac{\gamma}{\beta\,\Delta t}c + \frac{1}{\beta(\Delta t)^2}m \tag{5.4.14}$$

and

$$\Delta\hat{p}_i = \Delta p_i + \left(\frac{1}{\beta\,\Delta t}m + \frac{\gamma}{\beta}c\right)\dot{u}_i + \left[\frac{1}{2\beta}m + \Delta t\left(\frac{\gamma}{2\beta} - 1\right)c\right]\ddot{u}_i \tag{5.4.15}$$

With $\hat{k}$ and $\Delta\hat{p}_i$ known from the system properties m, k, and c, algorithm parameters γ and β, and the $\dot{u}_i$ and $\ddot{u}_i$ at the beginning of the time step, the incremental displacement is computed from

$$\Delta u_i = \frac{\Delta\hat{p}_i}{\hat{k}} \tag{5.4.16}$$

Once Δu_i is known, $\Delta\dot{u}_i$ and $\Delta\ddot{u}_i$ can be computed from Eqs. (5.4.11) and (5.4.10), respectively, and u_{i+1}, $\dot{u}_{i+1}$, and $\ddot{u}_{i+1}$ from Eq. (5.4.7).

The acceleration can also be obtained from the equation of motion at t_{i+1}:

$$\ddot{u}_{i+1} = \frac{p_{i+1} - c\dot{u}_{i+1} - ku_{i+1}}{m} \tag{5.4.17}$$

rather than by Eqs. (5.4.10) and (5.4.7). Equation (5.4.17) is needed to obtain $\ddot{u}_0$ to start the computations [see Eq. (5.3.10)].

In Newmark's method the solution at time $i + 1$ is determined from Eq. (5.4.12), which is equivalent to the use of the equilibrium condition, Eq. (5.1.4), at time $i + 1$. Such methods are called *implicit methods.*

Table 5.4.2 summarizes the time-stepping solution using Newmark's method as it might be implemented on the computer.

TABLE 5.4.2 NEWMARK'S METHOD: LINEAR SYSTEMS

Special cases

(1) Average acceleration method ($\gamma = \frac{1}{2}$, $\beta = \frac{1}{4}$)

(2) Linear acceleration method ($\gamma = \frac{1}{2}$, $\beta = \frac{1}{6}$)

1.0 *Initial calculations*

1.1 $\ddot{u}_0 = \dfrac{p_0 - c\dot{u}_0 - ku_0}{m}$.

1.2 Select Δt.

1.3 $\hat{k} = k + \dfrac{\gamma}{\beta\,\Delta t}c + \dfrac{1}{\beta(\Delta t)^2}m$.

1.4 $a = \dfrac{1}{\beta\,\Delta t}m + \dfrac{\gamma}{\beta}c$; and $b = \dfrac{1}{2\beta}m + \Delta t\left(\dfrac{\gamma}{2\beta} - 1\right)c$.

2.0 *Calculations for each time step, i*

2.1 $\Delta\hat{p}_i = \Delta p_i + a\dot{u}_i + b\ddot{u}_i$.

2.2 $\Delta u_i = \dfrac{\Delta\hat{p}_i}{\hat{k}}$.

2.3 $\Delta\dot{u}_i = \dfrac{\gamma}{\beta\,\Delta t}\Delta u_i - \dfrac{\gamma}{\beta}\dot{u}_i + \Delta t\left(1 - \dfrac{\gamma}{2\beta}\right)\ddot{u}_i$.

2.4 $\Delta\ddot{u}_i = \dfrac{1}{\beta(\Delta t)^2}\Delta u_i - \dfrac{1}{\beta\Delta t}\dot{u}_i - \dfrac{1}{2\beta}\ddot{u}_i$.

2.5 $u_{i+1} = u_i + \Delta u_i$, $\dot{u}_{i+1} = \dot{u}_i + \Delta\dot{u}_i$, $\ddot{u}_{i+1} = \ddot{u}_i + \Delta\ddot{u}_i$.

3.0 *Repetition for the next time step.* Replace i by $i + 1$ and implement steps 2.1 to 2.5 for the next time step.

Newmark's method is stable if

$$\frac{\Delta t}{T_n} \leq \frac{1}{\pi\sqrt{2}}\frac{1}{\sqrt{\gamma - 2\beta}} \tag{5.4.18}$$

For $\gamma = \frac{1}{2}$ and $\beta = \frac{1}{4}$ this condition becomes

$$\frac{\Delta t}{T_n} < \infty$$

This implies that the average acceleration method is stable for any Δt, no matter how large; however, it is accurate only if Δt is small enough. For $\gamma = \frac{1}{2}$ and $\beta = \frac{1}{6}$, Eq. (5.4.18) indicates that the linear acceleration method is stable if

$$\frac{\Delta t}{T_n} \leq 0.551$$

However, as in the case of the central difference method, this condition has little significance in the analysis of SDF systems because a much shorter time step than $0.551T_n$ must be used to obtain an accurate representation of the excitation and response.

Example 5.3

Solve Example 5.1 by the average acceleration method using $\Delta t = 0.1$ sec.

Solution

1.0 *Initial calculations*

$$m = 0.2533 \qquad k = 10 \qquad c = 0.1592$$

$$u_0 = 0 \qquad \dot{u}_0 = 0 \qquad p_0 = 0$$

1.1 $\ddot{u}_0 = \dfrac{p_0 - c\dot{u}_0 - ku_0}{m} = 0.$

1.2 $\Delta t = 0.1.$

1.3 $\hat{k} = k + \dfrac{2}{\Delta t}c + \dfrac{4}{(\Delta t)^2}m = 114.5.$

1.4 $a = \dfrac{4}{\Delta t}m + 2c = 10.45$; and $b = 2m = 0.5066.$

2.0 *Calculations for each time step*

2.1 $\Delta\hat{p}_i = \Delta p_i + a\dot{u}_i + b\ddot{u}_i = \Delta p_i + 10.45\dot{u}_i + 0.5066\ddot{u}_i.$

2.2 $\Delta u_i = \dfrac{\Delta\hat{p}_i}{\hat{k}} = \dfrac{\Delta\hat{p}_i}{114.5}.$

2.3 $\Delta\dot{u}_i = \dfrac{2}{\Delta t}\Delta u_i - 2\dot{u}_i = 20\Delta u_i - 2\dot{u}_i.$

2.4 $\Delta\ddot{u}_i = \dfrac{4}{(\Delta t)^2}(\Delta u_i - \Delta t\dot{u}_i) - 2\ddot{u}_i = 400(\Delta u_i - 0.1\dot{u}_i) - 2\ddot{u}_i.$

2.5 $u_{i+1} = u_i + \Delta u_i,\ \dot{u}_{i+1} = \dot{u}_i + \Delta\dot{u}_i,\ \ddot{u}_{i+1} = \ddot{u}_i + \Delta\ddot{u}_i.$

3.0 *Repetition for the next time step.* Steps 2.1 to 2.5 are repeated for successive time steps and are summarized in Table E5.3, where the theoretical result (from Table E5.1a) is also included.

Example 5.4

Solve Example 5.1 by the linear acceleration method using $\Delta t = 0.1$ sec.

Solution

1.0 *Initial calculations*

$$m = 0.2533 \qquad k = 10 \qquad c = 0.1592$$

$$u_0 = 0 \qquad \dot{u}_0 = 0 \qquad p_0 = 0$$

1.1 $\ddot{u}_0 = \dfrac{p_0 - c\dot{u}_0 - ku_0}{m} = 0.$

1.2 $\Delta t = 0.1.$

1.3 $\hat{k} = k + \dfrac{3}{\Delta t}c + \dfrac{6}{(\Delta t)^2}m = 166.8.$

1.4 $a = \dfrac{6}{\Delta t}m + 3c = 15.68$; and $b = 3m + \dfrac{\Delta t}{2}c = 0.7679.$

TABLE E5.3 NUMERICAL SOLUTION BY AVERAGE ACCELERATION METHOD

t_i	p_i	$\ddot{u}_i$ (Step 2.5)	Δp_i	$\Delta \hat{p}_i$ (Step 2.1)	Δu_i (Step 2.2)	$\Delta \dot{u}_i$ (Step 2.3)	$\Delta \ddot{u}_i$ (Step 2.4)	$\dot{u}_i$ (Step 2.5)	u_i (Step 2.5)	Theoretical u_i
0.0	0.0000	0.0000	5.0000	5.0000	0.0437	0.8733	17.4666	0.0000	0.0000	0.0000
0.1	5.0000	17.4666	3.6603	21.6356	0.1890	2.0323	5.7137	0.8733	0.0437	0.0328
0.2	8.6602	23.1803	1.3398	43.4485	0.3794	1.7776	−10.8078	2.9057	0.2326	0.2332
0.3	10.0000	12.3724	−1.3397	53.8708	0.4705	0.0428	−23.8893	4.6833	0.6121	0.6487
0.4	8.6603	−11.5169	−3.6602	39.8948	0.3484	−2.4839	−26.6442	4.7261	1.0825	1.1605
0.5	5.0000	−38.1611	−5.0000	−0.9009	−0.0079	−4.6417	−16.5122	2.2422	1.4309	1.5241
0.6	0.0000	−54.6733	0.0000	−52.7740	−0.4609	−4.4187	20.9716	−2.3995	1.4231	1.4814
0.7	0.0000	−33.7017	0.0000	−88.3275	−0.7714	−1.7912	31.5787	−6.8183	0.9622	0.9245
0.8	0.0000	−2.1229	0.0000	−91.0486	−0.7952	1.3159	30.5646	−8.6095	0.1908	0.0593
0.9	0.0000	28.4417	0.0000	−61.8123	−0.5398	3.7907	18.9297	−7.2936	−0.6044	−0.7751
1.0	0.0000	47.3714						−3.5029	−1.1442	−1.2718

2.0 *Calculations for each time step*

2.1 $\Delta\hat{p}_i = \Delta p_i + a\dot{u}_i + b\ddot{u}_i = \Delta p_i + 15.68\dot{u}_i + 0.7679\ddot{u}_i.$

2.2 $\Delta u_i = \dfrac{\Delta\hat{p}_i}{\hat{k}} = \dfrac{\Delta\hat{p}_i}{166.8}.$

2.3 $\Delta\dot{u}_i = \dfrac{3}{\Delta t}\Delta u_i - 3\dot{u}_i - \dfrac{\Delta t}{2}\ddot{u}_i = 30\Delta u_i - 3\dot{u}_i - 0.05\ddot{u}_i.$

2.4 $\Delta\ddot{u}_i = \dfrac{6}{(\Delta t)^2}(\Delta u_i - \Delta t\dot{u}_i) - 3\ddot{u}_i = 600(\Delta u_i - 0.1\dot{u}_i) - 3\ddot{u}_i.$

2.5 $u_{i+1} = u_i + \Delta u_i,\ \dot{u}_{i+1} = \dot{u}_i + \Delta\dot{u}_i,\ \ddot{u}_{i+1} = \ddot{u}_i + \Delta\ddot{u}_i.$

3.0 *Repetition for the next time step.* Steps 2.1 to 2.5 are repeated for successive time steps and are summarized in Table E5.4, where the theoretical result (from Table E5.1a) is also included.

5.5 STABILITY AND COMPUTATIONAL ERROR

5.5.1 Stability

Numerical procedures that lead to bounded solutions if the time step is shorter than some stability limit are called *conditionally stable* procedures. Procedures that lead to bounded solutions regardless of the time-step length are called *unconditionally stable* procedures. The average acceleration method is unconditionally stable. The linear acceleration method is stable if $\Delta t/T_n < 0.551$, and the central difference method is stable if $\Delta t/T_n < 1/\pi$.

The stability criteria are not restrictive (i.e., they do not dictate the choice of time step) in the analysis of SDF systems because $\Delta t/T_n$ must be considerably smaller than the stability limit (say, 0.1 or less) to ensure adequate accuracy in the numerical results. Stability of the numerical method is important, however, in the analysis of MDF systems, where it is often necessary to use unconditionally stable methods (Chapter 15).

5.5.2 Computational Error

Error is inherent in any numerical solution of the equation of motion. We not discuss error analysis from a mathematical point of view. Rather, we examine two of the important characteristics of numerical solutions to develop a feel for the nature of the errors, and then mention a simple, useful way of managing error.

Consider the free vibration problem

$$m\ddot{u} + ku = 0 \qquad u(0) = 1 \quad \text{and} \quad \dot{u}(0) = 0$$

for which the theoretical solution is

$$u(t) = \cos\omega_n t \tag{5.5.1}$$

TABLE E5.4 NUMERICAL SOLUTION BY LINEAR ACCELERATION METHOD

t_i	p_i	$\ddot{u}_i$ (Step 2.5)	Δp_i	$\Delta \hat{p}_i$ (Step 2.1)	Δu_i (Step 2.2)	$\Delta \dot{u}_i$ (Step 2.3)	$\Delta \ddot{u}_i$ (Step 2.4)	$\dot{u}_i$ (Step 2.5)	u_i (Step 2.5)	Theoretical u_i
0.0	0.0000	0.0000	5.0000	5.0000	0.0300	0.8995	17.9903	0.0000	0.0000	0.0000
0.1	5.0000	17.9903	3.6603	31.5749	0.1893	2.0824	5.6666	0.8995	0.0300	0.0328
0.2	8.6602	23.6569	1.3398	66.2479	0.3973	1.7897	−11.5191	2.9819	0.2193	0.2332
0.3	10.0000	12.1378	−1.3397	82.7784	0.4964	−0.0296	−24.8677	4.7716	0.6166	0.6487
0.4	8.6603	−12.7299	−3.6602	60.8987	0.3652	−2.6336	−27.2127	4.7420	1.1130	1.1605
0.5	5.0000	−39.9426	−5.0000	−2.6205	−0.0157	−4.7994	−16.1033	2.1084	1.4782	1.5241
0.6	0.0000	−56.0459	0.0000	−85.2198	−0.5110	−4.4558	22.9749	−2.6911	1.4625	1.4814
0.7	0.0000	−33.0710	0.0000	−137.4264	−0.8241	−1.6292	33.5584	−7.1469	0.9514	0.9245
0.8	0.0000	0.4874	0.0000	−137.1965	−0.8227	1.6218	31.4613	−8.7761	0.1273	0.0593
0.9	0.0000	31.9487	0.0000	−87.6156	−0.5254	4.1031	18.1644	−7.1543	−0.6954	−0.7751
1.0	0.0000	50.1130						−3.0512	−1.2208	−1.2718

This problem is solved by four numerical methods: central difference method, average acceleration method, linear acceleration method, and Wilson's method. The last of these methods is presented in Chapter 15. The numerical results obtained using $\Delta t = 0.1T_n$ are compared with the theoretical solution in Fig. 5.5.1. This comparison shows that some numerical methods may predict that the displacement amplitude decays with time, although the system is undamped, and that the natural period is elongated or shortened.

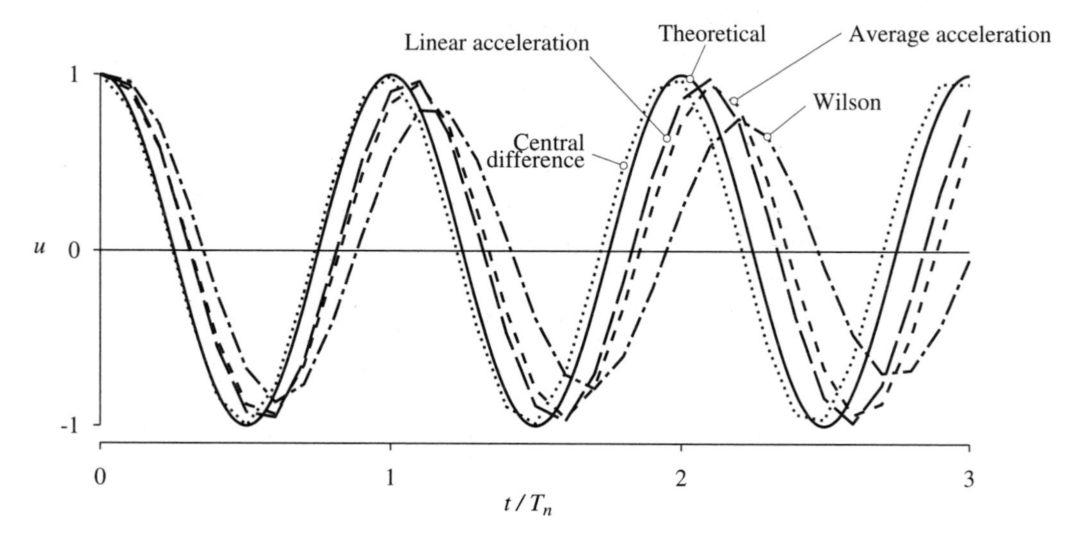

Figure 5.5.1 Free vibration solution by four numerical methods ($\Delta t/T_n = 0.1$) and the theoretical solution.

Figure 5.5.2 shows the amplitude decay AD and period elongation PE in the four numerical methods as a function of $\Delta t/T_n$; AD and PE are defined in parts (b) and (c) of the figure, respectively. The mathematical analyses that led to these data are not presented, however. Three of the methods predict no decay of displacement amplitude. Wilson's method contains decay of amplitude, however, implying that this method introduces *numerical damping* in the system; the equivalent viscous damping ratio $\bar{\zeta}$ is shown in part (a) of the figure. Observe the rapid increase in the period error in the central difference method near $\Delta t/T_n = 1/\pi$, the stability limit for the method. The central difference method introduces the largest period error. In this sense it is the least accurate of the methods considered. For $\Delta t/T_n$ less than its stability limit, the linear acceleration method gives the least period elongation. This property, combined with no amplitude decay, make this method the most suitable method (of the methods presented) for SDF systems. However, we shall arrive at a different conclusion for MDF systems because of stability requirements (Chapter 15).

The choice of time step also depends on the time variation of the dynamic excitation, in addition to the natural vibration period of the system. Figure 5.5.2 suggests that $\Delta t = 0.1T_n$ would give reasonably accurate results. The time step should also be

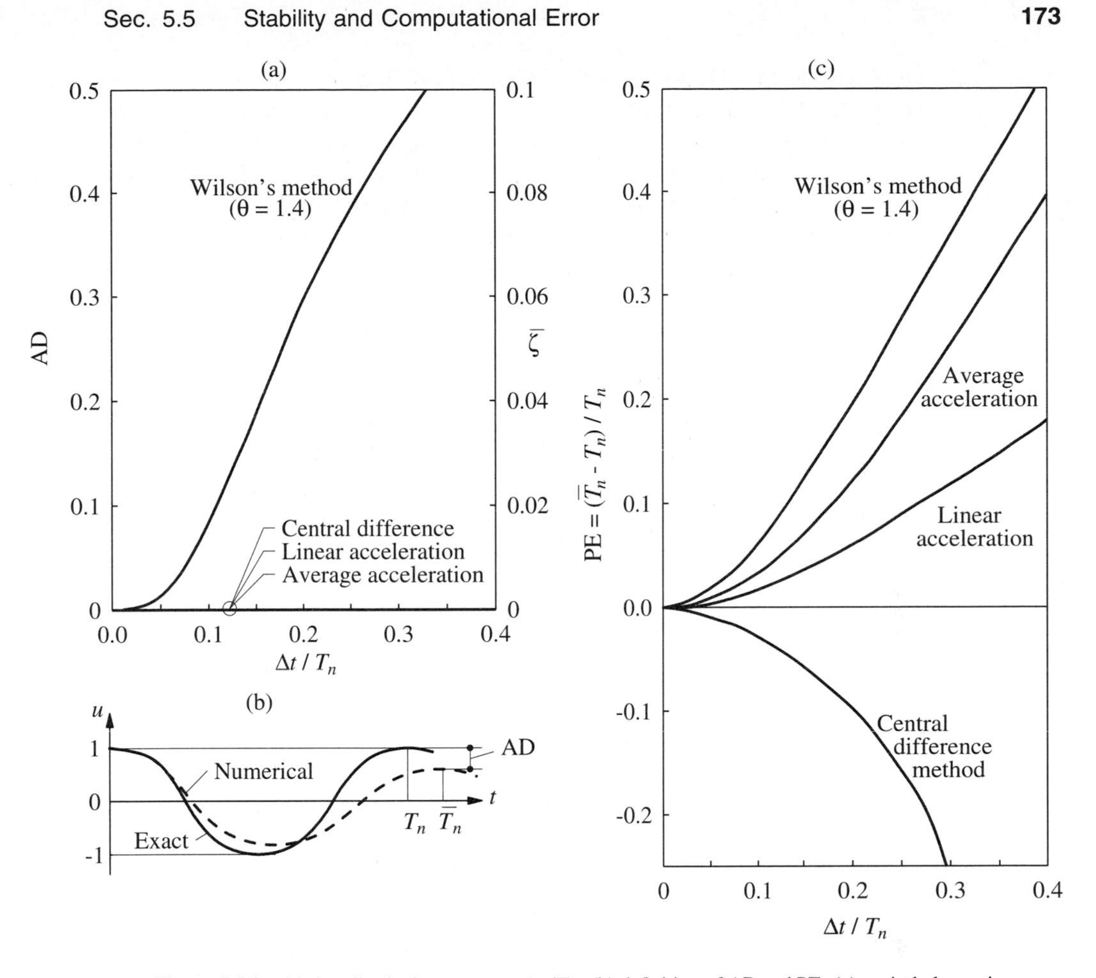

Figure 5.5.2 (a) Amplitude decay versus $\Delta t/T_n$; (b) definition of AD and PE; (c) period elongation versus $\Delta t/T_n$.

short enough to keep the distortion of the excitation function to a minimum. A very fine time step is necessary to describe numerically the highly irregular earthquake ground acceleration recorded during earthquakes; typically, $\Delta t = 0.02$ sec is chosen and this dictates a maximum time step for computing the response of a structure to earthquake excitation.

One useful, although unsophisticated technique for selecting the time step is to solve the problem with a time step that seems reasonable, then repeat the solution with a slightly smaller time step and compare the results, continuing the process until two successive solutions are close enough.

The preceding discussion of stability and accuracy applies strictly to linear systems. The reader should consult other references for how these issues affect nonlinear response analysis.

5.6 ANALYSIS OF NONLINEAR RESPONSE: CENTRAL DIFFERENCE METHOD

The dynamic response of a system beyond its linearly elastic range is generally not amenable to analytical solution even if the time variation of the excitation is described by a simple function. Numerical methods are therefore essential in the analysis of nonlinear systems. The central difference method can easily be adapted for solving the nonlinear equation of motion, Eq. (5.1.3), at time i. Substituting Eqs. (5.3.1), the central difference approximation for velocity and acceleration, gives Eq. (5.3.2) with ku_i replaced by $(f_S)_i$, which can be rewritten to obtain the following expression for response at time $i+1$:

$$\hat{k}u_{i+1} = \hat{p}_i \tag{5.6.1}$$

where

$$\hat{k} = \frac{m}{(\Delta t)^2} + \frac{c}{2\Delta t} \tag{5.6.2}$$

and

$$\hat{p}_i = p_i - \left[\frac{m}{(\Delta t)^2} - \frac{c}{2\Delta t}\right]u_{i-1} - (f_S)_i + \frac{2m}{(\Delta t)^2}u_i \tag{5.6.3}$$

Comparing these equations with those for linear systems, it is seen that the only difference is in the definition for $\hat{p}_i$. With this modification Table 5.3.1 also applies to nonlinear systems.

The resisting force $(f_S)_i$ appears *explicitly* as it depends only on the response at time i, not on the unknown response at time $i+1$. Thus it is easily calculated, making the central difference method perhaps the simplest procedure for nonlinear systems. Although attractive in this regard, the method is not popular for practical or research applications because more effective methods are available.

5.7 ANALYSIS OF NONLINEAR RESPONSE: NEWMARK'S METHOD

In this section Newmark's method, described in Section 5.4 for linear systems, is extended to nonlinear systems. Although not as simple as the central difference method, it is perhaps the most popular method because of its superior accuracy.

The difference between Eqs. (5.1.3) and (5.1.4) gives an incremental equilibrium equation:

$$m\,\Delta\ddot{u}_i + c\,\Delta\dot{u}_i + (\Delta f_S)_i = \Delta p_i \tag{5.7.1}$$

The incremental resisting force

$$(\Delta f_S)_i = (k_i)_{\text{sec}}\,\Delta u_i \tag{5.7.2}$$

where the secant stiffness $(k_i)_{\text{sec}}$, shown in Fig. 5.7.1, cannot be determined because u_{i+1} is *not* known. If we make the assumption that over a small time step Δt, the secant

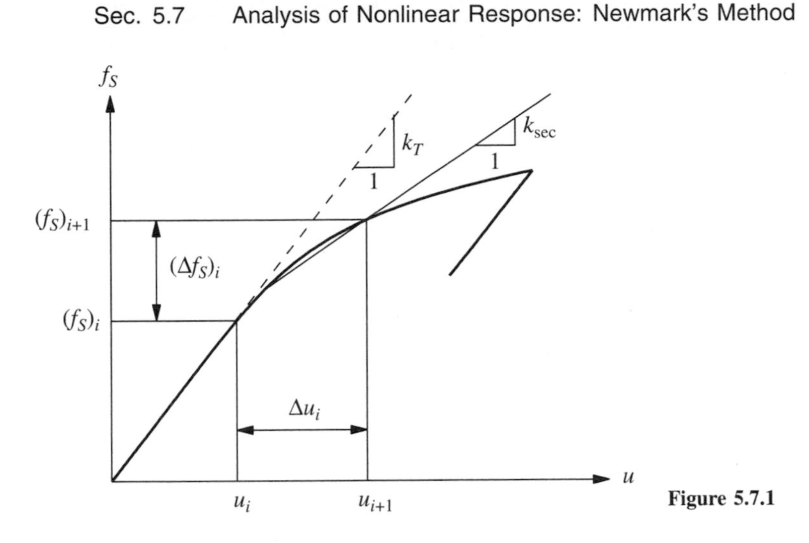

Figure 5.7.1

stiffness $(k_i)_{sec}$ could be replaced by the tangent stiffness $(k_i)_T$ shown in Fig. 5.7.1, then Eq. (5.7.2) could be approximated by

$$(\Delta f_S)_i \simeq (k_i)_T \, \Delta u_i \tag{5.7.3}$$

Dropping the subscript T from $(k_i)_T$ in Eq. (5.7.3) and substituting it in Eq. (5.7.1) gives

$$m \, \Delta \ddot{u}_i + c \, \Delta \dot{u}_i + k_i \, \Delta u_i = \Delta p_i \tag{5.7.4}$$

The similarity between this equation and the corresponding equation for linear systems, Eq. (5.4.12), suggests that the noniterative formulation of Newmark's method presented earlier for linear systems may also be used in the analysis of nonlinear response. All that needs to be done is to replace k in Eq. (5.4.14) by the tangent stiffness k_i to be evaluated at the beginning of each time step. This change implies that step 1.3 of Table 5.4.2 should follow step 2.1. For nonlinear systems step 2.5 and Eq. (5.4.17) would give different values of $\ddot{u}_{i+1}$ and the latter value is preferable because it satisfies equilibrium at time $i + 1$.

This procedure with a constant time step Δt can lead to unacceptably inaccurate results. Significant errors arise for two reasons: (1) the tangent stiffness was used instead of the secant stiffness, and (2) use of a constant time step delays detection of the transitions in the force–deformation relationship.

First, we consider the second source of error, illustrated by the force–deformation relation of Fig. 5.7.2a. Suppose that the displacement at time i, the beginning of a time step, is u_i and the velocity $\dot{u}_i$ is positive (i.e., the displacement is increasing); this is shown by point a. Application of the previously described numerical procedure for the time step results in displacement u_{i+1} and velocity $\dot{u}_{i+1}$ at time $i + 1$; this is shown by point b. If $\dot{u}_{i+1}$ is negative, then at some point b' during the time step, the velocity became zero, changed sign, and the displacement started decreasing. In the numerical procedure, if we do not bother to locate b', continue with the computations by starting

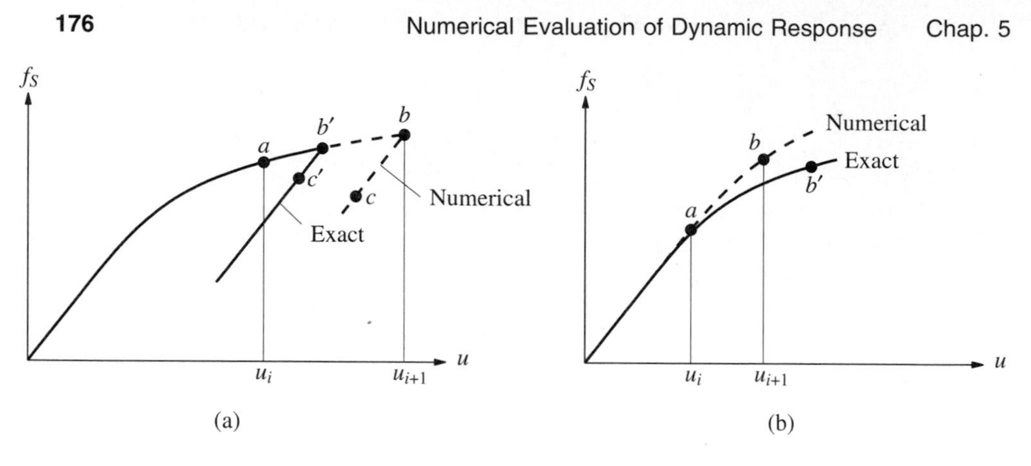

Figure 5.7.2

the next time step at point b, and use the tangent stiffness associated with the unloading branch of the force–deformation diagram, this procedure locates the point c at the end of the next time step with displacement u_{i+2} and negative velocity. On the other hand, if the time instant associated with b'—when the velocity actually became zero—could be determined, computations for the next time step would start with the state of the system at b' and determine the displacement and velocity at the end of the time step, identified as c'. Not locating b' has the effect of overshooting to b and not following the exact path on the force–deformation diagram. These departures from the exact path would occur at each reversal of velocity, leading to errors in the numerical results. A similar problem arises at sharp corners in the force–deformation relationship, as in elastoplastic systems.

These errors could be avoided by locating b' accurately. This could be achieved by retracing the integration over the time interval t_i to t_{i+1} with a smaller time step, say, $\Delta t/4$. Alternatively, an iterative process may be used in which integration is resumed from time i with a step smaller than the full time step, whose size is progressively adjusted so that at the end of such an adjusted time step, the velocity is close to zero.

Now, we return to the first source of error that is associated with the use of tangent stiffness instead of the unknown secant stiffness, and is illustrated by the force–deformation relation of Fig. 5.7.2b. The displacement at time i, the beginning of a time step, is shown as point a. Using the tangent stiffness at a, numerical integration from time i to time $i + 1$ leads to the displacement u_{i+1}, identified as point b. If we were able to follow the curve exactly, the result may have been the displacement at b'. This discrepancy accumulating over a series of time steps may introduce significant errors.

These errors can be minimized by using an iterative procedure. The key equation that is solved at each time step in Newmark's method is Eq. (5.4.13), which, modified for nonlinear systems, becomes

$$\hat{k}_i \, \Delta u_i = \Delta \hat{p}_i \tag{5.7.5}$$

where $\Delta \hat{p}_i$ is given by Eq. (5.4.15) and

$$\hat{k}_i = k_i + \frac{\gamma}{\beta\,\Delta t}c + \frac{1}{\beta(\Delta t)^2}m \tag{5.7.6}$$

For convenience in notation we drop the subscript i in k_i and replace it by T to emphasize that this is the tangent stiffness; also, the subscript i is dropped from Δu_i and $\Delta \hat{p}_i$. Equations (5.7.5) and (5.7.6) then become

$$\hat{k}_T\,\Delta u = \Delta \hat{p} \tag{5.7.7}$$

and

$$\hat{k}_T = k_T + \frac{\gamma}{\beta\,\Delta t}c + \frac{1}{\beta(\Delta t)^2}m \tag{5.7.8}$$

Figure 5.7.3a shows a schematic plot of Eq. (5.7.7). The relationship is nonlinear because the tangent stiffness k_T depends on the displacement u and hence the slope $\hat{k}_T$ is not constant. In static analysis of a nonlinear system, $\hat{k}_T = k_T$ and the nonlinearity in $\hat{k}_T$ is the same as in k_T. In dynamic analysis the presence of mass and damping terms in $\hat{k}_T$ decreases the nonlinearity because the constant term $m/\beta(\Delta t)^2$ for typical values of Δt is usually much larger than k_T.

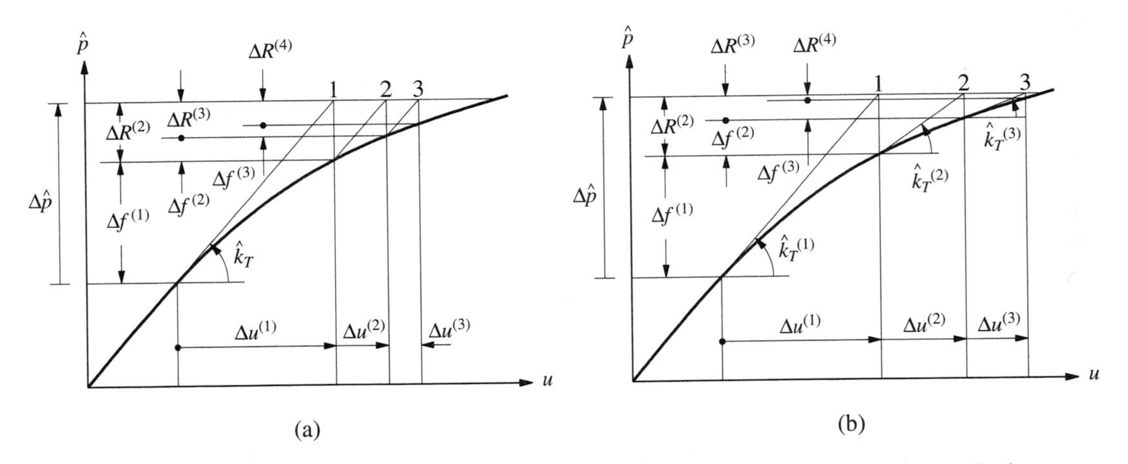

Figure 5.7.3 Iteration within a time step for nonlinear systems: (a) modified Newton–Raphson iteration; (b) Newton–Raphson iteration.

The iterative procedure is described next with reference to Fig. 5.7.3a. The first iterative step is the application of Eq. (5.7.7) in the procedure described previously:

$$\hat{k}_T\,\Delta u^{(1)} = \Delta \hat{p} \tag{5.7.9}$$

to determine $\Delta u^{(1)}$ (corresponding to point b in Fig. 5.7.2b), the first approximation to the final Δu (corresponding to point b' in Fig. 5.7.2b). Associated with $\Delta u^{(1)}$ is the true force $\Delta f^{(1)}$, which is less than $\Delta \hat{p}$, and a residual force is defined: $\Delta R^{(2)} = \Delta \hat{p} - \Delta f^{(1)}$.

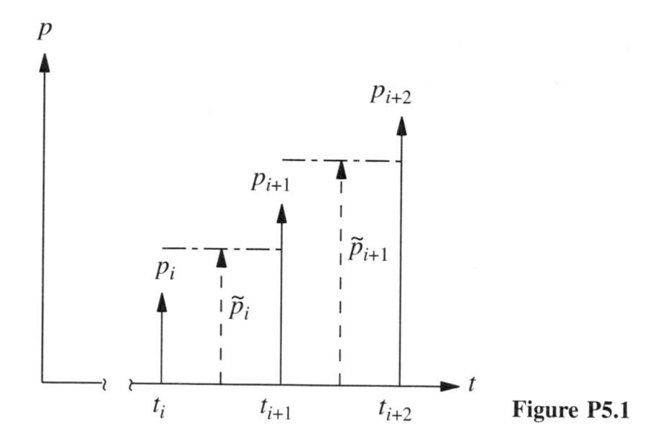

Figure P5.1

*5.2 Solve Example 5.1 using the piecewise-constant approximation of the forcing function; neglect damping in the SDF system.

*5.3 Solve the problem in Example 5.1 by the central difference method, implemented by a computer program in a language of your choice, using $\Delta t = 0.1$ sec. Note that this problem was solved as Example 5.2 and that the results were presented in Table E5.2.

*5.4 Repeat Problem 5.3 using $\Delta t = 0.05$ sec. How does the time step affect the accuracy of the solution?

*5.5 An SDF system has the same mass and stiffness as in Example 5.1, but the damping ratio is $\zeta = 20\%$. Determine the response of this system to the excitation of Example 5.1 by the central difference method using $\Delta t = 0.05$ sec. Plot the response as a function of time, compare with the solution of Problem 5.3, and comment on how damping affects the peak response.

*5.6 Solve the problem in Example 5.1 by the central difference method using $\Delta t = \frac{1}{3}$ sec. Carry out your solution to 2 sec, and comment on what happens to the solution and why.

*5.7 Solve the problem in Example 5.1 by the average acceleration method, implemented by a computer program in a language of your choice, using $\Delta t = 0.1$ sec. Note that this problem was solved as Example 5.3, and the results are presented in Table E5.3. Compare these results with those of Example 5.2, and comment on the relative accuracy of the average acceleration and central difference methods.

*5.8 Repeat Problem 5.7 using $\Delta t = 0.05$ sec. How does the time step affect the accuracy of the solution?

*5.9 Solve the problem in Example 5.1 by the average acceleration method using $\Delta t = \frac{1}{3}$ sec. Carry out the solution to 2 sec, and comment on the accuracy and stability of the solution.

*5.10 Solve the problem of Example 5.1 by the linear acceleration method, implemented by a computer program in a language of your choice, using $\Delta t = 0.1$ sec. Note that this problem was solved as Example 5.4 and that the results are presented in Table E5.4. Compare with the solution of Example 5.3, and comment on the relative accuracy of the average acceleration and linear acceleration methods.

*5.11 Repeat Problem 5.10 using $\Delta t = 0.05$ sec. How does the time step affect the accuracy of the solution?

*Denotes that a computer is necessary to solve this problem.

*__5.12__ Solve the problem of Example 5.5 by the central difference method, implemented by a computer program in a language of your choice, using $\Delta t = 0.05$ sec.

*__5.13__ Solve Example 5.5 by the average acceleration method without iteration. For this purpose implement the method using a language of your choice.

*__5.14__ Solve Example 5.6 by the average acceleration method with modified Newton–Raphson iteration within each time step. For this purpose implement the method by a computer program in a language of your choice.

*Denotes that a computer is necessary to solve this problem.

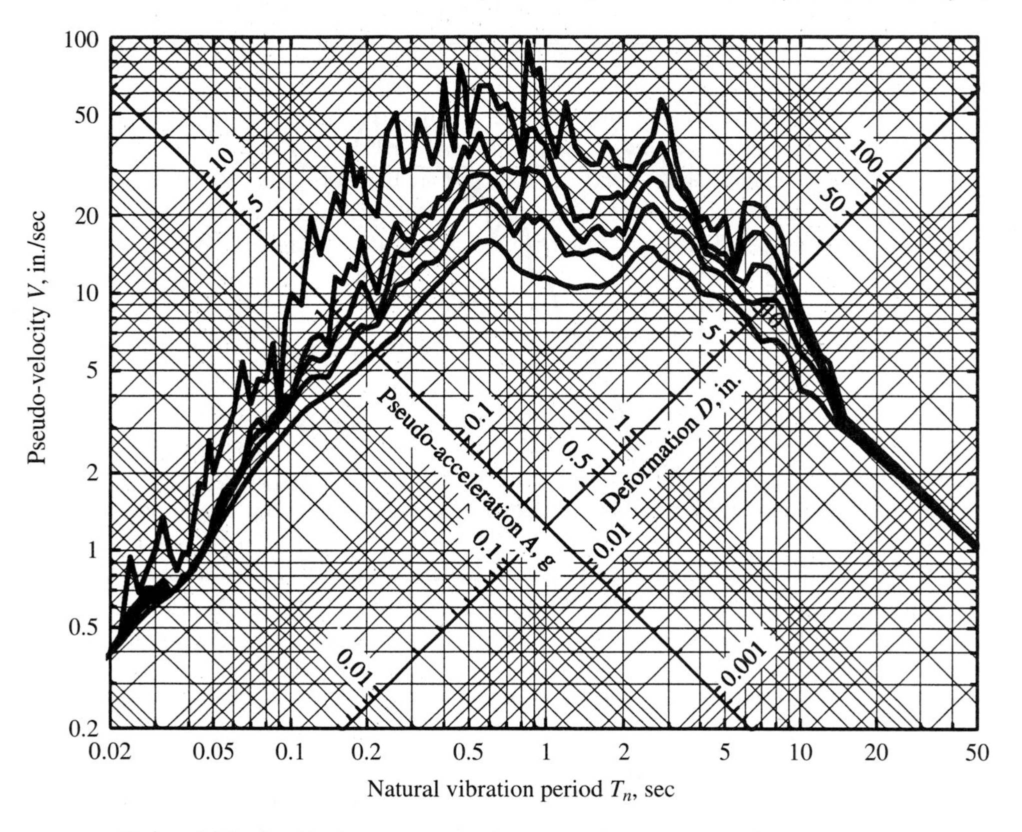

Figure 6.6.4 Combined *D–V–A* response spectrum for El Centro ground motion; $\zeta =$ 0, 2, 5, 10, and 20%.

earthquakes, and how response spectra are affected by distance to the causative fault, local soil conditions, and regional geology.

6.6.5 Construction of Response Spectrum

The response spectrum for a given ground motion component $\ddot{u}_g(t)$ can be developed by implementation of the following steps:

1. Numerically define the ground acceleration $\ddot{u}_g(t)$; typically, the ground motion ordinates are defined every 0.02 sec.
2. Select the natural vibration period T_n and damping ratio ζ of a SDF system.
3. Compute the deformation response $u(t)$ of this SDF system due to the ground motion $\ddot{u}_g(t)$ by any of the numerical methods described in Chapter 5. [In obtaining the responses shown in Fig. 6.4.1, the exact solution of Eq. (6.2.1) for ground motion assumed to be piecewise linear over every $\Delta t = 0.02$ sec was used; see Section 5.2.]

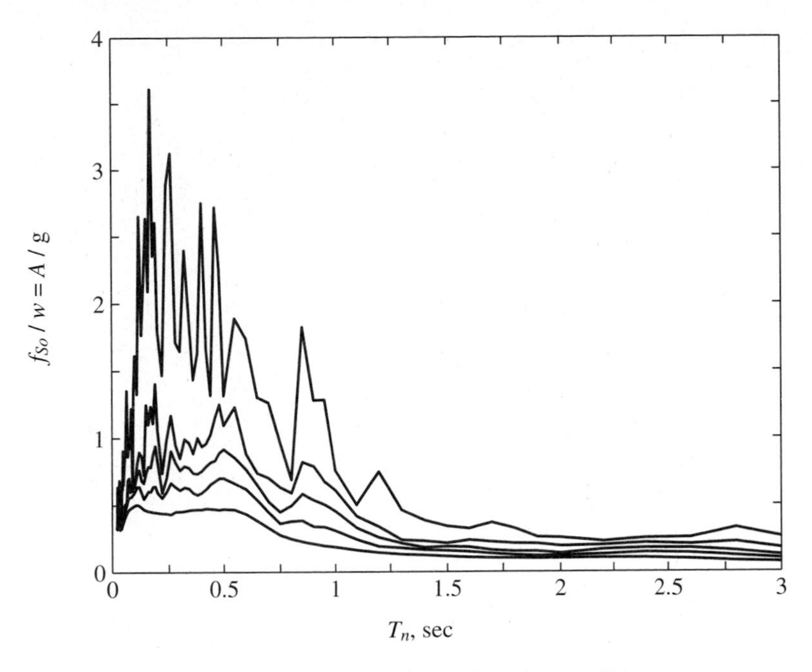

Figure 6.6.5 Normalized pseudo-acceleration, or base shear coefficient, response spectrum for El Centro ground motion; $\zeta = 0, 2, 5, 10$, and 20%.

4. Determine u_o, the peak value of $u(t)$.
5. The spectral ordinates are $D = u_o$, $V = (2\pi/T_n)D$, and $A = (2\pi/T_n)^2 D$.
6. Repeat steps 2 to 5 for a range of T_n and ζ values covering all possible systems of engineering interest.
7. Present the results of steps 2 to 6 graphically to produce three separate spectra like those in Fig. 6.6.2 or a combined spectrum like the one in Fig. 6.6.4.

Considerable computational effort is required to generate an earthquake response spectrum. A complete dynamic analysis to determine the time variation (or history) of the deformation of an SDF system provides the data for one point on the spectrum corresponding to the T_n and ζ of the system. Each curve in the response spectrum of Fig. 6.6.4 was produced from such data for 112 values of T_n unevenly spaced over the range $T_n = 0.02$ to 50 sec.

Example 6.1

Derive equations for and plot deformation, pseudo-velocity, and pseudo-acceleration response spectra for ground acceleration $\ddot{u}_g(t) = \dot{u}_{go}\delta(t)$, where $\delta(t)$ is the Dirac delta function and $\dot{u}_{go}$ is the increment in velocity, or the magnitude of the acceleration impulse. Only consider systems without damping.

Solution

1. *Determine the response history.* The response of an SDF system to $p(t) = \delta(t - \tau)$ is available in Eq. (4.1.6). Adapting that solution to $p_{\text{eff}}(t) = -m\ddot{u}_g(t) =$

$-m\dot{u}_{go}\delta(t)$ gives

$$u(t) = -\frac{\dot{u}_{go}}{\omega_n} \sin \omega_n (t - \tau) \tag{a}$$

The peak value of $u(t)$ is

$$u_o = \frac{\dot{u}_{go}}{\omega_n} \tag{b}$$

2. *Determine the spectral values.*

$$D \equiv u_o = \frac{\dot{u}_{go}}{\omega_n} = \frac{\dot{u}_{go}}{2\pi} T_n \tag{c}$$

$$V = \omega_n D = \dot{u}_{go} \qquad A = \omega_n^2 D = \frac{2\pi \dot{u}_{go}}{T_n} \tag{d}$$

Two of these response spectra are plotted in Fig. E6.1.

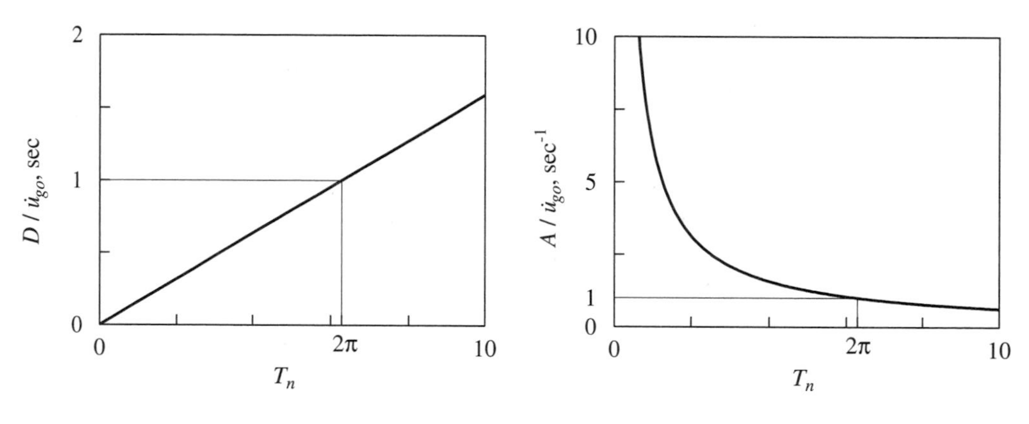

Figure E6.1

6.7 PEAK STRUCTURAL RESPONSE FROM THE RESPONSE SPECTRUM

If the response spectrum for a given ground motion component is available, the peak value of deformation or of an internal force in any linear SDF system can be determined readily. This is the case because the computationally intensive dynamic analyses summarized in Section 6.6.5 have already been completed in generating the response spectrum. Corresponding to the natural vibration period T_n and damping ratio ζ of the system, the values of D, V, or A are read from the spectrum, such as Fig. 6.6.4 or 6.6.5. Now all response quantities of interest can be expressed in terms of D, V, or A and the mass or stiffness properties of the system. In particular, the peak deformation of the system is

$$u_o = D = \frac{T_n}{2\pi} V = \left(\frac{T_n}{2\pi}\right)^2 A \tag{6.7.1}$$

and the peak value of the equivalent static force f_{So} is [from Eqs. (6.6.4) and (6.6.3)]

$$f_{So} = kD = mA \tag{6.7.2}$$

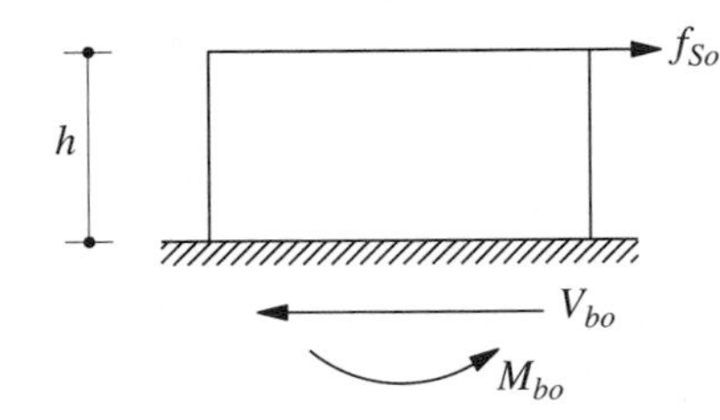

Figure 6.7.1 Peak value of equivalent static force.

Static analysis of the one-story frame subjected to lateral force f_{So} (Fig. 6.7.1) provides the internal forces (e.g., shears and moments in columns and beams). This involves application of well-known procedures of static structural analysis, as will be illustrated later by examples. We emphasize again that no further dynamic analysis is required beyond that necessary to determine $u(t)$. In particular, the peak values of shear and overturning moment at the base of the one-story structure are

$$V_{bo} = kD = mA \qquad M_{bo} = hV_{bo} \tag{6.7.3}$$

We note that only one of these response spectra—deformation, pseudo-velocity, or pseudo-acceleration—is sufficient for computing the peak deformations and forces required in structural design. For such applications the velocity or acceleration spectra (defined in Section 6.5) are not required, but for completeness we discuss these spectra briefly at the end of this chapter.

Example 6.2

A 12-ft-long vertical cantilever, a 4-in.-nominal-diameter standard steel pipe, supports a 5200-lb weight attached at the tip as shown in Fig. E6.2. The properties of the pipe are: outside diameter, $d_o = 4.500$ in., inside diameter $d_i = 4.026$ in., thickness $t = 0.237$ in., and second moment of cross-sectional area, $I = 7.23$ in^4, elastic modulus $E = 29,000$ ksi, and weight = 10.79 lb/foot length. Determine the peak deformation and bending stress in the cantilever due to the El Centro ground motion. Assume that $\zeta = 2\%$.

Solution The lateral stiffness of this SDF system is

$$k = \frac{3EI}{L^3} = \frac{3(29 \times 10^3)7.23}{(12 \times 12)^3} = 0.211 \text{ kip/in.}$$

The total weight of the pipe is $10.79 \times 12 = 129.5$ lb, which may be neglected relative to the lumped weight of 5200 lb. Thus

$$m = \frac{w}{g} = \frac{5.20}{386} = 0.01347 \text{ kip-sec}^2\text{/in.}$$

The natural vibration frequency and period of the system are

$$\omega_n = \sqrt{\frac{k}{m}} = \sqrt{\frac{0.211}{0.01347}} = 3.958 \text{ rad/sec} \qquad T_n = 1.59 \text{ sec}$$

From the response spectrum curve for $\zeta = 2\%$ (Fig. E6.2b), for $T_n = 1.59$ sec, $D = 5.0$ in. and $A = 0.20$ g. The peak deformation is

$$u_o = D = 5.0 \text{ in.}$$

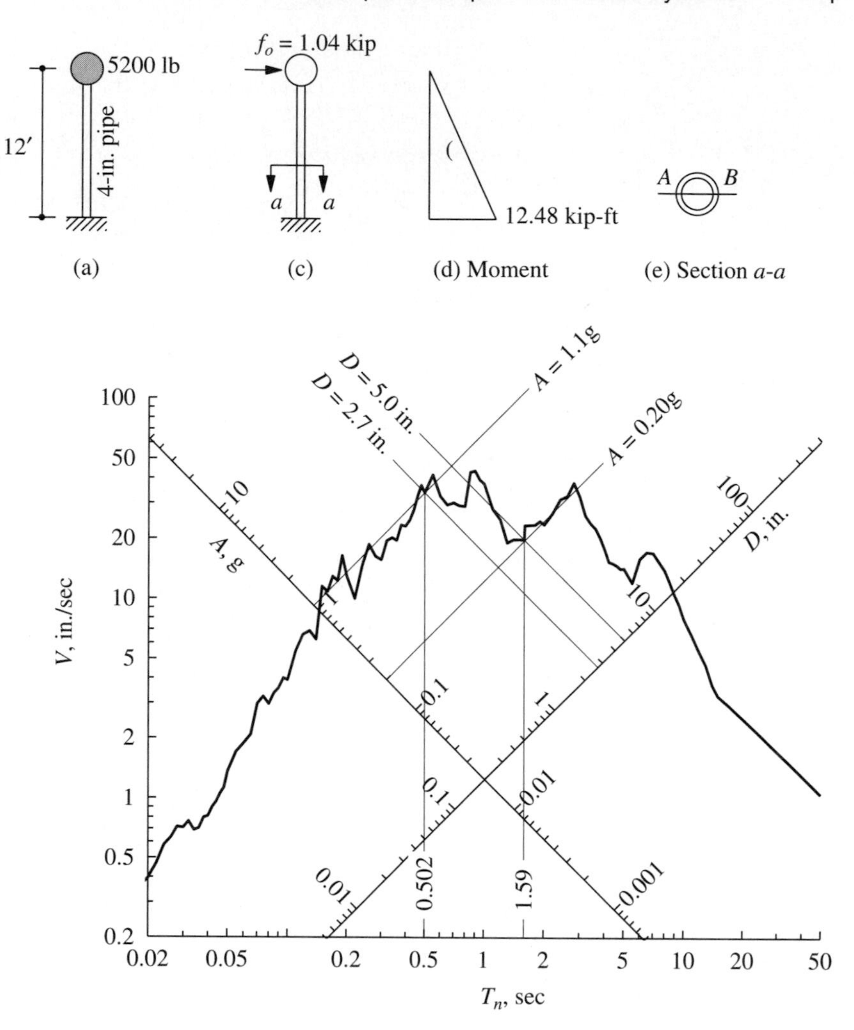

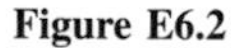
Figure E6.2

The peak value of the equivalent static force is

$$f_{So} = \frac{A}{g} w = 0.20 \times 5.2 = 1.04 \text{ kips}$$

The bending moment diagram is shown in Fig. E6.2c with the maximum moment at the base = 12.48 kip-ft. Points A and B shown in Fig. E6.2d are the locations of maximum bending stress:

$$\sigma_{\max} = \frac{Mc}{I} = \frac{(12.48 \times 12)(4.5/2)}{7.23} = 46.5 \text{ ksi}$$

As shown, $\sigma = +46.5$ ksi at A and $\sigma = -46.5$ ksi at B, where + denotes tension. The algebraic signs of these stresses are irrelevant because the direction of the peak force is not known, as the pseudo-acceleration spectrum is, by definition, positive.

Example 6.3

The stress computed in Example 6.2 exceeded the allowable stress and the designer decided to increase the size of the pipe to an 8-in.-nominal standard steel pipe. Its properties are $d_o = 8.625$ in., $d_i = 7.981$ in., $t = 0.322$ in., and $I = 72.5$ in^4. Comment on the advantages and disadvantages of using the bigger pipe.

Solution

$$k = \frac{3(29 \times 10^3)72.5}{(12 \times 12)^3} = 2.112 \text{ kips/in.}$$

$$\omega_n = \sqrt{\frac{2.112}{0.01347}} = 12.52 \text{ rad/sec} \qquad T_n = 0.502 \text{ sec}$$

From the response spectrum (Fig. E6.2b): $D = 2.7$ in. and $A = 1.1g$. Therefore,

$$u_o = D = 2.7 \text{ in.}$$

$$f_{So} = 1.1 \times 5.2 = 5.72 \text{ kips}$$

$$M_{\text{base}} = 5.72 \times 12 = 68.64 \text{ kip-ft}$$

$$\sigma_{\max} = \frac{(68.64 \times 12)(8.625/2)}{72.5} = 49.0 \text{ ksi}$$

Using the 8-in.-diameter pipe decreases the deformation from 5.0 in. to 2.7 in. However, contrary to the designer's objective, the bending stress increases slightly.

This example points out an important difference between the response of structures to earthquake excitation and to a fixed value of static force. In the latter case, the stress would decrease, obviously, by increasing the member size. In the case of earthquake excitation, the increase in pipe diameter shortens the natural vibration period from 1.59 sec to 0.50 sec, which for this response spectrum has the effect of increasing the equivalent static force f_{So}. Whether the bending stress decreases or increases by increasing the pipe diameter depends on the increase in section modulus, I/c, and the increase or decrease in f_{So}, depending on the response spectrum.

Example 6.4

A small one-story reinforced concrete building is idealized for purposes of structural analysis as a massless frame supporting a total dead load of 10 kips at the beam level (Fig. E6.4a). The frame is 24 ft wide and 12 ft high. Each column and the beam has a 10-in.-square cross section. Assume that the Young's modulus of concrete is 3×10^3 ksi and the damping ratio for the building is estimated as 5%. Determine the peak response of this frame to the El Centro ground motion. In particular, determine the peak lateral deformation at the beam level and plot the diagram of bending moments at the instant of peak response.

Solution The lateral stiffness of such a frame was calculated in Chapter 1: $k = 96EI/7h^3$, where EI is the flexural rigidity of the beam and columns and h is the height of the frame. For this particular frame,

$$k = \frac{96(3 \times 10^3)(10^4/12)}{7(12 \times 12)^3} = 11.48 \text{ kips/in.}$$

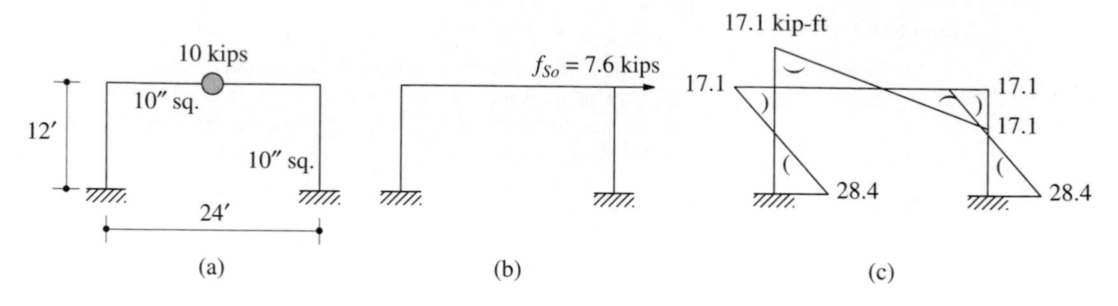

Figure E6.4 (a) Frame; (b) equivalent static force; (c) bending moment diagram.

The natural vibration period is

$$T_n = \frac{2\pi}{\sqrt{k/m}} = 2\pi\sqrt{\frac{10/386}{11.48}} = 0.30\,\text{sec}$$

For $T_n = 0.3$ and $\zeta = 0.05$, we read from the response spectrum of Fig. 6.6.4: $D = 0.67$ in. and $A = 0.76$g. Peak deformation: $u_o = D = 0.67$ in. Equivalent static force: $f_{So} = (A/\text{g})w = 0.76 \times 10 = 7.6$ kips. Static analysis of the frame for this lateral force, shown in Fig. E6.4b, gives the bending moments that are plotted in Fig. E6.4c.

Example 6.5

The frame of Example 6.4 is modified for use in a building to be located on sloping ground (Fig. E6.5). The beam is now made much stiffer than the columns and can be assumed to be rigid. The cross sections of the two columns are 10 in. square, as before, but their lengths are 12 ft and 24 ft, respectively. Determine the base shears in the two columns at the instant of peak response due to the El Centro ground motion. Assume the damping ratio to be 5%.

Solution

1. *Compute the natural vibration period.*

$$k = \frac{12(3 \times 10^3)(10^4/12)}{(12 \times 12)^3} + \frac{12(3 \times 10^3)(10^4/12)}{(24 \times 12)^3}$$

$$= 10.05 + 1.26 = 11.31\ \text{kips/in.}$$

$$T_n = 2\pi\sqrt{\frac{10/386}{11.31}} = 0.30\ \text{sec}$$

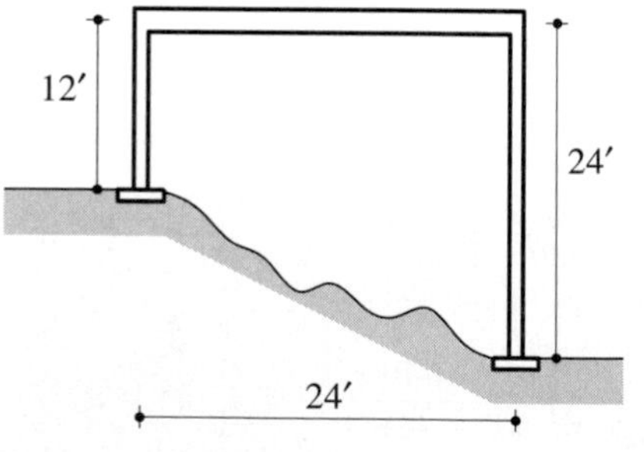

12′

24′

24′

Figure E6.5

2. *Compute the shear force at the base of the short and long columns.*

$$u_o = D = 0.67 \text{ in.}, \qquad A = 0.76\text{g}$$

$$V_{\text{short}} = k_{\text{short}} u_o = (10.05)0.67 = 6.73 \text{ kips}$$

$$V_{\text{long}} = k_{\text{long}} u_o = (1.26)0.67 = 0.84 \text{ kip}$$

Observe that both columns go through equal deformation. Undergoing equal deformations, the stiffer column carries a greater force than the flexible column; the lateral force is distributed to the elements in proportion to their relative stiffnesses. Sometimes this basic principle has, inadvertently, not been recognized in building design, leading to unanticipated damage of the stiffer elements.

6.8 RESPONSE SPECTRUM CHARACTERISTICS

We now study the important properties of earthquake response spectra. Figure 6.8.1 shows the response spectrum for El Centro ground motion together with u_{go}, $\dot{u}_{go}$, and $\ddot{u}_{go}$, the peak values of ground acceleration, ground velocity, and ground displacement, respectively, identified in Fig. 6.1.4. To show more directly the relationship between

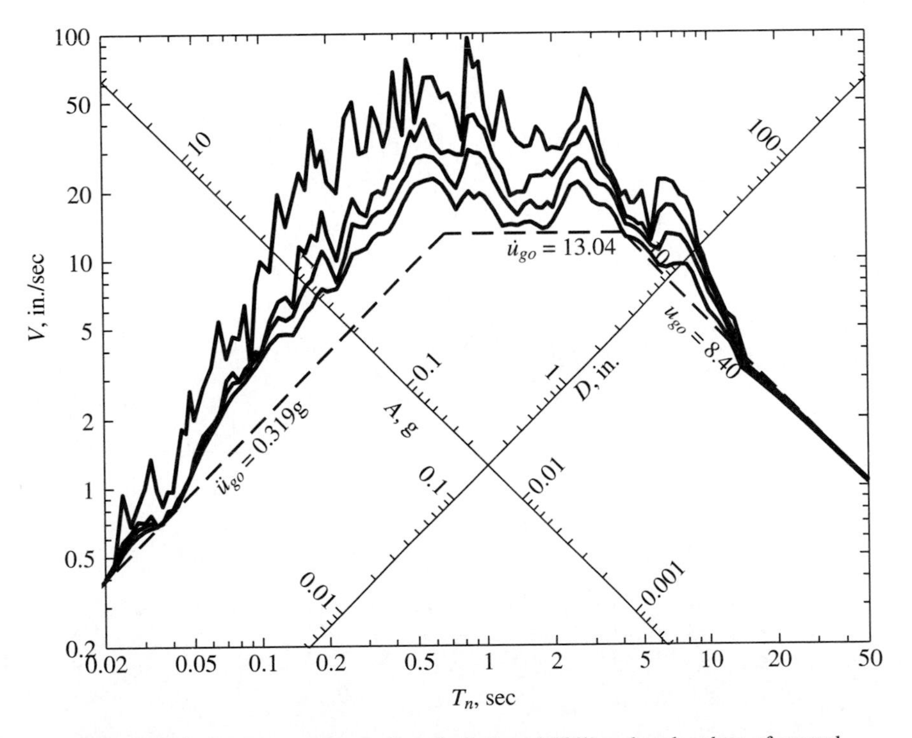

Figure 6.8.1 Response spectrum (ζ = 0, 2, 5, and 10%) and peak values of ground acceleration, ground velocity, and ground displacement for El Centro ground motion.

the response spectrum and the ground motion parameters, the data of Fig. 6.8.1 have been presented again in Fig. 6.8.2 using normalized scales: D/u_{go}, $V/\dot{u}_{go}$, and $A/\ddot{u}_{go}$. Figure 6.8.3 shows one of the spectrum curves of Fig. 6.8.2, the one for 5% damping, together with an idealized version shown in dashed lines; the latter will provide a basis for constructing smooth design spectra directly from the peak ground motion parameters (see Section 6.9). Based on Figs. 6.8.1 to 6.8.3, we first study the properties of the response spectrum over various ranges of the natural vibration period of the system separated by the period values at a, b, c, d, e, and f: $T_a = 0.035$ sec, $T_b = 0.125$, $T_c = 0.5$, $T_d = 3.0$, $T_e = 10$, and $T_f = 15$ sec. Subsequently, we identify the effects of damping on spectrum ordinates.

For systems with very short period, say $T_n < T_a = 0.035$ sec, the pseudo-acceleration A for all damping values approaches $\ddot{u}_{go}$ and D is very small. This trend can be understood based on physical reasoning. For a fixed mass, a very short period system is extremely stiff or essentially rigid. Such a system would be expected to undergo very little deformation and its mass would move rigidly with the ground; its peak acceleration should be approximately equal to $\ddot{u}_{go}$ (Fig. 6.8.4d). This expectation is confirmed by Fig. 6.8.4, where the ground acceleration is presented in part (a), the total acceleration $\ddot{u}^t(t)$ of a system with $T_n = 0.02$ sec and $\zeta = 2\%$ in part (b), and the pseudo-acceleration

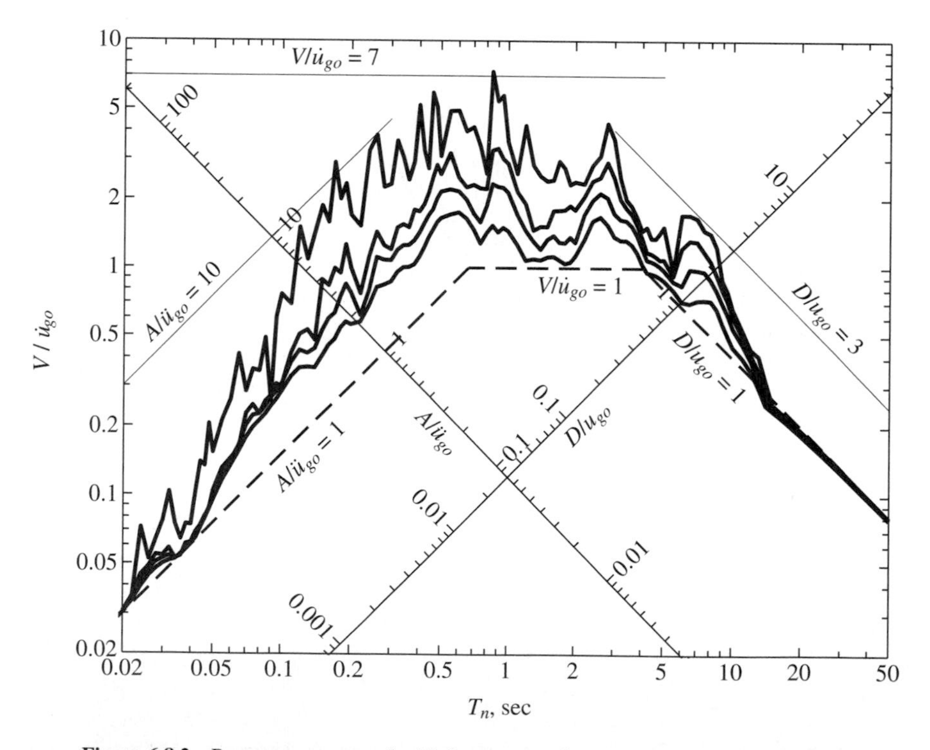

Figure 6.8.2 Response spectrum for El Centro ground motion plotted with normalized scales $A/\ddot{u}_{go}$, $V/\dot{u}_{go}$, and D/u_{go}; $\zeta = 0$, 2, 5, and 10%.

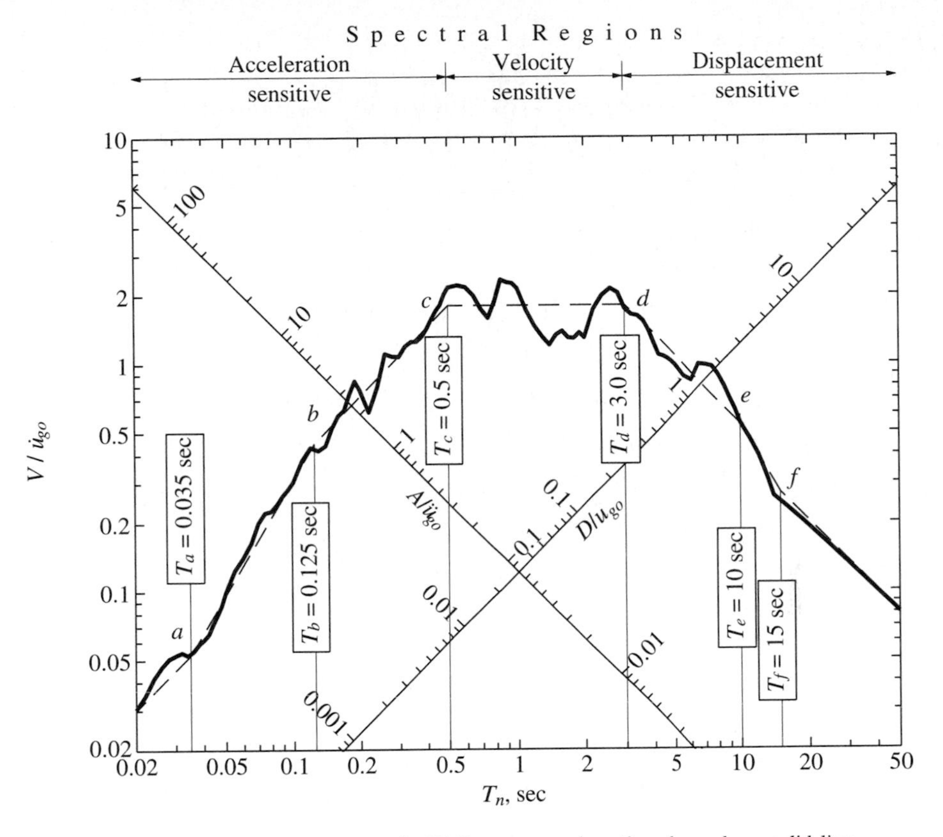

Figure 6.8.3 Response spectrum for El Centro ground motion shown by a solid line together with an idealized version shown by a dashed line; $\zeta = 5\%$.

$A(t)$ for the same system in part (c). Observe that $\ddot{u}^t(t)$ and $\ddot{u}_g(t)$ are almost identical functions. Furthermore, for such short-period systems $\ddot{u}^t(t) \simeq -A(t)$ and the peak acceleration $\ddot{u}^t_o$ of the mass is almost identical to the peak pseudo-acceleration A.

For systems with very long period, say $T_n > T_f = 15$ sec, D for all damping values approaches u_{go} and A is very small; thus the forces in the structure, which are related to mA, would be very small. This trend can again be explained by relying on physical reasoning. For a fixed mass, a very-long-period system is extremely flexible. The mass would be expected to remain essentially stationary while the ground below moves (Fig. 6.8.5c). Thus $\ddot{u}^t(t) \simeq 0$, implying that $A(t) \simeq 0$ (see Section 6.12.2); and $u(t) \simeq -u_g(t)$, implying that $D \simeq u_{go}$. This expectation is confirmed by Fig. 6.8.5, where the deformation response $u(t)$ of a system with $T_n = 30$ sec and $\zeta = 2\%$ to the El Centro ground motion is compared with the ground displacement $u_g(t)$. Observe that the peak values for u_o and u_{go} are close and the time variation of $u(t)$ is similar to that of $-u_g(t)$, but for rotation of the baseline. The discrepancy between the two arises, in part, from the loss of the initial portion of the recorded ground motion prior to triggering of the recording accelerograph.

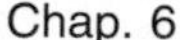

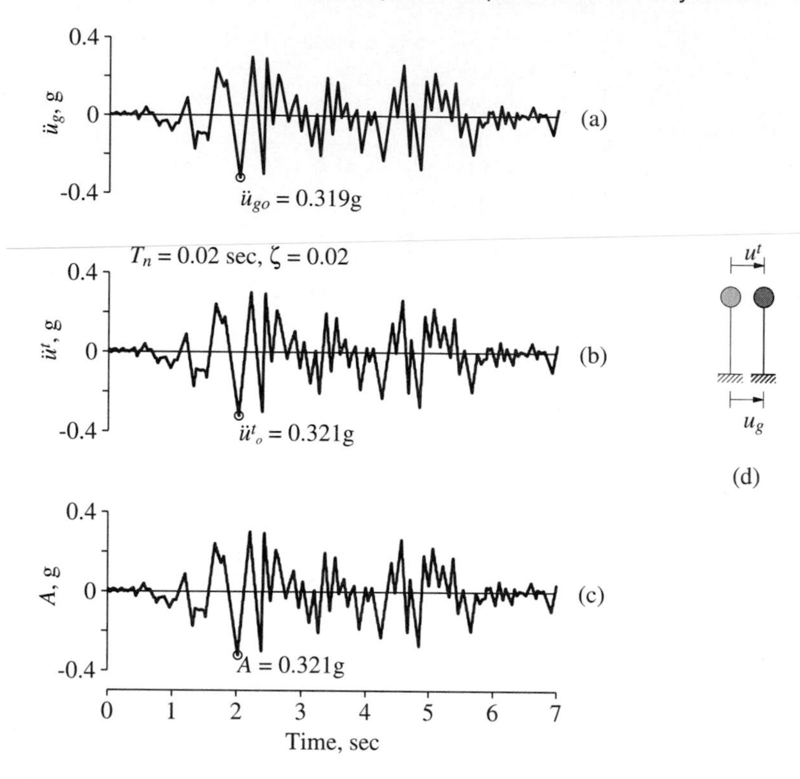

Figure 6.8.4 (a) El Centro ground acceleration; (b) total acceleration response of an SDF system with $T_n = 0.02$ sec and $\zeta = 2\%$; (c) pseudo-acceleration response of the same system; (d) rigid system.

For short-period systems with T_n between $T_a = 0.035$ sec and $T_c = 0.50$ sec, A exceeds $\ddot{u}_{go}$, with the amplification depending on T_n and ζ. Over a portion of this period range, $T_b = 0.125$ sec to $T_c = 0.5$ sec, A may be idealized as constant at a value equal to $\ddot{u}_{go}$ amplified by a factor depending on ζ.

For long-period systems with T_n between $T_d = 3$ sec and $T_f = 15$ sec, D generally exceeds u_{go}, with the amplification depending on T_n and ζ. Over a portion of this period range, $T_d = 3.0$ sec to $T_e = 10$ sec, D may be idealized as constant at a value equal to u_{go} amplified by a factor depending on ζ.

For intermediate-period systems with T_n between $T_c = 0.5$ sec and $T_d = 3.0$ sec, V exceeds $\dot{u}_{go}$. Over this period range, V may be idealized as constant at a value equal to $\dot{u}_{go}$, amplified by a factor depending on ζ.

Based on these observations, it is logical to divide the spectrum into three period ranges (Fig. 6.8.3). The long-period region to the right of point d, $T_n > T_d$, is called the *displacement-sensitive region* because structural response is most directly related to ground displacement. The short-period region to the left of point c, $T_n < T_c$, is called the *acceleration-sensitive region* because structural response is most directly related to ground acceleration. The intermediate period region between points c and d, $T_c < T_n < T_d$,

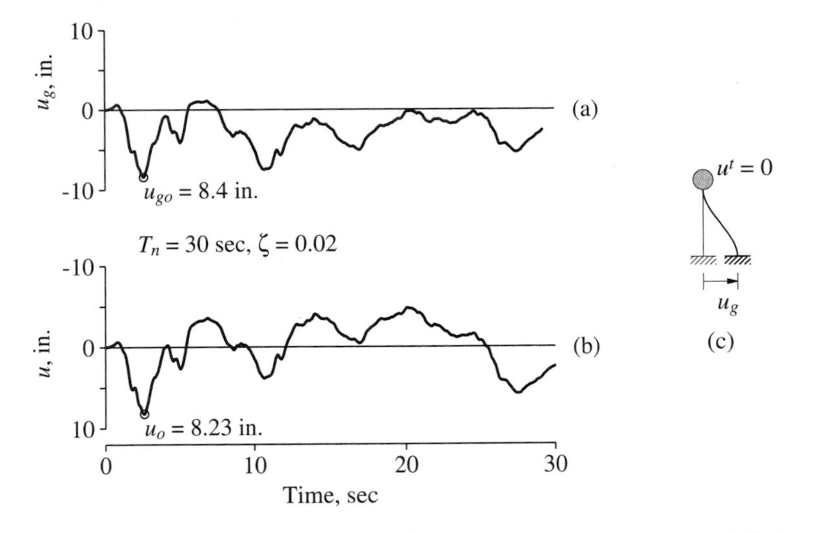

Figure 6.8.5 (a) El Centro ground displacement; (b) deformation response of SDF system with $T_n = 30$ sec and $\zeta = 2\%$; (c) very flexible system.

is called the *velocity-sensitive region* because structural response appears to be better related to ground velocity than to other ground motion parameters. For a particular ground motion, the periods T_a, T_b, T_e, and T_f on the idealized spectrum are independent of damping, but T_c and T_d vary with damping.

The preceding observations and discussion have brought out the usefulness of the four-way logarithmic plot of the combined deformation, pseudo-velocity, and pseudo-acceleration response spectra. These observations would be difficult to glean from the three individual spectra.

Idealizing the spectrum by a series of straight lines a–b–c–d–e–f in the four-way logarithmic plot is obviously not a precise process. For a given ground motion, the period values associated with the points a, b, c, d, e, and f and the amplification factors for the segments b–c, c–d, and d–e are somewhat judgmental in the way we have approached them. However, formal curve-fitting techniques can be used to replace the actual spectrum by an idealized spectrum of a selected shape. In any case, the idealized spectrum in Fig. 6.8.3 is not a close approximation to the actual spectrum. This may not be visually apparent but becomes obvious when we note that the scales are logarithmic. As we shall see in the next section, the greatest benefit of the idealized spectrum is in constructing a design spectrum representative of many ground motions.

The period values associated with the points T_a, T_b, T_c, T_d, T_e, and T_f and the amplification factors for the segments b–c, c–d, and d–e are not unique in the sense that they vary from one ground motion to the next. Some of the variation in these parameters reflects the inherent probabilistic differences that exist among ground motions even if they are recorded under similar conditions: magnitude of the earthquake, the distance of the site from the earthquake source, the source mechanism for the earthquake, and the local soil conditions at the site. Greater variation exists in the response spectrum parameters among ground motions recorded under dissimilar conditions. However, researchers have demonstrated that response trends identified earlier from the three regions of one response spectrum are generally valid for the corresponding spectral regions of other ground motions.

We now turn to damping, which has significant influence on the earthquake response spectrum (Figs. 6.6.4 and 6.6.5). The zero damping curve is marked by abrupt jaggedness, which indicates that the response is very sensitive to small differences in the natural vibration period. The introduction of damping makes the response much less sensitive to the period.

Damping reduces the response of a structure, as expected, and the reduction achieved with a given amount of damping is different in the three spectral regions. In the limit as $T_n \to 0$ damping does not affect the response because the structure moves rigidly with the ground. In the other limit as $T_n \to \infty$, damping again does not affect the response because the structural mass stays still while the ground underneath moves. Among the three period regions defined earlier, the effect of damping tends to be greatest in the velocity-sensitive region of the spectrum. In this spectral region the effect of damping depends on the ground motion characteristics. If the ground motion is nearly harmonic over many cycles (e.g., the record from Mexico City shown in Fig. 6.1.3), the effect of damping would be especially large for systems near "resonance" (Chapter 3). If the ground motion is short in duration with only a few major cycles (e.g., the record from Parkfield, California, shown in Fig. 6.1.3) the influence of damping would be small, as in the case of impulsive excitations (Chapter 4).

Figure 6.8.6 shows the peak pseudo-acceleration $A(\zeta)$, normalized relative to $A(\zeta = 0)$, plotted as a function of ζ for several T_n values. This is some of the data from

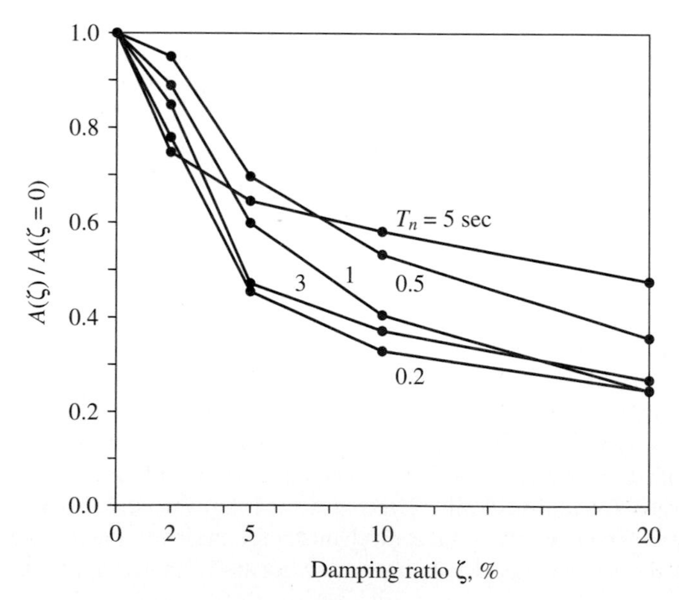

Figure 6.8.6 Variation of peak pseudo-acceleration with damping for systems with $T_n = 0.2$, 0.5, 1, 3, and 5 sec; El Centro ground motion.

the response spectrum of Figs. 6.6.4 and 6.6.5 replotted in a different format. Observe that the effect of damping is stronger for smaller damping values. This means that if the damping ratio is increased from 0 to 2%, the reduction in response is greater compared with the response reduction due to an increase in damping from 10% to 12%. The effect of damping in reducing the response depends on the period T_n of the system, but there is no clear trend from Fig. 6.8.6. This is yet another indication of the complexity of structural response to earthquakes.

The motion of a structure and the associated forces could be reduced by increasing the effective damping of the structure. The addition of dampers achieves this goal without significantly changing the natural vibration periods of the structure. Viscoelastic dampers have been introduced in many structures; for example, 10,000 dampers have been installed throughout the height of each tower of the World Trade Center in New York City to reduce wind-induced motion to within a comfortable range for the occupants. In recent years there is a growing interest in developing dampers suitable for structures in earthquake-prone regions. Because the inherent damping in most structures is relatively small, their earthquake response can be reduced significantly by the addition of dampers. These can be especially useful in improving the seismic safety of an existing structure. A viscoelastic damper such as the one shown in Fig. 6.8.7a could be installed in the diagonal bracing of a structure as indicated in Fig. 6.8.7b.

6.9 ELASTIC DESIGN SPECTRUM

In this section we introduce the concept of earthquake design spectrum for elastic systems and present a procedure to construct it from estimated peak values for ground acceleration, ground velocity, and ground displacement.

The design spectrum should satisfy certain requirements because it is intended for the design of new structures, or the seismic safety evaluation of existing structures, to resist future earthquakes. For this purpose the response spectrum for a ground motion recorded during a past earthquake is inappropriate. The jaggedness in the response spectrum, as seen in Fig. 6.6.4, is characteristic of that one excitation. The response spectrum for another ground motion recorded at the same site during a different earthquake is also jagged, but the peaks and valleys are not necessarily at the same periods. This is apparent from Fig. 6.9.1, where the response spectra for ground motions recorded at the same site during three past earthquakes are plotted. Similarly, it is not possible to predict the jagged response spectrum in all its detail for a ground motion that may occur in the future. Thus the design spectrum should consist of a set of smooth curves or a series of straight lines with one curve for each level of damping.

The design spectrum should, in a general sense, be representative of ground motions recorded at the site during past earthquakes. If none have been recorded at the site, the design spectrum should be based on ground motions recorded at other sites under similar conditions. The factors that one tries to match in the selection include the magnitude of the earthquake, the distance of the site from the earthquake fault, the fault mechanism, the geology of the travel path of seismic waves from the source to the site, and the

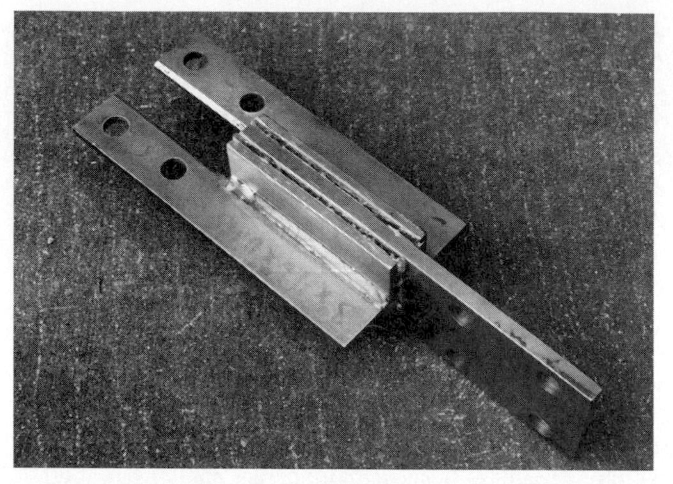

(a)

(b)

Figure 6.8.7 (a) Viscoelastic shear damper, and (b) diagonal bracing configuration with viscoelastic dampers. (Courtesy of I. D. Aiken.)

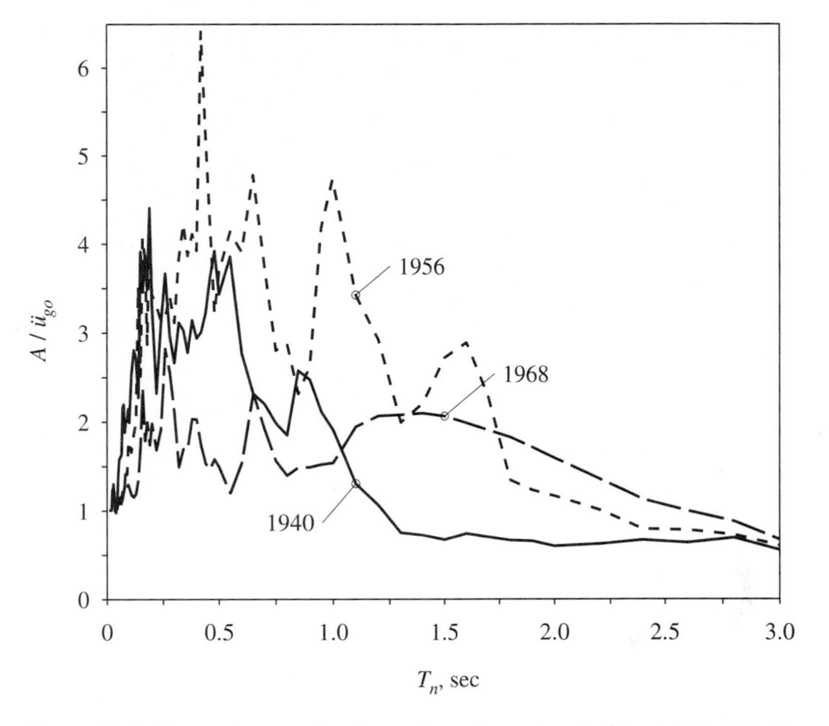

Figure 6.9.1 Response spectra for the north–south component of ground motions recorded at the Imperial Valley Irrigation District substation, El Centro, California, during earthquakes of May 18, 1940; February 9, 1956; and April 8, 1968. $\zeta = 2\%$.

local soil conditions at the site. While this approach is feasible for some parts of the world, such as California and Japan, where numerous ground motion records are available, in many other regions it is hampered by the lack of a sufficient number of such records. In such situations compromises in the approach are necessary by considering ground motion records that were recorded for conditions different from those at the site. Detailed discussion of these issues is beyond the scope of this book. The presentation here is focused on the narrow question of how to develop the design spectrum that is representative of an available ensemble (or set) of recorded ground motions.

The design spectrum is based on statistical analysis of the response spectra for the ensemble of ground motions. Suppose that the response spectrum for each ground motion is computed by the procedures described in Section 6.6, and plotted in normalized form, as in Fig. 6.8.2. At a particular natural period, the ordinates of the response spectrum for the ith ground motion in the ensemble are D^i/u^i_{go}, $V^i/\dot{u}^i_{go}$, and $A^i/\ddot{u}^i_{go}$, where D^i, V^i, and A^i are the deformation, pseudo-velocity, and pseudo-acceleration spectral ordinates; and u^i_{go}, $\dot{u}^i_{go}$, and $\ddot{u}^i_{go}$ are the peak displacement, velocity, and acceleration of the ground motion. Thus at each natural period there are as many spectral values as the number I of ground motion records in the ensemble (e.g.,

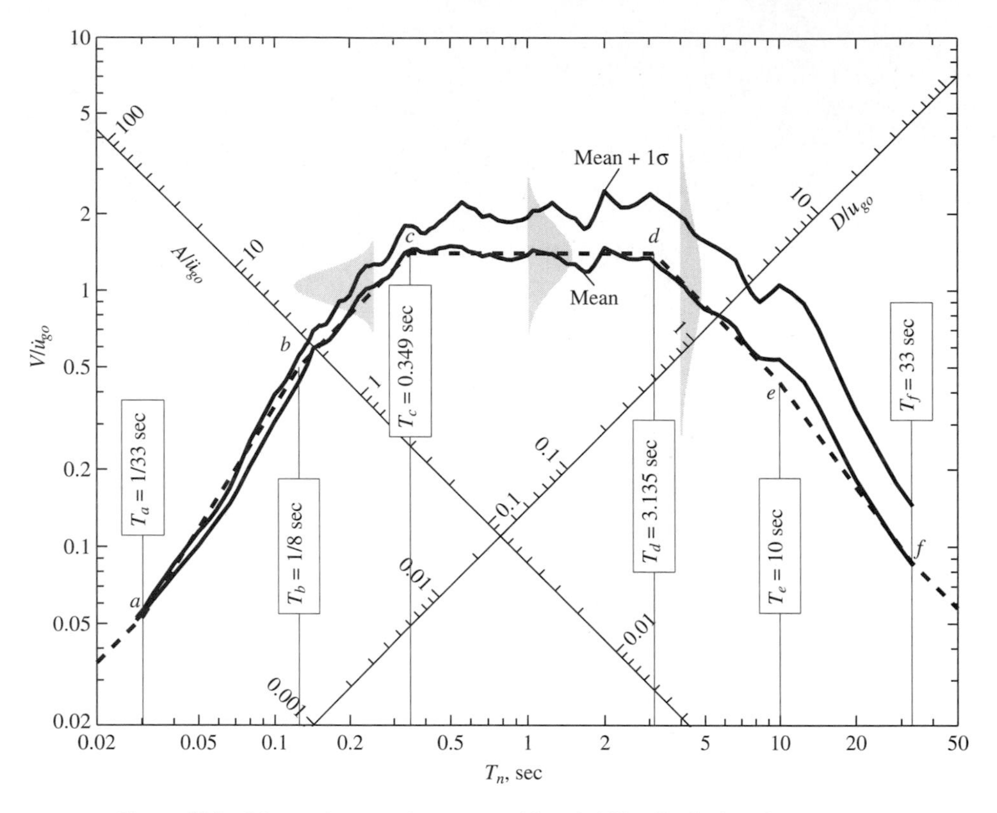

Figure 6.9.2 Mean and mean $+1\sigma$ spectra with probability distributions for V at $T_n =$ 0.25, 1, and 4 sec; $\zeta = 5\%$. Dashed lines show an idealized design spectrum. (Based on numerical data from R. Riddell and N. Newmark, 1979.)

$A^i/\ddot{u}^i_{go}$, $i = 1, 2, \ldots, I$). Such data were generated for a group of 10 earthquake records, and selected aspects of the results are presented in Fig. 6.9.2. Statistical analysis of these data provide the probability distribution for the spectral ordinate, its mean value, and its standard deviation at each period T_n. The probability distributions are shown schematically at three selected T_n values, indicating that the coefficient of variation (= standard deviation ÷ mean value) varies with T_n. Connecting all the mean values gives the *mean response spectrum* in normalized form. Similarly connecting all the mean-plus-one-standard-deviation values gives the *mean-plus-one-standard-deviation response spectrum.* Observe that these two response spectra are much smoother than the response spectrum for an individual ground motion (Fig. 6.6.4). As shown in Fig. 6.9.2, such a smooth spectrum curve lends itself to idealization by a series of straight lines much better than the spectrum for an individual ground motion (Fig. 6.8.3).

Researchers have developed procedures to construct such design spectra from ground motion parameters. One such procedure is illustrated in Fig. 6.9.3. The recommended period values T_a, T_b, T_e, and T_f and the amplification factors for the three

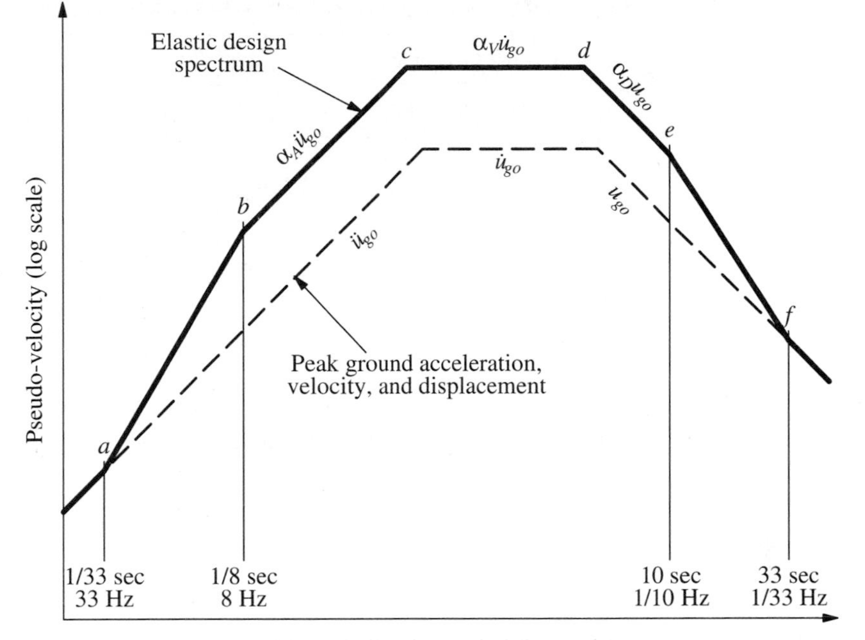

Figure 6.9.3 Construction of elastic design spectrum.

spectral regions were developed by the preceding analysis of a larger ensemble of ground motions recorded on firm ground (rock, soft rock, and competent sediments). The amplification factors for two different nonexceedance probabilities, 50% and 84.1%, are given in Table 6.9.1 for several values of damping and in Table 6.9.2 as a function of damping ratio. The 50% nonexceedance probability represents the median value of the

TABLE 6.9.1 AMPLIFICATION FACTORS: ELASTIC DESIGN SPECTRA

Damping, ζ (%)	Median (50 percentile)			One Sigma (84.1 percentile)		
	α_A	α_V	α_D	α_A	α_V	α_D
1	3.21	2.31	1.82	4.38	3.38	2.73
2	2.74	2.03	1.63	3.66	2.92	2.42
5	2.12	1.65	1.59	2.71	2.30	2.01
10	1.64	1.37	1.20	1.99	1.84	1.69
20	1.17	1.08	1.01	1.26	1.37	1.38

Source: N. M. Newmark and W. J. Hall, *Earthquake Spectra and Design*, Earthquake Engineering Research Institute, Berkeley, Calif., 1982, pp. 35 and 36.

TABLE 6.9.2 AMPLIFICATION FACTORS: ELASTIC DESIGN SPECTRA[a]

	Median (50 percentile)	One Sigma (84.1 percentile)
α_A	$3.21 - 0.68 \ln \zeta$	$4.38 - 1.04 \ln \zeta$
α_V	$2.31 - 0.41 \ln \zeta$	$3.38 - 0.67 \ln \zeta$
α_D	$1.82 - 0.27 \ln \zeta$	$2.73 - 0.45 \ln \zeta$

Source: N. M. Newmark and W. J. Hall, *Earthquake Spectra and Design*, Earthquake Engineering Research Institute, Berkeley, Calif., 1982, pp. 35 and 36.

[a]Damping ratio in percent.

spectral ordinates and the 84.1% represents the median-plus-one-standard-deviation value assuming lognormal probability distribution for the spectral ordinates.

Summary. A procedure to construct a design spectrum is now summarized with reference to Fig. 6.9.3:

1. Plot the three dashed lines corresponding to the peak values of ground acceleration $\ddot{u}_{go}$, velocity $\dot{u}_{go}$, and displacement u_{go} for the design ground motion.
2. Obtain from Table 6.9.1 or 6.9.2 the values for α_A, α_V, and α_D for the ζ selected.
3. Multiply $\ddot{u}_{go}$ by the amplification factor α_A to obtain the straight line *b–c* representing a constant value of pseudo-acceleration A.
4. Multiply $\dot{u}_{go}$ by the amplification factor α_V to obtain the straight line *c–d* representing a constant value of pseudo-velocity V.
5. Multiply u_{go} by the amplification factor α_D to obtain the straight line *d–e* representing a constant value of deformation D.
6. Draw the line $A = \ddot{u}_{go}$ for periods shorter than T_a and the line $D = u_{go}$ for periods longer than T_f.
7. The transition lines *a–b* and *e–f* complete the spectrum.

Observe that the period values associated with points a, b, e, and f on the spectrum are fixed; the values in Fig. 6.9.3 are for firm ground. Points c and d are located at intersections of the constant-A, constant-V, and constant-D branches of the spectrum. The locations of these intersection points vary with damping ratio ζ because they depend on the amplification factors α_A, α_V, and α_D.

We now illustrate use of this procedure by constructing the median-plus-one-standard-deviation design spectrum for systems with 5% damping. For convenience, a peak ground acceleration $\ddot{u}_{go} = 1\text{g}$ is selected; the resulting spectrum can be scaled by η to obtain the design spectrum corresponding to $\ddot{u}_{go} = \eta\text{g}$. Consider also that no specific estimates for peak ground velocity $\dot{u}_{go}$ and displacement u_{go} are provided; thus typical values $\dot{u}_{go}/\ddot{u}_{go} = 48$ in./sec/g and $\ddot{u}_{go} \times u_{go}/\dot{u}_{go}^2 = 6$, recommended for firm ground, are used. For $\ddot{u}_{go} = 1\text{g}$, these ratios give $\dot{u}_{go} = 48$ in./sec and $u_{go} = 36$ in.

The design spectrum shown in Fig. 6.9.4 is determined by the following steps:

1. The peak parameters for the ground motion: $\ddot{u}_{go} = 1\text{g}$, $\dot{u}_{go} = 48$ in./sec, and $u_{go} = 36$ in. are plotted.
2. From Table 6.9.1, the amplification factors for median-plus-one-standard-deviation spectrum and 5% damping are obtained: $\alpha_A = 2.71$, $\alpha_V = 2.30$, and $\alpha_D = 2.01$.

3–5. The ordinate for the constant-A branch is $A = 1\text{g} \times 2.71 = 2.71\text{g}$, for the constant-$V$ branch: $V = 48 \times 2.30 = 110.4$, and for the constant-$D$ branch: $D = 36 \times 2.01 = 72.4$. The three branches are drawn as shown; they intersect at $T_c = 0.66$ sec and $T_d = 4.12$ sec.

6. The line $A = 1\text{g}$ is plotted for $T_n < \frac{1}{33}$ sec and $D = 36$ in. for $T_n > 33$ sec.
7. The transition line b–a is drawn to connect the point $A = 2.71\text{g}$ at $T_n = \frac{1}{8}$ sec to $\ddot{u}_{go} = 1\text{g}$ at $T_n = \frac{1}{33}$ sec. Similarly, the transition line e–f is drawn to connect the point $D = 72.4$ at $T_n = 10$ sec to $u_{go} = 36$ in. at $T_n = 33$ sec.

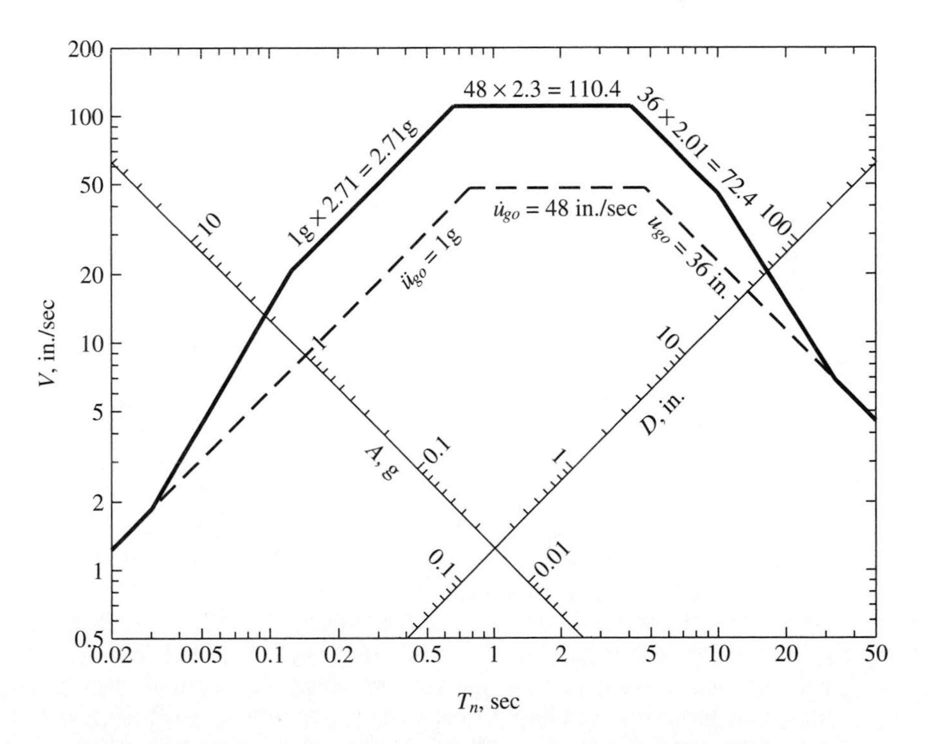

Figure 6.9.4 Construction of elastic design spectrum for ground motions with $\ddot{u}_{go} = 1\text{g}$, $\dot{u}_{go} = 48$ in./sec, and $u_{go} = 36$ in.; $\zeta = 5\%$.

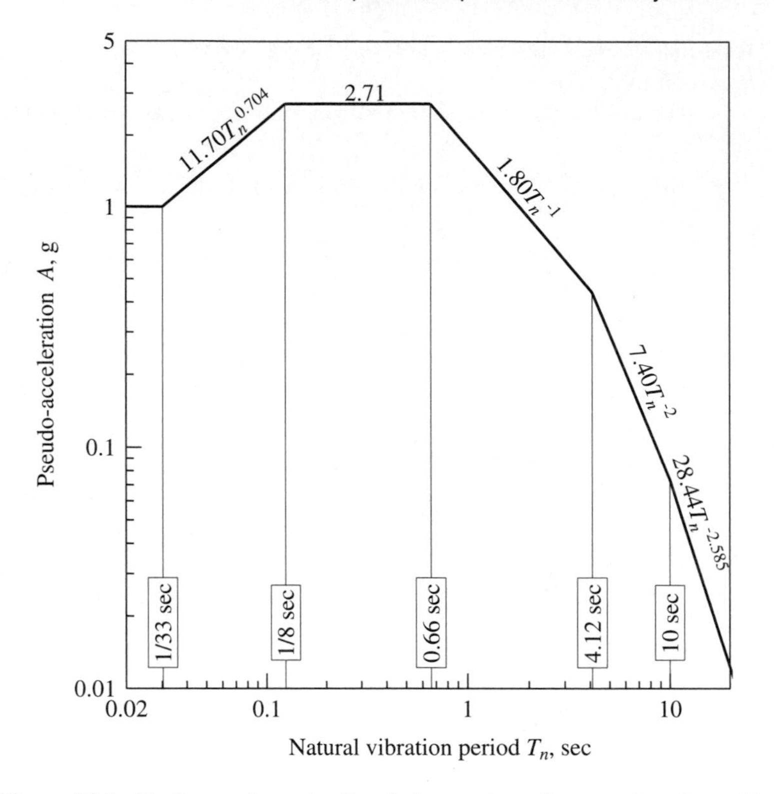

Figure 6.9.5 Elastic pseudo-acceleration design spectrum for ground motions with $\ddot{u}_{go} = 1\text{g}$, $\dot{u}_{go} = 48$ in./sec, and $u_{go} = 36$ in.; $\zeta = 5\%$.

The resulting design spectrum of Fig. 6.9.4 is replotted in Fig. 6.9.5 as a pseudo-acceleration design spectrum.

The elastic design spectrum provides a basis for calculating the design force and deformation for SDF systems to be designed to remain elastic. For this purpose the design spectrum is used in the same way as the response spectrum was used to compute peak response; see Examples 6.2 to 6.5. The errors in reading spectral ordinates from a four-way logarithmic plot can be avoided, however, because simple functions of T_n define various branches of the spectrum in Figs. 6.9.4 and 6.9.5.

Parameters that enter into construction of the elastic design spectrum should be selected considering the factors that influence ground motion mentioned previously. Thus the selection of design ground motion parameter $\ddot{u}_{go}$, $\dot{u}_{go}$, and u_{go} should be based on earthquake magnitude, distance to the earthquake fault, fault mechanism, wave-travel-path geology, and local soil conditions. Results of research on these factors and related issues are available; they are used to determine site-dependent design spectra for important projects. Similarly, numerical values for the amplification factors α_A, α_V, and α_D should be chosen consistent with the expected frequency content of the ground motion.

6.10 COMPARISON OF DESIGN AND RESPONSE SPECTRA

The design spectrum presented in Fig. 6.9.4 is for ground motions with $\ddot{u}_{go} = 1\text{g}$, $\dot{u}_{go} = 48$ in./sec, and $u_{go} = 36$ in., consistent with $\dot{u}_{go}/\ddot{u}_{go} = 48$ in./sec/g and $\ddot{u}_{go} \times u_{go}/\dot{u}_{go}^2 = 6$. These ratios are considered representative of ground motions on firm ground. The resulting spectrum shape is widely used in engineering practice, by scaling the spectrum of Fig. 6.9.4 to conform to the peak ground acceleration estimated for the site. Thus if this estimate is 0.4g, the spectrum of Fig. 6.9.4 multiplied by 0.4 gives the design spectrum for the site.

It is instructive to compare this "standard" design spectrum for firm ground with an actual response spectrum for similar soil conditions. Figure 6.10.1 shows a standard design spectrum for $\ddot{u}_{go} = 0.319\text{g}$, the peak acceleration for the El Centro ground motion; the implied values for $\dot{u}_{go}$ and u_{go} are 15.3 in./sec and 11.5 in., respectively, based on the standard ratios mentioned in the preceding paragraph. Also shown in Fig. 6.10.1 is the response spectrum for the El Centro ground motion; recall that the actual peak values for this motion are $\dot{u}_{go} = 13.04$ in./sec and $u_{go} = 8.40$ in. The El Centro response spectrum agrees well with the design spectrum in the acceleration-sensitive region, largely because the peak accelerations for the two are matched. However, the two spectra are considerably different in the velocity-sensitive region because of the differences (15.3 in./sec versus 13.04 in./sec) in the peak ground

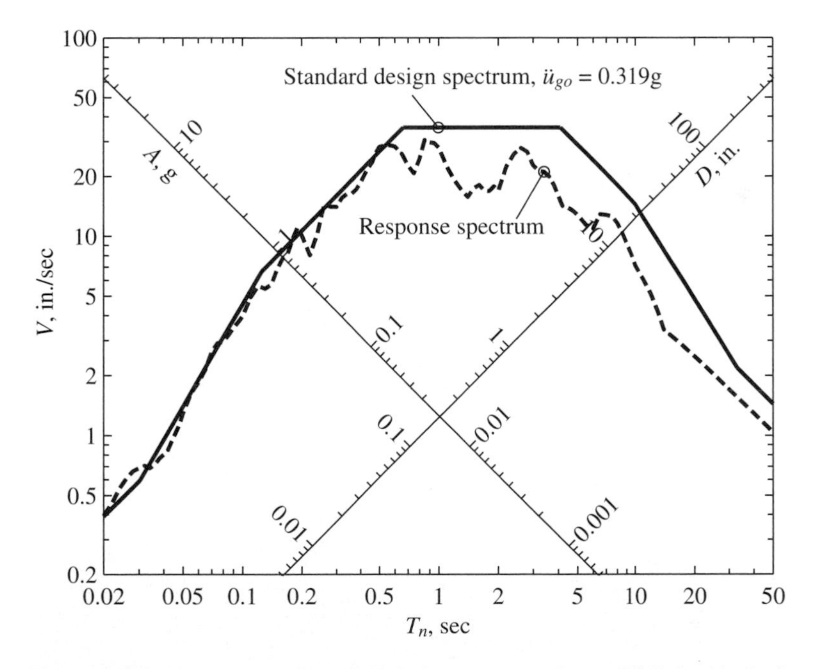

Figure 6.10.1 Comparison of standard design spectrum ($\ddot{u}_{go} = 0.319\text{g}$) with elastic response spectrum for El Centro ground motion; $\zeta = 5\%$.

velocity. Similarly, they are even more different in the displacement-sensitive region because of the larger differences (11.5 in. versus 8.4 in.) in the peak ground displacement.

The response spectrum for an individual ground motion differs from the design spectrum even if the peak values $\ddot{u}_{go}$, $\dot{u}_{go}$, and u_{go} for the two spectra are matched. In Fig. 6.10.2 the response spectrum for the El Centro ground motion is compared with the design spectrum for ground motion parameters $\ddot{u}_{go} = 0.319\text{g}$, $\dot{u}_{go} = 13.04$ in./sec, and $u_{go} = 8.40$ in.—the same as for the El Centro ground motion. Two design spectra are included: the median spectrum and median-plus-one-standard-deviation spectrum. The agreement between the response and design spectra is now better because the ground motion parameters are matched. However, significant differences remain: over the acceleration-sensitive region the response spectrum is close to the median-plus-one-standard-deviation design spectrum; over the velocity- and displacement-sensitive regions the response spectrum is between the two design spectra for some periods and below the median design spectrum for other periods.

Such differences are to be expected because the design spectrum is not intended to match the response spectrum for any particular ground motion but is constructed to represent the average characteristics of many ground motions. These differences are due to the inherent variability in ground motions as reflected in the probability distributions of the amplification factors and responses.

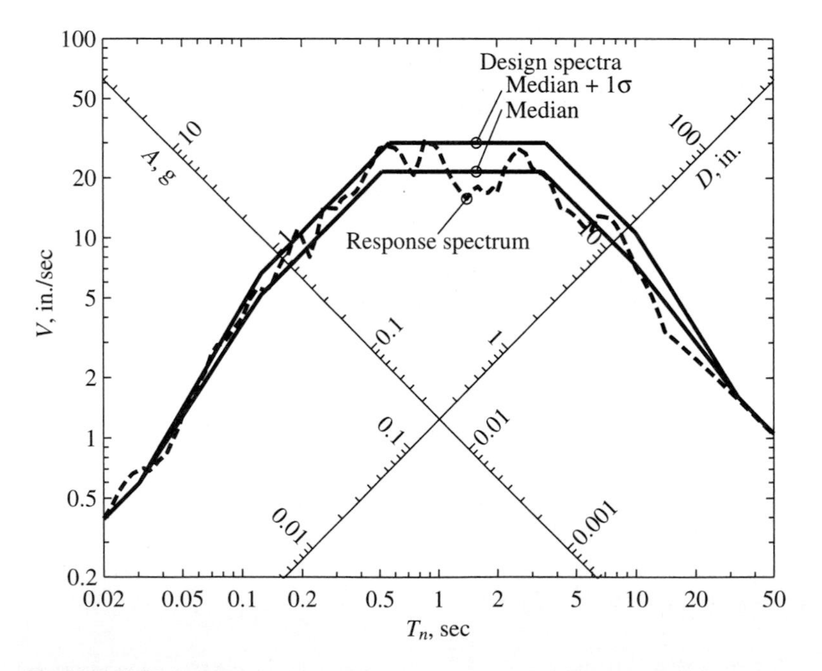

Figure 6.10.2 Comparison of design spectra ($\ddot{u}_{go} = 0.319\text{g}$, $\dot{u}_{go} = 13.04$ in./sec, $u_{go} = 8.40$ in.) with elastic response spectrum for El Centro ground motion; $\zeta = 5\%$.

6.11 DISTINCTION BETWEEN DESIGN AND RESPONSE SPECTRA

A design spectrum differs conceptually from a response spectrum in two important ways. First, the jagged response spectrum is a plot of the peak response of all possible SDF systems and hence is a description of a particular ground motion. The smooth design spectrum, however, is a specification of the level of seismic design force, or deformation, as a function of natural vibration period and damping ratio. This conceptual difference between the two spectra should be recognized, although in some situations, their shapes may be similar. Such is the case when the design spectrum is determined by statistical analysis of several comparable response spectra.

Second, for some sites a design spectrum is the envelope of two different elastic design spectra. Consider a site in southern California that could be affected by two different types of earthquakes: a Magnitude 6.5 earthquake originating on a nearby fault and a Magnitude 8.5 earthquake on the distant San Andreas fault. The design spectrum for each earthquake could be determined by the procedure developed in Section 6.9. The ordinates and shapes of the two design spectra would differ, as shown schematically in Fig. 6.11.1, because of the differences in earthquake magnitude and distance of the site from the earthquake fault. The design spectrum for this site is defined as the envelope of the design spectra for the two different types of earthquakes. Note that the short-period portion of the design spectrum is governed by the nearby earthquake, while the long-period portion of the design spectrum is controlled by the distant earthquake.

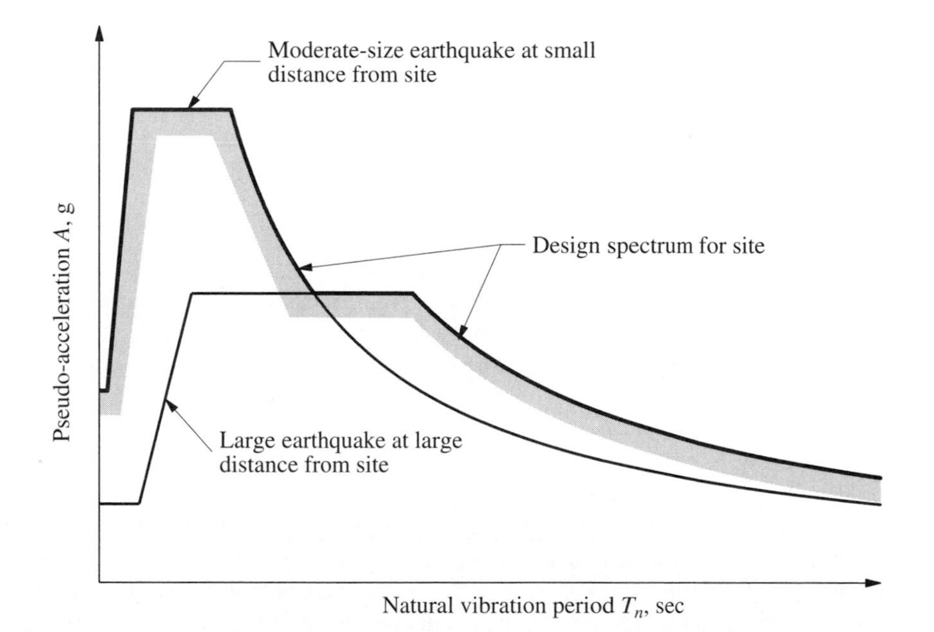

Figure 6.11.1 Design spectrum defined as the envelope of design spectra for earthquakes originating on two different faults.

6.12 VELOCITY AND ACCELERATION RESPONSE SPECTRA

We now return to the relative velocity response spectrum and the acceleration response spectrum that were introduced in Section 6.5. In one sense there is little motivation to study these "true" spectra because they are not needed to determine the peak deformations and forces in a system; for this purpose the pseudo-acceleration (or pseudo-velocity or deformation) response spectrum is sufficient. A brief discussion of these "true" spectra is included, however, because the distinction between them and "pseudo" spectra has not always been made in the early publications, and the two have sometimes been used interchangeably.

To study the relationship between these spectra, we write them in mathematical form. The deformation response of a linear SDF system to an arbitrary ground motion with zero initial conditions is given by Duhamel's integral, Eq. (4.2.3), with $p(t)$ replaced by $p_{\text{eff}}(t) = -m\ddot{u}_g(t)$:

$$u(t) = -\frac{1}{\omega_D}\int_0^t \ddot{u}_g(\tau)e^{-\zeta\omega_n(t-\tau)}\sin[\omega_D(t-\tau)]\,d\tau \tag{6.12.1}$$

Using theorems from calculus to differentiate under the integral sign leads to

$$\dot{u}(t) = -\zeta\omega_n u(t) - \int_0^t \ddot{u}_g(\tau)e^{-\zeta\omega_n(t-\tau)}\cos[\omega_D(t-\tau)]\,d\tau \tag{6.12.2}$$

An equation for the acceleration $\ddot{u}^t(t)$ of the mass can be obtained by differentiating Eq. (6.12.2) and adding the ground acceleration $\ddot{u}_g(t)$. However, the equation of motion for the system provides a more convenient alternative:

$$\ddot{u}^t(t) = -\omega_n^2 u(t) - 2\zeta\omega_n\dot{u}(t) \tag{6.12.3}$$

As defined earlier, the relative-velocity spectrum and acceleration spectrum are plots of $\dot{u}_o$ and $\ddot{u}_o^t$, the peak values of $\dot{u}(t)$ and $\ddot{u}^t(t)$, respectively, as functions of T_n.

6.12.1 Pseudo-velocity and Relative-Velocity Spectra

In Fig. 6.12.1a the relative-velocity response spectrum is compared with the pseudo-velocity response spectrum, both for El Centro ground motion and systems with $\zeta = 10\%$. The latter spectrum is simply one of the curves of Fig. 6.6.4 presented in a different form. Each point on the relative-velocity response spectrum represents the peak velocity of an SDF system obtained from $\dot{u}(t)$ determined by the numerical methods of Chapter 5. The differences between the two spectra depend on the natural period of the system. For long-period systems, V is less than $\dot{u}_o$ and the differences between the two are large. This can be understood by recognizing that as T_n becomes very long, the mass of the system stays still while the ground underneath moves. Thus, as $T_n \to \infty$, $D \to u_{go}$ (see Section 6.8 and Fig. 6.8.5) and $V \to \dot{u}_{go}$. Now $D \to u_{go}$ implies that $V \to 0$ because of Eq. (6.6.1). These trends are confirmed by the results presented in Fig. 6.12.1a. For short-period systems V exceeds $\dot{u}_o$, with the differences increasing as T_n becomes shorter. For medium-period systems, the differences between V and $\dot{u}_o$ are small over a wide range of T_n.

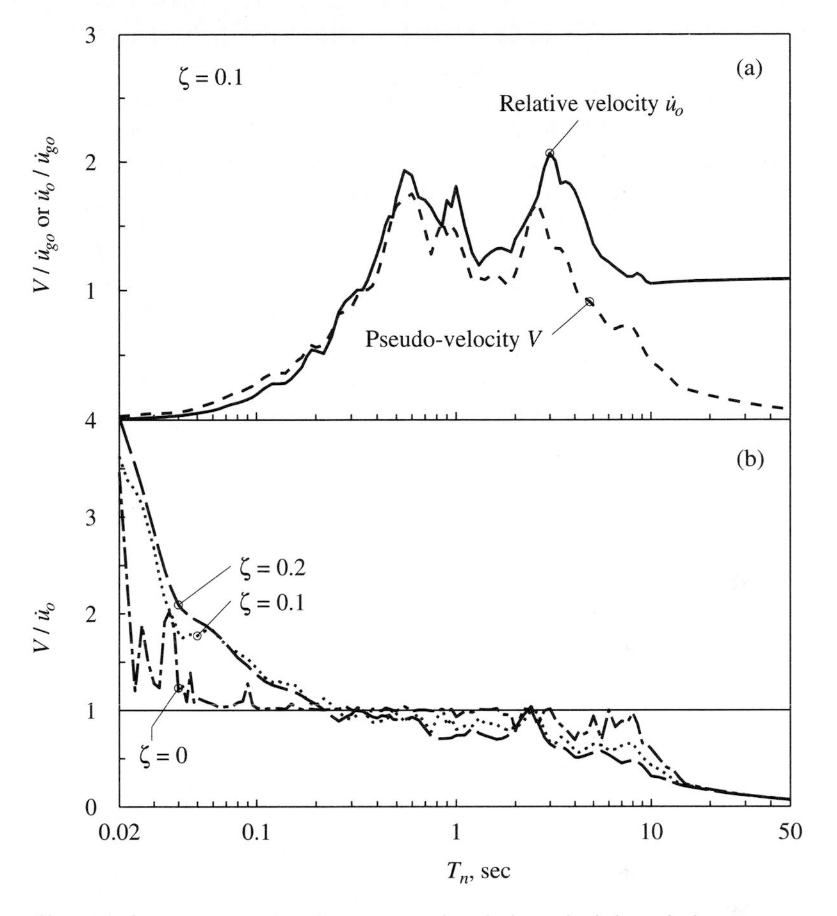

Figure 6.12.1 (a) Comparison between pseudo-velocity and relative-velocity response spectra; $\zeta = 10\%$; (b) ratio $V/\dot{u}_o$ for $\zeta = 0$, 10, and 20%.

In Fig. 6.12.1b the ratio $V/\dot{u}_o$ is plotted for three damping values, $\zeta = 0$, 10, and 20%. The differences between the two spectra, as indicated by how much the ratio $V/\dot{u}_o$ differs from unity, are smallest for undamped systems and increase with damping. This can be explained from Eqs. (6.12.1) and (6.12.2) by observing that for $\zeta = 0$, $\dot{u}(t)$ and $\omega_n u(t)$ are the same except for the sine and cosine terms in the integrand. With damping, the first term in Eq. (6.12.2) contributes to $\dot{u}(t)$, suggesting that $\dot{u}(t)$ would differ from $\omega_n u(t)$ to a greater degree. Over the medium-period range V can be taken as an approximation to $\dot{u}_o$ for the practical range of damping.

6.12.2 Pseudo-acceleration and Acceleration Spectra

The pseudo-acceleration and acceleration response spectra are identical for systems without damping. This is apparent from Eq. (6.12.3), which for undamped systems specializes to

$$\ddot{u}^t(t) = -\omega_n^2 u(t) \tag{6.12.4}$$

The peak values of the two sides are therefore equal, that is,

$$\ddot{u}_o^t = \omega_n^2 u_o = \omega_n^2 D = A \tag{6.12.5}$$

With damping, Eq. (6.12.4) is not valid at all times, but only at the time instants when $\dot{u}(t) = 0$, in particular when $u(t)$ attains its peak u_o. At this instant, $-\omega_n^2 u$ represents the true acceleration of the mass. The peak value $\ddot{u}_o^t$ of $\ddot{u}^t(t)$ does not occur at the same instant, however, unless $\zeta = 0$. The peak values $\ddot{u}_o^t$ and A occur at the same time and are equal only for $\zeta = 0$.

Equation (6.12.3) suggests that the differences between A and $\ddot{u}_o^t$ are expected to increase as the damping increases. This expectation is confirmed by the data presented in Fig. 6.12.2, where the pseudo-acceleration and the acceleration spectra for the El

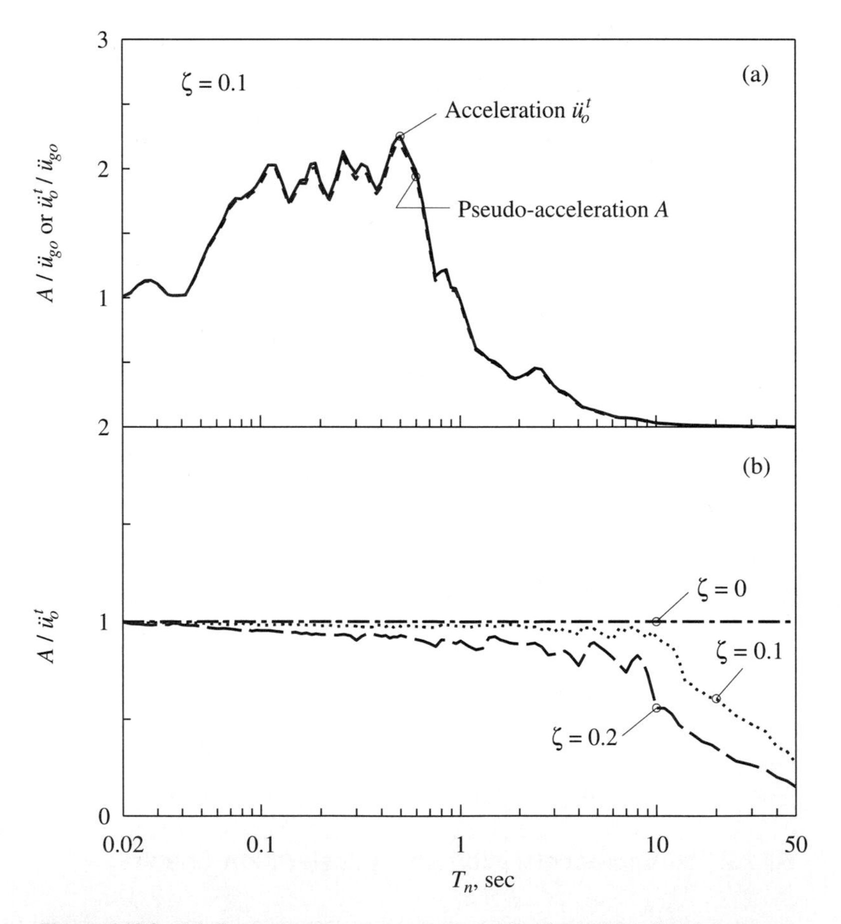

Figure 6.12.2 (a) Comparison between pseudo-acceleration and acceleration response spectra; $\zeta = 10\%$; (b) ratio $A/\ddot{u}_o^t$ for $\zeta = 0$, 10, and 20%.

Centro ground motion are plotted for $\zeta = 10\%$, and the ratio $A/\ddot{u}_o^t$ is presented for three damping values. The difference between the two spectra is small for short-period systems and is of some significance only for long-period systems with large values of damping. Thus for a wide range of conditions the pseudo-acceleration may be treated as an approximation to the true acceleration.

As the natural vibration period T_n of a system approaches infinity, the mass of the system stays still while the ground underneath moves. Thus, as $T_n \to \infty$, $\ddot{u}_o^t \to 0$ and $D \to u_{go}$; the latter implies that $A \to 0$ because of Eq. (6.6.3). Both A and $(\ddot{u}_t)_o \to 0$ as $T_n \to \infty$, but at different rates, as evident from the ratio $A/\ddot{u}_o^t$ plotted as a function of T_n; $A \to 0$ at a much faster rate because of T_n^2 in the denominator of Eq. (6.6.3).

Another way of looking at the differences between the two spectra is by recalling that mA is equal to the peak value of the elastic-resisting force. In contrast, $m\ddot{u}_o^t$ is equal to the peak value of the sum of elastic and damping forces. As seen in Fig. 6.12.2b, the pseudo-acceleration is smaller than the true acceleration, because it is that part of the true acceleration which gives the elastic force.

Parenthetically, we note that the widespread adoption of the prefix *pseudo* is in one sense misleading. The literal meaning of *pseudo* (false) is not really appropriate since we are dealing with approximation rather than with concepts that are in any sense false or inappropriate. In fact, there is rarely the need to use the "pseudo"-spectra as approximations to the "true" spectra because the latter can be computed by the same numerical procedures as those used for the former. Furthermore, as emphasized earlier, the pseudo quantities provide the exact values of the desired deformation and forces.

FURTHER READING

Benioff, H., "The Physical Evaluation of Seismic Destructiveness," *Bulletin of the Seismological Society of America*, **24**, 1934, pp. 398–403.

Biot, M. A., "Theory of Elastic Systems under Transient Loading with an Application to Earthquake Proof Buildings," *Proceedings, National Academy of Sciences*, **19**, 1933, pp. 262–268.

Biot, M. A., "A Mechanical Analyzer for the Prediction of Earthquake Stresses," *Bulletin of the Seismological Society of America*, **31**, 1941, pp. 151–171.

Bolt, B. A., *Earthquakes*, W.H. Freeman, New York, 1993, Chapters 1–7.

Clough, R. W., and Penzien, J., *Dynamics of Structures*, McGraw-Hill, New York, 1993, pp. 586–597.

Housner, G. W., "Calculating the Response of an Oscillator to Arbitrary Ground Motion," *Bulletin of the Seismological Society of America*, **31**, 1941, pp. 143–149.

Housner, G. W., and Jennings, P. C., *Earthquake Design Criteria*, Earthquake Engineering Research Institute, Berkeley, Calif., 1982, pp. 19–41 and 58–88.

Hudson, D. E., "Response Spectrum Techniques in Engineering Seismology," *Proceedings of the First World Conference in Earthquake Engineering*, Berkeley, Calif., 1956, pp. 4–1 to 4–12.

Hudson, D. E., "A History of Earthquake Engineering," *Proceedings of the IDNDR International Symposium on Earthquake Disaster Reduction Technology—30th Anniversary of IISEE*, Tsukuba, Japan, 1992, pp. 3–13.

Hudson, D. E., *Reading and Interpreting Strong Motion Accelerograms*, Earthquake Engineering Research Institute, Berkeley, Calif., 1979, pp. 22–70 and 95–97.

Mohraz, B., and Elghadamsi, F. E., "Earthquake Ground Motion and Response Spectra," Chapter 2 in *The Seismic Design Handbook* (ed. F. Naeim), Van Nostrand Reinhold, New York, 1989.

Newmark, N. M., and Hall, W. J., *Earthquake Spectra and Design*, Earthquake Engineering Research Institute, Berkeley, Calif., 1982, pp. 29–37.

Newmark, N. M., and Rosenblueth, E., *Fundamentals of Earthquake Engineering*, Prentice Hall, Englewood Cliffs, N.J., 1971, Chapter 7.

Riddell, R., and Newmark, N. M., "Statistical Analysis of the Response of Nonlinear Systems Subjected to Earthquakes," *Structural Research Series No. 468*, University of Illinois at Urbana-Champaign, Urbana, Ill., August 1979.

Rosenblueth, E., "Characteristics of Earthquakes," Chapter 1 in *Design of Earthquake Resistant Structures* (ed. E. Rosenblueth), Pentech Press, London, 1980.

Seed, H. B., and Idriss, I. M., *Ground Motions and Soil Liquefaction during Earthquakes*, Earthquake Engineering Research Institute, Berkeley, Calif., 1982, pp. 21–56.

Veletsos, A. S., "Maximum Deformation of Certain Nonlinear Systems," *Proceedings of the 4th World Conference on Earthquake Engineering*, Santiago, Chile, Vol. 1, 1969, pp. 155–170.

Veletsos, A. S., and Newmark, N. M., "Response Spectra for Single-Degree-of-Freedom Elastic and Inelastic Systems," *Report No. RTD-TDR-63-3096*, Vol. III, Air Force Weapons Laboratory, Albuquerque, N.Mex., June 1964.

Veletsos, A. S., Newmark, N. M., and Chelapati, C. V., "Deformation Spectra for Elastic and Elastoplastic Systems Subjected to Ground Shock and Earthquake Motion," *Proceedings of the 3rd World Conference on Earthquake Engineering*, New Zealand, Vol. II, 1965, pp. 663–682.

APPENDIX 6: EL CENTRO, 1940 GROUND MOTION

The north–south component of the ground motion recorded at a site in El Centro, California during the Imperial Valley, California earthquake of May 18, 1940 is shown in Fig. 6.1.4. This particular version of this record is used throughout this book, and is required in solving some of the end-of-chapter problems. Numerical values for the ground acceleration in units of g, the acceleration due to gravity, are presented in Table A6.1. This includes 1559 data points at equal time spacings of 0.02 sec, to be read row by row; the first value is at $t = 0$. These data are also available electronically from the National Information Service for Earthquake Engineering, University of California at Berkeley, via anonymous ftp if one has access to internet. To access these data, proceed as follows:

```
ftp nisee.ce.berkeley.edu
(128.32.43.154)
login: anonymous
passwd: your_ident
cd pub/a.k.chopra
get el_centro_data
quit
```

TABLE A6.1 GROUND ACCELERATION DATA

0.00630	0.00364	0.00099	0.00428	0.00758	0.01087	0.00682	0.00277
−0.00128	0.00368	0.00864	0.01360	0.00727	0.00094	0.00420	0.00221
0.00021	0.00444	0.00867	0.01290	0.01713	−0.00343	−0.02400	−0.00992
0.00416	0.00528	0.01653	0.02779	0.03904	0.02449	0.00995	0.00961
0.00926	0.00892	−0.00486	−0.01864	−0.03242	−0.03365	−0.05723	−0.04534
−0.03346	−0.03201	−0.03056	−0.02911	−0.02766	−0.04116	−0.05466	−0.06816
−0.08166	−0.06846	−0.05527	−0.04208	−0.04259	−0.04311	−0.02428	−0.00545
0.01338	0.03221	0.05104	0.06987	0.08870	0.04524	0.00179	−0.04167
−0.08513	−0.12858	−0.17204	−0.12908	−0.08613	−0.08902	−0.09192	−0.09482
−0.09324	−0.09166	−0.09478	−0.09789	−0.12902	−0.07652	−0.02401	0.02849
0.08099	0.13350	0.18600	0.23850	0.21993	0.20135	0.18277	0.16420
0.14562	0.16143	0.17725	0.13215	0.08705	0.04196	−0.00314	−0.04824
−0.09334	−0.13843	−0.18353	−0.22863	−0.27372	−0.31882	−0.25024	−0.18166
−0.11309	−0.04451	0.02407	0.09265	0.16123	0.22981	0.29839	0.23197
0.16554	0.09912	0.03270	−0.03372	−0.10014	−0.16656	−0.23299	−0.29941
−0.00421	0.29099	0.22380	0.15662	0.08943	0.02224	−0.04495	0.01834
0.08163	0.14491	0.20820	0.18973	0.17125	0.13759	0.10393	0.07027
0.03661	0.00295	−0.03071	−0.00561	0.01948	0.04458	0.06468	0.08478
0.10487	0.05895	0.01303	−0.03289	−0.07882	−0.03556	0.00771	0.05097
0.01013	−0.03071	−0.07156	−0.11240	−0.15324	−0.11314	−0.07304	−0.03294
0.00715	−0.06350	−0.13415	−0.20480	−0.12482	−0.04485	0.03513	0.11510
0.19508	0.12301	0.05094	−0.02113	−0.09320	−0.02663	0.03995	0.10653
0.17311	0.11283	0.05255	−0.00772	0.01064	0.02900	0.04737	0.06573
0.02021	−0.02530	−0.07081	−0.04107	−0.01133	0.00288	0.01709	0.03131
−0.02278	−0.07686	−0.13095	−0.18504	−0.14347	−0.10190	−0.06034	−0.01877
0.02280	−0.00996	−0.04272	−0.02147	−0.00021	0.02104	−0.01459	−0.05022
−0.08585	−0.12148	−0.15711	−0.19274	−0.22837	−0.18145	−0.13453	−0.08761
−0.04069	0.00623	0.05316	0.10008	0.14700	0.09754	0.04808	−0.00138
0.05141	0.10420	0.15699	0.20979	0.26258	0.16996	0.07734	−0.01527
−0.10789	−0.20051	−0.06786	0.06479	0.01671	−0.03137	−0.07945	−0.12753
−0.17561	−0.22369	−0.27177	−0.15851	−0.04525	0.06802	0.18128	0.14464
0.10800	0.07137	0.03473	0.09666	0.15860	0.22053	0.18296	0.14538
0.10780	0.07023	0.03265	0.06649	0.10033	0.13417	0.10337	0.07257
0.04177	0.01097	−0.01983	0.04438	0.10860	0.17281	0.10416	0.03551
−0.03315	−0.10180	−0.07262	−0.04344	−0.01426	0.01492	−0.02025	−0.05543
−0.09060	−0.12578	−0.16095	−0.19613	−0.14784	−0.09955	−0.05127	−0.00298
−0.01952	−0.03605	−0.05259	−0.04182	−0.03106	−0.02903	−0.02699	0.02515
0.01770	0.02213	0.02656	0.00419	−0.01819	−0.04057	−0.06294	−0.02417
0.01460	0.05337	0.02428	−0.00480	−0.03389	−0.00557	0.02274	0.00679
−0.00915	−0.02509	−0.04103	−0.05698	−0.01826	0.02046	0.00454	−0.01138
−0.00215	0.00708	0.00496	0.00285	0.00074	−0.00534	−0.01141	0.00361
0.01863	0.03365	0.04867	0.03040	0.01213	−0.00614	−0.02441	0.01375
0.01099	0.00823	0.00547	0.00812	0.01077	−0.00692	−0.02461	−0.04230
−0.05999	−0.07768	−0.09538	−0.06209	−0.02880	0.00448	0.03777	0.01773
−0.00231	−0.02235	0.01791	0.05816	0.03738	0.01660	−0.00418	−0.02496
−0.04574	−0.02071	0.00432	0.02935	0.01526	0.01806	0.02086	0.00793
−0.00501	−0.01795	−0.03089	−0.01841	−0.00593	0.00655	−0.02519	−0.05693
−0.04045	−0.02398	−0.00750	0.00897	0.00384	−0.00129	−0.00642	−0.01156
−0.02619	−0.04082	−0.05545	−0.04366	−0.03188	−0.06964	−0.05634	−0.04303
−0.02972	−0.01642	−0.00311	0.01020	0.02350	0.03681	0.05011	0.02436

TABLE A6.1 GROUND ACCELERATION DATA (CONTINUED)

−0.00139	−0.02714	−0.00309	0.02096	0.04501	0.06906	0.05773	0.04640
0.03507	0.03357	0.03207	0.03057	0.03250	0.03444	0.03637	0.01348
−0.00942	−0.03231	−0.02997	−0.03095	−0.03192	−0.02588	−0.01984	−0.01379
−0.00775	−0.01449	−0.02123	0.01523	0.05170	0.08816	0.12463	0.16109
0.12987	0.09864	0.06741	0.03618	0.00495	0.00420	0.00345	0.00269
−0.05922	−0.12112	−0.18303	−0.12043	−0.05782	0.00479	0.06740	0.13001
0.08373	0.03745	0.06979	0.10213	−0.03517	−0.17247	−0.13763	−0.10278
−0.06794	−0.03310	−0.03647	−0.03984	−0.00517	0.02950	0.06417	0.09883
0.13350	0.05924	−0.01503	−0.08929	−0.16355	−0.06096	0.04164	0.01551
−0.01061	−0.03674	−0.06287	−0.08899	−0.05430	−0.01961	0.01508	0.04977
0.08446	0.05023	0.01600	−0.01823	−0.05246	−0.08669	−0.06769	−0.04870
−0.02970	−0.01071	0.00829	−0.00314	0.02966	0.06246	−0.00234	−0.06714
−0.04051	−0.01388	0.01274	0.00805	0.03024	0.05243	0.02351	−0.00541
−0.03432	−0.06324	−0.09215	−0.12107	−0.08450	−0.04794	−0.01137	0.02520
0.06177	0.04028	0.01880	0.04456	0.07032	0.09608	0.12184	0.06350
0.00517	−0.05317	−0.03124	−0.00930	0.01263	0.03457	0.03283	0.03109
0.02935	0.04511	0.06087	0.07663	0.09239	0.05742	0.02245	−0.01252
0.00680	0.02611	0.04543	0.01571	−0.01402	−0.04374	−0.07347	−0.03990
−0.00633	0.02724	0.06080	0.03669	0.01258	−0.01153	−0.03564	−0.00677
0.02210	0.05098	0.07985	0.06915	0.05845	0.04775	0.03706	0.02636
0.05822	0.09009	0.12196	0.10069	0.07943	0.05816	0.03689	0.01563
−0.00564	−0.02690	−0.04817	−0.06944	−0.09070	−0.11197	−0.11521	−0.11846
−0.12170	−0.12494	−0.16500	−0.20505	−0.15713	−0.10921	−0.06129	−0.01337
0.03455	0.08247	0.07576	0.06906	0.06236	0.08735	0.11235	0.13734
0.12175	0.10616	0.09057	0.07498	0.08011	0.08524	0.09037	0.06208
0.03378	0.00549	−0.02281	−0.05444	−0.04030	−0.02615	−0.01201	−0.02028
−0.02855	−0.06243	−0.03524	−0.00805	−0.04948	−0.03643	−0.02337	−0.03368
−0.01879	−0.00389	0.01100	0.02589	0.01446	0.00303	−0.00840	0.00463
0.01766	0.03069	0.04372	0.02165	−0.00042	−0.02249	−0.04456	−0.03638
−0.02819	−0.02001	−0.01182	−0.02445	−0.03707	−0.04969	−0.05882	−0.06795
−0.07707	−0.08620	−0.09533	−0.06276	−0.03018	0.00239	0.03496	0.04399
0.05301	0.03176	0.01051	−0.01073	−0.03198	−0.05323	0.00186	0.05696
0.01985	−0.01726	−0.05438	−0.01204	0.03031	0.07265	0.11499	0.07237
0.02975	−0.01288	0.01212	0.03711	0.03517	0.03323	0.01853	0.00383
0.00342	−0.02181	−0.04704	−0.07227	−0.09750	−0.12273	−0.08317	−0.04362
−0.00407	0.03549	0.07504	0.11460	0.07769	0.04078	0.00387	0.00284
0.00182	−0.05513	0.04732	0.05223	0.05715	0.06206	0.06698	0.07189
0.02705	−0.01779	−0.06263	−0.10747	−0.15232	−0.12591	−0.09950	−0.07309
−0.04668	−0.02027	0.00614	0.03255	0.00859	−0.01537	−0.03932	−0.06328
−0.03322	−0.00315	0.02691	0.01196	−0.00300	0.00335	0.00970	0.01605
0.02239	0.04215	0.06191	0.08167	0.03477	−0.01212	−0.01309	−0.01407
−0.05274	−0.02544	0.00186	0.02916	0.05646	0.08376	0.01754	−0.04869
−0.02074	0.00722	0.03517	−0.00528	−0.04572	−0.08617	−0.06960	−0.05303
−0.03646	−0.01989	−0.00332	0.01325	0.02982	0.01101	−0.00781	−0.02662
−0.00563	0.01536	0.03635	0.05734	0.03159	0.00584	−0.01992	−0.00201
0.01589	−0.01024	−0.03636	−0.06249	−0.04780	−0.03311	−0.04941	−0.06570
−0.08200	−0.04980	−0.01760	0.01460	0.04680	0.07900	0.04750	0.01600
−0.01550	−0.00102	0.01347	0.02795	0.04244	0.05692	0.03781	0.01870
−0.00041	−0.01952	−0.00427	0.01098	0.02623	0.04148	0.01821	−0.00506
−0.00874	−0.03726	−0.06579	−0.02600	0.01380	0.05359	0.09338	0.05883

TABLE A6.1 GROUND ACCELERATION DATA (CONTINUED)

0.02429	−0.01026	−0.04480	−0.01083	−0.01869	−0.02655	−0.03441	−0.02503
−0.01564	−0.00626	−0.01009	−0.01392	0.01490	0.04372	0.03463	0.02098
0.00733	−0.00632	−0.01997	0.00767	0.03532	0.03409	0.03287	0.03164
0.02403	0.01642	0.00982	0.00322	−0.00339	0.02202	−0.01941	−0.06085
−0.10228	−0.07847	−0.05466	−0.03084	−0.00703	0.01678	0.01946	0.02214
0.02483	0.01809	−0.00202	−0.02213	−0.00278	0.01656	0.03590	0.05525
0.07459	0.06203	0.04948	0.03692	−0.00145	0.04599	0.04079	0.03558
0.03037	0.03626	0.04215	0.04803	0.05392	0.04947	0.04502	0.04056
0.03611	0.03166	0.00614	−0.01937	−0.04489	−0.07040	−0.09592	−0.07745
−0.05899	−0.04052	−0.02206	−0.00359	0.01487	0.01005	0.00523	0.00041
−0.00441	−0.00923	−0.01189	−0.01523	−0.01856	−0.02190	−0.00983	0.00224
0.01431	0.00335	−0.00760	−0.01856	−0.00737	0.00383	0.01502	0.02622
0.01016	−0.00590	−0.02196	−0.00121	0.01953	0.04027	0.02826	0.01625
0.00424	0.00196	−0.00031	−0.00258	−0.00486	−0.00713	−0.00941	−0.01168
−0.01396	−0.01750	−0.02104	−0.02458	−0.02813	−0.03167	−0.03521	−0.04205
−0.04889	−0.03559	−0.02229	−0.00899	0.00431	0.01762	0.00714	−0.00334
−0.01383	0.01314	0.04011	0.06708	0.04820	0.02932	0.01043	−0.00845
−0.02733	−0.04621	−0.03155	−0.01688	−0.00222	0.01244	0.02683	0.04121
0.05559	0.03253	0.00946	−0.01360	−0.01432	−0.01504	−0.01576	−0.04209
−0.02685	−0.01161	0.00363	0.01887	0.03411	0.03115	0.02819	0.02917
0.03015	0.03113	0.00388	−0.02337	−0.05062	−0.03820	−0.02579	−0.01337
−0.00095	0.01146	0.02388	0.03629	0.01047	−0.01535	−0.04117	−0.06699
−0.05207	−0.03715	−0.02222	−0.00730	0.00762	0.02254	0.03747	0.04001
0.04256	0.04507	0.04759	0.05010	0.04545	0.04080	0.02876	0.01671
0.00467	−0.00738	−0.00116	0.00506	0.01128	0.01750	−0.00211	−0.02173
−0.04135	−0.06096	−0.08058	−0.06995	−0.05931	−0.04868	−0.03805	−0.02557
−0.01310	−0.00063	0.01185	0.02432	0.03680	0.04927	0.02974	0.01021
−0.00932	−0.02884	−0.04837	−0.06790	−0.04862	−0.02934	−0.01006	0.00922
0.02851	0.04779	0.02456	0.00133	−0.02190	−0.04513	−0.06836	−0.04978
−0.03120	−0.01262	0.00596	0.02453	0.04311	0.06169	0.08027	0.09885
0.06452	0.03019	−0.00414	−0.03848	−0.07281	−0.05999	−0.04717	−0.03435
−0.03231	−0.03028	−0.02824	−0.00396	0.02032	0.00313	−0.01406	−0.03124
−0.04843	−0.06562	−0.05132	−0.03702	−0.02272	−0.00843	0.00587	0.02017
0.02698	0.03379	0.04061	0.04742	0.05423	0.03535	0.01647	0.01622
0.01598	0.01574	0.00747	−0.00080	−0.00907	0.00072	0.01051	0.02030
0.03009	0.03989	0.03478	0.02967	0.02457	0.03075	0.03694	0.04313
0.04931	0.05550	0.06168	−0.00526	−0.07220	−0.06336	−0.05451	−0.04566
−0.03681	−0.03678	−0.03675	−0.03672	−0.01765	0.00143	0.02051	0.03958
0.05866	0.03556	0.01245	−0.01066	−0.03376	−0.05687	−0.04502	−0.03317
−0.02131	−0.00946	0.00239	−0.00208	−0.00654	−0.01101	−0.01548	−0.01200
−0.00851	−0.00503	−0.00154	0.00195	0.00051	−0.00092	0.01135	0.02363
0.03590	0.04818	0.06045	0.07273	0.02847	−0.01579	−0.06004	−0.05069
−0.04134	−0.03199	−0.03135	−0.03071	−0.03007	−0.01863	−0.00719	0.00425
0.01570	0.02714	0.03858	0.02975	0.02092	0.02334	0.02576	0.02819
0.03061	0.03304	0.01371	−0.00561	−0.02494	−0.02208	−0.01923	−0.01638
−0.01353	−0.01261	−0.01170	−0.00169	0.00833	0.01834	0.02835	0.03836
0.04838	0.03749	0.02660	0.01571	0.00482	−0.00607	−0.01696	−0.00780
0.00136	0.01052	0.01968	0.02884	−0.00504	−0.03893	−0.02342	−0.00791
0.00759	0.02310	0.00707	−0.00895	−0.02498	−0.04100	−0.05703	−0.02920
−0.00137	0.02645	0.05428	0.03587	0.01746	−0.00096	−0.01937	−0.03778

TABLE A6.1 GROUND ACCELERATION DATA (CONTINUED)

−0.02281	−0.00784	0.00713	0.02210	0.03707	0.05204	0.06701	0.08198
0.03085	−0.02027	−0.07140	−0.12253	−0.08644	−0.05035	−0.01426	0.02183
0.05792	0.09400	0.13009	0.03611	−0.05787	−0.04802	−0.03817	−0.02832
0.02970	0.03993	0.05017	0.06041	0.07065	0.08089	−0.00192	−0.08473
−0.01846	−0.00861	−0.03652	−0.06444	−0.06169	−0.05894	−0.05618	−0.06073
−0.06528	−0.04628	−0.02728	−0.00829	0.01071	0.02970	0.03138	0.03306
0.03474	0.03642	0.04574	0.05506	0.06439	0.07371	0.08303	0.03605
−0.01092	−0.05790	−0.04696	−0.03602	−0.02508	−0.01414	−0.03561	−0.05708
−0.07855	−0.06304	−0.04753	−0.03203	−0.01652	−0.00102	0.00922	0.01946
−0.07032	−0.05590	−0.04148	−0.05296	−0.06443	−0.07590	−0.08738	−0.09885
−0.06798	−0.03710	−0.00623	0.02465	0.05553	0.08640	0.11728	0.14815
0.08715	0.02615	−0.03485	−0.09584	−0.07100	−0.04616	−0.02132	0.00353
0.02837	0.05321	−0.00469	−0.06258	−0.12048	−0.09960	−0.07872	−0.05784
−0.03696	−0.01608	0.00480	0.02568	0.04656	0.06744	0.08832	0.10920
0.13008	0.10995	0.08982	0.06969	0.04955	0.04006	0.03056	0.02107
0.01158	0.00780	0.00402	0.00024	−0.00354	−0.00732	−0.01110	−0.00780
−0.00450	−0.00120	0.00210	0.00540	−0.00831	−0.02203	−0.03575	−0.04947
−0.06319	−0.05046	−0.03773	−0.02500	−0.01227	0.00046	0.00482	0.00919
0.01355	0.01791	0.02228	0.00883	−0.00462	−0.01807	−0.03152	−0.02276
−0.01401	−0.00526	0.00350	0.01225	0.02101	0.01437	0.00773	0.00110
0.00823	0.01537	0.02251	0.01713	0.01175	0.00637	0.01376	0.02114
0.02852	0.03591	0.04329	0.03458	0.02587	0.01715	0.00844	−0.00027
−0.00898	−0.00126	0.00645	0.01417	0.02039	0.02661	0.03283	0.03905
0.04527	0.03639	0.02750	0.01862	0.00974	0.00086	−0.01333	−0.02752
−0.04171	−0.02812	−0.01453	−0.00094	0.01264	0.02623	0.01690	0.00756
−0.00177	−0.01111	−0.02044	−0.02977	−0.03911	−0.02442	−0.00973	0.00496
0.01965	0.03434	0.02054	0.00674	−0.00706	−0.02086	−0.03466	−0.02663
−0.01860	−0.01057	−0.00254	−0.00063	0.00128	0.00319	0.00510	0.00999
0.01488	0.00791	0.00093	−0.00605	0.00342	0.01288	0.02235	0.03181
0.04128	0.02707	0.01287	−0.00134	−0.01554	−0.02975	−0.04395	−0.03612
−0.02828	−0.02044	−0.01260	−0.00476	0.00307	0.01091	0.00984	0.00876
0.00768	0.00661	0.01234	0.01807	0.02380	0.02953	0.03526	0.02784
0.02042	0.01300	−0.03415	−0.00628	−0.00621	−0.00615	−0.00609	−0.00602
−0.00596	−0.00590	−0.00583	−0.00577	−0.00571	−0.00564	−0.00558	−0.00552
−0.00545	−0.00539	−0.00532	−0.00526	−0.00520	−0.00513	−0.00507	−0.00501
−0.00494	−0.00488	−0.00482	−0.00475	−0.00469	−0.00463	−0.00456	−0.00450
−0.00444	−0.00437	−0.00431	−0.00425	−0.00418	−0.00412	−0.00406	−0.00399
−0.00393	−0.00387	−0.00380	−0.00374	−0.00368	−0.00361	−0.00355	−0.00349
−0.00342	−0.00336	−0.00330	−0.00323	−0.00317	−0.00311	−0.00304	−0.00298
−0.00292	−0.00285	−0.00279	−0.00273	−0.00266	−0.00260	−0.00254	−0.00247
−0.00241	−0.00235	−0.00228	−0.00222	−0.00216	−0.00209	−0.00203	−0.00197
−0.00190	−0.00184	−0.00178	−0.00171	−0.00165	−0.00158	−0.00152	−0.00146
−0.00139	−0.00133	−0.00127	−0.00120	−0.00114	−0.00108	−0.00101	−0.00095
−0.00089	−0.00082	−0.00076	−0.00070	−0.00063	−0.00057	−0.00051	−0.00044
−0.00038	−0.00032	−0.00025	−0.00019	−0.00013	−0.00006	0.00000	

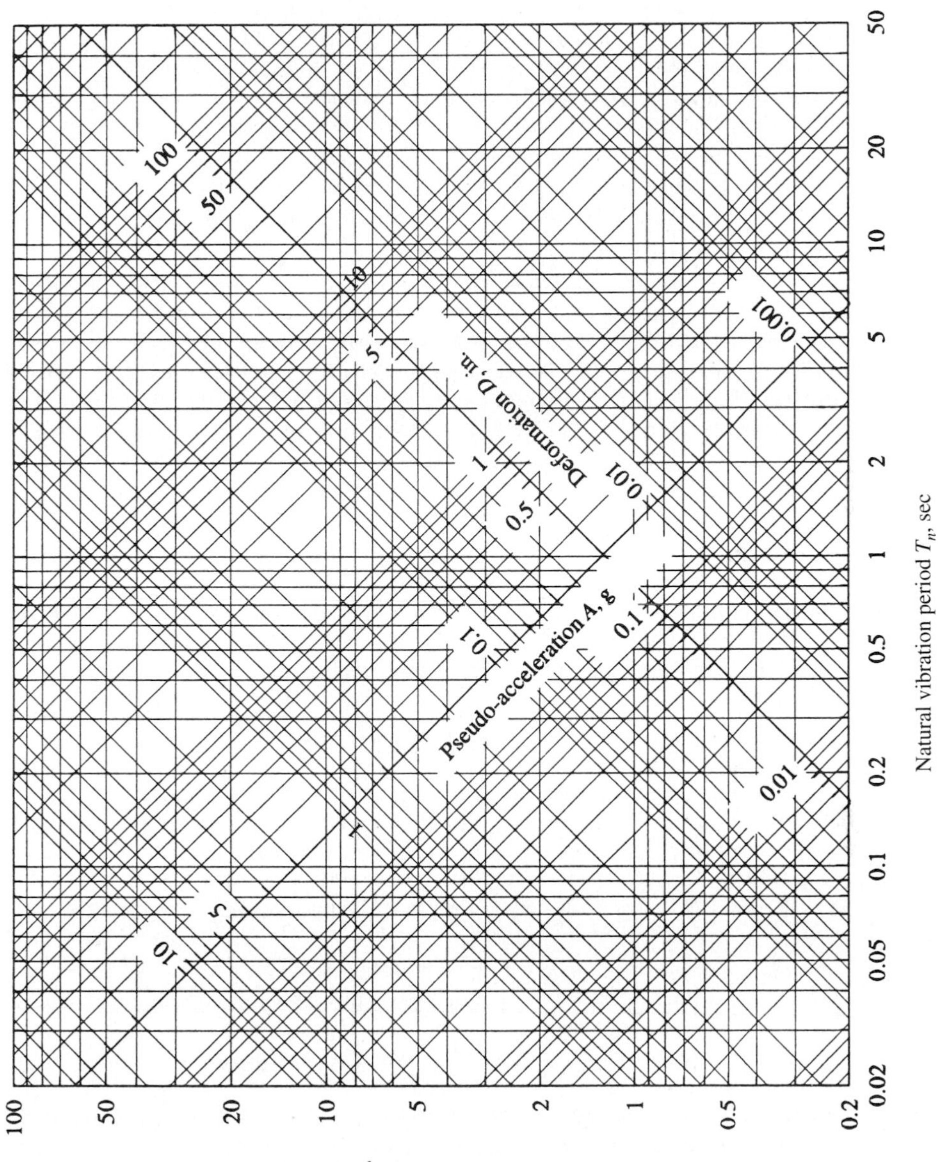

Figure A6.1 Graph paper with four-way logarithmic scales.

PROBLEMS

6.1 Determine the deformation response $u(t)$ for $0 \leq t \leq 10$ sec for an SDF system with natural period $T_n = 2$ sec and damping ratio $\zeta = 5\%$ to El Centro 1940 ground motion. The ground acceleration values are available at every $\Delta t = 0.02$ sec in Appendix 6. Use any existing computer program available to you or implement one of the numerical time-stepping algorithms of Chapter 5. Plot $u(t)$ and compare it with Fig. 6.4.1.

6.2 Derive equations for the deformation, pseudo-velocity, and pseudo-acceleration response spectra for ground acceleration $\ddot{u}_g(t) = \dot{u}_{go}\delta(t)$, where $\delta(t)$ is the Dirac delta function and $\dot{u}_{go}$ is the increment in velocity or the magnitude of the acceleration impulse. Plot the spectra for $\zeta = 0$ and 10%.

6.3 Consider harmonic ground motion $\ddot{u}_g(t) = \ddot{u}_{go} \sin(2\pi t/T)$, an unrealistic assumption.
(a) Derive equations for A and for $\ddot{u}_o^t$ in terms of the natural vibration period T_n and the damping ratio ζ of the SDF system. A is the peak value for the pseudo-acceleration, and $\ddot{u}_o^t$ is the peak value of the true acceleration. Consider only the steady-state response.
(b) Show that A and $\ddot{u}_o^t$ are identical for undamped systems but different for damped systems.
(c) Graphically display the two response spectra by plotting the normalized values $A/\ddot{u}_{go}$ and $\ddot{u}_o^t/\ddot{u}_{go}$ against T_n/T, the ratio of the natural vibration period of the system and the period of the excitation.

6.4 A 10-ft-long vertical cantilever made of a 6-in.-nominal-diameter standard steel pipe supports a 3000-lb weight attached at the tip, as shown in Fig. P6.4. The properties of the pipe are: outside diameter = 6.625 in., inside diameter = 6.065 in., thickness = 0.280 in., second moment of cross-sectional area $I = 28.1$ in^4, Young's modulus $E = 29{,}000$ ksi, and weight = 18.97 lb/ft length. Determine the peak deformation and the bending stress in the cantilever due to the El Centro ground motion; assume that $\zeta = 5\%$.

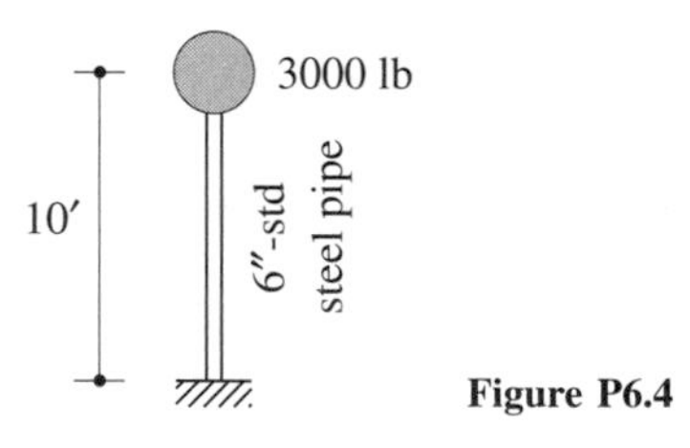

Figure P6.4

6.5 **(a)** A full water tank is supported on an 80-ft-high cantilever tower. It is idealized as an SDF system with weight $w = 100$ kips, lateral stiffness $k = 4$ kips/in., and damping ratio $\zeta = 5\%$. The tower supporting the tank is to be designed for ground motion characterized by the design spectrum of Fig. 6.9.4 scaled to 0.5g peak ground acceleration. Determine the design values of lateral deformation and base shear.
(b) The deformation computed for the system in part (a) seemed excessive to the structural designer, who decided to stiffen the tower by increasing its size. Determine the design values of deformation and base shear for the modified system if its lateral stiffness is 8 kips/in.; assume that the damping ratio is still 5%. Comment on how stiffening the system has affected the design requirements. What is the disadvantage of stiffening the system?
(c) If the stiffened tower were to support a tank weighing 200 kips, determine the design requirements; assume for purposes of this example that the damping ratio is still 5%. Comment on how the increased weight has affected the design requirements.

6.6 A one-story reinforced-concrete building is idealized for structural analysis as a massless frame supporting a dead load of 10 kips at the beam level. The frame is 24 ft wide and 12 ft high. Each column, clamped at the base, has a 10-in.-square cross section. The Young's modulus of concrete is 3×10^3 ksi, and the damping ratio of the building is estimated as 5%. If the building is to be designed for the design spectrum of Fig. 6.9.5 scaled to a peak ground acceleration of 0.5g, determine the design values of lateral deformation and bending moments in the columns for two conditions:

(a) The cross section of the beam is much larger than that of the columns, so the beam may be assumed as rigid in flexure.

(b) The beam cross section is much smaller than the columns, so the beam stiffness can be ignored. Comment on the influence of beam stiffness on the design quantities.

6.7 The columns of the frame of Problem 6.6 with condition (a) (i.e., rigid beam) are hinged at the base. For the same design earthquake, determine the design values of lateral deformation and bending moments on the columns. Comment on the influence of base fixity on the design deformation and bending moments.

6.8 A one-story steel frame of 24-ft span and 12-ft height has the following properties: The second moments of cross-sectional area for beam and columns are $I_b = 160$ in^4 and $I_c = 320$ in^4, respectively; the elastic modulus for steel is 30×10^3 ksi. For purposes of dynamic analysis the frame is considered massless with a mass of 100 kips lumped at the beam level; the damping ratio is estimated at 5%. Determine the peak values of lateral displacement at the beam level and bending moments throughout the frame due to the design spectrum of Fig. 6.9.5 scaled to a peak ground acceleration of 0.5g.

6.9 For the design earthquake at a site, the peak values of ground acceleration, velocity, and displacement have been estimated: $\ddot{u}_{go} = 0.5$g, $\dot{u}_{go} = 24$ in./sec, and $u_{go} = 18$ in. For systems with 2% damping ratio, construct the median design spectrum and median-plus-one-standard-deviation design spectrum.

(a) Plot both spectra, together, on four-way log paper.

(b) Plot the median-plus-one-standard-deviation spectrum for pseudo-acceleration on log-log paper, and determine the equations for $A(T_n)$ for each branch of the spectrum and the period values at the intersections of the branches.

(c) Plot the spectrum of part (b) on a linear-linear graph (the T_n scale should cover the range of 0 to 5 sec).

7

Earthquake Response of Inelastic Systems

PREVIEW

We have shown that the peak base shear induced in a linearly elastic system by ground motion is $V_b = (A/g)w$, where w is the weight of the system and A is the pseudo-acceleration spectrum ordinate corresponding to the natural vibration period and damping of the system (Chapter 6). Most buildings are designed, however, for base shear smaller than the elastic base shear associated with the strongest shaking that can occur at the site. This becomes clear from Fig. 7.1, wherein the base shear coefficient A/g

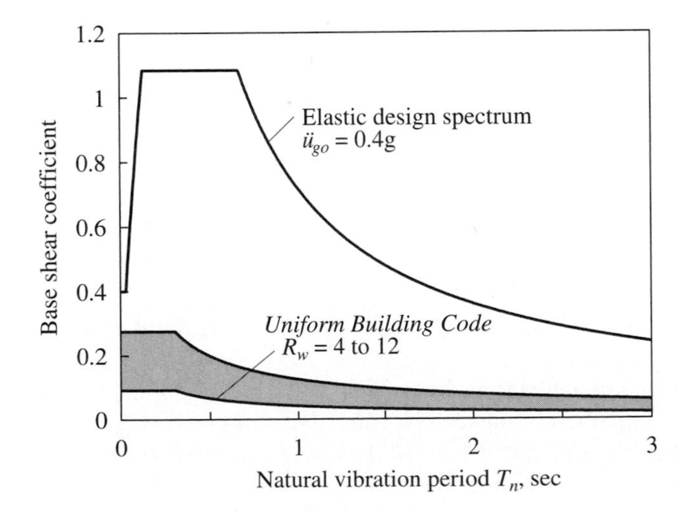

Figure 7.1 Comparison of base shear coefficients from elastic design spectrum and *Uniform Building Code*.

from the design spectrum of Fig. 6.9.5, scaled by 0.4 to correspond to peak ground acceleration of 0.4g, is compared with the base shear coefficient specified in the 1994 *Uniform Building Code*. This disparity implies that buildings designed for the code forces would be deformed beyond the limit of linearly elastic behavior when subjected to ground motions represented by the 0.4g design spectrum. Thus it should not be surprising that buildings suffer damage during intense ground shaking. The challenge to the engineer is to design the structure so that the damage is controlled to an acceptable degree. If an earthquake causes damage that is too severe to be repaired economically (Fig. 7.2) or it causes a building to collapse (Fig. 7.3), the design was obviously not successful.

The response of structures deforming into their inelastic range during intense ground shaking is therefore of central importance in earthquake engineering. This chapter is concerned with this important subject. After introducing the elastoplastic system and the parameters describing the system, the equation of motion is presented and the various parameters describing the system and excitation are identified. Then the earthquake response of elastic and inelastic systems is compared with the objective of understanding how yielding influences structural response. This is followed by a procedure to determine the response spectrum for yield force associated with specified values of the ductility factor, together with a discussion of how the spectrum can be used to determine the design force and deformation for inelastic systems. The chapter closes with a procedure to determine the design spectrum for inelastic systems from the elastic design spectrum, followed by a discussion of the important distinction between design and response spectra.

7.1 FORCE–DEFORMATION RELATIONS

7.1.1 Laboratory Tests

Since the 1960s hundreds of laboratory tests have been conducted to determine the force–deformation behavior of structures for earthquake conditions. During an earthquake structures undergo oscillatory motion with reversal of deformation. Cyclic tests simulating this condition have been conducted on structural members, assemblages of members, reduced-scale models of structures, and on small full-scale structures. The experimental results indicate that the cyclic force–deformation behavior of a structure depends on the structural material (Fig. 7.1.1) and on the structural system. The force–deformation plots show hysteresis loops under cyclic deformations because of inelastic behavior. The shapes of these loops depend on the structural system and materials.

Since the 1960s many computer simulation studies have focused on the earthquake response of SDF systems with their force–deformation behavior defined by idealized versions of experimental curves, such as in Fig. 7.1.1. For this chapter, the simplest of such idealized force–deformation behavior is chosen.

(a)

(b)

Figure 7.2 The six-story Imperial County Services Building was overstrained by the Imperial Valley, California earthquake of October 15, 1979. The building is located in El Centro, California, 9 km from the causative fault of the Magnitude 6.5 earthquake; the peak ground acceleration near the building was 0.23g. The first-story reinforced-concrete columns were overstrained top and bottom with partial hinging. The four columns at the right end were shattered at ground level, which dropped the end of the building about 6 in.; see detail. (Courtesy of G. W. Housner.)

(a)

(b)

Figure 7.3 Psychiatric Day Care Center (a) before and (b) after the San Fernando, California earthquake of February 9, 1971. This Magnitude 6.4 earthquake caused very strong shaking at this site which disintegrated the first story of this building. (Courtesy of G. W. Housner.)

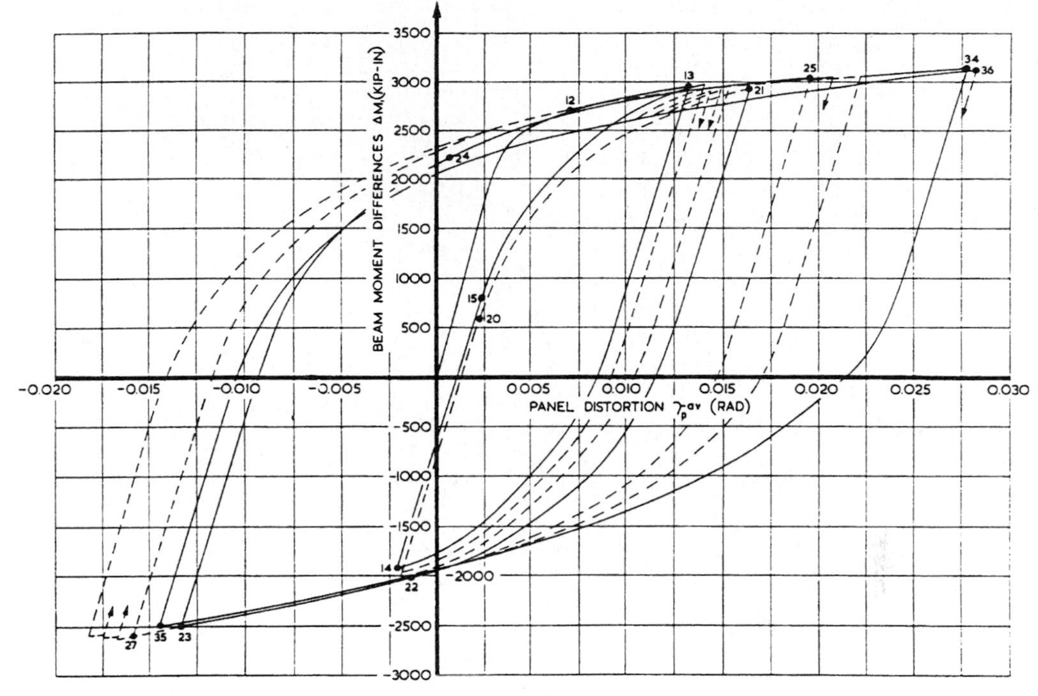

(a)

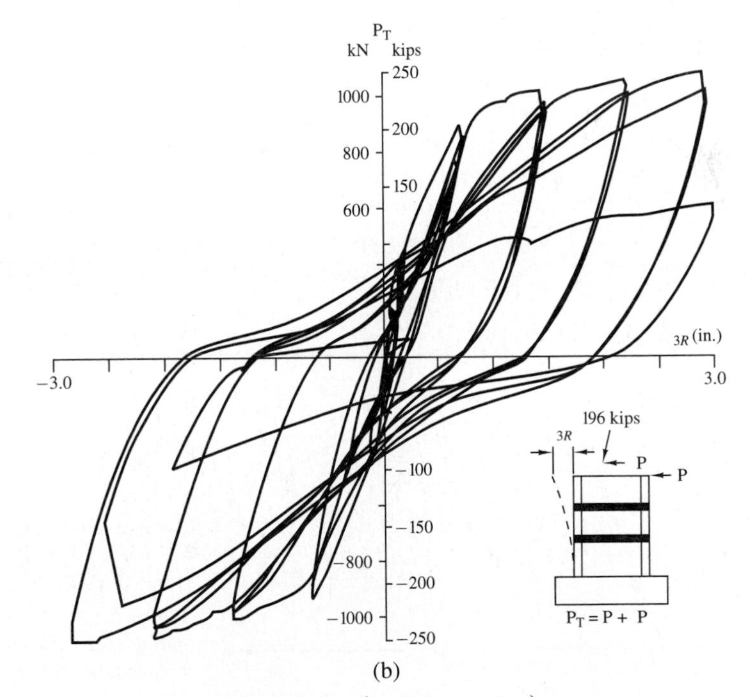

(b)

Figure 7.1.1 (*continues overleaf*).

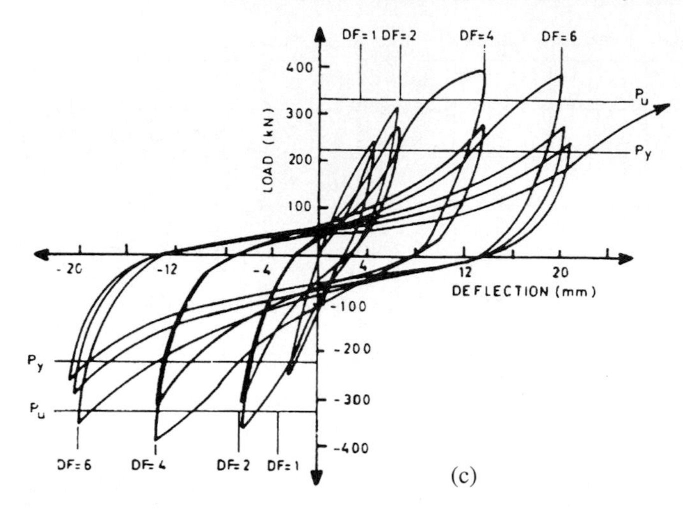

Figure 7.1.1 Force–deformation relations for structural components in different materials: (a) structural steel (from H. Krawinkler, V. V. Bertero, and E. P. Popov, "Inelastic Behavior of Steel Beam-to Column Subassemblages," *Report No. EERC 71-7*, University of California, Berkeley, Calif., 1971); (b) reinforced concrete [from E. P. Popov and V. V. Bertero, "On Seismic Behavior of Two R/C Structural Systems for Tall Buildings," in *Structural and Geotechnical Mechanics* (ed. W. J. Hall), Prentice Hall, Englewood Cliffs, N.J., 1977]; (c) masonry [from M. J. N. Priestley, "Masonry," in *Design of Earthquake Resistant Structures* (ed. Emilio Rosenblueth), Pentech Press, Plymouth, U.K., 1980].

7.1.2 Elastoplastic Idealization

Consider the force–deformation relation for a structure during its initial loading shown in Fig. 7.1.2. It is convenient to idealize this curve by an *elastic–perfectly plastic* (or *elastoplastic* for brevity) force–deformation relation because this approximation permits, as we will see later, the development of response spectra in a manner similar to linearly elastic systems. The elastoplastic approximation to the actual force–deformation curve is

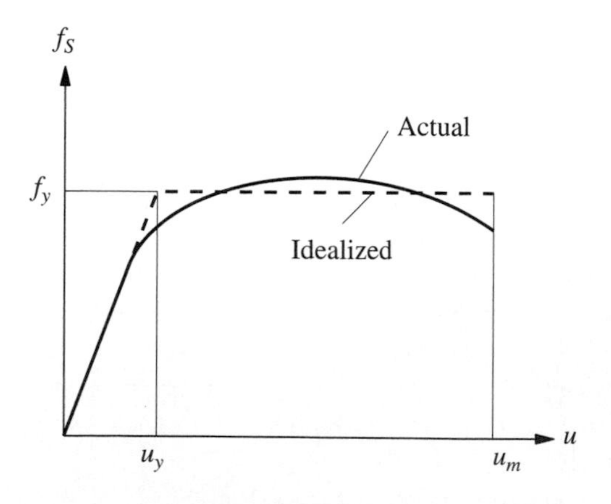

Figure 7.1.2 Force–deformation curve during initial loading: actual and elastoplastic idealization.

drawn, as shown in Fig. 7.1.2, so that the areas under the two curves are the same at the selected value of the maximum displacement u_m. On initial loading this idealized system is linearly elastic with stiffness k as long as the force does not exceed f_y. Yielding begins when the force reaches f_y, the *yield strength*. The deformation at which yielding begins is u_y, the *yield deformation*. Yielding takes place at constant force (i.e., the stiffness is zero).

Figure 7.1.3 shows a typical cycle of loading, unloading, and reloading for an elastoplastic system. The yield force is the same in the two directions of deformation. Unloading from a point of maximum deformation takes place along a path parallel to the initial elastic branch. Similarly, reloading from a point of minimum deformation takes place along a path parallel to the initial elastic branch. The maximum and minimum values of the resisting force for deformations in excess of the yield deformation are f_y. The force–deformation relation is no longer single-valued if the system is unloading or reloading; for deformation u at time t the resisting force f_S depends on the prior history of motion of the system and whether the deformation is currently increasing (velocity $\dot{u} > 0$) or decreasing ($\dot{u} < 0$).

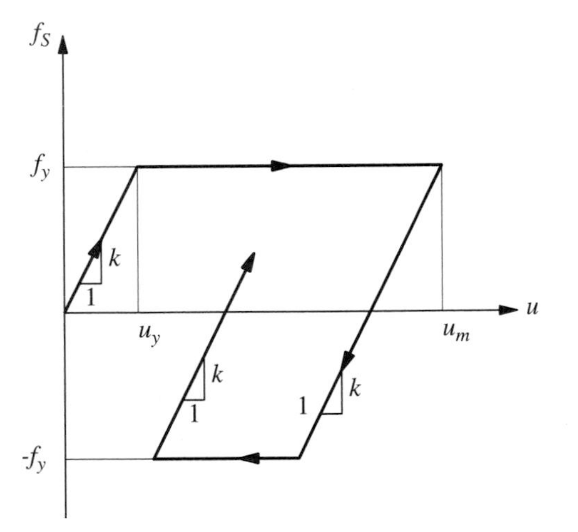

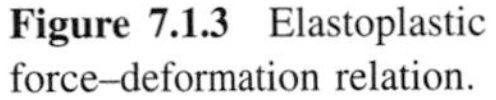
Figure 7.1.3 Elastoplastic force–deformation relation.

7.1.3 Corresponding Linear System

It is desired to evaluate the peak deformation of an elastoplastic system due to earthquake ground motion and to compare this deformation to the peak deformation caused by the same excitation in the *corresponding linear system*. This elastic system is defined to have the same stiffness as the stiffness of the elastoplastic system during its initial loading; see Fig 7.1.4. Both systems have the same mass and damping. Therefore, the natural vibration period of the corresponding linear system is the same as the period of the elastoplastic system undergoing small ($u \leq u_y$) oscillations.

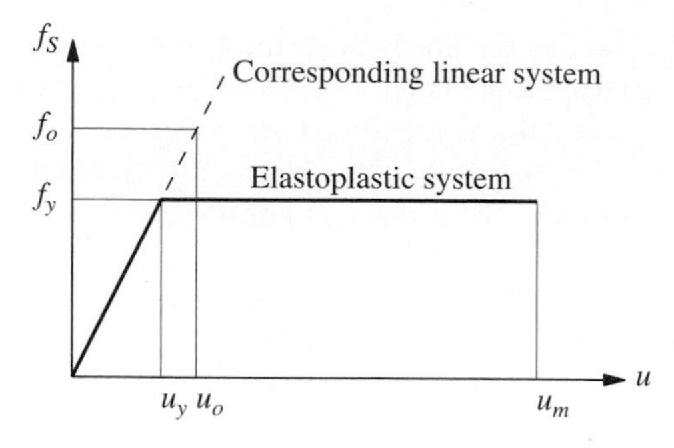

Figure 7.1.4 Elastoplastic system and its corresponding linear system.

7.2 NORMALIZED YIELD STRENGTH, YIELD REDUCTION FACTOR, AND DUCTILITY FACTOR

The *normalized yield strength* $\overline{f}_y$ of an elastoplastic system is defined as

$$\overline{f}_y = \frac{f_y}{f_o} = \frac{u_y}{u_o} \tag{7.2.1}$$

where f_o and u_o are the peak values of the earthquake-induced resisting force and deformation, respectively, in the corresponding linear system. (For brevity the notation f_o has been used instead of f_{So} employed in preceding chapters.) We may interpret f_o as the strength required for the structure to remain within its linearly elastic limit during the ground motion. The second part of Eq. (7.2.1) is obvious because $f_y = ku_y$ and $f_o = ku_o$. If the normalized strength of a system is less than unity, the system will deform beyond its linearly elastic limit (e.g., $\overline{f}_y = 0.5$ implies that the yield strength of the system is one-half of the strength required for the system to remain elastic during the ground motion). The normalized strength of a system that does not deform beyond its linearly elastic limit is equal to unity because such a system can be interpreted as an elastoplastic system with $f_y = f_o$.

Alternatively, f_y can be related to f_o through a *yield reduction factor* R_y defined by

$$R_y = \frac{f_o}{f_y} = \frac{u_o}{u_y} \tag{7.2.2}$$

Obviously, R_y is the reciprocal of $\overline{f}_y$; R_y is equal to 1 for linear systems and is greater then 1 for a system that deforms into the inelastic range. For example, $R_y = 2$ implies that the yield strength of the system is the strength required for the system to remain elastic divided by 2.

The peak, or absolute (without regard to algebraic sign) maximum, deformation of the elastoplastic system due to the ground motion is denoted by u_m. It is meaningful to normalize u_m relative to the yield deformation of the system

$$\mu = \frac{u_m}{u_y} \tag{7.2.3}$$

This dimensionless ratio is called the *ductility factor*. For systems deforming into the inelastic range, by definition, u_m exceeds u_y and the ductility factor is greater than unity.

For the corresponding linear system, the ductility factor is unity if this system is interpreted as an elastoplastic system with $f_y = f_o$. Later, we relate the peak deformations u_m and u_o of the elastoplastic and corresponding linear systems. Their ratio can be expressed as

$$\frac{u_m}{u_o} = \mu \bar{f}_y = \frac{\mu}{R_y} \tag{7.2.4}$$

This equation follows directly from Eqs. (7.2.1) to (7.2.3).

7.3 EQUATION OF MOTION AND CONTROLLING PARAMETERS

The governing equation for an inelastic system, Eq. (1.7.5), is repeated here for convenience:

$$m\ddot{u} + c\dot{u} + f_S(u, \dot{u}) = -m\ddot{u}_g(t) \tag{7.3.1}$$

where the resisting force $f_S(u, \dot{u})$ for an elastoplastic system is shown in Fig. 7.1.3. Equation (7.3.1) will be solved numerically using the procedures of Chapter 5 to determine $u(t)$. The response results presented in this chapter were obtained by the average acceleration method using a time step $\Delta t = 0.02$ sec, which was further subdivided to detect the transition from elastic to plastic branches, and vice versa, in the force–deformation relation (Section 5.7).

To identify the system parameters that influence the deformation response $u(t)$, Eq. (7.3.1) is divided by m to obtain

$$\ddot{u} + 2\zeta\omega_n \dot{u} + \omega_n^2 u_y \tilde{f}_S(u, \dot{u}) = -\ddot{u}_g(t) \tag{7.3.2}$$

where

$$\omega_n = \sqrt{\frac{k}{m}} \qquad \zeta = \frac{c}{2m\,\omega_n} \qquad \tilde{f}_S(u, \dot{u}) = \frac{f_S(u, \dot{u})}{f_y} \tag{7.3.3}$$

The quantity ω_n is the natural frequency ($T_n = 2\pi/\omega_n$ is the natural period) of the inelastic system vibrating within its linearly elastic range (i.e., $u \le u_y$). It is also the natural frequency of the corresponding linear system. Similarly, ζ is the damping ratio of the system based on the critical damping $2m\omega_n$ of the inelastic system vibrating within its linearly elastic range. It is also the damping ratio of the corresponding linear system. The function $\tilde{f}_S(u, \dot{u})$ describes the force–deformation relation in partially dimensionless form, as shown in Fig. 7.3.1a. Equation (7.3.2) indicates that for a given $\ddot{u}_g(t)$, $u(t)$ depends on three system parameters: ω_n, ζ, and u_y, in addition to the form of the force–deformation relation; here the elastoplastic form has been selected.

Equation (7.3.2) is rewritten in terms of $\mu(t) \equiv u(t)/u_y$ in order to identify the parameters that influence the ductility factor μ, Eq. (7.2.3), the peak value of $\mu(t)$. Substituting $u(t) = u_y\,\mu(t)$, $\dot{u}(t) = u_y\,\dot{\mu}(t)$, and $\ddot{u}(t) = u_y\,\ddot{\mu}(t)$ in Eq. (7.3.2) and dividing by u_y gives

$$\ddot{\mu} + 2\zeta\omega_n \dot{\mu} + \omega_n^2 \tilde{f}_S(\mu, \dot{\mu}) = -\omega_n^2 \frac{\ddot{u}_g(t)}{a_y} \tag{7.3.4}$$

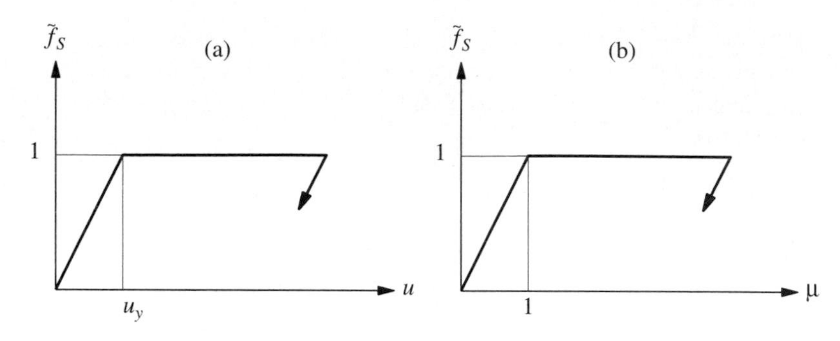

Figure 7.3.1 Force–deformation relations in normalized form.

where $a_y = f_y/m$ may be interpreted as the acceleration of the mass necessary to produce the yield force f_y, and $\tilde{f}_S(\mu, \dot{\mu})$ is shown in Fig. 7.3.1b. The acceleration ratio $\ddot{u}_g(t)/a_y$ is the ratio between the ground acceleration and a measure of the yield strength of the structure. Equation (7.3.4) indicates that doubling the ground accelerations $\ddot{u}_g(t)$ will produce the same response $\mu(t)$ as if the yield strength had been halved.

For a given $\ddot{u}_g(t)$, the ductility factor μ depends on three system parameters: ω_n, ζ, and $\overline{f}_y$; recall that $\overline{f}_y$ is the normalized strength of the elastoplastic system. This can be demonstrated by observing from Eq. (7.3.4) that for a given $\ddot{u}_g(t)$ and form for $\tilde{f}_S(\mu, \dot{\mu})$, say elastoplastic, $\mu(t)$ depends on ω_n, ζ, and a_y. In turn, a_y depends on ω_n, ζ, and $\overline{f}_y$; this can be shown by substituting Eq. (7.2.1) in the definition of $a_y = f_y/m$ to obtain $a_y = \omega_n^2 u_o \overline{f}_y$, and noting that the peak deformation u_o of the corresponding linear system depends on ω_n and ζ.

7.4 EFFECTS OF YIELDING

To understand how the response of SDF systems is affected by inelastic action or yielding, in this section we compare the response of an elastoplastic system to that of its corresponding linear system. The excitation selected is the El Centro ground motion shown in Fig. 6.1.4.

7.4.1 Response History

Figure 7.4.1 shows the response of a linearly elastic system with weight w, natural vibration period $T_n = 0.5$ sec, and no damping. The time variation of deformation shows that the system oscillates about its undeformed equilibrium position and the peak deformation, $u_o = 3.34$ in.; this is also the deformation response spectrum ordinate for $T_n = 0.5$ sec and $\zeta = 0$. Also shown is the time variation of the elastic resisting force f_S; the peak value of this force f_o is given by $f_o/w = 1.37$. In passing, note from Eq. (7.3.1) that for undamped systems, $f_S(t)/w = -\ddot{u}^t(t)/g$; recall that $\ddot{u}^t$ is the total acceleration of the mass. Thus the peak value of this acceleration is $\ddot{u}_o^t = 1.37g$; this is also the acceleration spectrum ordinate for $T_n = 0.5$ sec and $\zeta = 0$.

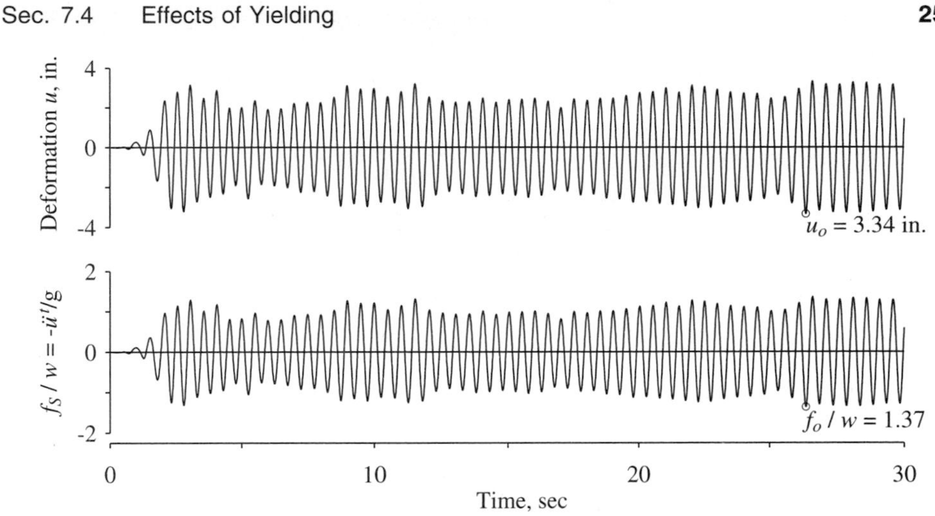

Figure 7.4.1 Response of linear system with $T_n = 0.5$ sec and $\zeta = 0$ to El Centro ground motion.

Figure 7.4.2 shows the response of an elastoplastic system having the same mass and initial stiffness as the linearly elastic system, with normalized strength $\overline{f}_y = 0.125$ (or yield reduction factor $R_y = 8$). The yield strength of this system is $f_y = 0.125 f_o$, where $f_o = 1.37w$ (Fig. 7.4.1); therefore, $f_y = 0.125(1.37w) = 0.171w$. To show more detail, only the first 10 sec of the response is shown in Fig. 7.4.2, which is organized in four parts: (a) shows the deformation $u(t)$; (b) shows the resisting force $f_S(t)$ and acceleration $\ddot{u}^t(t)$; (c) identifies the time intervals during which the system is yielding; and (d) shows the force–deformation relation for one cycle of motion. In the beginning, up to point b, the deformation is small, $f_S < f_y$, and the system is vibrating within its linearly elastic range. We now follow in detail a vibration cycle starting at point a when u and f_S are both zero. At this point the system is linearly elastic and remains so until point b. When the deformation reaches the yield deformation for the first time, identified as b, yielding begins. From b to c the system is yielding (Fig. c), the force is constant at f_y (Fig. b), and the system is on the plastic branch b–c of the force–deformation relation (Fig. d). At c a local maximum of deformation, the velocity is zero, and the deformation begins to reverse (Fig. a); the system begins to unload elastically along c–d (Fig. d) and is not yielding during this time (Fig. c). Unloading continues until point d (Fig. d), when the resisting force reaches zero. Then the system begins to load in the opposite direction and this continues until f_S reaches $-f_y$ at point e (Figs. b and d). Now yielding begins in the opposite direction and continues until point f (Fig. c); $f_S = -f_y$ during this time span (Fig. b) and the system is moving along the plastic branch e–f (Fig. d). At f a local minimum for deformation, the velocity is zero, and the deformation begins to reverse (Fig. a); the system begins to reload elastically along f–g (Fig. d) and is not yielding during this time (Fig. c). Reloading brings the resisting force in the system to zero at g, and it continues along this elastic branch until the resisting force reaches $+f_y$.

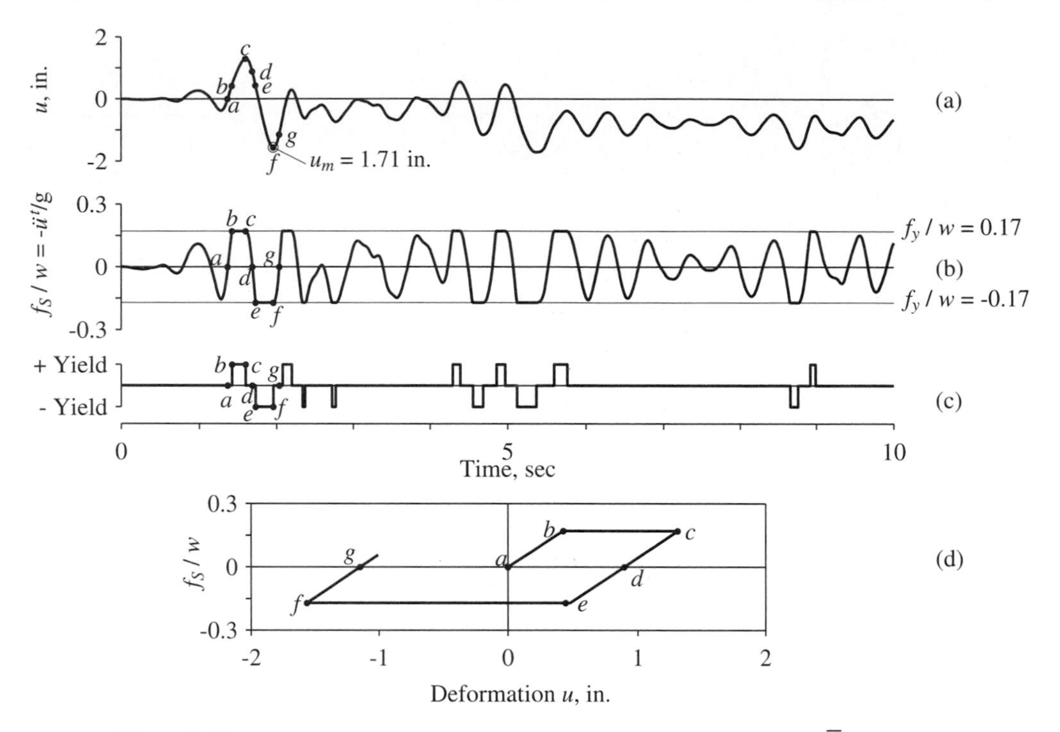

Figure 7.4.2 Response of elastoplastic system with $T_n = 0.5$ sec, $\zeta = 0$, and $\overline{f}_y = 0.125$ to El Centro ground motion: (a) deformation; (b) resisting force and acceleration; (c) time intervals of yielding; (d) force–deformation relation.

The time variation of deformation of the yielding system differs from that of the elastic system. Unlike the elastic system (Fig. 7.4.1), the inelastic system after it has yielded does not oscillate about its initial equilibrium position. Yielding causes the system to drift from its initial equilibrium position, and the system oscillates around a new equilibrium position until this gets shifted by another episode of yielding. Therefore, after the ground has stopped shaking, the system will come to rest at a position different from its initial equilibrium position (i.e., permanent deformation remains). Thus a structure that has undergone significant yielding during an earthquake may not stand exactly straight at the end of the motion. For example, the roof of the pergola shown in Fig. 1.1.1 was displaced by 9 in. relative to its original position at the end of the Caracas, Venezuela earthquake of July 29, 1967; this permanent displacement resulted from yielding of the pipe columns. In contrast, a linear system returns to its initial equilibrium position following the decay of free vibration after the ground has stopped shaking. The peak deformation, 1.71 in., of the elastoplastic system is different from the peak deformation, 3.34 in., of the corresponding linear system (Figs. 7.4.1 and 7.4.2); also, these peak values are reached at different times in the two cases.

We next examine how the response of an elastoplastic system is affected by its yield strength. Consider four SDF systems all with identical properties in their linearly elastic range: $T_n = 0.5$ sec and $\zeta = 5\%$, but they differ in their yield strength: $\overline{f}_y = 1, 0.5, 0.25$, and 0.125. $\overline{f}_y = 1$ implies a linearly elastic system; it is the corresponding linear system for the other three elastoplastic systems. Decreasing values of $\overline{f}_y$ indicate smaller yield strength f_y. The deformation response of these four systems to the El Centro ground motion is presented in Fig. 7.4.3. The linearly elastic system ($\overline{f}_y = 1$) oscillates around its equilibrium position and its peak deformation $u_o = 2.25$ in. The corresponding peak value of the resisting force is $f_o = ku_o$, the force required for a system with $T_n = 0.5$ and $\zeta = 5\%$ to remain elastic during the selected ground motion. The other three systems with yield strength $f_y = 0.5f_o$, $0.25f_o$, and $0.125f_o$, respectively, are therefore expected to deform into the inelastic range. This expectation is confirmed by Fig. 7.4.3, where the time intervals of yielding of these systems are identified. As might be intuitively expected, systems with lower yield strength yield more frequently and for longer

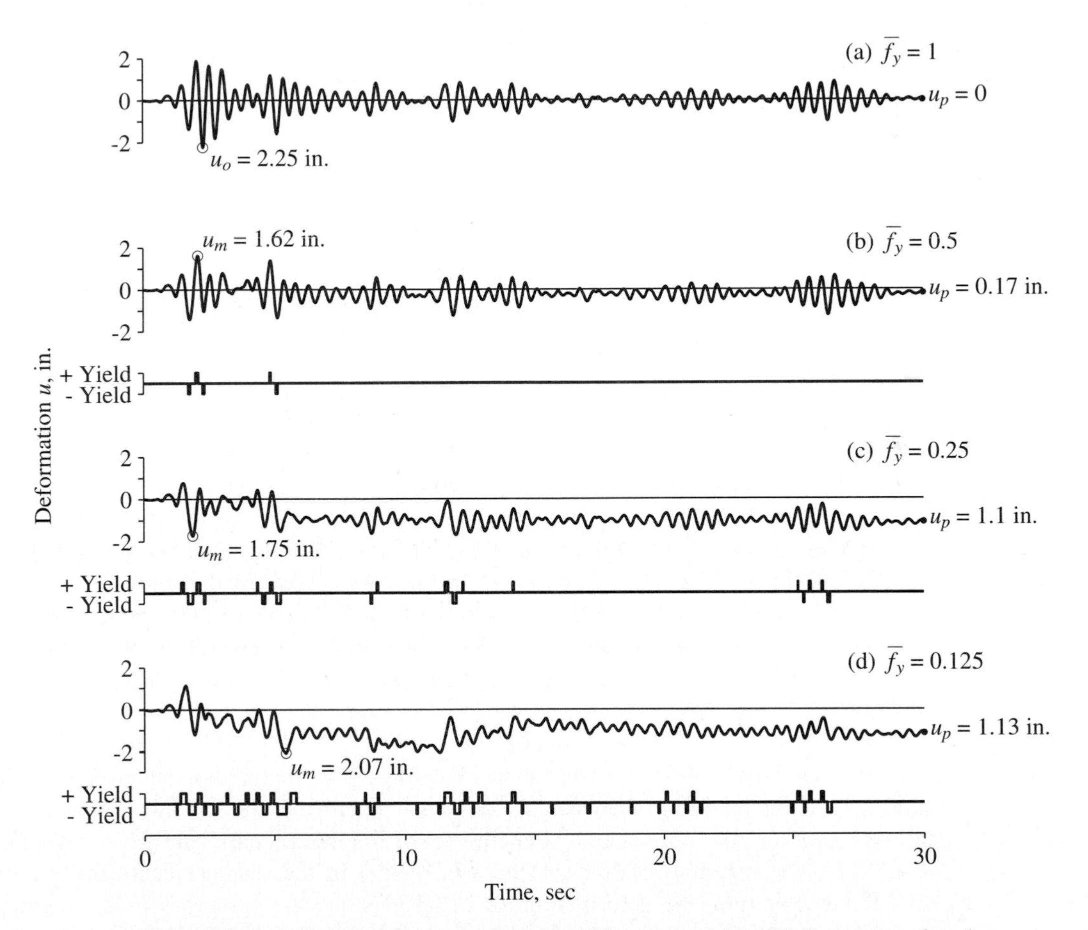

Figure 7.4.3 Deformation response and yielding of four systems due to El Centro ground motion; $T_n = 0.5$ sec, $\zeta = 5\%$; and $\overline{f}_y = 1, 0.5, 0.25$, and 0.125.

intervals. With more yielding, the permanent deformation u_p of the structure after the ground stops shaking tends to increase, but this trend may not be perfect. For the values of T_n and ζ selected, the peak deformations u_m of the three elastoplastic systems are smaller than the peak deformation u_o of the corresponding linear system. This is not always the case, however, because the relative values of u_m and u_o depend on the natural vibration period T_n of the system and the characteristics of the ground motion, and to a lesser degree on the damping in the system.

The ductility factor for an elastoplastic system can be computed using Eq. (7.2.4). For example, the peak deformations of an elastoplastic system with $\bar{f}_y = 0.25$ and the corresponding linear system are $u_m = 1.75$ in. and $u_o = 2.25$ in., respectively. Substituting for u_m, u_o, and $\bar{f}_y$ in Eq. (7.2.4) gives the ductility factor: $\mu = (1.75/2.25)(1/0.25) = 3.11$. This is the *ductility demand* imposed on this elastoplastic system by the ground motion. It represents a requirement on the design of the system in the sense that its *ductility capacity* (i.e., the ability to deform beyond the elastic limit) should exceed the ductility demand.

7.4.2 Ductility Demand, Peak Deformations, and Normalized Yield Strength

In this section we examine how the natural vibration period T_n influences (1) the ductility demand μ on the elastoplastic system; (2) the relative values of the peak deformations u_m and u_o of the elastoplastic system and the corresponding linear system, respectively; and (3) the relative values of the yield strength f_y of the elastoplastic system and the peak force f_o imposed on the elastic system, as defined by the normalized yield strength $\bar{f}_y$ of the elastoplastic system. Figure 7.4.4 is a plot of u_m as a function of T_n for four values of $\bar{f}_y = 1$, 0.5, 0.25, and 0.125; u_o is the same as u_m for $\bar{f}_y = 1$. (Note that u_o and u_m have been divided by the peak ground displacement $u_{go} = 8.4$ in.) In Fig. 7.4.5 the ductility factor μ is plotted versus T_n for the same four values of $\bar{f}_y$; $\mu = 1$ if $\bar{f}_y = 1$. The response histories presented in Fig. 7.4.3 for systems with $T_n = 0.5$ sec and $\zeta = 5\%$ provide the value for $u_o = 2.25$ in., and $u_m = 1.62$, 1.75, and 2.07 in. for $\bar{f}_y = 0.5$, 0.25, and 0.125, respectively. Two of these four data points are identified in Fig. 7.4.4. The ductility demands μ for the three elastoplastic systems are 1.44, 3.11 (computed earlier), and 7.36, respectively. These three data points are identified in Fig. 7.4.5. Also identified in these plots are the period values T_a, T_b, T_c, T_d, T_e, and T_f that define the various spectral regions; these were introduced in Section 6.8.

Figures 7.4.4 and 7.4.5 demonstrate that for a given excitation, the ductility demand and the relationship between u_m and u_o depend on the natural vibration period T_n and on the normalized strength $\bar{f}_y$ or its reciprocal, the yield reduction factor R_y. For very-long-period systems ($T_n > T_f$) in the displacement-sensitive region of the spectrum, the deformation u_m of an elastoplastic system is independent of $\bar{f}_y$ and is essentially equal to the peak deformation of the corresponding linear system. For a fixed mass, such a system is very flexible and, as mentioned in Sec-

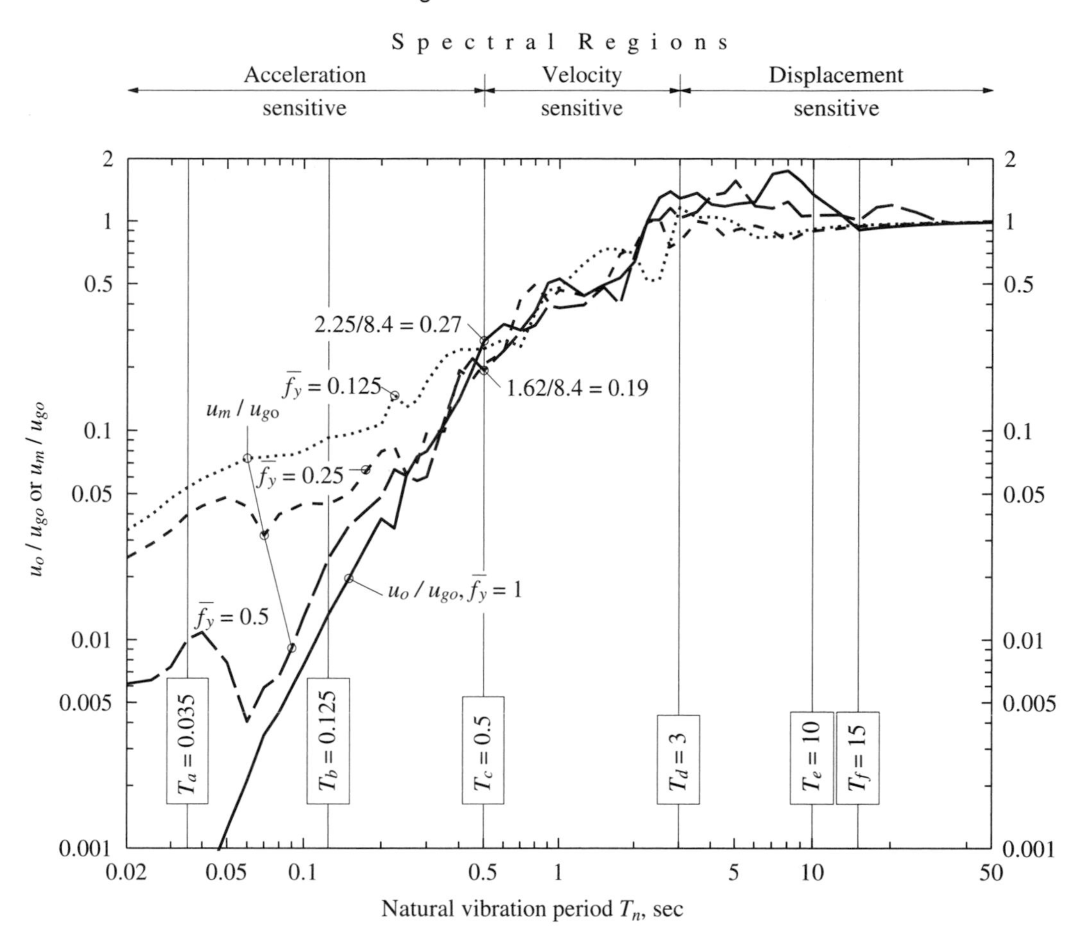

Figure 7.4.4 Peak deformation of elastoplastic systems and corresponding linear system due to El Centro ground motion; T_n is varied; $\zeta = 5\%$ and $\overline{f}_y = 1$, 0.5, 0.25, and 0.125.

tion 6.8, its mass stays still while the ground beneath moves. It experiences a peak deformation equal to the peak ground displacement, independent of $\overline{f}_y$. Thus $u_m \simeq u_o \simeq u_{go}$ and Eq. (7.2.4) gives $\mu \simeq 1/\overline{f}_y$ or $\mu \simeq R_y$, a result confirmed by Fig. 7.4.5. This implies that for a given μ, the design yield strength for an elastoplastic system with $T_n > T_f$ is $1/\mu$ times the strength required for the system to remain elastic.

For systems with T_n in the velocity-sensitive region of the spectrum, u_m may be larger or smaller than u_o; both are affected irregularly by variations in $\overline{f}_y$; the ductility demand μ may be larger or smaller than R_y; and the influence of $\overline{f}_y$, although small, is not negligible.

For systems in the acceleration-sensitive region of the spectrum, u_m is greater than u_o, and u_m increases with decreasing $\overline{f}_y$ (i.e., decreasing yield strength) and decreasing

apparent considering the response results of Fig. 7.4.3 for four systems, all having the same $T_n = 0.5$ sec and $\zeta = 5\%$, but different yield strengths, as defined by the normalized strength $\overline{f}_y = 1$, 0.5, 0.25, and 0.125. The ductility factors for these four systems are 1, 1.44, 3.11, and 7.36 (Section 7.4.2). Clearly, these results do not provide the $\overline{f}_y$ value corresponding to a specified ductility factor, say, 4.

These results provide the basis, however, to obtain the desired information. They lead to a plot showing $\overline{f}_y$ (or R_y) as functions of μ for a fixed T_n and ζ. The solid lines in Fig. 7.5.1 show such plots for several values of T_n and $\zeta = 5\%$. In the plot for $T_n = 0.5$ sec the four pairs of $\overline{f}_y$ and μ values mentioned in the preceding paragraph are identified. To develop some insight into the trends, for each $\overline{f}_y$ two values of the ductility factor are shown: u_m^+/u_y, where u_m^+ is the maximum deformation in the positive direction, and u_m^-/u_y, where u_m^- is the absolute value of the largest deformation in the negative direction. The solid line represents μ, the larger of the two values of the ductility factor.

Contrary to intuition, the ductility factor μ does not always increase monotonically as the normalized strength $\overline{f}_y$ decreases. In particular, more than one yield strength is possible corresponding to a given μ. For example, the plot for $T_n = 2$ sec features two values of $\overline{f}_y$ corresponding to $\mu = 5$. This peculiar phenomenon occurs where the u_m^+/u_y

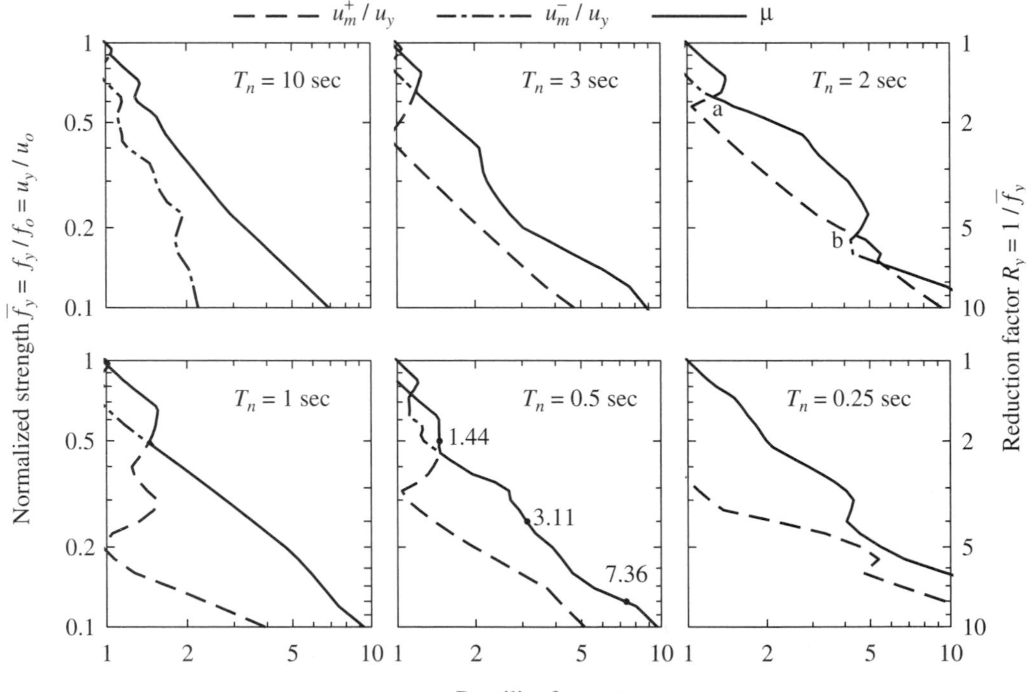

Figure 7.5.1 Relationship between normalized strength (or reduction factor) and ductility factor due to El Centro ground motion; $\zeta = 5\%$.

and u_m^-/u_y curves cross (e.g., points a or b in Fig. 7.5.1). Such a point often corresponds to a local minimum of the ductility factor, which permits more than one value of $\overline{f}_y$ for a slightly larger value of μ. For each μ value, it is the largest $\overline{f}_y$, or the largest yield strength, that is relevant for design.

The yield strength f_y of an elastoplastic system for a specified ductility factor μ can be obtained using the corresponding $\overline{f}_y$ value and Eq. (7.2.1). To ensure accuracy in this $\overline{f}_y$ value it is obtained by an iterative procedure, not from a plot like Fig. 7.5.1. From the available data pairs $(\overline{f}_y, \mu)$ interpolation assuming a linear relation between $\log(\overline{f}_y)$ and $\log(\mu)$ leads to $\overline{f}_y$ corresponding to the specified μ. The response history of the system with this $\overline{f}_y$ is computed to determine the ductility factor. If this is close enough to, say within 1% of, the μ specified, the $\overline{f}_y$ value is considered satisfactory; otherwise, it is modified until satisfactory agreement is reached.

7.5.3 Construction of Constant-Ductility Spectrum

The procedure to construct the response spectrum for elastoplastic systems corresponding to specified levels of ductility factor is summarized as a sequence of steps:

1. Numerically define the ground motion $\ddot{u}_g(t)$.
2. Select and fix the damping ratio ζ for which the spectrum is to be plotted.
3. Select a value for T_n.
4. Determine the response $u(t)$ of the linear system with T_n and ζ equal to the values selected. From $u(t)$ determine the peak deformation u_o and the peak force $f_o = ku_o$. Such results for $T_n = 0.5$ sec are shown in Fig. 7.4.3a.
5. Determine the response $u(t)$ of an elastoplastic system with the same T_n and ζ and yield force $f_y = \overline{f}_y f_o$, with a selected $\overline{f}_y < 1$. From $u(t)$ determine the peak deformation u_m and the associated ductility factor from Eq. (7.2.4). Repeat such an analysis for enough values of $\overline{f}_y$ to develop data points $(\overline{f}_y, \mu)$ covering the ductility range of interest. Such results are shown in Fig. 7.4.3 for $\overline{f}_y =$ 0.5, 0.25, and 0.125, which provide three data points for the $T_n = 0.5$ case in Fig. 7.5.1.
6. **a.** For a selected μ determine the $\overline{f}_y$ value from the results of step 5 using the interpolative procedure described in Section 7.5.2. If more than one $\overline{f}_y$ value corresponds to a particular value of μ, the largest value of $\overline{f}_y$ is chosen.
   b. Determine the spectral ordinates corresponding to the value of $\overline{f}_y$ determined in step 6a. Equation (7.2.1) gives u_y from which D_y, V_y, and A_y can be determined using Eq. (7.5.1). These data provide one point on the response spectrum plots of Figs. 7.5.2 and 7.5.3.
7. Repeat steps 3 to 6 for a range of T_n resulting in the spectrum valid for the μ value chosen in step 6(a).
8. Repeat steps 3 to 7 for several values of μ.

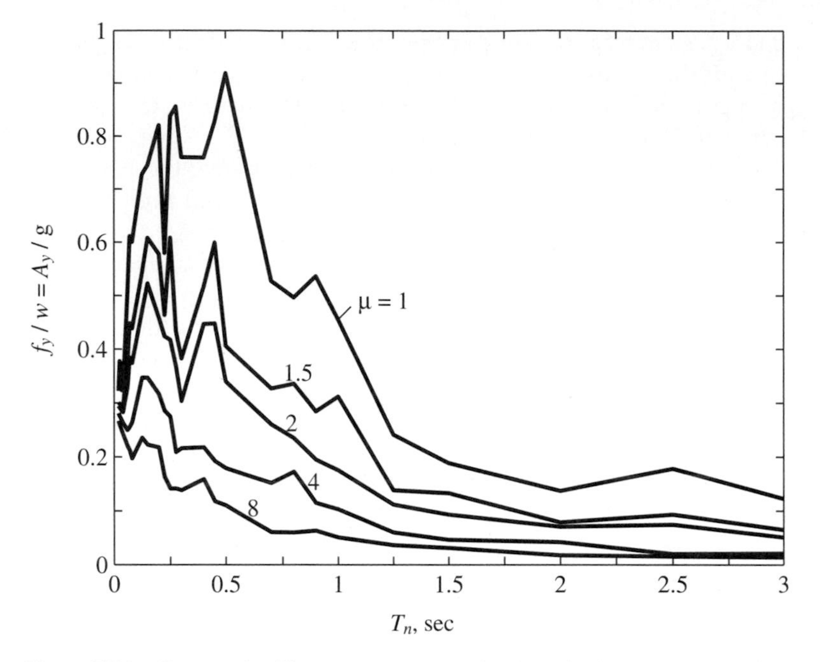

Figure 7.5.2 Constant-ductility response spectrum for elastoplastic systems and El Centro ground motion; $\mu = 1$, 1.5, 2, 4, and 8; $\zeta = 5\%$.

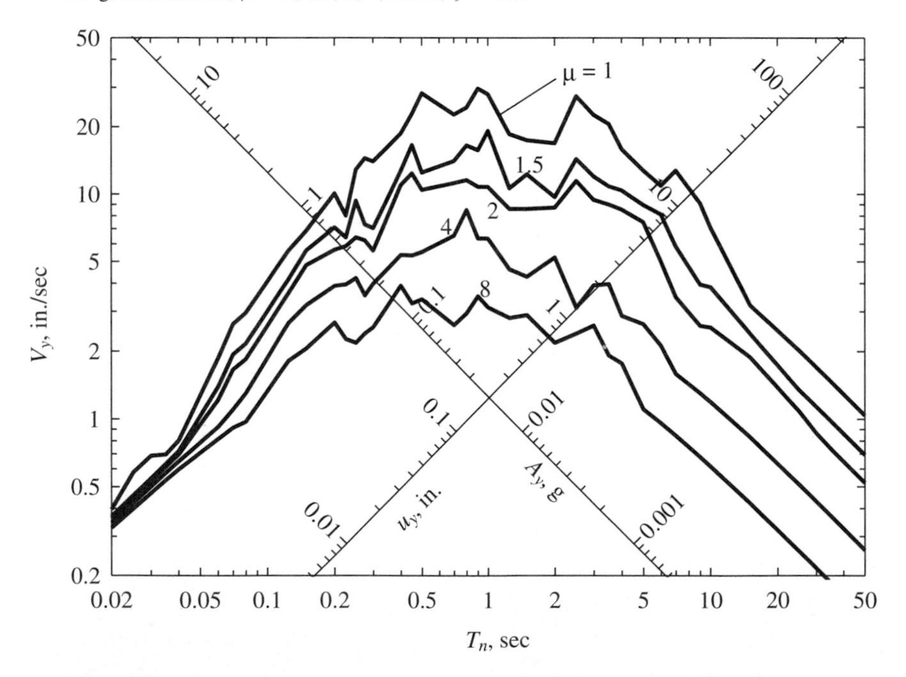

Figure 7.5.3 Constant-ductility response spectrum for elastoplastic systems and El Centro ground motion; $\mu = 1$, 1.5, 2, 4, and 8; $\zeta = 5\%$.

Constructed by this procedure, the response spectrum for elastoplastic systems with $\zeta = 5\%$ subjected to the El Centro ground motion is presented for $\mu = 1$, 1.5, 2, 4, and 8 in two different forms: linear plot of A_y/g versus T_n (Fig. 7.5.2) and a four-way logarithmic plot showing D_y, V_y, and A_y (Fig. 7.5.3).

7.6 DESIGN STRENGTH AND DEFORMATION FROM THE RESPONSE SPECTRUM

Consider an SDF system to be designed for an *allowable ductility* μ, which has been decided based on the allowable deformation and on the ductility capacity that can be achieved for the materials and design details selected. It is desired to determine the design yield strength and the design deformation for the system corresponding to a given excitation, say the El Centro ground motion. Corresponding to the allowable μ and the known values of T_n and ζ, the value of A_y/g is read from the spectrum of Fig. 7.5.2 or 7.5.3. Equation (7.5.3) gives the yield strength f_y necessary to limit the ductility demand to the allowable ductility. The peak deformation

$$u_m = \mu\, u_y \tag{7.6.1}$$

where $u_y = f_y/k = A_y/\omega_n^2$.

The ductility factor μ and the peak deformation u_m represent design requirements associated with the design force f_y. Thus the designer should design and detail the structure to possess the required ductility capacity and deformation capacity. Alternatively, we may consider f_y to be the strength demand imposed by the ground motion on a system with ductility capacity μ, and the designer should provide the strength required.

7.7 DESIGN YIELD STRENGTH

The design yield strength f_y for an SDF system permitted to undergo inelastic deformation is less than the strength required for the structure to remain elastic. Figure 7.5.2 shows that the design yield strength is reduced with increasing values of the ductility factor. Even small amounts of inelastic deformation, corresponding to $\mu = 1.5$, produce a significant reduction in the design force. Additional reductions are achieved with increasing values of μ but at a slower rate.

To study these reductions quantitatively, Fig. 7.7.1 shows the normalized yield strength $\bar{f}_y$ of elastoplastic systems as a function of T_n for four values of μ. This is simply the data of Fig. 7.5.2 (or Fig. 7.5.3) plotted in a different form. From Fig. 7.5.2, for each value of T_n, the $\mu = 1$ curve gives f_o/w and the curve for another μ gives the corresponding f_y/w. The normalized strength $\bar{f}_y$ is then computed from Eq. (7.2.1). For example, consider systems with $T_n = 0.5$ sec; $f_o = 0.919w$ and $f_y = 0.179w$ for $\mu = 4$; the corresponding $\bar{f}_y = 0.195$. Such computations for $\mu = 1$, 1.5, 2, 4, and 8 give $\bar{f}_y = 1$, 0.442, 0.370, 0.195, and 0.120 (or 100, 44.2, 37.0, 19.5, and 12.0%), respectively; three of these data points are identified in Fig. 7.7.1. Repeating

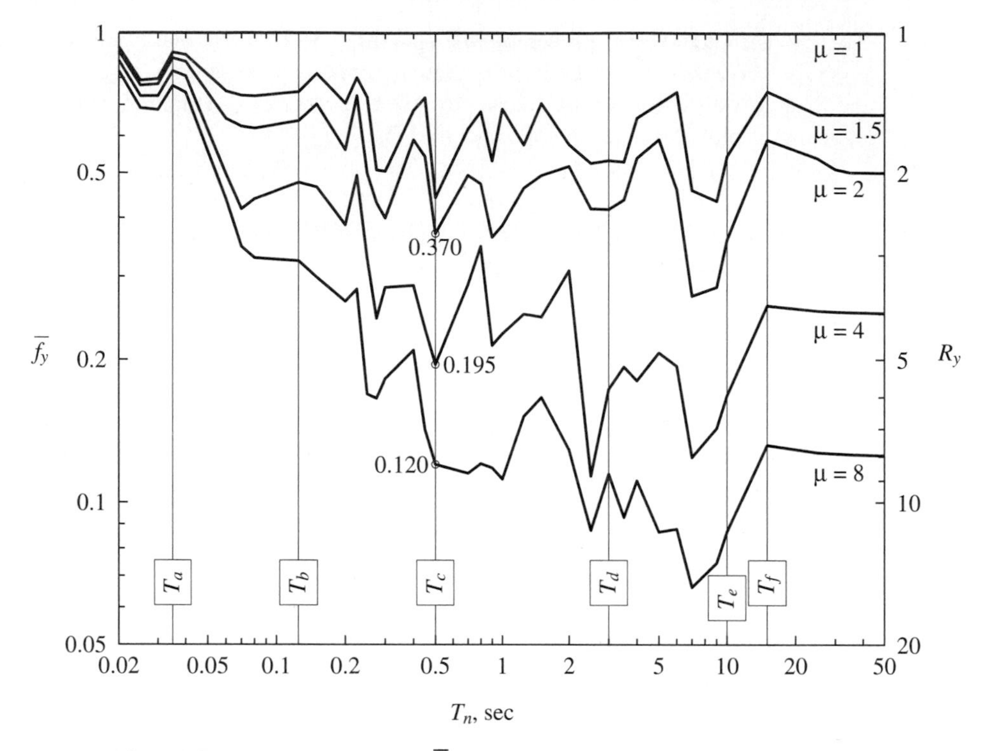

Figure 7.7.1 Normalized strength $\overline{f_y}$ of elastoplastic systems as a function of natural vibration period T_n for $\mu = 1$, 1.5, 2, 4, and 8; $\zeta = 5\%$; El Centro ground motion.

such computations for a range of T_n leads to Fig. 7.7.1, wherein the period values T_a, T_b, T_c, T_d, T_e, and T_f that define the various spectral regions are identified; these were introduced in Section 7.4.

The practical implication of these results is that a structure may be designed for earthquake resistance by making it strong, by making it ductile, or by designing it for economic combinations of both properties. Consider again an SDF system with $T_n =$ 0.5 sec and $\zeta = 5\%$ to be designed for the El Centro ground motion. If this system is designed for a strength $f_o = 0.919w$ or larger, it will remain within its linearly elastic range during this excitation; therefore, it need not be ductile. On the other hand, if it can develop a ductility factor of 8, it need be designed for only 12% of the strength f_o required for elastic behavior. Alternatively, it may be designed for strength equal to 37% of f_o and a ductility capacity of 2; or strength equal to 19.5% of f_o and a ductility capacity of 4. For some types of materials and structural members, ductility is difficult to achieve, and economy dictates designing for large lateral forces; for others, providing ductility is much easier than providing lateral strength and the design practice reflects this.

The strength reduction permitted for a specified allowable ductility varies with T_n. As shown in Fig. 7.7.1, the normalized strength $\overline{f_y}$ tends to 1, implying no reduction, at the short-period end of the spectrum; and to $\overline{f}_y = 1/\mu$ at the long-period end of the spectrum. In between, $\overline{f}_y$ determined for a single ground motion varies in an irregular manner. However, smooth curves can be developed for design purposes (Section 7.10).

The normalized strength for a specified ductility factor also depends on the damping ratio ζ, but this dependence is not strong. It is usually ignored, therefore, in design applications.

7.8 RELATIVE EFFECTS OF YIELDING AND DAMPING

Figure 7.8.1 shows the response spectra for linearly elastic systems for three values of viscous damping: $\zeta = 2$, 5, and 10%. For the same three damping values, response spectra for elastoplastic systems are presented for two different ductility factors: $\mu = 4$ and $\mu = 8$. From these results the relative effects of yielding and damping are identified in this section.

The effects of yielding and viscous damping are similar in one sense but different in another. They are similar in the sense that both mechanisms reduce the pseudo-acceleration A_y and hence the peak value of the lateral force for which the system should be designed. The relative effectiveness of yielding and damping is quite different,

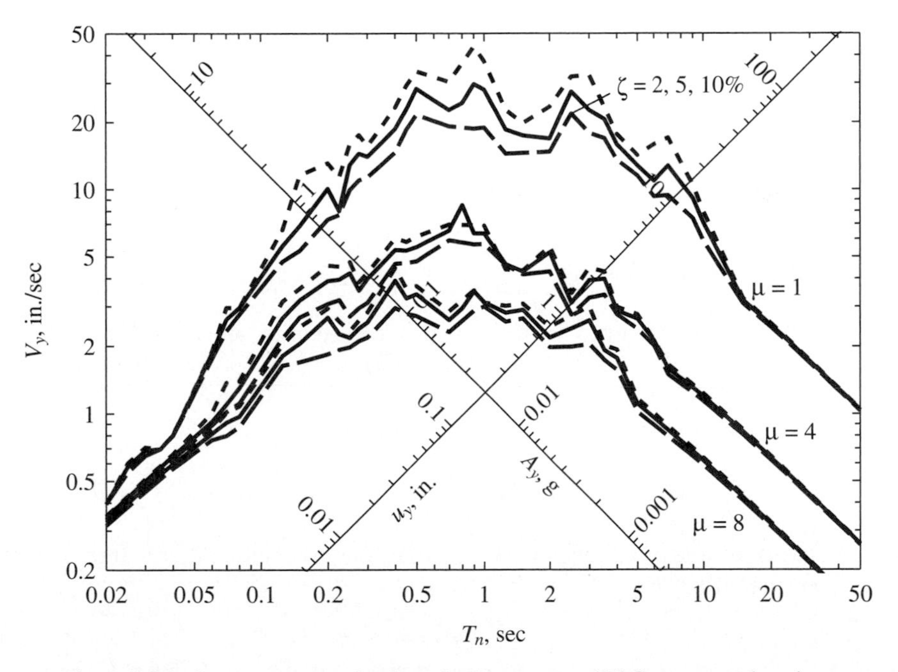

Figure 7.8.1 Response spectra for elastoplastic systems and El Centro ground motion; $\zeta = 2$, 5, and 10% and $\mu = 1$, 4, and 8.

however, in the various spectral regions:

1. Damping has negligible influence on the response of systems with $T_n > T_f$ in the displacement-sensitive region of the spectrum, whereas for such systems the effects of yielding on the design force are very important, but on the peak deformation u_m they are negligible (Fig. 7.4.4).

2. Damping has negligible influence in the response of systems with $T_n < T_a$ in the acceleration-sensitive region of the spectrum, whereas for such systems the effects of yielding on the peak deformation and ductility demand are very important, but on the design force they are small (Figs. 7.4.4 and 7.4.5). In the limit as T_n tends to zero, the pseudo-acceleration A or A_y will approach the peak ground acceleration, implying that this response parameter is unaffected by damping or yielding.

3. Damping is most effective in reducing the response of systems with T_n in the velocity-sensitive region of the spectrum, where yielding is even more effective.

Thus, in general, the effects of yielding cannot be considered in terms of a fixed amount of equivalent viscous damping. If this were possible, the peak response of inelastic systems could be determined directly from the response spectrum for linearly elastic systems, which would have been convenient.

The effectiveness of damping in reducing the response is smaller for inelastic systems and decreases as inelastic deformations increase (Fig. 7.8.1). For example, averaged over the velocity-sensitive spectral region, the percentage response reduction resulting from increasing the damping ratio from 2% to 10% for systems with $\mu = 4$ is about one-half of the reduction for linearly elastic systems. Thus the added viscoelastic dampers mentioned in Section 6.8 may be less beneficial in reducing the response of inelastic systems compared to elastic systems.

7.9 DISSIPATED ENERGY

The input energy imparted to an inelastic system by an earthquake is dissipated by both viscous damping and yielding. These energy quantities are defined and discussed in this section. The various energy terms can be defined by integrating the equation of motion of an inelastic system, Eq. (7.3.1), as follows:

$$\int_0^u m\ddot{u}(t)\, du + \int_0^u c\dot{u}(t)\, du + \int_0^u f_S(u, \dot{u})\, du = -\int_0^u m\ddot{u}_g(t)\, du \qquad (7.9.1)$$

The right side of this equation is the total energy input to the structure since the earthquake excitation began:

$$E_I(t) = -\int_0^u m\ddot{u}_g(t)\, du \qquad (7.9.2)$$

This is clear by noting that as the structure moves through an increment of displacement

du, the energy supplied to the structure by the effective force $p_{\text{eff}}(t) = -m\ddot{u}_g(t)$ is

$$dE_I = -m\ddot{u}_g(t)\,du$$

The first term on the left side of Eq. (7.9.1) is the kinetic energy of the mass associated with its motion relative to the ground:

$$E_K(t) = \int_0^u m\ddot{u}(t)\,du = \int_0^{\dot{u}} m\dot{u}(t)\,d\dot{u} = \frac{m\dot{u}^2}{2} \tag{7.9.3}$$

The second term on the left side of Eq. (7.9.1) is the energy dissipated by viscous damping, defined earlier in Section 3.8:

$$E_D(t) = \int_0^u f_D(t)\,du = \int_0^u c\dot{u}(t)\,du \tag{7.9.4}$$

The third term on the left side of Eq. (7.9.1) is the sum of the energy dissipated by yielding and the recoverable strain energy of the system:

$$E_S(t) = \frac{[f_S(t)]^2}{2k} \tag{7.9.5}$$

where k is the initial stiffness of the inelastic system. Thus the energy dissipated by yielding is

$$E_Y(t) = \int_0^u f_S(u,\,\dot{u})\,du - E_S(t) \tag{7.9.6}$$

Based on these energy quantities, Eq. (7.9.1) is a statement of energy balance for the system:

$$E_I(t) = E_K(t) + E_D(t) + E_S(t) + E_Y(t) \tag{7.9.7}$$

Concurrent with the earthquake response analysis of a system these energy quantities can be computed conveniently by rewriting the integrals with respect to time. Thus

$$\begin{aligned} E_D(t) &= \int_0^t c[\dot{u}(t)]^2\,dt \\ E_Y(t) &= \left[\int_0^t \dot{u}\,f_S(u,\,\dot{u})\,dt\right] - E_S(t) \end{aligned} \tag{7.9.8}$$

The kinetic energy E_K and strain energy E_S at any time t can be computed conveniently from Eqs. (7.9.3) and (7.9.5), respectively.

The foregoing energy analysis is for a structure whose mass is acted upon by a force $-m\ddot{u}_g(t)$, not for a structure whose base is excited by acceleration $\ddot{u}_g(t)$. Therefore, the kinetic energy terms in Eq. (7.9.1) represent the energy of motion relative to the base rather than that due to the total motion. As it is the relative displacements and velocities that cause forces in a structure, an energy equation expressed in terms of the relative motion is more meaningful than one expressed in terms of absolute velocities and displacement. The energy dissipated in viscous damping or yielding depends only on the relative motion, however.

Shown in Fig. 7.9.1 is the variation of these energy quantities with time for two SDF systems subjected to the El Centro ground motion. The results presented are for a linearly elastic system with natural period $T_n = 0.5$ sec and damping ratio $\zeta = 0.05$, and for an elastoplastic system with the same properties in the elastic range and normalized strength $\bar{f}_y = 0.25$. Recall that the deformation response of these two systems was presented in Fig. 7.4.3.

The results of Fig. 7.9.1 show that eventually the structure dissipates by viscous damping and yielding all the energy supplied to it. This is indicated by the fact that

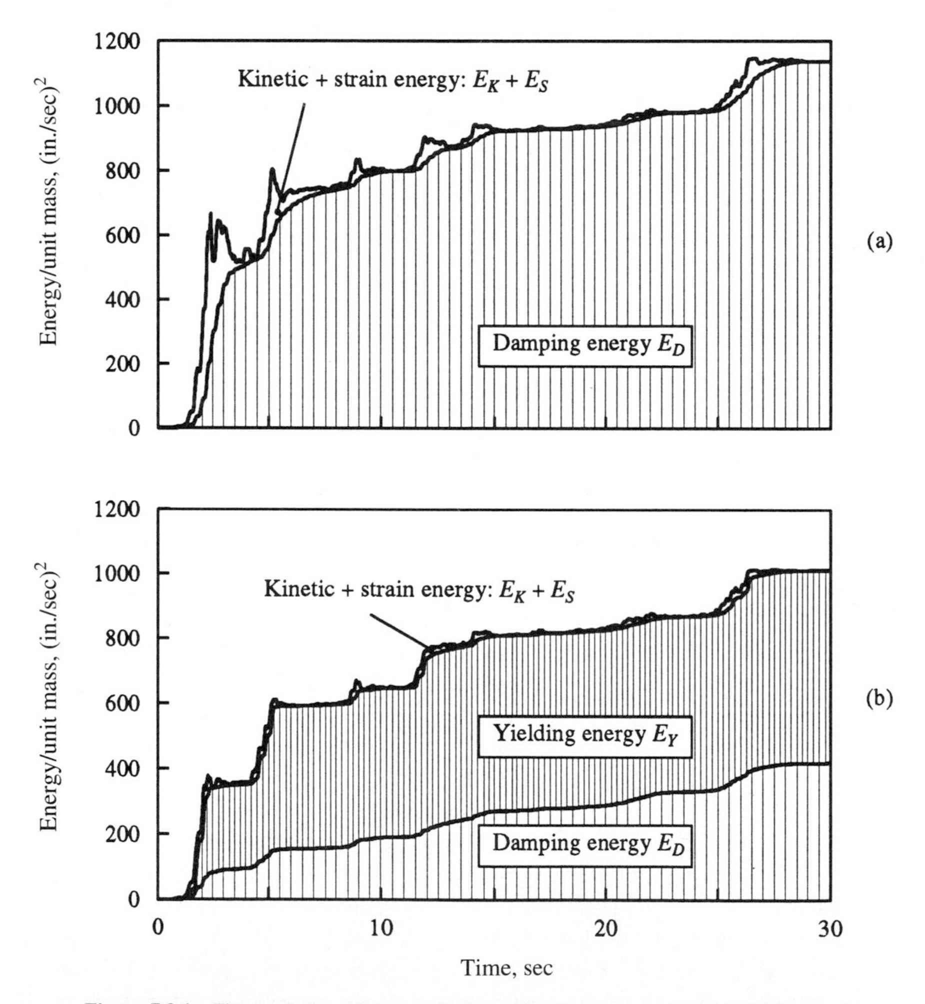

Figure 7.9.1 Time variation of energy dissipated by viscous damping and yielding, and of kinetic plus strain energy; (a) linear system, $T_n = 0.5$ sec, $\zeta = 5\%$; (b) elastoplastic system, $T_n = 0.5$ sec, $\zeta = 5\%$, $\overline{f_y} = 0.25$.

the kinetic energy and recoverable strain energy diminish near the end of the ground shaking. Viscous damping dissipates less energy from the inelastic system, implying smaller velocities relative to the elastic system. Figure 7.9.1 also indicates that the energy input to a linear system and to an inelastic system, both with the same T_n and ζ, is not the same. Furthermore, the input energy varies with T_n for both systems.

The yielding energy shown in Fig. 7.9.1b indicates a demand imposed on the structure. If this much energy can be dissipated through yielding of the structure, it needs to be designed only for $\bar{f}_y = 0.25$ (i.e., one-fourth the force developed in the corresponding linear system). The repeated yielding that dissipates energy causes damage to the structure, however, and leaves it in a permanently deformed condition at the end of the earthquake.

If part of this energy could be dissipated through special devices, which can easily be replaced, as necessary, after an earthquake, the structural damage could be reduced. Various types of energy-dissipating devices, utilizing friction as means of energy dissipation, have been developed and tested by researchers. They increase the capacity of the structure to dissipate energy but do not change the natural periods significantly. One of these devices is the slotted bolted connection (SBC). Figure 7.9.2 includes a schematic diagram of an SBC, the resulting almost rectangular hysteresis loop, the SBC connected to the top of Chevron braces, and a test structure with 12 SBCs. Such devices may be useful in the design of new buildings to reduce construction costs. They can also be added to existing buildings that are deficient in their earthquake resistance.

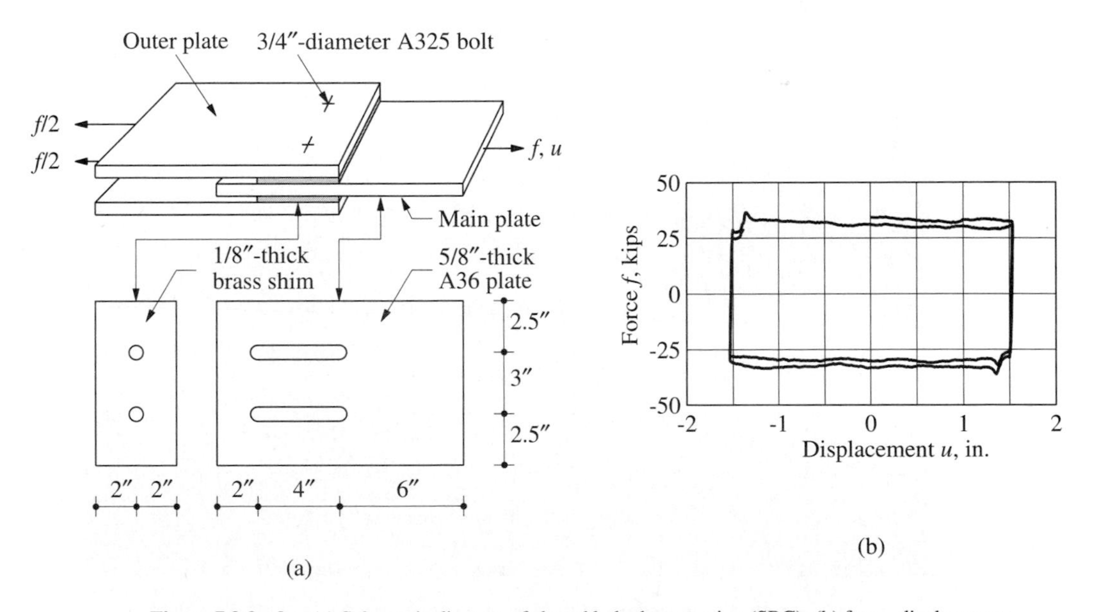

Figure 7.9.2a, b (a) Schematic diagram of slotted bolted connection (SBC); (b) force–displacement diagram of an SBC. (Adapted from C. E. Grigorian and E. P. Popov, 1994.)

(c)

(d)

Figure 7.9.2c, d (c) SBC at top of Chevron brace in test structure; (d) test structure with 12 SBCs on the shaking table at the University of California at Berkeley. (Courtesy of C. E. Grigorian and E. P. Popov.)

7.10 INELASTIC DESIGN SPECTRUM

In this section a procedure is presented for constructing the design spectrum for elastoplastic systems for specified ductility factors. This could be achieved by constructing the constant-ductility response spectrum (Section 7.5.3) for many plausible ground motions for the site and, based on these data, the design spectrum associated with an exceedance probability could be established. This procedure is computationally too demanding to be appropriate for practical application except where it is warranted by the exceptional importance and safety requirements of a structure.

A simpler approach is to develop a constant-ductility design spectrum by multiplying the elastic design spectrum (from Section 6.9) by the normalized strength $\overline{f}_y$ of the elastoplastic system. The variation of $\overline{f}_y$ with T_n for one ground motion was presented in Fig. 7.7.1, and researchers have obtained such results for many ground motions. Based on statistical analysis of these data, several different proposals exist for the variation of $\overline{f}_y$ with T_n for different values of μ. One of the early simpler proposals is shown in Fig. 7.10.1[†]:

$$\overline{f}_y = \begin{cases} 1 & T_n < T_a \\ (2\mu - 1)^{-1/2} & T_b < T_n < T_c \\ \mu^{-1} & T_n > T_c \end{cases} \tag{7.10.1}$$

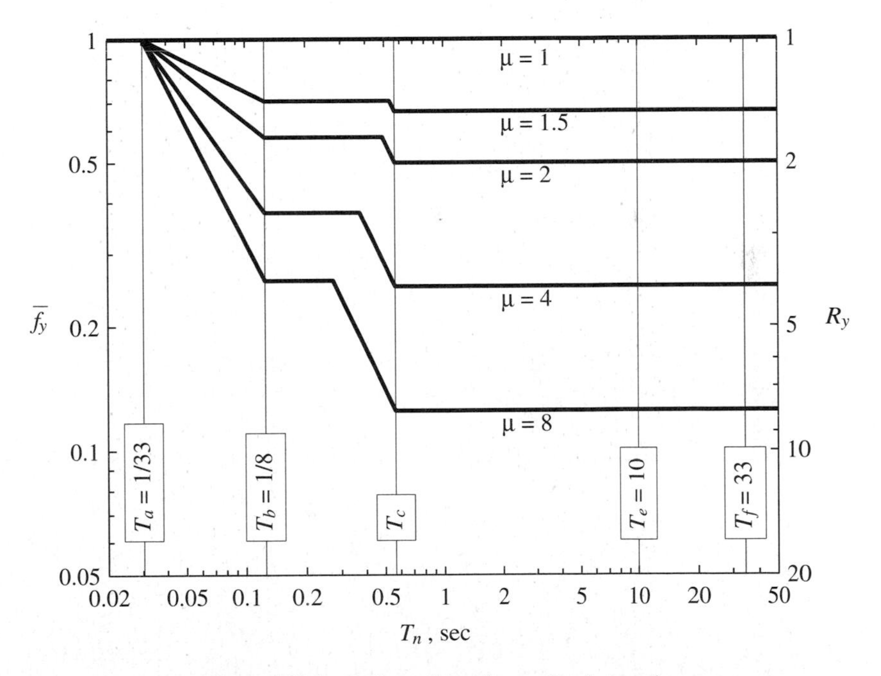

Figure 7.10.1 Design values of normalized strength.

[†]Although there is little rational basis for doing so, Eq. (7.10.1) for $T_b < T_n < T_c$ can be derived by equating the areas under the force–deformation relations for elastic and elastoplastic systems (Fig. 7.1.4).

The sloping straight lines shown in Fig. 7.10.1 provide transition among the three constant segments. For ground motions on firm ground, $T_a = \frac{1}{33}$ sec, $T_b = \frac{1}{8}$ sec, $T_e = 10$ sec, and $T_f = 33$ sec (Fig. 6.9.3); T_c depends on ζ.

It is presumed that the elastic design spectrum a–b–c–d–e–f shown in Fig. 7.10.2 has been determined by the procedure described in Section 6.9. From this elastic design spectrum, for a chosen value of ductility factor μ, the inelastic design spectrum a'–b'–c'–d'–e'–f' shown in Fig. 7.10.2 is obtained as follows:

1. Multiply the constant-A ordinate of segment b–c by $\overline{f}_y = (2\mu - 1)^{-1/2}$ to locate the segment b'–c'.
2. Multiply the constant-V ordinate of segment c–d by $\overline{f}_y = \mu^{-1}$ to locate the segment c'–d'.
3. Multiply the constant-D ordinate of segment d–e by $\overline{f}_y = \mu^{-1}$ to locate the segment d'–e'.
4. Multiply the ordinate at f by $\overline{f}_y = \mu^{-1}$ to locate f'. Join points f' and e'. Draw $D_y = \mu^{-1} u_{go}$ for $T_n > 33$ sec.
5. Take the ordinate a' of the inelastic spectrum at $T_n = \frac{1}{33}$ sec as equal to that of point a of the elastic spectrum. This is equivalent to $\overline{f}_y = 1$. Join points a' and

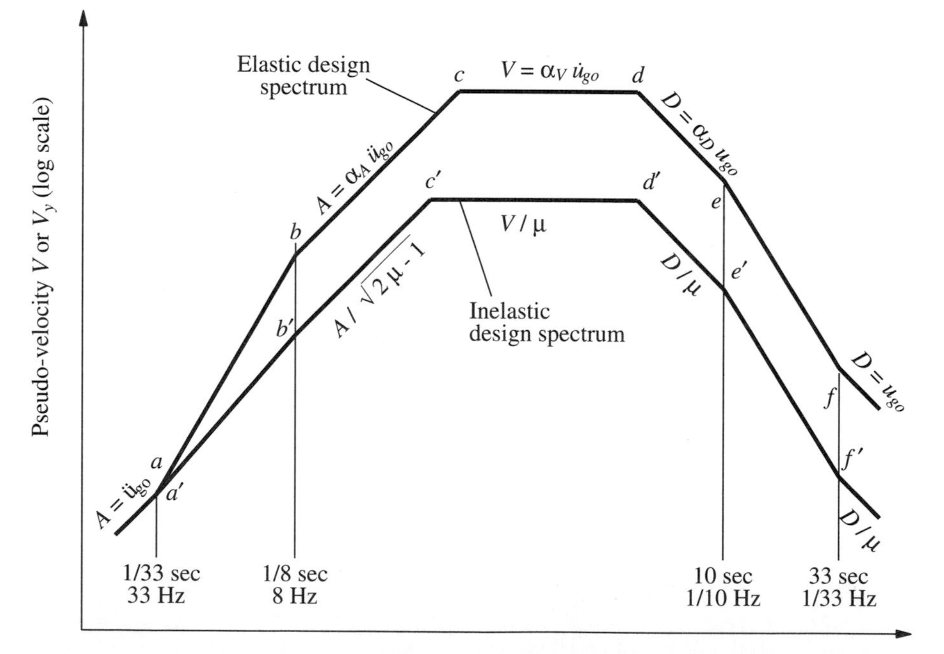

Figure 7.10.2 Construction of inelastic design spectrum.

b'. For a large value of the ductility factor the A_y ordinate for point b' may turn out to be lower than the peak ground acceleration $\ddot{u}_{go}$ (point a). In this case it is more appropriate to join a' and c' directly.

6. Draw $A_y = \ddot{u}_{go}$ for $T_n < \frac{1}{33}$ sec.

The period values associated with points a', b', e', and f' are fixed, as shown in Fig. 7.10.2, at the same values as the corresponding points of the elastic spectrum. The period values associated with points c and d of the elastic spectrum are determined by the amplification factors α_A, α_V, and α_D, which depend on damping. The period values for points c' and d' depend on the $\overline{f}_y$ values used to reduce the segments b–c, c–d, and d–e of the elastic design spectrum. With the selected $\overline{f}_y$ values of Eq. (7.10.1) for the three spectral regions, respectively, the period value for point d' is the same as for d but the period value for point c' is different than for c.

Consider ground motions on firm ground with peak acceleration $\ddot{u}_{go} = 1$g, peak velocity $\dot{u}_{go} = 48$ in./sec, and peak displacement $u_{go} = 36$ in. The median-plus-1σ spectrum is desired for elastoplastic systems with damping ratio $\zeta = 5\%$ for ductility factors of $\mu = 2$ and 8. The design spectrum for elastic systems with $\zeta = 5\%$ and the ground motion selected was presented in Fig. 6.9.4 and is reproduced in Fig. 7.10.3. The

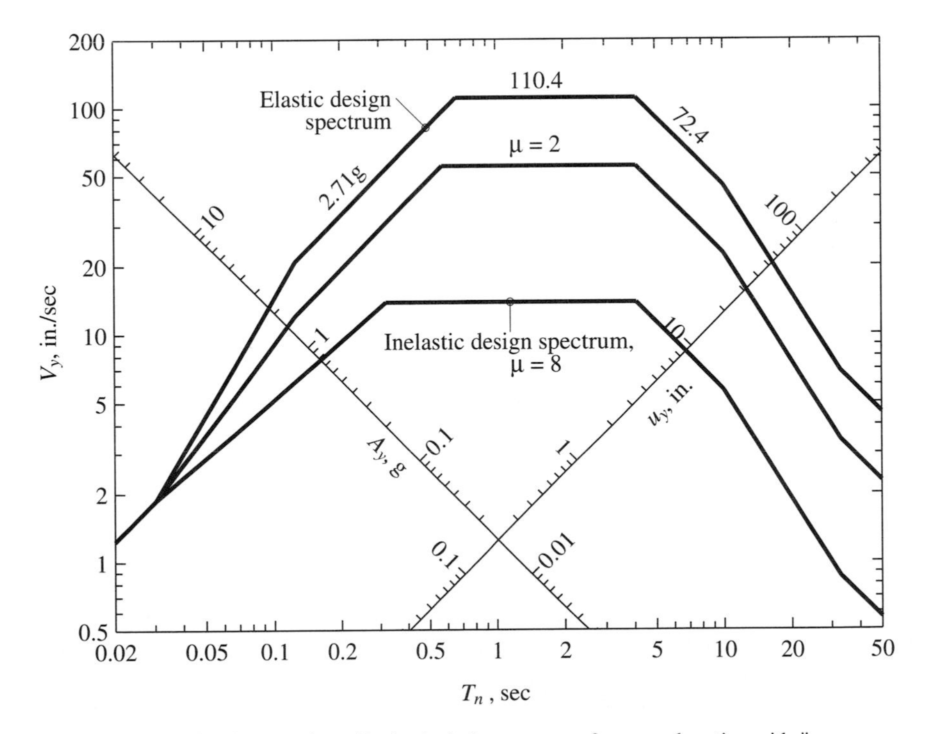

Figure 7.10.3 Construction of inelastic design spectrum for ground motion with $\ddot{u}_{go} = 1$g, $\dot{u}_{go} = 48$ in./sec, and $u_{go} = 36$ in.; $\mu = 2$ and 8; $\zeta = 5\%$.

Solution The system with $T_n = 0.25$ sec is on the constant-A_y branch of the design spectrum for which Eqs. (7.10.3) apply. For an elastic system with $T_n = 0.25$ sec, $A = (2.71\text{g})\,0.5 = 1.355\text{g}$. Then

$$f_o = \frac{A}{g}w = 1.355w \qquad u_o = \frac{A}{\omega_n^2} = 0.83\text{ in.}$$

Substituting for f_o and u_o in Eq. (7.10.3) gives the following results for $\mu = 1$, 4, and 8.

μ	f_y/w	u_m (in.)
1	1.355	0.83
4	0.512	1.25
8	0.350	1.71

7.11 COMPARISON OF DESIGN AND RESPONSE SPECTRA

In this section the design spectrum presented in Section 7.10 is compared with the response spectrum for elastoplastic systems. Such a comparison for elastic systems was presented in Section 6.10, and the data presented in Figs. 6.10.1 and 6.10.2 are reproduced in Figs. 7.11.1 and 7.11.2. Also included now are (1) the inelastic design spectrum for

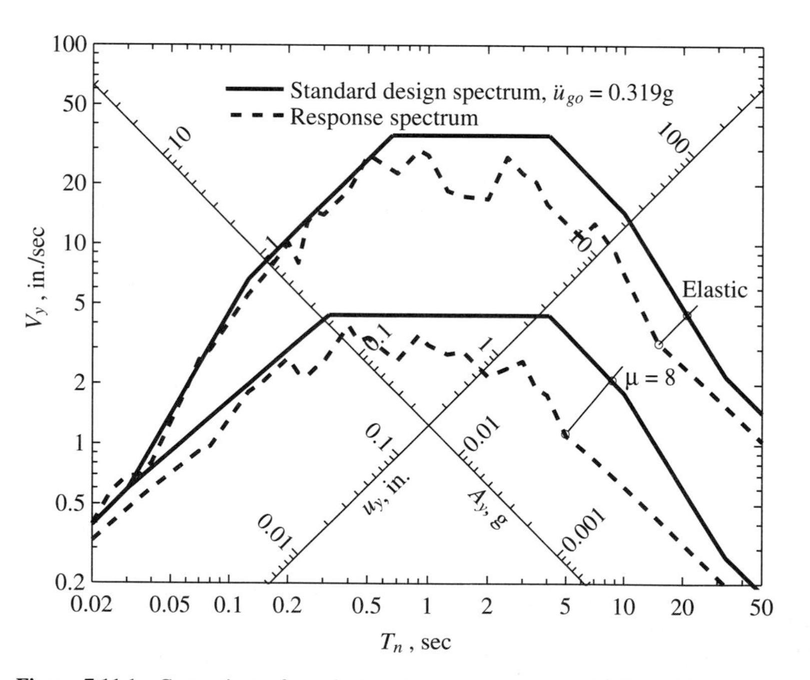

Figure 7.11.1 Comparison of standard design spectrum ($\ddot{u}_{go} = 0.319$g) with response spectrum for El Centro ground motion; $\mu = 1$ and 8; $\zeta = 5\%$.

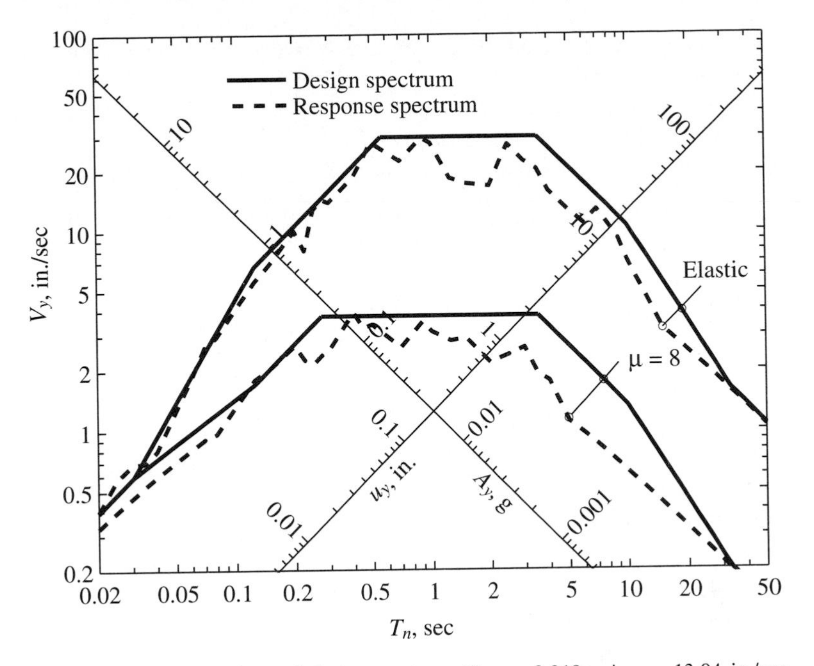

Figure 7.11.2 Comparison of design spectrum ($\ddot{u}_{go} = 0.319$g, $\dot{u}_{go} = 13.04$ in./sec, $u_{go} = 8.40$ in.) with response spectrum for El Centro ground motion; $\mu = 1$ and 8; $\zeta = 5\%$.

$\mu = 8$ determined from the elastic design spectrum using the procedure described in Section 7.10, and (2) the actual spectrum for the El Centro ground motion for $\mu = 8$, reproduced from Fig. 7.5.3.

Observe that the differences between the design and response spectra for elastoplastic systems are greater than between the two spectra for elastic systems. For the latter case, the reasons underlying these differences were discussed in Section 6.10. Additional discrepancies arise in the two spectra for elastoplastic systems because the jagged variation of $\overline{f}_y$ with T_n (Fig. 7.7.1) was approximated by simple functions (Fig. 7.10.1). The errors introduced by this simplification are responsible for the additional discrepancies in the velocity- and displacement-sensitive regions of the spectrum.

FURTHER READING

Grigorian, C. E., and Popov, E. P., "Energy Dissipation with Slotted Bolted Connections," *Report No. UCB/EERC-94/02*, Earthquake Engineering Research Center, University of California at Berkeley, February 1994.

Mohraz, B., and Elghadamsi, F. E., "Earthquake Ground Motion and Response Spectra," Chapter 2 in *The Seismic Design Handbook* (ed. F. Naeim), Van Nostrand Reinhold, New York, 1989.

Newmark, N. M., and Hall, W. J., *Earthquake Spectra and Design*, Earthquake Engineering Research Institute, Berkeley, Calif., 1982, pp. 29–37.

Newmark, N. M., and Rosenblueth, E., *Fundamentals of Earthquake Engineering*, Prentice Hall, Englewood Cliffs, N.J., 1971, Chapter 11.

"Passive Energy Dissipation," *Earthquake Spectra,* **9**, 1993, pp. 319–636.

Riddell, R., and Newmark, N. M., "Statistical Analysis of the Response of Nonlinear Systems Subjected to Earthquakes," *Structural Research Series No. 468*, University of Illinois at Urbana-Champaign, Urbana, Ill., August 1979.

Veletsos, A. S., "Maximum Deformation of Certain Nonlinear Systems," *Proceedings of the 4th World Conference on Earthquake Engineering*, Santiago, Chile, Vol. 1, 1969, pp. 155–170.

Veletsos, A. S., and Newmark, N. M., "Effect of Inelastic Behavior on the Response of Simple System to Earthquake Motions," *Proceedings of the 2nd World Conference on Earthquake Engineering*, Japan, Vol. 2, 1960, pp. 395–912.

Veletsos, A. S., and Newmark, N. M., "Response Spectra for Single-Degree-of-Freedom Elastic and Inelastic Systems," *Report No. RTD-TDR-63-3096*, Vol. III, Air Force Weapons Laboratory, Albuquerque, N.Mex., June 1964.

Veletsos, A. S., Newmark, N. M., and Chelapati, C. V., "Deformation Spectra for Elastic and Elastoplastic Systems Subjected to Ground Shock and Earthquake Motion," *Proceedings of the 3rd World Conference on Earthquake Engineering*, New Zealand, Vol. II, 1965, pp. 663–682.

PROBLEMS

7.1 From the response results presented in Fig. 7.4.3, compute the ductility demands for $\bar{f}_y = 0.5$, 0.25, and 0.125.

7.2 Consider a vertical cantilever tower that supports a lumped weight w at the top; assume that the tower mass is negligible, $\zeta = 5\%$, and that the force–deformation relation is elastoplastic. The design earthquake has a peak acceleration of 0.5g, and its elastic design spectrum is given by Fig. 6.9.4 multiplied by 0.5. For three different values of the natural vibration period in the linearly elastic range, $T_n = 0.02$, 0.2, and 2 sec, determine the lateral deformation and lateral force (in terms of w) for which the tower should be designed if (i) the system is required to remain elastic, and (ii) the allowable ductility factor is 2, 4, or 8. Comment on how the design deformation and design force are affected by structural yielding.

7.3 For the design earthquake at a site the peak values of ground acceleration, velocity, and displacement have been estimated: $\ddot{u}_{go} = 0.5\text{g}$, $\dot{u}_{go} = 24$ in./sec, and $u_{go} = 18$ in. For systems with a 2% damping ratio and allowable ductility of 3, construct the median-plus-one standard deviation design spectrum. Plot the elastic and inelastic spectra together on **(a)** four-way log paper, **(b)** log-log paper showing pseudo-acceleration versus natural vibration period, T_n, and **(c)** linear-linear paper showing pseudo-acceleration versus T_n from 0 to 5 sec. Determine equations $A(T_n)$ for each branch of the inelastic spectrum and the period values at intersections of branches.

Generalized Single-Degree-of-Freedom Systems

PREVIEW

So far in this book we have considered single-degree-of-freedom systems involving a single point mass (Figs. 1.2.1 and 1.6.1) or the translation of a rigid distributed mass (Fig. 1.1.3a) which is exactly equivalent to a mass lumped at a single point. Once the stiffness of the system was determined, the equation of motion was readily formulated, and solution procedures were presented in Chapters 2 to 7.

In this chapter we develop the analysis of more complex systems treated as SDF system, hence called generalized SDF systems. The analysis provides exact results for an assemblage of rigid bodies supported such that it can deflect in only one shape, but only approximate results for systems with distributed mass and flexibility. In the latter case, the approximate natural frequency is shown to depend on the assumed deflected shape. The same frequency estimate is also determined by the classical Rayleigh's method, based on the principle of conservation of energy; this method also provides insight into the error in the estimated natural frequency.

8.1 GENERALIZED SDF SYSTEMS

Consider, for example, the system of Fig. 8.1.1a, consisting of a rigid, massless bar supported by a hinge at the left end with two lumped masses, a spring and a damper, attached to it, subjected to a time-varying external force $p(t)$. Because the bar is rigid, its deflections can be related to a single *generalized displacement* $z(t)$ as shown, and can

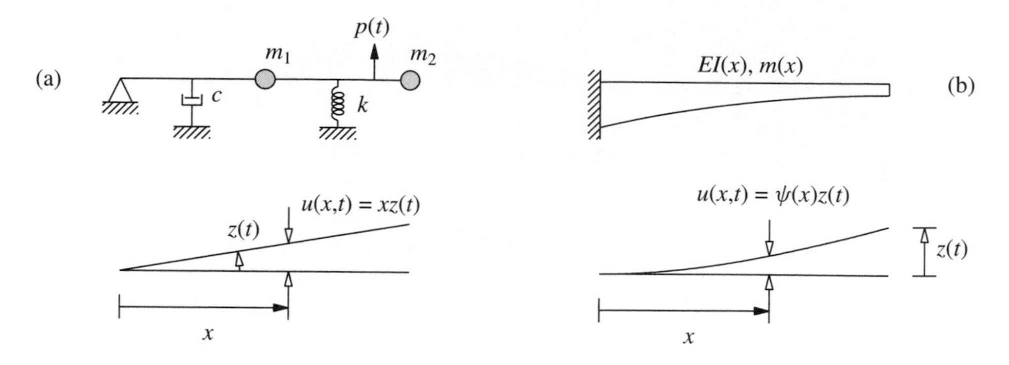

Figure 8.1.1 Generalized SDF systems.

be expressed as

$$u(x, t) = \psi(x)z(t) \tag{8.1.1}$$

We have some latitude in choosing the displacement coordinate and, quite arbitrarily, we have chosen the rotation z of the bar. For this system the *shape function* $\psi(x) = x$ is known exactly from the configuration of the system and how it is constrained by a hinged support. This is an SDF system, but it is difficult to replace the two masses by an equivalent mass lumped at a single point.

Next consider, for example, the system of Fig. 8.1.1b consisting of a cantilever beam with distributed mass. This system can deflect in a infinite variety of shapes, and for exact analysis it must be treated as an infinite-degree-of-freedom system. Such exact analysis, developed in Chapter 16, shows that the system, unlike an SDF system, possesses an infinite number of natural vibration frequencies, each paired with a natural mode of vibration. It is possible to obtain approximate results that are accurate to a useful degree for the lowest (also known as *fundamental*) natural frequency, however, by restricting the deflections of the beam to a single shape function $\psi(x)$ that approximates the fundamental vibration mode. The deflections of the beam are then given by Eq. (8.1.1), where the generalized coordinate $z(t)$ is the deflection of the cantilever beam at a selected location—say the free end, as shown in Fig. 8.1.1b.

The two systems of Fig. 8.1.1 are called *generalized SDF systems* because in each case the displacements at all locations are defined in terms of the generalized coordinate $z(t)$ through the shape function $\psi(x)$. We will show that the equation of motion for a generalized SDF system is of the form

$$\tilde{m}\ddot{z} + \tilde{c}\dot{z} + \tilde{k}z = \tilde{p}(t) \tag{8.1.2}$$

where $\tilde{m}$, $\tilde{c}$, $\tilde{k}$, and $\tilde{p}(t)$ are defined as the *generalized mass*, *generalized damping*, *generalized stiffness*, and *generalized force* of the system; these generalized properties are associated with the selected generalized displacement $z(t)$. Equation (8.1.2) is of the same form as the standard equation formulated in Chapter 1 for an SDF system with a single lumped mass. Thus the analysis procedures and response results presented in Chapters 2 to 7 can readily be adapted to determine the response $z(t)$ of generalized SDF systems. With $z(t)$ known, the displacements at all locations of the system are determined from Eq. (8.1.1). This analysis procedure leads to the exact results for the system of Fig. 8.1.1a because the shape function $\psi(x)$ could be determined exactly but provides only approximate results for the system of Fig. 8.1.1b because they are based on an assumed shape function.

The key step in the analysis outlined above that is new is the evaluation of the generalized properties $\tilde{m}$, $\tilde{c}$, $\tilde{k}$, and $\tilde{p}(t)$ for a given system. Procedures are developed to determine these properties for (1) assemblages of rigid bodies that permit exact evaluation of the deflected shape (Section 8.2), and (2) multi-degree-of-freedom systems with distributed mass or several lumped masses which require that a shape function be assumed that satisfies the displacement boundary conditions (Sections 8.3 and 8.4).

8.2 RIGID-BODY ASSEMBLAGES

In this section the equation of motion is formulated for an assemblage of rigid bodies with distributed mass supported by discrete springs and dampers subjected to time-varying forces. In formulating the equation of motion for such generalized SDF systems, application of Newton's second law of motion can be cumbersome, and it is simpler to use D'Alembert's principle and include inertia forces in the free-body diagram. The distributed inertia forces for a rigid body with distributed mass can be expressed in terms of the inertia force resultants at the center of gravity using the total mass and the moment of inertia of the body. These properties for rigid plates of three configurations are presented in Appendix 8.

Example 8.1

The system shown in Fig. E8.1a consists of a rigid bar supported by a fulcrum at O, with an attached spring and damper subjected to force $p(t)$. The mass m_1 of the part OB of the bar is distributed uniformly along its length. The portions OA and BC of the bar are massless, but a uniform circular plate of mass m_2 is attached at the midpoint of BC. Selecting the counterclockwise rotation about the fulcrum as the generalized displacement and considering small displacements, formulate the equation of motion for this generalized SDF system, determine the natural vibration frequency and damping ratio, and evaluate the dynamic response of the system without damping subjected to a suddenly applied force p_o. How would the equation of motion change with an axial force on the horizontal bar; what is the buckling load?

Solution

1. *Determine the shape function.* The L-shaped bar rotates about the fulcrum at O. Assuming small deflections, the deflected shape is shown in Fig. E8.1b.

2. *Draw the free-body diagram and write the equilibrium equation.* Figure. E8.1c shows the forces in the spring and damper associated with the displacements of Fig. E8.1b, together with the inertia forces. Setting the sum of the moments of all forces about O to zero gives

$$I_1\ddot{\theta}+\left(m_1\frac{L}{2}\ddot{\theta}\right)\frac{L}{2}+I_2\ddot{\theta}+(m_2L\ddot{\theta})L+\left(m_2\frac{L}{4}\ddot{\theta}\right)\frac{L}{4}+\left(c\frac{L}{2}\dot{\theta}\right)\frac{L}{2}+\left(k\frac{3L}{4}\theta\right)\frac{3L}{4}=p(t)\frac{L}{2}$$

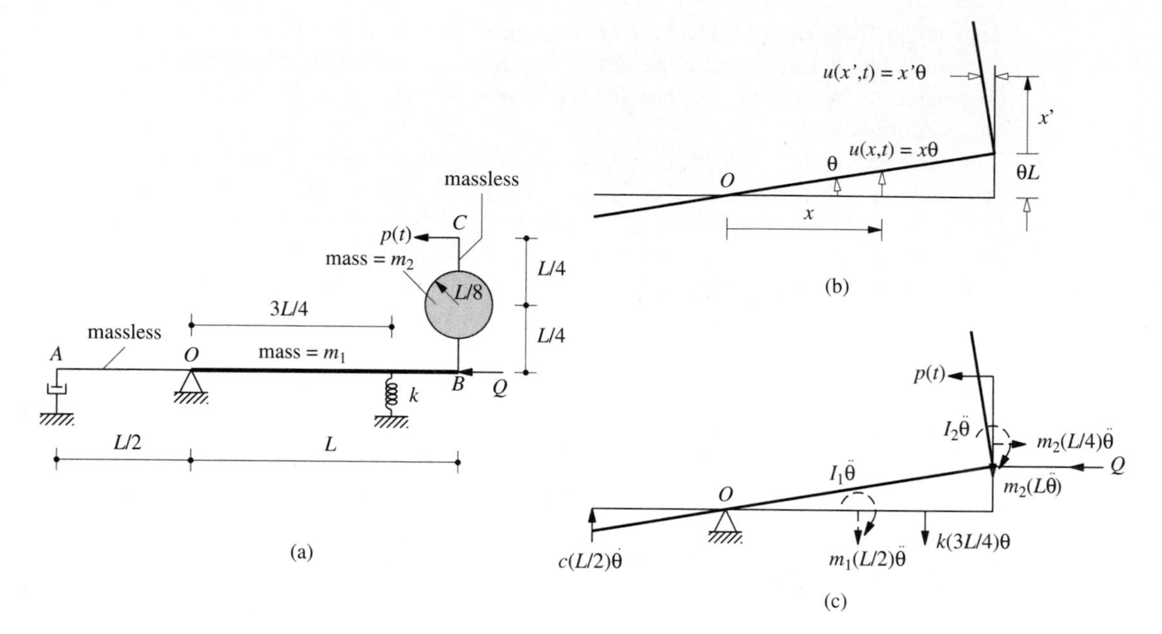

Figure E8.1

Substituting $I_1 = m_1 L^2/12$ and $I_2 = m_2(L/8)^2/2 = m_2 L^2/128$ (see Appendix 8) gives

$$\left(\frac{m_1 L^2}{3} + \frac{137}{128} m_2 L^2\right)\ddot{\theta} + \frac{cL^2}{4}\dot{\theta} + \frac{9kL^2}{16}\theta = p(t)\frac{L}{2} \tag{a}$$

The equation of motion is

$$\tilde{m}\ddot{\theta} + \tilde{c}\dot{\theta} + \tilde{k}\theta = \tilde{p}(t) \tag{b}$$

where

$$\tilde{m} = \left(\frac{m_1}{3} + \frac{137}{128} m_2\right)L^2 \qquad \tilde{c} = \frac{cL^2}{4} \qquad \tilde{k} = \frac{9kL^2}{16} \qquad \tilde{p}(t) = p(t)\frac{L}{2} \tag{c}$$

3. *Determine the natural frequency and damping ratio.*

$$\omega_n = \sqrt{\frac{\tilde{k}}{\tilde{m}}} \qquad \zeta = \frac{\tilde{c}}{2\sqrt{\tilde{k}\tilde{m}}} \tag{d}$$

4. *Solve the equation of motion.*

$$\tilde{p}(t) = \frac{p(t)L}{2} = \frac{p_o L}{2} \equiv \tilde{p}_o$$

By adapting Eq. (4.3.2), the solution of Eq. (b) with $c = 0$ is

$$\theta(t) = \frac{\tilde{p}_o}{\tilde{k}}(1 - \cos\omega_n t) = \frac{8p_o}{9kL}(1 - \cos\omega_n t) \tag{e}$$

5. *Determine the displacements.*

$$u(x, t) = x\theta(t) \qquad u(x', t) = x'\theta(t) \tag{f}$$

where $\theta(t)$ is given by Eq. (e).

6. *Include the axial force.* In the displaced position of the bar, the axial force Q introduces a counterclockwise moment $= QL\theta$. Thus Eq. (b) becomes

$$\tilde{m}\ddot{\theta} + \tilde{c}\dot{\theta} + (\tilde{k} - QL)\theta = \tilde{p}(t) \tag{g}$$

A compressive axial force decreases the stiffness of the system and hence its natural vibration frequency. These become zero if the axial force is

$$Q_{\text{cr}} = \frac{\tilde{k}}{L} = \frac{9kL}{16}$$

This is the critical or buckling axial load for the system.

8.3 SYSTEMS WITH DISTRIBUTED MASS AND ELASTICITY

As an illustration of approximating a system having an infinite number of degrees of freedom by a generalized SDF system, consider the cantilever tower shown in Fig. 8.3.1. This tower has mass $m(x)$ per unit length and flexural rigidity $EI(x)$, and the excitation is earthquake ground motion $u_g(t)$. In this section, first the equation of motion for this system without damping is formulated; damping is usually expressed by a damping ratio estimated based on experimental data from similar structures. Then the equation of motion is solved to determine displacements and a procedure is developed to determine the internal forces in the tower. Finally, this procedure is applied to evaluation of the peak response of the system to earthquake ground motion.

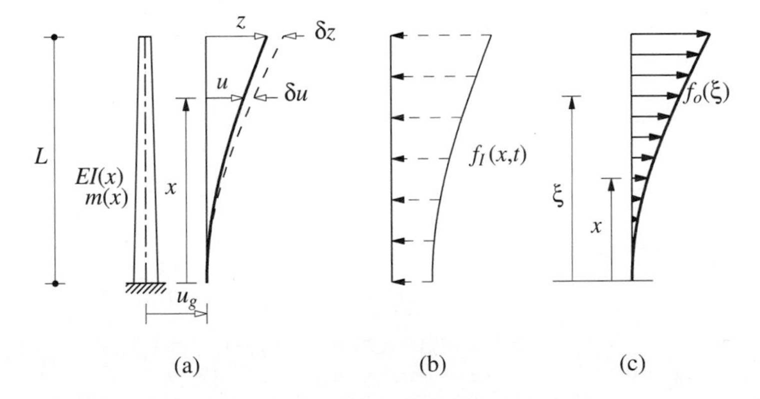

Figure 8.3.1 (a) Tower deflections and virtual displacements; (b) inertia forces; (c) equivalent static forces.

8.3.1 Assumed Shape Function

We assume that the displacement relative to the ground can be expressed by Eq. (8.1.1). The total displacement of the tower is

$$u^t(x,t) = u(x,t) + u_g(t) \tag{8.3.1}$$

The shape function $\psi(x)$ in Eq. (8.1.1) must satisfy the displacement boundary conditions (Fig. 8.3.1a). For this tower, these conditions at the base of the tower are $\psi(0) = 0$ and $\psi'(0) = 0$. Within these constraints a variety of shape function could be chosen. One possibility is to determine the shape function as the deflections of the tower due to some static forces. For example, the deflections of a uniform tower with flexural rigidity EI due to a unit lateral force at the top are $u(x) = (3Lx^2 - x^3)/6EI$. If we select the generalized coordinate as the deflection of some convenient reference point, say the top of the tower, then $z = u(L) = L^3/3EI$, and

$$u(x) = \psi(x)z \qquad \psi(x) = \frac{3}{2}\frac{x^2}{L^2} - \frac{1}{2}\frac{x^3}{L^3} \tag{8.3.2a}$$

This $\psi(x)$ automatically satisfies the displacement boundary conditions at $x = 0$ because it was determined from static analysis of the system. The $\psi(x)$ of Eq. (8.3.2a) may also be used as the shape function for a nonuniform tower, although it was determined for a uniform tower. It is not necessary to select the shape function based on deflections due to static forces, and it could be assumed directly; possibilities are

$$\psi(x) = \frac{x^2}{L^2} \quad \text{and} \quad \psi(x) = 1 - \cos\frac{\pi x}{2L} \tag{8.3.2b}$$

The three shape functions above have $\psi(L) = 1$, although this is not necessary. The accuracy of the generalized SDF system formulation depends on the assumed shape function $\psi(x)$ in which the structure is constrained to vibrate. This issue will be discussed later together with how to select the shape function.

8.3.2 Equation of Motion

We now proceed to formulate the equation of motion for the tower. At each time instant the system is in equilibrium under the action of the internal resisting bending moments and the fictitious inertia forces (Fig. 8.3.1b), which by D'Alembert's principle are

$$f_I(x,t) = -m(x)\ddot{u}^t(x,t)$$

Substituting Eq. (8.3.1) for u^t gives

$$f_I(x,t) = -m(x)[\ddot{u}(x,t) + \ddot{u}_g(t)] \tag{8.3.3}$$

The equation of dynamic equilibrium of this generalized SDF system can be formulated conveniently only by work or energy principles. We prefer to use the principle of virtual displacements. This principle states that if the system in equilibrium is subjected to virtual displacements $\delta u(x)$, the external virtual work δW_E is equal to the internal virtual work δW_I:

$$\delta W_I = \delta W_E \tag{8.3.4}$$

The external virtual work is due to the forces $f_I(x, t)$ acting through the virtual displacements $\delta u(x)$:

$$\delta W_E = \int_0^L f_I(x, t)\delta u(x)\, dx$$

which after substituting Eq. (8.3.3) becomes

$$\delta W_E = -\int_0^L m(x)\ddot{u}(x, t)\delta u(x)\, dx - \ddot{u}_g(t)\int_0^L m(x)\delta u(x)\, dx \tag{8.3.5}$$

The internal virtual work is due to the bending moments $\mathcal{M}(x, t)$ acting through the curvature $\delta\kappa(x)$ associated with the virtual displacements:

$$\delta W_I = \int_0^L \mathcal{M}(x, t)\delta\kappa(x)\, dx$$

Substituting

$$\mathcal{M}(x, t) = EI(x)u''(x, t) \qquad \delta\kappa(x) = \delta[u''(x)]$$

where $u'' \equiv \partial^2 u/\partial x^2$ gives

$$\delta W_I = \int_0^L EI(x)u''(x, t)\delta[u''(x)]\, dx \tag{8.3.6}$$

The internal and external virtual work is expressed next in terms of the generalized coordinate z and shape function $\psi(x)$. For this purpose, from Eq. (8.1.1) we obtain

$$u''(x, t) = \psi''(x)z(t) \qquad \ddot{u}(x, t) = \psi(x)\ddot{z}(t) \tag{8.3.7}$$

The virtual displacement is selected consistent with the assumed shape function (Fig. 8.3.1a), giving Eq. (8.3.8a) and the virtual curvature is defined by Eq. (8.3.8b):

$$\delta u(x) = \psi(x)\delta z \qquad \delta[u''(x)] = \psi''(x)\delta z \tag{8.3.8}$$

Substituting Eq. (8.3.7b) and Eq. (8.3.8a) in Eq. (8.3.5) gives

$$\delta W_E = -\delta z\left[\ddot{z}\int_0^L m(x)[\psi(x)]^2\, dx + \ddot{u}_g(t)\int_0^L m(x)\psi(x)\, dx\right] \tag{8.3.9}$$

Substituting Eqs. (8.3.7a) and Eq. (8.3.8b) in Eq. (8.3.6) gives

$$\delta W_I = \delta z\left[z\int_0^L EI(x)[\psi''(x)]^2\, dx\right] \tag{8.3.10}$$

Having obtained the final expressions for δW_E and δW_I, Eq. (8.3.4) gives

$$\delta z\left[\tilde{m}\ddot{z} + \tilde{k}z + \tilde{L}\ddot{u}_g(t)\right] = 0 \tag{8.3.11}$$

where

$$\tilde{m} = \int_0^L m(x)[\psi(x)]^2\,dx$$

$$\tilde{k} = \int_0^L EI(x)[\psi''(x)]^2\,dx \tag{8.3.12}$$

$$\tilde{L} = \int_0^L m(x)\psi(x)\,dx$$

Because Eq. (8.3.11) is valid for every virtual displacement δz, we conclude that

$$\tilde{m}\ddot{z} + \tilde{k}z = -\tilde{L}\ddot{u}_g(t) \tag{8.3.13a}$$

This is the equation of motion for the tower assumed to deflect according to the shape function $\psi(x)$. For this generalized SDF system, the generalized mass $\tilde{m}$, generalized stiffness $\tilde{k}$, and generalized excitation $-\tilde{L}\ddot{u}_g(t)$ are defined by Eq. (8.3.12). Dividing by $\tilde{m}$ gives

$$\ddot{z} + \omega_n^2 z = -\tilde{\Gamma}\ddot{u}_g(t) \tag{8.3.13b}$$

This equation is the same as Eq. (6.2.1) for an SDF system, except for the factor

$$\tilde{\Gamma} = \frac{\tilde{L}}{\tilde{m}} \tag{8.3.14}$$

A damping term using an estimated modal damping ratio ζ should be included in Eq. (8.3.13b).

8.3.3 Response Analysis

Once the generalized properties $\tilde{m}$ and $\tilde{k}$ are determined, the system can be analyzed by the methods developed in preceding chapters. In particular, the natural vibration frequency of the system is given by

$$\omega_n^2 = \frac{\tilde{k}}{\tilde{m}} = \frac{\int_0^L EI(x)[\psi''(x)]^2\,dx}{\int_0^L m(x)[\psi(x)]^2\,dx} \tag{8.3.15}$$

The generalized coordinate response $z(t)$ of the system to specified ground acceleration can be determined by the methods presented in Chapters 5 and 6. Equation (8.1.1) then gives the displacements $u(x,t)$ of the tower relative to the base.

The next step is to compute the internal forces—bending moments and shears—in the tower associated with the displacements $u(x,t)$. The second of the two methods described in Section 1.8 is used if we are working with deflected shape $\psi(x)$ that is assumed and not exact, as for generalized SDF systems. In this method internal forces are computed by static analysis of the structure subjected to *equivalent static forces*. Denoted by $f_S(x)$, these forces are defined as external forces that would cause displacements $u(x)$. Elementary beam theory gives

$$f_S(x) = \left[EI(x)u''(x)\right]'' \tag{8.3.16}$$

Because u varies with time, so will f_S; thus

$$f_S(x,t) = \left[EI(x)u''(x,t)\right]'' \tag{8.3.17}$$

which after substituting Eq. (8.1.1) becomes

$$f_S(x,t) = \left[EI(x)\psi''(x)\right]'' z(t) \tag{8.3.18}$$

These external forces, which depend on derivatives of the assumed shape function, will give internal forces that are usually less accurate than the displacements, because the derivatives of the assumed shape function are poorer approximations than the shape function itself.

The best estimate, best within the constraints of the assumed shape function, for equivalent static forces is

$$f_S(x,t) = \omega_n^2 m(x)\psi(x)z(t) \tag{8.3.19}$$

This is identical to Eq. (8.3.18) if the assumed shape function is exact, as we shall see in Chapter 16. With an approximate function, the two sets of forces given by Eqs. (8.3.18) and (8.3.19) are not the same locally at all points along the length of the structure, but the two are globally equivalent (see Derivation 8.1). Furthermore, Eq. (8.3.19) does not involve the derivatives of the assumed $\psi(x)$, and is therefore a better approximation, relative to Eq. (8.3.18).

8.3.4 Peak Earthquake Response

The equivalent static forces of Eq. (8.3.19) are used to determine the peak values of the response of a tower to earthquake ground motion. Comparing Eq. (8.3.13b) to Eq. (6.2.1) for an SDF system and using the procedure of Section 6.7 gives the peak value of $z(t)$:

$$z_o = \tilde{\Gamma} D = \frac{\tilde{\Gamma}}{\omega_n^2} A \tag{8.3.20}$$

where D and A are the deformation and pseudo-acceleration ordinates, respectively, of the design spectrum at period $T_n = 2\pi/\omega_n$ for damping ratio ζ. In Eqs. (8.1.1) and (8.3.19), $z(t)$ is replaced by z_o of Eq. (8.3.20) to obtain the peak values of displacements and equivalent static forces:

$$u_o(x) = \tilde{\Gamma} D\psi(x) \qquad f_o(x) = \tilde{\Gamma} m(x)\psi(x)A \tag{8.3.21}$$

where the conventional subscript s has been dropped from f_{So} for brevity.

The internal forces—bending moments and shears—in the cantilever tower are obtained by static analysis of the structure subjected to the forces $f_o(x)$; see Fig. 8.3.1c. Thus the shear and bending moment at height x above the base are

$$\mathcal{V}_o(x) = \int_x^L f_o(\xi)\,d\xi = \tilde{\Gamma} A \int_x^L m(\xi)\psi(\xi)\,d\xi \tag{8.3.22a}$$

$$\mathcal{M}_o(x) = \int_x^L (\xi - x) f_o(\xi)\,d\xi = \tilde{\Gamma} A \int_x^L (\xi - x) m(\xi)\psi(\xi)\,d\xi \tag{8.3.22b}$$

In particular the shear and bending moment at the base of the tower are

$$\mathcal{V}_{bo} = \mathcal{V}_o(0) = \tilde{L}\tilde{\Gamma}A \qquad \mathcal{M}_{bo} = \mathcal{M}_o(0) = \tilde{L}^\theta\tilde{\Gamma}A \tag{8.3.23}$$

where $\tilde{L}$ was defined in Eq. (8.3.12) and

$$\tilde{L}^\theta = \int_0^L xm(x)\psi(x)\,dx \tag{8.3.24}$$

This completes the approximate evaluation of the earthquake response of a system with distributed mass and flexibility based on an assumed shape function $\psi(x)$.

8.3.5 Applied Force Excitation

If the excitation were external forces $p(x,t)$ instead of ground motion $\ddot{u}_g(t)$, the equation of motion could be derived following the methods of Section 8.3.2, leading to

$$\tilde{m}\ddot{z} + \tilde{k}z = \tilde{p}(t) \tag{8.3.25}$$

where the generalized force

$$\tilde{p}(t) = \int_0^L p(x,t)\psi(x)\,dx \tag{8.3.26}$$

Observe that the only difference in the two equations (8.3.25) and (8.3.13a) is in the excitation term.

Derivation 8.1

The equivalent static forces from elementary beam theory, Eq. (8.3.17), are written as

$$f_S(x,t) = \mathcal{M}''(x,t) \tag{a}$$

where the internal bending moments

$$\mathcal{M}(x,t) = EI(x)u''(x,t) \tag{b}$$

We seek lateral forces $\tilde{f}_S(x,t)$ that do not involve the derivatives of $\mathcal{M}(x,t)$ and at each time instant are in equilibrium with the internal bending moments; equilibrium is satisfied globally for the system (but not at every location x). Using the principle of virtual displacements the external work done by the unknown forces $\tilde{f}_S(x,t)$ in acting through the virtual displacement $\delta u(x)$ equals the internal work done by the bending moments acting through the curvatures $\delta\kappa(x)$ associated with the virtual displacements:

$$\int_0^L \tilde{f}_S(x,t)\delta u(x)\,dx = \int_0^L \mathcal{M}(x,t)\delta\kappa(x)\,dx \tag{c}$$

This equation is rewritten by substituting Eq. (8.3.8a) for $\delta u(x)$ in the left side and by using Eq. (8.3.10) for the integral on the right side; thus

$$\delta z \int_0^L \tilde{f}_S(x,t)\psi(x)\,dx = \delta z \left[z(t) \int_0^L EI(x)\left[\psi''(x)\right]^2 dx \right] \tag{d}$$

Utilizing Eq. (8.3.15) and dropping δz, Eq. (d) can be rewritten as

$$\int_0^L \left[\tilde{f}_S(x,t) - \omega_n^2 m(x)\psi(x)z(t)\right]\psi(x)\,dx = 0 \tag{e}$$

Setting the quantity in brackets to zero gives

$$f_S(x,t) = \omega_n^2 m(x)\psi(x)z(t) \tag{f}$$

where the tilde above f_S has now been dropped. This completes the derivation of Eq. (8.3.19).

Example 8.2

A uniform cantilever tower of length L has mass per unit length $= m$ and flexural rigidity EI (Fig. E8.2). Assuming the shape function $\psi(x) = 1 - \cos(\pi x/2L)$, formulate the equation of motion for the system excited by ground motion, and determine its natural frequency.

Solution

1. *Determine the generalized properties.*

$$\tilde{m} = m\int_0^L \left(1 - \cos\frac{\pi x}{2L}\right)^2 dx = 0.227\text{ mL} \tag{a}$$

$$\tilde{k} = EI\int_0^L \frac{\pi^2}{4L^2}\cos^2\frac{\pi x}{2L}\,dx = 3.04EI/L^3 \tag{b}$$

$$\tilde{L} = m\int_0^L \left(1 - \cos\frac{\pi x}{2L}\right) dx = 0.363\text{ mL} \tag{c}$$

The computed $\tilde{k}$ is close to the stiffness of the tower under a concentrated lateral force at the top.

2. *Determine the natural vibration frequency.*

$$\omega_n = \sqrt{\frac{\tilde{k}}{\tilde{m}}} = \frac{3.66}{L^2}\sqrt{\frac{EI}{m}} \tag{d}$$

This approximate result is close to the exact natural frequency, $w_{\text{exact}} = (3.516/L^2)\sqrt{EI/m}$, determined in Chapter 16. The error is only 4%.

3. *Formulate the equation of motion.* Substituting $\tilde{L}$ and $\tilde{m}$ in Eq. (8.3.14) gives $\tilde{\Gamma} = 1.6$ and Eq. (8.3.13b) becomes

$$\ddot{z} + \omega_n^2 z = -1.6\ddot{u}_g(t) \tag{e}$$

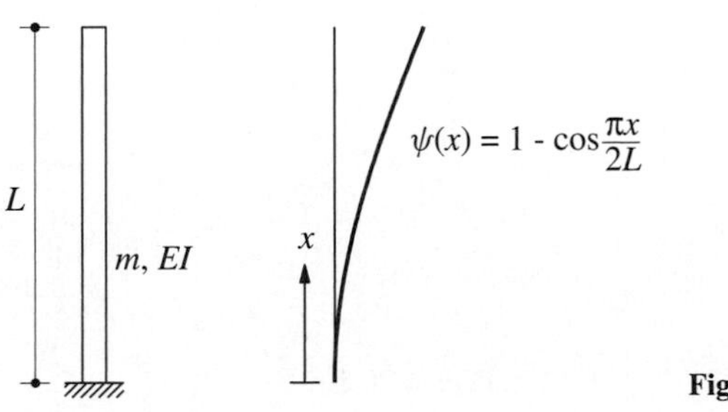

Figure E8.2

Example 8.3

A reinforced-concrete chimney, 600 ft high, has a uniform hollow circular cross section with outside diameter 50 ft and wall thickness 2 ft 6 in. (Fig. E8.3a). For purposes of preliminary earthquake analysis, the chimney is assumed clamped at the base, the mass and flexural rigidity are computed from the gross area of the concrete (neglecting the reinforcing steel), and the damping is estimated as 5%. The unit weight of concrete is 150 lb/ft^3 and its elastic modulus $E_c = 3600$ ksi.

Assuming the shape function as in Example 8.2, estimate the peak displacements, shear forces, and bending moments for the chimney due to ground motion characterized by the design spectrum of Fig. 6.9.5 scaled to a peak acceleration 0.25g.

Solution

1. *Determine the chimney properties.*

Length: $L = 600$ ft

Cross-sectional area: $A = \pi(25^2 - 22.5^2) = 373.1 \text{ ft}^2$

Mass/foot length: $m = \dfrac{150 \times 373.1}{32.2} = 1.738 \text{ kip-sec}^2/\text{ft}^2$

Second moment of area: $I = \dfrac{\pi}{4}(25^4 - 22.5^4) = 105{,}507 \text{ ft}^4$

Flexural rigidity: $EI = 5.469 \times 10^{10} \text{ kip-ft}^2$

2. *Determine the natural period.* From Example 8.2,

$$\omega_n = \frac{3.66}{L^2}\sqrt{\frac{EI}{m}} = 1.80 \text{ rad/sec}$$

$$T_n = \frac{2\pi}{\omega_n} = 3.49 \text{ sec}$$

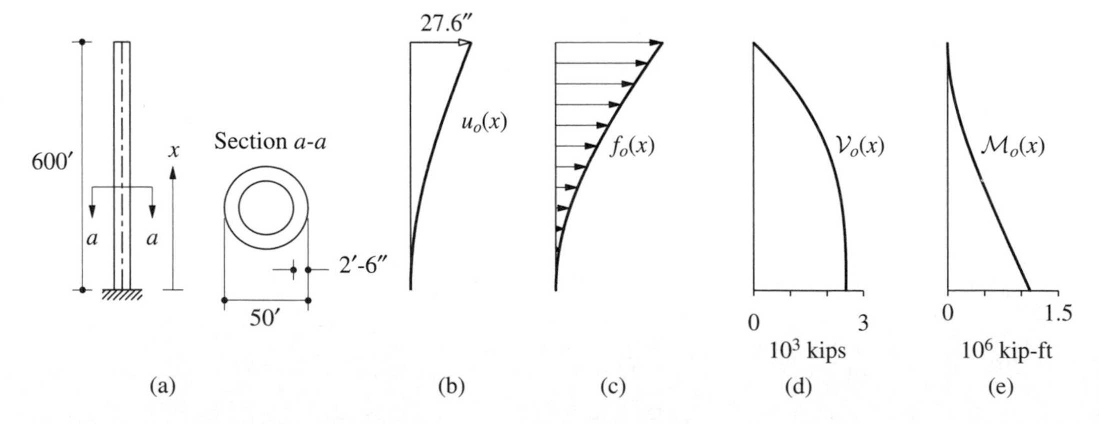

Figure E8.3

3. *Determine the peak value of $z(t)$.* For $T_n = 3.49$ sec and $\zeta = 0.05$, the design spectrum gives $A/g = 0.25(1.80/3.49) = 0.129$. The corresponding deformation is $D = A/\omega_n^2 = 15.3$ in. Equation (8.3.20) gives the peak value of $z(t)$:

$$z_o = 1.6D = 1.6 \times 15.3 = 27.6 \text{ in.}$$

4. *Determine the peak displacements $u_o(x)$ of the tower (Fig. E8.3b).*

$$u_o(x) = \psi(x) z_o = 27.6\left(1 - \cos\frac{\pi x}{2L}\right) \text{ in.}$$

5. *Determine the equivalent static forces.*

$$f_o(x) = \tilde{\Gamma} m(x) \psi(x) A = (1.6)(1.738)\left(1 - \cos\frac{\pi x}{2L}\right) 0.129\text{g}$$

$$= 11.58\left(1 - \cos\frac{\pi x}{2L}\right) \text{ kips/ft}$$

These forces are shown in Fig. E8.3c.

6. *Compute the shears and bending moments.* Static analysis of the chimney subjected to external forces $f_o(x)$ gives the shear forces and bending moments. The results using Eq. (8.3.22) are presented in Fig. E8.3d and e. If we were interested only in the forces at the base of the chimney, they could be computed directly from Eq. (8.3.23). In particular, the base shear is

$$V_{bo} = \tilde{L}\tilde{\Gamma} A = (0.363mL)(1.6)0.129\text{g}$$

$$= 0.0749mL\text{g} = 2518 \text{ kips}$$

This is 7.49% of the total weight of the chimney.

Example 8.4

A simply supported bridge with a single span of L feet has a deck of uniform cross section with mass m per foot length and flexural rigidity EI. A single wheel load p_o travels across the bridge at a uniform velocity of v, as shown in Fig. E8.4. Neglecting damping and assuming the shape function as $\psi(x) = \sin(\pi x/L)$, determine an equation for the deflection at midspan as a function of time. The properties of a prestressed-concrete box-girder elevated-freeway connector are $L = 200$ ft, $m = 11$ kips/g per foot, $I = 700$ ft^4, and $E = 576{,}000$ kips/ft^2. If $v = 55$ mph, determine the impact factor defined as the ratio of maximum deflection at midspan and the static deflection.

Solution We assume that the mass of the wheel load is small compared to the bridge mass, and it can be neglected.

1. *Determine the generalized mass, generalized stiffness, and natural frequency.*

$$\psi(x) = \frac{\sin \pi x}{L} \qquad \psi''(x) = -\frac{\pi^2}{L^2}\sin\frac{\pi x}{L}$$

$$\tilde{m} = \int_0^L m \sin^2\frac{\pi x}{L}\,dx = \frac{mL}{2} \tag{a}$$

$$\tilde{k} = \int_0^L EI\left(\frac{\pi^2}{L^2}\right)^2 \sin^2\frac{\pi x}{L}\,dx = \frac{\pi^4 EI}{2L^3} \tag{b}$$

$$\omega_n = \frac{\tilde{k}}{\tilde{m}} = \frac{\pi^2}{L^2}\sqrt{\frac{EI}{m}} \tag{c}$$

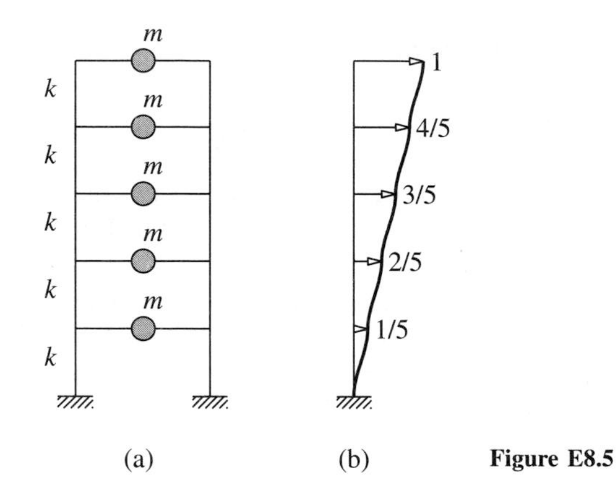

Figure E8.5

Solution

1. *Determine the generalized properties.*

$$\tilde{m} = \sum_{j=1}^{5} m_j \psi_j^2 = m \frac{1^2 + 2^2 + 3^2 + 4^2 + 5^2}{5^2} = \frac{11}{5} m$$

$$\tilde{k} = \sum_{j=1}^{5} k_j (\psi_j - \psi_{j-1})^2 = k \frac{1^2 + 1^2 + 1^2 + 1^2 + 1^2}{5^2} = \frac{k}{5}$$

$$\tilde{L} = \sum_{j=1}^{5} m_j \psi_j = m \frac{1 + 2 + 3 + 4 + 5}{5} = 3m$$

2. *Formulate the equation of motion.* Substituting for $\tilde{m}$ and $\tilde{L}$ in Eq. (8.3.14) gives $\tilde{\Gamma} = \frac{15}{11}$ and Eq. (8.3.13b) becomes

$$\ddot{z} + \omega_n^2 z = -\tfrac{15}{11} \ddot{u}_g(t)$$

where z is the lateral displacement at the location where $\psi_j = 1$, in this case the top of the frame.

3. *Determine the natural vibration frequency.*

$$\omega_n = \sqrt{\frac{k/5}{11m/5}} = 0.302 \sqrt{\frac{k}{m}}$$

This is about 6% higher than $\omega_n = 0.285\sqrt{k/m}$, the exact frequency of the system determined in Chapter 12.

Example 8.6

Determine the peak displacements, story shears, and floor overturning moments for the frame of Example 8.5 with $m = 100$ kips/g, $k = 31.54$ kips/in., and $h = 12$ ft (Fig. E8.6a) due to the ground motion characterized by the design spectrum of Fig. 6.9.5 scaled to a peak ground acceleration of 0.25g.

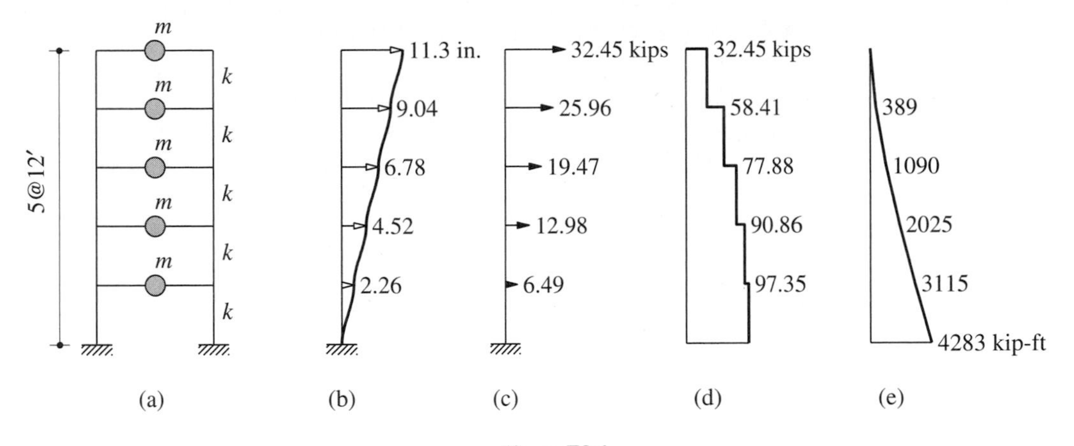

Figure E8.6

Solution

1. *Compute the natural period.*

$$\omega_n = 0.302\sqrt{\frac{31.54}{100/386}} = 3.332$$

$$T_n = \frac{2\pi}{3.332} = 1.89 \text{ sec}$$

2. *Determine the peak value of $z(t)$.* For $T_n = 1.89$ sec and $\zeta = 0.05$, the design spectrum gives $A/g = 0.25(1.80/1.89) = 0.238$ and $D = A/\omega_n^2 = 8.28$ in. The peak value of $z(t)$ is

$$z_o = \tfrac{15}{11} D = \tfrac{15}{11} 8.28 = 11.3 \text{ in.}$$

3. *Determine the peak values u_{jo} of floor displacements.*

$$u_{jo} = \psi_j z_o \qquad \psi_j = \frac{j}{5}$$

Therefore, $u_{1o} = 2.26$, $u_{2o} = 4.52$, $u_{3o} = 6.78$, $u_{4o} = 9.04$, and $u_{5o} = 11.3$, all in inches (Fig. E8.6b).

4. *Determine the equivalent static forces.*

$$f_{jo} = \tilde{\Gamma} m_j \psi_j A = \tfrac{15}{11} m \psi_j (0.238\text{g}) = 32.45\psi_j \text{ kips}$$

These forces are shown in Fig. E8.6c.

5. *Compute the story shears and overturning moments.* Static analysis of the structure subjected to external floor forces f_{jo}, Eq. (8.4.16), gives the story shears (Fig. E8.6d) and overturning moments (Fig. E8.6e). If we were interested only in the forces at the base, they could be computed directly from Eq. (8.4.17). In particular, the base shear is

$$V_{bo} = \tilde{L}\tilde{\Gamma} A = (3m)\tfrac{15}{11} 0.238\text{g}$$

$$= 0.195(5m\text{g}) = 97.35 \text{ kips}$$

This is 19.5% of the total weight of the building.

8.5 NATURAL VIBRATION FREQUENCY BY RAYLEIGH'S METHOD

Although the principle of virtual displacements provides an approximate result for the natural vibration frequency of any structure, it is instructive to obtain the same result by another approach, developed by Lord Rayleigh. Based on the principle of conservation of energy, Rayleigh's method was published in 1873. In this section this method is applied to a mass–spring system, a distributed-mass system, and a lumped-mass system.

8.5.1 Mass–Spring System

When an SDF system with lumped mass m and stiffness k is disturbed from its equilibrium position, it oscillates at its natural vibration frequency ω_n, and it was shown in Section 2.1 by solving the equation of motion that $\omega_n = \sqrt{k/m}$. Now we will obtain the same result using the principle of conservation of energy.

The simple harmonic motion of a freely vibrating mass–spring system, Eq. (2.1.3), can be described conveniently by redefining the origin for the time variable t' as shown in Fig. 8.5.1a:

$$u(t') = u_o \sin \omega_n t' \tag{8.5.1}$$

where the frequency ω_n is to be determined and the amplitude u_o of the motion is given by Eq. (2.1.9). The velocity of the mass, shown in Fig. 8.5.1b, is

$$\dot{u}(t') = \omega_n u_o \cos \omega_n t' \tag{8.5.2}$$

The potential energy of the system is the strain energy in the spring, which is proportional to the square of the spring deformation u (Eq. 2.3.2). Therefore, the strain energy is maximum at $t' = T_n/4$ (also at $t' = 3T_n/4$, $5T_n/4$, ...) when $u(t) = u_o$ and is

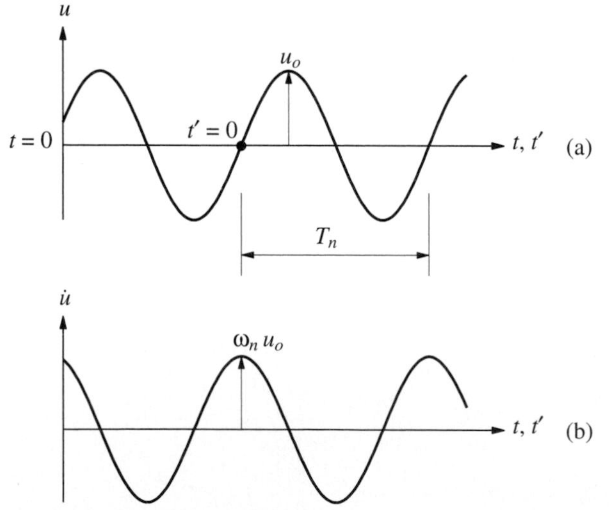

Figure 8.5.1 Simple harmonic motion of a freely vibrating system: (a) displacement; (b) velocity.

given by

$$E_{So} = \tfrac{1}{2} k u_o^2 \tag{8.5.3}$$

This is also the total energy of the system because at this t' the velocity is zero (Fig. 8.5.1b), implying that the kinetic energy is zero.

The kinetic energy of the system is maximum at $t' = 0$ (also at $t' = T_n/2$, $3T_n/2, \ldots$) when the velocity $\dot{u}(t) = \omega_n u_o$ and is given by

$$E_{Ko} = \tfrac{1}{2} m \omega_n^2 u_o^2 \tag{8.5.4}$$

This is also the total energy of the system because at this t', the deformation is zero (Fig. 8.5.1a), implying that the strain energy is zero.

The principle of conservation of energy states that the total energy in a freely vibrating system without damping is constant (i.e., it does not vary with time), as shown by Eq. (2.3.5). Thus the two alternative expressions, E_{Ko} and E_{So}, for the total energy must be equal, leading to the important result:

$$\text{maximum kinetic energy, } E_{Ko} = \text{maximum potential energy, } E_{So} \tag{8.5.5}$$

Substituting Eqs. (8.5.3) and (8.5.4) gives

$$\omega_n = \sqrt{\frac{k}{m}} \tag{8.5.6}$$

This is the same result for the natural vibration frequency as Eq. (2.1.4) obtained by solving the equation of motion.

Rayleigh's method does not provide any significant advantage in obtaining the natural vibration frequency of a mass–spring system, but the underlying concept of energy conservation is useful for complex systems, as shown in the next two sections.

8.5.2 Systems with Distributed Mass and Elasticity

As an illustration of such a system, consider the cantilever tower of Fig. 8.3.1 vibrating freely in simple harmonic motion:

$$u(x,t) = z_o \psi(x) \sin \omega_n t' \tag{8.5.7}$$

where $\psi(x)$ is an assumed shape function that defines the form of deflections, z_o is the amplitude of the generalized coordinate $z(t)$, and the frequency ω_n is to be determined. The velocity of the tower is

$$\dot{u}(x,t) = \omega_n z_o \psi(x) \cos \omega_n t' \tag{8.5.8}$$

The maximum potential energy of the system over a vibration cycle is equal to its strain energy associated with the maximum displacement $u_o(x)$:

$$E_{So} = \int_0^L \tfrac{1}{2} EI(x) [u_o''(x)]^2 \, dx \tag{8.5.9}$$

The maximum kinetic energy of the system over a vibration cycle is associated with the maximum velocity $\dot{u}_o(x)$:

$$E_{Ko} = \int_0^L \tfrac{1}{2} m(x)[\dot{u}_o(x)]^2 \, dx \tag{8.5.10}$$

From Eqs. (8.5.7) and (8.5.8), $u_o(x) = z_o\psi(x)$ and $\dot{u}_o(x) = \omega_n z_o \psi(x)$. Substituting these in Eqs. (8.5.9) and (8.5.10) and equating E_{Ko} to E_{So} gives

$$\omega_n^2 = \frac{\int_0^L EI(x)[\psi''(x)]^2 \, dx}{\int_0^L m(x)[\psi(x)]^2 \, dx} \tag{8.5.11}$$

This is known as *Rayleigh's quotient* for a system with distributed mass and elasticity; recall that the same result, Eq. (8.3.15), was obtained using the principle of virtual displacements. Rayleigh's quotient is valid for any natural vibration frequency of a multi-degree-of-freedom system, although its greatest utility is in determining the lowest or fundamental frequency.

8.5.3 Systems with Lumped Masses

As an illustration of such a system, consider the shear building of Fig. 8.4.1 vibrating freely in simple harmonic motion,

$$\mathbf{u}(t) = z_o \sin \omega_n t' \boldsymbol{\psi} \tag{8.5.12}$$

where the vector $\boldsymbol{\psi}$ is an assumed shape vector that defines the form of deflections, z_o is the amplitude of the generalized coordinate $z(t)$, and the natural vibration frequency ω_n is to be determined. The velocities of the lumped masses of the system are given by the vector

$$\dot{\mathbf{u}}(t) = \omega_n z_o \cos \omega_n t' \boldsymbol{\psi} \tag{8.5.13}$$

The maximum potential energy of the system over a vibration cycle is equal to its strain energy associated with the maximum displacements, $\mathbf{u}_o = \langle u_{1o} \quad u_{2o} \quad \cdots \quad u_{No} \rangle^T$:

$$E_{So} = \sum_{j=1}^{N} \tfrac{1}{2} k_j \left(u_{jo} - u_{j-1,o}\right)^2 \tag{8.5.14}$$

The maximum kinetic energy of the system over a vibration cycle is associated with the maximum velocities $\dot{u}_{jo}$:

$$E_{Ko} = \sum_{j=1}^{N} \tfrac{1}{2} m_j \dot{u}_{jo}^2 \tag{8.5.15}$$

From Eqs. (8.5.12) and (8.5.13), $u_{jo} = z_o\psi_j$ and $\dot{u}_{jo} = \omega_n z_o \psi_j$. Substituting these in Eqs. (8.5.14) and (8.5.15) and equating E_{Ko} to E_{So} gives

$$\omega_n^2 = \frac{\sum_{j=1}^{N} k_j (\psi_j - \psi_{j-1})^2}{\sum_{j=1}^{N} m_j \psi_j^2} \tag{8.5.16a}$$

Rewriting this in matrix notation gives

$$\omega_n^2 = \frac{\boldsymbol{\psi}^T \mathbf{k} \boldsymbol{\psi}}{\boldsymbol{\psi}^T \mathbf{m} \boldsymbol{\psi}} \tag{8.5.16b}$$

This is *Rayleigh's quotient* for the shear building with N lumped masses; recall that the same result, Eq. (8.4.13), was obtained using the principle of virtual displacements.

8.5.4 Properties of Rayleigh's Quotient

What makes Rayleigh's method especially useful for estimating a natural vibration frequency of a system are the properties of Rayleigh's quotient: First, the approximate frequency obtained from an assumed shape function is always greater than the exact value of the fundamental natural frequency—the smallest among all the natural frequencies—of the system. Second, Rayleigh's quotient provides excellent estimates of the fundamental frequency even with a not-so-good shape function.

Let us examine these properties in the context of a specific system, the cantilever tower considered in Example 8.2. Its fundamental frequency can be expressed as $\omega_n = \alpha_n\sqrt{EI/mL^4}$ [see Eq. (d) of Example 8.2]. Three different estimates of α_n using three different shape functions are summarized in Table 8.5.1. The second frequency estimate comes from Example 8.2. The same procedure leads to the results using the other two shape functions. The percentage error shown is relative to the exact value of $\alpha_n = 3.516$ (Chapter 16).

TABLE 8.5.1 NATURAL FREQUENCY ESTIMATES FOR A UNIFORM CANTILEVER

$\psi(x)$	α_n	% Error
$3x^2/2L^2 - x^3/2L^3$	3.57	1.5
$1 - \cos(\pi x/2L)$	3.66	4
x^2/L^2	4.47	27

Consistent with the properties of Rayleigh's quotient, the three estimates of the natural frequency are higher than its exact value. Even if we did not know the exact value, as would be the case for complex systems, we could say that the smallest value, $\alpha_n = 3.57$, is the best among the three estimates for the natural frequency. This concept can be used to determine the *exact* frequency of a two-DOF system by minimizing the Rayleigh's quotient over a shape function parameter.

Why such a large error in the third case of Table 8.5.1? The shape function $\psi(x) = x^2/L^2$ satisfies the displacement boundary conditions at the base of the tower but implies a constant bending moment over the height of the tower. A bending moment at the free end of a cantilever is unrealistic unless there is a mass at the free end with a moment of inertia. Thus a shape function that satisfies the geometric boundary conditions does not always ensure an accurate result for the natural frequency.

An estimate of the natural vibration frequency of a system obtained using Rayleigh's quotient can be improved by iterative methods. Such methods are developed in Chapter 10.

8.6 SELECTION OF SHAPE FUNCTION

The accuracy of the natural vibration frequency estimated using Rayleigh's quotient depends entirely on the shape function that is assumed to approximate the exact mode shape. In principle any shape may be selected that satisfies the displacement boundary conditions at the supports. In this section we address the question of how a reasonable shape function can be selected to ensure good results.

For this purpose it is useful to identify the properties of the exact mode shape. In free vibration the displacements are given by Eq. (8.5.7) and the associated inertia forces are

$$f_I(x,t) = -m(x)\ddot{u}(x,t) = \omega_n^2 z_o m(x)\psi(x) \sin \omega_n t'$$

If $\psi(x)$ were the exact mode shape, static application of these inertia forces at each time instant will produce deflections given by Eq. (8.5.7), a result that will become evident in Chapter 16. This concept is not helpful in evaluating the exact mode shape $\psi(x)$ because the inertia forces involve this unknown shape. However, it suggests that an approximate shape function $\psi(x)$ may be determined as the deflected shape due to static forces $p(x) = m(x)\tilde{\psi}(x)$, where $\tilde{\psi}(x)$ is any reasonable approximation of the exact mode shape.

In general, this procedure to select the shape function involves more computational effort than is necessary because, as mentioned earlier, Rayleigh's method gives good accuracy even if the shape function is not so good. However, the preceding discussion does support the concept of determining the shape function from deflections due to a selected set of static forces. One common selection for these forces is the weight of the structure applied in an appropriate direction. For the cantilever tower it is the lateral direction (Fig. 8.6.1a). This selection is equivalent to taking $\tilde{\psi}(x) = 1$ in $p(x) = m(x)\tilde{\psi}(x)$. Another selection includes several concentrated forces as shown in Fig. 8.6.1b.

The displacement boundary conditions are satisfied automatically if the shape function is determined from the static deflections due to a selected set of forces. This choice of shape function has the additional advantage that the strain energy can be calculated as the work done by the static forces in producing the deflections, an approach that is usually simpler than Eq. (8.5.9). Thus the maximum strain energy of the system associated with the forces of Fig. 8.6.1a is

$$E_{So} = \tfrac{1}{2} z_o g \int_0^L m(x)\psi(x)\,dx$$

Equating this E_{So} to E_{Ko} of Eq. (8.5.10) gives

$$\omega_n^2 = \frac{g}{z_o} \frac{\int_0^L m(x)\psi(x)\,dx}{\int_0^L m(x)[\psi(x)]^2\,dx} = g\frac{\int_0^L m(x)u(x)\,dx}{\int_0^L m(x)[u(x)]^2\,dx} \tag{8.6.1}$$

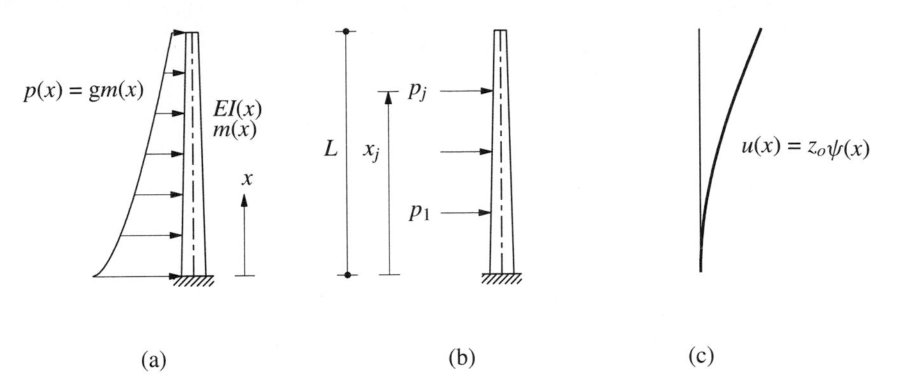

Figure 8.6.1 Shape function from deflections due to static forces.

Similarly, the maximum strain energy of the system associated with deflections $u(x)$ due to the forces of Fig. 8.6.1b is

$$E_{So} = \tfrac{1}{2} z_o \sum_j p_j \psi(x_j)$$

Equating this E_{So} to E_{Ko} of Eq. (8.5.10) gives

$$\omega_n^2 = \frac{1}{z_o} \frac{\sum p_j \psi(x_j)}{\int_0^L m(x)[\psi(x)]^2\,dx} = \frac{\sum p_j u(x_j)}{\int_0^L m(x)[u(x)]^2\,dx} \tag{8.6.2}$$

Although attractive in principle, selecting the shape function as the static deflections due to a set of forces can be cumbersome for a nonuniform beam. The reader is reminded, however, against doing complicated analysis to determine deflected shapes in the interest of obtaining an extremely accurate natural frequency. The principal attraction of Rayleigh's method lies in its ability to provide a useful estimate of the natural frequency from any reasonable assumption on the shape function that satisfies the displacement boundary conditions.

Example 8.7

Estimate the natural frequency of a uniform cantilever beam assuming the shape function obtained from static deflections due to a load p at the free end.

Solution

1. *Determine the deflections.* With the origin at the clamped end,

$$u(x) = \frac{p}{6EI}(3Lx^2 - x^3) \tag{a}$$

2. *Determine the natural frequency from* Eq. (8.6.2).

$$\sum_j p_j u(x_j) = pu(L) = p^2 \frac{L^3}{3EI} \tag{b}$$

$$\int_0^L m(x)[u(x)]^2\,dx = m\frac{p^2}{(6EI)^2}\int_0^L (3Lx^2 - x^3)^2\,dx = \frac{11p^2}{420}\frac{mL^7}{(EI)^2} \tag{c}$$

Substituting Eqs. (b) and (c) in Eq. (8.6.2) gives

$$\omega_n = \frac{3.57}{L^2}\sqrt{\frac{EI}{m}}$$

This is the first frequency estimate in Table 8.5.1.

The concept of using the shape function as the static deflections due to a selected set of forces is also useful for lumped-mass systems. Three sets of forces that may be used for a multistory building frame are shown in Fig. 8.6.2. The maximum strain energy of the system associated with deflections u_j in the three cases is

$$E_{So} = \tfrac{1}{2} p_N u_N \qquad \tfrac{1}{2}\sum_{j=1}^{N} p_j u_j \quad \text{and} \quad \tfrac{1}{2} g \sum_{j=1}^{N} m_j u_j$$

respectively. Equating these E_{So} to E_{Ko} of Eq. (8.5.15) with $\dot{u}_{jo} = \omega_n z_o \psi_j$ and simplifying gives

$$\omega_n^2 = \frac{p_N u_N}{\sum m_j u_j^2} \qquad \frac{\sum p_j u_j}{\sum m_j u_j^2} \quad \text{and} \quad \frac{g \sum m_j u_j}{\sum m_j u_j^2} \tag{8.6.3}$$

respectively. The second of these expressions appears in the *Uniform Building Code* to estimate the fundamental natural frequency of a building. Unlike Eq. (8.5.16a), these results are not restricted to a shear building as long as the deflections are calculated using the actual stiffness properties of the frame.

It is important to recognize that the success of Rayleigh's method for estimating the fundamental natural frequency of a structure depends on the ability to visualize the corresponding natural mode of vibration that the shape function is intended to approximate. The fundamental mode of a multistory building or of a single-span beam is easy to visualize because the deflections in this mode are all of the same sign. However, the mode shape of more complex systems may not be easy to visualize, and even a shape function calculated from the static deflections due to the self-weight of the structure may

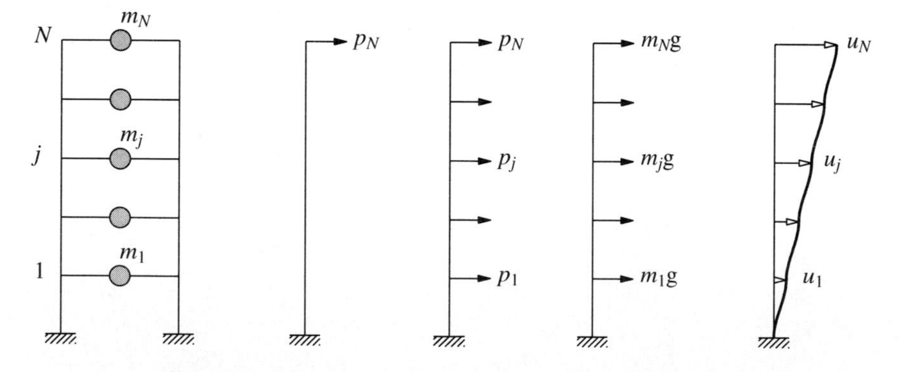

Figure 8.6.2 Shape function from deflections due to static forces.

not be appropriate. Consider, for example, a two-span continuous beam. Its symmetric deflected shape under its own weight, shown in Fig. 8.6.3a, is not appropriate for computing the lowest natural frequency because this frequency is associated with the antisymmetric mode shown in Fig. 8.6.3b. If this mode shape can be visualized, we can find ways of selecting the shape function (e.g., the static deflections due to the self-weight of the beam applied downward in one span and upward in the other span).

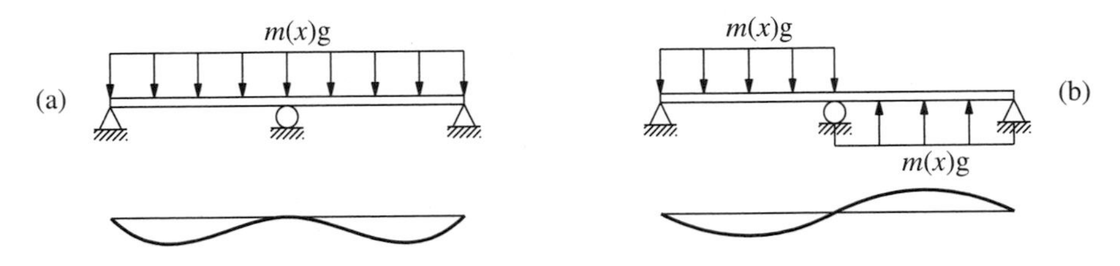

Figure 8.6.3 Shape functions resulting from self-weight applied in appropriate directions.

Example 8.8

Estimate the fundamental natural frequency of the five-story frame in Fig. E8.8 assuming the shape function obtained from static deflections due to lateral forces equal to floor weights $w = mg$.

Solution

1. *Determine the deflections due to applied forces.* The static deflections are determined as shown in Fig. E8.8 by calculating the story shears and the resulting story drifts; and adding these drifts from the bottom to the top to obtain

$$\mathbf{u}^T = \frac{w}{k}\langle 5 \quad 9 \quad 12 \quad 14 \quad 15\rangle^T$$

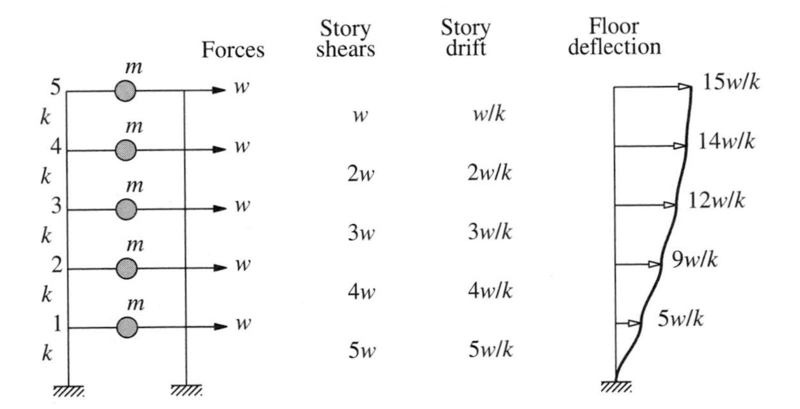

Figure E8.8

2. *Determine the natural frequency from Eq. (8.6.3c).*

$$\omega_n^2 = g\frac{(w/k)(5+9+12+14+15)}{(w/k)^2(25+81+144+196+225)} = \frac{55}{671}\frac{k}{m}$$

$$\omega_n = 0.286\sqrt{\frac{k}{m}}$$

This estimate is very close to the exact value, $\omega_{\text{exact}} = 0.285\sqrt{k/m}$, and better than from a linear shape function (Example 8.5).

FURTHER READING

Rayleigh, J. W. S., *Theory of Sound*, Dover, New York, 1945; originally published in 1894.

APPENDIX 8: INERTIA FORCES FOR RIGID BODIES

The inertia forces for a rigid bar and rectangular and circular rigid plates associated with accelerations $\ddot{u}_x$, $\ddot{u}_y$, and $\ddot{\theta}$ of the center of mass O (or center of gravity) are shown in Fig. A8.1. Each rigid body is of uniform thickness, and its total mass m is uniformly

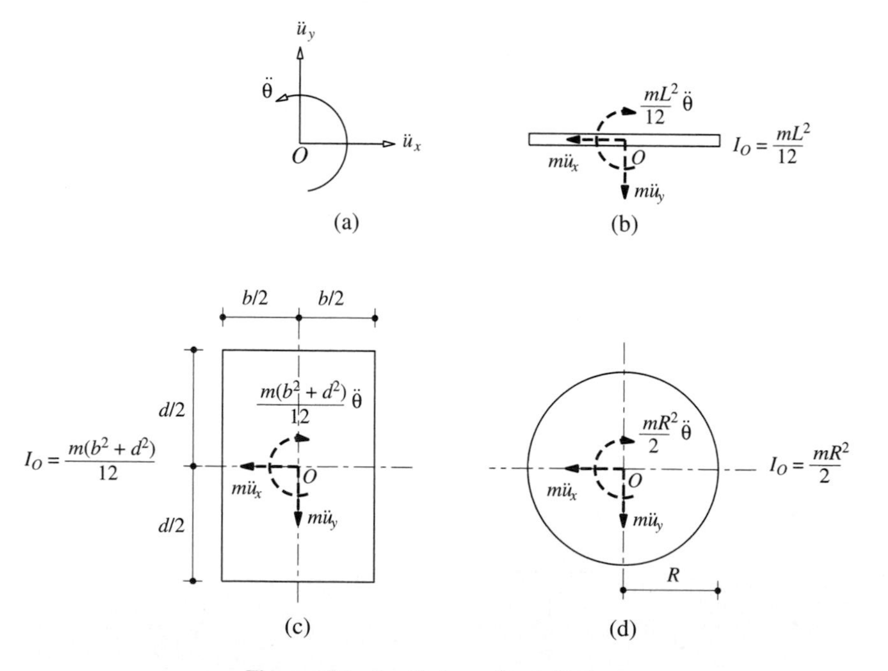

Figure A8.1 Inertia forces for rigid plates.

distributed; the moment of inertia I_O about the axis normal to the bar or plate and passing through O is as noted in the figure.

PROBLEMS

8.1 Repeat parts (a), (b), and (c) of Example 8.1 with one change: Use the horizontal displacement at C as the generalized coordinate. Show that the natural frequency, damping ratio, and displacement response are independent of the choice of generalized displacement.

8.2 For the rigid-body system shown in Fig. P8.2:
(a) Formulate the equation of motion governing the rotation at O.
(b) Determine the natural frequency and damping ratio.
(c) Determine the displacement response $u(x, t)$ to $p(t) = \delta(t)$, the Dirac delta function.

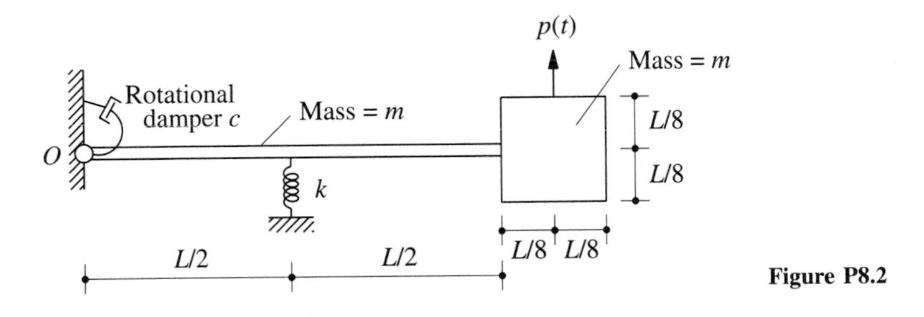

Figure P8.2

8.3 Solve Problem 8.2 with one change: Use the vertical displacement at the center of gravity of the square plate as the generalized displacement. Show that the results are independent of the choice of generalized displacement.

8.4 The rigid bar in Fig. P8.4 with a hinge at the center is bonded to a viscoelastic foundation, which can be modeled by stiffness k and damping coefficient c per unit of length. Using the rotation of the bar as the generalized coordinate:
(a) Formulate the equation of motion.
(b) Determine the natural vibration frequency and damping ratio.

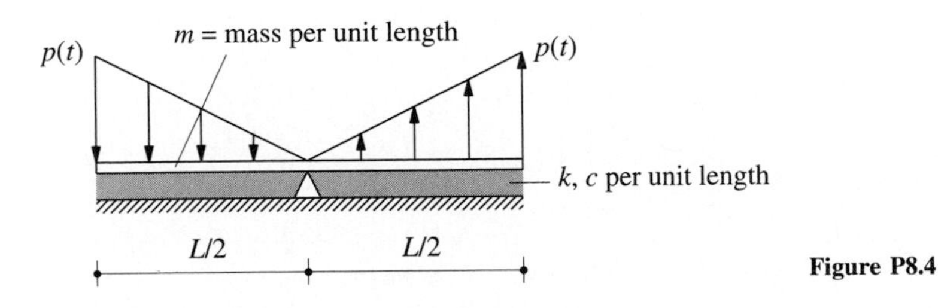

Figure P8.4

8.5 For the rigid-body system shown in Fig. P8.5:
(a) Choose a generalized coordinate.
(b) Formulate the equation of motion.
(c) Determine the natural vibration frequency and damping ratio.

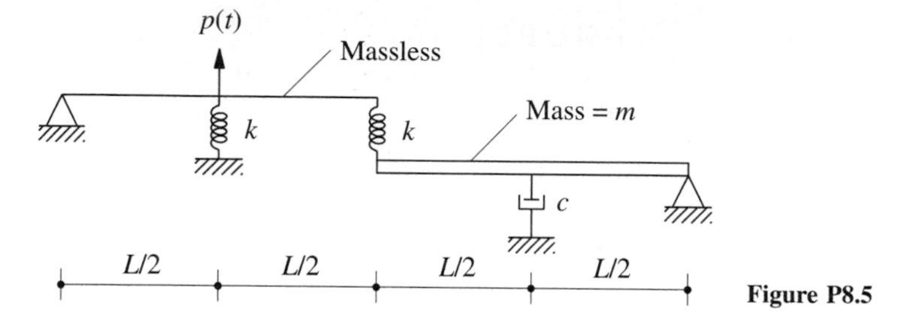

Figure P8.5

8.6 Solve Example 8.3 assuming the deflected shape function due to lateral force at the top:

$$\psi(x) = \frac{3}{2}\frac{x^2}{L^2} - \frac{1}{2}\frac{x^3}{L^3}$$

The shear forces and bending moments need to be calculated only at the base and mid-height. (Note that these forces were determined in Example 8.3 throughout the height of the chimney.)

8.7 A reinforced-concrete chimney 600 ft high has a hollow circular cross section with outside diameter 50 ft at the base and 25 ft at the top; the wall thickness is 2 ft 6 in., uniform over the height (Fig. P8.7). Using the approximation that the wall thickness is small compared to the radius, the mass and flexural stiffness properties are computed from the gross area of concrete (neglecting reinforcing steel). The chimney is assumed to be clamped at the base, and its damping ratio is estimated to be 5%. The unit weight of concrete is 150 lb/ft^3, and its elastic modulus $E_c = 3600$ ksi. Assuming that the shape function is

$$\psi(x) = 1 - \cos\frac{\pi x}{2L}$$

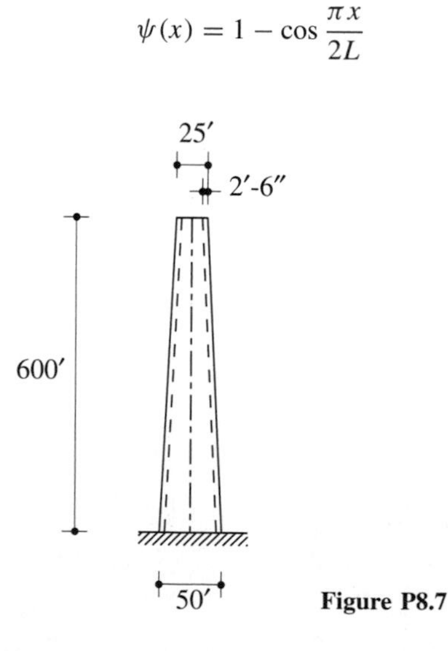

Figure P8.7

where L is the length of the chimney and x is measured from the base, calculate the following quantities: **(a)** the shear forces and bending moments at the base and at the midheight, and **(b)** the top deflection due to ground motion defined by the design spectrum of Fig. 6.9.5, which is scaled to a peak acceleration of 0.25g.

8.8 Repeat Problem 8.7 for a different excitation: a blast force varying linearly over height from zero at the base to $p(t)$ at the top, where $p(t)$ is given in Fig. P8.8.

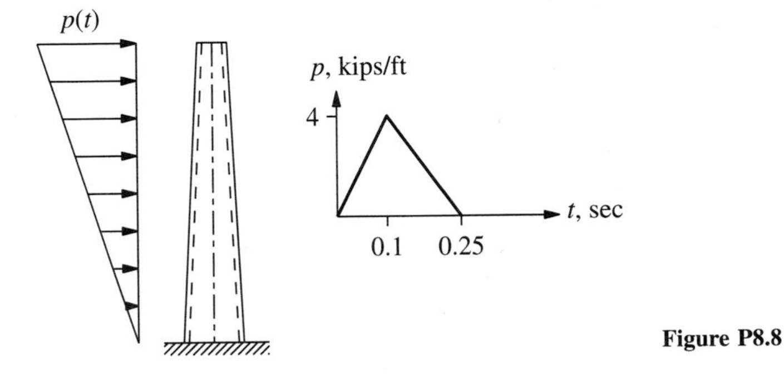

Figure P8.8

8.9 A three-story shear frame (rigid beams and flexible columns) in structural steel (E = 29,000 ksi) is shown in Fig. P8.9; w = 100 kips; I = 1400 in^4; and the modal damping ratios ζ_n are 5% for all modes. Assuming that the shape function is given by deflections due to lateral forces that are equal to the floor weights, determine the floor displacements, story shears, and overturning moments at the floors and base due to the ground motion characterized by the design spectrum of Fig. 6.9.5, which has been scaled to a peak ground acceleration of 0.25g.

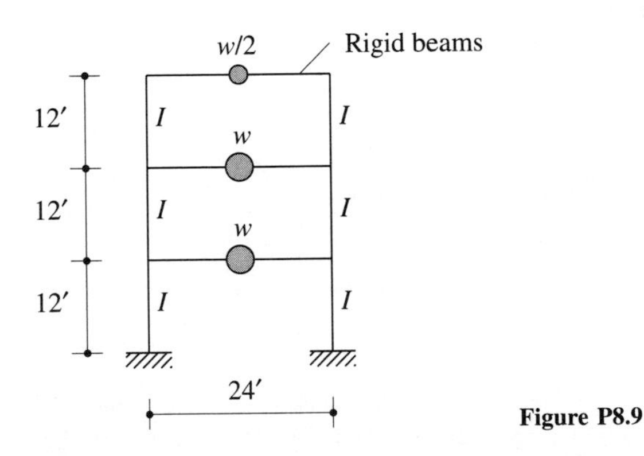

Figure P8.9

8.10 Solve Problem 8.9 using the shape function given by deflections due to a lateral force at the roof level.

8.11 Determine the natural vibration frequency of the inverted L-shaped frame shown in Fig. P8.11 using the shape function given by the deflections due to a vertical force at the free end. Neglect deformations due to shear and axial force.

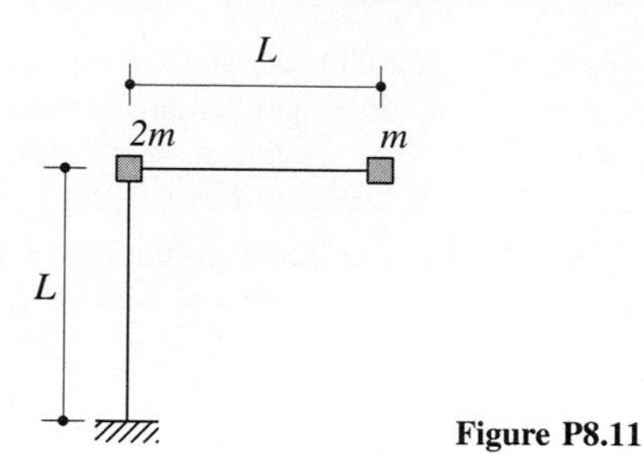

Figure P8.11

8.12 **(a)** By Rayleigh's method determine the natural vibration frequency of the rigid bar on two springs (Fig. P8.12) using the shape function shown. Note that the result involves the unknown ψ_r. Plot the value of ω_n^2 as a function of ψ_r.

(b) Using the properties of Rayleigh's quotient, determine the exact values of the two vibration frequencies and the corresponding vibration shapes.

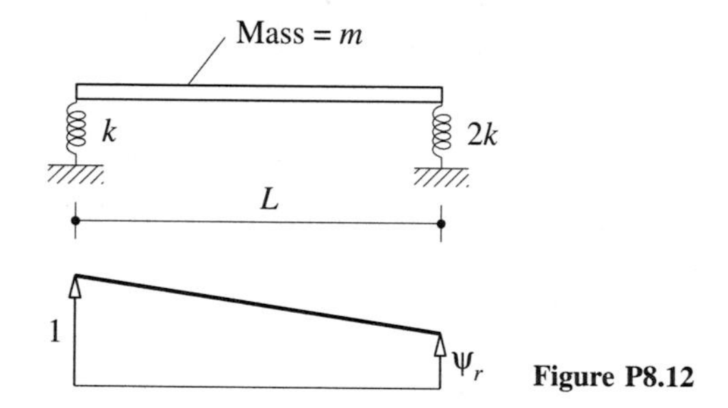

Figure P8.12

8.13 The umbrella structure shown in Fig. P8.13 consists of a uniform column of flexural rigidity EI supporting a uniform slab of radius R and mass m. By Rayleigh's method determine the natural vibration frequency of the structure. Neglect the mass of the column and the effect of axial force on column stiffness.

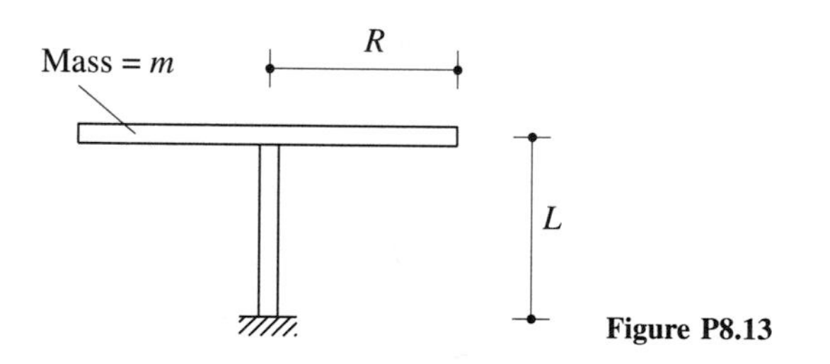

Figure P8.13

8.14 By Rayleigh's method determine the natural vibration frequency of the uniform beam shown in Fig. P8.14. Assume that the shape function is given by the deflections due to a force applied at the free end.

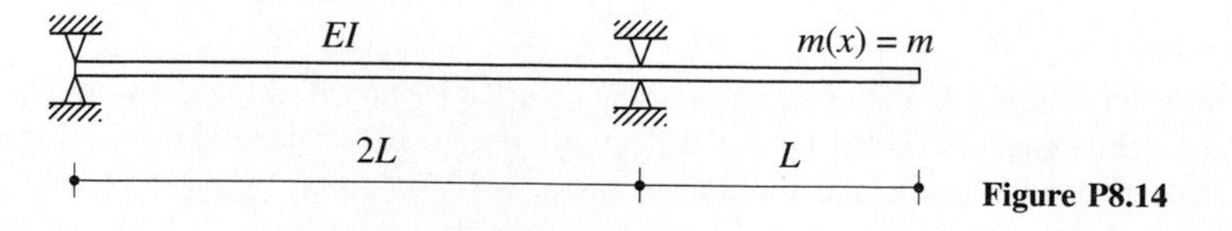

Figure P8.14

PART II

Multi-Degree-of-Freedom Systems

Equations of Motion, Problem Statement, and Solution Methods

PREVIEW

In this opening chapter of Part II the structural dynamics problem is formulated for structures discretized as systems with a finite number of degrees of freedom. The equations of motion are developed first for a simple multi-degree-of-freedom (MDF) system; a two-story shear frame is selected to permit easy visualization of elastic, damping, and inertia forces. Subsequently, a general formulation is presented for MDF systems subjected to external forces or earthquake-induced ground motion. This general formulation is then applied to develop the equations of motion for multistory buildings, first for symmetric-plan buildings and then for unsymmetric-plan buildings. Subsequently, the formulation for earthquake response analysis is extended to systems subjected to spatially varying ground motion and to inelastic systems. The chapter ends with an overview of methods for solving the differential equations governing the motion of the structure and of how our study of dynamic analysis of MDF systems is organized.

9.1 SIMPLE SYSTEM: TWO-STORY SHEAR BUILDING

We first formulate the equations of motion for the simplest possible MDF system, a highly idealized two-story frame subjected to external forces $p_1(t)$ and $p_2(t)$ (Fig. 9.1.1a). In this idealization the beams and floor systems are rigid (infinitely stiff) in flexure, and several factors are neglected: axial deformation of the beams and columns, and the effect of axial force on the stiffness of the columns. This shear-frame or shear-building idealization, although unrealistic, is convenient for illustrating how the equations of motion for an MDF system are developed. Later, we extend the formulation to more

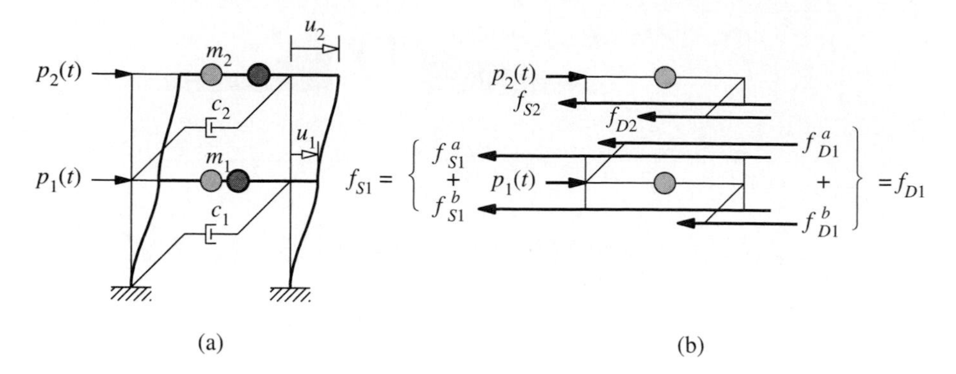

Figure 9.1.1 (a) Two-story shear frame; (b) forces acting on the two masses.

realistic idealizations of buildings that consider beam flexure and joint rotations, and to structures other than buildings.

The mass is distributed throughout the building, but we will idealize it as concentrated at the floor levels. This assumption is generally appropriate for multistory buildings because most of the building mass is indeed at the floor levels.

Just as in the case of SDF systems (Chapter 1), we assume that a linear viscous damping mechanism represents the energy dissipation in a structure. If energy dissipation is associated with the deformational motions of each story, the viscous dampers may be visualized as shown.

The number of independent displacements required to define the displaced positions of all the masses relative to their original equilibrium position is called the number of degrees of freedom. The two-story frame of Fig. 9.1.1a, with lumped mass at each floor level, has two DOFs: the lateral displacements u_1 and u_2 of the two floors in the direction of the x-axis.

9.1.1 Using Newton's Second Law of Motion

The forces acting on each floor mass m_j are shown in Fig. 9.1.1b. These include the external force $p_j(t)$, the elastic (or inelastic) resisting force f_{Sj}, and the damping force f_{Dj}. The external force is taken to be positive along the positive direction of the x-axis. The elastic and damping forces are shown acting in the opposite direction because they are internal forces that resist the motions.

Newton's second law of motion then gives for each mass:

$$p_j - f_{Sj} - f_{Dj} = m_j\ddot{u}_j \quad \text{or} \quad m_j\ddot{u}_j + f_{Dj} + f_{Sj} = p_j(t) \tag{9.1.1}$$

Equation (9.1.1) contains two equations for $j = 1$ and 2, and these can be written in matrix form:

$$\begin{bmatrix} m_1 & 0 \\ 0 & m_2 \end{bmatrix} \begin{Bmatrix} \ddot{u}_1 \\ \ddot{u}_2 \end{Bmatrix} + \begin{Bmatrix} f_{D1} \\ f_{D2} \end{Bmatrix} + \begin{Bmatrix} f_{S1} \\ f_{S2} \end{Bmatrix} = \begin{Bmatrix} p_1(t) \\ p_2(t) \end{Bmatrix} \tag{9.1.2}$$

Equation (9.1.2) can be written compactly as

$$\mathbf{m}\ddot{\mathbf{u}} + \mathbf{f}_D + \mathbf{f}_S = \mathbf{p}(t) \tag{9.1.3}$$

by introducing the following notation:

$$\mathbf{u} = \begin{Bmatrix} u_1 \\ u_2 \end{Bmatrix} \qquad \mathbf{m} = \begin{bmatrix} m_1 & 0 \\ 0 & m_2 \end{bmatrix} \qquad \mathbf{f}_D = \begin{Bmatrix} f_{D1} \\ f_{D2} \end{Bmatrix} \qquad \mathbf{f}_S = \begin{Bmatrix} f_{S1} \\ f_{S2} \end{Bmatrix} \qquad \mathbf{p} = \begin{Bmatrix} p_1 \\ p_2 \end{Bmatrix}$$

where $\mathbf{m}$ is the *mass matrix* for the two-story shear frame.

Assuming linear behavior, the elastic resisting forces $\mathbf{f}_S$ are next related to the floor displacements $\mathbf{u}$. For this purpose we introduce the lateral stiffness k_j of the jth story; it relates the story shear V_j to the story deformation or drift, $\Delta_j = u_j - u_{j-1}$, by

$$V_j = k_j \, \Delta_j \tag{9.1.4}$$

The story stiffness is the sum of the lateral stiffnesses of all columns in the story. For a story of height h and a column with modulus E and second moment of area I_c, the lateral stiffness of a column with fixed ends, implied by the shear-building idealization, is $12EI_c/h^3$. Thus the story stiffness is

$$k_j = \sum_{\text{columns}} \frac{12EI_c}{h^3} \tag{9.1.5}$$

With the story stiffnesses defined, we can relate the elastic resisting forces f_{S1} and f_{S2} to the floor displacements, u_1 and u_2. The force f_{S1} at the first floor is made up of two contributions: f_{S1}^a from the story above, and f_{S1}^b from the story below. Thus

$$f_{S1} = f_{S1}^b + f_{S1}^a$$

which, after substituting Eq. (9.1.4) and noting that $\Delta_1 = u_1$ and $\Delta_2 = u_2 - u_1$, becomes

$$f_{S1} = k_1 u_1 + k_2(u_1 - u_2) \tag{9.1.6a}$$

The force f_{S2} at the second floor is

$$f_{S2} = k_2(u_2 - u_1) \tag{9.1.6b}$$

Observe that f_{S1}^a and f_{S2} are equal in magnitude and opposite in direction because both represent the shear in the second story. In matrix form Eqs. (9.1.6a) and (9.1.6b) are

$$\begin{Bmatrix} f_{S1} \\ f_{S2} \end{Bmatrix} = \begin{bmatrix} k_1 + k_2 & -k_2 \\ -k_2 & k_2 \end{bmatrix} \begin{Bmatrix} u_1 \\ u_2 \end{Bmatrix} \quad \text{or} \quad \mathbf{f}_S = \mathbf{k}\mathbf{u} \tag{9.1.7}$$

Thus the elastic resisting force vector $\mathbf{f}_S$ and the displacement vector $\mathbf{u}$ are related through the *stiffness matrix* $\mathbf{k}$ for the two-story shear building.

The damping forces f_{D1} and f_{D2} are next related to the floor velocities $\dot{u}_1$ and $\dot{u}_2$. The jth story damping coefficient c_j relates the story shear V_j due to damping effects to the velocity $\dot{\Delta}_j$ associated with the story deformation by

$$V_j = c_j \, \dot{\Delta}_j \tag{9.1.8}$$

In a manner similar to Eq. (9.1.6), we can derive

$$f_{D1} = c_1\dot{u}_1 + c_2(\dot{u}_1 - \dot{u}_2) \qquad f_{D2} = c_2(\dot{u}_2 - \dot{u}_1) \tag{9.1.9}$$

In matrix form Eq. (9.1.9) is

$$\begin{Bmatrix} f_{D1} \\ f_{D2} \end{Bmatrix} = \begin{bmatrix} c_1 + c_2 & -c_2 \\ -c_2 & c_2 \end{bmatrix} \begin{Bmatrix} \dot{u}_1 \\ \dot{u}_2 \end{Bmatrix} \quad \text{or} \quad \mathbf{f}_D = \mathbf{c}\dot{\mathbf{u}} \tag{9.1.10}$$

The damping resisting force vector $\mathbf{f}_D$ and the velocity vector $\dot{\mathbf{u}}$ are related through the *damping matrix* **c** for the two-story shear building.

We now substitute Eqs. (9.1.7) and (9.1.10) into Eq. (9.1.3) to obtain

$$\mathbf{m}\ddot{\mathbf{u}} + \mathbf{c}\dot{\mathbf{u}} + \mathbf{k}\mathbf{u} = \mathbf{p}(t) \tag{9.1.11}$$

This matrix equation represents two ordinary differential equations governing the displacements $u_1(t)$ and $u_2(t)$ of the two-story frame subjected to external dynamic forces $p_1(t)$ and $p_2(t)$. Each equation contains both unknowns u_1 and u_2. The two equations are therefore coupled and in their present form must be solved simultaneously.

9.1.2 Dynamic Equilibrium

According to D'Alembert's principle (Chapter 1), with inertia forces included, a dynamic system is in equilibrium at each time instant. For the two masses in the system of Fig. 9.1.1a, Fig. 9.1.2 shows their free-body diagrams, including the inertia forces. Each inertia force is equal to the product of the mass times its acceleration and acts opposite to the direction of acceleration. From the free-body diagrams the condition of dynamic equilibrium also gives Eq. (9.1.11).

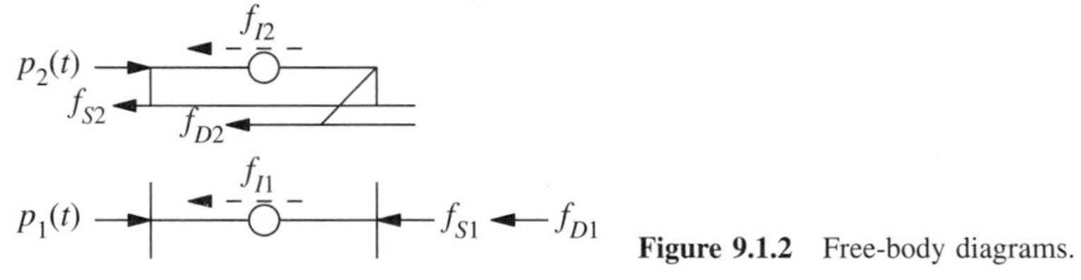

Figure 9.1.2 Free-body diagrams.

9.1.3 Mass–Spring–Damper System

We have introduced the linear two-DOF system by idealizing a two-story frame—an approach that should appeal to structural engineering students. However, the classic two-DOF system, shown in Fig. 9.1.3a, consists of two masses connected by linear springs and linear viscous dampers subjected to external forces $p_1(t)$ and $p_2(t)$. At any instant of time the forces acting on the two masses are as shown in their free-body diagrams (Fig. 9.1.3b). The resulting conditions of dynamic equilibrium also lead to Eq. (9.1.11) with **u**, **m**, **c**, **k**, and $\mathbf{p}(t)$ as defined earlier.

Example 9.1

Formulate the equations of motion for the two-story shear frame shown in Fig. E9.1.

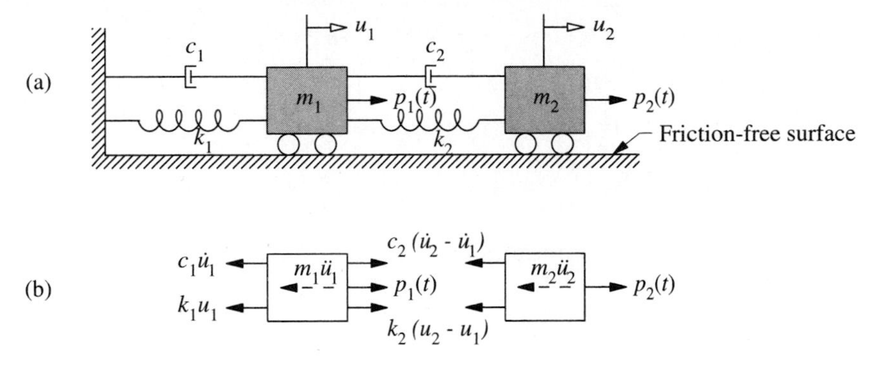

Figure 9.1.3 (a) Two-degree-of-freedom system; (b) free-body diagrams.

Solution Equation (9.1.11) is specialized for this system to obtain its equation of motion. To do so, we note that

$$m_1 = 2m \qquad m_2 = m$$

$$k_1 = 2\frac{12(2EI_c)}{h^3} = \frac{48EI_c}{h^3} \qquad k_2 = 2\frac{12(EI_c)}{h^3} = \frac{24EI_c}{h^3}$$

Substituting these data in Eqs. (9.1.2) and (9.1.7) gives the mass and stiffness matrices:

$$\mathbf{m} = m\begin{bmatrix} 2 & 0 \\ 0 & 1 \end{bmatrix} \qquad \mathbf{k} = \frac{24EI_c}{h^3}\begin{bmatrix} 3 & -1 \\ -1 & 1 \end{bmatrix}$$

Substituting these $\mathbf{m}$ and $\mathbf{k}$ in Eq. (9.1.11) gives the governing equations for this system without damping:

$$m\begin{bmatrix} 2 & 0 \\ 0 & 1 \end{bmatrix}\begin{Bmatrix} \ddot{u}_1 \\ \ddot{u}_2 \end{Bmatrix} + 24\frac{EI_c}{h^3}\begin{bmatrix} 3 & -1 \\ -1 & 1 \end{bmatrix}\begin{Bmatrix} u_1 \\ u_2 \end{Bmatrix} = \begin{Bmatrix} p_1(t) \\ p_2(t) \end{Bmatrix}$$

Observe that the stiffness matrix is nondiagonal, implying that the two equations are coupled.

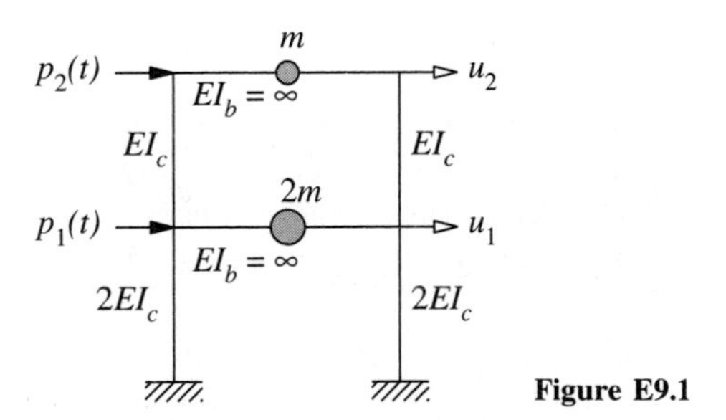

Figure E9.1

9.1.4 Stiffness, Damping, and Mass Components

In this section the governing equations for the two-story shear frame are formulated based on an alternative viewpoint. Under the action of external forces $p_1(t)$ and $p_2(t)$ the state of the system at any time instant is described by displacements $u_j(t)$, velocities $\dot{u}_j(t)$, and accelerations $\ddot{u}_j(t)$; see Fig. 9.1.4a. Now visualize this system as the combination

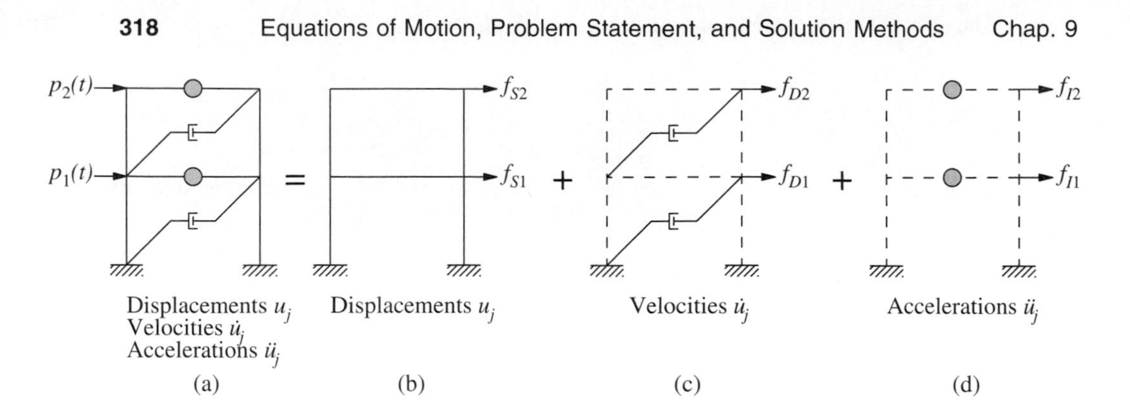

Figure 9.1.4 (a) System; (b) stiffness component; (c) damping component; (d) mass component.

of three pure components: (1) stiffness component: the frame without damping or mass (Fig. 9.1.4b); (2) damping component: the frame with its damping property but no stiffness or mass (Fig. 9.1.4c); and (3) mass component: the floor masses without the stiffness or damping of the frame (Fig. 9.1.4d). The external forces f_{Sj} on the stiffness component are related to the displacements by Eq. (9.1.7). Similarly, the external forces f_{Dj} on the damping component are related to the velocities by Eq. (9.1.10). Finally, the external forces f_{Ij} on the mass component are related to the accelerations by $\mathbf{f}_I = \mathbf{m}\ddot{\mathbf{u}}$. The external forces $\mathbf{p}(t)$ on the system may therefore by visualized as distributed among the three components of the structure. Thus $\mathbf{f}_S + \mathbf{f}_D + \mathbf{f}_I$ must equal the applied forces $\mathbf{p}(t)$, leading to Eq. (9.1.3). This alternative viewpoint may seem unnecessary for the two-story shear frame, but it can be useful in visualizing the formulation of the equations of motion for complex MDF systems (Section 9.2).

9.2 GENERAL APPROACH FOR LINEAR SYSTEMS

The formulation of the equations of motion in the preceding sections, while easy to visualize for a shear building and other simple systems, is not suitable for complex structures. For this purpose a more general approach is presented in this section. We first define the three types of forces—inertia forces, elastic forces, and damping forces—and then use the line of reasoning presented in Section 9.1.4 to develop the equations of motion.

9.2.1 Discretization

A frame structure can be idealized as an assemblage of elements—beams, columns, walls—interconnected at nodal points or nodes (Fig. 9.2.1a). The displacements of the nodes are the degrees of freedom. In general, a node in a planar two-dimensional frame has three DOFs—two translations and one rotation. A node in a three-dimensional frame has six DOFs—three translations (the x, y, and z components) and three rotations (about the x, y, and z axes). For example, a two-story, two-bay planar frame has six nodes and 18 DOFs (Fig. 9.2.1a). Axial deformations of beams can be neglected in analyzing most buildings, and axial deformations of columns need not be considered for low-

librium method is feasible to implement such calculations for simple structures with a few DOFs; it is not practical, however, for complex structures or for computer implementation. The most commonly used method is the direct stiffness method wherein the stiffness matrices of individual elements are assembled to obtain the structural stiffness matrix. This and other methods should be familiar to the reader. Therefore, these methods will not be developed in this book; we will use the simplest method appropriate for the problem to be solved.

9.2.3 Damping Forces

As mentioned in Section 1.4, the mechanisms by which the energy of a vibrating structure is dissipated can usually be idealized by equivalent viscous damping. With this assumption we relate the external forces f_{Dj} acting on the damping component of the structure to the velocities $\dot{u}_j$ (Fig. 9.2.4). We impart a unit velocity along DOF j, while the velocities in all other DOFs are kept zero. These velocities will generate internal damping forces that resist the velocities, and external forces would be necessary to equilibrate these forces. The *damping influence coefficient* c_{ij} is the external force in DOF i due to unit velocity in DOF j. The force f_{Di} at DOF i associated with velocities $\dot{u}_j$, $j = 1$ to N (Fig. 9.2.4), is obtained by superposition:

$$f_{Di} = c_{i1}\dot{u}_1 + c_{i2}\dot{u}_2 + \cdots + c_{ij}\dot{u}_j + \cdots + c_{iN}\dot{u}_N \tag{9.2.4}$$

Collecting all such equations for $i = 1$ to N and writing them in matrix form gives

$$\begin{bmatrix} f_{D1} \\ f_{D2} \\ \vdots \\ f_{DN} \end{bmatrix} = \begin{bmatrix} c_{11} & c_{12} & \cdots & c_{1j} & \cdots & c_{1N} \\ c_{21} & c_{22} & \cdots & c_{2j} & \cdots & c_{2N} \\ \vdots & \vdots & & \vdots & & \vdots \\ c_{N1} & c_{N2} & \cdots & c_{Nj} & \cdots & c_{NN} \end{bmatrix} \begin{Bmatrix} \dot{u}_1 \\ \dot{u}_2 \\ \vdots \\ \dot{u}_N \end{Bmatrix} \tag{9.2.5}$$

or

$$\mathbf{f}_D = \mathbf{c}\dot{\mathbf{u}} \tag{9.2.6}$$

where **c** is the *damping matrix* for the structure.

It is impractical to compute the coefficients c_{ij} of the damping matrix directly from the dimensions of the structure and the sizes of the structural elements. Therefore,

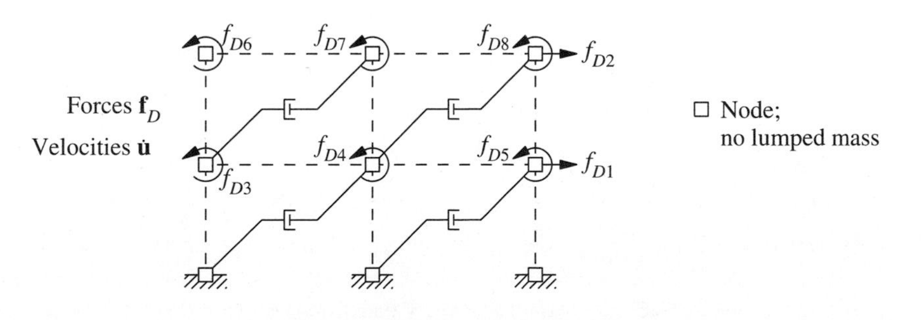

Figure 9.2.4 Damping component of frame.

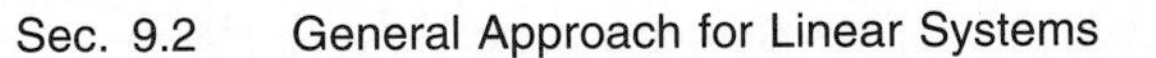

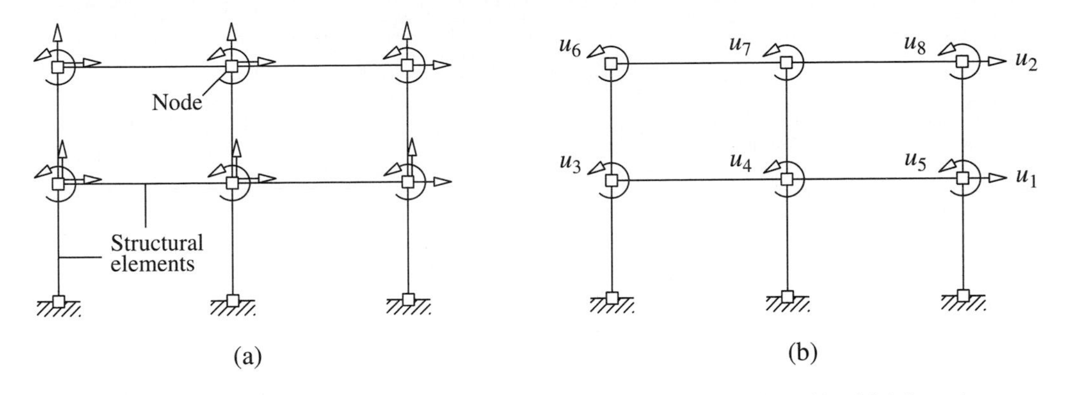

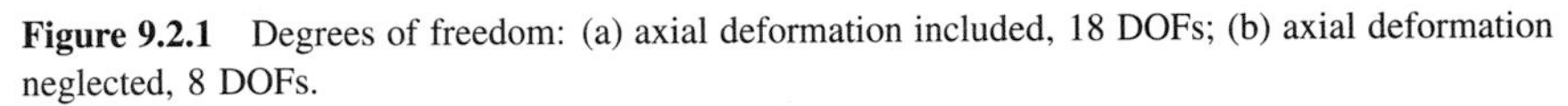

Figure 9.2.1 Degrees of freedom: (a) axial deformation included, 18 DOFs; (b) axial deformation neglected, 8 DOFs.

rise buildings. With these assumptions the two-story, two-bay frame has eight DOFs (Fig. 9.2.1b). This is the structural idealization we use for illustration. The external dynamic forces are applied at the nodes (Fig. 9.2.2). The moments $p_3(t)$ to $p_8(t)$ are zero in most (if not all) practical cases.

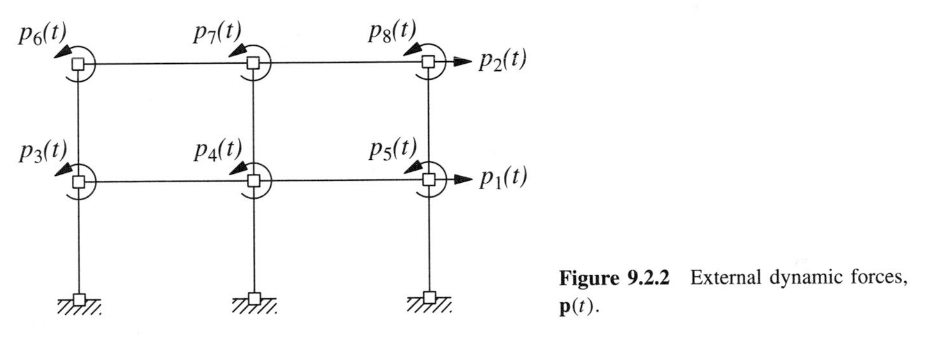

Figure 9.2.2 External dynamic forces, $\mathbf{p}(t)$.

9.2.2 Elastic Forces

We will relate the external forces f_{Sj} on the stiffness component of the structure to the resulting displacements u_j (Fig. 9.2.3a). For linear systems this relationship can be obtained by the method of superposition and the concept of stiffness influence coefficients.

We apply a unit displacement along DOF j, holding all other displacements to zero as shown; to maintain these displacements forces must be applied along all DOFs. The *stiffness influence coefficient* k_{ij} is the force required along DOF i due to unit displacement at DOF j. In particular, the forces k_{i1} $(i = 1, 2, \ldots, 8)$ shown in Fig. 9.2.3b are required to maintain the deflected shape associated with $u_1 = 1$ and all other $u_j = 0$. Similarly, the forces k_{i4} $(i = 1, 2, \ldots, 8)$ shown in Fig. 9.2.3c are required to maintain the deflected shape associated with $u_4 = 1$ and all other $u_j = 0$. All forces in Fig. 9.2.3 are shown with their positive signs, but some of them may be negative to be consistent with the imposed deformations.

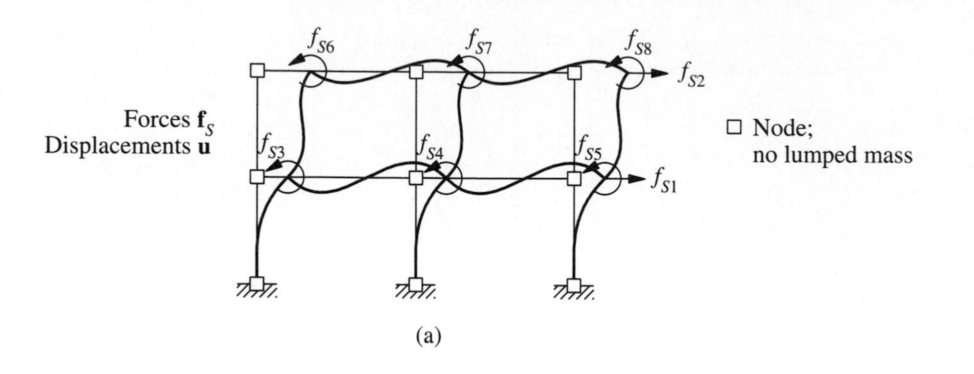

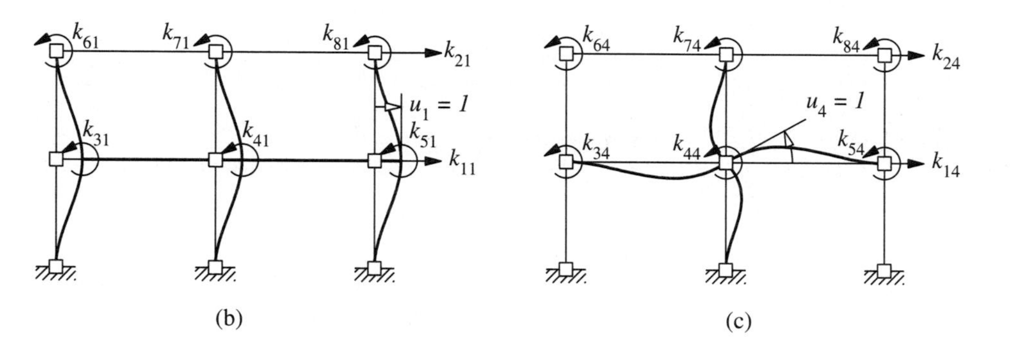

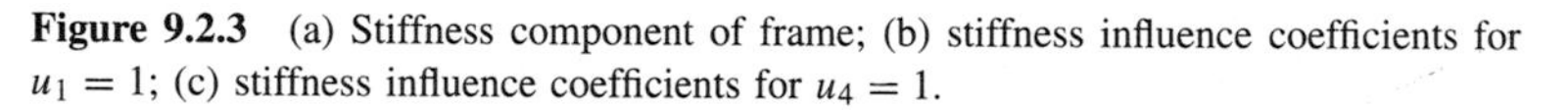

Figure 9.2.3 (a) Stiffness component of frame; (b) stiffness influence coefficients for $u_1 = 1$; (c) stiffness influence coefficients for $u_4 = 1$.

The force f_{Si} at DOF i associated with displacements u_j, $j = 1$ to N (Fig. 9.2.3a), is obtained by superposition:

$$f_{Si} = k_{i1}u_1 + k_{i2}u_2 + \cdots + k_{ij}u_j + \cdots + k_{iN}u_N \tag{9.2.1}$$

One such equation exists for each $i = 1$ to N. The set of N equations can be written in matrix form:

$$\begin{bmatrix} f_{S1} \\ f_{S2} \\ \vdots \\ f_{SN} \end{bmatrix} = \begin{bmatrix} k_{11} & k_{12} & \cdots & k_{1j} & \cdots & k_{1N} \\ k_{21} & k_{22} & \cdots & k_{2j} & \cdots & k_{2N} \\ \vdots & \vdots & & \vdots & & \vdots \\ k_{N1} & k_{N2} & \cdots & k_{Nj} & \cdots & k_{NN} \end{bmatrix} \begin{Bmatrix} u_1 \\ u_2 \\ \vdots \\ u_N \end{Bmatrix} \tag{9.2.2}$$

or

$$\mathbf{f}_S = \mathbf{k}\mathbf{u} \tag{9.2.3}$$

where $\mathbf{k}$ is the *stiffness matrix* of the structure; it is a symmetric matrix (i.e., $k_{ij} = k_{ji}$).

The stiffness matrix $\mathbf{k}$ for a discretized system can be determined by any one of several methods. The jth column of $\mathbf{k}$ can be obtained by calculating the forces k_{ij} $(i = 1, 2, \ldots, N)$ required to produce $u_j = 1$ (with all other $u_i = 0$). The direct equi-

damping for MDF systems is generally specified by numerical values for the damping ratios, as for SDF systems, based on experimental data for similar structures (Chapter 11). Methods are available to construct the damping matrix from known damping ratios (Chapter 11).

9.2.4 Inertia Forces

We will relate the external forces f_{Ij} acting on the mass component of the structure to the accelerations $\ddot{u}_j$ (Fig. 9.2.5a). We apply a unit acceleration along DOF j, while the accelerations in all other DOFs are kept zero. According to D'Alembert's principle, the fictitious inertia forces oppose these accelerations; therefore, external forces will be necessary to equilibrate these inertia forces. The *mass influence coefficient* m_{ij} is the external force in DOF i due to unit acceleration along DOF j. In particular, the forces m_{i1} $(i = 1, 2, \ldots, 8)$ shown in Fig. 9.2.5b are required in the various DOF to equilibrate the inertia forces associated with $\ddot{u}_1 = 1$ and all other $\ddot{u}_j = 0$. Similarly, the forces m_{i4} $(i = 1, 2, \ldots, 8)$ shown in Fig. 9.2.5c are necessary to cause acceleration $\ddot{u}_4 = 1$ and all other $\ddot{u}_j = 0$. The force f_{Ii} at DOF i associated with accelerations $\ddot{u}_j$, $j = 1$ to N

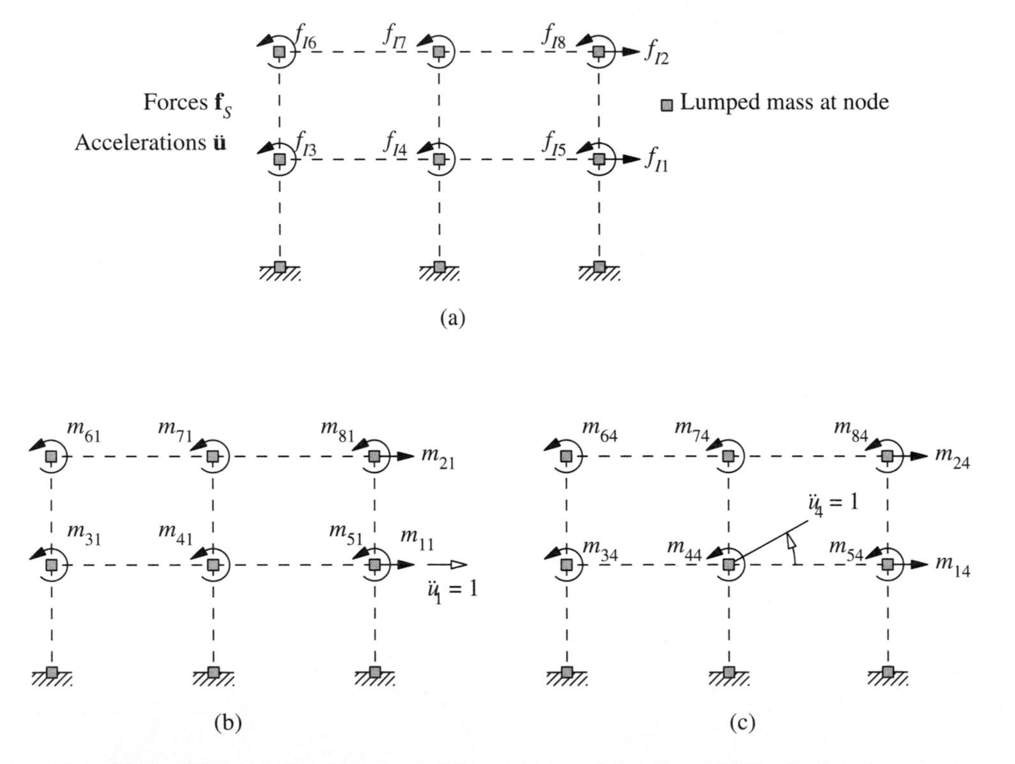

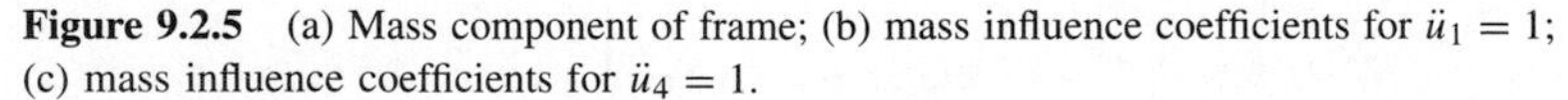

Figure 9.2.5 (a) Mass component of frame; (b) mass influence coefficients for $\ddot{u}_1 = 1$; (c) mass influence coefficients for $\ddot{u}_4 = 1$.

(Fig. 9.2.5a), is obtained by superposition:

$$f_{Ii} = m_{i1}\ddot{u}_1 + m_{i2}\ddot{u}_2 + \cdots + m_{ij}\ddot{u}_j + \cdots + m_{iN}\ddot{u}_N \tag{9.2.7}$$

One such equation exists for each $i = 1$ to N. The set of N equations can be written in matrix form:

$$\begin{bmatrix} f_{I1} \\ f_{I2} \\ \vdots \\ f_{IN} \end{bmatrix} = \begin{bmatrix} m_{11} & m_{12} & \cdots & m_{1j} & \cdots & m_{1N} \\ m_{21} & m_{22} & \cdots & m_{2j} & \cdots & m_{2N} \\ \vdots & \vdots & & \vdots & & \vdots \\ m_{N1} & m_{N2} & \cdots & m_{Nj} & \cdots & m_{NN} \end{bmatrix} \begin{Bmatrix} \ddot{u}_1 \\ \ddot{u}_2 \\ \vdots \\ \ddot{u}_N \end{Bmatrix} \tag{9.2.8}$$

or

$$\mathbf{f}_I = \mathbf{m}\ddot{\mathbf{u}} \tag{9.2.9}$$

where **m** is the *mass matrix.* Just like the stiffness matrix, the mass matrix is symmetric (i.e., $m_{ij} = m_{ji}$).

The mass is distributed throughout an actual structure, but it can be idealized as lumped or concentrated at the nodes of the discretized structure; usually, such a lumped-mass idealization is satisfactory. The lumped mass at a node is determined from the portion of the weight that can reasonably be assigned to the node. Each structural element is replaced by point masses at its two nodes, with the distribution of the two masses being determined by statics. The lumped mass at a node of the structure is the sum of the mass contributions of all the structural elements connected to the node. This procedure is illustrated schematically in Fig. 9.2.6 for a two-story, two-bay frame where the beam mass includes the floor–slab mass it supports. The lumped masses m_a, m_b, and so on, at the various nodes are identified.

Once the lumped masses at the nodes have been calculated, the mass matrix for the structure can readily be formulated. Consider again the two-story, two-bay frame of Fig. 9.2.1b. The external forces associated with acceleration $\ddot{u}_1 = 1$ (Fig. 9.2.5b) are $m_{11} = m_1$, where $m_1 = m_a + m_b + m_c$ (Fig. 9.2.6c), and $m_{i1} = 0$ for $i = 2, 3, \ldots, 8$. Similarly, the external forces m_{i4} associated with $\ddot{u}_4 = 1$ (Fig. 9.2.5c) are zero for all i, except possibly for $i = 4$. The coefficient m_{44} is equal to the rotational inertia of the

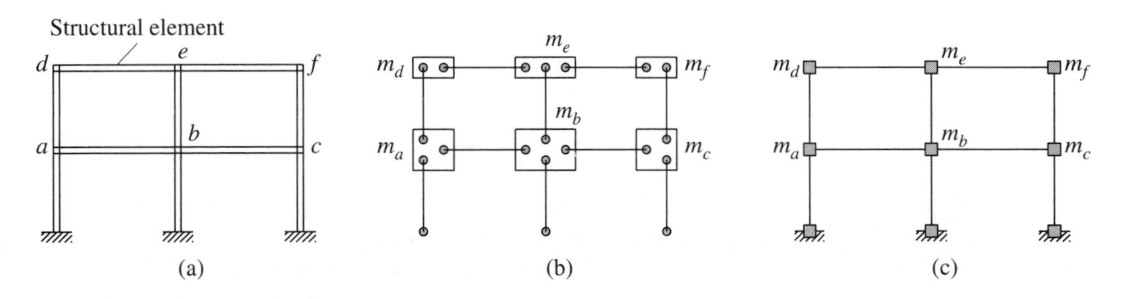

Figure 9.2.6 Lumping of mass at structural nodes.

mass lumped at the middle node at the first floor. This rotational inertia has negligible influence on the dynamics of practical structures; thus $m_{44} = 0$.

In general, then, for a lumped-mass idealization, the mass matrix is diagonal:

$$m_{ij} = 0 \qquad i \neq j \qquad m_{jj} = m_j \quad \text{or} \quad 0 \tag{9.2.10}$$

where m_j is the lumped mass associated with the jth translational DOF, and $m_{jj} = 0$ for a rotational DOF. The mass lumped at a node is associated with all the translational degrees of freedom of that node: (1) the horizontal (x) and vertical (z) DOFs for a two-dimensional frame, and (2) all three (x, y, and z) translational DOFs for a three-dimensional frame.

The mass representation can be simplified for multistory buildings because of the constraining effects of the floor slabs or floor diaphragms. Each floor diaphragm is usually assumed to be rigid in its own plane but is flexible in bending in the vertical direction, which is a reasonable representation of the true behavior of several types of floor systems (e.g., cast-in-place concrete). Introducing this assumption implies that both (x and y) horizontal DOFs of all the nodes at a floor level are related to the three rigid-body DOFs of the floor diaphragm in its own plane. For the jth floor diaphragm these three DOFs, defined at the center of mass, are translations u_{jx} and u_{jy} in the x and y directions and rotation $u_{j\theta}$ about the vertical axis (Fig. 9.2.7). Therefore, the mass needs to be defined only in these DOFs and need not be identified separately for each node. The diaphragm mass gives the mass associated with DOFs u_{jx} and u_{jy}, and the moment of inertia of the diaphragm about the vertical axis through O gives the mass associated with DOF $u_{j\theta}$. The diaphragm mass should include the contributions of the dead load and live load on the diaphragm and of the structural elements—columns, walls, etc.—and of the nonstructural elements—partition walls, architectural finishes, etc.—between floors.

The mass idealization for a multistory building becomes complicated if the floor diaphragm cannot be assumed as rigid in its own plane (e.g., floor system with wood framing and plywood sheathing). The diaphragm mass should then be assigned to individual nodes. The distributed dead and live loads at a floor level are assigned to the nodes at that floor in accordance with their respective tributary areas (Fig. 9.2.8). Similarly, the distributed weights of the structural and nonstructural elements between floors should be distributed to the nodes at the top and bottom of the story according to statics. The diaphragm flexibility must also be recognized in formulating the stiffness properties

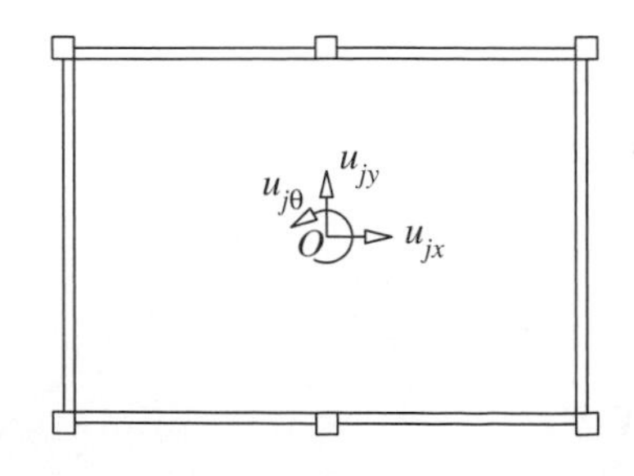

Figure 9.2.7 Degrees of freedom for in-plane-rigid floor diaphragm with distributed mass.

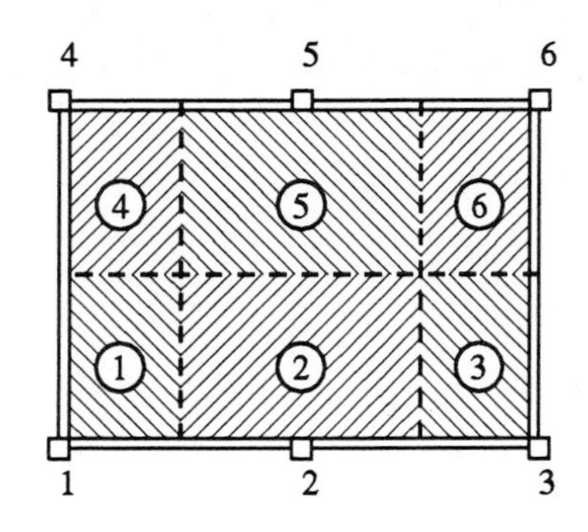

Figure 9.2.8 Tributary areas for distributing diaphragm mass to nodes.

of the structure; the finite element method (Chapter 17) is effective in idealizing flexible diaphragms for this purpose.

9.2.5 Equations of Motion: External Forces

We now write the equations of motion for an MDF system subjected to external dynamic forces $p_j(t)$, $j = 1$ to N. The dynamic response of the structure to this excitation is defined by the displacements $u_j(t)$, velocities $\dot{u}_j(t)$, and accelerations $\ddot{u}_j(t)$, $j = 1$ to N. As mentioned in Section 9.1.4, the external forces $\mathbf{p}(t)$ may be visualized as distributed among the three components of the structure: $\mathbf{f}_S(t)$ to the stiffness components (Fig. 9.2.3a), $\mathbf{f}_D(t)$ to the damping component (Fig. 9.2.4), and $\mathbf{f}_I(t)$ to the mass component (Fig. 9.2.5a). Thus

$$\mathbf{f}_I + \mathbf{f}_D + \mathbf{f}_S = \mathbf{p}(t) \qquad (9.2.11)$$

Substituting Eqs. (9.2.3), (9.2.6), and (9.2.9) into Eq. (9.2.11) gives

$$\mathbf{m}\ddot{\mathbf{u}} + \mathbf{c}\dot{\mathbf{u}} + \mathbf{k}\mathbf{u} = \mathbf{p}(t) \qquad (9.2.12)$$

This is a system of N ordinary differential equations governing the displacements $\mathbf{u}(t)$ due to applied forces $\mathbf{p}(t)$. Equation (9.2.12) is the MDF equivalent of Eq. (1.5.2) for an SDF system; each scalar term in the SDF equation has become a vector or a matrix of order N, the number of DOFs in the MDF system.

Coupling of equations. The off-diagonal terms in the coefficient matrices $\mathbf{m}$, $\mathbf{c}$, and $\mathbf{k}$ are known as *coupling terms.* In general, the equations have mass, damping, and stiffness coupling; however, the coupling in a system depends on the choice of degrees of freedom used to describe the motion. This is illustrated in Examples 9.2 and 9.3 for the same physical system with two different choices for the DOFs.

Example 9.2

A uniform rigid bar of total mass m is supported on two springs k_1 and k_2 at the two ends and subjected to dynamic forces shown in Fig. E9.2a. The bar is constrained so that it can move only vertically in the plane of the paper; with this constraint the system has two DOFs. Formulate the equations of motion with respect to displacements u_1 and u_2 of the two ends as the two DOFs.

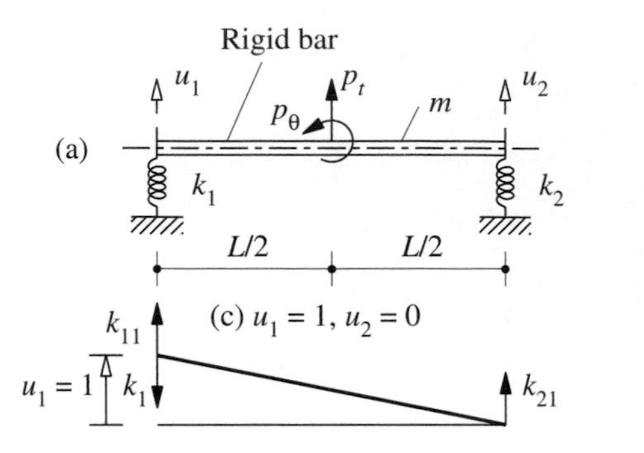

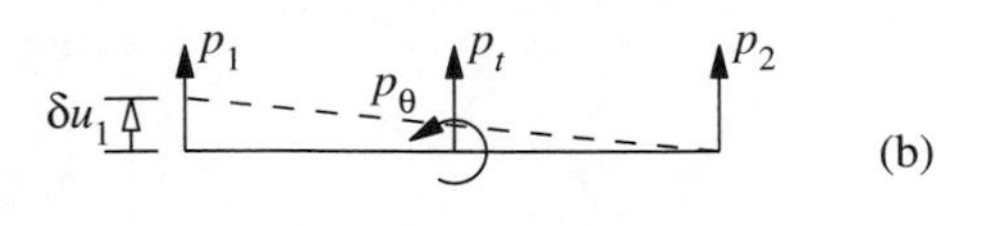

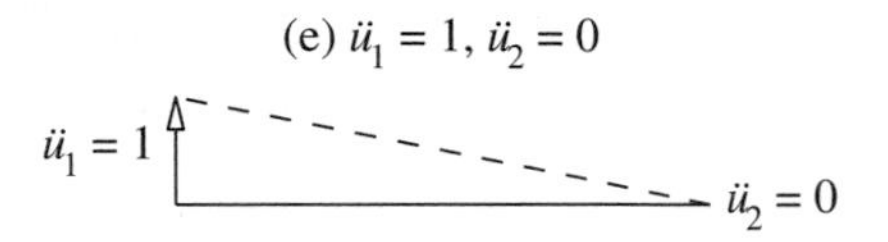

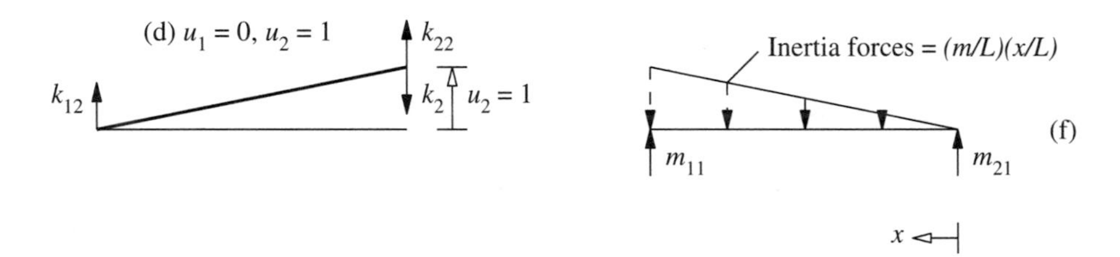

Figure E9.2

Solution

1. *Determine the applied forces.* The external forces do not act along the DOFs and should therefore be converted to equivalent forces p_1 and p_2 along the DOFs (Fig. E9.2b) using equilibrium equations. This can also be achieved by the principle of virtual displacements. Thus if we introduce a virtual displacement δu_1 along DOF 1, the work done by the applied forces is

$$\delta W = p_t \frac{\delta u_1}{2} - p_\theta \frac{\delta u_1}{L} \tag{a}$$

Similarly, the work done by the equivalent forces is

$$\delta W = p_1 \delta u_1 + p_2(0) \tag{b}$$

Because the work done by the two sets of forces should be the same, we equate Eqs. (a) and (b) and obtain

$$p_1 = \frac{p_t}{2} - \frac{p_\theta}{L} \tag{c}$$

In a similar manner, by introducing a virtual displacement δu_2, we obtain

$$p_2 = \frac{p_t}{2} + \frac{p_\theta}{L} \tag{d}$$

2. *Determine the stiffness matrix.* Apply a unit displacement $u_1 = 1$ with $u_2 = 0$ and identify the resulting elastic forces and the stiffness influence coefficients k_{11} and k_{21} (Fig. E9.2c). By statics, $k_{11} = k_1$ and $k_{21} = 0$. Now apply a unit displacement $u_2 = 1$ with $u_1 = 0$ and identify the resulting elastic forces and the stiffness influence coefficients

(Fig. E9.2d). By statics, $k_{12} = 0$ and $k_{22} = k_2$. Thus the stiffness matrix is

$$\mathbf{k} = \begin{bmatrix} k_1 & 0 \\ 0 & k_2 \end{bmatrix} \tag{e}$$

In this case the stiffness matrix is diagonal (i.e., there are no coupling terms) because the two DOFs are defined at the locations of the springs.

3. *Determine the mass matrix.* Impart a unit acceleration $\ddot{u}_1 = 1$ with $\ddot{u}_2 = 0$, determine the distribution of accelerations of (Fig. E9.2e) and the associated inertia forces, and identify mass influence coefficients (Fig. E9.2f). By statics, $m_{11} = m/3$ and $m_{21} = m/6$. Similarly, imparting a unit acceleration $\ddot{u}_2 = 1$ with $\ddot{u}_1 = 0$, defining the inertia forces and mass influence coefficients, and applying statics gives $m_{12} = m/6$ and $m_{22} = m/3$. Thus the mass matrix is

$$\mathbf{m} = \frac{m}{6}\begin{bmatrix} 2 & 1 \\ 1 & 2 \end{bmatrix} \tag{f}$$

The mass matrix is coupled, as indicated by the off-diagonal terms, because the mass is distributed and not lumped at the locations where the DOFs are defined.

4. *Determine the equations of motion.* Substituting Eqs. (c)–(f) in Eq. (9.2.12) with $\mathbf{c} = \mathbf{0}$ gives

$$\frac{m}{6}\begin{bmatrix} 2 & 1 \\ 1 & 2 \end{bmatrix}\begin{bmatrix} \ddot{u}_1 \\ \ddot{u}_2 \end{bmatrix} + \begin{bmatrix} k_1 & 0 \\ 0 & k_2 \end{bmatrix}\begin{bmatrix} u_1 \\ u_2 \end{bmatrix} = \begin{bmatrix} (p_t/2) - (p_\theta/L) \\ (p_t/2) + (p_\theta/L) \end{bmatrix}$$

The two differential equations are coupled because of mass coupling due to the off-diagonal terms in the mass matrix.

Example 9.3

Formulate the equations of motion of the system of Fig. E9.2a with the two DOFs defined at the center of mass O of the rigid bar: translation u_t and rotation u_θ (Fig. E9.3a).

Solution

1. *Determine the stiffness matrix.* Apply a unit displacement $u_t = 1$ with $u_\theta = 0$ and identify the resulting elastic forces and k_{tt} and $k_{\theta t}$ (Fig. E9.3b). By statics, $k_{tt} = k_1 + k_2$ and $k_{\theta t} = (k_2 - k_1)L/2$. Now, apply a unit rotation $u_\theta = 1$ with $u_t = 0$ and identify the resulting elastic forces and $k_{t\theta}$ and $k_{\theta\theta}$ (Fig. E9.3c). By statics, $k_{t\theta} = (k_2 - k_1)L/2$ and $k_{\theta\theta} = (k_1 + k_2)L^2/4$. Thus the stiffness matrix is

$$\bar{\mathbf{k}} = \begin{bmatrix} k_1 + k_2 & (k_2 - k_1)L/2 \\ (k_2 - k_1)L/2 & (k_1 + k_2)L^2/4 \end{bmatrix} \tag{a}$$

Observe that now the stiffness matrix has coupling terms because the chosen DOFs are not the displacements at the locations of the springs.

2. *Determine the mass matrix.* Impart a unit acceleration $\ddot{u}_t = 1$ with $\ddot{u}_\theta = 0$, determine the acceleration distribution (Fig. E9.3d) and the associated inertia forces, and identify m_{tt} and $m_{\theta t}$ (Fig. E9.3e). By statics, $m_{tt} = m$ and $m_{\theta t} = 0$. Now impart a unit rotational acceleration $\ddot{u}_\theta = 1$ with $\ddot{u}_t = 0$, determine the resulting accelerations (Fig. E9.3f) and the associated inertia forces, and identify $m_{t\theta}$ and $m_{\theta\theta}$ (Fig. E9.3g). By statics, $m_{t\theta} = 0$ and $m_{\theta\theta} = mL^2/12$. Note that $m_{\theta\theta} = I_O$, the moment of inertia of the bar about an axis that passes through O and is perpendicular to the plane of rotation. Thus the mass

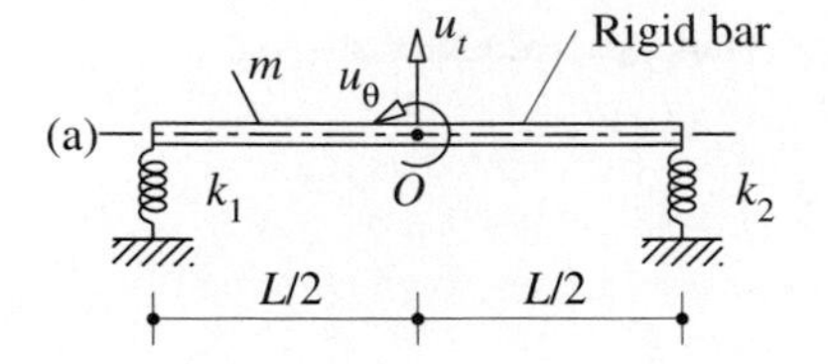

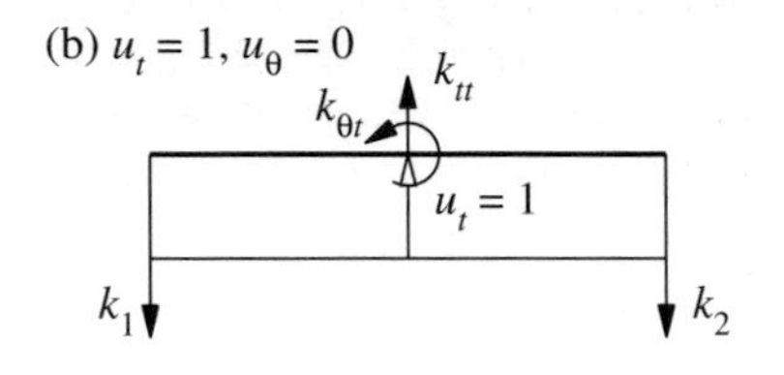

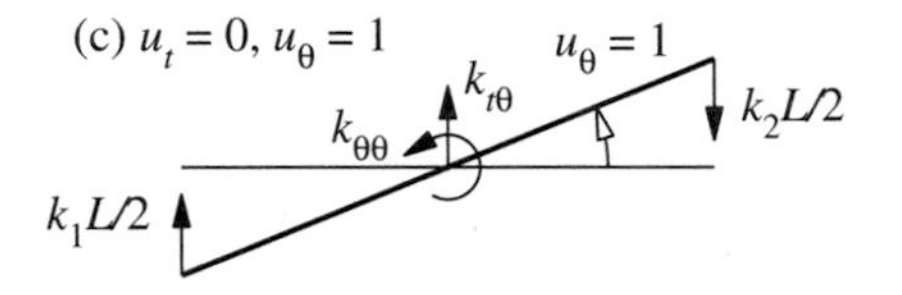

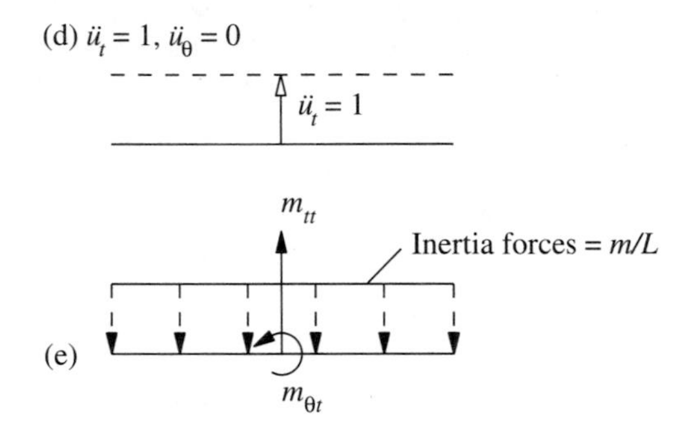

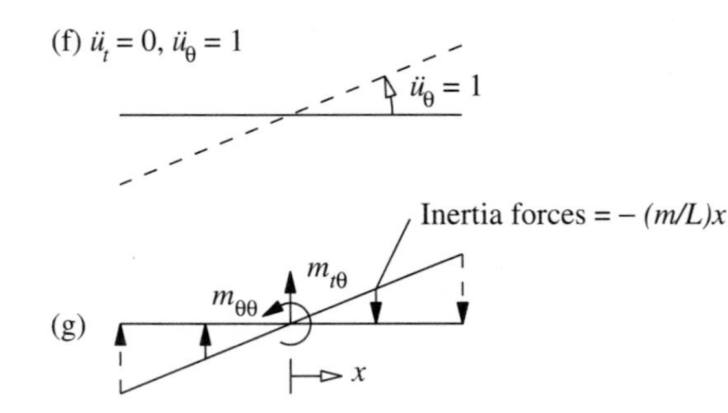

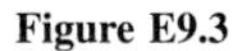
Figure E9.3

matrix is

$$\bar{\mathbf{m}} = \begin{bmatrix} m & 0 \\ 0 & mL^2/12 \end{bmatrix} \tag{b}$$

Now the mass matrix is diagonal (i.e., it has no coupling terms) because the DOFs of this rigid bar are defined at the mass center.

3. *Determine the equations of motion.* Substituting $\mathbf{u} = \langle u_t \quad u_\theta \rangle^T$, $\mathbf{p} = \langle p_t \quad p_\theta \rangle^T$, and Eqs. (a) and (b) in Eq. (9.2.12) gives

$$\begin{bmatrix} m & 0 \\ 0 & mL^2/12 \end{bmatrix} \begin{Bmatrix} \ddot{u}_t \\ \ddot{u}_\theta \end{Bmatrix} + \begin{bmatrix} k_1 + k_2 & (k_2 - k_1)L/2 \\ (k_2 - k_1)L/2 & (k_1 + k_2)L^2/4 \end{bmatrix} \begin{Bmatrix} u_t \\ u_\theta \end{Bmatrix} = \begin{Bmatrix} p_t \\ p_\theta \end{Bmatrix} \tag{c}$$

The two differential equations are now coupled through the stiffness matrix.

We should note that if the equations of motion for a system are available in one set of DOFs, they can be transformed to a different choice of DOF. This concept is illustrated for the system of Fig. E9.2a. Suppose that the mass and stiffness matrices and the applied force vector for the system are available for the first choice of DOF, $\mathbf{u} = \langle u_1 \quad u_2 \rangle^T$. These displacements are related to the second set of DOF, $\bar{\mathbf{u}} = \langle u_t \quad u_\theta \rangle^T$, by

$$\begin{Bmatrix} u_1 \\ u_2 \end{Bmatrix} = \begin{bmatrix} 1 & -L/2 \\ 1 & L/2 \end{bmatrix} \{ u_t \quad u_\theta \} \quad \text{or} \quad \mathbf{u} = \mathbf{a}\bar{\mathbf{u}} \tag{d}$$

where $\mathbf{a}$ denotes the coordinate transformation matrix. The stiffness and mass matrices and the applied force vector for the $\bar{\mathbf{u}}$ DOFs are given by

$$\bar{\mathbf{k}} = \mathbf{a}^T\mathbf{k}\mathbf{a} \qquad \bar{\mathbf{m}} = \mathbf{a}^T\mathbf{m}\mathbf{a} \qquad \bar{\mathbf{p}} = \mathbf{a}^T\mathbf{p} \tag{e}$$

Substituting for $\mathbf{a}$ from Eq. (d) and for $\mathbf{k}$, $\mathbf{m}$, and $\mathbf{p}$ from Example 9.2 into Eq. (e) leads to $\bar{\mathbf{k}}$ and $\bar{\mathbf{m}}$, which are identical to Eqs. (a) and (b) and to the $\bar{\mathbf{p}}$ in Eq. (c).

Example 9.4

A massless cantilever beam of length L supports two lumped masses $mL/2$ and $mL/4$ at the midpoint and free end as shown in Fig. E9.4a. The flexural rigidity of the uniform beam is EI. With the four DOFs chosen as shown in Fig. E9.4b and the applied forces $p_1(t)$ and $p_2(t)$, formulate the equations of motion of the system. Axial and shear deformations in the beam are neglected.

Solution

The beam consists of two beam elements and three nodes. The left node is constrained and each of the other two nodes has two DOFs (Fig. E9.4b).

1. *Determine the mass matrix.* With the DOFs defined at the locations of the lumped masses, the diagonal mass matrix is given by Eq. (9.2.10):

$$\mathbf{m} = \begin{bmatrix} mL/4 & & & \\ & mL/2 & & \\ & & 0 & \\ & & & 0 \end{bmatrix} \tag{a}$$

2. *Determine the stiffness matrix.* Several methods are available to determine the stiffness matrix. Here we use the direct equilibrium method based on the definition of stiffness influence coefficients (Appendix 1).

To obtain the first column of the stiffness matrix, we impose $u_1 = 1$ and $u_2 = u_3 = u_4 = 0$. The stiffness influence coefficients are k_{i1} (Fig. E9.4c). The forces necessary at the nodes of each beam element to maintain the deflected shape are determined from the beam stiffness coefficients (Fig. E9.4d). The two sets of forces in figures (c) and (d) are one and the same. Thus $k_{11} = 96EI/L^3$, $k_{21} = -96EI/L^3$, $k_{31} = -24EI/L^2$, and $k_{41} = -24EI/L^2$.

The second column of the stiffness matrix is obtained in a similar manner by imposing $u_2 = 1$ with $u_1 = u_3 = u_4 = 0$. The stiffness influence coefficients are k_{i2} (Fig. E9.4e) and the forces on each beam element necessary to maintain the imposed displacements are shown in Fig. E9.4f. The two sets of forces in figures (e) and (f) are one and the same. Thus $k_{12} = -96EI/L^3$, $k_{32} = 24EI/L^2$, $k_{22} = 96EI/L^3 + 96EI/L^3 = 192EI/L^3$, and $k_{42} = -24EI/L^2 + 24EI/L^2 = 0$.

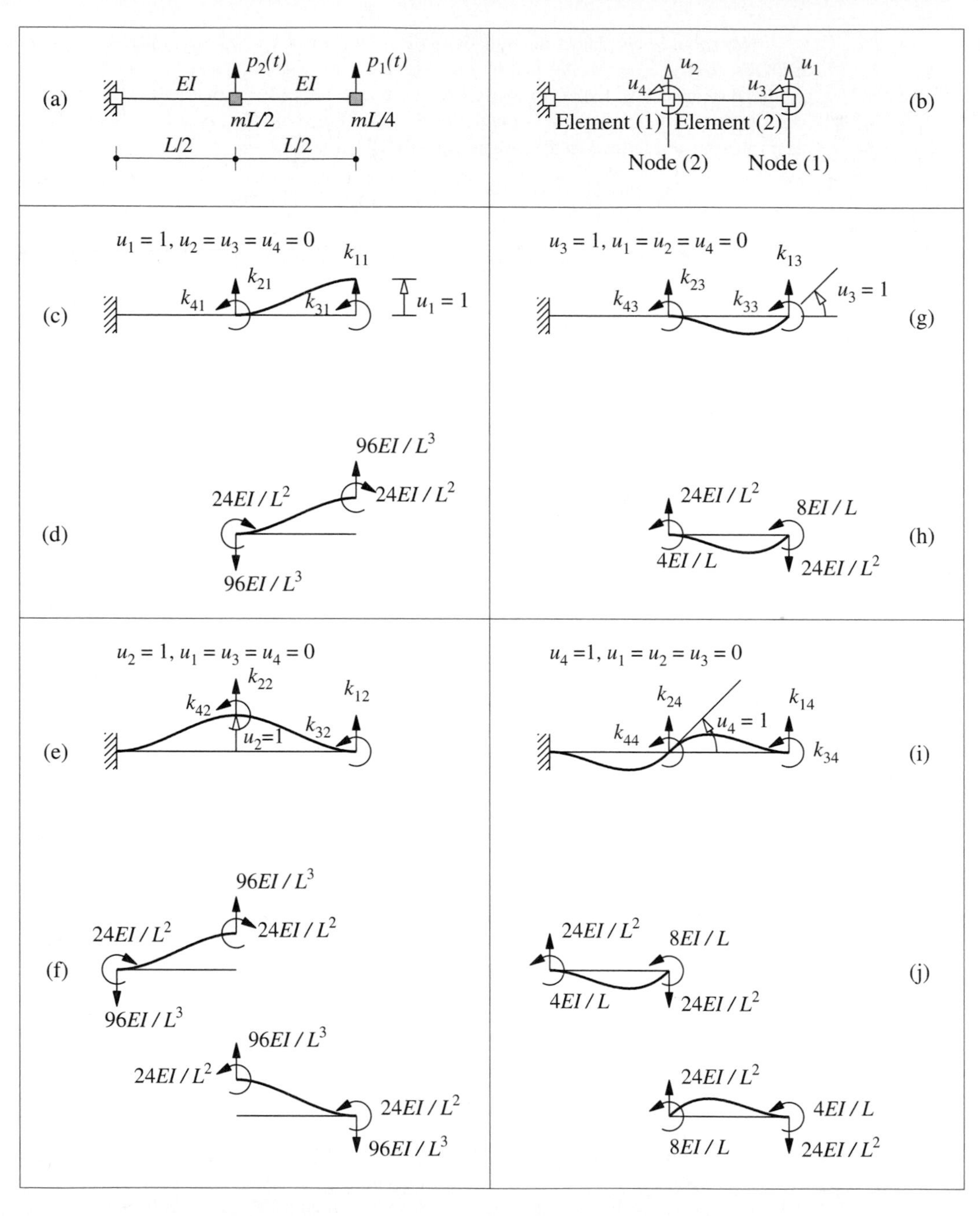
(a)
$p_2(t)$
$p_1(t)$
EI
EI
mL/2
mL/4
L/2
L/2
(b)
u_2
u_1
u_4
u_3
Element (1)
Element (2)
Node (2)
Node (1)
(c)
$u_1 = 1, u_2 = u_3 = u_4 = 0$
k_{11}
k_{21}
k_{41}
k_{31}
$u_1 = 1$
(g)
$u_3 = 1, u_1 = u_2 = u_4 = 0$
k_{13}
k_{23}
k_{43}
k_{33}
$u_3 = 1$
(d)
$96EI / L^3$
$24EI / L^2$
$24EI / L^2$
$96EI / L^3$
(h)
$24EI / L^2$
$8EI / L$
$4EI / L$
$24EI / L^2$
(e)
$u_2 = 1, u_1 = u_3 = u_4 = 0$
k_{22}
k_{12}
k_{42}
k_{32}
$u_2=1$
(i)
$u_4 =1, u_1 = u_2 = u_3 = 0$
k_{24}
k_{14}
k_{44}
$u_4 = 1$
k_{34}
(f)
$96EI / L^3$
$24EI / L^2$
$24EI / L^2$
$96EI / L^3$
$96EI / L^3$
$24EI / L^2$
$24EI / L^2$
$96EI / L^3$
(j)
$24EI / L^2$
$8EI / L$
$4EI / L$
$24EI / L^2$
$24EI / L^2$
$4EI / L$
$8EI / L$
$24EI / L^2$

Figure E9.4

The third column of the stiffness matrix is obtained in a similar manner by imposing $u_3 = 1$ with $u_1 = u_2 = u_4 = 0$. The stiffness influence coefficients k_{i3} are shown in Fig. E9.4g and the nodal forces in Fig. E9.4h. Thus $k_{13} = -24EI/L^2$, $k_{23} = 24EI/L^2$, $k_{33} = 8EI/L$, and $k_{43} = 4EI/L$.

The fourth column of the stiffness matrix is obtained in a similar manner by imposing $u_4 = 1$ with $u_1 = u_2 = u_3 = 0$. The stiffness influence coefficients k_{i4} are shown in Fig. E9.4i, and the nodal forces in Fig. E9.4j. Thus $k_{14} = -24EI/L^2$, $k_{34} = 4EI/L$, $k_{24} = -24EI/L^2 + 24EI/L^2 = 0$, and $k_{44} = 8EI/L + 8EI/L = 16EI/L$.

With all the stiffness influence coefficients determined, the stiffness matrix is

$$\mathbf{k} = \frac{8EI}{L^3}\begin{bmatrix} 12 & -12 & -3L & -3L \\ -12 & 24 & 3L & 0 \\ -3L & 3L & L^2 & L^2/2 \\ -3L & 0 & L^2/2 & 2L^2 \end{bmatrix} \tag{b}$$

3. *Determine the equations of motion.* The governing equations are

$$\mathbf{m}\ddot{\mathbf{u}} + \mathbf{k}\mathbf{u} = \mathbf{p}(t) \tag{c}$$

where $\mathbf{u} = \langle u_1 \quad u_2 \quad u_3 \quad u_4 \rangle^T$, $\mathbf{m}$ and $\mathbf{k}$ are given by Eqs. (a) and (b), and $\mathbf{p}(t) = \langle p_1(t) \quad p_2(t) \quad 0 \quad 0 \rangle^T$.

Example 9.5

Derive the equations of motion of the beam of Example 9.4 (also shown in Fig. E9.5a) expressed in terms of the displacements u_1 and u_2 of the masses (Fig. E9.5b).

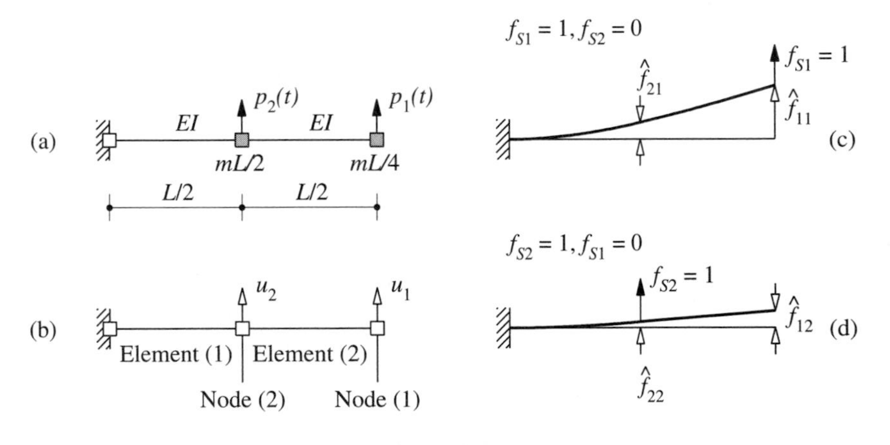

Figure E9.5

Solution This system is the same as that in Example 9.4, but its equations of motion will be formulated considering only the translational DOFs u_1 and u_2 (i.e., the rotational DOFs u_3 and u_4 will be excluded).

1. *Determine the stiffness matrix.* In a statically determinate structure such as the one in Fig. E9.5a, it is usually easier to calculate first the flexibility matrix and invert it to obtain the stiffness matrix. The flexibility influence coefficient $\hat{f}_{ij}$ is the displacement in

DOF i due to unit force applied in DOF j (Fig. E9.4c and d). The deflections are computed by standard procedures of structural analysis to obtain the flexibility matrix:

$$\hat{\mathbf{f}} = \frac{L^3}{48EI}\begin{bmatrix} 16 & 5 \\ 5 & 2 \end{bmatrix}$$

The off-diagonal elements $\hat{f}_{12}$ and $\hat{f}_{21}$ are equal, as expected, because of Maxwell's theorem of reciprocal deflections. By inverting $\hat{\mathbf{f}}$, the stiffness matrix is obtained:

$$\mathbf{k} = \frac{48EI}{7L^3}\begin{bmatrix} 2 & -5 \\ -5 & 16 \end{bmatrix} \tag{a}$$

2. *Determine the mass matrix.* This is a diagonal matrix because the lumped masses are located where the DOFs are defined:

$$\mathbf{m} = \begin{bmatrix} mL/4 & \\ & mL/2 \end{bmatrix} \tag{b}$$

3. *Determine the equations of motion.* Substituting $\mathbf{m}$, $\mathbf{k}$, and $\mathbf{p}(t) = \langle\, p_1(t) \quad p_2(t)\,\rangle^T$ in Eq. (9.2.12) with $\mathbf{c} = \mathbf{0}$ gives

$$\begin{bmatrix} mL/4 & \\ & mL/2 \end{bmatrix}\begin{Bmatrix} \ddot{u}_1 \\ \ddot{u}_2 \end{Bmatrix} + \frac{48EI}{7L^3}\begin{bmatrix} 2 & -5 \\ -5 & 16 \end{bmatrix}\begin{Bmatrix} u_1 \\ u_2 \end{Bmatrix} = \begin{Bmatrix} p_1(t) \\ p_2(t) \end{Bmatrix} \tag{c}$$

Example 9.6

Formulate the free vibration equations for the two-element frame of Fig. E9.6a. For both elements the flexural stiffness is EI, and axial deformations are to be neglected. The frame is massless with lumped masses at the two nodes as shown.

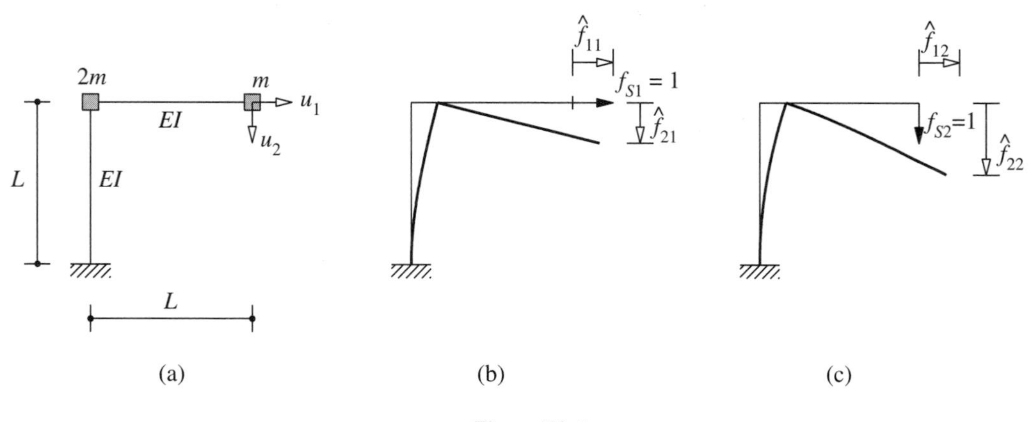

Figure E9.6

Solution The two degrees of freedom of the frame are shown. The mass matrix is

$$\mathbf{m} = \begin{bmatrix} 3m & \\ & m \end{bmatrix} \tag{a}$$

Note that the mass corresponding to $\ddot{u}_1 = 1$ is $2m+m = 3m$ because both masses will undergo the same acceleration since the beam connecting the two masses is axially inextensible.

The stiffness matrix is formulated by first evaluating the flexibility matrix and then inverting it. The flexibility influence coefficients are identified in Fig. E9.6b and c, and the deflections are computed by standard procedures of structural analysis to obtain the flexibility matrix

$$\hat{\mathbf{f}} = \frac{L^3}{6EI}\begin{bmatrix} 2 & 3 \\ 3 & 8 \end{bmatrix}$$

This matrix is inverted to determine the stiffness matrix:

$$\mathbf{k} = \frac{6EI}{7L^3}\begin{bmatrix} 8 & -3 \\ -3 & 2 \end{bmatrix}$$

Thus the equations in free vibration of the system (without damping) are

$$\begin{bmatrix} 3m & \\ & m \end{bmatrix}\begin{Bmatrix} \ddot{u}_1 \\ \ddot{u}_2 \end{Bmatrix} + \frac{6EI}{7L^3}\begin{bmatrix} 8 & -3 \\ -3 & 2 \end{bmatrix}\begin{Bmatrix} u_1 \\ u_2 \end{Bmatrix} = \begin{Bmatrix} 0 \\ 0 \end{Bmatrix}$$

Example 9.7

Formulate the equations of motion for the two-story frame in Fig. E9.7a. The flexural rigidity of the beams and columns and the lumped masses at the floor levels are as noted. The dynamic excitation consists of lateral forces $p_1(t)$ and $p_2(t)$ at the two floor levels. The story height is h and the bay width $2h$. Neglect axial deformations in the beams and the columns.

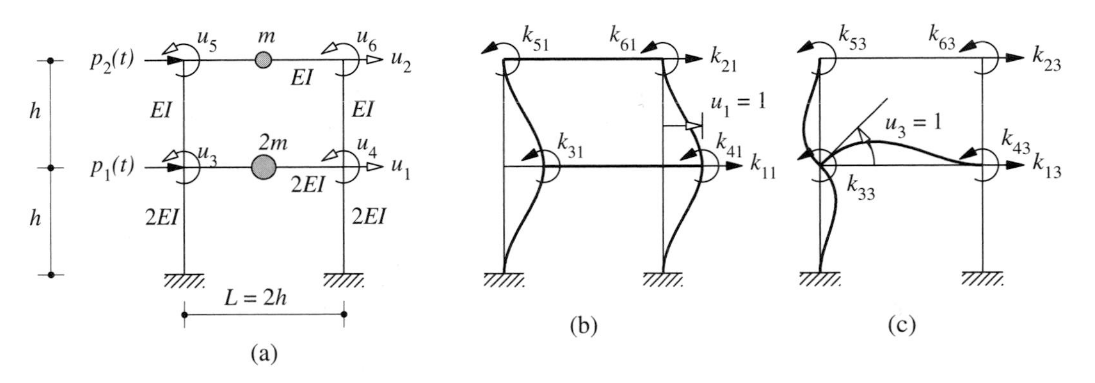

Figure E9.7

Solution The system has six degrees of freedom shown in Fig. E9.7a: lateral displacements u_1 and u_2 of the floors and joint rotations u_3, u_4, u_5, and u_6. The displacement vector is

$$\mathbf{u} = \langle u_1 \quad u_2 \quad u_3 \quad u_4 \quad u_5 \quad u_6 \rangle^T \tag{a}$$

The mass matrix is given by Eq. (9.2.10):

$$\mathbf{m} = m\begin{bmatrix} 2 & & & & & \\ & 1 & & & & \\ & & 0 & & & \\ & & & 0 & & \\ & & & & 0 & \\ & & & & & 0 \end{bmatrix} \tag{b}$$

The stiffness influence coefficients are evaluated following the procedure of Example 9.4. A unit displacement is imposed, one at a time, in each DOF while constraining the other five DOFs, and the stiffness influence coefficients (e.g., shown in Fig. E9.7b and c for $u_1 = 1$ and $u_3 = 1$, respectively) are calculated by statics from the nodal forces associated with the imposed displacements. These nodal forces are determined from the beam stiffness coefficients (Appendix 1). The result is

$$\mathbf{k} = \frac{EI}{h^3}\begin{bmatrix} 72 & -24 & 6h & 6h & -6h & -6h \\ -24 & 24 & 6h & 6h & 6h & 6h \\ 6h & 6h & 16h^2 & 2h^2 & 2h^2 & 0 \\ 6h & 6h & 2h^2 & 16h^2 & 0 & 2h^2 \\ -6h & 6h & 2h^2 & 0 & 6h^2 & h^2 \\ -6h & 6h & 0 & 2h^2 & h^2 & 6h^2 \end{bmatrix} \tag{c}$$

The dynamic forces applied are lateral forces $p_1(t)$ and $p_2(t)$ at the two floors without any moments at the nodes. Thus the applied force vector is

$$\mathbf{p}(t) = \langle\, p_1(t) \quad p_2(t) \quad 0 \quad 0 \quad 0 \quad 0 \rangle^T \tag{d}$$

The equations of motion are

$$\mathbf{m}\ddot{\mathbf{u}} + \mathbf{k}\mathbf{u} = \mathbf{p}(t)$$

where $\mathbf{u}$, $\mathbf{m}$, $\mathbf{k}$, and $\mathbf{p}(t)$ are given by Eqs. (a), (b), (c), and (d), respectively.

9.3 STATIC CONDENSATION

The static condensation method is used to eliminate from dynamic analysis those DOFs of a structure to which zero mass is assigned; however, all the DOFs are included in the static analysis. Consider the two-bay, two-story frame shown in Fig. 9.3.1. With axial deformations in structural elements neglected, the system has eight DOFs for formulating its stiffness matrix (Fig. 9.3.1a). As discussed in Section 9.2.4, typically the mass of the structure is idealized as concentrated in point lumps at the nodes (Fig. 9.3.1b), and the mass matrix contains zero diagonal elements in the rotational DOFs (see also Example 9.7). These are the DOFs that can be eliminated from the dynamic analysis of the structure provided that the dynamic excitation does not include any external forces in the rotational DOFs, as in the case of earthquake excitation (Section 9.4). Even if included in

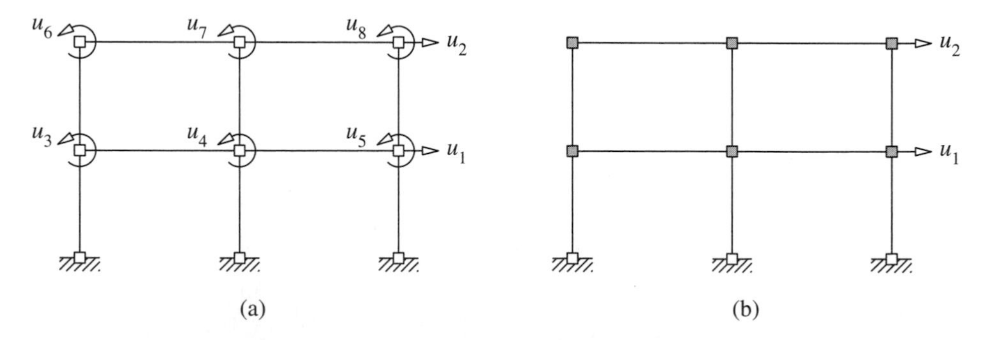

Figure 9.3.1 (a) Degrees of freedom (DOFs) for elastic forces—axial deformations neglected; (b) DOFs for inertia forces.

formulating the stiffness matrix, the vertical DOFs of the building can also be eliminated from dynamic analysis—because the inertial effects associated with the vertical DOFs of building frames are usually small—provided that the dynamic excitation does not include vertical forces at the nodes, as in the case of horizontal ground motion (Section 9.4).

The equations of motion for a system excluding damping [Eq. (9.2.12)] are written in partitioned form:

$$\begin{bmatrix} \mathbf{m}_{tt} & \mathbf{0} \\ \mathbf{0} & \mathbf{0} \end{bmatrix} \begin{Bmatrix} \ddot{\mathbf{u}}_t \\ \ddot{\mathbf{u}}_0 \end{Bmatrix} + \begin{bmatrix} \mathbf{k}_{tt} & \mathbf{k}_{t0} \\ \mathbf{k}_{0t} & \mathbf{k}_{00} \end{bmatrix} \begin{Bmatrix} \mathbf{u}_t \\ \mathbf{u}_0 \end{Bmatrix} = \begin{Bmatrix} \mathbf{p}_t(t) \\ \mathbf{0} \end{Bmatrix} \tag{9.3.1}$$

where $\mathbf{u}_0$ denotes the DOFs with zero mass and $\mathbf{u}_t$ the remaining DOFs. The two partitioned equations are:

$$\mathbf{m}_{tt}\ddot{\mathbf{u}}_t + \mathbf{k}_{tt}\mathbf{u}_t + \mathbf{k}_{t0}\mathbf{u}_0 = \mathbf{p}_t(t) \qquad \mathbf{k}_{0t}\mathbf{u}_t + \mathbf{k}_{00}\mathbf{u}_0 = \mathbf{0} \tag{9.3.2}$$

Because no inertia terms or external forces are associated with $\mathbf{u}_0$, Eq. (9.3.2b) permits a static relationship between $\mathbf{u}_0$ and $\mathbf{u}_t$:

$$\mathbf{u}_0 = -\mathbf{k}_{00}^{-1}\mathbf{k}_{0t}\mathbf{u}_t \tag{9.3.3}$$

Substituting Eq. (9.3.3) in Eq. (9.3.2a) gives

$$\mathbf{m}_{tt}\ddot{\mathbf{u}}_t + \hat{\mathbf{k}}_{tt}\mathbf{u}_t = \mathbf{p}_t(t) \tag{9.3.4}$$

where $\hat{\mathbf{k}}_{tt}$ is the *condensed stiffness matrix* given by

$$\hat{\mathbf{k}}_{tt} = \mathbf{k}_{tt} - \mathbf{k}_{0t}^T\mathbf{k}_{00}^{-1}\mathbf{k}_{0t} \tag{9.3.5}$$

Solution of Eq. (9.3.4) provides the displacements $\mathbf{u}_t(t)$ in the dynamic DOFs, and at each instant of time the displacements $\mathbf{u}_0(t)$ in the condensed DOFs are determined from Eq. (9.3.3). Henceforth, for notational convenience, Eq. (9.2.12) will also denote the equations of motion governing the dynamic DOFs [Eq. (9.3.4)], and it will be understood that static condensation to retain only the dynamic DOFs has already taken place before formulating these equations.

Example 9.8

Examples 9.4 and 9.5 were concerned with formulating the equations of motion for a cantilever beam with two lumped masses. The degrees of freedom chosen in Example 9.5 were the translational displacements u_1 and u_2 at the lumped masses; in Example 9.4 the four DOFs were u_1, u_2, and node rotations u_3 and u_4. Starting with the equations governing these four DOFs, derive the equations of motion in the two translational DOFs.

Solution The vector of four DOFs is partitioned in two parts: $\mathbf{u}_t = \langle u_1 \quad u_2 \rangle$ and $\mathbf{u}_0 = \langle u_3 \quad u_4 \rangle$. The equations of motion governing $\mathbf{u}_t$ are given by Eq. (9.3.4), where

$$\mathbf{m}_{tt} = \begin{bmatrix} mL/4 & \\ & mL/2 \end{bmatrix} \qquad \mathbf{p}_t(t) = \langle p_1(t) \quad p_2(t) \rangle \tag{a}$$

To determine $\hat{\mathbf{k}}_{tt}$, the 4×4 stiffness matrix determined in Example 9.4 is partitioned:

$$\mathbf{k} = \begin{bmatrix} \mathbf{k}_{tt} & \mathbf{k}_{t0} \\ \mathbf{k}_{0t} & \mathbf{k}_{00} \end{bmatrix} = \frac{8EI}{L^3} \left[\begin{array}{cc|cc} 12 & -12 & -3L & -3L \\ -12 & 24 & 3L & 0 \\ \hline -3L & 3L & L^2 & L^2/2 \\ -3L & 0 & L^2/2 & 2L^2 \end{array}\right] \tag{b}$$

Substituting these submatrices in Eq. (9.3.5) gives the condensed stiffness matrix:

$$\hat{\mathbf{k}}_{tt} = \frac{48EI}{7L^3}\begin{bmatrix} 2 & -5 \\ -5 & 16 \end{bmatrix} \tag{c}$$

This stiffness matrix of Eq. (c) is the same as that obtained in Example 9.5 by inverting the flexibility matrix corresponding to the two translational DOFs.

Substituting the stiffness submatrices in Eq. (9.3.3) gives the relation between the condensed DOF $\mathbf{u}_0$ and the dynamic DOF $\mathbf{u}_t$:

$$\mathbf{u}_0 = \mathbf{T}\mathbf{u}_t \qquad \mathbf{T} = \frac{1}{L}\begin{bmatrix} 2.57 & -3.43 \\ 0.857 & 0.857 \end{bmatrix} \tag{d}$$

The equations of motion are given by Eq. (9.3.4), where $\mathbf{m}_{tt}$ and $\mathbf{p}_t(t)$ are defined in Eq. (a) and $\hat{\mathbf{k}}_{tt}$ in Eq. (c). These are the same as Eq. (c) of Example 9.5.

Example 9.9

Formulate the equations of motion for the two-story frame of Example 9.7 governing the lateral floor displacements u_1 and u_2.

Solution The equations of motion for this system were formulated in Example 9.7 considering six DOFs which are partitioned into $\mathbf{u}_t = \langle u_1 \quad u_2 \rangle$ and $\mathbf{u}_0 = \langle u_3 \quad u_4 \quad u_5 \quad u_6 \rangle$.

The equations governing $\mathbf{u}_t$ are given by Eq. (9.3.4), where

$$\mathbf{m}_{tt} = m\begin{bmatrix} 2 & \\ & 1 \end{bmatrix} \qquad \mathbf{p}_t(t) = \langle p_1(t) \quad p_2(t) \rangle \tag{a}$$

To determine $\mathbf{k}_{tt}$, the 6×6 stiffness matrix determined in Example 9.7 is partitioned:

$$\mathbf{k} = \begin{bmatrix} \mathbf{k}_{tt} & \mathbf{k}_{t0} \\ \mathbf{k}_{0t} & \mathbf{k}_{00} \end{bmatrix} = \frac{EI}{h^3}\left[\begin{array}{cc:cccc} 72 & -24 & 6h & 6h & -6h & -6h \\ -24 & 24 & 6h & 6h & 6h & 6h \\ \hdashline 6h & 6h & 16h^2 & 2h^2 & 2h^2 & 0 \\ 6h & 6h & 2h^2 & 16h^2 & 0 & 2h^2 \\ -6h & 6h & 2h^2 & 0 & 6h^2 & h^2 \\ -6h & 6h & 0 & 2h^2 & h^2 & 6h^2 \end{array}\right] \tag{b}$$

Substituting these submatrices in Eq. (9.3.5) gives the condensed stiffness matrix:

$$\hat{\mathbf{k}}_{tt} = \frac{EI}{h^3}\begin{bmatrix} 54.88 & -17.51 \\ -17.51 & 11.61 \end{bmatrix} \tag{c}$$

This is called the *lateral stiffness matrix* because the DOFs are the lateral displacements of the floors. It enters into the earthquake analysis of buildings (Section 9.4).

Substituting the stiffness submatrices in Eq. (9.3.3) gives the relation between the condensed DOF $\mathbf{u}_0$ and the translational DOF $\mathbf{u}_t$:

$$\mathbf{u}_0 = \mathbf{T}\mathbf{u}_t \qquad \mathbf{T} = \frac{1}{h}\begin{bmatrix} -0.4426 & -0.2459 \\ -0.4426 & -0.2459 \\ 0.9836 & -0.7869 \\ 0.9836 & -0.7869 \end{bmatrix} \tag{d}$$

The equations of motion are given by Eq. (9.3.4), where $\mathbf{m}_{tt}$ and $\mathbf{p}_t$ are defined in Eq. (a) and $\hat{\mathbf{k}}_{tt}$ in Eq. (c):

$$m\begin{bmatrix} 2 & \\ & 1 \end{bmatrix}\begin{Bmatrix} \ddot{u}_1 \\ \ddot{u}_2 \end{Bmatrix} + \frac{EI}{h^3}\begin{bmatrix} 54.88 & -17.51 \\ -17.51 & 11.61 \end{bmatrix}\begin{Bmatrix} u_1 \\ u_2 \end{Bmatrix} = \begin{Bmatrix} p_1(t) \\ p_2(t) \end{Bmatrix} \tag{e}$$

9.4 PLANAR OR SYMMETRIC-PLAN SYSTEMS: GROUND MOTION

One of the important applications of structural dynamics is in predicting how structures respond to earthquake-induced motion of the base of the structure. In this and following sections equations of motion for MDF systems subjected to earthquake excitation are formulated. Planar systems subjected to translational and rotational ground motions are considered in Sections 9.4.1 and 9.4.3, symmetric-plan buildings subjected to translational and torsional excitations in Sections 9.4.2 and 9.6, and unsymmetric-plan buildings subjected to translational ground motion in Section 9.5. Systems excited by different prescribed motions at their multiple supports are the subject of Section 9.7.

9.4.1 Planar Systems: Translational Ground Motion

We start with the simplest case where all the dynamic degrees of freedom are displacements in the same direction as the ground motion. Two such structures—a tower and a building frame—are shown in Fig. 9.4.1. The displacement of the ground is denoted by u_g, the total (or absolute) displacement of the mass m_j by u_j^t, and the relative displacement between this mass and the ground by u_j. At each instant of time these displacements are related by

$$u_j^t(t) = u_j(t) + u_g(t) \tag{9.4.1a}$$

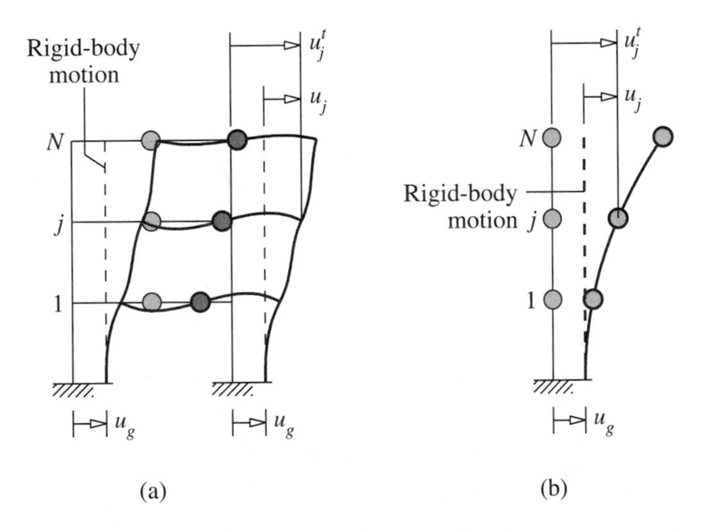

Figure 9.4.1 (a) Building frame; (b) tower.

Such equations for all the N masses can be combined in vector form:

$$\mathbf{u}^t(t) = \mathbf{u}(t) + u_g(t)\mathbf{1} \tag{9.4.1b}$$

where $\mathbf{1}$ is a vector of order N with each element equal to unity.

The equation of dynamic equilibrium, Eq. (9.2.11), developed earlier is still valid, except that $\mathbf{p}(t) = \mathbf{0}$ because no external dynamic forces are applied. Thus

$$\mathbf{f}_I + \mathbf{f}_D + \mathbf{f}_S = \mathbf{0} \tag{9.4.2}$$

Only the relative motions $\mathbf{u}$ between the masses and the base due to structural deformations produce elastic and damping forces (i.e., the rigid-body component of the displacement of the structure produces no internal forces). Thus for a linear system, Eqs. (9.2.3) and (9.2.6) are still valid. However, the inertia forces $\mathbf{f}_I$ are related to the total accelerations $\ddot{\mathbf{u}}^t$ of the masses, and Eq. (9.2.9) becomes

$$\mathbf{f}_I = \mathbf{m}\ddot{\mathbf{u}}^t \tag{9.4.3}$$

Substituting Eqs. (9.2.3), (9.2.6), and (9.4.3) in Eq. (9.4.2) and using Eq. (9.4.1b) gives

$$\mathbf{m}\ddot{\mathbf{u}} + \mathbf{c}\dot{\mathbf{u}} + \mathbf{k}\mathbf{u} = -\mathbf{m}\mathbf{1}\ddot{u}_g(t) \tag{9.4.4}$$

Equation (9.4.4) contains N differential equations governing the relative displacements $u_j(t)$ of a linearly elastic MDF system subjected to ground acceleration $\ddot{u}_g(t)$. The stiffness matrix in Eq. (9.4.4) refers to the horizontal displacements u_j and is obtained by the static condensation method (Section 9.3) to eliminate the rotational and vertical DOF of the nodes; hence this $\mathbf{k}$ is known as the *lateral stiffness matrix.*

Comparison of Eqs. (9.4.4) with Eq. (9.2.12) shows that the equations of motion for the structure subjected to two separate excitations—ground acceleration = $\ddot{u}_g(t)$ and external forces = $-m_j\ddot{u}_g(t)$—are one and the same. As shown in Fig. 9.4.2, the ground motion can therefore be replaced by the *effective earthquake forces:*

$$\mathbf{p}_{\text{eff}}(t) = -\mathbf{m}\mathbf{1}\ddot{u}_g(t) \tag{9.4.5}$$

A generalization of the preceding derivation is useful if all the DOFs of the system are not in the direction of the ground motion (later in this section), or if the earthquake excitation is not identical at all the structural supports (Section 9.7). In this general

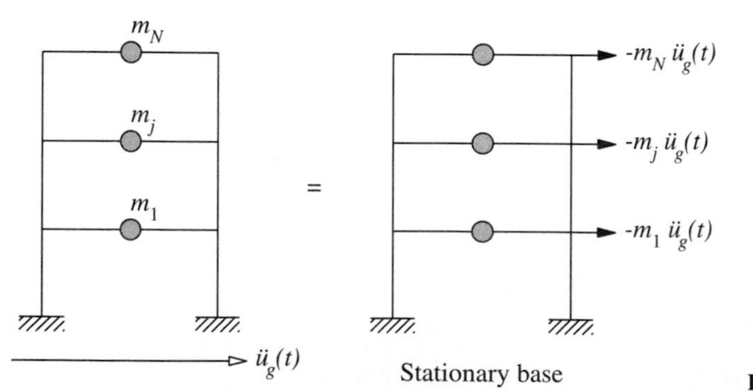

Figure 9.4.2 Effective earthquake forces.

approach the total displacement of each mass is expressed as its displacement u_j^s due to static application of the ground motion plus the dynamic displacement u_j relative to the quasi-static displacement:

$$u_j^t(t) = u_j(t) + u_j^s(t) \quad \text{or} \quad \mathbf{u}^t(t) = \mathbf{u} + \mathbf{u}^s(t) \tag{9.4.6}$$

The quasi-static displacements can be expressed as $\mathbf{u}^s(t) = \boldsymbol{\iota} u_g(t)$, where the *influence vector* $\boldsymbol{\iota}$ represents the displacements of the masses resulting from static application of a unit ground displacement; thus Eq. (9.4.6b) becomes

$$\mathbf{u}^t(t) = \mathbf{u}(t) + \boldsymbol{\iota} u_g(t) \tag{9.4.7}$$

The equations of motion are obtained as before, except that Eq. (9.4.7) is used instead of Eq. (9.4.1b):

$$\mathbf{m}\ddot{\mathbf{u}} + \mathbf{c}\dot{\mathbf{u}} + \mathbf{k}\mathbf{u} = -\mathbf{m}\boldsymbol{\iota}\ddot{u}_g(t) \tag{9.4.8}$$

Now the effective earthquake forces are

$$\mathbf{p}_{\text{eff}}(t) = -\mathbf{m}\boldsymbol{\iota}\ddot{u}_g(t) \tag{9.4.9}$$

This generalization is of no special benefit in deriving the governing equations for the systems of Fig. 9.4.1. Static application of $u_g = 1$ to these systems gives $u_j = 1$ for all j (i.e., $\boldsymbol{\iota} = \mathbf{1}$), as shown in Fig. 9.4.3, where the masses are blank to emphasize that the displacements are static. Thus Eqs. (9.4.8) and (9.4.9) become identical to Eqs. (9.4.4) and (9.4.5), respectively.

We next consider systems with not all the dynamic DOFs in the direction of the ground motion. An example is shown in Fig. 9.4.4a, where an inverted L-shaped frame with lumped masses is subjected to horizontal ground motion. Assuming the elements to be axially rigid, the three DOFs are as shown. Static application of $u_g = 1$ results in the displacements shown in Fig. 9.4.4b. Thus $\boldsymbol{\iota} = \langle 1 \quad 1 \quad 0 \rangle^T$ in Eq. (9.4.8), and Eq. (9.4.9) becomes

$$\mathbf{p}_{\text{eff}}(t) = -\mathbf{m}\boldsymbol{\iota}\ddot{u}_g(t) = -\ddot{u}_g(t) \begin{bmatrix} m_1 \\ m_2 + m_3 \\ 0 \end{bmatrix} \tag{9.4.10}$$

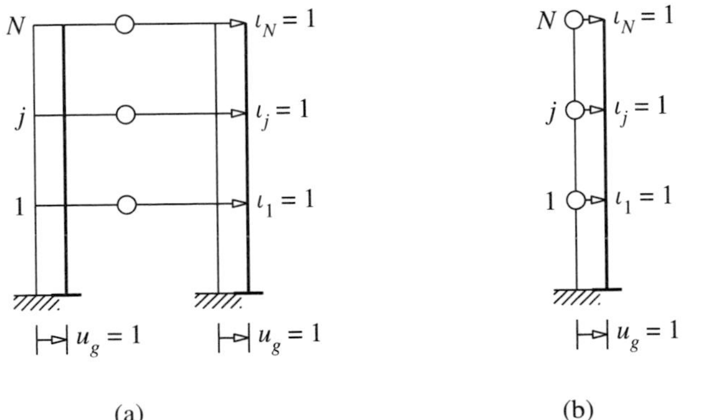

Figure 9.4.3 Influence vector $\boldsymbol{\iota}$: static displacements due to $u_g = 1$.

9.5.4 Multistory One-Way Unsymmetric System

It is apparent from the preceding sections that the simplest system that responds in coupled lateral-torsional motions is a one-story system symmetric about the x-axis but unsymmetric about the y-axis when subjected to the y-component of ground motion. In this section equations of motion are developed for a similar multistory system. Figure 9.5.3 shows such a system, which consists of some frames oriented in the y-direction and others in the x-direction. The framing plan is symmetric about the x-axis and the properties of the two symmetrically located frames are identical; the centers of mass O of all floor diaphragms lie on the same vertical axis.

Each floor diaphragm, assumed to be rigid in its own plane, has three DOFs defined at the center of mass (Fig. 9.5.3a). The DOFs for the jth floor are: translation u_{jx} along the x-axis, translation u_{jy} along the y-axis, and torsional rotation $u_{j\theta}$ about the

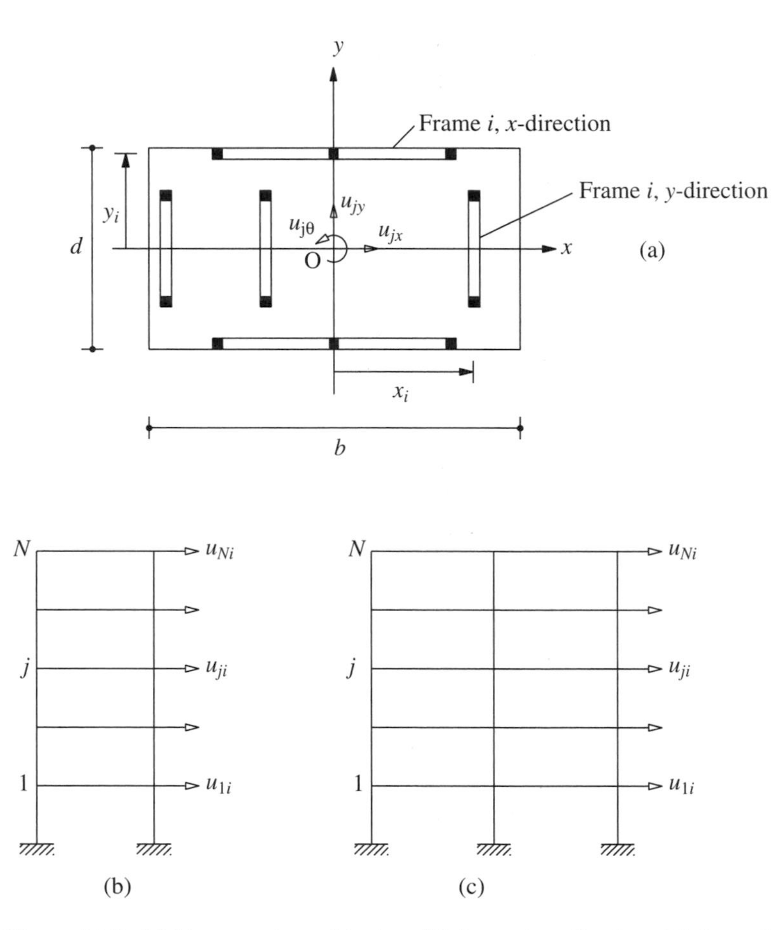

Figure 9.5.3 Multistory system: (a) plan; (b) frame i, y-direction; (c) frame i, x-direction.

vertical axis; u_{jx} and u_{jy} are defined relative to the ground. As suggested by the earlier formulation for a one-story system, the x-translational motion of the building due to the x-component of ground motion can be determined by planar analysis of the building in the x-direction, a system with N degrees of freedom: u_{jx}, $j = 1, 2, \ldots, N$. The equations governing such motion were presented in Section 9.4.2.

In this section the equations governing the response of the system to ground motion in the y-direction are formulated. As suggested by the earlier formulation for a one-story system, the building would undergo coupled lateral-torsional motion described by $2N$ degrees of freedom: u_{jy} and $u_{j\theta}$, $j = 1, 2, \ldots, N$. The displacement vector $\mathbf{u}$ of size $2N \times 1$ for the system is defined by

$$\mathbf{u}^T = \langle\, \mathbf{u}_y^T \quad \mathbf{u}_\theta^T \,\rangle$$

where

$$\mathbf{u}_y^T = \langle\, u_{1y} \quad u_{2y} \quad \cdots \quad u_{Ny} \,\rangle \qquad \mathbf{u}_\theta^T = \langle\, u_{1\theta} \quad u_{2\theta} \quad \cdots \quad u_{N\theta} \,\rangle$$

The stiffness matrix of this system with respect to the global DOF $\mathbf{u}$ is formulated by the direct stiffness method by implementing four major steps [similar to Eqs. (9.5.5) to (9.5.10) for a one-story frame].

Step 1. Determine the lateral stiffness matrix for each frame. For the ith frame it is determined by the following steps: (a) Define the DOF for the ith frame: lateral displacements at floor levels, $\mathbf{u}_i = \langle\, u_{1i} \quad u_{2i} \quad \cdots \quad u_{Ni} \,\rangle^T$ (Fig. 9.5.3b and c), and vertical displacement and rotation of each node. (b) Obtain the complete stiffness matrix for the ith frame with reference to the frame DOF. (c) Statically condense all the rotational and vertical DOF to obtain the $N \times N$ lateral stiffness matrix of the ith frame, denoted by $\mathbf{k}_{xi}$ if the frame is oriented in the x-direction, or by $\mathbf{k}_{yi}$ if the frame is parallel to the y-axis.

Step 2. Determine the displacement transformation matrix relating the lateral DOF $\mathbf{u}_i$ defined in step 1(a) for the ith frame to the global DOF $\mathbf{u}$ for the building. This $N \times 2N$ matrix is denoted by $\mathbf{a}_{xi}$ if the frame is oriented in the x-direction or $\mathbf{a}_{yi}$ if in the y-direction. Thus

$$\mathbf{u}_i = \mathbf{a}_{xi}\mathbf{u} \quad \text{or} \quad \mathbf{u}_i = \mathbf{a}_{yi}\mathbf{u} \tag{9.5.21}$$

These transformation matrices are

$$\mathbf{a}_{xi} = [\,\mathbf{O} \quad -y_i\mathbf{I}\,] \quad \text{or} \quad \mathbf{a}_{yi} = [\,\mathbf{I} \quad x_i\mathbf{I}\,] \tag{9.5.22}$$

where x_i and y_i define the location of the ith frame (Fig. 9.5.3a) oriented in the y and x directions, respectively, $\mathbf{I}$ is an identity matrix of order N, and $\mathbf{O}$ is a square matrix of order N with all elements equal to zero.

Step 3. Transform the lateral stiffness matrix for the ith frame to the building DOF $\mathbf{u}$ to obtain

$$\mathbf{k}_i = \mathbf{a}_{xi}^T\mathbf{k}_{xi}\mathbf{a}_{xi} \quad \text{or} \quad \mathbf{k}_i = \mathbf{a}_{yi}^T\mathbf{k}_{yi}\mathbf{a}_{yi} \tag{9.5.23}$$

The $2N \times 2N$ matrix $\mathbf{k}_i$ is the contribution of the ith frame to the building stiffness matrix.

Step 4. Add the stiffness matrices for all frames to obtain the stiffness matrix for the building:

$$\mathbf{k} = \sum_i \mathbf{k}_i \tag{9.5.24}$$

Substituting Eq. (9.5.22) into Eq. (9.5.23) and the latter into Eq. (9.5.24) leads to

$$\mathbf{k} = \begin{bmatrix} \mathbf{k}_{yy} & \mathbf{k}_{y\theta} \\ \mathbf{k}_{\theta y} & \mathbf{k}_{\theta\theta} \end{bmatrix} \tag{9.5.25}$$

where

$$\mathbf{k}_{yy} = \sum_i \mathbf{k}_{yi} \qquad \mathbf{k}_{y\theta} = \mathbf{k}_{\theta y}^T = \sum_i x_i \mathbf{k}_{yi} \qquad \mathbf{k}_{\theta\theta} = \sum_i (x_i^2 \mathbf{k}_{yi} + y_i^2 \mathbf{k}_{xi}) \tag{9.5.26}$$

The equations of undamped motion of the building subjected to ground acceleration $\ddot{u}_{gy}(t)$ along the y-axis can be developed as shown earlier for a one-story system:

$$\begin{bmatrix} \mathbf{m} & \\ & \mathbf{I}_O \end{bmatrix} \begin{Bmatrix} \ddot{\mathbf{u}}_y \\ \ddot{\mathbf{u}}_\theta \end{Bmatrix} + \begin{bmatrix} \mathbf{k}_{yy} & \mathbf{k}_{y\theta} \\ \mathbf{k}_{\theta y} & \mathbf{k}_{\theta\theta} \end{bmatrix} \begin{Bmatrix} \mathbf{u}_y \\ \mathbf{u}_\theta \end{Bmatrix} = - \begin{bmatrix} \mathbf{m} & \\ & \mathbf{I}_O \end{bmatrix} \begin{Bmatrix} \mathbf{1} \\ \mathbf{0} \end{Bmatrix} \ddot{u}_{gy}(t) \tag{9.5.27}$$

where $\mathbf{m}$ is a diagonal matrix of order N, with $m_{jj} = m_j$, the mass lumped at the jth floor diaphragm; $\mathbf{I}_O$ is a diagonal matrix of order N with $I_{jj} = I_{Oj}$, the moment of inertia of the jth floor diaphragm about the vertical axis through the center of mass; and $\mathbf{1}$ and $\mathbf{0}$ are vectors of dimension N with all elements equal to 1 and zero, respectively. Thus the ground motion in the y-direction can be replaced by effective earthquake forces—$m_j \ddot{u}_{gy}(t)$ in the y-lateral direction; the effective torques are zero. If all floor diaphragms have the same radius of gyration (i.e., $I_{Oj} = m_j r^2$), Eq. (9.5.27) can be rewritten as

$$\begin{bmatrix} \mathbf{m} & \mathbf{0} \\ \mathbf{0} & r^2\mathbf{m} \end{bmatrix} \begin{Bmatrix} \ddot{\mathbf{u}}_y \\ \ddot{\mathbf{u}}_\theta \end{Bmatrix} + \begin{bmatrix} \mathbf{k}_{yy} & \mathbf{k}_{y\theta} \\ \mathbf{k}_{\theta y} & \mathbf{k}_{\theta\theta} \end{bmatrix} \begin{Bmatrix} \mathbf{u}_y \\ \mathbf{u}_\theta \end{Bmatrix} = - \begin{bmatrix} \mathbf{m} & 0 \\ 0 & r^2\mathbf{m} \end{bmatrix} \begin{Bmatrix} \mathbf{1} \\ \mathbf{0} \end{Bmatrix} \ddot{u}_{gy}(t) \tag{9.5.28}$$

9.6 SYMMETRIC-PLAN BUILDINGS: TORSIONAL EXCITATION

Consider a multistory building with its floor plan symmetric about the x and y axes with its base undergoing rotational acceleration $\ddot{u}_{g\theta}(t)$ about a vertical axis. This excitation would cause only torsion of the building without any lateral motion, as demonstrated in Section 9.5.3 for a one-story symmetric system. The equations governing this torsional motion of a multistory building can be written by modifying Eq. (9.5.28): $\mathbf{k}_{\theta y}$ and $\mathbf{k}_{y\theta}$ vanish for symmetric-plan systems, and the translational excitation is replaced by rotational excitation. Without presenting the derivation details, the final equation is

$$r^2\mathbf{m}\ddot{\mathbf{u}}_\theta + \mathbf{k}_{\theta\theta}\mathbf{u}_\theta = -r^2\mathbf{m}\mathbf{1}\ddot{u}_{g\theta}(t) \tag{9.6.1}$$

which is the MDF counterpart of the third equation in Eq. (9.5.20). The rotational acceleration $\ddot{u}_{g\theta}(t)$ of the base of a building is not measured directly during an earthquake but can be calculated from the translational accelerations recorded in the same direction at two locations on the base (Section 13.4).

9.7 MULTIPLE SUPPORT EXCITATION

So far, we have assumed that all supports where the structure is connected to the ground undergo identical motion that is prescribed. In this section we generalize the previous formulation of the equations of motion to allow for different—possibly even multicomponent—prescribed motions at the various supports. Such multiple-support excitation may arise in several situations. First, consider the earthquake analysis of extended structures such as the Golden Gate Bridge, shown in Fig. 2.1.2. The ground motion generated by an earthquake on the nearby San Andreas fault is expected to vary significantly over the 6450-ft length of the structure. Therefore, different motions should be prescribed at the four supports: the base of the two towers and two ends of the bridge. Second, consider the dynamic analysis of piping in nuclear power plants. Although the piping may not be especially long, its ends are connected to different locations of the main structure and would therefore experience different motions during an earthquake.

For the analysis of such systems the formulation of Section 9.4 is extended to include the degrees of freedom at the supports (Fig. 9.7.1). The displacement vector now contains two parts: (1) $\mathbf{u}^t$ includes the N DOFs of the superstructure, where the superscript t denotes that these are total displacements; and (2) $\mathbf{u}_g$ contains the N_g components of support displacements. The equation of dynamic equilibrium for all the DOFs is written in partitioned form:

$$\begin{bmatrix} \mathbf{m} & \mathbf{m}_g \\ \mathbf{m}_g^T & \mathbf{m}_{gg} \end{bmatrix} \begin{Bmatrix} \ddot{\mathbf{u}}^t \\ \ddot{\mathbf{u}}_g \end{Bmatrix} + \begin{bmatrix} \mathbf{c} & \mathbf{c}_g \\ \mathbf{c}_g^T & \mathbf{c}_{gg} \end{bmatrix} \begin{Bmatrix} \dot{\mathbf{u}}^t \\ \dot{\mathbf{u}}_g \end{Bmatrix} + \begin{bmatrix} \mathbf{k} & \mathbf{k}_g \\ \mathbf{k}_g^T & \mathbf{k}_{gg} \end{bmatrix} \begin{Bmatrix} \mathbf{u}^t \\ \mathbf{u}_g \end{Bmatrix} = \begin{Bmatrix} \mathbf{0} \\ \mathbf{p}_g(t) \end{Bmatrix} \tag{9.7.1}$$

Observe that no external forces are applied along the superstructure DOFs. In Eq. (9.7.1) the mass, damping, and stiffness matrices can be determined from the properties of the

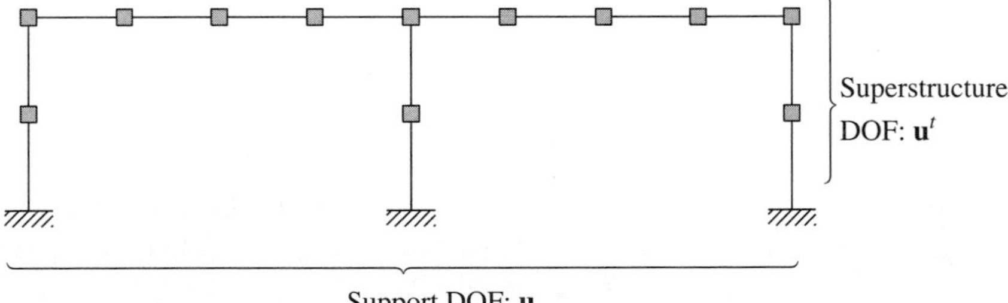

Figure 9.7.1 Definition of superstructure and support DOFs.

structure using the procedures presented earlier in this chapter, while the support motions $\mathbf{u}_g(t)$, $\dot{\mathbf{u}}_g(t)$, and $\ddot{\mathbf{u}}_g(t)$ must be specified. It is desired to determine the displacements $\mathbf{u}^t$ in the superstructure DOF and the support forces $\mathbf{p}_g$.

To write the governing equations in a form familiar from the earlier formulation for a single excitation, we separate the displacements into two parts:

$$\begin{Bmatrix} \mathbf{u}^t \\ \mathbf{u}_g \end{Bmatrix} = \begin{Bmatrix} \mathbf{u}^s \\ \mathbf{u}_g \end{Bmatrix} + \begin{Bmatrix} \mathbf{u} \\ \mathbf{0} \end{Bmatrix} \tag{9.7.2}$$

In this equation $\mathbf{u}^s$ is the vector of structural displacements due to static application of the prescribed support displacements $\mathbf{u}_g$ at each time instant. The two are related through

$$\begin{bmatrix} \mathbf{k} & \mathbf{k}_g \\ \mathbf{k}_g^T & \mathbf{k}_{gg} \end{bmatrix} \begin{Bmatrix} \mathbf{u}^s \\ \mathbf{u}_g \end{Bmatrix} = \begin{Bmatrix} \mathbf{0} \\ \mathbf{p}_g^s \end{Bmatrix} \tag{9.7.3}$$

where $\mathbf{p}_g^s$ are the support forces necessary to statically impose displacements $\mathbf{u}_g$ that vary with time; obviously, $\mathbf{u}^s$ varies with time and is therefore known as the vector of quasi-static displacements. Observe that $\mathbf{p}_g^s = \mathbf{0}$ if the structure is statically determinate or if the support system undergoes rigid-body motion; for the latter condition an obvious example is identical horizontal motion of all supports. The remainder $\mathbf{u}$ of the structural displacements are known as dynamic displacements because a dynamic analysis is necessary to evaluate them.

With the total structural displacements split into quasi-static and dynamic displacements, Eq. (9.7.2), we return to the first of the two partitioned equations (9.7.1):

$$\mathbf{m}\ddot{\mathbf{u}}^t + \mathbf{m}_g\ddot{\mathbf{u}}_g + \mathbf{c}\dot{\mathbf{u}}^t + \mathbf{c}_g\dot{\mathbf{u}}_g + \mathbf{k}\mathbf{u}^t + \mathbf{k}_g\mathbf{u}_g = \mathbf{0} \tag{9.7.4}$$

Substituting Eq. (9.7.2) and transferring all terms involving $\mathbf{u}_g$ and $\mathbf{u}^s$ to the right side leads to

$$\mathbf{m}\ddot{\mathbf{u}} + \mathbf{c}\dot{\mathbf{u}} + \mathbf{k}\mathbf{u} = \mathbf{p}_{\text{eff}}(t) \tag{9.7.5}$$

where the vector of effective earthquake forces is

$$\mathbf{p}_{\text{eff}}(t) = -(\mathbf{m}\ddot{\mathbf{u}}^s + \mathbf{m}_g\ddot{\mathbf{u}}_g) - (\mathbf{c}\dot{\mathbf{u}}^s + \mathbf{c}_g\dot{\mathbf{u}}_g) - (\mathbf{k}\mathbf{u}^s + \mathbf{k}_g\mathbf{u}_g) \tag{9.7.6}$$

This effective force vector can be rewritten in a more useful form. The last term drops out because Eq. (9.7.3) gives

$$\mathbf{k}\mathbf{u}^s + \mathbf{k}_g\mathbf{u}_g = \mathbf{0} \tag{9.7.7}$$

This relation also enables us to express the quasi-static displacements $\mathbf{u}^s$ in terms of the specified support displacements $\mathbf{u}_g$:

$$\mathbf{u}^s = \iota\mathbf{u}_g \qquad \iota = -\mathbf{k}^{-1}\mathbf{k}_g \tag{9.7.8}$$

We call ι the *influence matrix* because it describes the influence of support displacements on the structural displacements. Later, we will find it useful to use a different form:

$$\mathbf{u}^s = \sum_{l=1}^{N_g} \iota_l u_{gl}(t) \tag{9.7.9}$$

where ι_l, the lth column of the influence matrix $\boldsymbol{\iota}$, is the influence vector associated with the support displacement u_{gl}. It is the vector of static displacements in the structural DOF due to $u_{gl} = 1$. Substituting Eqs. (9.7.8) and (9.7.7) in Eq. (9.7.6) gives

$$\mathbf{p}_{\text{eff}}(t) = -(\mathbf{m}\boldsymbol{\iota} + \mathbf{m}_g)\ddot{\mathbf{u}}_g(t) - (\mathbf{c}\boldsymbol{\iota} + \mathbf{c}_g)\dot{\mathbf{u}}_g(t) \tag{9.7.10}$$

If the ground (or support) accelerations $\ddot{\mathbf{u}}_g(t)$ and velocities $\dot{\mathbf{u}}_g(t)$ are prescribed, $\mathbf{p}_{\text{eff}}(t)$ is known from Eq. (9.7.10), and this completes the formulation of the governing equation [Eq. (9.7.5)].

Simplification of $\mathbf{p}_{\text{eff}}(t)$. For many practical applications, further simplification of the effective force vector is possible on two counts. First, the damping term is zero if the damping matrices are proportional to the stiffness matrices (i.e., $\mathbf{c} = a_1\mathbf{k}$ and $\mathbf{c}_g = a_1\mathbf{k}_g$) because of Eq. (9.7.7); this stiffness-proportional damping will be shown in Chapter 11 to be unrealistic, however. While the damping term in Eq. (9.7.10) is not zero for arbitrary forms of damping, it is usually small relative to the inertia term and can therefore be dropped. Second, for structures with mass idealized as lumped at the DOF, the mass matrix is diagonal, implying that $\mathbf{m}_g$ is a null matrix and $\mathbf{m}$ is diagonal. With these simplifications Eq. (9.7.10) reduces to

$$\mathbf{p}_{\text{eff}}(t) = -\mathbf{m}\boldsymbol{\iota}\ddot{\mathbf{u}}_g(t) \tag{9.7.11}$$

Observe that this equation for the effective earthquake forces associated with multiple-support excitation is a generalization of Eq. (9.4.9) for structures with single support and for structures with identical motion at multiple supports. The $N \times N_g$ influence matrix $\boldsymbol{\iota}$ was previously an $N \times 1$ vector, and the $N_g \times 1$ vector $\ddot{\mathbf{u}}_g(t)$ of support motions was a scalar $\ddot{u}_g(t)$.

Interpretation of $\mathbf{p}_{\text{eff}}(t)$. By using Eqs. (9.7.8) and (9.7.9), the effective force vector, Eq. (9.7.11), can be expressed as

$$\mathbf{p}_{\text{eff}}(t) = -\sum_{l=1}^{N_g} \mathbf{m}\boldsymbol{\iota}_l\ddot{u}_{gl}(t) \tag{9.7.12}$$

The lth term in Eq. (9.7.12) that denotes the effective earthquake forces due to acceleration in the lth support DOF is of the same form as Eq. (9.4.9) for structures with single support (and for structures with identical motion at multiple supports). The two cases differ in an important sense, however: In the latter case, the influence vector can be determined by kinematics, but N algebraic equations [Eq. (9.7.7)] are solved to determine each influence vector $\boldsymbol{\iota}_l$ for multiple-support excitations.

Example 9.10

A uniform two-span continuous bridge with flexural stiffness EI is idealized as a lumped-mass system (Fig. E9.10a). Formulate the equations of motion for the bridge subjected to vertical motions u_{g1}, u_{g2}, and u_{g3} of the three supports. Consider only the translational degrees of freedom. Neglect damping.

This general equation replaces Eq. (9.2.3) and Eq. (9.4.8) becomes

$$\mathbf{m}\ddot{\mathbf{u}} + \mathbf{c}\dot{\mathbf{u}} + \mathbf{f}_S(\mathbf{u}, \dot{\mathbf{u}}) = -\mathbf{m}\iota\ddot{u}_g(t) \tag{9.8.2}$$

These are the equations of motion for inelastic MDF systems subjected to ground acceleration $\ddot{u}_g(t)$, the same at all support points.

Following the approach outlined in Section 1.4 for SDF systems, the damping matrix that models the energy dissipation arising from dynamic effects within the linearly elastic range of deformations (see Chapter 11) is also assumed to represent this damping mechanism in the inelastic range of deformations. The additional energy dissipated due to inelastic behavior at larger deformations is accounted for by the inelastic force–deformation relation used in the numerical time-stepping procedures for solving the equations of motion (Chapter 15).

These numerical procedures are based on linearizing the equations of motion over a time step t_i to $t_i + \Delta t$. The structural stiffness matrix at t_i is formulated by direct assembly of the element stiffness matrices. For each structural element—column, beam, or wall, etc.—the element stiffness matrix is determined for the state—displacements and velocities—of the system at t_i and the prescribed yielding mechanism of the material. The element stiffness matrices are then assembled. These procedures are not presented in this structural dynamics text because the reader is expected to be familiar with static analysis of inelastic systems. However, we discuss this issue briefly in Chapter 19 in the context of nonlinear analysis of simple idealizations of multistory buildings.

9.9 PROBLEM STATEMENT

Given the mass matrix $\mathbf{m}$, the stiffness matrix $\mathbf{k}$ of a linearly elastic system or the force–deformation relations $\mathbf{f}_S(\mathbf{u}, \dot{\mathbf{u}})$ for an inelastic system, the damping matrix $\mathbf{c}$, and the dynamic excitation—which may be external forces $\mathbf{p}(t)$ or ground acceleration $\ddot{u}_g(t)$—a fundamental problem in structural dynamics is to determine the response of the MDF structure.

The term *response* denotes any response quantity, such as displacement, velocity, and acceleration of each mass, and also an internal force or internal stress in the structural elements. When the excitation is a set of external forces, the displacements $\mathbf{u}(t)$, velocities $\dot{\mathbf{u}}(t)$, and accelerations $\ddot{\mathbf{u}}(t)$ are of interest. For earthquake excitations the response quantities relative to the ground—$\mathbf{u}$, $\dot{\mathbf{u}}$, and $\ddot{\mathbf{u}}$—as well as the total responses—$\mathbf{u}^t$, $\dot{\mathbf{u}}^t$, and $\ddot{\mathbf{u}}^t$—may be needed. The relative displacements $\mathbf{u}(t)$ associated with deformations of the structure are the most important since the internal forces in the structure are directly related to $\mathbf{u}(t)$.

9.10 ELEMENT FORCES

Once the relative displacements $\mathbf{u}(t)$ have been determined by dynamic analysis, the element forces and stresses needed for structural design can be determined by static analysis of the structure at each time instant (i.e., no additional dynamic analysis is necessary). The static analysis of an MDF system can be visualized in one of two ways:

1. At each time instant the nodal displacements are known from $\mathbf{u}(t)$; if $\mathbf{u}(t)$ includes only the dynamic DOF, the displacements in the condensed DOF are given by Eq. (9.3.3). From the known displacements and rotations of the nodes of each structural element (beam and column), the element forces (bending moments and shears) can be determined through the element stiffness properties (Appendix 1), and stresses can be determined from the element forces.

2. The second approach is to introduce *equivalent static forces;* at any instant of time t these forces $\mathbf{f}_S$ are the external forces that will produce the displacements $\mathbf{u}$ at the same t in the stiffness component of the structure. Thus

$$\mathbf{f}_S(t) = \mathbf{k}\mathbf{u}(t) \tag{9.10.1}$$

Element forces or stresses can be determined at each time instant by static analysis of the structure subjected to the forces $\mathbf{f}_S$. The repeated static analyses at many time instants can be implemented efficiently as described in Chapter 13.

For inelastic systems the element forces can be determined by appropriate modifications of these procedures to recognize that such systems are analyzed by incremental time-stepping procedures (Chapter 15). The change $\Delta\mathbf{u}_i$ in displacements $\mathbf{u}$ over a small time step t_i to $t_i + \Delta t$ is determined by dynamic analysis. The element forces associated with displacements $\Delta\mathbf{u}_i$ are computed from the linear force–deformation using the secant stiffness valid over the time step by implementing the first of the two approaches mentioned in the preceding paragraph. The displacements and forces at time t_i are added to their increments over the time step to determine their values at $t_i + \Delta t$.

To keep the preceding problem statement simple, we have excluded systems subjected to spatially varying multiple-support excitations (Section 9.7). Such dynamic response analyses involve additional considerations that are discussed in Section 13.5.

9.11 METHODS FOR SOLVING THE EQUATIONS OF MOTION: OVERVIEW

The dynamic response of linear systems with classical damping that is a reasonable model for many structures can be determined by classical modal analysis. Classical natural frequencies and modes of vibration exist for such systems (Chapter 10), and their equations of motion, when transformed to modal coordinates, become uncoupled (Chapters 12 and 13). Thus the response in each natural vibration mode can be computed independently of the others, and the modal responses can be combined to determine the total response. Each mode responds with its own particular pattern of deformation, the mode shape; with its own frequency, the natural frequency; and with its own damping. Each modal response can be computed as a function of time by analysis of an SDF system with the vibration properties—natural frequency and damping—of the particular mode. These SDF equations can be solved in closed form for excitations that can be described analytically (Chapters 3 and 4), or they can be solved by time-stepping methods for complicated excitations that are defined numerically (Chapter 5).

Classical modal analysis is not applicable to a structure consisting of subsystems with very different levels of damping. For such systems the classical damping model may not be appropriate, classical vibration modes do not exist, and the equations of motion cannot be uncoupled by transforming to modal coordinates of the system without damping. Such systems can be analyzed by (1) transforming the equations of motion to the eigenvectors of the complex eigenvalue problem that includes the damping matrix (Chapter 12); or (2) direct solution of the coupled system of differential equations (Chapter 15). The latter approach requires numerical methods because closed-form analytical solutions are not possible even if the dynamic excitation is a simple, analytically described function of time and also, of course, if the dynamic excitation is described numerically.

Classical modal analysis is also not applicable to inelastic systems irrespective of the damping model, classical or nonclassical. The standard approach is to solve directly the coupled equations in the original nodal displacements by numerical methods (Chapter 15). The overview of analysis procedures presented in this section is summarized in Fig. 9.11.1.

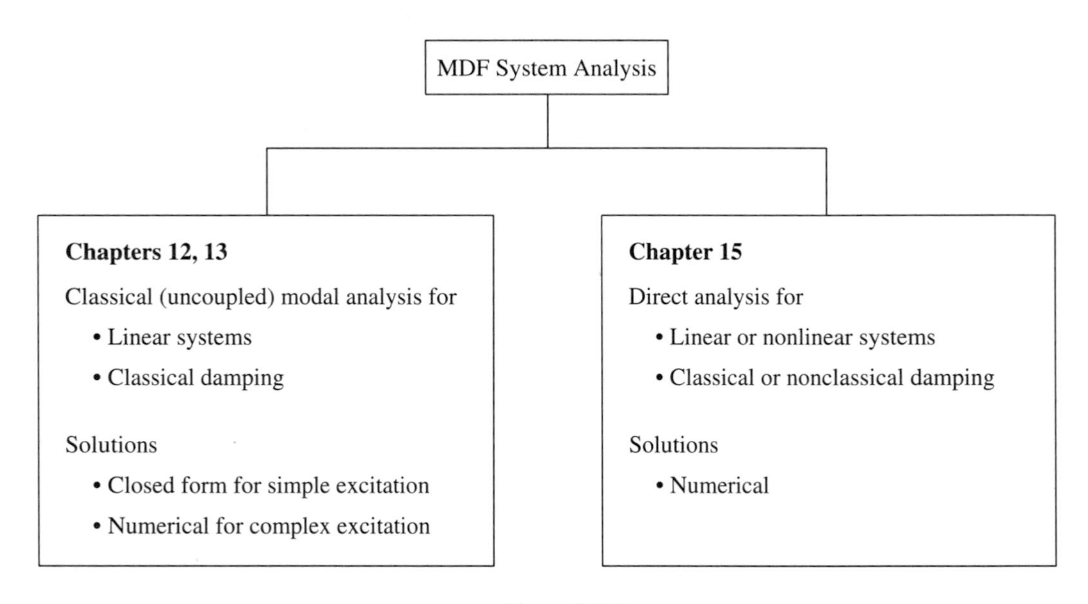

Figure 9.11.1

FURTHER READING

Clough, R. W., and Penzien, J., *Dynamics of Structures,* McGraw-Hill, New York, 1993, Chapters 9 and 10.

Craig, R. R., Jr., *Structural Dynamics,* Wiley, New York, 1981, Chapter 11.

Humar, J. L., *Dynamics of Structures,* Prentice Hall, Englewood Cliffs, N.J., 1990, Chapter 3.

PROBLEMS

9.1 A uniform rigid bar of total mass m issupported on two springs k_1 and k_2 at the two ends and subjected to dynamic forces as shown in Fig. P9.1. The bar is constrained so that it can move only vertically in the plane of the paper. (*Note*: This is the system of Example 9.2.) Formulate the equations of motion with respect to the two DOFs defined at the left end of the bar.

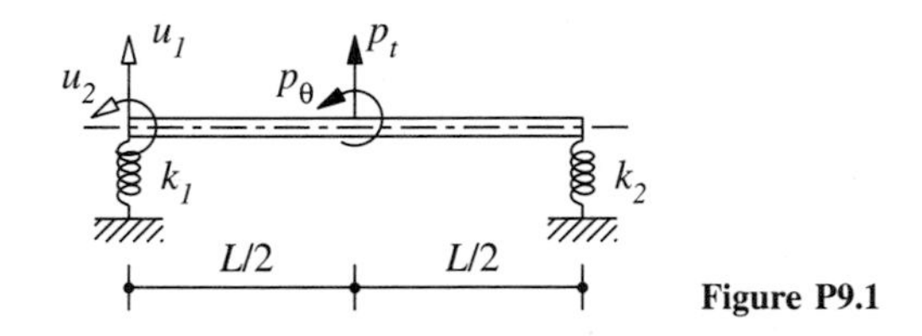

Figure P9.1

***9.2** A uniform simply supported beam of length L, flexural rigidity EI, and mass m per unit length has been idealized as the lumped-mass system shown in Fig. P9.2. The applied forces are also shown.

(a) Identify the DOFs for representing the elastic properties and determine the stiffness matrix. Neglect the axial deformations of the beam.

(b) Identify the DOFs for representing the inertial properties and determine the mass matrix.

(c) Formulate the equations governing the translational motion of the beam.

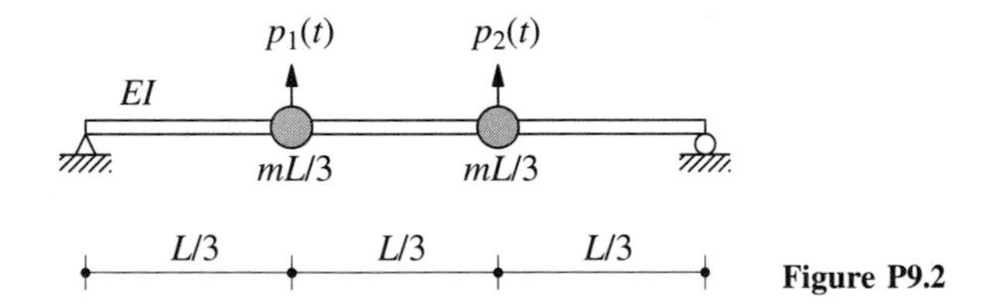

Figure P9.2

9.3 Derive the equations of motion of the beam of Fig. P9.2 governing the translational displacements u_1 and u_2 by starting directly with these two DOFs only.

9.4 A rigid bar is supported by a weightless column as shown in Fig. P9.4. Evaluate the mass, flexibility, and stiffness matrices of the system defined for the two DOFs shown. Do not use a lumped-mass approximation.

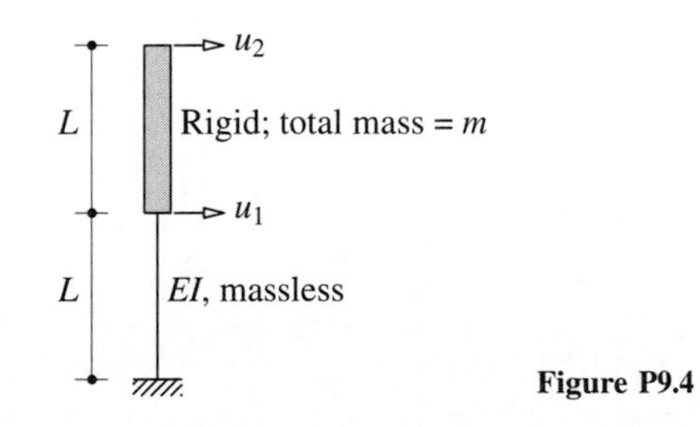

Figure P9.4

*Denotes that a computer is necessary to solve this problem.

*9.5 An umbrella structure has been idealizedas an assemblage of three flexural elements with lumped masses at the nodes as shown in Fig. P9.5.

(a) Identify the DOFs for representing the elastic properties and determine the stiffness matrix. Neglect axial deformations in all members.

(b) Identify the DOFs for representing the inertial properties and determine the mass matrix.

(c) Formulate the equations of motion governing the DOFs in part (b) when the excitation is (i) horizontal ground motion and (ii) vertical ground motion.

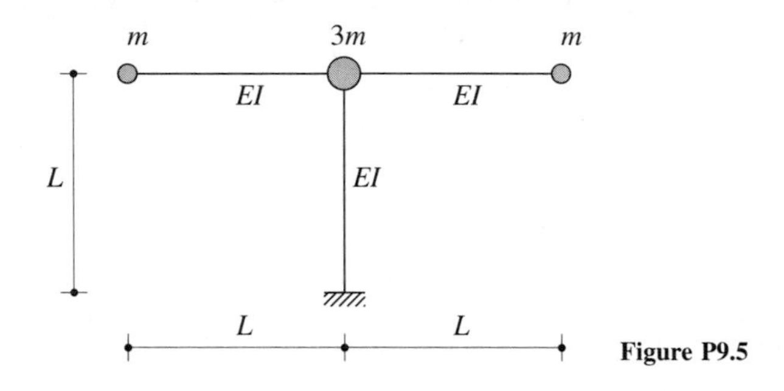

Figure P9.5

9.6 Using the definition of stiffness and mass influence coefficients, formulate the equations of motion for the two-story shear frame with lumped masses shown in Fig. P9.6. The beams are rigid and the flexural rigidity of the columns is EI. Neglect axial deformations in all elements.

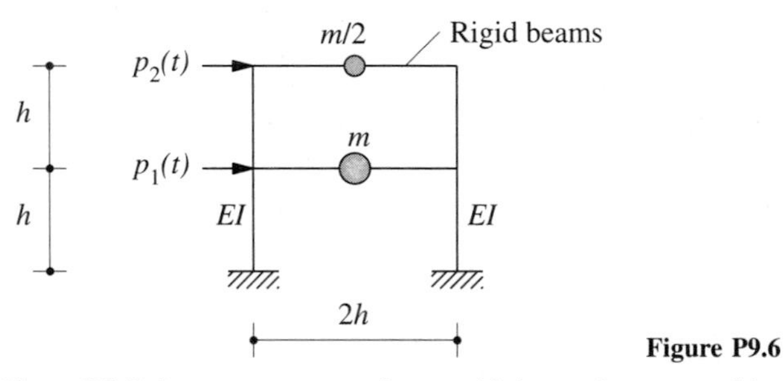

Figure P9.6

*9.7 Figure P9.7 shows a two-story frame with lumped masses subjected to lateral forces, together with some of its properties; in addition, the flexural rigidity is EI for columns and beams.

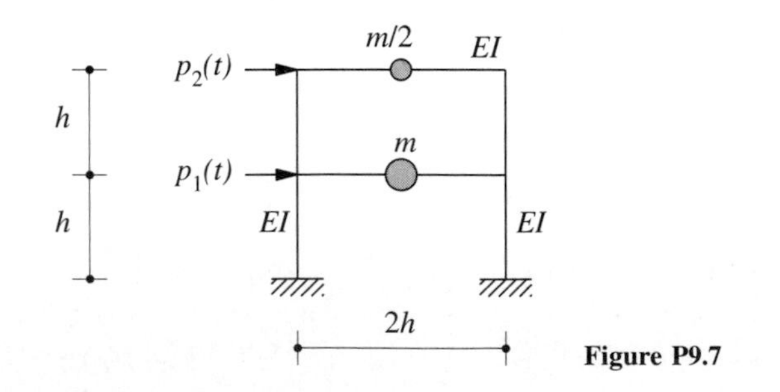

Figure P9.7

*Denotes that a computer is necessary to solve this problem.

(a) Identify the DOFs for representing the elastic properties and determine the stiffness matrix. Neglect the axial deformation of the members.
(b) Identify the DOFs for representing the inertial properties and determine the mass matrix. Assume the members to be massless and neglect their rotational inertia.
(c) Formulate the equations governing the motion of the frame in the DOFs in part (b).

9.8 Using the definition of stiffness and mass influence coefficients, formulate the equations of motion for the three-story shear frame with lumped masses shown in Fig. P9.8. The beams are rigid in flexure, and the flexural rigidity of the columns is EI. Neglect axial deformations in all elements.

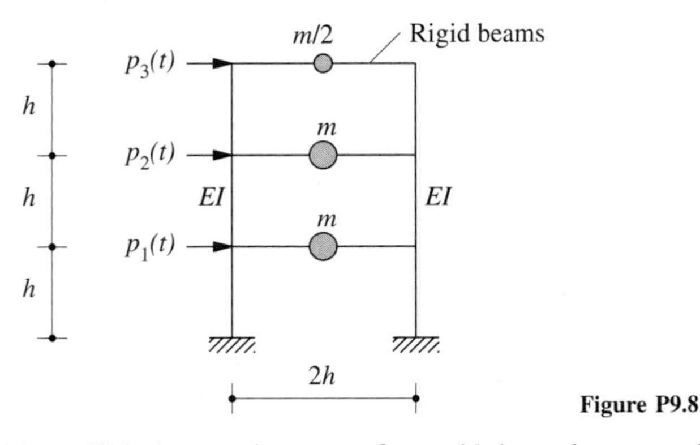

Figure P9.8

***9.9** Figure P9.9 shows a three-story framewith lumped masses subjected to lateral forces, together with its properties; in addition, the flexural rigidity is EI for columns and beams.
(a) Identify the DOFs for representing the elastic properties and determine the stiffness matrix. Neglect the axial deformation of the members.
(b) Identify the DOFs for representing the inertial properties and determine the mass matrix. Assume the members to be massless and neglect their rotational inertia.
(c) Formulate the equations governing the motion of the frame in the DOFs in part (b).

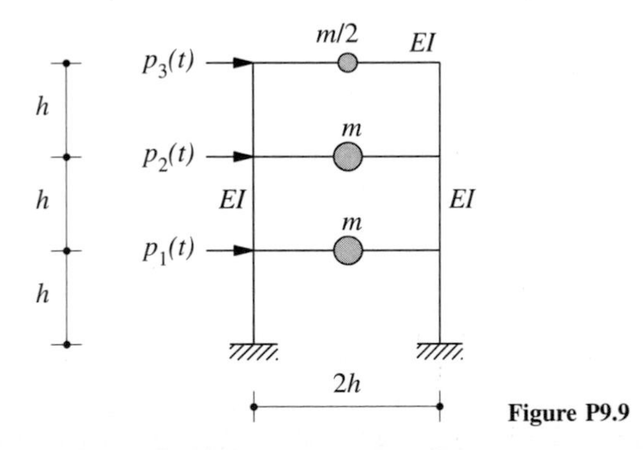

Figure P9.9

9.10 Figure P9.10 shows the plan view of a uniform slab supported on four columns rigidly attached to the slab and clamped at the base. The slab has a total mass m and is rigid in plane and out of plane. Each column is of circular cross section, and its second moment of

*Denotes that a computer is necessary to solve this problem.

cross-sectional area about any diametrical axis is as noted. With the DOFs selected as u_x, u_y, and u_θ at the center of the slab, and using influence coefficients:
(a) Formulate the mass and stiffness matrices in terms of m and the lateral stiffness $k = 12EI/h^3$ of the smaller column; h is the height.
(b) Formulate the equations of motion for ground motion in (i) the x-direction, (ii) the y-direction, and (iii) the direction d–b.

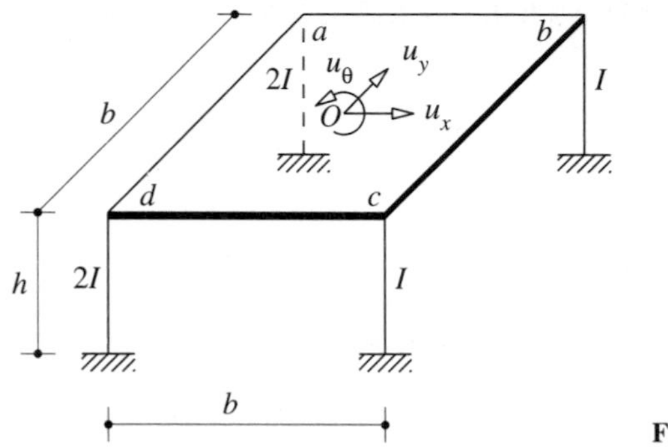

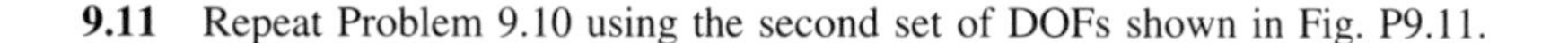

Figure P9.10

9.11 Repeat Problem 9.10 using the second set of DOFs shown in Fig. P9.11.

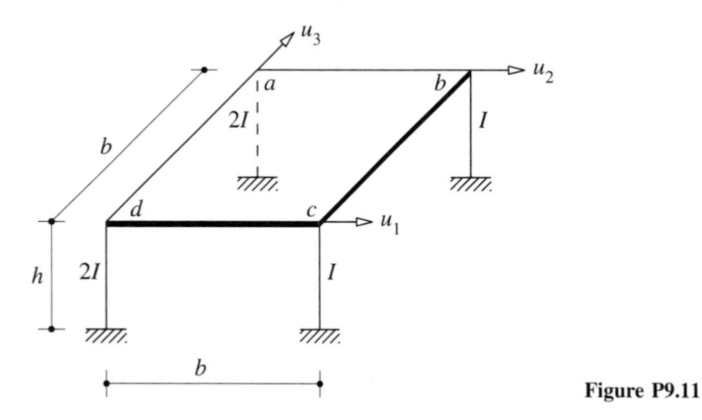

Figure P9.11

9.12 Formulate the equations of motion for the system shown in Fig. P9.12 subjected to support displacements $u_{g1}(t)$ and $u_{g2}(t)$. These equations governing the dynamic components of

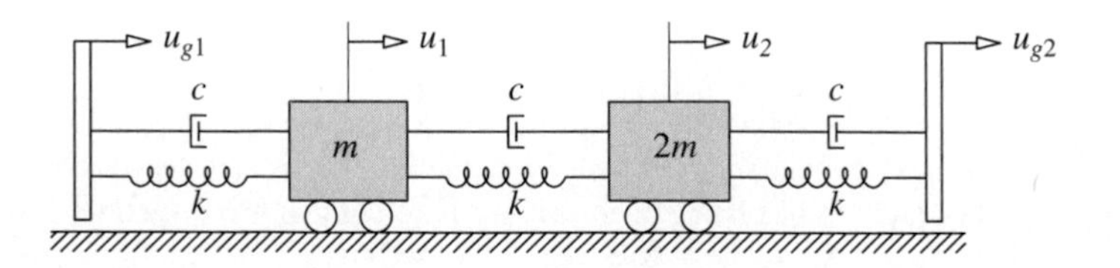

Figure P9.12

displacements u_1 and u_2 (total displacements minus quasistatic displacements) should be expressed in terms of m, k, $\ddot{u}_{g1}(t)$, and $\ddot{u}_{g2}(t)$.

9.13 Figure P9.13 shows a simply supported massless beam with a lumped mass at the center subjected to motions $u_{g1}(t)$ and $u_{g2}(t)$ at the two supports. Formulate the equation of motion governing the dynamic component of displacement u (total displacement minus quasi-static displacement) of the lumped mass. Express this equation in terms of m, EI, L, $\ddot{u}_{g1}(t)$, and $\ddot{u}_{g2}(t)$.

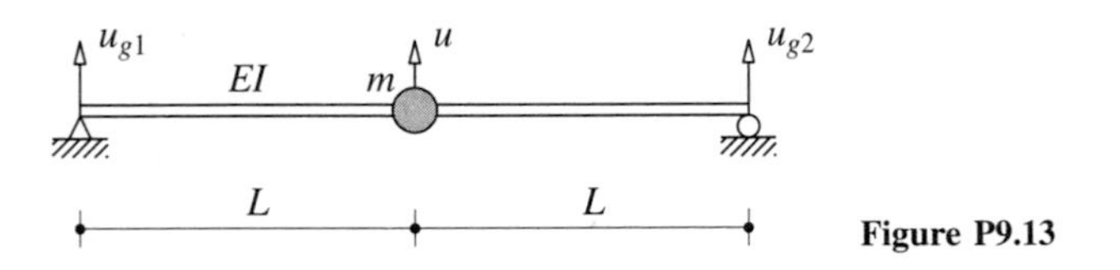

Figure P9.13

*__9.14__ An intake-outlet tower fixed at thebase is partially submerged in water and is accessible from the edge of the reservoir by a foot bridge that is axially rigid and pin-connected to the tower (Fig. P9.14). (In practice, sliding is usually permitted at the connection. The pin connection has been used here only for this hypothetical problem.) The 200-ft-high uniform tower has a hollow reinforced-concrete cross section with outside diameter = 25 ft and wall thickness = 1 ft 3 in. An approximate value of the flexural stiffness EI may be computed from the gross properties of the concrete section without the reinforcement; the elastic modulus of concrete $E = 3.6 \times 10^3$ ksi. For purposes of preliminary analysis the mass of the tower is lumped as shown at two equally spaced locations, where m is the mass per unit length and L the total length of the tower; the unit weight of concrete is 150 lb/ft^3. (The added mass of the surrounding water may be neglected here, but it should be considered in practical analysis.) It is desired to analyze the response of this structure to support motions $u_{g1}(t)$ and $u_{g2}(t)$. Formulate the equations of motion governing the dynamic components of displacements u_1 and u_2 (dynamic component = total displacement − quasi-static component).

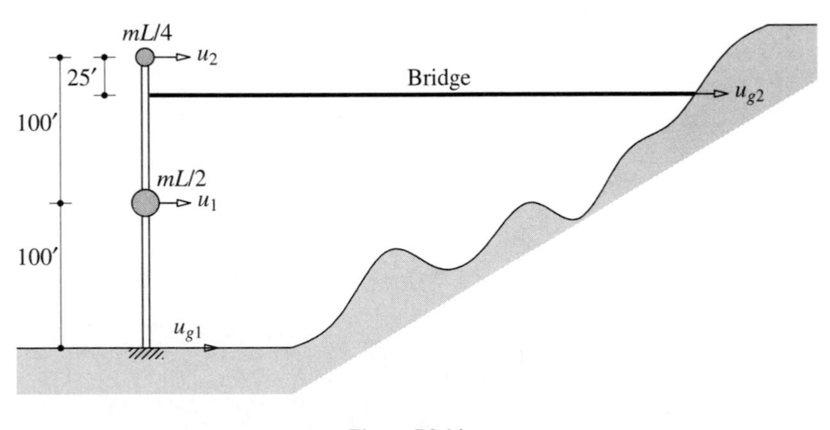

Figure P9.14

*Denotes that a computer is necessary to solve this problem.

Free Vibration

PREVIEW

By *free vibration* we mean the motion of a structure without any dynamic excitation—external forces or support motion. Free vibration is initiated by disturbing the structure from its equilibrium position by some initial displacements and/or by imparting some initial velocities.

This chapter on free vibration of MDF systems is divided into three parts. In Part A we develop the notion of natural frequencies and natural modes of vibration of a structure; these concepts play a central role in the dynamic and earthquake analysis of linear systems (Chapters 12 and 13).

In Part B we describe the use of these vibration properties to determine the free vibration response of systems. Undamped systems are analyzed first. We then discuss the differences in the free vibration response of systems with classical damping and of systems with nonclassical damping. The analysis procedure is extended to systems with classical damping, recognizing that such systems possess the same natural modes as the undamped system.

Part C is concerned with numerical solution of the eigenvalue problem to determine the natural frequencies and modes of vibration. Vector iteration methods are effective in structural engineering applications, and we restrict this presentation to such methods. Only the basic ideas of vector iteration are included without getting into subspace iteration or the Lanczos method. While this limited treatment would suffice for many practical problems and research applications, the reader should recognize that a wealth of knowledge exists on the subject.

PART A: NATURAL VIBRATION FREQUENCIES AND MODES

10.1 SYSTEMS WITHOUT DAMPING

Free vibration of linear MDF systems is governed by Eq. (9.2.12) with $\mathbf{p}(t) = \mathbf{0}$, which for systems without damping is

$$\mathbf{m}\ddot{\mathbf{u}} + \mathbf{k}\mathbf{u} = 0 \tag{10.1.1}$$

Equation (10.1.1) represents N homogeneous differential equations that are coupled through the mass matrix, the stiffness matrix, or both matrices; N is the number of DOFs. It is desired to find the solution $\mathbf{u}(t)$ of Eq. (10.1.1) that satisfies the initial conditions

$$\mathbf{u} = \mathbf{u}(0) \qquad \dot{\mathbf{u}} = \dot{\mathbf{u}}(0) \tag{10.1.2}$$

at $t = 0$. A general procedure to obtain the desired solution for any MDF system is developed in Section 10.8. In this section the solution is presented in graphical form that enables us to understand free vibration of an MDF system in qualitative terms.

Figure 10.1.1 shows the free vibration of a two-story shear frame. The story stiffnesses and lumped masses at the floors are noted, and the free vibration is initiated by the deflections shown by curve a in Fig. 10.1.1b. The resulting motion u_j of the two masses is plotted in Fig. 10.1.1d as a function of the time parameter t/T_1, where T_1 is a natural vibration period of the structure, which will be defined later.

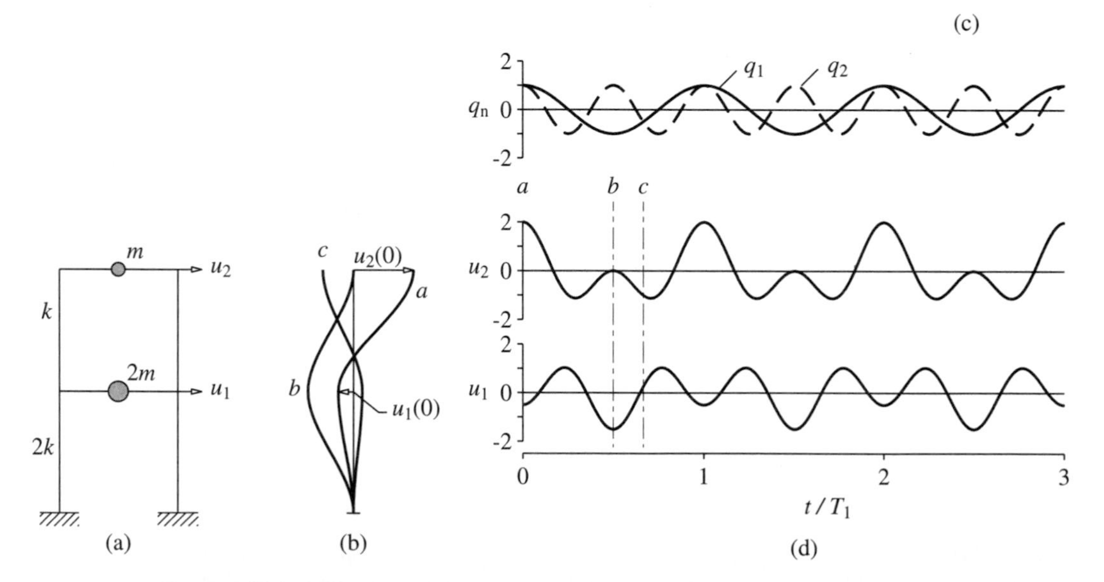

Figure 10.1.1 Free vibration of an undamped system due to arbitrary initial displacement: (a) two-story frame; (b) deflected shapes at time instants a, b, and c; (c) modal coordinates $q_n(t)$; (d) displacement history.

The deflected shapes of the structure at selected time instants a, b, and c are also shown; the $q_n(t)$ plotted in Fig.10.1.1c are discussed in Example 10.11. The displacement–time plot for the jth floor starts with the initial conditions $u_j(0)$ and $\dot{u}_j(0)$; the $u_j(0)$ are identified in Fig. 10.1.1b and $\dot{u}_j(0) = 0$ for both floors. Contrary to what we observed in Fig. 2.1.1 for SDF systems, the motion of each mass (or floor) is not a simple harmonic motion and the frequency of the motion cannot be defined. Furthermore, the deflected shape (i.e., the ratio u_1/u_2) varies with time, as is evident from the differing deflected shapes b and c, which are in turn different from the initial deflected shape a.

An undamped structure would undergo simple harmonic motion without change of deflected shape, however, if free vibration is initiated by appropriate distributions of displacements in the various DOFs. As shown in Figs. 10.1.2 and 10.1.3, two characteristic deflected shapes exist for this two-DOF system such that if it is displaced in one of these shapes and released, it will vibrate in simple harmonic motion, maintaining the initial deflected shape. Both floors reach their extreme displacements at the same time and pass through the equilibrium position at the same time. Observe that the displacements of both floors are in the same direction in the first characteristic deflected shape but in opposite directions in the second characteristic shape. The point of zero displacement, called a *node,*† does not move at all (Fig. 10.1.3); as the mode number n increases, the number of nodes increases accordingly (see Fig. 12.8.2). Each characteristic deflected shape is called a *natural mode of vibration* of an MDF system.

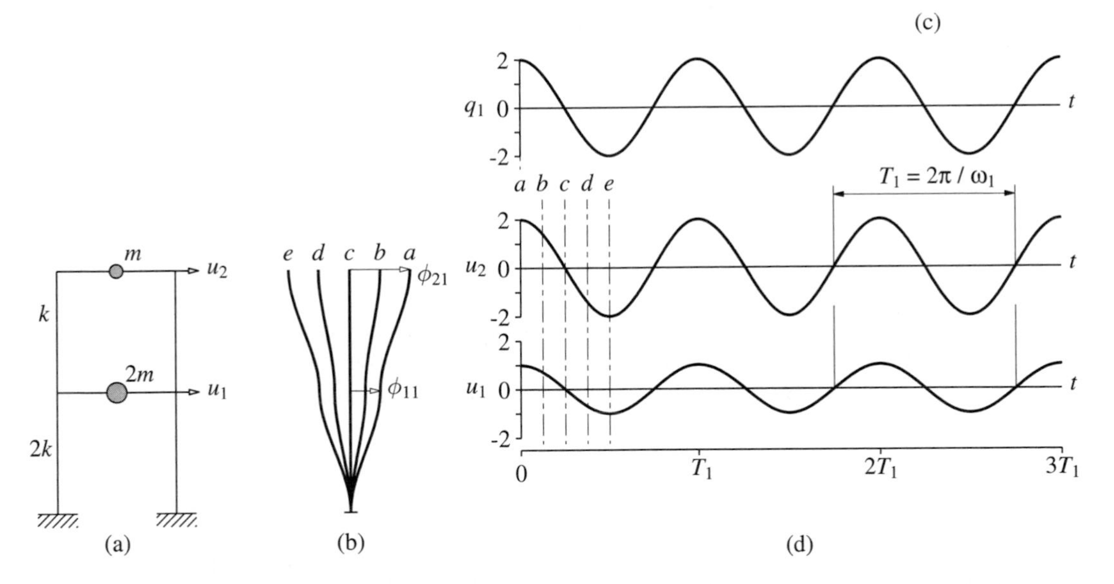

Figure 10.1.2 Free vibration of an undamped system in its first natural mode of vibration: (a) two-story frame; (b) deflected shapes at time instants a, b, c, d, and e; (c) modal coordinate $q_1(t)$; (d) displacement history.

†Recall that we have already used the term *node* for nodal points in the structural idealization; the two different uses of *node* should be clear from the context.

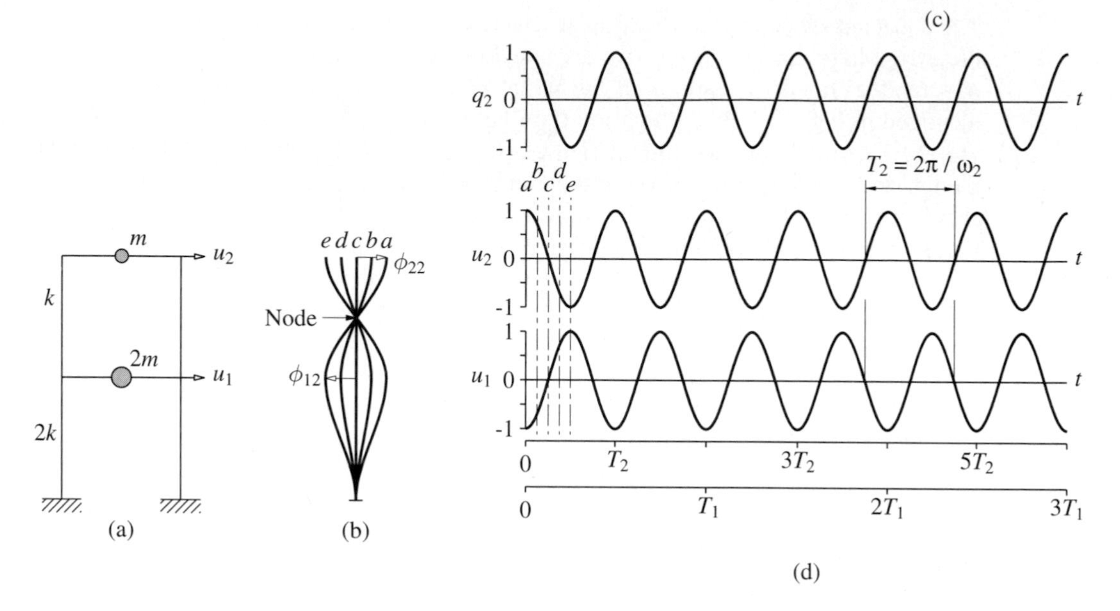

Figure 10.1.3 Free vibration of an undamped system in its second natural mode of vibration: (a) two-story frame; (b) deflected shapes at the time instants a, b, c, d, and e; (c) modal coordinate $q_2(t)$; (d) displacement history.

A *natural period of vibration* T_n of an MDF system is the time required for one cycle of the simple harmonic motion in one of these natural modes. The corresponding *natural circular frequency of vibration* is ω_n and the *natural cyclic frequency of vibration* is f_n, where

$$T_n = \frac{2\pi}{\omega_n} \qquad f_n = \frac{1}{T_n} \tag{10.1.3}$$

Figures 10.1.2 and 10.1.3 show the two natural periods T_n and natural frequencies ω_n $(n = 1, 2)$ of the two-story building vibrating in its natural modes $\boldsymbol{\phi}_n = \langle \phi_{1n} \quad \phi_{2n} \rangle^T$. The smaller of the two natural vibration frequencies is denoted by ω_1, and the larger by ω_2. Correspondingly, the longer of the two natural vibration periods is denoted by T_1 and the shorter one as T_2.

10.2 NATURAL VIBRATION FREQUENCIES AND MODES

In this section we introduce the eigenvalue problem whose solution gives the natural frequencies and modes of a system. The free vibration of an undamped system in one of its natural vibration modes, graphically displayed in Figs. 10.1.2 and 10.1.3 for a two-DOF system, can be described mathematically by

$$\mathbf{u}(t) = q_n(t)\boldsymbol{\phi}_n \tag{10.2.1}$$

where the deflected shape ϕ_n does not vary with time. The time variation of the displacements is described by the simple harmonic function

$$q_n(t) = A_n \cos \omega_n t + B_n \sin \omega_n t \tag{10.2.2}$$

where A_n and B_n are constants of integration that can be determined from the initial conditions that initiate the motion. Combining Eqs. (10.2.1) and (10.2.2) gives

$$\mathbf{u}(t) = \phi_n (A_n \cos \omega_n t + B_n \sin \omega_n t) \tag{10.2.3}$$

where ω_n and ϕ_n are unknown.

Substituting this form of $\mathbf{u}(t)$ in Eq. (10.1.1) gives

$$\left[-\omega_n^2 \mathbf{m}\phi_n + \mathbf{k}\phi_n\right] q_n(t) = \mathbf{0}$$

This equation can be satisfied in one of two ways. Either $q_n(t) = 0$, which implies that $\mathbf{u}(t) = \mathbf{0}$ and there is no motion of the system (this is the so-called trivial solution), or the natural frequencies ω_n and modes ϕ_n must satisfy the following algebraic equation:

$$\mathbf{k}\phi_n = \omega_n^2 \mathbf{m}\phi_n \tag{10.2.4}$$

which provides a useful condition. This algebraic problem is called the *matrix eigenvalue problem.* When necessary it is called the real eigenvalue problem to distinguish it from the complex eigenvalue problem mentioned in Section 12.14 for systems with damping. The stiffness and mass matrices **k** and **m** are known; the problem is to determine the scalar ω_n^2 and vector ϕ_n.

To indicate the formal solution to Eq. (10.2.4), it is rewritten as

$$\left[\mathbf{k} - \omega_n^2 \mathbf{m}\right] \phi_n = \mathbf{0} \tag{10.2.5}$$

which can be interpreted as a set of N homogeneous algebraic equations for the N elements ϕ_{jn} $(j = 1, 2, \ldots, N)$. This set always has the trivial solution $\phi_n = \mathbf{0}$, which is not useful because it implies no motion. It has nontrivial solutions if

$$\det\left[\mathbf{k} - \omega_n^2 \mathbf{m}\right] = 0 \tag{10.2.6}$$

When the determinant is expanded, a polynomial of order N in ω_n^2 is obtained. Equation (10.2.6) is known as the *characteristic equation* or *frequency equation.* This equation has N real and positive roots for ω_n^2 because **m** and **k**, the structural mass and stiffness matrices, are symmetric and positive definite. The positive definite property of **k** is assured for all structures supported in a way that prevents rigid-body motion. Such is the case for civil engineering structures of interest to us, but not for unrestrained structures such as aircraft in flight—these are beyond the scope of this book. The positive definite property of **m** is also assured because the lumped masses are nonzero in all DOFs retained in the analysis after the DOFs with zero lumped mass have been eliminated by static condensation (Section 9.3).

The N roots of Eq. (10.2.6) determine the N natural frequencies ω_n $(n = 1, 2, \ldots, N)$ of vibration. These roots of the characteristic equation are also known as *eigenvalues, characteristic values,* or *normal values.* When a natural frequency ω_n is known, Eq. (10.2.5) can be solved for the corresponding vector ϕ_n to within a multiplicative constant. The eigenvalue problem does not fix the absolute amplitude of the vectors ϕ_n, only the shape of the vector given by the relative values of the N displacements ϕ_{jn} ($j = 1, 2, \ldots, N$). Corresponding to the N natural vibration frequencies ω_n of an N-DOF system, there are N independent vectors ϕ_n which are known as *natural modes of vibration,* or *natural mode shapes of vibration.* These vectors are also known as *eigenvectors, characteristic vectors,* or *normal modes.*

In summary, a vibrating system with N DOFs has N natural vibration frequencies ω_n ($n = 1, 2, \ldots, N$), arranged in sequence from smallest to largest ($\omega_1 < \omega_2 < \cdots < \omega_N$); corresponding natural periods T_n; and natural modes ϕ_n. The term *natural* is used to qualify each of these vibration properties to emphasize the fact that these are natural properties of the structure in free vibration, and they depend only on its mass and stiffness properties. The subscript n denotes the mode number and the first mode ($n = 1$) is also known as the fundamental mode.

10.3 MODAL AND SPECTRAL MATRICES

The N eigenvalues, N natural frequencies, and N natural modes can be assembled compactly into matrices. Let the natural mode ϕ_n corresponding to the natural frequency ω_n have elements ϕ_{jn}, where j indicates the DOFs. The N eigenvectors then can be displayed in a single square matrix, each column of which is a natural mode:

$$\boldsymbol{\Phi} = \left[\phi_{jn}\right] = \begin{bmatrix} \phi_{11} & \phi_{12} & \cdots & \phi_{1N} \\ \phi_{21} & \phi_{22} & \cdots & \phi_{2N} \\ \vdots & \vdots & \ddots & \vdots \\ \phi_{N1} & \phi_{N2} & \cdots & \phi_{NN} \end{bmatrix}$$

The matrix $\boldsymbol{\Phi}$ is called the *modal matrix* for the eigenvalue problem, Eq. (10.2.4). The N eigenvalues ω_n^2 can be assembled into a diagonal matrix $\boldsymbol{\Omega}^2$, which is known as the *spectral matrix* of the eigenvalue problem, Eq. (10.2.4):

$$\boldsymbol{\Omega}^2 = \begin{bmatrix} \omega_1^2 & & & \\ & \omega_2^2 & & \\ & & \ddots & \\ & & & \omega_N^2 \end{bmatrix}$$

Each eigenvalue and eigenvector satisfies Eq. (10.2.4), which can be rewritten as the relation

$$\mathbf{k}\phi_n = \mathbf{m}\phi_n\omega_n^2 \tag{10.3.1}$$

By using the modal and spectral matrices, it is possible to assemble all of these relations into a single matrix equation:

$$\mathbf{k}\boldsymbol{\Phi} = \mathbf{m}\boldsymbol{\Phi}\boldsymbol{\Omega}^2 \tag{10.3.2}$$

Equation (10.3.2) provides a compact presentation of the equations relating all eigenvalues and eigenvectors.

10.4 ORTHOGONALITY OF MODES

The natural modes corresponding to different natural frequencies can be shown to satisfy the following orthogonality conditions. When $\omega_n \neq \omega_r$,

$$\boldsymbol{\phi}_n^T \mathbf{k} \boldsymbol{\phi}_r = 0 \qquad \boldsymbol{\phi}_n^T \mathbf{m} \boldsymbol{\phi}_r = 0 \tag{10.4.1}$$

These properties can be proven as follows: The nth natural frequency and mode satisfy Eq. (10.2.4); premultiplying it by $\boldsymbol{\phi}_r^T$, the transpose of $\boldsymbol{\phi}_r$, gives

$$\boldsymbol{\phi}_r^T \mathbf{k} \boldsymbol{\phi}_n = \omega_n^2 \boldsymbol{\phi}_r^T \mathbf{m} \boldsymbol{\phi}_n \tag{10.4.2}$$

Similarly, the rth natural frequency and mode satisfy Eq. (10.2.4); thus $\mathbf{k}\boldsymbol{\phi}_r = \omega_r^2 \mathbf{m} \boldsymbol{\phi}_r$. Premultiplying by $\boldsymbol{\phi}_n^T$ gives

$$\boldsymbol{\phi}_n^T \mathbf{k} \boldsymbol{\phi}_r = \omega_r^2 \boldsymbol{\phi}_n^T \mathbf{m} \boldsymbol{\phi}_r \tag{10.4.3}$$

The transpose of the matrix on the left side of Eq. (10.4.2) will equal the transpose of the matrix on the right side of the equation; thus

$$\boldsymbol{\phi}_n^T \mathbf{k} \boldsymbol{\phi}_r = \omega_n^2 \boldsymbol{\phi}_n^T \mathbf{m} \boldsymbol{\phi}_r \tag{10.4.4}$$

wherein we have utilized the symmetry property of the mass and stiffness matrices. Subtracting Eq. (10.4.3) from (10.4.4) gives

$$(\omega_n^2 - \omega_r^2) \boldsymbol{\phi}_n^T \mathbf{m} \boldsymbol{\phi}_r = 0$$

Thus Eq. (10.4.1b) is true when $\omega_n^2 \neq \omega_r^2$ which for systems with positive natural frequencies implies that $\omega_n \neq \omega_r$. Substituting Eq. (10.4.1b) in (10.4.3) indicates that Eq. (10.4.1a) is true when $\omega_n \neq \omega_r$. This completes a proof for the orthogonality relations of Eq. (10.4.1).

We have established the orthogonality relations between modes with distinct frequencies (i.e., $\omega_n \neq \omega_r$). If the frequency equation (10.2.4) has a j-fold multiple root (i.e., the system has one frequency repeated j times) it is always possible to find j modes associated with this frequency that satisfy Eq. (10.4.1). If these j modes are included with the modes corresponding to the other frequencies, a set of N modes is obtained which satisfies Eq. (10.4.1) for $n \neq r$.

The orthogonality of natural modes implies that the following square matrices are diagonal:

$$\mathbf{K} \equiv \boldsymbol{\Phi}^T \mathbf{k} \boldsymbol{\Phi} \qquad \mathbf{M} \equiv \boldsymbol{\Phi}^T \mathbf{m} \boldsymbol{\Phi} \tag{10.4.5}$$

where the diagonal elements are

$$K_n = \boldsymbol{\phi}_n^T \mathbf{k} \boldsymbol{\phi}_n \qquad M_n = \boldsymbol{\phi}_n^T \mathbf{m} \boldsymbol{\phi}_n \tag{10.4.6}$$

Since $\mathbf{m}$ and $\mathbf{k}$ are positive definite, the diagonal elements of $\mathbf{K}$ and $\mathbf{M}$ are positive. They are related by

$$K_n = \omega_n^2 M_n \tag{10.4.7}$$

This can be demonstrated from the definitions of K_n and M_n as follows: Substituting Eq. (10.2.4) in (10.4.6a) gives

$$K_n = \phi_n^T(\omega_n^2 \mathbf{m}\phi_n) = \omega_n^2(\phi_n^T \mathbf{m}\phi_n) = \omega_n^2 M_n$$

10.5 INTERPRETATION OF MODAL ORTHOGONALITY

In this section we develop physically motivated interpretations of the orthogonality properties of natural modes. One implication of modal orthogonality is that the work done by the nth mode inertia forces in going through the rth mode displacements is zero. To demonstrate this result, consider a structure vibrating in the nth mode with displacements

$$\mathbf{u}_n(t) = q_n(t)\phi_n \tag{10.5.1}$$

The corresponding accelerations are $\ddot{\mathbf{u}}_n(t) = \ddot{q}_n(t)\phi_n$ and the associated inertia forces are

$$(\mathbf{f}_I)_n = -\mathbf{m}\ddot{\mathbf{u}}_n(t) = -\mathbf{m}\phi_n\ddot{q}_n(t) \tag{10.5.2}$$

Next, consider displacements of the structure in its rth natural mode:

$$\mathbf{u}_r(t) = q_r(t)\phi_r \tag{10.5.3}$$

The work done by the inertia forces of Eq. (10.5.2) in going through the displacements of Eq. (10.5.3) is

$$(\mathbf{f}_I)_n^T \mathbf{u}_r = -\left(\phi_n^T \mathbf{m}\phi_r\right)\ddot{q}_n(t)q_r(t) \tag{10.5.4}$$

which is zero because of the modal orthogonality relation of Eq. (10.4.1b). This completes the proof.

Another implication of the modal orthogonality properties is that the work done by the equivalent static forces associated with displacements in the nth mode in going through the rth mode displacements is zero. These forces are

$$(\mathbf{f}_S)_n = \mathbf{k}\mathbf{u}_n(t) = \mathbf{k}\phi_n q_n(t)$$

and the work they do in going through the displacements of Eq. (10.5.3) is

$$(\mathbf{f}_S)_n^T \mathbf{u}_r = \left(\phi_n^T \mathbf{k}\phi_r\right) q_n(t)q_r(t)$$

which is zero because of the modal orthogonality relation of Eq. (10.4.1a). This completes the proof.

10.6 NORMALIZATION OF MODES

As mentioned earlier, the eigenvalue problem, Eq. (10.2.4), determines the natural modes to only within a multiplicative factor. If the vector ϕ_n is a natural mode, any vector proportional to ϕ_n is essentially the same natural mode because it also satisfies Eq. (10.2.4). Scale factors are sometimes applied to natural modes to standardize their elements associated with amplitudes in various DOFs. This process is called *normalization*. Sometimes it is convenient to normalize each mode so that its largest element is unity. Other times it may be advantageous to normalize each mode so that the element corresponding to a particular DOF, say the top floor of a multistory building, is unity. In theoretical discussions and computer programs it is common to normalize modes so that the M_n have unit values. In this case

$$M_n = \phi_n^T \mathbf{m} \phi_n = 1 \qquad \mathbf{\Phi}^T \mathbf{m} \mathbf{\Phi} = \mathbf{I} \tag{10.5.5}$$

where **I** is the identity matrix, a diagonal matrix with unit values along the main diagonal. Equation (10.5.5) states that the natural modes are not only orthogonal but are normalized with respect to **m**. They are then called a mass *orthonormal set.* When the modes are normalized in this manner, Eqs. (10.4.6a) and (10.4.5a) become

$$K_n = \phi_n^T \mathbf{k} \phi_n = \omega_n^2 M_n = \omega_n^2 \qquad \mathbf{K} = \mathbf{\Phi}^T \mathbf{k} \mathbf{\Phi} = \mathbf{\Omega}^2 \tag{10.5.6}$$

Example 10.1

(**a**) Determine the natural vibration frequencies and modes of the system of Fig. E10.1a using the first set of DOFs shown. (**b**) Repeat part (a) using the second set of DOFs in Fig. E10.1b. (**c**) Show that the natural frequencies and modes determined using the two sets of DOFs are the same.

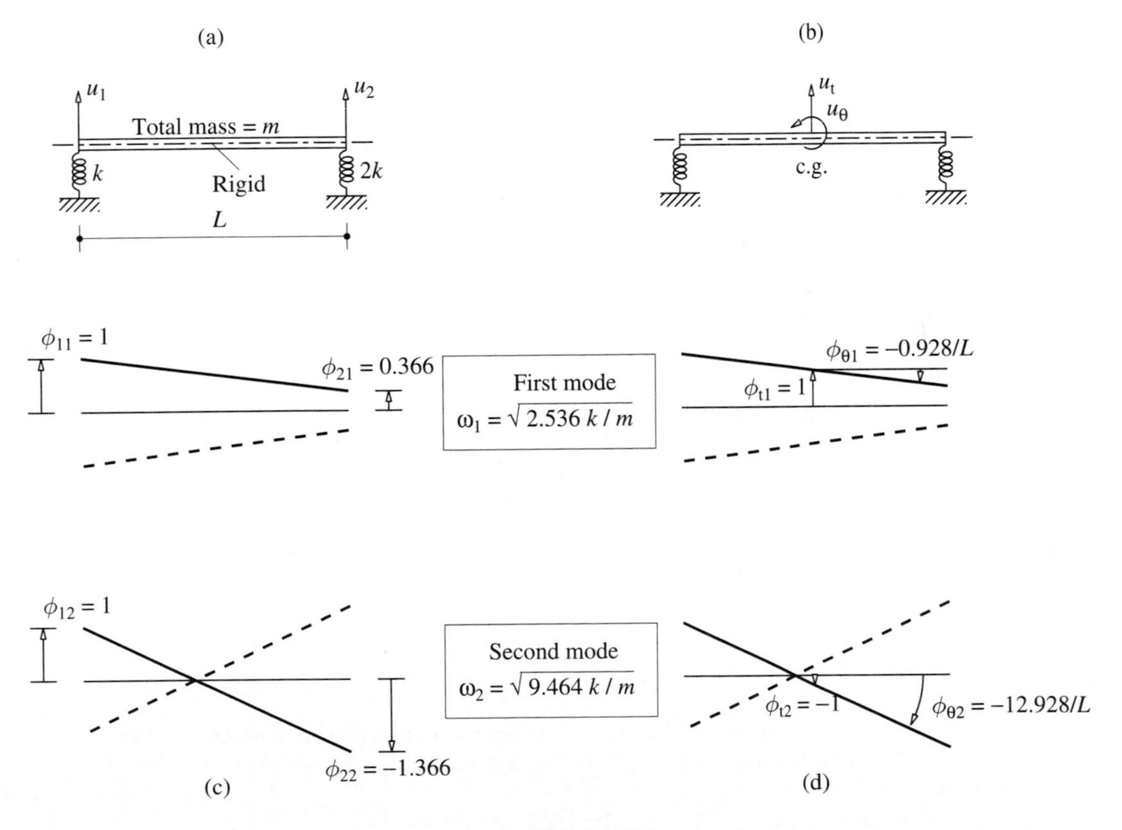

Figure E10.1

Solution **(a)** The mass and stiffness matrices for the system with the first set of DOFs were determined in Example 9.2:

$$\mathbf{m} = \frac{m}{6}\begin{bmatrix} 2 & 1 \\ 1 & 2 \end{bmatrix} \qquad \mathbf{k} = k\begin{bmatrix} 1 & 0 \\ 0 & 2 \end{bmatrix}$$

Then

$$\mathbf{k} - \omega_n^2\mathbf{m} = \begin{bmatrix} k - m\omega_n^2/3 & -m\omega_n^2/6 \\ -m\omega_n^2/6 & 2k - m\omega_n^2/3 \end{bmatrix} \tag{a}$$

is substituted in Eq. (10.2.6) to obtain the frequency equation:

$$m^2\omega_n^4 - 12km\omega_n^2 + 24k^2 = 0$$

This is a quadratic equation in ω_n^2 that has the solutions

$$\omega_1^2 = \left(6 - 2\sqrt{3}\right)\frac{k}{m} = 2.536\frac{k}{m} \qquad \omega_2^2 = \left(6 + 2\sqrt{3}\right)\frac{k}{m} = 9.464\frac{k}{m} \tag{b}$$

Taking the square root of Eq. (b) gives the natural frequencies ω_1 and ω_2.

The natural modes are determined by substituting $\omega_n^2 = \omega_1^2$ in Eq. (a), and then Eq. (10.2.5) gives

$$k\begin{bmatrix} 0.155 & -0.423 \\ -0.423 & 1.165 \end{bmatrix}\begin{Bmatrix} \phi_{11} \\ \phi_{21} \end{Bmatrix} = \begin{Bmatrix} 0 \\ 0 \end{Bmatrix} \tag{c}$$

Now select any value for one unknown, say $\phi_{11} = 1$. Then the first or second of the two equations gives $\phi_{21} = 0.366$. Substituting $\omega_n^2 = \omega_2^2$ in Eq. (10.2.5) gives

$$k\begin{bmatrix} -2.155 & -1.577 \\ -1.577 & -1.155 \end{bmatrix}\begin{Bmatrix} \phi_{12} \\ \phi_{22} \end{Bmatrix} = \begin{Bmatrix} 0 \\ 0 \end{Bmatrix} \tag{d}$$

Selecting $\phi_{12} = 1$, either of these equations gives $\phi_{22} = -1.366$. In summary, the two modes plotted in Fig. E10.1c are

$$\boldsymbol{\phi}_1 = \begin{Bmatrix} 1 \\ 0.366 \end{Bmatrix} \qquad \boldsymbol{\phi}_2 = \begin{Bmatrix} 1 \\ -1.366 \end{Bmatrix} \tag{e}$$

(b) The mass and stiffness matrices of the system described by the second set of DOF were developed in Example 9.3:

$$\mathbf{m} = \begin{bmatrix} m & 0 \\ 0 & mL^2/12 \end{bmatrix} \qquad \mathbf{k} = \begin{bmatrix} 3k & kL/2 \\ kL/2 & 3kL^2/4 \end{bmatrix} \tag{f}$$

Then

$$\mathbf{k} - \omega_n^2\mathbf{m} = \begin{bmatrix} 3k - m\omega_n^2 & kL/2 \\ kL/2 & (9k - m\omega_n^2)L^2/12 \end{bmatrix} \tag{g}$$

is substituted in Eq. (10.2.6) to obtain

$$m^2\omega_n^4 - 12km\omega_n^2 + 24k^2 = 0$$

This frequency equation is the same as obtained in part (a); obviously, it gives the ω_1 and ω_2 of Eq. (b).

To determine the nth mode we go back to either of the two equations of Eq. (10.2.5) with $\left[\mathbf{k} - \omega^2\mathbf{m}\right]$ given by Eq. (g). The first equation gives

$$\left(3k - m\omega_n^2\right)\phi_{tn} + \frac{kL}{2}\phi_{\theta n} = 0 \quad \text{or} \quad \phi_{\theta n} = -\frac{3k - m\omega_n^2}{kL/2}\phi_{tn} \tag{h}$$

Substituting for $\omega_1^2 = 2.536k/m$ and $\omega_2^2 = 9.464k/m$ in Eq. (h) gives

$$\frac{L}{2}\phi_{\theta 1} = -0.464\phi_{t1} \qquad \frac{L}{2}\phi_{\theta 2} = 6.464\phi_{t2}$$

If $\phi_{t1} = 1$, then $\phi_{\theta 1} = -0.928/L$, and if $\phi_{t2} = -1$, then $\phi_{\theta 2} = -12.928/L$. In summary, the two modes plotted in Fig. E10.1d are

$$\phi_1 = \begin{Bmatrix} 1 \\ -0.928/L \end{Bmatrix} \qquad \phi_2 = \begin{Bmatrix} -1 \\ -12.928/L \end{Bmatrix} \tag{i}$$

(c) The same natural frequencies were obtained using the two sets of DOFs. The mode shapes are given by Eqs. (e) and (i) for the two sets of DOFs. These two sets of results are plotted in Fig. E10.1c and d and can be shown to be equivalent on a graphical basis. Alternatively, the equivalence can be demonstrated by using the coordinate transformation from one set of DOFs to the other. The displacements $\mathbf{u} = \langle u_1 \quad u_2 \rangle^T$ are related to the second set of DOFs, $\bar{\mathbf{u}} = \langle u_t \quad u_\theta \rangle^T$ by

$$\begin{Bmatrix} u_1 \\ u_2 \end{Bmatrix} = \begin{bmatrix} 1 & -L/2 \\ 1 & L/2 \end{bmatrix} \begin{Bmatrix} u_t \\ u_\theta \end{Bmatrix} \qquad \text{or} \qquad \mathbf{u} = \mathbf{a}\bar{\mathbf{u}} \tag{j}$$

The displacements $\bar{\mathbf{u}}$ in the first two modes are given by Eq. (i). Substituting the first mode in Eq. (j) leads to $\mathbf{u} = \langle 1.464 \quad 0.536 \rangle^T$. Normalizing the vector yields $\mathbf{u} = \langle 1 \quad 0.366 \rangle^T$, which is identical to ϕ_1 of Eq. (e). Similarly, substituting the second mode from Eq. (i) in Eq. (j) gives $\mathbf{u} = \langle 1 \quad -1.366 \rangle$, which is identical to ϕ_2 of Eq. (e).

Example 10.2

Determine the natural frequencies and modes of vibration of the system shown in Fig. E10.2a and defined in Example 9.5. Show that the modes satisfy the orthogonality properties.

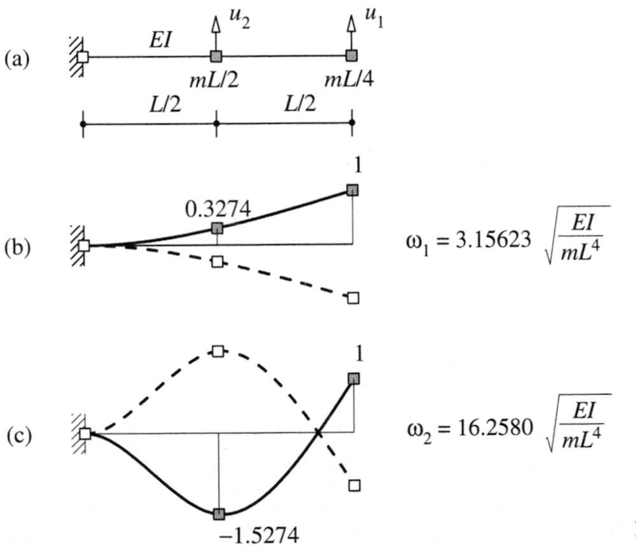

Figure E10.2

Solution The stiffness and mass matrices were determined in Example 9.5 with reference to the translational DOFs u_1 and u_2:

$$\mathbf{m} = \begin{bmatrix} mL/4 & \\ & mL/2 \end{bmatrix} \qquad \mathbf{k} = \frac{48EI}{7L^3}\begin{bmatrix} 2 & -5 \\ -5 & 16 \end{bmatrix}$$

Then

$$\mathbf{k} - \omega^2\mathbf{m} = \frac{48EI}{7L^3}\begin{bmatrix} 2-\lambda & -5 \\ -5 & 16-2\lambda \end{bmatrix} \tag{a}$$

where

$$\lambda = \frac{7mL^4}{192EI}\omega^2 \tag{b}$$

Substituting Eq. (a) in (10.2.6) gives the frequency equation

$$2\lambda^2 - 20\lambda + 7 = 0$$

which has two solutions: $\lambda_1 = 0.36319$ and $\lambda_2 = 9.6368$. The natural frequencies corresponding to the two values of λ are obtained from Eq. (b)†

$$\omega_1 = 3.15623\sqrt{\frac{EI}{mL^4}} \qquad \omega_2 = 16.2580\sqrt{\frac{EI}{mL^4}} \tag{c}$$

The natural modes are determined from Eq. (10.2.5) following the procedure shown in Example 10.1 to obtain

$$\phi_1 = \begin{Bmatrix} 1 \\ 0.3274 \end{Bmatrix} \qquad \phi_2 = \begin{Bmatrix} 1 \\ -1.5274 \end{Bmatrix} \tag{d}$$

These natural modes are plotted in Fig. E10.2b and c.

With the modes known we compute the left side of Eq. (10.4.1):

$$\phi_1^T\mathbf{m}\phi_2 = \frac{mL}{4}\langle 1 \quad 0.3274\rangle \begin{bmatrix} 1 & \\ & 2 \end{bmatrix}\begin{Bmatrix} 1 \\ -1.5274 \end{Bmatrix} = 0$$

$$\phi_1^T\mathbf{k}\phi_2 = \frac{48EI}{7L^3}\langle 1 \quad 0.3274\rangle \begin{bmatrix} 2 & -5 \\ -5 & 16 \end{bmatrix}\begin{Bmatrix} 1 \\ -1.5274 \end{Bmatrix} = 0$$

This verifies that the natural modes computed for the system are orthogonal.

Example 10.3

Determine the natural frequencies and modes of vibration of the system shown in Fig. E10.3a and defined in Example 9.6. Normalize the modes to have unit vertical deflection at the free end.

Solution The stiffness and mass matrices were determined in Example 9.6 with reference to DOFs u_1 and u_2:

$$\mathbf{m} = \begin{bmatrix} 3m & \\ & m \end{bmatrix} \qquad \mathbf{k} = \frac{6EI}{7L^3}\begin{bmatrix} 8 & -3 \\ -3 & 2 \end{bmatrix}$$

†Six significant digits are included so as to compare with the continuum model of a beam in Chapter 16.

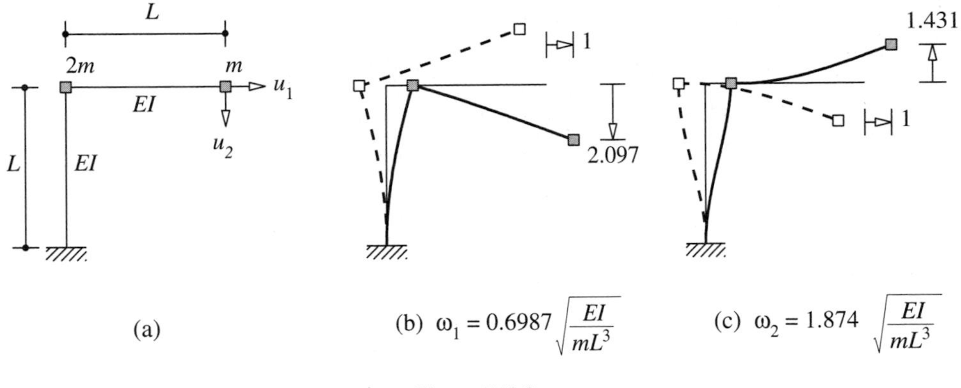

Figure E10.3

The frequency equation is Eq. (10.2.6), which, after substituting for **m** and **k**, evaluating the determinant, and defining

$$\lambda = \frac{7mL^3}{6EI}\omega^2 \tag{a}$$

can be written as

$$3\lambda^2 - 14\lambda + 7 = 0$$

The two roots are $\lambda_1 = 0.5695$ and $\lambda_2 = 4.0972$. The natural frequencies corresponding to the two values of λ are obtained from Eq. (a):

$$\omega_1 = 0.6987\sqrt{\frac{EI}{mL^3}} \qquad \omega_2 = 1.874\sqrt{\frac{EI}{mL^3}} \tag{b}$$

The natural modes are determined from Eq. (10.2.5) following the procedure used in Example 10.1 to obtain

$$\phi_1 = \begin{Bmatrix} 1 \\ 2.097 \end{Bmatrix} \qquad \phi_2 = \begin{Bmatrix} 1 \\ -1.431 \end{Bmatrix} \tag{c}$$

These modes are plotted in Fig. E10.3b and c.

In computing the natural modes the mode shape value for the first DOF had been arbitrarily set as unity. The resulting mode is normalized to unit value in DOF u_2 by dividing ϕ_1 in Eq. (c) by 2.097. Similarly, the second mode is normalized by dividing ϕ_2 in Eq. (c) by -1.431. Thus the normalized modes are

$$\phi_1 = \begin{Bmatrix} 0.4769 \\ 1 \end{Bmatrix} \qquad \phi_2 = \begin{Bmatrix} -0.6988 \\ 1 \end{Bmatrix} \tag{d}$$

Example 10.4

Determine the natural frequencies and modes of the system shown in Fig. E10.4a and defined in Example E9.1, a two-story frame idealized as a shear building. Normalize the modes so that $M_n = 1$.

Solution The mass and stiffness matrices of the system, determined in Example 9.1, are

$$\mathbf{m} = \begin{bmatrix} 2m & \\ & m \end{bmatrix} \qquad \mathbf{k} = \begin{bmatrix} 3k & -k \\ -k & k \end{bmatrix} \tag{a}$$

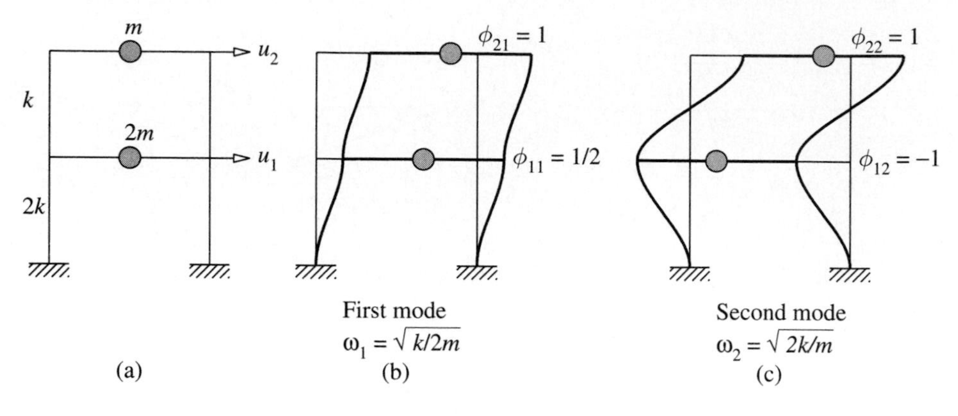

Figure E10.4

where $k = 24EI_c/h^3$. The frequency equation is Eq. (10.2.6), which, after substituting for **m** and **k** and evaluating the determinant, can be written as

$$(2m^2)\omega^4 + (-5km)\omega^2 + 2k^2 = 0 \tag{b}$$

The two roots are $\omega_1^2 = k/2m$ and $\omega_2^2 = 2k/m$, and the two natural frequencies are

$$\omega_1 = \sqrt{\frac{k}{2m}} \qquad \omega_2 = \sqrt{\frac{2k}{m}} \tag{c}$$

Substituting for k gives

$$\omega_1 = 3.464\sqrt{\frac{EI_c}{mh^3}} \qquad \omega_2 = 6.928\sqrt{\frac{EI_c}{mh^3}} \tag{d}$$

The natural modes are determined from Eq. (10.2.5) following the procedure used in Example 10.1 to obtain

$$\phi_1 = \begin{Bmatrix} \frac{1}{2} \\ 1 \end{Bmatrix} \qquad \phi_2 = \begin{Bmatrix} -1 \\ 1 \end{Bmatrix} \tag{e}$$

These natural modes are shown in Fig. E10.4b, and c.

To normalize the first mode, M_1 is calculated using Eq. (10.4.6), with ϕ_1 given by Eq. (e):

$$M_1 = \phi_1^T \mathbf{m} \phi_1 = m \left\langle \tfrac{1}{2} \quad 1 \right\rangle \begin{bmatrix} 2 & \\ & 1 \end{bmatrix} \begin{Bmatrix} \frac{1}{2} \\ 1 \end{Bmatrix} = \frac{3}{2}m$$

To make $M_1 = 1$, divide ϕ_1 of Eq. (e) by $\sqrt{3m/2}$ to obtain the normalized mode

$$\phi_1 = \frac{1}{\sqrt{6m}} \begin{Bmatrix} 1 \\ 2 \end{Bmatrix}$$

For this ϕ_1 it can be verified that $M_1 = 1$. The second mode can be normalized similarly.

Example 10.5

Determine the natural frequencies and modes of the system shown in Fig. E10.5a and defined earlier in Example 9.9. The story height $h = 10$ ft.

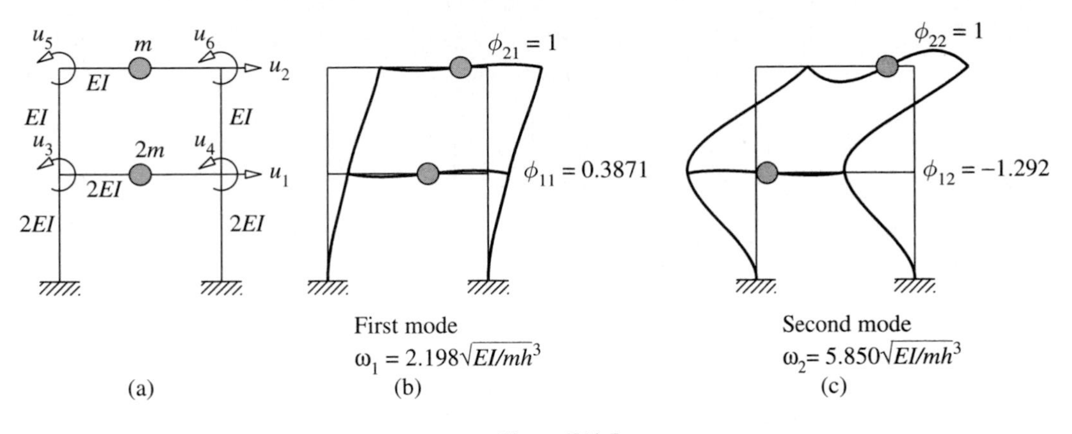

Figure E10.5

Solution With reference to the lateral displacements u_1 and u_2 of the two floors as the two DOFs, the mass matrix and the condensed stiffness matrix were determined in Example 9.9:

$$\mathbf{m}_{tt} = m\begin{bmatrix} 2 & \\ & 1 \end{bmatrix} \qquad \hat{\mathbf{k}}_{tt} = \frac{EI}{h^3}\begin{bmatrix} 54.88 & -17.51 \\ -17.51 & 11.61 \end{bmatrix} \tag{a}$$

The frequency equation is

$$\det(\hat{\mathbf{k}}_{tt} - \omega^2 \mathbf{m}_{tt}) = 0 \tag{b}$$

Substituting for $\mathbf{m}_{tt}$ and $\hat{\mathbf{k}}_{tt}$, evaluating the determinant, and obtaining the two roots just as in Example 10.4 leads to

$$\omega_1 = 2.198\sqrt{\frac{EI}{mh^3}} \qquad \omega_2 = 5.850\sqrt{\frac{EI}{mh^3}} \tag{c}$$

It is of interest to compare these frequencies for a frame with flexible beams with those for the frame with flexurally rigid beams, determined in Example 10.4. It is clear that beam flexibility has the effect of lowering the frequencies, consistent with intuition.

The natural modes are determined by solving

$$(\hat{\mathbf{k}}_{tt} - \omega_n^2 \mathbf{m}_{tt})\phi_n = \mathbf{0} \tag{d}$$

with ω_1 and ω_2 substituted successively from Eq. (c) to obtain

$$\phi_1 = \begin{Bmatrix} 0.3871 \\ 1 \end{Bmatrix} \qquad \phi_2 = \begin{Bmatrix} -1.292 \\ 1 \end{Bmatrix} \tag{e}$$

These vectors define the lateral displacements of each floor. They are shown in Fig. E10.5b and c together with the joint rotations. The joint rotations associated with the first mode are determined by substituting $\mathbf{u}_t = \phi_1$ from Eq. (e) in Eq. (d) of Example 9.9:

$$\begin{Bmatrix} u_3 \\ u_4 \\ u_5 \\ u_6 \end{Bmatrix} = \frac{1}{h}\begin{bmatrix} -0.4426 & -0.2459 \\ -0.4426 & -0.2459 \\ 0.9836 & -0.7869 \\ 0.9836 & -0.7869 \end{bmatrix}\begin{Bmatrix} 0.3871 \\ 1.0000 \end{Bmatrix} = \frac{1}{h}\begin{Bmatrix} -0.4172 \\ -0.4172 \\ -0.4061 \\ -0.4061 \end{Bmatrix} \tag{f}$$

Similarly, the joint rotations associated with the second mode are obtained by substituting $\mathbf{u}_t = \phi_2$ from Eq. (e) in Eq. (d) of Example 9.9:

$$\begin{Bmatrix} u_3 \\ u_4 \\ u_5 \\ u_6 \end{Bmatrix} = \frac{1}{h} \begin{Bmatrix} 0.3258 \\ 0.3258 \\ -2.0573 \\ -2.0573 \end{Bmatrix} \tag{g}$$

Example 10.6

Figure 9.5.1 shows the plan view of a one-story building. The structure consists of a roof, idealized as a rigid diaphragm, supported on three frames, A, B, and C, as shown. The roof weight is uniformly distributed and has a magnitude of 100 lb/ft.^{2} The lateral stiffnesses of the frames are $k_y = 75$ kips/ft for frame A, and $k_x = 40$ kips/ft for frames B and C. The plan dimensions are $b = 30$ ft and $d = 20$ ft, the eccentricity is $e = 1.5$ ft, and the height of the building is 12 ft. Determine the natural periods and modes of vibration of the structure.

Solution

$$\text{Weight of roof slab: } w = 30 \times 20 \times 100 \text{ lb} = 60 \text{ kips}$$

$$\text{Mass: } m = w/g = 1.863 \text{ kips-sec}^2/\text{ft}$$

$$\text{Moment of inertia: } I_O = \frac{m(b^2 + d^2)}{12} = 201.863 \text{ kips-ft-sec}^2$$

Lateral motion of the roof diaphragm in the x-direction is governed by Eq. (9.5.18):

$$m\ddot{u}_x + 2k_x u_x = 0 \tag{a}$$

Thus the natural frequency of x-lateral vibration is

$$\omega_x = \sqrt{\frac{2kx}{m}} = \sqrt{\frac{2(40)}{1.863}} = 6.553 \text{ rad/sec}$$

The corresponding natural mode is shown in Fig. E10.6c.

The coupled lateral (u_y)-torsional (u_θ) motion of the roof diaphragm is governed by Eq. (9.5.19). Substituting for m and I_O gives

$$\mathbf{m} = \begin{bmatrix} 1.863 & \\ & 201.863 \end{bmatrix}$$

From Eqs. (9.5.16) and (9.5.19) the stiffness matrix has four elements:

$$k_{yy} = k_y = 75 \text{ kips/ft}$$

$$k_{y\theta} = k_{\theta y} = ek_y = 1.5 \times 75 = 112.5 \text{ kips}$$

$$k_{\theta\theta} = e^2 k_y + \frac{d^2}{2} k_x = 8168.75 \text{ kips-ft}$$

Hence,

$$\mathbf{k} = \begin{bmatrix} 75.00 & 112.50 \\ 112.50 & 8168.75 \end{bmatrix}$$

With $\mathbf{k}$ and $\mathbf{m}$ known, the eigenvalue problem for this two-DOF system is solved by standard procedures to obtain:

$$\text{Natural frequencies (rad/sec): } \omega_1 = 5.878,\ \omega_2 = 6.794$$

$$\text{Natural modes: } \phi_1 = \begin{Bmatrix} -0.5228 \\ 0.0493 \end{Bmatrix},\ \phi_2 = \begin{Bmatrix} -0.5131 \\ -0.0502 \end{Bmatrix}$$

These mode shapes are plotted in Fig. E10.6a and b. The motion of the structure in each mode consists of translation of the rigid diaphragm coupled with torsion about the vertical axis through the center of mass.

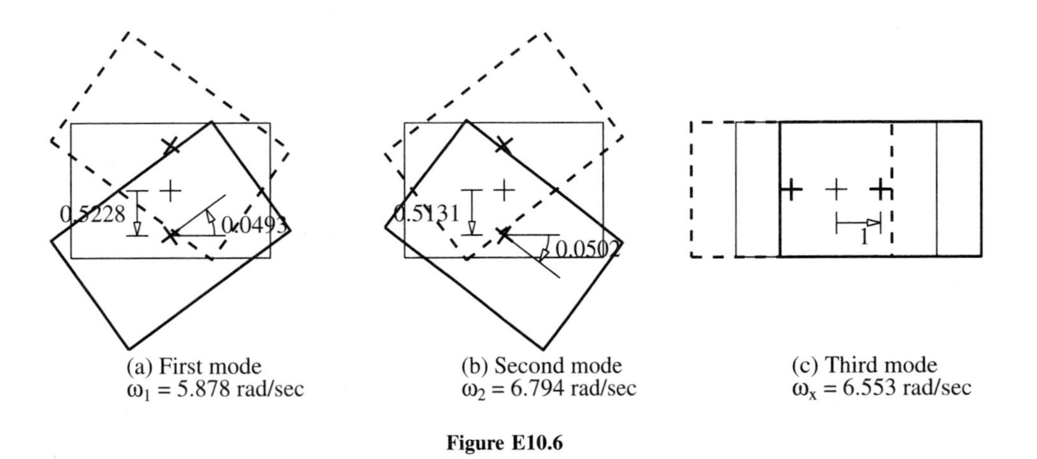

(a) First mode
$\omega_1 = 5.878$ rad/sec

(b) Second mode
$\omega_2 = 6.794$ rad/sec

(c) Third mode
$\omega_x = 6.553$ rad/sec

Figure E10.6

Example 10.7

Consider a special case of the system of Example 10.6 in which frame *A* is located at the center of mass (i.e., $e = 0$). Determine the natural frequencies and modes of this system.

Solution Equation (9.5.20) specialized for free vibration of this system gives three equations of motion:

$$m\ddot{u}_x + 2k_x u_x = 0 \qquad m\ddot{u}_y + k_y u_y = 0 \qquad I_O \ddot{u}_\theta + \frac{d^2}{2} k_x u_\theta = 0 \tag{a}$$

The first equation of motion indicates that translational motion in the x-direction would occur at the natural frequency

$$\omega_x = \sqrt{\frac{2k_x}{m}} = \sqrt{\frac{2(40)}{1.863}} = 6.553 \text{ rad/sec}$$

This motion is independent of lateral motion u_y or torsional motion u_θ (Fig. E10.7c). The second equation of motion indicates that translational motion in the y-direction would occur at the natural frequency

$$\omega_y = \sqrt{\frac{k_y}{m}} = \sqrt{\frac{75}{1.863}} = 6.344 \text{ rad/sec}$$

This motion is independent of the lateral motion u_x or torsional motion u_θ (Fig. E10.7b). The third equation of motion indicates that torsional motion would occur at the natural frequency

$$\omega_\theta = \sqrt{\frac{d^2 k_x}{2 I_O}} = \sqrt{\frac{(20)^2 40}{2(201.863)}} = 6.295 \text{ rad/sec}$$

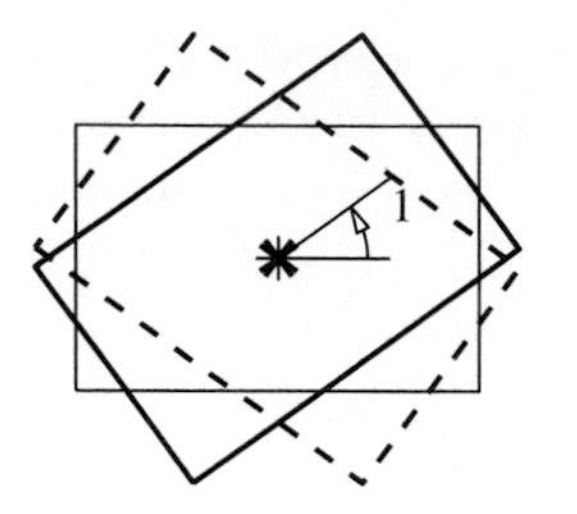

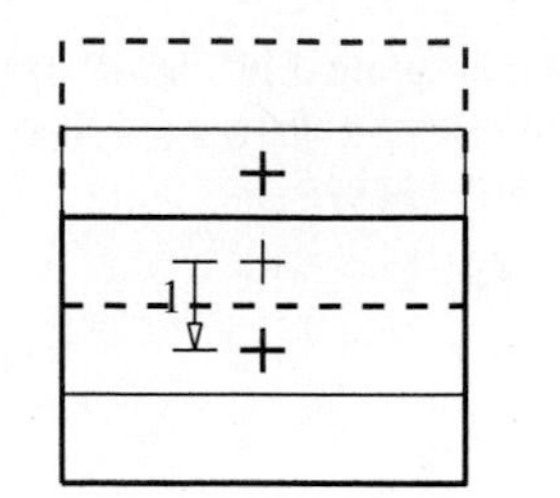

(a) First mode
$\omega_\theta = 6.295$ rad/sec

(b) Second mode
$\omega_y = 6.344$ rad/sec

(c) Third mode
$\omega_x = 6.553$ rad/sec

Figure E10.7

The roof diaphragm would rotate about the vertical axis through its center of mass without any translation of this point in the x or y directions (Fig. E10.7a).

Observe that the natural frequencies ω_1 and ω_2 of the unsymmetric-plan system (Example 10.6) are different from and more separated than the natural frequencies ω_y and ω_θ of the symmetric-plan system (Example 10.7).

10.7 MODAL EXPANSION OF DISPLACEMENTS

Any set of N independent vectors can be used as a basis for representing any other vector of order N. In the following sections the natural modes are used as such a basis. Thus, a modal expansion of any displacement vector $\mathbf{u}$ has the form

$$\mathbf{u} = \sum_{r=1}^{N} \phi_r q_r = \mathbf{\Phi q} \tag{10.7.1}$$

where q_r are scalar multipliers called *modal coordinates* or *normal coordinates*. When the ϕ_r are known, for a given $\mathbf{u}$ it is possible to evaluate the q_r by multiplying both sides of Eq. (10.7.1) by $\phi_n^T m$:

$$\phi_n^T \mathbf{mu} = \sum_{r=1}^{N} (\phi_n^T \mathbf{m} \phi_r) q_r$$

Because of the orthogonality relation of Eq. (10.4.1b), all terms in the summation above vanish except the $r = n$ term; thus

$$\phi_n^T \mathbf{mu} = (\phi_n^T \mathbf{m} \phi_n) q_n$$

The matrix products on both sides are scalars. Therefore,

$$q_n = \frac{\phi_n^T \mathbf{mu}}{\phi_n^T \mathbf{m} \phi_n} = \frac{\phi_n^T \mathbf{mu}}{M_n} \tag{10.7.2}$$

The modal expansion of the displacement vector **u**, Eq. (10.7.1), is employed in Section 10.8 to obtain solutions for the free vibration response of undamped systems. It also plays a central role in the analysis of forced vibration response and earthquake response of MDF systems (Chapters 12 and 13).

Example 10.8

For the two-story shear frame of Example 10.4, determine the modal expansion of the displacement vector $\mathbf{u} = \langle 1 \quad 1 \rangle^T$.

Solution The displacement **u** is substituted in Eq. (10.7.2) together with $\phi_1 = \langle \frac{1}{2} \quad 1 \rangle^T$ and $\phi_2 = \langle -1 \quad 1 \rangle^T$, from Example 10.4, to obtain

$$q_1 = \frac{\langle \frac{1}{2} \quad 1 \rangle \begin{bmatrix} 2m & \\ & m \end{bmatrix} \begin{Bmatrix} 1 \\ 1 \end{Bmatrix}}{\langle \frac{1}{2} \quad 1 \rangle \begin{bmatrix} 2m & \\ & m \end{bmatrix} \begin{Bmatrix} \frac{1}{2} \\ 1 \end{Bmatrix}} = \frac{2m}{3m/2} = \frac{4}{3}$$

$$q_2 = \frac{\langle -1 \quad 1 \rangle \begin{bmatrix} 2m & \\ & m \end{bmatrix} \begin{Bmatrix} 1 \\ 1 \end{Bmatrix}}{\langle -1 \quad 1 \rangle \begin{bmatrix} 2m & \\ & m \end{bmatrix} \begin{Bmatrix} -1 \\ 1 \end{Bmatrix}} = \frac{-m}{3m} = -\frac{1}{3}$$

Substituting q_n in Eq. (10.7.1) gives the desired modal expansion, which is shown in Fig. E10.8.

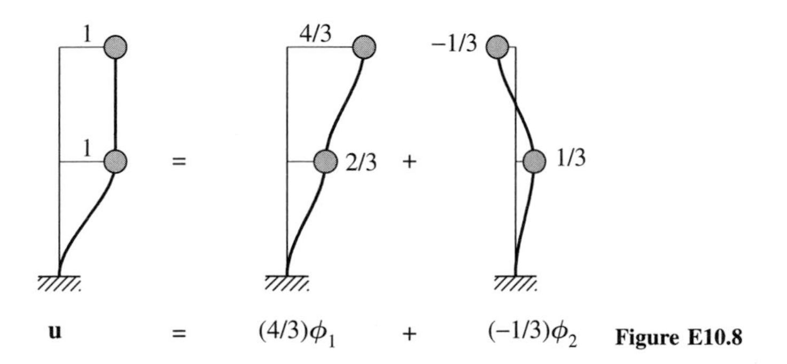

Figure E10.8

PART B: FREE VIBRATION RESPONSE

10.8 SOLUTION OF FREE VIBRATION EQUATIONS: UNDAMPED SYSTEMS

We now return to the problem posed by Eqs. (10.1.1) and (10.1.2) and find its solution. The differential equation (10.1.1) to be solved had led to the matrix eigenvalue problem of Eq. (10.2.4). Assuming that the eigenvalue problem has been solved for the natural frequencies and modes, the general solution of Eq. (10.1.1) is given by a superposition of the response in individual modes given by Eq. (10.2.3). Thus

$$\mathbf{u}(t) = \sum_{n=1}^{N} \boldsymbol{\phi}_n (A_n \cos \omega_n t + B_n \sin \omega_n t) \tag{10.8.1}$$

where A_n and B_n are $2N$ constants of integration. To determine these constants, we will also need the equation for the velocity vector, which is

$$\dot{\mathbf{u}}(t) = \sum_{n=1}^{N} \boldsymbol{\phi}_n \omega_n (-A_n \sin \omega_n t + B_n \cos \omega_n t) \tag{10.8.2}$$

Setting $t = 0$ in Eqs. (10.8.1) and (10.8.2) gives

$$\mathbf{u}(0) = \sum_{n=1}^{N} \boldsymbol{\phi}_n A_n \qquad \dot{\mathbf{u}}(0) = \sum_{n=1}^{N} \boldsymbol{\phi}_n \omega_n B_n \tag{10.8.3}$$

With the initial displacements $\mathbf{u}(0)$ and initial velocities $\dot{\mathbf{u}}(0)$ known, each of these two equation sets represents N algebraic equations in the unknowns A_n and B_n, respectively. Simultaneous solution of these equations is not necessary because they can be interpreted as a modal expansion of the vectors $\mathbf{u}(0)$ and $\dot{\mathbf{u}}(0)$. Following Eq. (10.7.1), we can write

$$\mathbf{u}(0) = \sum_{n=1}^{N} \boldsymbol{\phi}_n q_n(0) \qquad \dot{\mathbf{u}}(0) = \sum_{n=1}^{N} \boldsymbol{\phi}_n \dot{q}_n(0) \tag{10.8.4}$$

where, analogous to Eq. (10.7.2), $q_n(0)$ and $\dot{q}_n(0)$ are given by

$$q_n(0) = \frac{\boldsymbol{\phi}_n^T \mathbf{m} \mathbf{u}(0)}{M_n} \qquad \dot{q}_n(0) = \frac{\boldsymbol{\phi}_n^T \mathbf{m} \dot{\mathbf{u}}(0)}{M_n} \tag{10.8.5}$$

Equations (10.8.3) and (10.8.4) are equivalent, implying that $A_n = q_n(0)$ and $B_n = \dot{q}_n(0)/\omega_n$. Substituting these in Eq. (10.8.1) gives

$$\mathbf{u}(t) = \sum_{n=1}^{N} \boldsymbol{\phi}_n \left[q_n(0) \cos \omega_n t + \frac{\dot{q}_n(0)}{\omega_n} \sin \omega_n t \right] \tag{10.8.6}$$

or, alternatively,

$$\mathbf{u}(t) = \sum_{n=1}^{N} \boldsymbol{\phi}_n q_n(t) \tag{10.8.7}$$

where

$$q_n(t) = q_n(0) \cos \omega_n t + \frac{\dot{q}_n(0)}{\omega_n} \sin \omega_n t \tag{10.8.8}$$

is the time variation of modal coordinates, which is analogous to the free vibration response of SDF systems [Eq. (2.1.3)]. Equation (10.8.6) is the solution of the free vibration problem. It provides the displacement $\mathbf{u}$ as a function of time due to initial displacement $\mathbf{u}(0)$ and velocity $\dot{\mathbf{u}}(0)$. Assuming that the natural frequencies ω_n and modes $\boldsymbol{\phi}_n$ are available, the right side of Eq. (10.8.6) is known with $q_n(0)$ and $\dot{q}_n(0)$ defined by Eq. (10.8.5).

Example 10.9

Determine the free vibration response of the two-story shear frame of Example 10.4 due to initial displacement $\mathbf{u}(0) = \langle 1 \quad 2 \rangle$.

Solution The initial displacement and velocity vectors are

$$\mathbf{u}(0) = \begin{Bmatrix} 1 \\ 2 \end{Bmatrix} \qquad \dot{\mathbf{u}}(0) = \begin{Bmatrix} 0 \\ 0 \end{Bmatrix}$$

For the given $\mathbf{u}(0)$, $q_n(0)$ are calculated following the procedure of Example 10.8 and using ϕ_n from Eq. (e) of Example 10.4; the results are $q_1(0) = 2$ and $q_2(0) = 0$. Because the initial velocity $\dot{\mathbf{u}}(0)$ is zero, $\dot{q}_1(0) = \dot{q}_2(0) = 0$. Inserting $q_n(0)$ and $\dot{q}_n(0)$ in Eq. (10.8.8) gives the solution for modal coordinates

$$q_1(t) = 2\cos\omega_1 t \qquad q_2(t) = 0$$

Substituting $q_n(t)$ and ϕ_n in Eq. (10.8.7) leads to

$$\begin{Bmatrix} u_1(t) \\ u_2(t) \end{Bmatrix} = \begin{Bmatrix} \frac{1}{2} \\ 1 \end{Bmatrix} 2\cos\omega_1 t = \begin{Bmatrix} 1 \\ 2 \end{Bmatrix} \cos\omega_1 t$$

where $\omega_1 = \sqrt{k/2m}$ from Example 10.4. These solutions for $q_1(t)$, $u_1(t)$, and $u_2(t)$ had been plotted in Fig. 10.1.2c and d. Note that $q_2(t) = 0$ implies that the second mode has no contribution to the response, which is all due to the first mode. Such is the case because the initial displacement is proportional to the first mode and hence orthogonal to the second mode.

Example 10.10

Determine the free vibration response of the two-story shear frame of Example 10.4 due to initial displacement $\mathbf{u}(0) = \langle -1 \quad 1 \rangle^T$.

Solution The calculations proceed as in Example 10.9, leading to $q_1(0) = 0$, $q_2(0) = -1$, and $q_n(0) = \dot{q}_n(0) = 0$. Inserting these in Eq. (10.8.8) gives the solutions for modal coordinates:

$$q_1(t) = 0 \qquad q_2(t) = 1\cos\omega_2 t$$

Substituting $q_n(t)$ and ϕ_n in Eq. (10.8.7) leads to

$$\begin{Bmatrix} u_1(t) \\ u_2(t) \end{Bmatrix} = \begin{Bmatrix} -1 \\ 1 \end{Bmatrix} \cos\omega_2 t$$

where $\omega_2 = \sqrt{2k/m}$ from Example 10.4. These solutions for $q_2(t)$, $u_1(t)$, and $u_2(t)$ had been plotted in Fig. 10.1.3c and d. Note that $q_1(t) = 0$ implies that the first mode has no contribution to the response and the response is due entirely to the second mode. Such is the case because the initial displacement is proportional to the second mode and hence orthogonal to the first mode.

Example 10.11

Determine the free vibration response of the two-story shear frame of Example 10.4 due to initial displacements $\mathbf{u}(0) = \langle -\frac{1}{2} \quad 2 \rangle^T$.

Solution Following Example 10.8, $q_n(0)$ and $\dot{q}_n(0)$ are evaluated: $q_1(0) = 1$, $q_2(0) = 1$, and $\dot{q}_1(0) = \dot{q}_2(0) = 0$. Substituting these in Eq. (10.8.8) gives the solution for modal coordinates

$$q_1(t) = 1\cos\omega_1 t \qquad q_2(t) = 1\cos\omega_2 t$$

Substituting $q_n(t)$ and ϕ_n in Eq. (10.8.7) leads to

$$\begin{Bmatrix} u_1(t) \\ u_2(t) \end{Bmatrix} = \begin{Bmatrix} \frac{1}{2} \\ 1 \end{Bmatrix} \cos \omega_1 t + \begin{Bmatrix} -1 \\ 1 \end{Bmatrix} \cos \omega_2 t$$

These solutions for $q_n(t)$ and $u_j(t)$ had been plotted in Fig. 10.1.1c and d. Observe that both natural modes contribute to the response due to these initial displacements.

10.9 FREE VIBRATION OF SYSTEMS WITH DAMPING

When damping is included, the free vibration response of the system is governed by Eq. (9.2.12) with $\mathbf{p}(t) = \mathbf{0}$:

$$\mathbf{m}\ddot{\mathbf{u}} + \mathbf{c}\dot{\mathbf{u}} + \mathbf{k}\mathbf{u} = \mathbf{0} \tag{10.9.1}$$

It is desired to find the solution $\mathbf{u}(t)$ of Eq. (10.9.1) that satisfies the initial conditions

$$\mathbf{u} = \mathbf{u}(0) \qquad \dot{\mathbf{u}} = \dot{\mathbf{u}}(0) \tag{10.9.2}$$

at $t = 0$. A procedure to obtain the desired solution will be developed in Section 10.10 for certain forms of damping that are reasonable models for many real structures. In this section the solution is presented in graphical form for a specific system that enables us to understand qualitatively the effects of damping on the free vibration of MDF systems.

For this purpose we express the displacement $\mathbf{u}$ in terms of the natural modes of the system without damping; thus we substitute Eq. (10.7.1) in Eq. (10.9.1):

$$\mathbf{m}\boldsymbol{\Phi}\ddot{\mathbf{q}} + \mathbf{c}\boldsymbol{\Phi}\dot{\mathbf{q}} + \mathbf{k}\boldsymbol{\Phi}\mathbf{q} = \mathbf{0}$$

Premultiplying by $\boldsymbol{\Phi}^T$ gives

$$\mathbf{M}\ddot{\mathbf{q}} + \mathbf{C}\dot{\mathbf{q}} + \mathbf{K}\mathbf{q} = \mathbf{0} \tag{10.9.3}$$

where the diagonal matrices $\mathbf{M}$ and $\mathbf{K}$ were defined in Eq. (10.4.5) and

$$\mathbf{C} = \boldsymbol{\Phi}^T \mathbf{c} \boldsymbol{\Phi} \tag{10.9.4}$$

The square matrix $\mathbf{C}$ may or may not be diagonal, depending on the distribution of damping in the system. If $\mathbf{C}$ is diagonal, Eq. (10.9.3) represents N uncoupled differential equations in modal coordinates q_n, and the system is said to have *classical damping* because classical modal analysis (Chapters 12 and 13) is applicable to such systems. These systems possess the same natural modes as those of the undamped system. The general form of the damping matrix that satisfies these properties is presented in Chapter 11. Systems with damping such that $\mathbf{C}$ is nondiagonal are said to have *nonclassical damping*. These systems are not amenable to classical modal analysis, and they do not possess the same natural modes as the undamped system.

10.9.1 Nonclassically Damped System

Figure 10.9.1a shows the two-story system of Fig. E10.4 but now with viscous dampers of coefficients c and $4c$ in the two stories. The equations of motion for this system are given by Eq. (10.9.1) with

$$\mathbf{m} = m\begin{bmatrix} 2 & \\ & 1 \end{bmatrix} \qquad \mathbf{k} = k\begin{bmatrix} 3 & -1 \\ -1 & 1 \end{bmatrix} \qquad \mathbf{c} = c\begin{bmatrix} 5 & -4 \\ -4 & 4 \end{bmatrix} \tag{10.9.5}$$

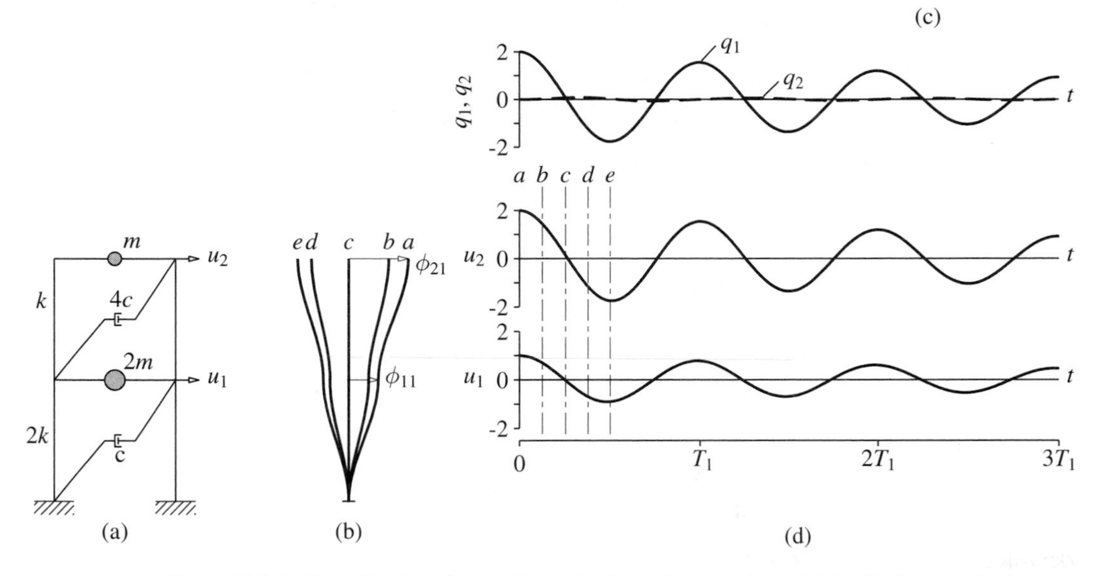

Figure 10.9.1 Free vibration of a nonclassically damped system due to initial displacement in the first natural mode of the undamped system: (a) two-story frame; (b) deflected shapes at time instants a, b, c, d, and e; (c) modal coordinates $q_n(t)$; (d) displacement history.

The **m** and **k** are available from Example 9.1 and **c** is obtained from Eq. (9.1.10) with $c_1 = c$ and $c_2 = 4c$. The natural modes of the system without damping were computed in Example 10.4:

$$\phi_1 = \begin{Bmatrix} \frac{1}{2} \\ 1 \end{Bmatrix} \qquad \phi_2 = \begin{Bmatrix} -1 \\ 1 \end{Bmatrix} \tag{10.9.6}$$

Substituting these **m**, **k**, **c**, and ϕ_n in Eqs. (10.4.5), (10.4.6), and (10.9.4) gives the **M**, **C**, and **K** which are inserted in Eq. (10.9.3) to obtain

$$m\begin{bmatrix} 1.5 & \\ & 3.0 \end{bmatrix}\begin{Bmatrix} \ddot{q}_1 \\ \ddot{q}_2 \end{Bmatrix} + c\begin{bmatrix} 2.25 & 4.50 \\ 4.50 & 18.00 \end{bmatrix}\begin{Bmatrix} \dot{q}_1 \\ \dot{q}_2 \end{Bmatrix} + k\begin{bmatrix} 0.75 & \\ & 6.00 \end{bmatrix}\begin{Bmatrix} q_1 \\ q_2 \end{Bmatrix} = \begin{Bmatrix} 0 \\ 0 \end{Bmatrix} \tag{10.9.7}$$

Note that **C** is not a diagonal matrix; thus the two differential equations in $q_1(t)$ and $q_2(t)$ are coupled.

Numerical methods are presented in Chapter 15 for solving such coupled differential equations. Using those methods, Eq. (10.9.7) was solved for the system considered, with $c = \sqrt{km/200}$ subjected to initial displacements $\mathbf{u}(0)$. The results for initial displacements $\mathbf{u}(0) = \langle 1 \quad 2 \rangle^T$ are presented in Fig. 10.9.1 and for $\mathbf{u}(0) = \langle -1 \quad 1 \rangle^T$ in Fig. 10.9.2. The solutions for $q_1(t)$ and $q_2(t)$ are presented in part (c) of these figures. Substituting $q_n(t)$ and ϕ_n in Eq. (10.7.1) gives the floor displacements $u_1(t)$ and $u_2(t)$, which are plotted in part (d) of these figures; the deflected

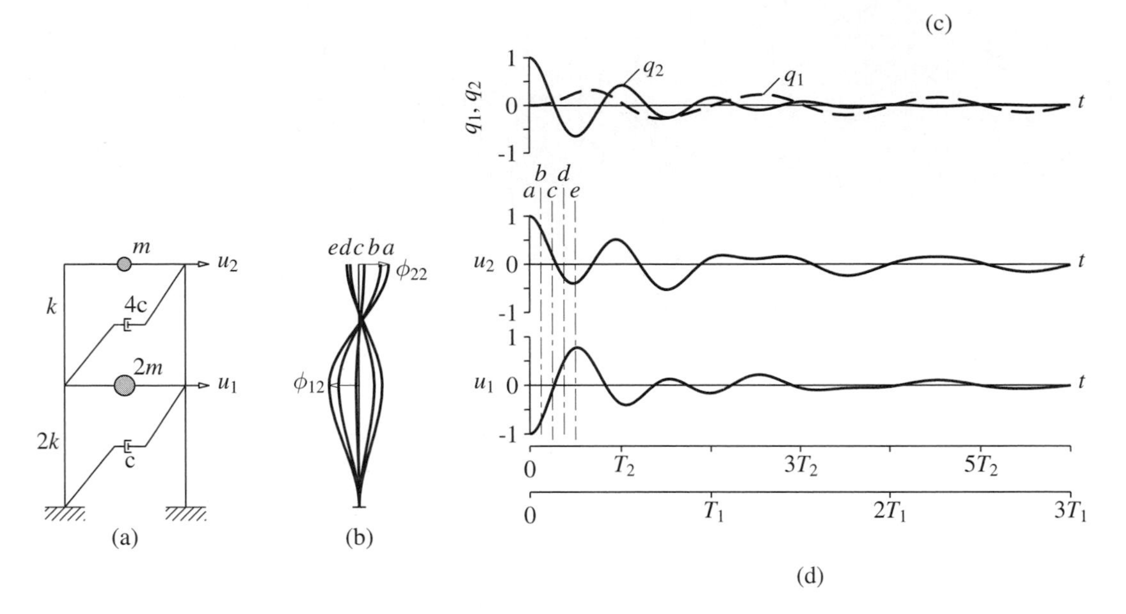

Figure 10.9.2 Free vibration of a nonclassically damped system due to initial displacement in the second natural mode of the undamped system: (a) two-story frame; (b) deflected shapes at time instants *a*, *b*, *c*, *d*, and *e*; (c) modal coordinates $q_n(t)$; (d) displacement history.

shapes at selected time instants—*a*, *b*, *c*, *d*, and *e*—are plotted in part (b) of these figures.

These results are for a system disturbed from its equilibrium position by imposing displacements—defined by vectors $\mathbf{u}(0)$ in the preceding paragraph—which are proportional to the two natural modes ϕ_n of the undamped system; the results permit three observations. First, contrary to what we observed for undamped systems (Figs. 10.1.2 and 10.1.3), $q_2(t) \neq 0$ in Fig. 10.9.1c and $q_1(t) \neq 0$ in Fig. 10.9.2c. Second, the initial deflected shape is not maintained in free vibration; this is noticeable in Fig. 10.9.2b. Third, unlike for SDF systems with damping (Fig. 2.2.2), the motion in each DOF is not a damped simple harmonic motion with a unique frequency.

10.9.2 Classically Damped System

Consider now the system shown in Fig. 10.9.3a, similar to the one considered previously but for a different distribution of damping. In this case the viscous damping coefficients for the first and second stories are $4c$ and $2c$, respectively, and $c = \sqrt{km}/200$. The equations of motion for this system are also given by Eq. (10.9.1), with $\mathbf{m}$ and $\mathbf{k}$ the same as in Eq. (10.9.5) and

$$\mathbf{c} = c\begin{bmatrix} 6 & -2 \\ -2 & 2 \end{bmatrix}$$

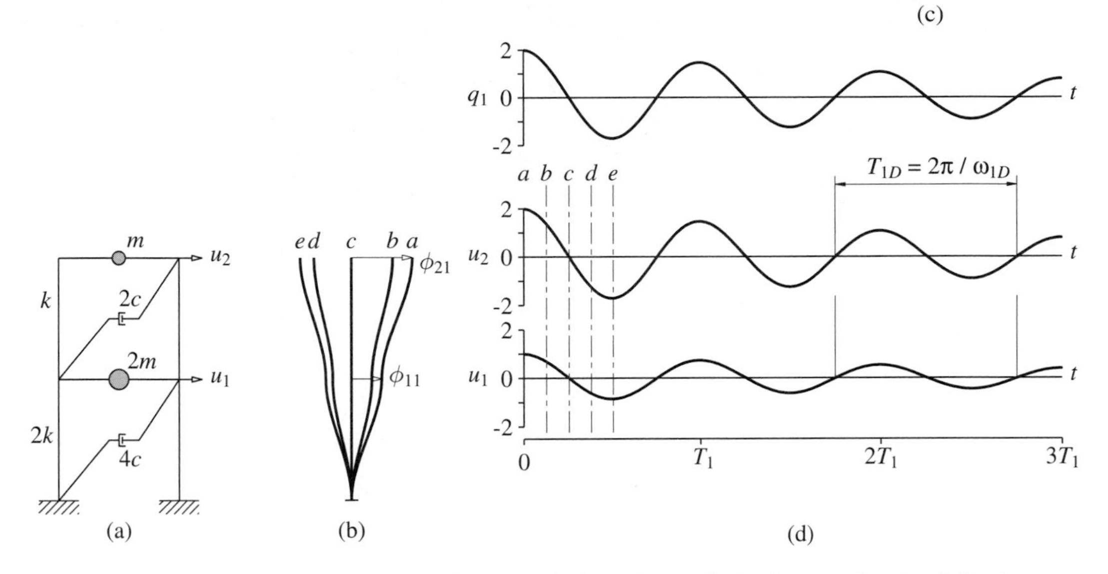

Figure 10.9.3 Free vibration of a classically damped system in the first natural mode of vibration: (a) two-story frame; (b) deflected shapes at time instants a, b, c, d, and e; (c) modal coordinate $q_1(t)$; (d) displacement history.

The equations governing the modal coordinates, Eq. (10.9.3), for this system are

$$m\begin{bmatrix} 1.5 & \\ & 3.0 \end{bmatrix}\begin{Bmatrix} \ddot{q}_1 \\ \ddot{q}_2 \end{Bmatrix} + c\begin{bmatrix} 1.5 & \\ & 12.0 \end{bmatrix}\begin{Bmatrix} \dot{q}_1 \\ \dot{q}_2 \end{Bmatrix} + k\begin{bmatrix} 0.75 & \\ & 6.00 \end{bmatrix}\begin{Bmatrix} q_1 \\ q_2 \end{Bmatrix} = \begin{Bmatrix} 0 \\ 0 \end{Bmatrix} \tag{10.9.8}$$

In this case **C** is a diagonal matrix and the two differential equations are uncoupled.

For an N-DOF system with classical damping, each of the N differential equations in modal coordinates is

$$M_n\ddot{q}_n + C_n\dot{q}_n + K_nq_n = 0 \tag{10.9.9}$$

where M_n and K_n were defined in Eq. (10.4.6) and

$$C_n = \boldsymbol{\phi}_n^T \mathbf{c} \boldsymbol{\phi}_n \tag{10.9.10}$$

Equation (10.9.9) is of the same form as Eq. (2.2.1a) for an SDF system with damping. Thus the damping ratio can be defined for each mode in a manner analogous to Eq. (2.2.2) for an SDF system:

$$\zeta_n = \frac{C_n}{2M_n\omega_n} \tag{10.9.11}$$

and Eq. (10.9.9) can be solved for $q_n(t)$ by the methods of Chapter 2.

These calculations are implemented for the system of Fig. 10.9.3a: $C_1 = 1.5c$, $M_1 = 1.5m$, $\omega_1 = \sqrt{k/2m}$, and $c = \sqrt{km/200}$, and Eq. (10.9.11) gives $\zeta_1 = 0.05$. Similar computations lead to $\zeta_2 = 0.10$. Solution of Eq. (10.9.9) for each mode gives $q_n(t)$, and this is substituted in Eq. (10.7.1) to obtain the displacements $u_j(t)$. The results are plotted for the system subject to the two sets of initial displacements defined earlier. The results for $\mathbf{u}(0) = \langle 1 \quad 2 \rangle^T$ are presented in Fig. 10.9.3 and for $\mathbf{u}(0) = \langle -1 \quad 1 \rangle^T$ in Fig. 10.9.4.

These results are for a system disturbed from its equilibrium position by imposing displacements—defined by vectors $\mathbf{u}(0)$ in the preceding paragraph—which are proportional to the natural modes ϕ_n of the undamped system; the results permit four observations. First, for the mode r other than the mode in which the initial displacement is introduced, $q_r(t) = 0$, implying no response due to the other mode. Second, the initial deflected shape is maintained during free vibration just as in the case of systems without damping (Figs. 10.1.2 and 10.1.3). Thus ϕ_n is also a natural mode of the damped system. Third, the motion of each mass is similar to that of the system without damping except that the amplitude of motion decreases with every vibration cycle because of damping. Fourth, the motion of each floor is a damped simple harmonic motion with a unique frequency, just as for SDF systems with damping (Fig. 2.2.2).

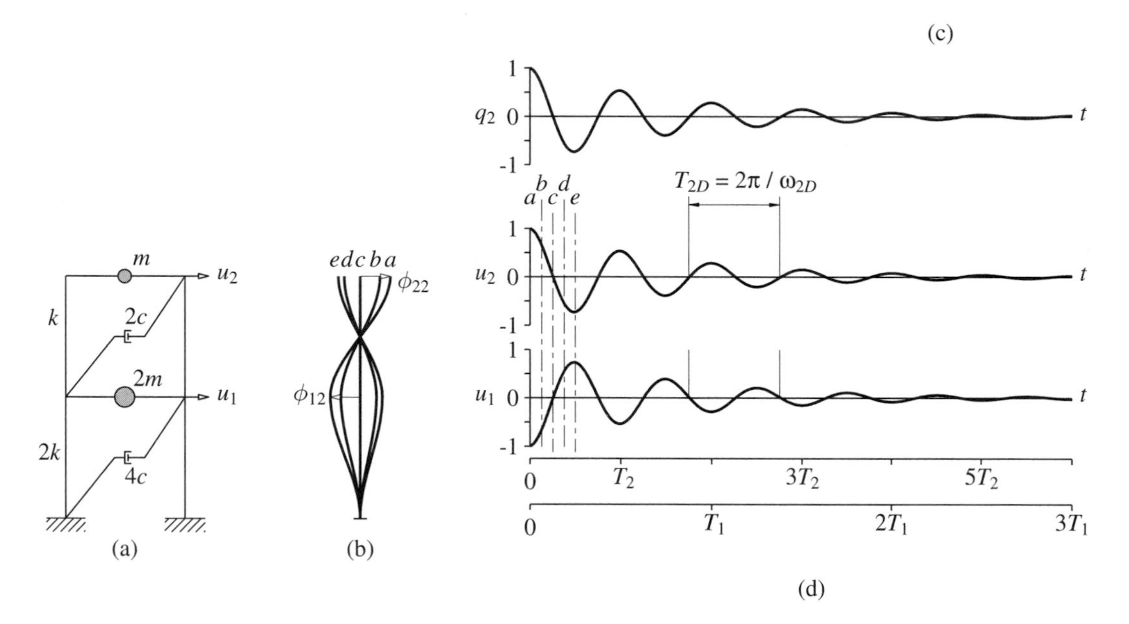

Figure 10.9.4 Free vibration of a classically damped system in the second natural mode of vibration: (a) two-story frame; (b) deflected shapes at time instants a, b, c, d, and e; (c) modal coordinate $q_2(t)$; (d) displacement history.

10.10 SOLUTION OF FREE VIBRATION EQUATIONS: CLASSICALLY DAMPED SYSTEMS

In this section a formal solution for free vibration of systems with classical damping due to initial displacements and/or initial velocities is presented. For this form of damping the natural modes are unaffected by damping. Therefore, the natural frequencies and modes of the system are first computed for the system without damping; the effect of damping on the natural frequencies is considered in the same manner as for an SDF system. This becomes apparent by dividing Eq. (10.9.9) governing $q_n(t)$ by M_n to obtain

$$\ddot{q}_n + 2\zeta_n\omega_n\dot{q}_n + \omega_n^2 q_n = 0 \tag{10.10.1}$$

This equation is of the same form as Eq. (2.2.1b) governing the free vibration of an SDF system with damping for which the solution is Eq. (2.2.4). Adapting this result, the solution for Eq. (10.10.1) is given by

$$q_n(t) = e^{-\zeta_n\omega_n t}\left[q_n(0)\cos\omega_{nD}t + \frac{\dot{q}_n(0) + \zeta_n\omega_n q_n(0)}{\omega_{nD}}\sin\omega_{nD}t\right] \tag{10.10.2}$$

where the nth natural frequency with damping

$$\omega_{nD} = \omega_n\sqrt{1 - \zeta_n^2} \tag{10.10.3}$$

The displacement response of the system is then obtained by substituting Eq. (10.10.2) for $q_n(t)$ in Eq. (10.8.7):

$$\mathbf{u}(t) = \sum_{n=1}^{N} \boldsymbol{\phi}_n e^{-\zeta_n\omega_n t}\left[q_n(0)\cos\omega_{nD}t + \frac{\dot{q}_n(0) + \zeta_n\omega_n q_n(0)}{\omega_{nD}}\sin\omega_{nD}t\right] \tag{10.10.4}$$

This is the solution of the free vibration problem for an MDF system with classical damping. It provides the displacement $\mathbf{u}$ as a function of time due to initial displacement $\mathbf{u}(0)$ and velocity $\dot{\mathbf{u}}(0)$. Assuming that the natural frequencies ω_n and modes ϕ_n of the system without damping are available together with the modal damping ratios ζ_n, the right side of Eq. (10.10.4) is known with $q_n(0)$ and $\dot{q}_n(0)$ defined by Eq. (10.8.5).

Damping influences the natural frequencies and periods of vibration of an MDF system according to Eq. (10.10.3), which is of the same form as Eq. (2.2.5) for an SDF system. Therefore, the effect of damping on the natural frequencies and periods of an MDF system is negligible for damping ratios ζ_n below 20% (Fig. 2.2.3), a range that includes most practical structures.

In an MDF system with classical damping undergoing free vibration in its nth natural mode, the displacement amplitude at any DOF decreases with each vibration cycle (Figs. 10.9.3 and 10.9.4). The rate of decay depends on the damping ratio ζ_n in that mode, in a manner similar to SDF systems. Thus the ratio of two response peaks separated by j cycles of vibration is related to the damping ratio by Eq. (2.2.12) with appropriate change in notation.

Consequently, the damping ratio in a natural mode of an MDF system can be determined, in principle, from a free vibration test following the procedure presented in Section 2.2.4 for SDF systems. In such a test the structure would be deformed by pulling on it with a cable that is then suddenly released, thus causing the structure to undergo free vibration about its static equilibrium position. A difficulty in such tests is to apply the pull and release in such a way that the structure will vibrate in only one of its natural modes. For this reason this test procedure is not an effective means to determine damping except possibly for the fundamental mode. After the response contributions of the higher modes have damped out, the free vibration is essentially in the fundamental mode, and the damping ratio for this mode can be computed from the decay rate of vibration amplitudes.

Example 10.12

Determine the free vibration response of the two-story shear frame of Fig. 10.9.3a with $c = \sqrt{km/200}$ due to initial displacement $\mathbf{u}(0) = \langle 1 \quad 2\rangle^T$.

Solution The $q_n(0)$ corresponding to this $\mathbf{u}(0)$ were determined in Example 10.9: $q_1(0) = 2$ and $q_2(0) = 0$; $\dot{q}_n(0) = 0$. The differential equations governing $q_n(t)$ are given by Eq. (10.10.1). Because $q_2(0)$ and $\dot{q}_2(0)$ are both zero, $q_2(t) = 0$ for all times. The response is given by the $n = 1$ term in Eq. (10.10.4). Substituting the aforementioned values for $q_1(0)$, $\dot{q}_1(0)$, and $\phi_1 = \langle \frac{1}{2} \quad 1\rangle$ gives

$$\begin{Bmatrix} u_1(t) \\ u_2(t) \end{Bmatrix} = \begin{Bmatrix} 1 \\ 2 \end{Bmatrix} e^{-\zeta_1\omega_1 t}\left(\cos\omega_{1D}t + \frac{\zeta_1}{\sqrt{1-\zeta_1^2}}\sin\omega_{1D}t\right)$$

where, as determined earlier, $\zeta_1 = 0.05$ and $\omega_1 = \sqrt{k/2m}$. This is the solution that has been plotted in Fig. 10.9.3d. Note that $q_2(t) = 0$ implies that the second mode has no contribution and the response is due entirely to the first mode. Such is the case because the initial displacement is proportional to the first mode and the system has classical damping.

Example 10.13

Determine the free vibration response of the two-story shear frame of Fig. 10.9.4a with $c = \sqrt{km/200}$ due to initial displacement $\mathbf{u}(0) = \langle -1 \quad 1\rangle^T$.

Solution The $q_n(0)$ corresponding to this $\mathbf{u}(0)$ were determined in Example 10.10: $q_1(0) = 0$ and $q_2(0) = 1$; $\dot{q}_n(0) = 0$. The differential equations governing $q_n(t)$ are given by Eq. (10.10.1). Because $q_1(0)$ and $\dot{q}_1(0)$ are both zero, $q_1(t) = 0$ at all times. The response is given by the $n = 2$ term in Eq. (10.10.4). Substituting for $q_2(0)$, $\dot{q}_2(0)$, and $\phi_2 = \langle -1 \quad 1\rangle$ gives

$$\begin{Bmatrix} u_1(t) \\ u_2(t) \end{Bmatrix} = \begin{Bmatrix} -1 \\ 1 \end{Bmatrix} e^{-\zeta_2\omega_2 t}\left(\cos\omega_{2D}t + \frac{\zeta_2}{\sqrt{1-\zeta_2^2}}\sin\omega_{2D}t\right)$$

where, as determined earlier, $\zeta_2 = 0.10$ and $\omega_2 = \sqrt{2k/m}$. This is the solution that has been plotted in Fig. 10.9.4d. Note that $q_1(t) = 0$ implies that the first mode has no contribution, and the response is due entirely to the second mode. Such is the case because the initial displacement is proportional to the second mode, and the system has classical damping.

PART C: COMPUTATION OF VIBRATION PROPERTIES

10.11 SOLUTION METHODS FOR THE EIGENVALUE PROBLEM

Finding the vibration properties—natural frequencies and modes—of a structure requires solution of the matrix eigenvalue problem of Eq. (10.2.4), which is repeated for convenience:

$$\mathbf{k}\boldsymbol{\phi} = \lambda\mathbf{m}\boldsymbol{\phi} \tag{10.11.1}$$

As mentioned earlier, the eigenvalues $\lambda_n \equiv \omega_n^2$ are the roots of the characteristic equation (10.2.6):

$$p(\lambda) = \det(\mathbf{k} - \lambda\mathbf{m}) = 0 \tag{10.11.2}$$

where $p(\lambda)$ is a polynomial of order N, the number of DOFs of the system. This is not a practical method, especially for large systems (i.e., a large number of DOFs), because evaluation of the N coefficients of the polynomial requires much computational effort and the roots of $p(\lambda)$ are sensitive to numerical round-off errors in the coefficients.

Finding reliable and efficient methods to solve the eigenvalue problem has been the subject of much research, especially since development of the digital computer. Most of the methods available can be classified into three broad categories depending on which basic property is used as the basis of the solution algorithm: (1) Vector iteration methods work directly with the property of Eq. (10.11.1). (2) Transformation methods use the orthogonality property of modes, Eqs. (10.4.1). (3) Polynomial iteration techniques work on the fact that $p(\lambda_n) = 0$. A number of solution algorithms have been developed within each of the foregoing three categories. Combination of two or more methods that belong to the same or to different categories have been developed to deal with large systems. Two examples of such combined procedures are the determinant search method and the subspace iteration method.

All solution methods for eigenvalue problems must be iterative in nature because, basically, solving the eigenvalue problem is equivalent to finding the roots of the polynomial $p(\lambda)$. No explicit formulas are available for these roots when N is larger than 4, thus requiring an iterative solution. To find an eigenpair (λ_n, ϕ_n), only one of them is calculated by iteration; the other can be obtained without further iteration. For example, if λ_n is obtained by iteration, then ϕ_n can be evaluated by solving the algebraic equations $(\mathbf{k} - \lambda_n\mathbf{m})\phi_n = \mathbf{0}$. On the other hand, if ϕ_n is determined by iteration, λ_n can be obtained by evaluating Rayleigh's quotient (Section 10.12). Is it most economical to solve first for λ_n and then calculate ϕ_n (or vice versa), or to solve for both simultaneously? The answer to this question and hence the choice among the three procedure categories mentioned above depends on the properties of the mass and stiffness matrices—size N, bandwidth of $\mathbf{k}$, and whether $\mathbf{m}$ is diagonal or banded—and on the number of eigenpairs required.

In structural engineering we are usually analyzing systems with narrowly banded $\mathbf{k}$ and diagonal or narrowly banded $\mathbf{m}$ subjected to excitations that excite primarily the lower few (relative to N) natural modes of vibration. Inverse vector iteration methods are usually effective (i.e., reliable in obtaining accurate solutions and computationally efficient) for such situations, and this presentation is restricted to such methods. Only the basic ideas of vector iteration are included, without getting into subspace iteration or the Lanczos method. Similarly, transformation methods and polynomial iteration techniques are excluded. In short, this is a limited treatment of solution methods for the eigenvalue problem arising in structural dynamics. This is sufficient for our purposes, but more comprehensive treatments are available in other books.

10.12 RAYLEIGH'S QUOTIENT

In this section Rayleigh's quotient and its properties are presented because it is needed in vector iteration methods. If Eq. (10.11.1) is premultiplied by ϕ^T, the following scalar equation is obtained:

$$\phi^T \mathbf{k}\phi = \omega^2 \phi^T \mathbf{m}\phi$$

The positive definiteness of $\mathbf{m}$ guarantees that $\phi^T \mathbf{m}\phi$ is nonzero, so that it is permissible to solve for ω^2:

$$\omega^2 = \frac{\phi^T \mathbf{k}\phi}{\phi^T \mathbf{m}\phi} \tag{10.12.1}$$

This quotient is called *Rayleigh's quotient.* It may also be derived by equating the maximum value of kinetic energy to the maximum value of potential energy under the assumption that the vibrating system is executing simple harmonic motion at frequency ω with the deflected shape given by ϕ (Section 8.5.3). Rayleigh's quotient has the following interesting properties: (1) When ϕ is an eigenvector ϕ_n of Eq. (10.11.1), Rayleigh's quotient is equal to the corresponding eigenvalue ω_n^2; (2) if ϕ is an approximation to ϕ_n with an error that is a first-order infinitesimal, Rayleigh's quotient is an approximation to ω_n^2 with an error which is a second-order infinitesimal (i.e., Rayleigh's quotient is *stationary* in the neighborhoods of the true eigenvectors); and (3) Rayleigh's quotient is bounded between ω_1^2 and ω_N^2, the smallest and largest eigenvalues.

A common engineering application of Rayleigh's quotient involves simply evaluating Eq. (10.12.1) for a trial vector ϕ which is selected on the basis of physical insight (Chapter 8). If the elements of an approximate eigenvector whose largest element is unity are correct to s decimal places, Rayleigh's quotient can be expected to be correct to about $2s$ decimal places. Several interesting numerical procedures for solving eigenvalue problems make use of the stationary property of Rayleigh's quotient.

10.13 INVERSE VECTOR ITERATION METHOD

10.13.1 Basic Concept and Procedure

We restrict this presentation to systems with a stiffness matrix $\mathbf{k}$ that is positive definite, whereas the mass matrix $\mathbf{m}$ may be a banded mass matrix, or it may be a diagonal matrix with or without zero diagonal elements. The fact that vector iteration methods can handle zero diagonal elements in the mass matrix implies that these methods can be applied without requiring static condensation of the stiffness matrix (Section 9.3).

Our goal is to satisfy Eq. (10.11.1) by operating on it directly. We assume a vector for ϕ, say $\mathbf{x}_1$, and evaluate the right-hand side of Eq. (10.11.1). This we can do except for the eigenvalue λ, which is unknown. Thus we drop λ, which is equivalent to saying that we set $\lambda = 1$. Because eigenvectors can be determined only within a scale factor, the choice of λ will not affect the final result. With $\lambda = 1$ the right-hand side of Eq. (10.11.1) can be computed:

$$\mathbf{R}_1 = \mathbf{m}\mathbf{x}_1 \tag{10.13.1}$$

Since $\mathbf{x}_1$ was an arbitrary choice, in general $\mathbf{k}\mathbf{x}_1 \neq \mathbf{R}_1$. (If by coincidence we find that $\mathbf{k}\mathbf{x}_1 = \mathbf{R}_1$, the chosen $\mathbf{x}_1$ is an eigenvector.) We now set up an equilibrium equation

$$\mathbf{k}\mathbf{x}_2 = \mathbf{R}_1 \tag{10.13.2}$$

where $\mathbf{x}_2$ is the displacement vector corresponding to forces $\mathbf{R}_1$ and $\mathbf{x}_2 \neq \mathbf{x}_1$. Since we are using iteration to solve for an eigenvector, intuition may suggest that $\mathbf{x}_2$, obtained after one cycle of iteration, may be a better approximation to ϕ than was $\mathbf{x}_1$. This is indeed the case, as we shall demonstrate later, and by repeating the iteration cycle, we obtain an increasingly better approximation to the eigenvector. A corresponding value for the eigenvalue can be computed using Rayleigh's quotient, and the iteration can be terminated when two successive estimates of the eigenvalue are close enough. As the number of iterations increases, $\mathbf{x}_{i+1}$ approaches ϕ_1 and the eigenvalue approaches λ_1.

Thus the procedure starts with the assumption of a starting iteration vector $\mathbf{x}_1$ and consists of the following steps to be repeated for $j = 1, 2, 3, \ldots$ until convergence:

1. Determine $\bar{\mathbf{x}}_{j+1}$ by solving the algebraic equations:

$$\mathbf{k}\bar{\mathbf{x}}_{j+1} = \mathbf{m}\mathbf{x}_j \tag{10.13.3}$$

2. Obtain an estimate of the eigenvalue by evaluating Rayleigh's quotient:

$$\lambda^{(j+1)} = \frac{\bar{\mathbf{x}}_{j+1}^T \mathbf{k}\bar{\mathbf{x}}_{j+1}}{\bar{\mathbf{x}}_{j+1}^T \mathbf{m}\bar{\mathbf{x}}_{j+1}} = \frac{\bar{\mathbf{x}}_{j+1}^T \mathbf{m}\mathbf{x}_j}{\bar{\mathbf{x}}_{j+1}^T \mathbf{m}\bar{\mathbf{x}}_{j+1}} \tag{10.13.4}$$

3. Check convergence by comparing two successive values of λ:

$$\frac{|\,\lambda^{(j+1)} - \lambda^{(j)}\,|}{\lambda^{(j+1)}} \leq \text{tolerance} \tag{10.13.5}$$

4. If the convergence criterion is not satisfied, normalize $\bar{\mathbf{x}}_{j+1}$:

$$\mathbf{x}_{j+1} = \frac{\bar{\mathbf{x}}_{j+1}}{(\bar{\mathbf{x}}_{j+1}^T \mathbf{m}\bar{\mathbf{x}}_{j+1})^{1/2}} \tag{10.13.6}$$

   and go back to the first step and carry out another iteration using the next j.

5. Let l be the last iteration [i.e., the iteration that satisfies Eq. (10.13.5)]. Then

$$\lambda_1 \doteq \lambda^{(l+1)} \qquad \phi_1 \doteq \frac{\bar{\mathbf{x}}_{l+1}}{(\bar{\mathbf{x}}_{l+1}^T \mathbf{m}\bar{\mathbf{x}}_{l+1})^{1/2}} \tag{10.13.7}$$

The basic step in the iteration is the solution of Eq. (10.13.3)—a set of N algebraic equations—which gives a better approximation to ϕ_1. The calculation in Eq. (10.13.4) gives an approximation to the eigenvalue λ_1 according to Rayleigh's quotient. It is this approximation to λ_1 that we use to determine convergence in the iteration.

Equation (10.13.6) simply assures that the new vector satisfies the mass orthonormality relation

$$\mathbf{x}_{j+1}^T \mathbf{m} \mathbf{x}_{j+1} = 1 \tag{10.13.8}$$

Such normalizing of the new vector does not affect convergence, and it is numerically useful. If such normalizing is not included, the elements of the iteration vectors grow (or decrease) in each step, and this may cause numerical problems. Normalizing keeps the element values similar from one iteration to the next. The tolerance is selected depending on the accuracy desired. It should be 10^{-2s} or smaller when λ_1 is required to $2s$-digit accuracy. Then the eigenvector will be accurate to about s or more digits.

The inverse vector iteration algorithm can be organized a little differently for convenience in computer implementation, but such computational issues are not included in this presentation.

Example 10.14

The floor masses and story stiffnesses of the three-story frame, idealized as a shear frame, are shown in Fig. E10.14, where $m = 100$ kips/g, and $k = 168$ kips/in. Determine the fundamental frequency ω_1 and mode shape ϕ_1 by inverse vector iteration.

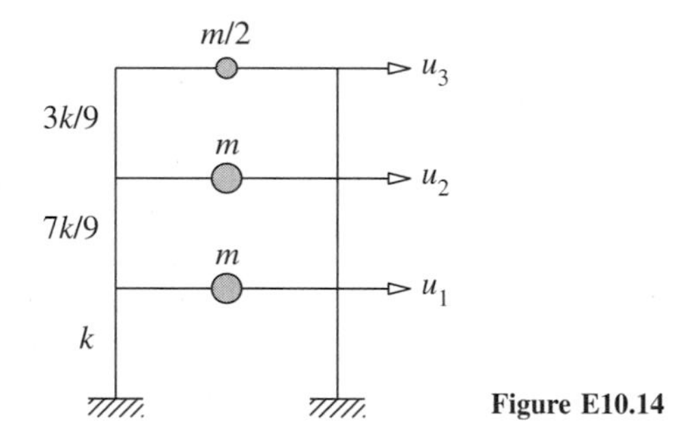

Figure E10.14

Solution The mass and stiffness matrices for the system are

$$\mathbf{m} = m \begin{bmatrix} 1 & & \\ & 1 & \\ & & \frac{1}{2} \end{bmatrix} \qquad \mathbf{k} = \frac{k}{9} \begin{bmatrix} 16 & -7 & 0 \\ -7 & 10 & -3 \\ 0 & -3 & 3 \end{bmatrix}$$

where $m = 0.259$ kip-sec^2/in. and $k = 168$ kips/in.

The inverse iteration algorithm of Eqs. (10.13.3) to (10.13.7) is implemented starting with an initial vector $\mathbf{x}_1 = \langle 1 \quad 1 \quad 1 \rangle^T$ leading to Table E10.14. The final result is $\omega_1 \doteq \sqrt{144.14} = 12.006$ and $\phi_1 \doteq \langle 0.6377 \quad 1.2752 \quad 1.9122 \rangle^T$.

10.13.2 Convergence of Iteration

In the preceding section we have merely presented the inverse iteration scheme and stated that it converges to the first eigenvector associated with the smallest eigenvalue. We now demonstrate this convergence because the proof is instructive, especially in suggesting how to modify the procedure to achieve convergence to a higher eigenvector.

TABLE E10.14 INVERSE VECTOR ITERATION FOR THE FIRST EIGENPAIR

Iteration	$\mathbf{x}_j$	$\bar{\mathbf{x}}_{j+1}$	$\lambda^{(j+1)}$	$\mathbf{x}_{j+1}$
1	$\begin{bmatrix} 1 \\ 1 \\ 1 \end{bmatrix}$	$\begin{bmatrix} 0.0039 \\ 0.0068 \\ 0.0091 \end{bmatrix}$	147.73	$\begin{bmatrix} 0.7454 \\ 1.3203 \\ 1.7676 \end{bmatrix}$
2	$\begin{bmatrix} 0.7454 \\ 1.3203 \\ 1.7676 \end{bmatrix}$	$\begin{bmatrix} 0.0045 \\ 0.0089 \\ 0.0130 \end{bmatrix}$	144.29	$\begin{bmatrix} 0.6574 \\ 1.2890 \\ 1.8800 \end{bmatrix}$
3	$\begin{bmatrix} 0.6574 \\ 1.2890 \\ 1.8800 \end{bmatrix}$	$\begin{bmatrix} 0.0044 \\ 0.0089 \\ 0.0132 \end{bmatrix}$	144.15	$\begin{bmatrix} 0.6415 \\ 1.2785 \\ 1.9052 \end{bmatrix}$
4	$\begin{bmatrix} 0.6415 \\ 1.2785 \\ 1.9052 \end{bmatrix}$	$\begin{bmatrix} 0.0044 \\ 0.0089 \\ 0.0133 \end{bmatrix}$	144.14	$\begin{bmatrix} 0.6384 \\ 1.2758 \\ 1.9109 \end{bmatrix}$
5	$\begin{bmatrix} 0.6384 \\ 1.2758 \\ 1.9109 \end{bmatrix}$	$\begin{bmatrix} 0.0044 \\ 0.0088 \\ 0.0133 \end{bmatrix}$	144.14	$\begin{bmatrix} 0.6377 \\ 1.2752 \\ 1.9122 \end{bmatrix}$

The modal expansion of vector $\mathbf{x}$ is [from Eqs. (10.7.1) and (10.7.2)]

$$\mathbf{x} = \sum_{n=1}^{N} \boldsymbol{\phi}_n q_n = \sum_{n=1}^{N} \boldsymbol{\phi}_n \frac{\boldsymbol{\phi}_n^T \mathbf{m}\mathbf{x}}{\boldsymbol{\phi}_n^T \mathbf{m}\boldsymbol{\phi}_n} \tag{10.13.9}$$

The nth term in this summation represents the nth modal component in $\mathbf{x}$.

The first iteration cycle involves solving the equilibrium equations (10.13.3) with $j = 1$: $\mathbf{k}\bar{\mathbf{x}}_2 = \mathbf{m}\mathbf{x}_1$, where $\mathbf{x}_1$ is the starting vector. This solution can be expressed as $\bar{\mathbf{x}}_2 = \mathbf{k}^{-1}\mathbf{m}\mathbf{x}_1$. Substituting the modal expansion of Eq. (10.13.9) for $\mathbf{x}_1$ gives

$$\bar{\mathbf{x}}_2 = \sum_{n=1}^{N} \mathbf{k}^{-1}\mathbf{m}\boldsymbol{\phi}_n q_n \tag{10.13.10}$$

By rewriting Eq. (10.11.1) for the nth eigenpair as $\mathbf{k}^{-1}\mathbf{m}\boldsymbol{\phi}_n = (1/\lambda_n)\boldsymbol{\phi}_n$ and substituting it in Eq. (10.13.10), we get

$$\bar{\mathbf{x}}_2 = \sum_{n=1}^{N} \frac{1}{\lambda_n} \boldsymbol{\phi}_n q_n = \frac{1}{\lambda_1} \sum_{n=1}^{N} \frac{\lambda_1}{\lambda_n} \boldsymbol{\phi}_n q_n \tag{10.13.11}$$

The second iteration cycle involves solving Eq. (10.13.3) with $j = 2$: $\bar{\mathbf{x}}_3 = \mathbf{k}^{-1}\mathbf{m}\bar{\mathbf{x}}_2$, wherein we have used the unnormalized vector $\bar{\mathbf{x}}_2$ instead of the normalized vector $\mathbf{x}_2$. This is acceptable for the present purpose because convergence is unaffected by normalization and eigenvectors are arbitrary within a multiplicative factor. Following the derivation of Eqs. (10.13.10) and (10.13.11), it can be shown that

$$\bar{\mathbf{x}}_3 = \frac{1}{\lambda_1^2} \sum_{n=1}^{N} \left(\frac{\lambda_1}{\lambda_n}\right)^2 \boldsymbol{\phi}_n q_n \tag{10.13.12}$$

Similarly, the vector $\bar{\mathbf{x}}_{j+1}$ after j iteration cycles can be expressed as

$$\bar{\mathbf{x}}_{j+1} = \frac{1}{\lambda_1^j} \sum_{n=1}^{N} \left(\frac{\lambda_1}{\lambda_n}\right)^j \phi_n q_n \tag{10.13.13}$$

Since $\lambda_1 < \lambda_n$ for $n > 1$, $(\lambda_1/\lambda_n)^j \to 0$ as $j \to \infty$, and only the $n = 1$ term in Eq. (10.13.13) remains significant, indicating that

$$\bar{\mathbf{x}}_{j+1} \to \frac{1}{\lambda_1^j} \phi_1 q_1 \quad \text{as} \quad j \to \infty \tag{10.13.14}$$

Thus $\bar{\mathbf{x}}_{j+1}$ converges to a vector proportional to ϕ_1. Furthermore, the normalized vector $\mathbf{x}_{j+1}$ of Eq. (10.13.6) converges to ϕ_1, which is mass orthonormal.

The rate of convergence depends on λ_1/λ_2, the ratio that appears in the second term in the summation of Eq. (10.13.13). The smaller this ratio is, the faster is the convergence; this implies that convergence is very slow when λ_2 is nearly equal to λ_1. For such situations the convergence rate can be improved by the procedures of Section 10.14. The initial trial vector $\mathbf{x}_1$ may be chosen arbitrarily and the iteration process should converge to the first eigenvector ϕ_1. If $\mathbf{x}_1$ were orthogonal to ϕ_1 [i.e., $q_1 = 0$ in Eq. (10.13.9)], theoretically the iteration will not converge to ϕ_1 but to some other eigenvector—the one with the next-higher eigenvalue which is contained in the modal expansion of $\mathbf{x}_1$. However, in practice this never occurs since the inevitable round-off errors in finite-precision arithmetic continuously reintroduce small components of ϕ_1 which the iteration process magnifies.

If only the first natural mode ϕ_1 and the associated natural frequency ω_1 are required, there is no need to proceed further. This is an advantage of the iteration method. It is unnecessary to solve the complete eigenvalue problem to obtain one or two of the modes.

10.13.3 Evaluation of Higher Modes

To continue the solution after ϕ_1 and λ_1 have been determined, the starting vector is modified to make the iteration procedure converge to the second eigenvector. The necessary modification is suggested by the proof presented in Section 10.13.2 to show that the iteration process converges to the first eigenvector. Observe that after each iteration cycle the other modal components are reduced relative to the first modal component because its eigenvalue λ_1 is smaller than all other eigenvalues λ_n. The iteration process converges to the first mode for the same reason because $(\lambda_1/\lambda_n)^j \to 0$ as $j \to \infty$. In general, the iteration procedure will converge to the mode with the lowest eigenvalue contained in a trial vector $\mathbf{x}$.

To make the iteration procedure converge to the second mode, a trial vector $\mathbf{x}$ should therefore be chosen so that it does not contain any first-mode component [i.e., q_1 should be zero in Eq. (10.13.9)] and $\mathbf{x}$ is said to be orthogonal to ϕ_1. It is not possible to start a priori with such an $\mathbf{x}$, however. We therefore start with an arbitrary $\mathbf{x}$ and make it orthogonal to ϕ_1 by the Gram–Schmidt orthogonalization process. This process can also be used to orthogonalize a trial vector with respect to the first n eigenvectors that have already been determined so that iteration on the purified trial vectors will converge to the $(n+1)$th mode, the mode with the next eigenvalue in ascending sequence.

In principle, the Gram–Schmidt orthogonalization process, combined with the inverse iteration procedure, provides a tool for computing the second and higher eigenvalues and eigenvectors. This tool is not effective, however, as a general computer method because the convergence of the iteration process becomes progressively slower for the higher modes. It is for this reason that this method is not developed in this book.

10.14 VECTOR ITERATION WITH SHIFTS: PREFERRED PROCEDURE

The inverse vector iteration procedure of Section 10.13, combined with the concept of "shifting" the eigenvalue spectrum (or scale), provides an effective means to improve the convergence rate of the iteration process and to make it converge to an eigenpair other than (λ_1, ϕ_1). Thus this is the preferred method, as it provides a practical tool for computing as many pairs of natural vibration frequencies and modes of a structure as desired.

10.14.1 Basic Concept and Procedure

The solutions of Eq. (10.11.1) are the eigenvalues λ_n and eigenvectors ϕ_n; the number of such pairs equals N, the order of $\mathbf{m}$ and $\mathbf{k}$. Figure 10.14.1a shows the eigenvalue spectrum (i.e., a plot of $\lambda_1, \lambda_2, \ldots$ along the eigenvalue axis). Introducing a shift μ in the origin of the eigenvalue axis (Fig. 10.14.1b) and defining $\check{\lambda}$ as the shifted eigenvalue measured from the shifted origin gives $\lambda = \check{\lambda} + \mu$. Substituting this in Eq. (10.11.1) leads to

$$\check{\mathbf{k}}\phi = \check{\lambda}\mathbf{m}\phi \tag{10.14.1}$$

where

$$\check{\mathbf{k}} = \mathbf{k} - \mu\mathbf{m} \qquad \check{\lambda} = \lambda - \mu \tag{10.14.2}$$

The eigenvectors of the two eigenvalue problems—original Eq. (10.11.1) and shifted Eq. (10.14.1)—are the same. This is obvious because if a ϕ satisfies one equation, it

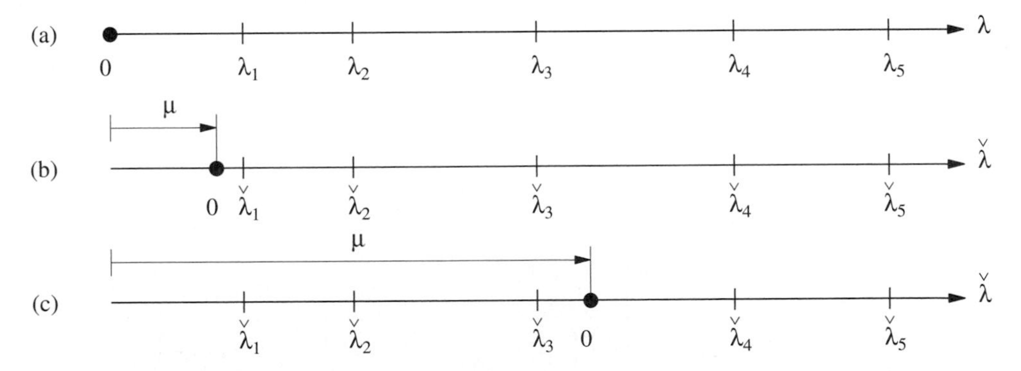

Figure 10.14.1 (a) Eigenvalue spectrum; (b) eigenvalue measured from a shifted origin; (c) location of shift point for convergence to λ_3.

will also satisfy the other. However, the eigenvalues $\check{\lambda}$ of the shifted problem differ from the eigenvalues λ of the original problem by the shift μ [Eq. (10.14.2)]. The spectrum of the shifted eigenvalues $\check{\lambda}$ is also shown in Fig. 10.14.1b with the origin at μ. If the inverse vector iteration method of Section 10.13 were applied to the eigenvalue problem of Eq. (10.14.1), it obviously will converge to the eigenvector having the smallest magnitude of the shifted eigenvalue $|\check{\lambda}_n|$ (i.e., the eigenvector with original eigenvalue λ_n closest to the shift value μ).

If μ were chosen as in Fig. 10.14.1b, the iteration will converge to the first eigenvector. The rate of convergence depends on the ratio $\check{\lambda}_1/\check{\lambda}_2 = (\lambda_1 - \mu)/(\lambda_2 - \mu)$. The convergence rate has improved because this ratio is smaller than the ratio λ_1/λ_2 of the original eigenvalue problem. If μ were chosen between λ_n and λ_{n+1}, and μ is closer to λ_n than λ_{n+1}, the iteration will converge to λ_n. On the other hand, if μ is closer to λ_{n+1} than λ_n, the iteration will converge to λ_{n+1}. Thus the "shifting" concept enables computation of any pair (λ_n, ϕ_n). In particular, if μ were chosen as in Fig. 10.14.1c, the iteration will converge to the third eigenvector.

The inverse iteration method with shifts converges rapidly if a shift is chosen near enough to the eigenvalue of interest. However, selection of an appropriate shift is difficult without knowledge of the eigenvalue. Many techniques have been developed to overcome this difficulty; one of these is presented in the next section.

Example 10.15

Determine the natural frequencies and modes of vibration of the system of Example 10.14 by inverse vector iteration with shifting.

Solution Equation (10.14.1) with a selected shift μ is solved by inverse vector iteration. Selecting the shift $\mu_1 = 100$, $\check{\mathbf{k}}$ is calculated from Eq. (10.14.2) and the inverse vector iteration algorithm of Eqs. (10.13.3) to (10.13.7) is implemented starting with an initial vector of $\mathbf{x}_1 = \langle 1 \quad 1 \quad 1 \rangle^T$ leading to Table E10.15a. The final result is $\omega_1 = \sqrt{144.14} = 12.006$

TABLE E10.15a VECTOR ITERATION WITH SHIFT: FIRST EIGENPAIR

Iteration	$\mathbf{x}_j$	μ	$\bar{\mathbf{x}}_{j+1}$	$\lambda^{(j+1)}$	$\mathbf{x}_{j+1}$
1	$\begin{bmatrix} 1 \\ 1 \\ 1 \end{bmatrix}$	100	$\begin{bmatrix} 0.0114 \\ 0.0218 \\ 0.0313 \end{bmatrix}$	144.60	$\begin{bmatrix} 0.6759 \\ 1.2933 \\ 1.8610 \end{bmatrix}$
2	$\begin{bmatrix} 0.6759 \\ 1.2933 \\ 1.8610 \end{bmatrix}$	100	$\begin{bmatrix} 0.0145 \\ 0.0289 \\ 0.0432 \end{bmatrix}$	144.15	$\begin{bmatrix} 0.6401 \\ 1.2769 \\ 1.9083 \end{bmatrix}$
3	$\begin{bmatrix} 0.6401 \\ 1.2769 \\ 1.9083 \end{bmatrix}$	100	$\begin{bmatrix} 0.0144 \\ 0.0289 \\ 0.0433 \end{bmatrix}$	144.14	$\begin{bmatrix} 0.6377 \\ 1.2752 \\ 1.9122 \end{bmatrix}$
4	$\begin{bmatrix} 0.6377 \\ 1.2752 \\ 1.9122 \end{bmatrix}$	100	$\begin{bmatrix} 0.0144 \\ 0.0289 \\ 0.0433 \end{bmatrix}$	144.14	$\begin{bmatrix} 0.6375 \\ 1.2750 \\ 1.9125 \end{bmatrix}$

and $\phi_1 = \langle 0.6375 \quad 1.2750 \quad 1.9125 \rangle^T$. This is obtained in one iteration cycle less than in the iteration without shift in Example 10.14.

Starting with the shift $\mu_1 = 600$ and the same $\mathbf{x}_1$, the inverse iteration algorithm leads to Table E10.15b. The final result is $\omega_2 = \sqrt{648.65} = 25.468$ and $\phi_2 = \langle 0.9827 \quad 0.9829 \quad -1.9642 \rangle^T$. Convergence is attained in four iteration cycles.

TABLE E10.15b VECTOR ITERATION WITH SHIFT: SECOND EIGENPAIR

Iteration	$\mathbf{x}_j$	μ	$\bar{\mathbf{x}}_{j+1}$	$\lambda^{(j+1)}$	$\mathbf{x}_{j+1}$
1	$\begin{bmatrix} 1 \\ 1 \\ 1 \end{bmatrix}$	600	$\begin{bmatrix} 0.0044 \\ 0.0028 \\ -0.0133 \end{bmatrix}$	605.11	$\begin{bmatrix} 0.8030 \\ 0.5189 \\ -2.4277 \end{bmatrix}$
2	$\begin{bmatrix} 0.8030 \\ 0.5189 \\ -2.4277 \end{bmatrix}$	600	$\begin{bmatrix} 0.0197 \\ 0.0201 \\ -0.0373 \end{bmatrix}$	648.10	$\begin{bmatrix} 1.0062 \\ 1.0221 \\ -1.8994 \end{bmatrix}$
3	$\begin{bmatrix} 1.0062 \\ 1.0221 \\ -1.8994 \end{bmatrix}$	600	$\begin{bmatrix} 0.0201 \\ 0.0201 \\ -0.0405 \end{bmatrix}$	648.64	$\begin{bmatrix} 0.9804 \\ 0.9778 \\ -1.9717 \end{bmatrix}$
4	$\begin{bmatrix} 0.9804 \\ 0.9778 \\ -1.9717 \end{bmatrix}$	600	$\begin{bmatrix} 0.0202 \\ 0.0202 \\ -0.0404 \end{bmatrix}$	648.65	$\begin{bmatrix} 0.9827 \\ 0.9829 \\ -1.9642 \end{bmatrix}$

Starting with the shift $\mu_1 = 1500$ and the same $\mathbf{x}_1$, the inverse iteration algorithm leads to Table E10.15c. The final result is $\omega_3 = \sqrt{1513.5} = 38.904$ and $\phi_3 = \langle 1.5778 \quad -1.1270 \quad 0.4508 \rangle^T$. Convergence is attained in three cycles.

TABLE E10.15c VECTOR ITERATION WITH SHIFT: THIRD EIGENPAIR

Iteration	$\mathbf{x}_j$	μ	$\bar{\mathbf{x}}_{j+1}$	$\lambda^{(j+1)}$	$\mathbf{x}_{j+1}$
1	$\begin{bmatrix} 1 \\ 1 \\ 1 \end{bmatrix}$	1500	$\begin{bmatrix} 0.0198 \\ -0.0156 \\ 0.0054 \end{bmatrix}$	1510.6	$\begin{bmatrix} 1.5264 \\ -1.2022 \\ 0.4148 \end{bmatrix}$
2	$\begin{bmatrix} 1.5264 \\ -1.2022 \\ 0.4148 \end{bmatrix}$	1500	$\begin{bmatrix} 0.1167 \\ -0.0832 \\ 0.0333 \end{bmatrix}$	1513.5	$\begin{bmatrix} 1.5784 \\ -1.1261 \\ 0.4509 \end{bmatrix}$
3	$\begin{bmatrix} 1.5784 \\ -1.1261 \\ 0.4509 \end{bmatrix}$	1500	$\begin{bmatrix} 0.1168 \\ -0.0834 \\ 0.0334 \end{bmatrix}$	1513.5	$\begin{bmatrix} 1.5778 \\ -1.1270 \\ 0.4508 \end{bmatrix}$

10.14.2 Rayleigh's Quotient Iteration

The Rayleigh quotient calculated by Eq. (10.13.4) to estimate the eigenvalue provides an appropriate shift value. It is not necessary to calculate and introduce a new shift at each iteration cycle. If this is done, however, the resulting procedure is called *Rayleigh's quotient iteration.*

This procedure starts with the assumption of a starting iteration vector $\mathbf{x}_1$ and starting shift $\lambda^{(1)}$ and consists of the following steps to be repeated for $j = 1, 2, 3, \ldots$ until convergence:

1. Determine $\bar{\mathbf{x}}_{j+1}$ by solving the algebraic equations:

$$[\mathbf{k} - \lambda^{(j)}\mathbf{m}]\bar{\mathbf{x}}_{j+1} = \mathbf{m}\mathbf{x}_j \tag{10.14.3}$$

2. Obtain an estimate of the eigenvalue and the shift for the next iteration from

$$\lambda^{(j+1)} = \frac{\bar{\mathbf{x}}_{j+1}^T\mathbf{m}\mathbf{x}_j}{\bar{\mathbf{x}}_{j+1}^T\mathbf{m}\bar{\mathbf{x}}_{j+1}} + \lambda^{(j)} \tag{10.14.4}$$

3. Normalize $\bar{\mathbf{x}}_{j+1}$:

$$\mathbf{x}_{j+1} = \frac{\bar{\mathbf{x}}_{j+1}}{(\bar{\mathbf{x}}_{j+1}^T\mathbf{m}\bar{\mathbf{x}}_{j+1})^{1/2}} \tag{10.14.5}$$

This iteration converges to a particular eigenpair (λ_n, ϕ_n) depending on the starting vector $\mathbf{x}_1$ and the initial shift $\lambda^{(1)}$. If $\mathbf{x}_1$ includes a strong contribution of the eigenvector ϕ_n and $\lambda^{(1)}$ is close enough to λ_n, the iteration converges to the eigenpair (λ_n, ϕ_n). The rate of convergence is faster than the standard vector iteration described in Section 10.14.1, but at the expense of additional computation because a new $[\mathbf{k} - \lambda^{(j)}\mathbf{m}]$ has to be factorized in each iteration.

Example 10.16

Determine all three natural frequencies and modes of the system of Example 10.14 by inverse vector iteration with the shift in each iteration cycle equal to Rayleigh's quotient from the previous cycle.

Solution The iteration procedure of Eqs. (10.14.3) to (10.14.5) is implemented with starting shifts of $\mu_1 = 100$, $\mu_2 = 600$, and $\mu_3 = 1500$, leading to Tables E10.16a, E10.16b, and E10.16c, respectively, where the final results are $\omega_1 = \sqrt{144.14} = 12.006$ and $\phi_1 = \langle 0.6375 \quad 1.2750 \quad 1.9125\rangle^T$, $\omega_2 = \sqrt{648.65} = 25.468$ and $\phi_2 = \langle 0.9825 \quad 0.9825 \quad -1.9649\rangle^T$, and $\omega_3 = \sqrt{1513.5} = 38.904$ and $\phi_3 = \langle 1.5778 \quad -1.1270 \quad 0.4508\rangle^T$.

Observe that convergence is faster when a new shift is used in each iteration cycle. Only two cycles are required instead of four (Example 10.15) for the first mode, and three instead of four for the second mode.

TABLE E10.16a RAYLEIGH'S QUOTIENT ITERATION: FIRST EIGENPAIR

Iteration	$\mathbf{x}_j$	μ	$\bar{\mathbf{x}}_{j+1}$	$\lambda^{(j+1)}$	$\mathbf{x}_{j+1}$
1	$\begin{bmatrix} 1 \\ 1 \\ 1 \end{bmatrix}$	100	$\begin{bmatrix} 0.0114 \\ 0.0218 \\ 0.0313 \end{bmatrix}$	144.60	$\begin{bmatrix} 0.6579 \\ 1.2933 \\ 1.8610 \end{bmatrix}$
2	$\begin{bmatrix} 0.6579 \\ 1.2933 \\ 1.8610 \end{bmatrix}$	144.60	$\begin{bmatrix} -1.3947 \\ -2.7895 \\ -4.1845 \end{bmatrix}$	144.14	$\begin{bmatrix} 0.6375 \\ 1.2750 \\ 1.9126 \end{bmatrix}$

TABLE E10.16b RAYLEIGH'S QUOTIENT ITERATION: SECOND EIGENPAIR

Iteration	$\mathbf{x}_j$	μ	$\bar{\mathbf{x}}_{j+1}$	λ^{j+1}	$\mathbf{x}_{j+1}$
1	$\begin{bmatrix} 1 \\ 1 \\ 1 \end{bmatrix}$	600	$\begin{bmatrix} 0.0044 \\ 0.0028 \\ -0.0133 \end{bmatrix}$	605.11	$\begin{bmatrix} 0.8030 \\ 0.5189 \\ -2.4277 \end{bmatrix}$
2	$\begin{bmatrix} 0.8030 \\ 0.5189 \\ -2.4277 \end{bmatrix}$	605.11	$\begin{bmatrix} 0.0220 \\ 0.0223 \\ -0.0418 \end{bmatrix}$	648.21	$\begin{bmatrix} 1.0036 \\ 1.0176 \\ -1.9070 \end{bmatrix}$
3	$\begin{bmatrix} 1.0036 \\ 1.0176 \\ -1.9070 \end{bmatrix}$	648.21	$\begin{bmatrix} 2.2624 \\ 2.2623 \\ -4.5249 \end{bmatrix}$	648.65	$\begin{bmatrix} 0.9825 \\ 0.9824 \\ -1.9650 \end{bmatrix}$
4	$\begin{bmatrix} 0.9825 \\ 0.9824 \\ -1.9650 \end{bmatrix}$	648.65	$\begin{bmatrix} 3.0372 \times 10^6 \\ 3.0372 \times 10^6 \\ -6.0745 \times 10^6 \end{bmatrix}$	648.65	$\begin{bmatrix} 0.9825 \\ 0.9825 \\ -1.9649 \end{bmatrix}$

TABLE E10.16c RAYLEIGH'S QUOTIENT ITERATION: THIRD EIGENPAIR

Iteration	$\mathbf{x}_j$	μ	$\bar{\mathbf{x}}_{j+1}$	λ^{j+1}	$\mathbf{x}_{j+1}$
1	$\begin{bmatrix} 1 \\ 1 \\ 1 \end{bmatrix}$	1500	$\begin{bmatrix} 0.0198 \\ -0.0156 \\ 0.0054 \end{bmatrix}$	1510.6	$\begin{bmatrix} 1.5264 \\ -1.2022 \\ 0.4148 \end{bmatrix}$
2	$\begin{bmatrix} 1.5264 \\ -1.2022 \\ 0.4148 \end{bmatrix}$	1510.6	$\begin{bmatrix} 0.5431 \\ -0.3879 \\ 0.1552 \end{bmatrix}$	1513.5	$\begin{bmatrix} 1.5779 \\ -1.1268 \\ 0.4508 \end{bmatrix}$
3	$\begin{bmatrix} 1.5779 \\ -1.1268 \\ 0.4508 \end{bmatrix}$	1513.5	$\begin{bmatrix} 9.7061 \times 10^4 \\ -6.9329 \times 10^4 \\ 2.7732 \times 10^4 \end{bmatrix}$	1513.5	$\begin{bmatrix} 1.5778 \\ -1.1270 \\ 0.4508 \end{bmatrix}$

Application to structural dynamics. In modal analysis of the dynamic response of structures, we are interested in the lower J natural frequencies and modes (Chapters 12 and 13); typically, J is much smaller than N, the number of degrees of freedom. Although Rayleigh's quotient iteration may appear to be an effective tool for the necessary computation, it may not always work. For example, with the starting vector $\mathbf{x}_1$ and starting shift $\lambda^{(1)} = 0$, Eq. (10.14.4) may provide a value for Rayleigh's quotient (which, according to Section 10.12, is always higher than the first eigenvalue) closer to the second eigenvalue than the first, resulting in the iteration converging to the second mode. Thus it is necessary to supplement Rayleigh's quotient iteration by another technique to assure convergence to the lowest eigenpair (λ_1, ϕ_1). One possibility is to use first the inverse iteration without shift, Eqs. (10.13.3) to (10.13.7), for a few cycles to obtain an iteration vector that is a good approximation (but has not converged) to ϕ_1, and then start with Rayleigh's quotient iteration.

Computer implementation of inverse vector iteration with shift should be reliable and efficient. By reliability we mean that it should give the desired eigenpair. Efficiency implies that with the fewest iterations and least computation, the method should provide results to the desired degree of accuracy. Both of these requirements are essential; otherwise, the computer program may skip a desired eigenpair, or the computations may be unnecessarily time consuming. The issues related to reliability and efficiency of computer methods for solving the eigenvalue problem are discussed further in other books.

10.15 TRANSFORMATION OF $\mathbf{k}\phi = \omega^2\mathbf{m}\phi$ TO THE STANDARD FORM

The standard eigenvalue problem $\mathbf{Ay} = \lambda\mathbf{y}$ arises in many situations in mathematics and in applications to problems in the physical sciences and engineering. It has therefore attracted much attention and many solution algorithms have been developed and are available in computer software libraries. These computer procedures could be used to solve the structural dynamics eigenvalue problem, $\mathbf{k}\phi = \omega^2\mathbf{m}\phi$, provided that it can be transformed to the standard form. Such a transformation is presented in this section.

We assume that $\mathbf{m}$ is positive definite; that is, it is either a diagonal matrix with nonzero masses or a banded matrix as in a consistent mass formulation (Chapter 17). If $\mathbf{m}$ is a diagonal matrix with zero mass in some degrees of freedom, these have been eliminated by static condensation (Section 9.3). Positive definiteness of $\mathbf{m}$ implies that $\mathbf{m}^{-1}$ can be calculated. Premultiplying the structural dynamics eigenvalue problem

$$\mathbf{k}\phi = \omega^2\mathbf{m}\phi \tag{10.15.1}$$

by $\mathbf{m}^{-1}$ gives the standard eigenvalue problem:

$$\mathbf{A}\phi = \lambda\phi \tag{10.15.2}$$

where

$$\mathbf{A} = \mathbf{m}^{-1}\mathbf{k} \qquad \lambda = \omega^2 \tag{10.15.3}$$

In general, $\mathbf{A}$ is not symmetric, although $\mathbf{m}$ and $\mathbf{k}$ are both symmetric matrices. Because the computational effort could be greatly reduced if $\mathbf{A}$ were symmetric, we seek methods that yield a symmetric $\mathbf{A}$. Consider that $\mathbf{m} = \text{diag}(m_j)$, a diagonal matrix with elements $m_{jj} = m_j$, and define $\mathbf{m}^{1/2} = \text{diag}(m_j^{1/2})$ and $\mathbf{m}^{-1/2} = \text{diag}(m_j^{-1/2})$. Then $\mathbf{m}$ and the identity matrix $\mathbf{I}$ can be expressed as

$$\mathbf{m} = \mathbf{m}^{1/2}\mathbf{m}^{1/2} \qquad \mathbf{I} = \mathbf{m}^{-1/2}\mathbf{m}^{1/2} \tag{10.15.4}$$

Using Eq. (10.15.4), Eq. (10.15.1) can be rewritten as

$$\mathbf{k}\mathbf{m}^{-1/2}\mathbf{m}^{1/2}\phi = \omega^2\mathbf{m}^{1/2}\mathbf{m}^{1/2}\phi$$

Premultiplying both sides by $\mathbf{m}^{-1/2}$ leads to

$$\mathbf{m}^{-1/2}\mathbf{k}\mathbf{m}^{-1/2}\mathbf{m}^{1/2}\phi = \omega^2\mathbf{m}^{-1/2}\mathbf{m}^{1/2}\mathbf{m}^{1/2}\phi$$

Utilizing Eq. (10.15.4b) to simplify the right-hand side of the equation above gives

$$\mathbf{Ay} = \lambda\mathbf{y} \tag{10.15.5}$$

where

$$\mathbf{A} = \mathbf{m}^{-1/2}\mathbf{k}\mathbf{m}^{-1/2} \qquad \mathbf{y} = \mathbf{m}^{1/2}\boldsymbol{\phi} \qquad \lambda = \omega^2 \tag{10.15.6}$$

Equation (10.15.5) is the standard eigenvalue problem and $\mathbf{A}$ is now symmetric.

Thus if a computer program to solve $\mathbf{Ay} = \lambda\mathbf{y}$ were available, it could be utilized to determine the natural frequencies ω_n and modes ϕ_n of a system for which $\mathbf{m}$ and $\mathbf{k}$ were known as follows:

1. Compute $\mathbf{A}$ from Eq. (10.15.6a).
2. Determine the eigenvalues λ_n and eigenvectors $\mathbf{y}_n$ of $\mathbf{A}$ by solving Eq. (10.15.5).
3. Determine the natural frequencies and modes by

$$\omega_n = \sqrt{\lambda_n} \qquad \phi_n = \mathbf{m}^{-1/2}\mathbf{y}_n \tag{10.15.7}$$

The transformation of Eq. (10.15.6) can be generalized to situations where the mass matrix is not diagonal but is banded like the stiffness matrix; such banding is typical of finite element formulations (Chapter 17). Then, $\mathbf{A}$ is a full matrix, although $\mathbf{k}$ and $\mathbf{m}$ are banded. This is a major computational disadvantage for large systems. For such situations the transformation of $\mathbf{k}\boldsymbol{\phi} = \omega^2\mathbf{m}\boldsymbol{\phi}$ to $\mathbf{Ay} = \lambda\mathbf{y}$ may not be an effective approach and the inverse iteration method, which works directly with $\mathbf{k}\boldsymbol{\phi} = \omega^2\mathbf{m}\boldsymbol{\phi}$ may be more efficient.

FURTHER READING

Bathe, K. J., *Finite Element Procedures in Engineering Analysis,* Prentice Hall, Englewood Cliffs, N.J., 1982, Chapters 10–12.

Crandall, S. H., and McCalley, R. B., Jr., "Matrix Methods of Analysis," Chapter 28 in *Shock and Vibration Handbook* (ed. C. M. Harris), McGraw-Hill, New York, 1988.

Humar, J. L., *Dynamics of Structures,* Prentice Hall, Englewood Cliffs, N.J., 1990, Chapter 11.

Parlett, B. N., *The Symmetric Eigenvalue Problem,* Prentice Hall, Englewood Cliffs, N.J., 1980.

PROBLEMS

Parts A and B

10.1 Determine the natural vibration frequencies and modes of the system of Fig. P9.1 with $k_1 = k$ and $k_2 = 2k$ in terms of the DOFs in the figure. Show that these results are equivalent to those presented in Fig. E10.1.

10.2 For the system defined in Problem 9.2:
(a) Determine the natural vibration frequencies and modes; express the frequencies in terms of m, EI, and L. Sketch the modes and identify the associated natural frequencies.

(**b**) Verify that the modes satisfy the orthogonality properties.

(**c**) Normalize each mode so that the modal mass M_n has unit value. Sketch these normalized modes. Compare these modes with those obtained in part (a) and comment on the differences.

10.3 Determine the free vibration response of the system of Problem 9.2 (and Problem 10.2) due to each of the three sets of initial displacements: (**a**) $u_1(0) = 1,\ u_2(0) = 0$; (**b**) $u_1(0) = 1,\ u_2(0) = 1$; (**c**) $u_1(0) = 1,\ u_2(0) = -1$. Comment on the relative contribution of the modes to the response in the three cases. Neglect damping in the system.

10.4 Repeat Problem 10.3(a) considering damping in the system. For each mode the damping ratio is $\zeta_n = 5\%$.

10.5 For the system defined in Problem 9.4:

(**a**) Determine the natural vibration frequencies and modes. Express the frequencies in terms of m, EI, and L, and sketch the modes.

(**b**) Determine the displacement response due to an initial velocity $\dot{u}_2(0)$ imparted to the top of the system.

10.6 (**a**) For the system in Problem 9.5 determine the natural vibration frequencies and modes. Express the frequencies in terms of m, EI, and L, and sketch the modes.

(**b**) The structure is pulled through a lateral displacement $u_1(0)$ and released. Determine the free vibration response.

10.7 For the two-story shear building shown in Problem 9.6:

(**a**) Determine the natural vibration frequencies and modes; express the frequencies in terms of m, EI, and h.

(**b**) Verify that the modes satisfy the orthogonality properties.

(**c**) Normalize each mode so that the roof displacement is unity. Sketch the modes and identify the associated natural frequencies.

(**d**) Normalize each mode so that the modal mass M_n has unit value. Compare these modes with those obtained in part (c) and comment on the differences.

10.8 The structure of Problem 9.6 is modified so that the columns are hinged at the base. Determine the natural vibration frequencies and modes of the modified system, and compare them with the vibration properties of the original structure determined in Problem 10.7. Comment on the effect of the column support condition on the vibration properties.

10.9 Determine the free vibration response of the structure of Problem 10.7 (and Problem 9.6) if it is displaced as shown in Fig. P10.9a and b and released. Comment on the relative contributions of the two vibration modes to the response that was produced by the two initial displacements. Neglect damping.

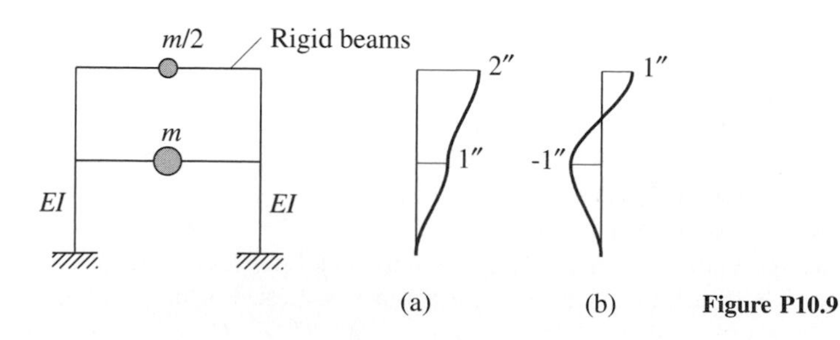

Figure P10.9

10.10 Repeat Problem 10.9 for the initial displacement of Fig. P10.9a, assuming that the damping ratio for each mode is 5%.

***10.11** Determine the natural vibration frequencies and modes of the system defined in Problem 9.7. Express the frequencies in terms of m, EI, and h and the joint rotations in terms of h. Normalize each mode to unit displacement at the roof and sketch it, identifying all DOFs.

10.12 For the three-story shear building shown in Problem 9.8:

(a) Determine the natural vibration frequencies and modes; express the frequencies in terms of m, EI, and h. Sketch the modes and identify the associated natural frequencies.

(b) Verify that the modes satisfy the orthogonality properties.

(c) Normalize each mode so that the modal mass M_n has unit value. Sketch these normalized modes. Compare these modes with those obtained in part (a) and comment on the differences.

10.13 The structure of Problem 9.8 is modified so that the columns are hinged at the base. Determine the natural vibration frequencies and modes of the modified system, and compare them with the vibration properties of the original structure determined in Problem 10.12. Comment on the effect of the column support condition on the vibration properties.

10.14 Determine the free vibration response of the structure of Problem 10.12 (and Problem 9.8) if it is displaced as shown in Fig. P10.14a, b, and c and released. Comment on the relative contributions of the three vibration modes to the response that was produced by the three initial displacements. Neglect damping.

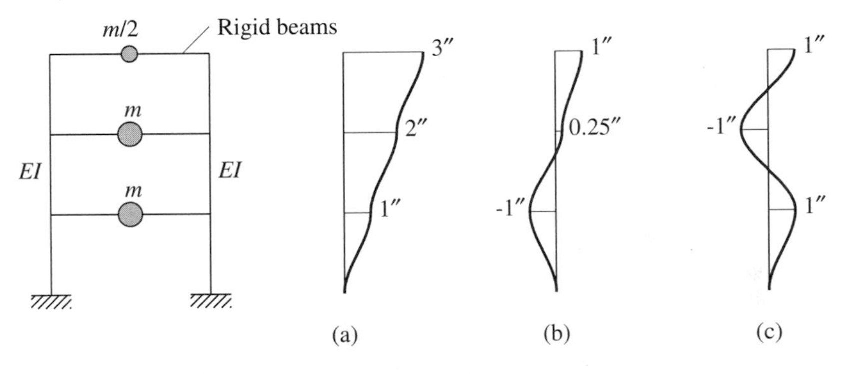

Figure P10.14

10.15 Repeat Problem 10.14 for the initial displacement of Fig. P10.14a, assuming that the damping ratio for each mode is 5%.

***10.16** Determine the natural vibration frequencies and modes of the system definedin Problem 9.9. Express the frequencies in terms of m, EI, and h and the joint rotations in terms of h. Normalize each mode to unit displacement at the roof and sketch it, including all DOFs.

10.17 For the system defined in Problem 9.10 $m = 90$ kips/g, $k = 1.5$ kips/in., and $b = 25$ ft.

(a) Determine the natural vibration frequencies and modes.

(b) Normalize each mode so that the modal mass M_n has unit value. Sketch these modes.

10.18 Repeat Problem 10.17 using a different set of DOFs—those defined in Problem 9.11. Show that the natural vibration frequencies and modes determined using the two sets of DOFs are the same.

*Denotes that a computer is necessary to solve this problem.

Part C

***10.19** The floor weights and story stiffnesses ofthe three-story shear frame are shown in Fig. P10.19, where $w = 100$ kips and $k = 326.32$ kips/in. Determine the fundamental natural vibration frequency ω_1 and mode ϕ_1 by inverse vector iteration.

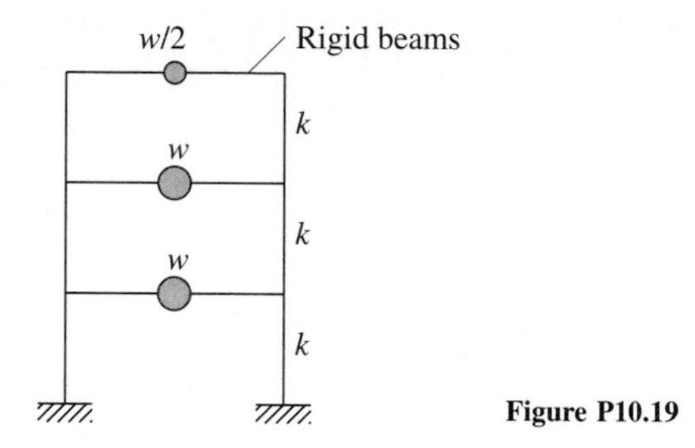

Figure P10.19

***10.20** For the system defined in Problem 10.19 there is concern for possible resonant vibrations due to rotating machinery mounted at the second-floor level. The operating speed of the motor is 430 rpm. Obtain the natural vibration frequency of the structure that is closest to the machine frequency.

***10.21** Determine the three natural vibration frequencies and modes of the system defined in Problem 10.19 by inverse vector iteration with shifting.

***10.22** Determine the three natural vibration frequencies and modes of the system defined in Problem 10.19 by inverse vector iteration with the shift in each iteration cycle equal to Rayleigh's quotient from the previous cycle.

*Denotes that a computer is necessary to solve this problem.

Damping in Structures

PREVIEW

Several issues that arise in defining the damping properties of structures are discussed in this chapter. It is impractical to determine the coefficients of the damping matrix directly from the structural dimensions, structural member sizes, and the damping properties of the structural materials used. Therefore, damping is generally specified by numerical values for the modal damping ratios and these are sufficient for analysis of linear systems with classical damping. The experimental data that provide a basis for estimating these damping ratios are discussed in Part A of this chapter, which ends with recommended values for modal damping ratios. The damping matrix is needed, however, for analysis of linear systems with nonclassical damping and for analysis of nonlinear structures. Two procedures for constructing the damping matrix for a structure from the modal damping ratios are presented in Part B; classically damped systems as well as nonclassically damped systems are considered.

PART A: EXPERIMENTAL DATA AND RECOMMENDED MODAL DAMPING RATIOS

11.1 VIBRATION PROPERTIES OF MILLIKAN LIBRARY BUILDING

The Robert A. Millikan Library building is a nine-story reinforced-concrete building constructed in 1966–1967 on the campus of the California Institute of Technology in Pasadena, California. Figure 11.1.1 is a photograph of the library building. It is 69

Figure 11.1.1 Millikan Library, California Institute of Technology, Pasadena. (Courtesy of G. W. Housner.)

by 75 ft in plan and extends 144 ft above grade and 158 ft above the basement level. This includes an enclosed roof that houses air-conditioning equipment. Lateral forces in the north–south direction are resisted mainly by the 12-in. reinforced-concrete shear walls located at the east and west ends of the building. In the east–west direction the 12-in. reinforced-concrete walls of the central core, which houses the elevator and the emergency stairway, provide most of the lateral resistance. Precast concrete window wall panels are bolted in place on the north and south walls. These were intended to be architectural but provide stiffness in the east–west direction for low levels of vibration.

The vibration properties—natural periods, natural modes, and modal damping ratios—of the Millikan Library have been determined by forced harmonic vibration tests using the vibration generator shown in Fig. 3.3.1. Such a test leads to a frequency-response curve that shows a resonant peak corresponding to each natural frequency of the structure (e.g., the frequency-response curve near the fundamental natural frequency of vibration in the east–west direction is shown in Fig. 11.1.2). From such data the natural frequency and damping ratio for the fundamental natural mode were determined by the methods of Section 3.4.2, and the results are presented in Table 11.1.1. The

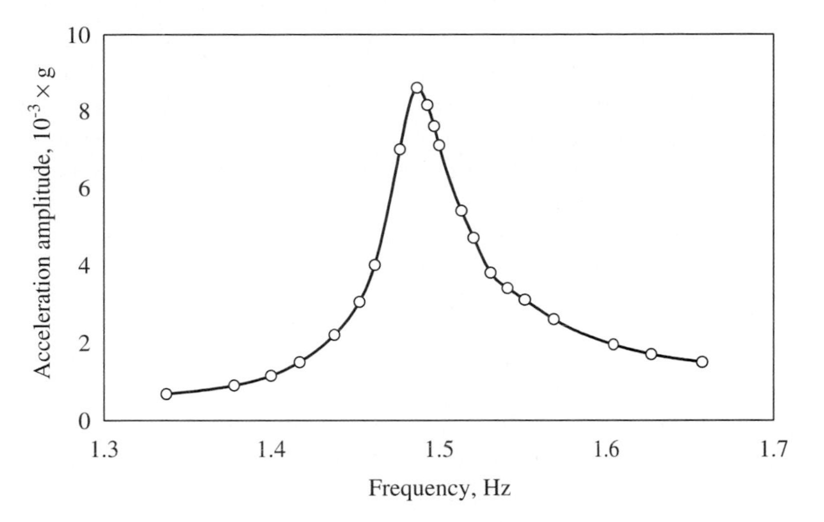

Figure 11.1.2 Frequency response curve for Millikan Library near its fundamental natural frequency of vibration in the east–west direction; acceleration measured is at the eighth floor. [Adapted from Jennings and Kuroiwa (1968).]

TABLE 11.1.1 NATURAL VIBRATION PERIODS AND MODAL DAMPING RATIOS OF MILLIKAN LIBRARY

Excitation	Roof Acceleration (g)	Fundamental Mode		Second Mode	
		Period (sec)	Damping (%)	Period (sec)	Damping (%)
North–South Direction					
Vibration generator	5×10^{-3} to 20×10^{-3}	0.51–0.53	1.2–1.8	[a]	[a]
Lytle Creek earthquake	0.05	0.52	2.9	0.12	1.0
San Fernando earthquake	0.312	0.62	6.4	0.13	4.7
East–West Direction					
Vibration generator	3×10^{-3} to 17×10^{-3}	0.66–0.68	0.7–1.5	[b]	[b]
Lytle Creek earthquake	0.035	0.71	2.2	0.18	3.6
San Fernando earthquake	0.348	0.98	7.0	0.20	5.9

[a]Not measured.

[b]Data not reliable.

natural period for this mode of vibration in the east–west direction was 0.66 sec. This value increased roughly 3% over the resonant amplitude range of testing: acceleration of 3×10^{-3}g to 17×10^{-3}g at the roof. The mode shape corresponding to this mode was found from measurements taken at various floors of the structure but is not presented here. In the vibration test the damping ratio in the fundamental east–west mode varied between 0.7 and 1.5%, increasing with the amplitude of response. In the north–south direction, the natural period of the fundamental mode was 0.51 sec, increasing roughly 4% over the resonant amplitude range of testing: acceleration of 5×10^{-3}g to 20×10^{-3}g at the roof. The damping ratio in this mode varied between 1.2 and 1.8%, again increasing with the amplitude of response.

The Millikan Library is located approximately 19 miles from the center of the Magnitude 6.4 San Fernando earthquake of February 9, 1971. The strong motion accelerographs installed in the basement and the roof of the building recorded three components (two horizontal and one vertical) of accelerations. The recorded accelerations in the north–south direction given in Fig. 11.1.3 show that the peak acceleration of 0.202g at the basement amplified to 0.312g at the roof. Figure 11.1.4 shows that in the east–west direction the peak acceleration at the basement and roof were 0.185g and 0.348g, respectively. The accelerations at the roof represent the total motion of the building, which is composed of the relative motions of the building with respect to the ground plus the motion of the ground. The total displacement at the roof of the building and the displacement of the basement were obtained by twice-integrating the recorded accelerations. The north–south and east–west components of the relative displacement of the roof, determined by subtracting the ground (basement) displacement from the total displacement at the roof, are presented in Fig. 11.1.5.

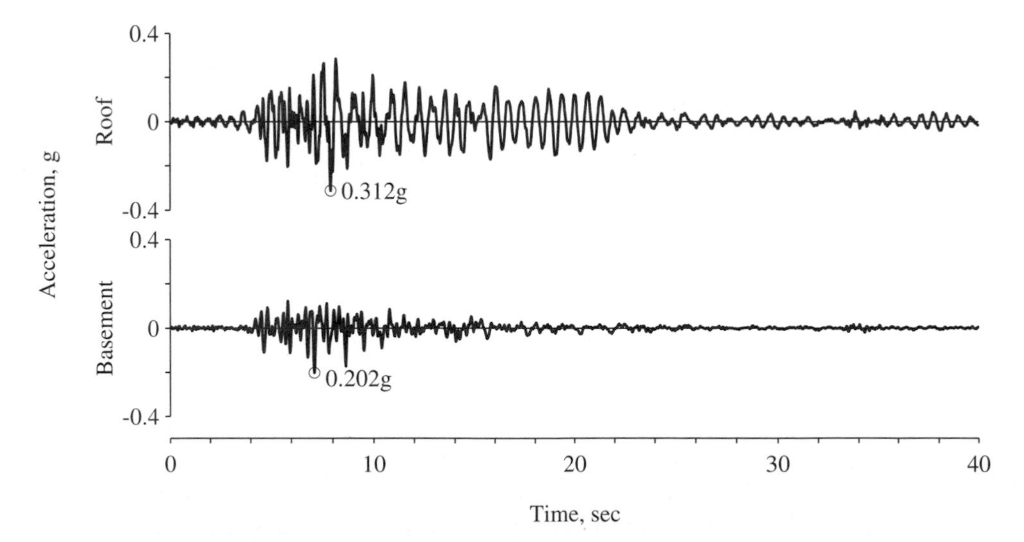

Figure 11.1.3 Accelerations in the north–south direction recorded at Millikan Library during the 1971 San Fernando earthquake.

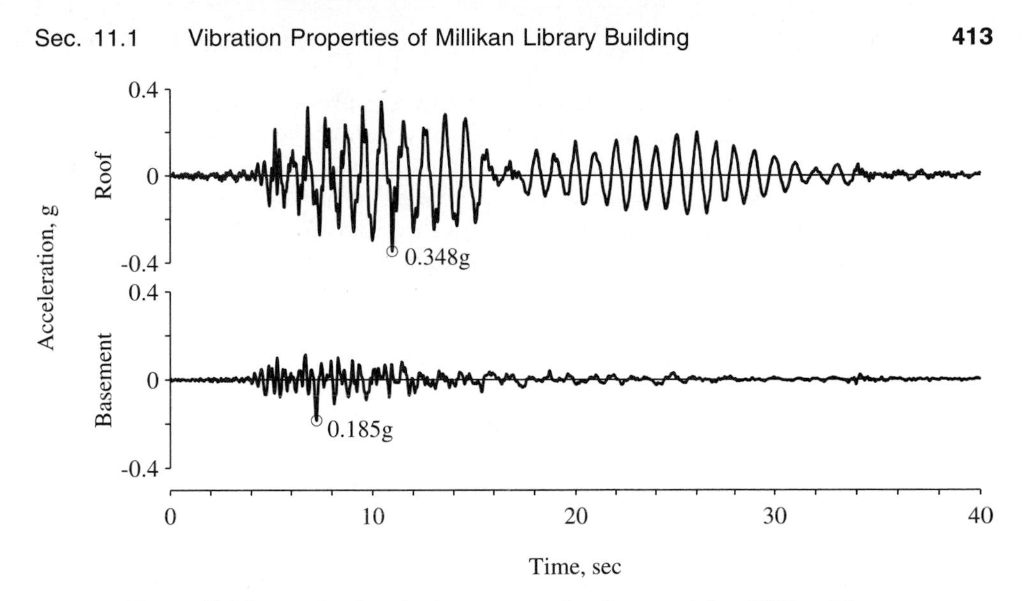

Figure 11.1.4 Accelerations in the east–west direction recorded at Millikan Library during the 1971 San Fernando earthquake.

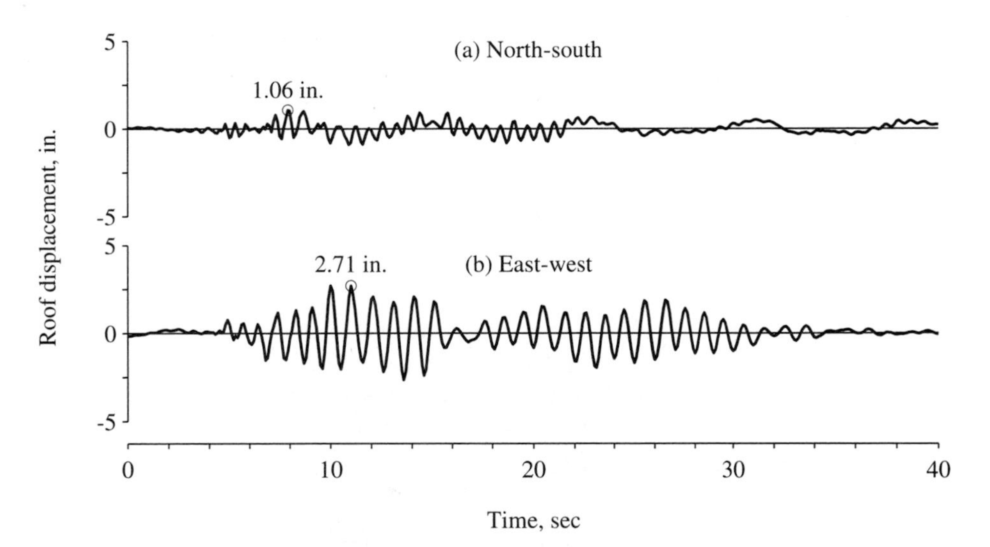

Figure 11.1.5 Relative displacement of the roof in (a) north–south direction; (b) east–west direction. [Adapted from Foutch, Housner, and Jennings (1975).]

It can be seen that the horizontal accelerations of the roof of the building are larger and their time variation is different from the ground (basement) accelerations. These differences arise because the building is flexible, not rigid. It is seen in the displacement plots that the displacement amplitude of the roof relative to the basement was 1.06 in. in the north–south direction and 2.71 in. in the east–west direction. The building vibrated in the north–south direction with a fundamental-mode period of approximately six-tenths of a second, while in the east–west direction this period was about 1 sec. These period values were estimated as the duration of a vibration cycle in Fig. 11.1.5. More accurate values for the first few natural periods and modal damping ratios can be determined from the recorded accelerations at the basement and the roof by system identification procedures (not presented in this book). The results for the first two modes in the north–south and east–west directions are presented in Table 11.1.1 for the Millikan Library building.

Acceleration records were also obtained in the basement and on the roof of this building from the Lytle Creek earthquake of September 12, 1970. The Magnitude 5.4 Lytle Creek earthquake, centered 40 miles from Millikan Library, produced a peak ground acceleration of approximately 0.02g and a roof acceleration of 0.05g in the building, fairly low levels for measured earthquake motion. System identification analysis of these records led to values for natural periods and damping ratios shown in Table 11.1.1. For the low-level vibrations due to the Lytle Creek earthquake, the fundamental periods of 0.52 and 0.71 sec in the north–south and east–west directions, respectively, were similar to—only slightly longer than—those determined in vibration generator tests. Similarly, the damping ratios were slightly increased relative to the vibration generator tests.

For the larger motions of the building during the San Fernando earthquake, the natural periods and damping ratios were increased significantly relative to the values from vibration generator tests. The fundamental period in the north–south direction increased from 0.51 to 0.62 sec and the damping ratio increased substantially, to 6.4%. In the east–west direction the building vibrated with a fundamental period of 0.98 sec, which is 50% longer than the period of 0.66 sec from vibration generator tests; the damping increased substantially, to 7.0%.

The lengthening of natural periods at the larger amplitudes of motion experienced by the building during the San Fernando earthquake implies a reduction in the stiffness of the structure. The stiffness in the east–west direction is reduced substantially, although except for the collapse of bookshelves and minor plaster cracking, the building suffered no observable damage. The apparent damage of the structure due to the earthquake is also the cause of the substantial increase in damping. Following the earthquake there is apparent recovery of the structural stiffness, as suggested by measured natural periods (not presented here) that are shorter than during the earthquake. Whether this recovery is complete or only partial appears to depend on how strongly the structure was excited by the earthquake. These are all indications of the complexity of the behavior of actual structures during earthquakes. We return to this issue in Chapter 13 after we have presented analytical procedures to compute the response of linearly elastic structures to specified ground motion.

11.2 ESTIMATING MODAL DAMPING RATIOS

It is usually not feasible to determine the damping properties or natural vibration periods of a structure to be analyzed in the way they were determined for the Millikan Library. If the seismic safety of an existing structure is to be evaluated, ideally we would like to determine experimentally the important properties of the structure, including its damping, but this is rarely done, for lack of budget and time. For a new building being designed, obviously its damping or other properties cannot be measured.

The modal damping ratios for a structure should therefore be estimated using measured data from similar structures. Although researchers have accumulated a substantial body of valuable data, it should be used with discretion because some of it is *not* directly applicable to earthquake analysis and design. It is clear from the Millikan Library data that the damping ratios determined from the low-amplitude forced vibration tests should not be used directly for the analysis of response to earthquakes that cause much larger motions of the structure, say, up to yielding of the structural materials. Modal damping ratios for such analysis should be based on data from recorded earthquake motions.

The data that are most useful but hard to come by are from structures shaken strongly but not deformed into the inelastic range. The damping ratios determined from structural motions that are small are not representative of the larger damping expected at higher amplitudes of structural motions. On the other hand, recorded motions of structures that have experienced significant yielding during an earthquake would provide damping ratios that also include the energy dissipation due to yielding. These damping ratios would not be useful in dynamic analysis because the energy dissipation in yielding is accounted for separately through nonlinear force–deformation relationships (see Section 5.7).

Useful data on damping are slow to accumulate because relatively few structures are installed with permanent accelerographs ready to record motions when an earthquake occurs and strong earthquakes are infrequent. The bulk of records of earthquake-induced structural motions in the United States are from multistory buildings in California: more than 50 buildings in the greater Los Angeles area during the 1971 San Fernando earthquake; over 40 buildings in the Monterey Bay and San Francisco Bay areas during the 1989 Loma Prieta earthquake; and over 100 buildings in the greater Los Angeles area during the 1994 Northridge earthquake. Furthermore, the recorded motions of only some of these buildings have been analyzed to determine their natural periods and modal damping ratios.

Ideally, we would like to have data on damping extracted from recorded earthquake motions of many structures of various types—buildings, bridges, dams, etc.—using different materials—steel, reinforced concrete, prestressed concrete, masonry, wood, etc. Such data would provide the basis for estimating the damping ratios for an existing structure to be evaluated for its seismic safety or for a new structure to be designed. Until we accumulate a sufficiently large data base, selection of damping ratios is based on whatever data are available and expert opinion. Recommended damping values are given in Table 11.2.1 for two levels of motion: working stress levels or stress levels no more than one-half the yield point, and stresses at or just below the yield point. For each stress level, a range of damping values is given; the higher values of damping are to be used for ordinary structures, and the lower values are for special structures to be designed more conservatively. In addition to Table 11.2.1, recommended damping values are 3% for unreinforced masonry structures and 7% for reinforced masonry construction.

Most building codes do not recognize the variation in damping with structural materials; and typically a 5% damping ratio is implicit in the code-specified earthquake forces and design spectrum.

TABLE 11.2.1 RECOMMENDED DAMPING VALUES

Stress Level	Type and Condition of Structure	Damping Ratio (%)
Working stress, no more than about $\frac{1}{2}$ yield point	Welded steel, prestressed concrete, well-reinforced concrete (only slight cracking)	2–3
	Reinforced concrete with considerable cracking	3–5
	Bolted and/or riveted steel, wood structures with nailed or bolted joints	5–7
At or just below yield point	Welded steel, prestressed concrete (without complete loss in prestress)	5–7
	Prestressed concrete with no prestress left	7–10
	Reinforced concrete	7–10
	Bolted and/or riveted steel, wood structures with bolted joints	10–15
	Wood structures with nailed joints	15–20

Source: N. M. Newmark, and W. J. Hall, *Earthquake Spectra and Design*, Earthquake Engineering Research Institute, Berkeley, Calif., 1982.

The recommended damping ratios can be used directly for the linearly elastic analysis of structures with classical damping. For such systems the equations of motion when transformed to natural vibration modes of the undamped system become uncoupled, and the estimated modal damping ratios are used directly in each modal equation. This concept was introduced in Section 10.9.2 and will be developed further in Chapters 12 and 13.

PART B: CONSTRUCTION OF DAMPING MATRIX

11.3 DAMPING MATRIX

When is the damping matrix needed? The damping matrix must be defined completely if classical modal analysis is not applicable. Such is the case for structures with nonclassical damping (see Section 11.5 for examples), even if our interest is confined to their linearly elastic response. Classical modal analysis is also not applicable to the analysis of nonlinear systems even if the damping is of classical form. One of the most important nonlinear problems of interest to us is calculating the response of structures beyond their linearly elastic range during earthquakes.

The damping matrix for practical structures should not be calculated from the structural dimensions, structural member sizes, and the damping of the structural materials used. One might think that it should be possible to determine the damping matrix for the structure from the damping properties of individual structural elements, just as the structural stiffness matrix is determined. However, it is impractical to determine the damping matrix in this manner because unlike the elastic modulus, which enters into the computation of stiffness, the damping properties of materials are not well established. Even if these properties were known, the resulting damping matrix would not account for a significant part of the energy dissipated in friction at steel connections, opening and closing of microcracks in concrete, stressing of nonstructural elements—partition walls, mechanical equipment, fireproofing, etc.—friction between the structure itself and nonstructural elements, and other similar mechanisms, some of which are even hard to identify.

Thus the damping matrix for a structure should be determined from its modal damping ratios, which account for all energy dissipating mechanisms. As discussed in Section 11.2, the modal damping ratios should be estimated from available data on similar structures shaken strongly during past earthquakes but not deformed into the inelastic range; lacking such data the values of Table 11.2.1 are recommended.

11.4 CLASSICAL DAMPING MATRIX

Classical damping is an appropriate idealization if similar damping mechanisms are distributed throughout the structure (e.g., a multistory building with a similar structural system and structural materials over its height). In this section we develop two procedures for constructing a classical damping matrix for a structure from modal damping ratios which have been estimated as described in Section 11.2. These two procedures are presented in the following two subsections.

11.4.1 Rayleigh Damping and Caughey Damping

Consider first mass-proportional damping and stiffness-proportional damping:

$$\mathbf{c} = a_0\mathbf{m} \quad \text{and} \quad \mathbf{c} = a_1\mathbf{k} \tag{11.4.1}$$

where the constants a_0 and a_1 have units of sec^{-1} and sec, respectively. For both of these damping matrices the matrix $\mathbf{C}$ of Eq. (10.9.4) is diagonal by virtue of the modal orthogonality properties of Eq. (10.4.1); therefore, these are classical damping matrices. Physically, they represent the damping models shown in Fig. 11.4.1 for a multistory building. The stiffness-proportional damping appeals to intuition because it can be interpreted to model the energy dissipation arising from story deformations. In contrast,

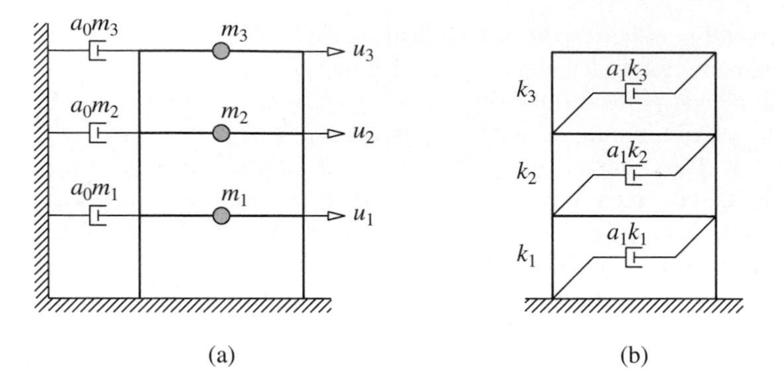

Figure 11.4.1 (a) Mass-proportional damping; (b) stiffness-proportional damping.

the mass-proportional damping is difficult to justify physically because the air damping it can be interpreted to model is negligibly small for most structures. Later we shall see that, by themselves, neither of the two damping models are appropriate for practical application.

We now relate the modal damping ratios for a system with mass-proportional damping to the coefficient a_0. The generalized damping for the nth mode, Eq. (10.9.10), is

$$C_n = a_0 M_n \tag{11.4.2}$$

and the modal damping ratio, Eq. (10.9.11), is

$$\zeta_n = \frac{a_0}{2}\frac{1}{\omega_n} \tag{11.4.3}$$

The damping ratio is inversely proportional to the natural frequency (Fig. 11.4.2a). The coefficient a_0 can be selected to obtain a specified value of damping ratio in any one mode, say ζ_i for the ith mode. Equation (11.4.3) then gives

$$a_0 = 2\zeta_i\omega_i \tag{11.4.4}$$

With a_0 determined, the damping matrix $\mathbf{c}$ is known from Eq. (11.4.1a), and the damping ratio in any other mode, say the nth mode, is given by Eq. (11.4.3).

Similarly, the modal damping ratios for a system with stiffness-proportional damping can be related to the coefficient a_1. In this case

$$C_n = a_1\omega_n^2 M_n \quad \text{and} \quad \zeta_n = \frac{a_1}{2}\omega_n \tag{11.4.5}$$

wherein Eq. (10.2.4) is used. The damping ratio increases linearly with the natural frequency (Fig. 11.4.2a). The coefficient a_1 can be selected to obtain a specified value of the damping ratio in any one mode, say ζ_j for the jth mode. Equation (11.4.5b) then gives

$$a_1 = \frac{2\zeta_j}{\omega_j} \tag{11.4.6}$$

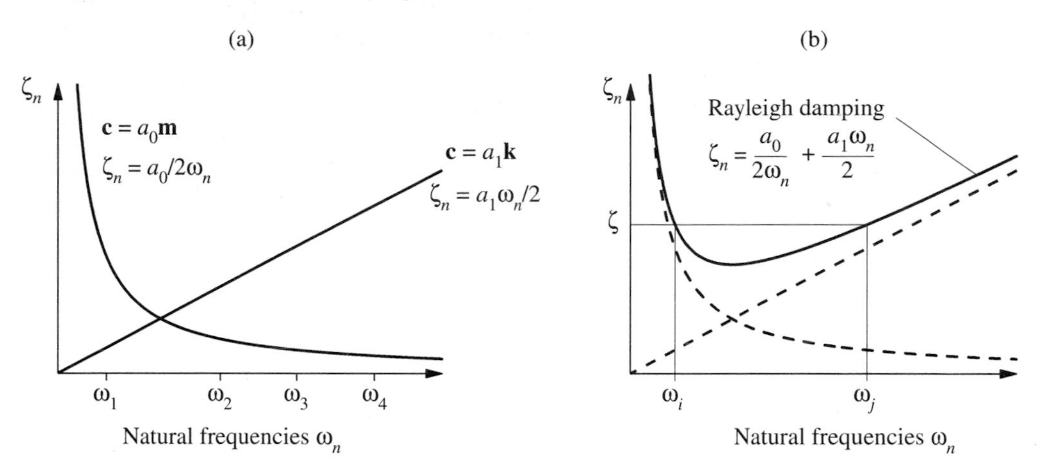

Figure 11.4.2 Variation of modal damping ratios with natural frequency: (a) mass-proportional damping and stiffness-proportional damping; (b) Rayleigh damping.

With a_1 determined, the damping matrix $\mathbf{c}$ is known from Eq. (11.4.1b), and the damping ratio in any other mode is given by Eq. (11.4.5b). Neither of the damping matrices defined by Eq. (11.4.1) are appropriate for practical analysis of MDF systems. The variations of modal damping ratios with natural frequencies they represent (Fig. 11.4.2a) are not consistent with experimental data that indicate roughly the same damping ratios for several vibration modes of a structure.

As a first step toward constructing a classical damping matrix consistent with experimental data, we consider *Rayleigh damping*:

$$\mathbf{c} = a_0\mathbf{m} + a_1\mathbf{k} \tag{11.4.7}$$

The damping ratio for the nth mode of such a system is

$$\zeta_n = \frac{a_0}{2}\frac{1}{\omega_n} + \frac{a_1}{2}\omega_n \tag{11.4.8}$$

The coefficients a_0 and a_1 can be determined from specified damping ratios ζ_i and ζ_j for the ith and jth modes, respectively. Expressing Eq. (11.4.8) for these two modes in matrix form leads to

$$\frac{1}{2}\begin{bmatrix} 1/\omega_i & \omega_i \\ 1/\omega_j & \omega_j \end{bmatrix} \begin{Bmatrix} a_0 \\ a_1 \end{Bmatrix} = \begin{Bmatrix} \zeta_i \\ \zeta_j \end{Bmatrix} \tag{11.4.9}$$

These two algebraic equations can be solved to determine the coefficients a_0 and a_1. If both modes are assumed to have the same damping ratio ζ, which is reasonable based on experimental data, then

$$a_0 = \zeta\frac{2\omega_i\omega_j}{\omega_i + \omega_j} \qquad a_1 = \zeta\frac{2}{\omega_i + \omega_j} \tag{11.4.10}$$

The damping matrix is then known from Eq. (11.4.7) and the damping ratio for any other mode, given by Eq. (11.4.8), varies with natural frequency as shown in Fig. 11.4.2b.

In applying this procedure to a practical problem, the modes i and j with specified damping ratios should be chosen to ensure reasonable values for the damping ratios in all the modes contributing significantly to the response. Consider, for example, that five modes are to be included in the response analysis and roughly the same damping ratio ζ is desired for all modes. This ζ should be specified for the first mode and possibly for the fourth mode. Then Fig. 11.4.2b suggests that the damping ratio for the second and third modes will be somewhat smaller than ζ and for the fifth mode it will be somewhat larger than ζ. The damping ratio for modes higher than the fifth will increase monotonically with frequency and the corresponding modal responses will be essentially eliminated because of their high damping.

Example 11.1

The properties of a three-story shear building are given in Fig. E11.1. These include the floor weights, story stiffnesses, natural frequencies, and modes. Derive a Rayleigh damping matrix such that the damping ratio is 5% for the first and second modes. Compute the damping ratio for the third mode.

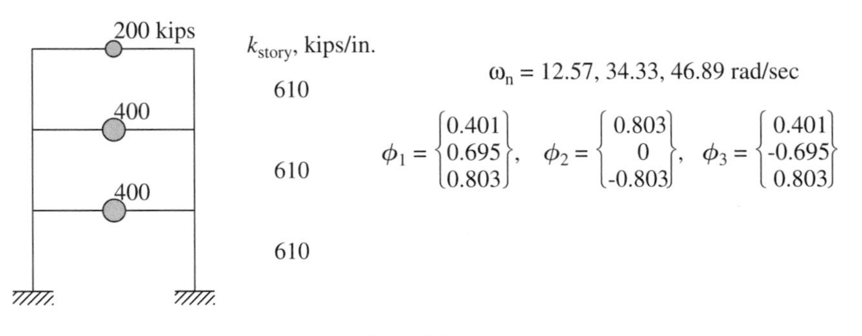

Figure E11.1

Solution

1. *Set up the mass and stiffness matrices.*

$$\mathbf{m} = \frac{1}{386}\begin{bmatrix} 400 & & \\ & 400 & \\ & & 200 \end{bmatrix} \qquad \mathbf{k} = 610\begin{bmatrix} 2 & -1 & 0 \\ -1 & 2 & -1 \\ 0 & -1 & 1 \end{bmatrix}$$

2. *Determine a_0 and a_1 from Eq.* (11.4.9).

$$\begin{bmatrix} 1/12.57 & 12.57 \\ 1/34.33 & 34.33 \end{bmatrix}\begin{Bmatrix} a_0 \\ a_1 \end{Bmatrix} = 2\begin{Bmatrix} 0.05 \\ 0.05 \end{Bmatrix}$$

These algebraic equations have the following solution:

$$a_0 = 0.9198 \qquad a_1 = 0.0021$$

3. *Evaluate the damping matrix.*

$$\mathbf{c} = a_0\mathbf{m} + a_1\mathbf{k} = \begin{bmatrix} 3.55 & -1.30 & 0 \\ & 3.55 & -1.30 \\ \text{(sym)} & & 1.78 \end{bmatrix}$$

4. *Compute ζ_3 from Eq.* (11.4.8).

$$\zeta_3 = \frac{0.9198}{2(46.89)} + \frac{0.0021(46.89)}{2} = 0.0593$$

If we wish to specify values for damping ratios in more than two modes, we need to consider the general form for a classical damping matrix (see Derivation 11.1), known as *Caughey damping*:

$$\mathbf{c} = \mathbf{m} \sum_{l=0}^{N-1} a_l [\mathbf{m}^{-1}\mathbf{k}]^l \tag{11.4.11}$$

where N is the number of degrees of freedom in the system and a_l are constants. The first three terms of the series are

$$a_0\mathbf{m}(\mathbf{m}^{-1}\mathbf{k})^0 = a_0\mathbf{m} \qquad a_1\mathbf{m}(\mathbf{m}^{-1}\mathbf{k})^1 = a_1\mathbf{k} \qquad a_2\mathbf{m}(\mathbf{m}^{-1}\mathbf{k})^2 = a_2\mathbf{k}\mathbf{m}^{-1}\mathbf{k} \tag{11.4.12}$$

Thus Eq. (11.4.11) with only the first two terms is the same as Rayleigh damping. Suppose that we wish to specify the damping ratios for J modes of an N-DOF system. Then J terms need to be included in the Caughey series and these could be any J of the N terms in Eq. (11.4.11). Typically, the first J terms are included:

$$\mathbf{c} = \mathbf{m} \sum_{l=0}^{J-1} a_l [\mathbf{m}^{-1}\mathbf{k}]^l \tag{11.4.13}$$

and the modal damping ratio ζ_n is given by (see Derivation 11.2)

$$\zeta_n = \frac{1}{2} \sum_{l=0}^{J-1} a_l \omega_n^{2l-1} \tag{11.4.14}$$

The coefficients a_l can be determined from the damping ratios specified in any J modes, say the first J modes, by solving the J algebraic equations (11.4.14) for the unknowns $a_l, l = 0$ to $J-1$. With a_l determined, the damping matrix **c** is known from Eq. (11.4.13), and the damping ratios for modes $n = J+1, J+2, \ldots, N$ are given by Eq. (11.4.14). It is recommended that these damping ratios be computed to ensure that their values are reasonable; in particular, they should not take on negative values, which may happen depending on the number of terms included in the Caughey series. Negative values of modal damping ratios are obviously unrealistic because they imply free vibration response that grows with time instead of decaying with time.

While the general classical damping matrix given by Eq. (11.4.13) makes it possible to specify the damping ratios in any number of modes, there are two problems associated with its use. First, the algebraic equations (11.4.14) are numerically ill conditioned because the coefficients ω_n^{-1}, ω_n, ω_n^3, ω_n^5, ... can differ by orders of magnitude. Second, if more than two terms are included in the Caughey series, **c** is a full matrix, although **k** is a banded matrix, and for a lumped mass system, **m** is a diagonal matrix. Since the cost of analysis of large systems is increased by a very significant amount if the damping matrix is not banded, Rayleigh damping is assumed in most practical analyses.

Example 11.2

For the system of Fig. E11.1, evaluate the classical damping matrix if the damping ratio is 5% for all three modes.

Solution

1. *Caughey series for a 3-DOF system:*

$$\mathbf{c} = a_0\mathbf{m} + a_1\mathbf{k} + a_2\mathbf{k}\mathbf{m}^{-1}\mathbf{k} \tag{a}$$

2. *Determine a_0, a_1, and a_2 from Eq.* (11.4.14):

$$\zeta_n = \frac{a_0}{2}\frac{1}{\omega_n} + \frac{a_1}{2}\omega_n + \frac{a_2}{2}\omega_n^3 \qquad n = 1, 2, 3 \tag{b}$$

or

$$\begin{bmatrix} 1/12.57 & 12.57 & (12.57)^3 \\ 1/34.33 & 34.33 & (34.33)^3 \\ 1/46.89 & 46.89 & (46.89)^3 \end{bmatrix} \begin{Bmatrix} a_0 \\ a_1 \\ a_2 \end{Bmatrix} = 2\begin{Bmatrix} 0.05 \\ 0.05 \\ 0.05 \end{Bmatrix} \tag{c}$$

These algebraic equations have the following solution:

$$a_0 = 0.8377 \qquad a_1 = 0.0027 \qquad a_2 = -4.416 \times 10^{-7} \tag{d}$$

3. *Evaluate* **c**. Substituting a_0, a_1, and a_2 from Eq. (d) in Eq. (a) gives

$$\mathbf{c} = \begin{bmatrix} 3.40 & -1.03 & -0.159 \\ & 3.08 & -1.03 \\ \text{(sym)} & & 1.62 \end{bmatrix} \tag{e}$$

Derivation 11.1

The natural frequencies ω_r and modes ϕ_r satisfy

$$\mathbf{k}\phi_r = \omega_r^2\mathbf{m}\phi_r \tag{a}$$

Premultiplying both sides by $\phi_n^T\mathbf{k}\mathbf{m}^{-1}$ gives

$$\phi_n^T\left[\mathbf{k}\mathbf{m}^{-1}\mathbf{k}\right]\phi_r = \omega_r^2\phi_n^T\mathbf{k}\phi_r = 0 \qquad n \neq r \tag{b}$$

wherein the second equality comes from the orthogonality equation (10.4.1a). Premultiplying both sides of Eq. (a) by $\phi_n^T(\mathbf{k}\mathbf{m}^{-1})^2$ gives

$$\begin{aligned} \phi_n^T\left[(\mathbf{k}\mathbf{m}^{-1})^2\mathbf{k}\right]\phi_r &= \omega_r^2\phi_n^T\left[\mathbf{k}\mathbf{m}^{-1}\mathbf{k}\mathbf{m}^{-1}\mathbf{m}\right]\phi_r \\ &= \omega_r^2\phi_n^T\left[\mathbf{k}\mathbf{m}^{-1}\mathbf{k}\right]\phi_r = 0 \qquad n \neq r \end{aligned} \tag{c}$$

wherein the second equality comes from Eq. (b). By repeated application of this procedure, a family of orthogonality relations can be obtained which can all be expressed in a compact form:

$$\phi_n^T\mathbf{c}_l\phi_r = 0 \qquad n \neq r \tag{d}$$

where

$$\mathbf{c}_l = \left[\mathbf{k}\mathbf{m}^{-1}\right]^l\mathbf{k} \qquad l = 0, 1, 2, 3, \ldots, \infty \tag{e}$$

The matrices $\mathbf{c}_l$ can be written in an alternative form by premultiplying Eq. (e) by the identity matrix, $\mathbf{I} = \mathbf{m}\mathbf{m}^{-1}$:

$$\begin{aligned}\mathbf{c}_l &= \mathbf{m}\mathbf{m}^{-1}\mathbf{k}\mathbf{m}^{-1}\mathbf{k}\mathbf{m}^{-1}\cdots\mathbf{k}\mathbf{m}^{-1}\mathbf{k} \\ &= \mathbf{m}\left[\mathbf{m}^{-1}\mathbf{k}\right]^l \qquad l = 0, 1, 2, 3, \ldots, \infty\end{aligned} \tag{f}$$

Premultiplying Eq. (a) by $\phi_n^T\mathbf{m}\mathbf{k}^{-1}$ and following the procedure above it can be shown that Eq. (d) is satisfied by another infinite sequence of matrices:

$$\mathbf{c}_l = \mathbf{m}\left[\mathbf{m}^{-1}\mathbf{k}\right]^l \qquad l = -1, -2, -3, \ldots, -\infty \tag{g}$$

Combining Eqs. (f) and (g) gives

$$\mathbf{c} = \mathbf{m}\sum_{l=-\infty}^{\infty} a_l\left[\mathbf{m}^{-1}\mathbf{k}\right]^l \tag{h}$$

It can be shown that only N terms in this infinite series are independent, leading to Eq. (11.4.11) as the general form of classical damping matrices.

Derivation 11.2

For the nth mode the generalized damping is

$$C_n = \phi_n^T\mathbf{c}\phi_n = \sum_{l=0}^{N-1}\phi_n^T\mathbf{c}_l\phi_n \tag{a}$$

where $\mathbf{c}_l$ is given by Eq. (f) of Derivation 11.1; and the various terms in this series are

$$l = 0{:}\quad \phi_n^T\mathbf{c}_o\phi_n = \phi_n^T(a_0\mathbf{m})\phi_n = a_o M_n$$

$$l = 1{:}\quad \phi_n^T\mathbf{c}_1\phi_n = \phi_n^T(a_1\mathbf{k})\phi_n = a_1\omega_n^2 M_n$$

$$l = 2{:}\quad \phi_n^T\mathbf{c}_2\phi_n = \phi_n^T(a_2\mathbf{k}\mathbf{m}^{-1}\mathbf{k})\phi_n = a_2\omega_n^2\phi_n^T\mathbf{k}\phi_n = a_2\omega_n^4 M_n$$

wherein Eq. (10.2.4) is used. Thus Eq. (a) becomes

$$C_n = \sum_{l=0}^{N-1} a_l\omega_n^{2l} M_n \tag{b}$$

The damping ratio for the nth mode, Eq. (10.9.11), is given by

$$\zeta_n = \frac{1}{2}\sum_{l=0}^{N-1} a_l\omega_n^{2l-1} \tag{c}$$

which is similar to Eq. (11.4.14).

11.4.2 Superposition of Modal Damping Matrices

An alternative procedure to determine a classical damping matrix from modal damping ratios can be derived starting with Eq. (10.9.4):

$$\mathbf{\Phi}^T\mathbf{c}\mathbf{\Phi} = \mathbf{C} \tag{11.4.15}$$

where $\mathbf{C}$ is a diagonal matrix with the nth diagonal element equal to the generalized modal damping:

$$C_n = \zeta_n(2M_n\omega_n) \tag{11.4.16}$$

With ζ_n estimated as described in Section 11.2, $\mathbf{C}$ is known from Eq. (11.4.16) and Eq. (11.4.15) can be rewritten as

$$\mathbf{c} = \left(\mathbf{\Phi}^T\right)^{-1}\mathbf{C}\mathbf{\Phi}^{-1} \tag{11.4.17}$$

Using this equation to compute $\mathbf{c}$ may appear to be an inefficient procedure because it seems to require the inversion of two matrices of order N, the number of DOFs. However, the inverse of the modal matrix $\mathbf{\Phi}$ and of $\mathbf{\Phi}^T$ can be determined with little computation because of the orthogonality property of modes.

Starting with the orthogonality relationship of Eq. (10.4.5),

$$\mathbf{\Phi}^T\mathbf{m}\mathbf{\Phi} = \mathbf{M} \tag{11.4.18}$$

it can be shown that

$$\mathbf{\Phi}^{-1} = \mathbf{M}^{-1}\mathbf{\Phi}^T\mathbf{m} \qquad \left(\mathbf{\Phi}^T\right)^{-1} = \mathbf{m}\mathbf{\Phi}\mathbf{M}^{-1} \tag{11.4.19}$$

Because $\mathbf{M}$ is a diagonal matrix of generalized modal masses M_n, $\mathbf{M}^{-1}$ is known immediately as a diagonal matrix with elements $= 1/M_n$. Thus $\mathbf{\Phi}$ and $\mathbf{\Phi}^{-1}$ can be computed efficiently from Eq. (11.4.19).

Substituting Eq. (11.4.19) in Eq. (11.4.17) leads to

$$\mathbf{c} = (\mathbf{m}\mathbf{\Phi}\mathbf{M}^{-1})\mathbf{C}(\mathbf{M}^{-1}\mathbf{\Phi}^T\mathbf{m}) \tag{11.4.20}$$

Since $\mathbf{M}$ and $\mathbf{C}$ are diagonal matrices, defined by Eqs. (11.4.18) and (11.4.15), respectively, Eq. (11.4.20) can be expressed as

$$\mathbf{c} = \mathbf{m}\left(\sum_{n=1}^{N}\frac{2\zeta_n\omega_n}{M_n}\phi_n\phi_n^T\right)\mathbf{m} \tag{11.4.21}$$

The nth term in this summation is the contribution of the nth mode with its damping ratio ζ_n to the damping matrix $\mathbf{c}$; if this term is not included, the resulting $\mathbf{c}$ implies a zero damping ratio in the nth mode. It is reasonable to include only the first J modes in Eq. (11.4.21) that are expected to contribute significantly to the response. The lack of damping in modes $J+1$ to N does not create numerical problems if an unconditionally stable time-stepping procedure is used to integrate the equations of motion; see Chapter 15.

Example 11.3

Determine a damping matrix for the system of Fig. E11.1 by superposing the damping matrices for the first two modes, each with $\zeta_n = 5\%$.

Solution

1. *Determine the individual terms in Eq.* (11.4.21).

$$\mathbf{c}_1 = \frac{2(0.05)(12.57)}{1.0}\mathbf{m}\phi_1\phi_1^T\mathbf{m} = \begin{bmatrix} 0.217 & 0.376 & 0.217 \\ & 0.651 & 0.376 \\ \text{(sym)} & & 0.217 \end{bmatrix} \qquad \mathbf{c}_2 = \frac{2(0.05)(34.33)}{1.0}\mathbf{m}\phi_2\phi_2^T\mathbf{m} = \begin{bmatrix} 2.37 & 0 & -1.19 \\ & 0 & 0 \\ \text{(sym)} & & 0.593 \end{bmatrix}$$

2. *Determine* **c**.

$$\mathbf{c} = \mathbf{c}_1 + \mathbf{c}_2 = \begin{bmatrix} 2.59 & 0.376 & -0.969 \\ & 0.651 & 0.376 \\ \text{(sym)} & & 0.810 \end{bmatrix}$$

Recall that this **c** implies a zero damping ratio for the third mode.

Example 11.4

Determine the damping matrix for the system of Fig. E11.1 by superposing the damping matrices for the three modes, each with $\zeta_n = 5\%$.

Solution

1. *Determine the individual terms in Eq.* (11.4.21). The first two terms, $\mathbf{c}_1$ and $\mathbf{c}_2$, are already computed in Example 11.3, and

$$\mathbf{c}_3 = \frac{2(0.05)(46.89)}{1.0}\mathbf{m}\phi_3\phi_3^T\mathbf{m} = \begin{bmatrix} 0.809 & -1.40 & 0.810 \\ & 2.43 & -1.40 \\ \text{(sym)} & & 0.811 \end{bmatrix}$$

2. *Determine* **c**.

$$\mathbf{c} = \sum_{n=1}^{3}\mathbf{c}_n = \begin{bmatrix} 3.40 & -1.03 & -0.159 \\ & 3.08 & -1.03 \\ \text{(sym)} & & 1.62 \end{bmatrix}$$

Note that this **c** is the same as in Example 11.2 because $\zeta_n = 5\%$ for all three modes in both examples.

11.5 NONCLASSICAL DAMPING MATRIX

The assumption of classical damping is not appropriate if the system to be analyzed consists of two or more parts with significantly different levels of damping. One such example is a structure–soil system. While the underlying soil can be assumed as rigid in the analysis of many structures, soil–structure interaction should be considered in the analysis of structures with very short natural periods, such as the nuclear containment structure of Fig. 1.10.1. The modal damping ratio for the soil system would typically be much different than the structure, say 15 to 20% for the soil region compared to 3 to 5% for the structure. Therefore, the assumption of classical damping would not be appropriate for the combined structure–soil system, although it may be reasonable for the structure and soil regions separately. Another example is a concrete dam with water impounded behind the dam (Fig. 1.10.2). The damping of the water is negligible relative to damping for the dam, and classical damping is not an appropriate model for the dam–water system. While substructure methods (not developed in this book) are especially effective for the analysis of structure–soil and structure–fluid systems, these systems are also analyzed by standard methods, requiring the damping matrix for the complete system.

The damping matrix for the complete system is constructed by directly assembling the damping matrices for the two subsystems—structure and soil in the first case, dam and water in the second. As shown in Fig. 11.5.1, the stiffness and mass matrices of the combined structure–soil system are assembled from the corresponding matrices for the two subsystems. The portion of these matrices associated with the common DOFs at the interface (I) between the two subsystems include contributions from both subsystems. Thus all that remains to be described is the procedure to construct damping matrices for the individual subsystems, assumed to be classically damped.

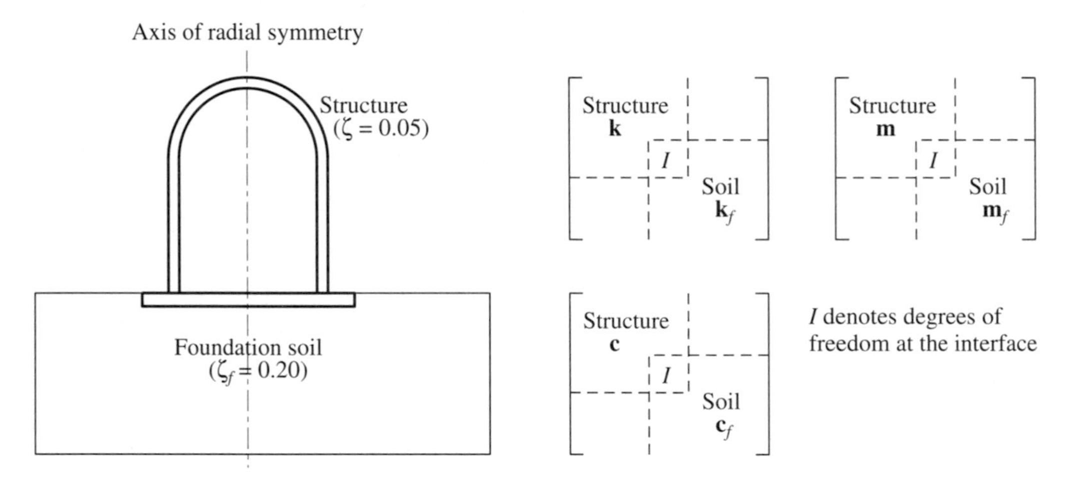

Figure 11.5.1 Assembly of subsystem matrices.

In principle, these subsystem damping matrices could be constructed by any of the procedures developed in Section 11.4, but Rayleigh damping is perhaps most convenient for practical analyses. Thus the damping matrices for the structure and the foundation soil (denoted by subscript f) are

$$\mathbf{c} = a_0\mathbf{m} + a_1\mathbf{k} \qquad \mathbf{c}_f = a_{0f}\mathbf{m}_f + a_{1f}\mathbf{k}_f$$

The coefficients a_0 and a_1 are given by Eq. (11.4.10) using an appropriate damping ratio for the structure, say $\zeta = 0.05$, where ω_i and ω_j are selected as the frequencies of the ith and jth natural vibration modes of the combined system without damping. The coefficients a_{0f} and a_{1f} are determined similarly and they would be four times larger if the damping ratio for the foundation soil region is estimated as $\zeta_f = 0.20$. As mentioned earlier, ω_i and ω_j may be taken as the first and fourth natural frequencies of the combined system if five modes are to be included in linear analysis by the procedure of Section 12.14.

The assumption of classical damping may not be appropriate also for structures with special energy dissipating devices (Section 6.8) or on a base isolation system (Chapter 20), even if the structure itself has classical damping. The nonclassical damping matrix for the system is constructed by first evaluating the damping matrix for the structure alone (without the special devices or isolators) from the damping ratios appropriate for the structure, using the procedures of Section 11.4. The damping contributions of the energy dissipating devices or the isolators are then included to obtain the damping matrix of the complete system.

FURTHER READING

Caughey, T. K., "Classical Normal Modes in Damped Linear Dynamic Systems," *Journal of Applied Mechanics, ASME,* **27**, 1960, pp 269–271.

Caughey, T. K., and O'Kelly, M. E. J., "Classical Normal Modes in Damped Linear Dynamic Systems," *Journal of Applied Mechanics, ASME,* **32**, 1965, pp. 583–588.

Foutch, D. A., Housner, G. W., and Jennings, P. C., "Dynamic Responses of Six Multistory Buildings during the San Fernando Earthquake," *Report No. EERL 75-02*, California Institute of Technology, Pasadena, Calif., October 1975.

Hart, G. C., and Vasudevan, R., "Earthquake Design of Buildings: Damping," *Journal of the Structural Division, ASCE,* **101**, 1975, pp. 11–30.

Hashimoto, P. S., Steele, L. K., Johnson, J. J., and Mensing, R. W., "Review of Structure Damping Values for Elastic Seismic Analysis of Nuclear Power Plants," *Report No. NUREG/CR-6011,* U.S. Nuclear Regulatory Commission, Washington, D.C., March 1993.

Jennings, P. C., and Kuroiwa, J. H., "Vibration and Soil–Structure Interaction Tests of a Nine-Story Reinforced Concrete Building," *Bulletin of the Seismological Society of America*, **58**, 1968, pp. 891–916.

McVerry, G. H., "Frequency Domain Identification of Structural Models from Earthquake Records," *Report No. EERL 79-02*, California Institute of Technology, Pasadena, Calif., October 1979.

Newmark, N. M., and Hall, W. J., *Earthquake Spectra and Design*, Earthquake Engineering Research Institute, Berkeley, Calif., 1982, pp. 53–54.

Rayleigh, Lord, *Theory of Sound*, Vol. 1, Dover Publications, New York, 1945.

Wilson, E. L., and Penzien, J., "Evaluation of Orthogonal Damping Matrices," *International Journal for Numerical Methods in Engineering*, **4**, 1972, pp. 5–10.

PROBLEMS

11.1 The properties of a three-story shear building are given in Fig. P11.1. These include the floor weights, story stiffnesses, natural vibration frequencies, and modes. Derive a Rayleigh damping matrix such that the damping ratio is 5% for the first and third modes. Compute the damping ratio for the second mode.

11.2 For the system of Fig. P11.1, using a Caughey series determine the classical damping matrix if the damping ratio is 5% for all three modes.

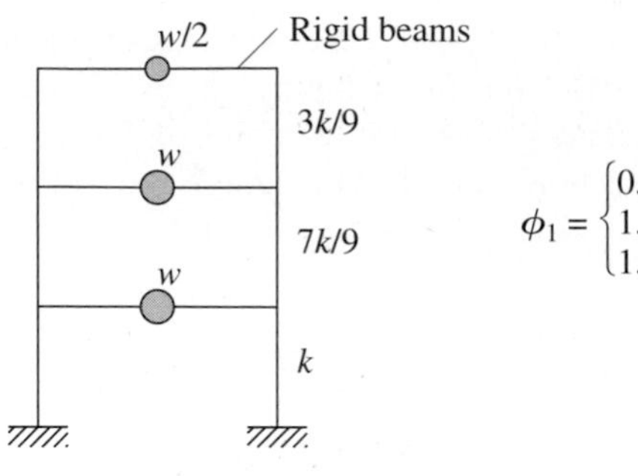

$w = 100$ kips $\quad k = 168$ kips/in.

$\omega_n = 12.01, 25.47, 38.90$ rad/sec

$$\phi_1 = \begin{Bmatrix} 0.6375 \\ 1.2750 \\ 1.9125 \end{Bmatrix}, \quad \phi_2 = \begin{Bmatrix} 0.9827 \\ 0.9829 \\ -1.9642 \end{Bmatrix}, \quad \phi_3 = \begin{Bmatrix} 1.5778 \\ -1.1270 \\ 0.4508 \end{Bmatrix}$$

Figure P11.1

11.3 Determine a damping matrix for the system of Fig. P11.1 by superimposing the damping matrices for the first and third modes, each with $\zeta_n = 5\%$. Verify that the resulting damping matrix gives no damping in the second mode.

11.4 Determine the classical damping matrix for the system of Fig. P11.1 by superimposing the damping matrices for the three modes, each with $\zeta_n = 5\%$.

Dynamic Analysis and Response of Linear Systems

PREVIEW

Now that we have developed the procedures to formulate the equations of motion for MDF systems subjected to dynamic forces (Chapters 9 and 11), we are ready to present the solution of these equations. In Part A of this chapter we show that the equations for a two-DOF system without damping subjected to harmonic forces can be solved analytically. Then we use these results to explain how a vibration absorber or tuned mass damper works to decrease or eliminate unwanted vibration. This simultaneous solution of the coupled equations of motion is not feasible in general, so in Part B we develop the classical modal analysis procedure. The equations of motion are transformed to modal coordinates, leading to an uncoupled set of modal equations; each modal equation is solved to determine the modal contributions to the response, and these modal responses are combined to obtain the total response. An understanding of the relative response contributions of the various modes is developed in Part C with the objective of deciding the number of modes to include in dynamic analysis. The chapter closes with Part D, which includes three analysis procedures useful in special situations; one of these is for systems with nonclassical damping.

PART A: TWO-DEGREE-OF-FREEDOM SYSTEMS

12.1 ANALYSIS OF TWO-DOF SYSTEMS WITHOUT DAMPING

Consider the two-DOF systems shown in Fig. 12.1.1 excited by a harmonic force $p_1(t) = p_o \sin \omega t$ applied to the mass m_1. For both systems the equations of motion are

$$\begin{bmatrix} m_1 & 0 \\ 0 & m_2 \end{bmatrix} \begin{Bmatrix} \ddot{u}_1 \\ \ddot{u}_2 \end{Bmatrix} + \begin{bmatrix} k_1 + k_2 & -k_2 \\ -k_2 & k_2 \end{bmatrix} \begin{Bmatrix} u_1 \\ u_2 \end{Bmatrix} = \begin{Bmatrix} p_o \\ 0 \end{Bmatrix} \sin \omega t \tag{12.1.1}$$

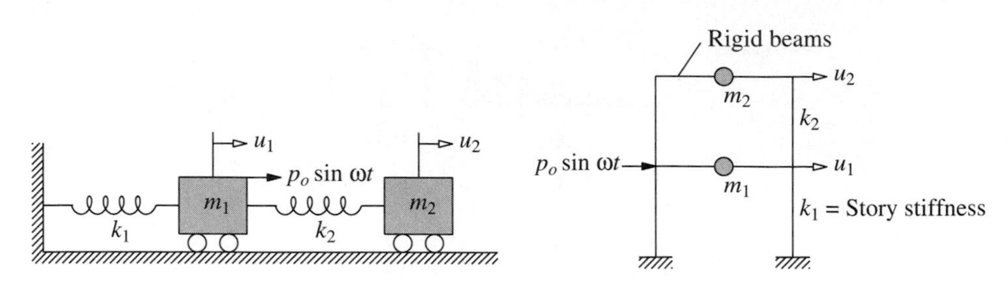

Figure 12.1.1 Two-degree-of-freedom systems.

Observe that the equations are coupled through the stiffness matrix. One equation cannot be solved independent of the other and both equations must be solved simultaneously. Because the system is undamped, the steady-state solution can be assumed as

$$\begin{Bmatrix} u_1 \\ u_2 \end{Bmatrix} = \begin{Bmatrix} u_{1o} \\ u_{2o} \end{Bmatrix} \sin \omega t$$

Substituting this into Eq. (12.1.1), we obtain

$$\begin{bmatrix} k_1 + k_2 - m_1\omega^2 & -k_2 \\ -k_2 & k_2 - m_2\,\omega^2 \end{bmatrix} \begin{Bmatrix} u_{1o} \\ u_{2o} \end{Bmatrix} = \begin{Bmatrix} p_o \\ 0 \end{Bmatrix} \tag{12.1.2}$$

or

$$\left[\mathbf{k} - \omega^2\mathbf{m}\right] \begin{Bmatrix} u_{1o} \\ u_{2o} \end{Bmatrix} = \begin{Bmatrix} p_o \\ 0 \end{Bmatrix}$$

Premultiplying by $[\mathbf{k} - \omega^2\mathbf{m}]^{-1}$ gives

$$\begin{Bmatrix} u_{1o} \\ u_{2o} \end{Bmatrix} = \left[\mathbf{k} - \omega^2\mathbf{m}\right]^{-1} \begin{Bmatrix} p_o \\ 0 \end{Bmatrix} = \frac{1}{\det\left[\mathbf{k} - \omega^2\mathbf{m}\right]} \operatorname{adj}\left[\mathbf{k} - \omega^2\mathbf{m}\right] \begin{Bmatrix} p_o \\ 0 \end{Bmatrix} \tag{12.1.3}$$

where det[·] and adj[·] denote the determinant and adjoint of the matrix[·], respectively. The frequency equation

$$\det[\mathbf{k} - \omega^2\mathbf{m}] = 0$$

can be solved for the natural frequencies ω_1 and ω_2 of the system. In terms of these frequencies, this determinant can be expressed as

$$\det[\mathbf{k} - \omega^2\mathbf{m}] = m_1 m_2(\omega^2 - \omega_1^2)(\omega^2 - \omega_2^2) \tag{12.1.4}$$

Thus Eq. (12.1.3) becomes

$$\begin{Bmatrix} u_{1o} \\ u_{2o} \end{Bmatrix} = \frac{1}{\det[\mathbf{k} - \omega^2\mathbf{m}]} \begin{bmatrix} k_2 - m_2\omega^2 & k_2 \\ k_2 & k_1 + k_2 - m_1\omega^2 \end{bmatrix} \begin{Bmatrix} p_o \\ 0 \end{Bmatrix} \tag{12.1.5}$$

or

$$u_{1o} = \frac{p_o(k_2 - m_2\omega^2)}{m_1 m_2(\omega^2 - \omega_1^2)(\omega^2 - \omega_2^2)} \qquad u_{2o} = \frac{p_o k_2}{m_1 m_2(\omega^2 - \omega_1^2)(\omega^2 - \omega_2^2)} \tag{12.1.6}$$

Example 12.1

Plot the frequency-response curve for the system shown in Fig. 12.1.1 with $m_1 = 2m$, $m_2 = m$, $k_1 = 2k$, and $k_2 = k$ subjected to harmonic force p_o applied on mass m_1.

Solution Substituting the given mass and stiffness values in Eq. (12.1.6) gives

$$u_{1o} = \frac{p_o(k - m\omega^2)}{2m^2(\omega^2 - \omega_1^2)(\omega^2 - \omega_2^2)} \qquad u_{2o} = \frac{p_o k}{2m^2(\omega^2 - \omega_1^2)(\omega^2 - \omega_2^2)} \tag{a}$$

where $\omega_1 = \sqrt{k/2m}$ and $\omega_2 = \sqrt{2k/m}$; these natural frequencies were obtained in Example 10.4. For given system parameters, Eq. (a) provides solutions for the response amplitudes u_{1o} and u_{2o}. It is instructive to rewrite them as

$$\frac{u_{1o}}{p_o/2k} = \frac{1 - \frac{1}{2}(\omega/\omega_1)^2}{\left[1 - (\omega/\omega_1)^2\right]\left[1 - (\omega/\omega_2)^2\right]} \qquad \frac{u_{2o}}{p_o/2k} = \frac{1}{\left[1 - (\omega/\omega_1)^2\right]\left[1 - (\omega/\omega_2)^2\right]} \tag{b}$$

In these equations the response amplitudes have been divided by $p_o/2k$ to obtain normalized or nondimensional responses which depend on frequency ratios ω/ω_1 and ω/ω_2, not separately on ω, ω_1, and ω_2.

Figure E12.1 shows the normalized response amplitudes u_{1o} and u_{2o} plotted against the frequency ratio ω/ω_1. These frequency-response curves show two resonance conditions

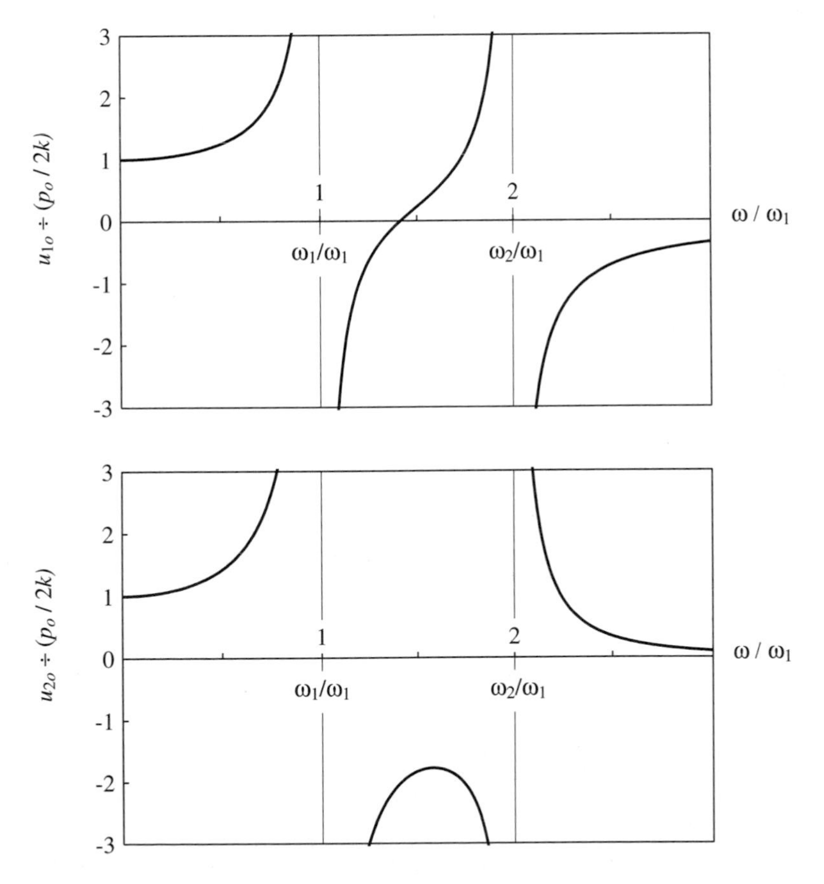

Figure E12.1

at $\omega = \omega_1$ and $\omega = \omega_2$; at these exciting frequencies the steady-state response is unbounded. At other exciting frequencies, the vibration is finite and could be calculated from Eq. (b). Note that there is an exciting frequency where the vibration of the first mass, where the exciting force is applied, is reduced to zero. This is the entire basis of the *dynamic vibration absorber* or *tuned mass damper* discussed next.

12.2 VIBRATION ABSORBER OR TUNED MASS DAMPER

The vibration absorber is a mechanical device used to decrease or eliminate unwanted vibration. The description *tuned mass damper* is often used in modern installation; this modern name has the advantage of showing its relationship to other types of dampers. Its greatest application is to synchronous machinery, operating at nearly constant frequency, for the vibration absorber is tuned to one particular frequency and is only effective over a narrow band of frequencies. However, absorbers are also used in situations where the excitation is not nearly harmonic. The dumbbell-shaped devices that hang from highest-voltage transmission lines are vibration absorbers used to mitigate the fatiguing effects of wind-induced vibration. Vibration absorbers have also been used to reduce the wind-induced vibration of tall buildings when the motions have reached annoying levels for the occupants. An example of this is the 59-story Citicorp Center in midtown Manhattan; completed in 1977, this building has a 820-kip block of concrete installed on the 59th floor in a movable platform connected to the building by large hydraulic arms. When the building sways more than 1 foot a second, the computer directs the arms to move the block in the other direction. This action reduces such oscillation by 40%, considerably easing the discomfort of the building's occupants during high winds.

In the brief presentation that follows, we restrict ourselves to the basic principle of a vibration absorber without getting into the many important aspects of its practical design. In its simplest form, a vibration absorber consists of one spring and a mass. Such an absorber system is attached to a SDF system, as shown in Fig. 12.2.1a. The equations of motion for the main mass m_1 and the absorber mass m_2 are the same as Eq. (12.1.1). For harmonic force applied to the main mass we already have the solution given by Eq. (12.1.6). Introducing the notation

$$\omega_1^* = \sqrt{\frac{k_1}{m_1}} \qquad \omega_2^* = \sqrt{\frac{k_2}{m_2}} \qquad \mu = \frac{m_2}{m_1} \tag{12.2.1}$$

the available solution can be rewritten as

$$u_{1o} = \frac{p_o}{k_1} \frac{1 - (\omega/\omega_2^*)^2}{\left[1 + \mu\,(\omega_2^*/\omega_1^*)^2 - (\omega/\omega_1^*)^2\right]\left[1 - (\omega/\omega_2^*)^2\right] - \mu\,(\omega_2^*/\omega_1^*)^2} \tag{12.2.2a}$$

$$u_{2o} = \frac{p_o}{k_1} \frac{1}{\left[1 + \mu\,(\omega_2^*/\omega_1^*)^2 - (\omega/\omega_1^*)^2\right]\left[1 - (\omega/\omega_2^*)^2\right] - \mu\,(\omega_2^*/\omega_1^*)^2} \tag{12.2.2b}$$

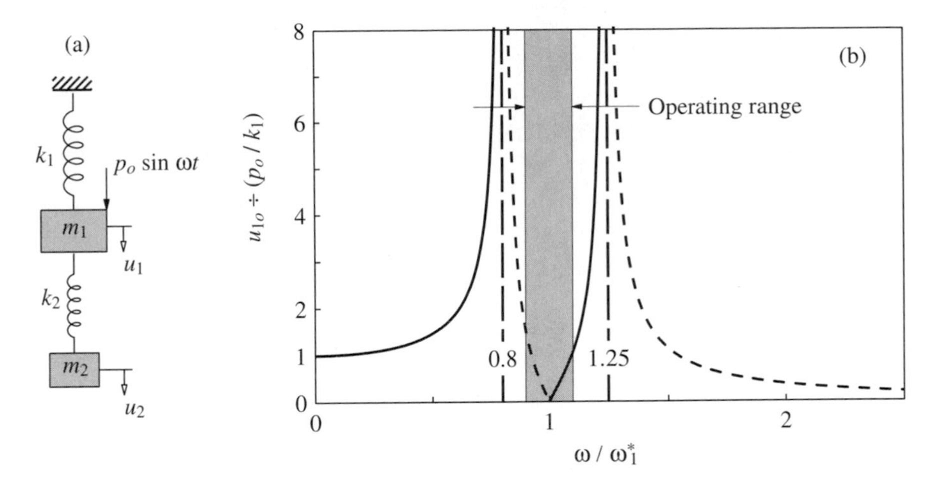

Figure 12.2.1 (a) Vibration absorber attached to an SDF system; (b) response amplitude versus exciting frequency (dashed curve indicates negative u_{1o} or phase opposite to excitation); $\mu = 0.2$ and $\omega_1^* = \omega_2^*$.

At $\omega = \omega_2^*$, Eq. (12.2.2a) indicates that the motion of the main mass m_1 does not simply diminish, it ceases altogether. Figure 12.2.1b shows a plot of response amplitude u_{1o} versus exciting frequency ω; for this example, the mass ratio $\mu = 0.2$ and $\omega_1^* = \omega_2^*$, the absorber being tuned to the natural frequency of the main system. The operating frequency range where $u_{1o} \div (p_o/k_1) < 1$ is shown. Because the system has two DOFs, two resonant frequencies exist, and the response is unbounded at those frequencies.

The usefulness of the vibration absorber becomes obvious if we compare the frequency-response function of Fig. 12.2.1b with the response of the main mass alone, without the absorber mass. At $\omega = \omega_1^*$ the response amplitude of the main mass alone is unbounded but is zero with the presence of the absorber mass. Thus if the exciting frequency ω is close to the natural frequency ω_1^* of the main system, and operating restrictions make it impossible to vary either one, the vibration absorber can be used to reduce the response amplitude of the main system to near zero.

What should be the size of the absorber mass? To answer this question, we use Eq. (12.2.2b) to determine the motion of the absorber mass at $\omega = \omega_2^*$:

$$u_{2o} = -\frac{p_o}{k_2} \tag{12.2.3}$$

The force acting on the absorber mass is

$$k_2 u_{2o} = \omega^2 m_2 u_{2o} = -p_o \tag{12.2.4}$$

This implies that the absorber system exerts a force equal and opposite to the exciting force. Thus the size of the absorber stiffness and mass, k_2 and m_2, depends on the allowable value of u_{2o}. There are other factors that affect the choice of the absorber mass. Obviously, a large absorber mass presents a practical problem. At the same time the smaller the mass ratio μ, the narrower will be the operating frequency range of the absorber.

PART B: MODAL ANALYSIS

12.3 MODAL EQUATIONS FOR UNDAMPED SYSTEMS

The equations of motion for a linear MDF system without damping were derived in Chapter 9 and are repeated here:

$$\mathbf{m}\ddot{\mathbf{u}} + \mathbf{k}\mathbf{u} = \mathbf{p}(t) \tag{12.3.1}$$

The simultaneous solution of these coupled equations of motion that we have illustrated in Section 12.1 for a two-DOF system subjected to harmonic excitation is not efficient for systems with more DOF, nor is it feasible for systems excited by other types of forces. Consequently, it is advantageous to transform these equations to modal coordinates, as we shall see next.

As mentioned in Section 10.7, the displacement $\mathbf{u}$ of an MDF system can be expanded in terms of modal contributions. Thus the dynamic response of a system can be expressed as

$$\mathbf{u}(t) = \sum_{r=1}^{N} \phi_r q_r(t) = \mathbf{\Phi}\mathbf{q}(t) \tag{12.3.2}$$

Using this equation, the coupled equations (12.3.1) in $u_j(t)$ can be transformed to a set of uncoupled equations with modal coordinates $q_n(t)$ as the unknowns. Substituting Eq. (12.3.2) in Eq. (12.3.1) gives

$$\sum_{r=1}^{N} \mathbf{m}\,\phi_r \ddot{q}_r(t) + \sum_{r=1}^{N} \mathbf{k}\,\phi_r q_r(t) = \mathbf{p}(t)$$

Premultiplying each term in this equation by ϕ_n^T gives

$$\sum_{r=1}^{N} \phi_n^T\,\mathbf{m}\,\phi_r\,\ddot{q}_r(t) + \sum_{r=1}^{N} \phi_n^T\,\mathbf{k}\,\phi_r\,q_r(t) = \phi_n^T\,\mathbf{p}(t)$$

Because of the orthogonality relations of Eq. (10.4.1), all terms in each of the summations vanish, except the $r = n$ term, reducing this equation to

$$(\phi_n^T\,\mathbf{m}\,\phi_n)\,\ddot{q}_n(t) + (\phi_n^T\,\mathbf{k}\,\phi_n)\,q_n(t) = \phi_n^T\,\mathbf{p}(t)$$

or

$$M_n\,\ddot{q}_n(t) + K_n\,q_n(t) = P_n(t) \tag{12.3.3}$$

where

$$M_n = \phi_n^T\,\mathbf{m}\,\phi_n \qquad K_n = \phi_n^T\,\mathbf{k}\,\phi_n \qquad P_n(t) = \phi_n^T\,\mathbf{p}(t) \tag{12.3.4}$$

Equation (12.3.3) may be interpreted as the equation governing the response $q_n(t)$ of the SDF system shown in Fig. 12.3.1 with mass M_n, stiffness K_n, and exciting force $P_n(t)$. Therefore, M_n is called the *generalized mass* for the nth natural mode, K_n the

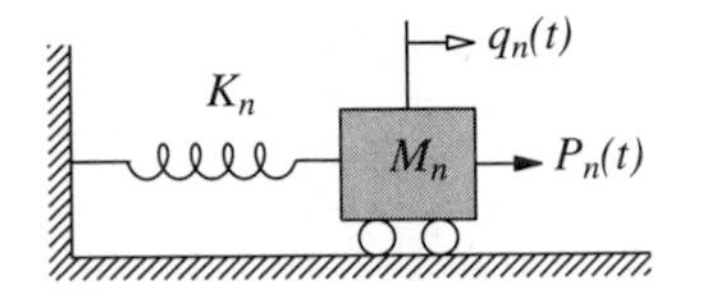

Figure 12.3.1 Generalized SDF system for the nth natural mode.

generalized stiffness for the nth mode, and $P_n(t)$ the *generalized force* for the nth mode. These parameters depend only on the nth-mode ϕ_n. Thus if we know only the nth mode, we can write the equation for q_n and solve it without even knowing the other modes. Dividing by M_n and using Eq. (10.4.7), Eq. (12.3.3) can be rewritten as

$$\ddot{q}_n + \omega_n^2 q_n = \frac{P_n(t)}{M_n} \tag{12.3.5}$$

Equation (12.3.3) or (12.3.5) governs the nth modal coordinate $q_n(t)$, the only unknown in the equation, and there are N such equations, one for each mode. Thus the set of N coupled differential equations (12.3.1) in nodal displacements $u_j(t)$—$j = 1, 2, \ldots, N$—has been transformed to the set of N uncoupled equations (12.3.3) in modal coordinates $q_n(t)$—$n = 1, 2, \ldots, N$. Written in matrix form the latter set of equations is

$$\mathbf{M}\ddot{\mathbf{q}} + \mathbf{K}\mathbf{q} = \mathbf{P}(t) \tag{12.3.6}$$

where $\mathbf{M}$ is a diagonal matrix of the generalized modal masses M_n, $\mathbf{K}$ is a diagonal matrix of the generalized modal stiffnesses K_n, and $\mathbf{P}(t)$ is a column vector of the generalized modal forces $P_n(t)$. Recall that $\mathbf{M}$ and $\mathbf{K}$ had been introduced in Section 10.4.

Example 12.2

Consider the systems and excitation of Example 12.1. By modal analysis determine the steady-state response of the system.

Solution The natural vibration frequencies and modes of this system were determined in Example 10.4, from which the generalized masses and stiffnesses are calculated using Eq. (12.3.4). These results are summarized next:

$$\omega_1 = \sqrt{\frac{k}{2m}} \qquad \omega_2 = \sqrt{\frac{2k}{m}}$$

$$\phi_1 = \langle \tfrac{1}{2} \quad 1 \rangle^T \qquad \phi_2 = \langle -1 \quad 1 \rangle^T$$

$$M_1 = \frac{3m}{2} \qquad M_2 = 3m$$

$$K_1 = \frac{3k}{4} \qquad K_2 = 6k$$

1. *Compute the generalized forces.*

$$P_1(t) = \phi_1^T \mathbf{p}(t) = \underbrace{(p_o/2)}_{P_{1o}} \sin \omega t \qquad P_2(t) = \phi_2^T \mathbf{p}(t) = \underbrace{-p_o}_{P_{2o}} \sin \omega t \tag{a}$$

2. *Set up the modal equations.*

$$M_n \ddot{q}_n + K_n q_n = P_{no} \sin \omega t \tag{b}$$

3. *Solve the modal equations.* To solve Eq. (b) we draw upon the solution presented in Eq. (3.1.7) for a SDF system subjected to harmonic force. The governing equation is

$$m\ddot{u} + ku = p_o \sin \omega t \tag{c}$$

and its steady-state solution is

$$u(t) = \frac{p_o}{k} C \sin \omega t \qquad C = \frac{1}{1 - (\omega/\omega_n)^2} \tag{d}$$

where $\omega_n = \sqrt{k/m}$. Comparing Eqs. (c) and (b), the solution for Eq. (b) is

$$q_n(t) = \frac{P_{no}}{K_n} C_n \sin \omega t \tag{e}$$

where C_n is given by Eq. (d) with ω_n interpreted as the natural frequency of the nth mode. Substituting for P_{no} and K_n for $n = 1$ and 2 gives

$$q_1(t) = \frac{2p_o}{3k} C_1 \sin \omega t \qquad q_2(t) = -\frac{p_o}{6k} C_2 \sin \omega t \tag{f}$$

4. *Determine the modal responses.* The nth mode contribution to displacements—from Eq. (12.3.2)—is $\mathbf{u}_n(t) = \phi_n q_n(t)$. Substituting Eq. (f) gives the displacement response due to the two modes:

$$\mathbf{u}_1(t) = \phi_1 \frac{2p_o}{3k} C_1 \sin \omega t \qquad \mathbf{u}_2(t) = \phi_2 \frac{-p_o}{6k} C_2 \sin \omega t \tag{g}$$

5. *Combine the modal responses.*

$$\mathbf{u}(t) = \mathbf{u}_1(t) + \mathbf{u}_2(t) \quad \text{or} \quad u_j(t) = u_{j1}(t) + u_{j2}(t) \qquad j = 1, 2 \tag{h}$$

Substituting Eq. (g) and for ϕ_1 and ϕ_2 gives

$$u_1(t) = \frac{p_o}{6k} (2C_1 + C_2) \sin \omega t \qquad u_2(t) = \frac{p_o}{6k} (4C_1 - C_2) \sin \omega t \tag{i}$$

These results are equivalent to those obtained in Example 12.1 by solving the coupled equations (12.3.1) of motion.

12.4 MODAL EQUATIONS FOR DAMPED SYSTEMS

When damping is included, the equations of motion for an MDF system are

$$\mathbf{m}\ddot{\mathbf{u}} + \mathbf{c}\dot{\mathbf{u}} + \mathbf{k}\mathbf{u} = \mathbf{p}(t) \tag{12.4.1}$$

Using the transformation of Eq. (12.3.2), where ϕ_r are the natural modes of the system without damping, these equations can be written in terms of the modal coordinates. Unlike the case of undamped systems (Section 12.3), these modal equations may be coupled through the damping terms. However, for certain forms of damping that are reasonable idealizations for many structures, the equations become uncoupled, just as for undamped systems. We shall demonstrate this next.

Substituting Eq. (12.3.2) in Eq. (12.4.1) gives

$$\sum_{r=1}^{N} \mathbf{m}\,\phi_r\, \ddot{q}_r + \sum_{r=1}^{N} \mathbf{c}\,\phi_r \dot{q}_r(t) + \sum_{r=1}^{N} \mathbf{k}\,\phi_r\, q_r = \mathbf{p}(t)$$

Premultiplying each term in this equation by ϕ_n^T gives

$$\sum_{r=1}^{N} \phi_n^T \mathbf{m} \phi_r \ddot{q}_r + \sum_{r=1}^{N} \phi_n^T \mathbf{c} \phi_r \dot{q}_r(t) + \sum_{r=1}^{N} \phi_n^T \mathbf{k} \phi_r q_r = \phi_n^T \mathbf{p}(t)$$

which can be rewritten as

$$M_n \ddot{q}_n + \sum_{r=1}^{N} C_{nr} \dot{q}_r + K_n q_n = P_n(t) \tag{12.4.2}$$

where M_n, K_n, and $P_n(t)$ were defined in Eq. (12.3.4) and

$$C_{nr} = \phi_n^T \mathbf{c} \phi_r \tag{12.4.3}$$

Equation (12.4.2) exists for each $n = 1$ to N and the set of N equations can be written in matrix form:

$$\mathbf{M}\ddot{\mathbf{q}} + \mathbf{C}\dot{\mathbf{q}} + \mathbf{K}\mathbf{q} = \mathbf{P}(t) \tag{12.4.4}$$

where **M**, **K**, and $\mathbf{P}(t)$ were introduced in Eq. (12.3.6) and **C** is a nondiagonal matrix of coefficients C_{nr}. These N equations in modal coordinates $q_n(t)$ are coupled through the damping terms because Eq. (12.4.2) contains more than one modal velocity.

The modal equations will be uncoupled if the system has classical damping. For such systems, as defined in Section 10.9, $C_{nr} = 0$ if $n \neq r$ and Eq. (12.4.2) reduces to

$$M_n \ddot{q}_n + C_n \dot{q}_n + K_n q_n = P_n(t) \tag{12.4.5}$$

where the generalized damping C_n is defined by Eq. (10.9.10). This equation governs the response of the SDF system shown in Fig. 12.4.1. Dividing Eq. (12.4.5) by M_n gives

$$\ddot{q}_n + 2\zeta_n \omega_n \dot{q}_n + \omega_n^2 q_n = \frac{P_n(t)}{M_n} \tag{12.4.6}$$

where ζ_n is the damping ratio for the nth mode. The damping ratio is usually not computed using Eq. (10.9.11) but is estimated based on experimental data for structures similar to the one being analyzed (Chapter 11). Equation (12.4.5) governs the nth modal coordinate $q_n(t)$, and the parameters M_n, K_n, C_n, and $P_n(t)$ depend only on the nth-mode ϕ_n, not on other modes. Thus we have N uncoupled equations like Eq. (12.4.5), one for each natural mode. In summary, the set of N coupled differential equations (12.4.1) in nodal displacements $u_j(t)$ has been transformed to the set of N uncoupled equations (12.4.5) in modal coordinates $q_n(t)$.

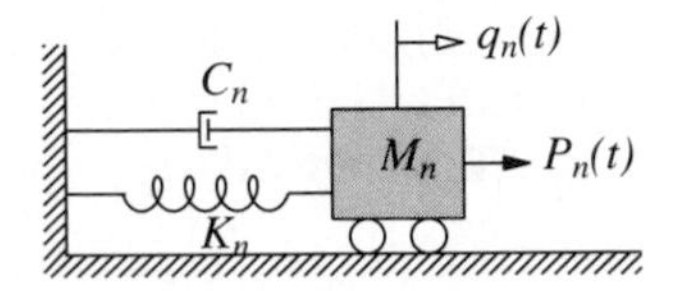

Figure 12.4.1 Generalized SDF system for the nth natural mode.

12.5 DISPLACEMENT RESPONSE

For given external dynamic forces defined by $\mathbf{p}(t)$, the dynamic response of an MDF system can be determined by solving Eq. (12.4.5) or (12.4.6) for the modal coordinate $q_n(t)$. Each modal equation is of the same form as the equation of motion for an SDF system. Thus the solution methods and results available for SDF systems (Chapters 3 to 5) can be adapted to obtain solutions $q_n(t)$ for the modal equations. Once the modal coordinates $q_n(t)$ have been determined, Eq. (12.3.2) indicates that the contribution of the nth mode to the displacement $\mathbf{u}(t)$ is

$$\mathbf{u}_n(t) = \phi_n \, q_n(t) \tag{12.5.1}$$

and combining these modal contributions gives the total displacement:

$$\mathbf{u}(t) = \sum_{n=1}^{N} \mathbf{u}_n(t) = \sum_{n=1}^{N} \phi_n \, q_n(t) \tag{12.5.2}$$

This procedure is known as *classical modal analysis* or the *classical mode superposition method* because individual (uncoupled) modal equations are solved to determine the modal coordinates $q_n(t)$ and the modal responses $\mathbf{u}_n(t)$, and the latter are combined to obtain the total response $\mathbf{u}(t)$. More precisely, this method is called the *classical mode displacement superposition method* because modal displacements are superposed. For brevity we usually refer to this procedure as *modal analysis*. This analysis method is restricted to linear systems with classical damping. Linearity of the system is implicit in using the principle of superposition, Eq. (12.3.2). Damping must be of the classical form in order to obtain modal equations that are uncoupled, a central feature of modal analysis.

12.6 ELEMENT FORCES

Two procedures described in Section 9.10 are available to determine the forces in various elements—beams, columns, walls, etc.—of the structure at time instant t from the displacement $\mathbf{u}(t)$ at the same time instant. In modal analysis it is instructive to determine the contributions of the individual modes to the element forces. In the first procedure, the nth-mode contribution $r_n(t)$ to an element force $r(t)$ is determined from modal displacements $\mathbf{u}_n(t)$ using element stiffness properties (Appendix 1). Then the element force considering contributions of all modes is

$$r(t) = \sum_{n=1}^{N} r_n(t) \tag{12.6.1}$$

In the second procedure, the equivalent static forces associated with the nth-mode response are defined using Eq. (9.10.1) with subscript s deleted: $\mathbf{f}_n(t) = \mathbf{k}\mathbf{u}_n(t)$. Substituting Eq. (12.5.1) and using Eq. (10.2.4) gives

$$\mathbf{f}_n(t) = \omega_n^2 \mathbf{m} \phi_n q_n(t) \tag{12.6.2}$$

Static analysis of the structure subjected to these external forces at each time instant gives the element force $r_n(t)$. Then the total force $r(t)$ is given by Eq. (12.6.1).

12.7 MODAL ANALYSIS: SUMMARY

The dynamic response of an MDF system to external forces $\mathbf{p}(t)$ can be computed by modal analysis, summarized next as a sequence of steps:

1. Define the structural properties.
   a. Determine the mass matrix $\mathbf{m}$ and stiffness matrix $\mathbf{k}$ (Chapter 9).
   b. Estimate the modal damping ratios ζ_n (Chapter 11).
2. Determine the natural frequencies ω_n and modes ϕ_n (Chapter 10).
3. Compute the response in each mode by the following steps:
   a. Set up Eq. (12.4.5) or (12.4.6) and solve for $q_n(t)$.
   b. Compute the nodal displacements $\mathbf{u}_n(t)$ from Eq. (12.5.1).
   c. Compute the element forces associated with the nodal displacements $\mathbf{u}_n(t)$ by implementing one of the two methods described in Section 12.6 for the desired values of t and the element forces of interest.
4. Combine the contributions of all the modes to determine the total response. In particular, the nodal displacements $\mathbf{u}(t)$ are given by Eq. (12.5.2) and element forces by Eq. (12.6.1).

Example 12.3

Consider the systems and excitation of Example 12.1. Determine the spring forces $V_j(t)$ for the system of Fig. 12.1.1a, or story shears $V_j(t)$ in the system of Fig. 12.1.1b, without introducing equivalent static forces. Consider only the steady-state response.

Solution Steps 1, 2, 3a, and 3b of the analysis summary of Section 12.7 have already been completed in Example 12.2.

Step 3c: The spring forces in the system of Fig. 12.1.1a or the story shears in the system of Fig. 12.1.1b are

$$V_{1n}(t) = k_1 u_{1n}(t) = k_1 \phi_{1n} q_n(t) \tag{a}$$

$$V_{2n}(t) = k_2 [u_{2n}(t) - u_{1n}(t)] = k_2 (\phi_{2n} - \phi_{1n}) q_n(t) \tag{b}$$

Substituting Eq. (f) of Example 12.2 in Eqs. (a) and (b) with $n = 1$, $k_1 = 2k$, $k_2 = k$, $\phi_{11} = \frac{1}{2}$, and $\phi_{21} = 1$ gives the forces due to the first mode:

$$V_{11}(t) = \frac{2p_o}{3} C_1 \sin \omega t \qquad V_{21}(t) = \frac{p_o}{3} C_1 \sin \omega t \tag{c}$$

Substituting Eq. (f) of Example 12.2 in Eqs. (a) and (b) with $n = 2$, $\phi_{12} = -1$, and $\phi_{22} = 1$ gives the second-mode forces:

$$V_{12}(t) = \frac{p_o}{3} C_2 \sin \omega t \qquad V_{22}(t) = -\frac{p_o}{3} C_2 \sin \omega t \tag{d}$$

Step 4b: Substituting Eqs. (c) and (d) in $V_j(t) = V_{j1}(t) + V_{j2}(t)$ gives

$$V_1(t) = \frac{p_o}{3}\,(2C_1 + C_2)\ \sin\,\omega t \qquad V_2(t) = \frac{p_o}{3}\,(C_1 - C_2)\ \sin\,\omega t \tag{e}$$

Equation (e) gives the time variation of spring forces and story shears. For a given p_o and ω and the ω_n already determined, all quantities on the right side of these equations are known; thus $V_j(t)$ can be computed.

Example 12.4

Repeat Example 12.3 using equivalent static forces.

Solution From Eq. (12.6.2), for a lumped mass system the equivalent static force in the jth DOF due to the nth mode is

$$f_{jn}(t) = \omega_n^2\, m_j\, \phi_{jn}\, q_n(t) \tag{a}$$

Step 3c: In Eq. (a) with $n = 1$, substitute $m_1 = 2m$, $m_2 = m$, $\phi_{11} = \frac{1}{2}$, $\phi_{21} = 1$, $\omega_1^2 = k/2m$, and $q_1(t)$ from Eq. (f) of Example 12.2 to obtain

$$f_{11}(t) = \frac{p_o}{3}\, C_1\ \sin\,\omega t \qquad f_{21}(t) = \frac{p_o}{3}\, C_1\ \sin\,\omega t \tag{b}$$

In Eq. (a) with $n = 2$, substituting $m_1 = 2m$, $m_2 = m$, $\phi_{12} = -1$, $\phi_{22} = 1$, $\omega_1^2 = 2k/m$, and $q_2(t)$ from Eq. (f) of Example 12.2 gives

$$f_{12}(t) = \frac{2p_o}{3}\, C_2\ \sin\,\omega t \qquad f_{22}(t) = -\frac{p_o}{3}\, C_2\ \sin\,\omega t \tag{c}$$

Static analysis of the systems of Fig. E12.4 subjected to forces $f_{jn}(t)$ gives the two spring forces and story shears due to the nth mode:

$$V_{1n}(t) = f_{1n}(t) + f_{2n}(t) \qquad V_{2n}(t) = f_{2n}(t) \tag{d}$$

Substituting Eq. (b) in Eq. (d) with $n = 1$ gives the first mode forces that are identical to Eq. (c) of Example 12.3. Similarly, substituting Eq. (c) in Eq. (d) with $n = 2$ gives the second-mode results that are identical to Eq. (d) of Example 12.3.

Step 4: Proceed as in step 4b of Example 12.3.

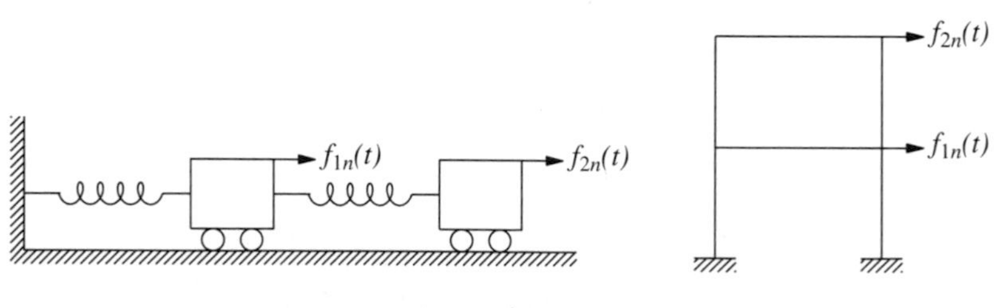

Figure E12.4

Example 12.5

Consider the system and excitation of Example 12.1 with modal damping ratios ζ_n. Determine the steady-state displacement amplitudes of the system.

Solution Steps 1 and 2 of the analysis summary have been completed in Example 12.2.

Step 3: The modal equations without damping were developed in Example 12.2. Now including damping they become

$$M_n \ddot{q}_n + C_n \dot{q}_n + K_n q_n = P_{no} \sin \omega t \tag{a}$$

where M_n, K_n, and P_{no} are available and C_n is known in terms of ζ_n.

To solve Eq. (a), we draw upon the solution presented in Eq. (3.2.3) for an SDF system with damping subjected to harmonic force. The governing equation is

$$m\ddot{u} + c\dot{u} + ku = p_o \sin \omega t \tag{b}$$

and its steady-state solution is

$$u(t) = \frac{p_o}{k} (\mathcal{C} \sin \omega t + \mathcal{D} \cos \omega t) \tag{c}$$

with

$$\mathcal{C} = \frac{1 - (\omega/\omega_n)^2}{[1 - (\omega/\omega_n)^2]^2 + (2\zeta \, \omega/\omega_n)^2} \qquad \mathcal{D} = \frac{-2\zeta \, \omega/\omega_n}{[1 - (\omega/\omega_n)^2]^2 + (2\zeta \, \omega/\omega_n)^2} \tag{d}$$

where $\omega_n = \sqrt{k/m}$ and $\zeta = c/2m\omega_n$.

Comparing Eqs. (b) and (a), the solution for the latter is

$$q_n(t) = \frac{P_{no}}{K_n} (\mathcal{C}_n \sin \omega t + \mathcal{D}_n \cos \omega t) \tag{e}$$

where $\mathcal{C}_n$ and $\mathcal{D}_n$ are given by Eq. (d) with ω_n interpreted as the natural frequency of the nth mode and $\zeta = \zeta_n$, the damping ratio for the nth mode. Substituting for P_{no} and K_n for $n = 1$ and 2 gives

$$q_1(t) = \frac{2p_o}{3k} (\mathcal{C}_1 \sin \omega t + \mathcal{D}_1 \cos \omega t) \tag{f}$$

$$q_2(t) = -\frac{p_o}{6k} (\mathcal{C}_2 \sin \omega t + \mathcal{D}_2 \cos \omega t) \tag{g}$$

Steps 3b and 4: Substituting ϕ_n in Eqs. (12.5.2) gives the nodal displacements:

$$u_1(t) = \tfrac{1}{2} q_1(t) - q_2(t) \qquad u_2(t) = q_1(t) + q_2(t)$$

Substituting Eqs. (f) and (g) for $q_n(t)$ gives

$$u_1(t) = \frac{p_o}{6k} [(2\mathcal{C}_1 + \mathcal{C}_2) \sin \omega t + (2\mathcal{D}_1 + \mathcal{D}_2) \cos \omega t] \tag{h}$$

$$u_2(t) = \frac{p_o}{6k} [(4\mathcal{C}_1 - \mathcal{C}_2) \sin \omega t + (4\mathcal{D}_1 - \mathcal{D}_2) \cos \omega t] \tag{i}$$

The displacement amplitudes are

$$u_{1o} = \frac{p_o}{6k} \sqrt{(2\mathcal{C}_1 + \mathcal{C}_2)^2 + (2\mathcal{D}_1 + \mathcal{D}_2)^2} \tag{j}$$

$$u_{2o} = \frac{p_o}{6k} \sqrt{(4\mathcal{C}_1 - \mathcal{C}_2)^2 + (4\mathcal{D}_1 - \mathcal{D}_2)^2} \tag{k}$$

These u_{jo} can be computed when the amplitude p_o and frequency ω of the exciting force are known together with system properties k, ω_n, and ζ_n.

It can be shown that Eqs. (h) and (i), specialized for $\zeta_n = 0$, are identical to the results for the system without damping obtained in Example 12.2.

Example 12.6

The dynamic response of the system of Fig. E12.6a to the excitation shown in Fig. E12.6b is desired. Determine **(a)** displacements $u_1(t)$ and $u_2(t)$; **(b)** bending moments and shears at sections a, b, c, and d as functions of time; **(c)** the shearing force and bending moment diagrams at $t = 0.18$ sec. The system and excitation parameters are $E = 29{,}000$ ksi, $I = 100$ in^4, $L = 120$ in., $mL = 0.1672$ kip-sec^2/in., and $p_o = 5$ kips. Neglect damping.

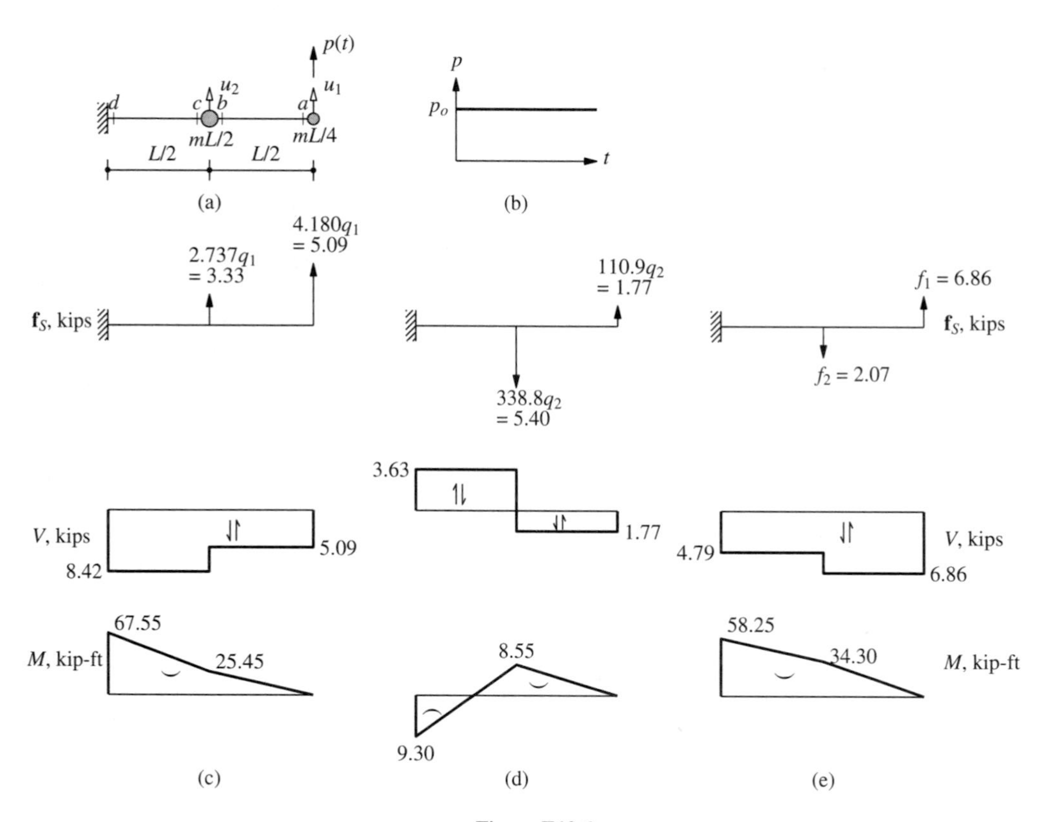

Figure E12.6

Solution The mass and stiffness matrices are available from Example 9.5. The natural frequencies and modes of this system were determined in Example 10.2. They are $\omega_1 = 3.156\sqrt{EI/mL^4}$ and $\omega_2 = 16.258\sqrt{EI/mL^4}$; $\phi_1 = \langle 1 \quad 0.3274 \rangle^T$ and $\phi_2 = \langle 1 \quad -1.5274 \rangle^T$. Substituting for E, I, m, and L gives $\omega_1 = 10.00$ and $\omega_2 = 51.51$ rad/sec.

1. *Set up the modal equations.*

$$M_1 = \phi_1^T \mathbf{m} \phi_1 = 0.0507 \qquad M_2 = \phi_2^T \mathbf{m} \phi_2 = 0.2368 \text{ kip-sec}^2\text{/in.}$$

$$P_1(t) = \phi_1^T \begin{Bmatrix} p_o \\ o \end{Bmatrix} = 5 \qquad P_2(t) = \phi_2^T \begin{Bmatrix} p_o \\ o \end{Bmatrix} = 5 \text{ kips}$$

The modal equations (12.4.6) are

$$\ddot{q}_1 + 10^2 q_1 = \frac{5}{0.0507} = 98.62 \qquad \ddot{q}_2 + (51.51)^2 q_2 = \frac{5}{0.2368} = 21.12 \tag{a}$$

2. *Solve the modal equations.* Adapting the SDF system result, Eq. (4.3.2), to Eq. (a) gives

$$\begin{aligned} q_1(t) &= \frac{98.62}{10^2}(1 - \cos 10t) = 0.986(1 - \cos 10t) \\ q_2(t) &= \frac{21.12}{(51.61)^2}(1 - \cos 51.51t) = 0.008(1 - \cos 51.51t) \end{aligned} \tag{b}$$

3. *Determine the displacement response.* Substituting for ϕ_1, ϕ_2, $q_1(t)$, and $q_2(t)$ in Eq. (12.5.2) gives

$$\begin{aligned} u_1(t) &= 0.994 - 0.986 \cos 10t - 0.008 \cos 51.51t \\ u_2(t) &= 0.311 - 0.323 \cos 10t + 0.012 \cos 51.51t \end{aligned} \tag{c}$$

4. *Determine the equivalent static forces.* Substituting for ω_1^2, $\mathbf{m}$, and ϕ_1 in Eq. (12.6.2) gives the forces shown in Fig. E12.6c:

$$\mathbf{f}_1(t) = \begin{bmatrix} f_1(t) \\ f_2(t) \end{bmatrix}_1 = 10^2 \begin{bmatrix} 0.0418 & \\ & 0.0836 \end{bmatrix} \begin{Bmatrix} 1 \\ 0.3274 \end{Bmatrix} q_1(t) = \begin{Bmatrix} 4.180 \\ 2.737 \end{Bmatrix} q_1(t) \tag{d}$$

Similarly substituting ω_2^2, $\mathbf{m}$, and ϕ_2 gives the forces shown in Fig. E12.6d:

$$\mathbf{f}_2(t) = \begin{bmatrix} f_1(t) \\ f_1(t) \end{bmatrix}_2 = \begin{bmatrix} 110.9 \\ -338.8 \end{bmatrix} q_2(t) \tag{e}$$

The combined forces (Fig. E12.6e) are

$$f_1(t) = 4.180\, q_1(t) + 110.9\, q_2(t) \qquad f_2(t) = 2.737\, q_1(t) - 338.8\, q_2(t) \tag{f}$$

5. *Determine the internal forces.* Static analysis of the cantilever beam of Fig. E12.1e gives the shearing forces and bending moments at the various sections a, b, c, and d:

$$V_a(t) = V_b(t) = f_1(t) \qquad V_c(t) = V_d(t) = f_1(t) + f_2(t) \tag{g}$$

$$M_a(t) = 0 \qquad M_b(t) = \frac{L}{2} f_1(t) \qquad M_d(t) = L f_1(t) + \frac{L}{2} f_2(t) \tag{h}$$

where $f_1(t)$ and $f_2(t)$ are known from Eqs. (f) and (b).

6. *Determine the internal forces at $t = 0.18$ sec.* At $t = 0.18$ sec, from Eq. (b), $q_1 = 1.217$ in. and $q_2 = 0.0159$ in. Substituting these in Eqs. (d) and (e) gives numerical values for the equivalent static forces shown in Fig. E12.6c and d, wherein the shearing forces and bending moments due to each mode are plotted. The combined values of these element forces are shown in Fig. E12.6e.

PART C: MODAL RESPONSE CONTRIBUTIONS

12.8 MODAL EXPANSION OF EXCITATION VECTOR $\mathbf{p}(t) = \mathbf{s}p(t)$

We now consider a common loading case in which the applied forces $p_j(t)$ have the same time variation $p(t)$, and their spatial distribution is defined by $\mathbf{s}$, independent of time. Thus

$$\mathbf{p}(t) = \mathbf{s}p(t) \tag{12.8.1}$$

We will find it instructive to expand the vector $\mathbf{s}$ as

$$\mathbf{s} = \sum_{r=1}^{N} \mathbf{s}_r = \sum_{r=1}^{N} \Gamma_r \, \mathbf{m} \, \boldsymbol{\phi}_r \tag{12.8.2}$$

Premultiplying both sides of Eq. (12.8.2) by $\boldsymbol{\phi}_n^T$ and utilizing the orthogonality property of modes gives

$$\Gamma_n = \frac{\boldsymbol{\phi}_n^T \mathbf{s}}{M_n} \tag{12.8.3}$$

The contribution of the nth mode to the excitation vector $\mathbf{s}$ is

$$\mathbf{s}_n = \Gamma_n \mathbf{m} \boldsymbol{\phi}_n \tag{12.8.4}$$

which is independent of how the modes are normalized. This should be clear from the structure of Eqs. (12.8.3) and (12.8.4).

Equation (12.8.2) may be viewed as an expansion of the applied force distribution $\mathbf{s}$ in terms of inertia force distributions $\mathbf{s}_n$ associated with natural modes. This interpretation becomes apparent by considering the structure vibrating in its nth mode with accelerations $\ddot{\mathbf{u}}_n(t) = \ddot{q}_n(t)\,\boldsymbol{\phi}_n$. The associated inertia forces are

$$(\mathbf{f}_I)_n = -\mathbf{m}\ddot{\mathbf{u}}_n(t) = -\mathbf{m}\,\boldsymbol{\phi}_n\,\ddot{q}_n(t)$$

and their spatial distribution, given by the vector $\mathbf{m}\boldsymbol{\phi}_n$, is the same as that of $\mathbf{s}_n$.

The expansion of Eq. (12.8.2) has the useful property that the force vector $\mathbf{s}_n p(t)$ produces response only in the nth mode but no response in any other mode. This property can be demonstrated from the generalized force for the rth mode:

$$P_r(t) = \boldsymbol{\phi}_r^T \, \mathbf{s}_n \, p(t) = \Gamma_n (\boldsymbol{\phi}_r^T \, \mathbf{m} \, \boldsymbol{\phi}_n) \, p(t) \tag{12.8.5}$$

Because of Eq. (10.4.1b), the orthogonality property of modes,

$$P_r(t) = 0 \qquad r \neq n \tag{12.8.6}$$

indicating that the excitation vector $\mathbf{s}_n p(t)$ produces no generalized force and hence no response in the nth mode, $r \neq n$. Equation (12.8.5) for $r = n$ is

$$P_n(t) = \Gamma_n \, M_n \, p(t) \tag{12.8.7}$$

The expansion of Eq. (12.8.2) has another useful property in that the dynamic response in the nth mode is due entirely to the partial force vector $\mathbf{s}_n p(t)$. This property becomes obvious by examining the generalized force for the nth mode associated with the total force vector:

$$P_n(t) = \phi_n^T \mathbf{s} p(t)$$

Substituting Eq. (12.8.2) for $\mathbf{s}$ gives

$$P_n(t) = \sum_{r=1}^{N} \Gamma_r \left(\phi_n^T \mathbf{m} \phi_r\right) p(t)$$

which, after utilizing the orthogonality property of modes, reduces to

$$P_n(t) = \Gamma_n M_n p(t) \tag{12.8.8}$$

This generalized force for the complete force vector $\mathbf{s}p(t)$ is the same as Eq. (12.8.7) associated with the partial force vector $\mathbf{s}_n p(t)$.

To study the modal expansion of the force vector $\mathbf{s}$ further, we consider the structure of Fig. 12.8.1: a five-story shear building (i.e., flexurally rigid floor beams and slabs) with lumped mass m at each floor, and same story stiffness k for all stories.

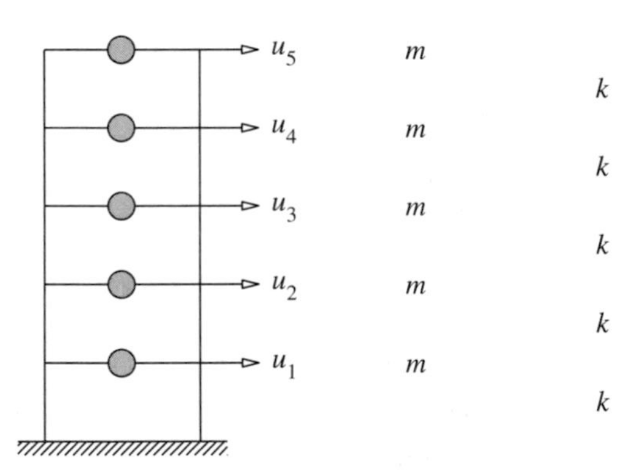

Figure 12.8.1 Uniform five-story shear building.

The mass and stiffness matrices of the structure are

$$\mathbf{m} = m \begin{bmatrix} 1 & & & & \\ & 1 & & & \\ & & 1 & & \\ & & & 1 & \\ & & & & 1 \end{bmatrix} \qquad \mathbf{k} = k \begin{bmatrix} 2 & -1 & & & \\ -1 & 2 & -1 & & \\ & -1 & 2 & -1 & \\ & & -1 & 2 & -1 \\ & & & -1 & 1 \end{bmatrix}$$

Determined by solving the eigenvalue problem, the natural frequencies are

$$\omega_n = \alpha_n \left(\frac{k}{m}\right)^{1/2}$$

where $\alpha_1 = 0.285$, $\alpha_2 = 0.831$, $\alpha_3 = 1.310$, $\alpha_4 = 1.682$, and $\alpha_5 = 1.919$. For a structure with $m = 100$ kips/g, the natural vibration modes, which have been normalized to obtain $M_n = 1$, are (Fig. 12.8.2)

$$\phi_1 = \begin{Bmatrix} 0.334 \\ 0.641 \\ 0.895 \\ 1.078 \\ 1.173 \end{Bmatrix} \quad \phi_2 = \begin{Bmatrix} -0.895 \\ -1.173 \\ -0.641 \\ 0.334 \\ 1.078 \end{Bmatrix} \quad \phi_3 = \begin{Bmatrix} 1.173 \\ 0.334 \\ -1.078 \\ -0.641 \\ 0.895 \end{Bmatrix} \quad \phi_4 = \begin{Bmatrix} -1.078 \\ 0.895 \\ 0.334 \\ -1.173 \\ 0.641 \end{Bmatrix} \quad \phi_5 = \begin{Bmatrix} 0.641 \\ -1.078 \\ 1.173 \\ -0.895 \\ 0.334 \end{Bmatrix}$$

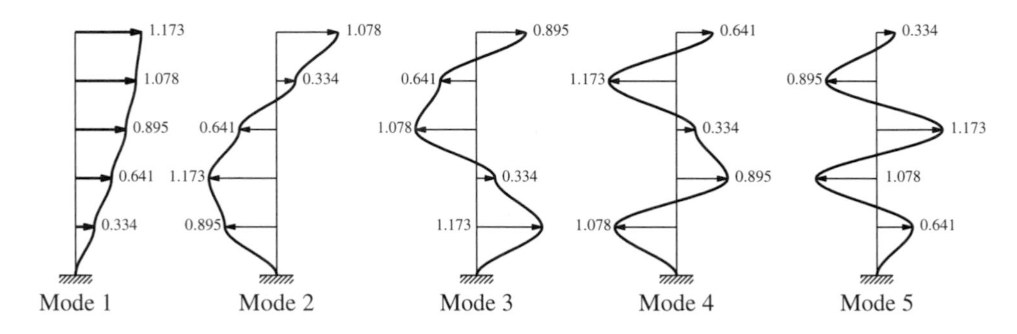

Figure 12.8.2 Natural modes of vibration of uniform five-story shear building.

Consider two different sets of applied forces: $\mathbf{p}(t) = \mathbf{s}_a p(t)$ and $\mathbf{p}(t) = \mathbf{s}_b p(t)$, where $\mathbf{s}_a^T = \langle 0 \quad 0 \quad 0 \quad 0 \quad 1 \rangle$ and $\mathbf{s}_b^T = \langle 0 \quad 0 \quad 0 \quad -1 \quad 2 \rangle$; note that the resultant force is unity in both cases (Fig. 12.8.3). Substituting for $\mathbf{m}$, ϕ_n, and $\mathbf{s} = \mathbf{s}_a$ in Eqs. (12.8.2) and (12.8.3) gives the modal contributions $\mathbf{s}_n$:

$$\mathbf{s}_1 = \begin{Bmatrix} 0.101 \\ 0.195 \\ 0.272 \\ 0.327 \\ 0.356 \end{Bmatrix} \quad \mathbf{s}_2 = \begin{Bmatrix} -0.250 \\ -0.327 \\ -0.179 \\ 0.093 \\ 0.301 \end{Bmatrix} \quad \mathbf{s}_3 = \begin{Bmatrix} 0.272 \\ 0.077 \\ -0.250 \\ -0.149 \\ 0.208 \end{Bmatrix} \quad \mathbf{s}_4 = \begin{Bmatrix} -0.179 \\ 0.149 \\ 0.055 \\ -0.195 \\ 0.106 \end{Bmatrix} \quad \mathbf{s}_5 = \begin{Bmatrix} 0.055 \\ -0.093 \\ 0.101 \\ -0.077 \\ 0.029 \end{Bmatrix}$$

Similarly, for $\mathbf{s} = \mathbf{s}_b$, the $\mathbf{s}_n$ vectors are

$$\mathbf{s}_1 = \begin{Bmatrix} 0.110 \\ 0.210 \\ 0.294 \\ 0.354 \\ 0.385 \end{Bmatrix} \quad \mathbf{s}_2 = \begin{Bmatrix} -0.423 \\ -0.553 \\ -0.302 \\ 0.157 \\ 0.508 \end{Bmatrix} \quad \mathbf{s}_3 = \begin{Bmatrix} 0.739 \\ 0.210 \\ -0.679 \\ -0.403 \\ 0.564 \end{Bmatrix} \quad \mathbf{s}_4 = \begin{Bmatrix} -0.685 \\ 0.569 \\ 0.212 \\ -0.746 \\ 0.407 \end{Bmatrix} \quad \mathbf{s}_5 = \begin{Bmatrix} 0.259 \\ -0.436 \\ 0.475 \\ -0.363 \\ 0.135 \end{Bmatrix}$$

These vectors are displayed in Fig. 12.8.3. The contributions of the higher modes, especially the second and third modes, to $\mathbf{s}$ are larger for $\mathbf{s}_b$ than for $\mathbf{s}_a$, suggesting that these modes may contribute more to the response if the force distribution is $\mathbf{s}_b$ than if it is $\mathbf{s}_a$. We will return to this observation in Section 12.11.

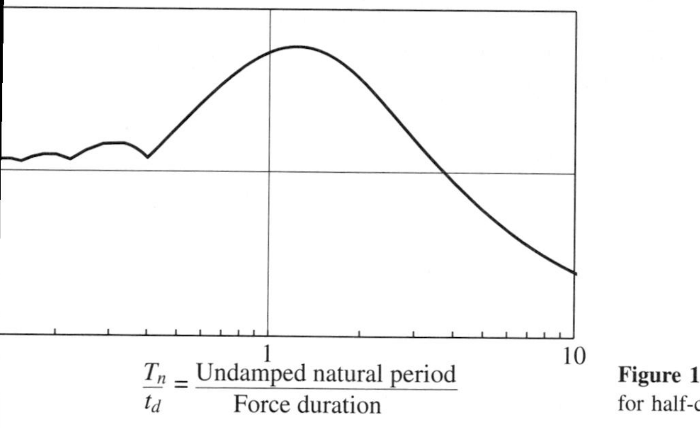

Figure 12.11.3 Dynamic resp
for half-cycle sine pulse force;

monic force because, as seen in Fig. 12.11.2, several modes would have simi
of R_{dn}. However, for lightly damped systems subjected to harmonic excitatio
especially large for modes with natural period T_n close to the forcing period
modes would contribute most to the response and are perhaps the only modes
be included in modal analysis unless the modal contribution factors $\bar{r}_n$ for the
are much smaller than for some other modes.

To explore these ideas further, consider the five-story shear frame (Fig. 12.
out damping subjected to harmonic forces $\mathbf{p}(t) = \mathbf{s}_b p(t)$, where $p(t) = p_o$ sin
Further consider four different values of the forcing period T relative to the fun
natural period of the system: $T_1/T = 0.75$, 2.75, 3.50, and 4.30. For each
forcing periods the R_{dn} values for the five natural periods of the system (define
tion 12.8) are identified in Fig. 12.11.4. These data permit the following obs
for each of the four cases:

1. $T_1/T = 0.75$ (Fig. 12.11.4a): R_{dn} is largest for the first mode, and R_d
nificantly larger than other R_{dn}. The larger R_{d1} combined with the larg
contribution factor $\bar{V}_{b1}$ for the first mode, compared to these factors $\bar{V}_{bn}$
modes (Table 12.11.1), will make the first-mode response largest. Obs
R_{dn} for higher modes are close to 1, indicating that the response in these
essentially static.
2. $T_1/T = 2.75$ (Fig. 12.11.4b): For this case $T_2/T = 0.943$, and therefor
8.89 is much larger than the other R_{dn}. Therefore, the second-mode
will dominate all the modal responses, even the first-mode response,
$\bar{V}_{b1} = 1.353$ is more than twice $\bar{V}_{b2} = -0.612$. The second mode alon
provide a reasonably accurate result.
3. $T_1/T = 3.5$ (Fig. 12.11.4c): R_{d2} and R_{d3} are similar in magnitude, muc
than R_{d1}, and significantly larger than R_{d4} and R_{d5}. This suggests that the
and third-mode contributions to the response would be the larger ones. The
mode response would exceed the third-mode response because the magn
$\bar{V}_{b2} = -0.612$ is larger than $\bar{V}_{b3} = 0.431$.

This important result can be proven by recognizing that $\mathbf{s} = \sum \mathbf{s}_n$ [Eq. (12.8.2)], which implies that $r^{\text{st}} = \sum r_n^{\text{st}}$. Dividing by r^{st} gives the desired result.

12.11 MODAL CONTRIBUTIONS TO RESPONSE

The peak value of $r_n(t)$, Eq. (12.10.1), is

$$r_{no} = r^{\text{st}}\,\bar{r}_n\,\omega_n^2 D_{no} \tag{12.11.1}$$

where D_{no} is the peak value of $D_n(t)$. We shall rewrite this equation in terms of a dynamic response factor introduced in Chapters 3 and 4. For the SDF system governed by Eq. (12.9.2), this factor is $R_{dn} = D_{no}/(D_{n,\text{st}})_o$, where $(D_{n,\text{st}})_o$ is the maximum value of $D_{n,\text{st}}(t)$, the static response. Obtained by dropping the $\dot{D}_n$ and $\ddot{D}_n$ terms in Eq. (12.9.2), $D_{n,\text{st}}(t) = p(t)/\omega_n^2$ and its maximum value is $(D_{n,\text{st}})_o = p_o/\omega_n^2$. Therefore, Eq. (12.11.1) becomes

$$r_{no} = p_o r^{\text{st}}\bar{r}_n R_{dn} \tag{12.11.2}$$

Equation (12.11.2) indicates that the maximum value of the contribution of the nth mode to the response r is the product of four quantities: (1) the dimensionless dynamic response factor R_{dn} for an SDF system with natural frequency ω_n and damping ratio ζ_n subjected to force $p(t)$; (2) the dimensionless modal contribution factor $\bar{r}_n$ for the response quantity r; (3) r^{st}, the static value of r due to the external forces $\mathbf{s}$; and (4) p_o, the maximum value of $p(t)$. The quantities r^{st} and $\bar{r}_n$ depend on the spatial distribution $\mathbf{s}$ of the applied forces but are independent of the time variation $p(t)$ of the forces; on the other hand, R_{dn} depends on $p(t)$, but is independent of $\mathbf{s}$. Next, we will study how the relative contributions of various modes to the response depend on the spatial distribution $\mathbf{s}$ of the applied forces and on their time variation $p(t)$.

12.11.1 Modal Contribution Factors

In this section we determine the modal contribution factors for the base shear V_b and roof (Nth floor) displacement u_N of a multistory building. Figure 12.11.1 shows the external forces $\mathbf{s}$ and $\mathbf{s}_n$. The latter is defined by Eq. (12.8.4); in particular the lateral force at the jth floor level is the jth element of $\mathbf{s}_n$:

$$s_{jn} = \Gamma_n m_j \phi_{jn} \tag{12.11.3}$$

where m_j is the lumped mass and ϕ_{jn} the nth-mode shape value at the jth floor. As a result of the static forces $\mathbf{s}_n$ (Fig. 12.11.1b), the base shear is

$$V_{bn}^{\text{st}} = \sum_{j=1}^{N} s_{jn} = \Gamma_n \sum_{j=1}^{N} m_j \phi_{jn} \tag{12.11.4}$$

and the floor displacements are $\mathbf{u}_n^{\text{st}} = \mathbf{k}^{-1}\mathbf{s}_n$. Inserting Eq. (12.8.4) and using Eq. (10.2.4)

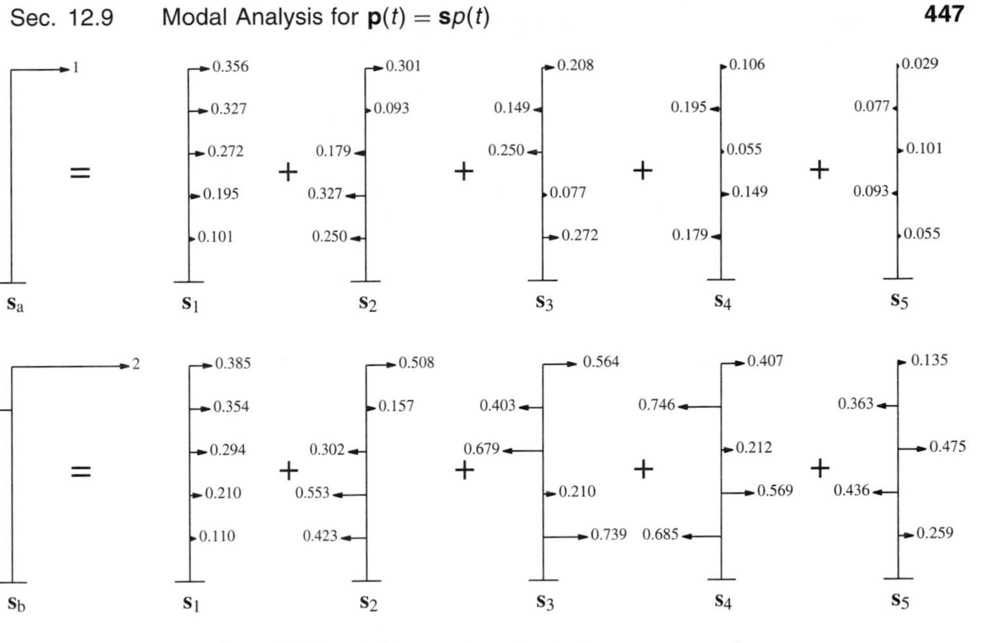

Figure 12.8.3 Modal expansion of excitation vectors $\mathbf{s}_a$ and $\mathbf{s}_b$.

12.9 MODAL ANALYSIS FOR $\mathbf{p}(t) = \mathbf{s}p(t)$

The dynamic analysis of an MDF system subjected to forces $\mathbf{p}(t)$ is specialized in this section for the excitation $\mathbf{p}(t) = \mathbf{s}\,p(t)$. The generalized force $P_n(t)$ for the nth mode, Eq. (12.8.7), is substituted in Eq. (12.4.6) to obtain the modal equation

$$\ddot{q}_n + 2\zeta_n\omega_n\dot{q}_n + \omega_n^2 q_n = \Gamma_n\, p(t) \tag{12.9.1}$$

The factor Γ_n that multiplies the force $p(t)$ is sometimes called a *modal participation factor*, implying that it is a measure of the degree to which the nth mode participates in the response. This is not a useful definition, however, because Γ_n is not independent of how the mode is normalized, nor a measure of the modal contribution to a response quantity. Both these drawbacks are overcome by modal contribution factors that will be introduced in the next section.

We will write the solution $q_n(t)$ in terms of the response of an SDF system. Consider such a system with unit mass, and vibration properties—natural frequency ω_n and damping ratio ζ_n—of the nth mode of the MDF system excited by the force $p(t)$. This nth-*mode SDF system* is governed by Eq. (1.5.2) with $m = 1$ and $\zeta = \zeta_n$, which is repeated here with u replaced by D_n to emphasize its connection with the nth mode:

$$\ddot{D}_n + 2\zeta_n\omega_n\dot{D}_n + \omega_n^2 D_n = p(t) \tag{12.9.2}$$

Comparing Eqs. (12.9.2) and (12.9.1) gives

$$q_n(t) = \Gamma_n D_n(t) \tag{12.9.3}$$

Thus $q_n(t)$ is readily available once Eq. (12.9.2) has been solved for $D_n(t)$, utilizing the available results for SDF systems subjected to, for example, harmonic, step, and impulsive forces (Chapters 3 and 4).

Then the contribution of the nth mode to nodal displacements $\mathbf{u}(t)$, Eq. (12.5.1), is

$$\mathbf{u}_n(t) = \Gamma_n \phi_n D_n(t) \tag{12.9.4}$$

Substituting Eq. (12.9.3) in (12.6.2) gives the equivalent static forces:

$$\mathbf{f}_n(t) = \mathbf{s}_n\left[\omega_n^2 D_n(t)\right] \tag{12.9.5}$$

The nth-mode contribution $r_n(t)$ to any response quantity $r(t)$ is determined by static analysis of the structure subjected to forces $\mathbf{f}_n(t)$. If r_n^{st} denotes the *modal static response*, the static value (indicated by superscript "st") of r due to external forces $\mathbf{s}_n$, then

$$r_n(t) = r_n^{st}\left[\omega_n^2 D_n(t)\right] \tag{12.9.6}$$

Thus the dynamic response $r_n(t)$ is the product of results from two analyses: (1) static analysis of the structure subjected to external forces $\mathbf{s}_n$, and (2) dynamic analysis of the nth-mode SDF system excited by the force $p(t)$.

Combining the response contributions of all the modes gives the total response:

$$r(t) = \sum_{n=1}^{N} r_n(t) = \sum_{n=1}^{N} r_n^{st}\left[\omega_n^2 D_n(t)\right] \tag{12.9.7}$$

The modal analysis procedure just presented, a special case of the one presented in Section 12.7, has the advantage in that it provides a basis to identify and understand the factors that influence the relative modal contributions to the response, as we shall see in Section 12.11.

12.10 MODAL CONTRIBUTION FACTORS

The contribution of the nth mode to response quantity r, Eq. (12.9.6), can be expressed as

$$r_n(t) = r^{st}\,\overline{r}_n\left[\omega_n^2 D_n(t)\right] \tag{12.10.1}$$

where r^{st} is the static value of r due to external forces $\mathbf{s}$, and the nth *modal contribution factor*:

$$\overline{r}_n = \frac{r_n^{st}}{r^{st}} \tag{12.10.2}$$

These modal contribution factors $\overline{r}_n$ have three important properties. First, by definition they are dimensionless. Second, they are independent of how the modes are normalized because r_n^{st} is the static effect of $\mathbf{s}_n$, which does not depend on the normalization, and the modal properties do not enter into r^{st}. Third, the sum of the modal contribution factors over all modes is unity, that is,

$$\sum_{n=1}^{N} \overline{r}_n = 1 \tag{12.10.3}$$

It is not necessary to repeat the analysis above for all response quan some of the key response quantities, especially those that are likely to l higher modes, should be identified for deciding the number of modes to modal analysis.

12.11.2 Dynamic Response Factor

We now study how the modal response contributions depend on the tin the excitation. The dynamic response to $p(t)$ is characterized by the dyn factor R_{dn} in Eq. (12.11.2). This factor was derived in Chapter 3 for harm (Fig. 3.2.6) and in Chapter 4 for various excitations, including the half-c (Fig. 4.8.3c). In Fig. 12.11.2, R_d for harmonic force of period T is plotte for SDF systems with natural period T_n and two damping ratios: $\zeta = 5$ for a half-cycle sine pulse force of duration t_d is plotted against T_n/t_d in F undamped SDF systems.

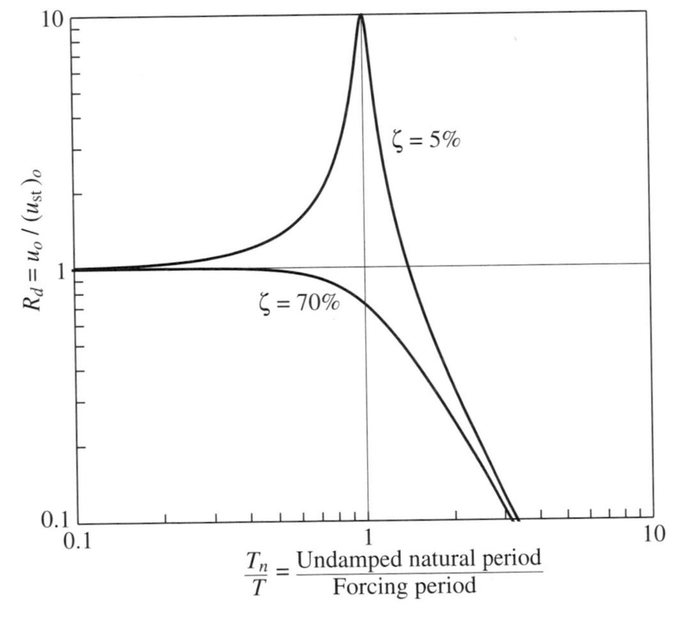

Figure 12.11.2 Dynami for harmonic force; $\zeta =$

How R_{dn}, the value of R_d for the nth mode, for a given excitati with n depends on where the natural periods T_n fall on the period scale. impulsive excitation, Fig. 12.11.3 shows that R_{dn} varies over a narrow ra range of T_n and could have similar values for several modes. Thus severa generally have to be included in modal analysis with their relative respons Eq. (12.11.2), determined primarily by the relative values of $\overline{r}_n$, the mod factors. The same conclusion also applies to highly damped systems su

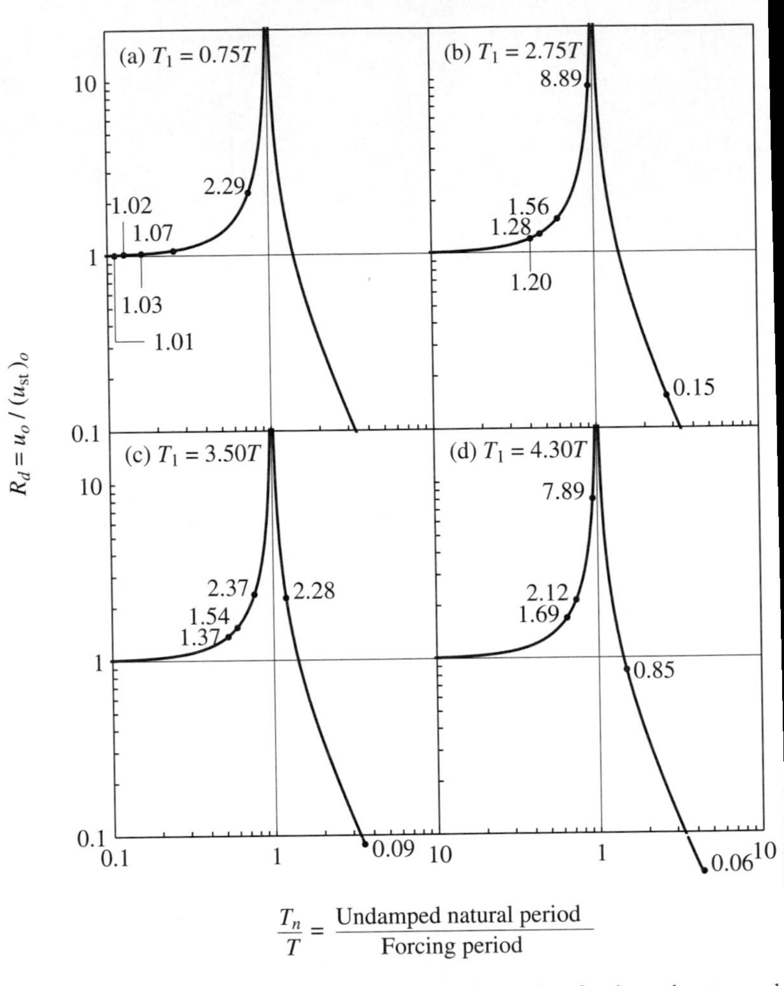

Figure 12.11.4 Dynamic response factors R_{dn} for five modes of undamped system and four forcing periods T: (a) $T_1 = 0.75T$; (b) $T_1 = 2.75T$; (c) $T_1 = 3.50T$; (d) $T_1 = 4.30T$.

4. $T_1/T = 4.3$ (Fig. 12.11.4d): For this case the third-mode period is cl the exciting period ($T_3/T = 0.935$), and therefore $R_{d3} = 7.89$ is muc than the other R_{dn}. Therefore, the third-mode response will dominate modal responses, even the first-mode response, although $\overline{V}_{b1} = 1.353$ $\overline{V}_{b3} = 0.431$ by a factor of over 3. The third mode alone would pr reasonably accurate result.

It is not necessary to compute all the natural periods of a system having number of DOFs in ascertaining which of the R_{dn} values are significant. Only few natural periods need to be calculated and located on the plot showing the response factor. Then the approximate locations of the higher natural periods

readily apparent, thus providing sufficient information to estimate the range of R_{dn} values and to make a preliminary decision on the modes that may contribute significant response. Precise values of R_{dn} can then be calculated for these modes to be included in modal analysis.

In judging the contribution of a natural mode to the dynamic response of a structure, it is necessary to consider the combined effects of the modal contribution factor $\bar{r}_n$ and the dynamic response factor R_{dn}. The two factors have been discussed separately in preceding sections because $\bar{r}_n$ depends on the spatial distribution **s** of the applied forces, whereas R_{dn} depends on the time variation $p(t)$ of the excitation. However, they both enter into the modal response, Eq. (12.11.2). By retaining in dynamic analysis only those natural modes with significant values of $\bar{r}_n$ or R_{dn}, or both, the computational effort can be reduced. This reduction may not be significant in analysis of systems with a few dynamic DOFs, such as the five-story shear frame considered here. However, substantial reduction in computation can be achieved for practical, complex structures that may require hundreds of DOFs for their idealization. Typically in analyzing such an N-DOF system, the first J modes are included, where J is much smaller than N, and the modal summation of Eq. (12.9.7) is truncated accordingly. Thus we need to compute the natural frequencies, natural modes, and the response $D_n(t)$ only for the first J modes.

PART D: SPECIAL ANALYSIS PROCEDURES

12.12 STATIC CORRECTION METHOD

In Section 12.11 we have shown that the dynamic response factor R_{dn} for some of the higher modes of a structure may be only slightly larger than unity. For the five-story shear frame subjected to harmonic excitation with $T_1/T = 0.75$, $R_{dn} = 1.07$, 1.03, 1.02, and 1.01 for the second, third, fourth, and fifth modes, respectively (Fig. 12.11.4a). Such is the case when the higher-mode period T_n is much shorter than the period T of the harmonic excitation or the duration t_d of an impulsive excitation (Fig. 12.11.3). The response in such a higher mode is essentially static and could be determined by the easier static analysis instead of dynamic analysis. This is the essence of the *static correction method*, developed next.

Suppose that we include all N modes in the analysis but divide them into two parts: (1) the first N_d modes with natural periods T_n such that the dynamic effects are significant, as indicated by R_{dn} being significantly larger than 1; and (2) modes $N_d + 1$ to N with natural periods T_n such that R_{dn} is close to 1. Then the modal contributions to the response can be divided into two parts:

$$r(t) = \sum_{n=1}^{N_d} r_n(t) + \sum_{n=N_d+1}^{N} r_n(t) \tag{12.12.1}$$

uncoupled modal equations for only the significant modes. Similarly, only these modes need to be considered in analysis of systems with nonclassical damping.

If only the first J modes contribute significantly to the response, the size of Eq. (12.4.4) can be reduced accordingly; $\mathbf{\Phi}$ is now an $N \times J$ matrix; $\mathbf{M}$, $\mathbf{C}$, and $\mathbf{K}$ are $J \times J$ matrices; and $\mathbf{P}(t)$ is a $J \times 1$ vector. Thus the problem reduces to solving these J coupled equations for $q_n(t)$, $n = 1, 2, \ldots, J$. Numerical methods are generally necessary for this solution even if $p(t) = 0$ or is a simple function because the equations are coupled (Chapter 15). Once $q_n(t)$ have been determined at each time instant, $\mathbf{u}(t)$ is computed from Eq. (12.3.2) with its summation truncated at $N = J$.

An alternative approach for analysis of linear systems with nonclassical damping may be interpreted as an extension of the classical modal analysis procedure with uncoupled modal equations. This approach requires solution of the eigenvalue problem including damping:

$$(\nu^2 \mathbf{m} + \nu \mathbf{c} + \mathbf{k})\, \phi = \mathbf{0} \tag{12.14.1}$$

This is a complex eigenvalue problem because the eigenvalue ν and the elements of the eigenvector ϕ are complex numbers. Its solution gives $2N$ eigenvalues and corresponding eigenvectors. The dynamic response can then be expressed as a linear combination of $2N$ "modal" solutions. Thus

$$\mathbf{u}(t) = \sum_{n=1}^{2N} q_n(t)\, \phi_n e^{\nu_n t} \tag{12.14.2}$$

where $q_n(t)$ is the modal coordinate response. This solution procedure is computationally demanding for two reasons. First, the solution of a complex eigenvalue problem requires about eight times the computations for the solution of a real eigenvalue problem of the system without damping. Second, the eigenvalues and eigenvectors are complex numbers that require dealing with complex arithmetic throughout the analysis, although the final result of Eq. (12.14.2) is real valued. For these reasons this solution is not developed in this textbook, and the reader is referred to the published literature.

FURTHER READING

Bisplinghoff, R. L., Ashley, H., and Halfman, R. L., *Aeroelasticity*, Addison-Wesley, Reading, Mass., 1955.

Craig, R. R., Jr., *Structural Dynamics: An Introduction to Computer Methods*, Wiley, New York, 1981, pp. 350–353.

Crandall, S. H., and McCalley, R. B., Jr., "Matrix Methods of Analysis," Chapter 28 in *Shock and Vibration Handbook* (ed. C. M. Haris), McGraw-Hill, New York, 1988.

Den Hartog, J. P., *Mechanical Vibrations*, McGraw-Hill, New York, 1956, pp. 87–105.

Humar, J. L., *Dynamics of Structures,* Prentice Hall, Englewood Cliffs, N.J., 1990, Chapter 10.

Veletsos, A. S., and Ventura, C.E., "Modal Analysis of Non-classically Damped Linear Systems," *Earthquake Engineering and Structural Dynamics*, **14**, 1986, pp. 217–243.

PROBLEMS

Part A

12.1 Figure P12.1 shows a shear frame (i.e., rigid beams) and its floor masses and story stiffnesses. This structure is subjected to harmonic horizontal force $p(t) = p_o \sin \omega t$ at the top floor.
(a) Derive equations for the steady-state displacements of the structure by two methods: (i) direct solution of coupled equations, and (ii) modal analysis.
(b) Show that both methods give equivalent results.
(c) Plot on the same graph the two displacement amplitudes u_{1o} and u_{2o} as functions of the excitation frequency. Use appropriate normalizations of the displacement and frequency scales. Neglect damping.

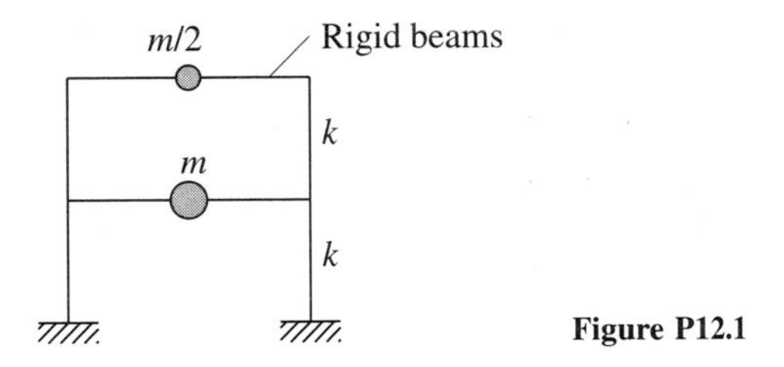

Figure P12.1

12.2 For the system and excitation of Problem 12.1 derive equations for story shears (considering steady-state response only) by two methods: **(a)** directly from displacements (without introducing equivalent static forces), and **(b)** using equivalent static forces. Show that the two methods give equivalent results.

Part B

12.3 Consider the system of Fig. P12.1 with modal damping ratios ζ_n subjected to the same excitation. Derive equations for the steady-state displacement amplitudes of the system.

12.4 The undamped system of Fig. P12.1 is subjected to an impulsive force at the first-floor mass: $p_1(t) = p_o\delta(t)$. Derive equations for the lateral floor displacements as functions of time.

12.5 The undamped system of Fig. P12.1 is subjected to a suddenly applied force at the first-floor mass: $p_1(t) = p_o$, $t \geq 0$. Derive equations for **(a)** the lateral floor displacements as functions of time, and **(b)** the story drift (or deformation) in the second story as a function of time.

12.6 The undamped system of Fig. P12.1 is subjected to a rectangular pulse force at the first floor. The pulse has an amplitude p_o and duration $t_d = T_1/2$, where T_1 is the fundamental vibration period of the system. Derive equations for the floor displacements as functions of time.

12.7 Figure P12.7 shows a shear frame (i.e., rigid beams) and its floor weights and story stiffnesses. This structure is subjected to harmonic force $p(t) = p_o \sin \omega t$ at the top floor.
(a) Determine the steady-state displacements as functions of ω by two methods: (i) direct solution of coupled equations, and (ii) modal analysis.

(b) Show that both methods give the same results.

(c) Plot on the same graph the three displacement amplitudes as a function of the excitation frequency over the frequency range 0 to $5\omega_1$. Use appropriate normalizations of the displacement and frequency scales.

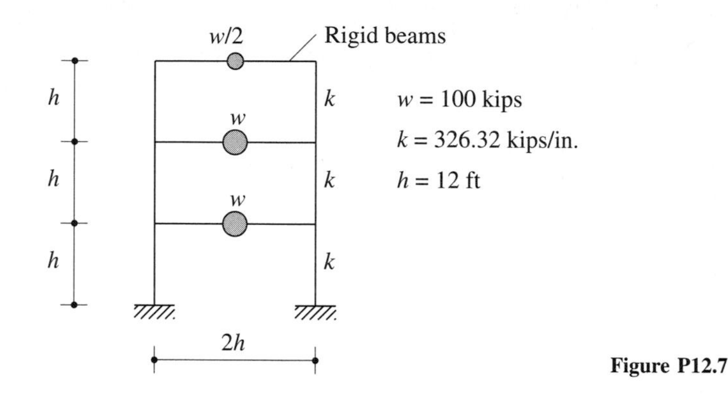

Figure P12.7

12.8 For the system and excitation of Problem 12.7 determine the story shears (considering steady-state response only) by two methods: **(a)** directly from displacements (without introducing equivalent static forces), and **(b)** using equivalent static forces. Show that the two methods give identical results.

12.9 An eccentric mass shaker is mounted on the roof of the system of Fig. P12.7. The shaker has two counterrotating weights, each 20 lb, at an eccentricity of 12 in. with respect to the vertical axis of rotation. Determine the steady-state amplitudes of displacement and acceleration at the roof as a function of excitation frequency. Plot the frequency response curves over the frequency range 0 to 15 Hz. Assume that the modal damping ratios ζ_n are 5%.

12.10 The undamped system of Fig. P12.7 is subjected to an impulsive force at the second-floor mass: $p_2(t) = p_o\delta(t)$, where $p_o = 20$ kips. Derive equations for the lateral floor displacements as functions of time.

12.11 The undamped system of Fig. P12.7 is subjected to a suddenly applied force at the first-floor mass: $p_1(t) = p_o$, $t \geq 0$, where $p_o = 200$ kips. Derive equations for **(a)** the lateral floor displacements as functions of time, and **(b)** the story drift (or deformation) in the second story as a function of time.

12.12 The undamped system of Fig. P12.7 is subjected to a rectangular pulse force at the third floor. The pulse has an amplitude $p_o = 200$ kips and duration $t_d = T_1/2$, where T_1 is the fundamental vibration period of the system. Derive equations for the floor displacements as functions of time.

12.13 Figure P12.13 shows a structural steel beam with $E = 30,000$ ksi, $I = 100$ in^4, $L = 150$ in, and $mL = 0.864$ kip-sec^2/in. Determine the displacement response of the system to an impulsive force $p_1(t) = p_o\delta(t)$ at the left mass, where $p_o = 10$ kips. Plot as functions of time the displacements u_j due to each vibration mode separately and combined.

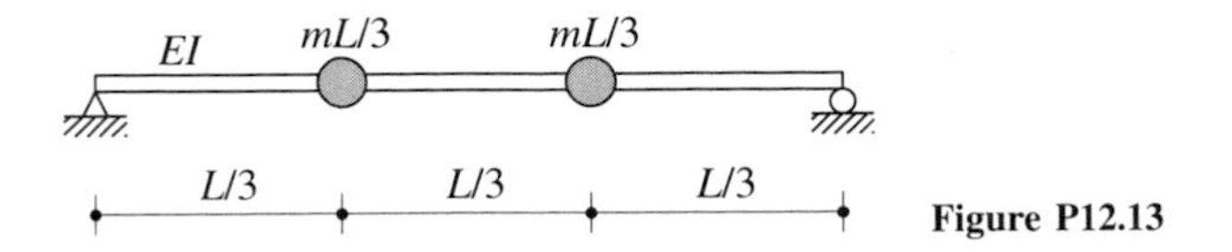

Figure P12.13

12.14 Determine the displacement response of the system of Fig. P12.13 to a suddenly applied force of 100 kips applied at the left mass. Plot as functions of time the displacements u_j due to each vibration mode separately and combined.

12.15 Determine the displacement response of the system of Fig. P12.13 to force $p(t)$, which is shown in Fig. P12.15 and applied at the left mass. Plot as functions of time the displacements $u_j(t)$ due to each vibration mode separately and combined.

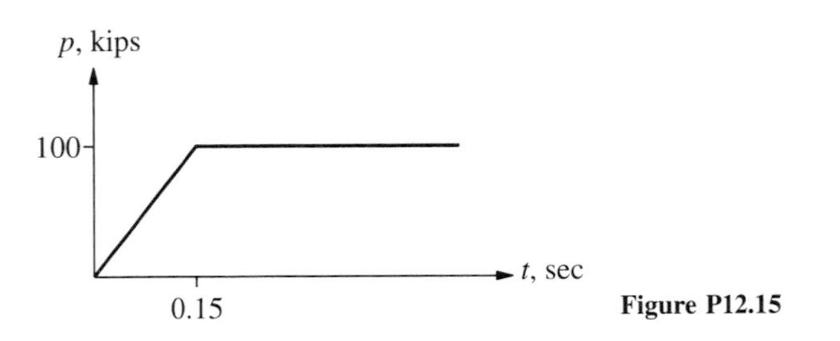

Figure P12.15

12.16 Determine the displacement response of the system of Fig. P12.13 to force $p(t)$, which is shown in Fig. P12.16 and applied at the right mass. Plot as functions of time the displacements $u_j(t)$ due to each vibration mode separately and combined.

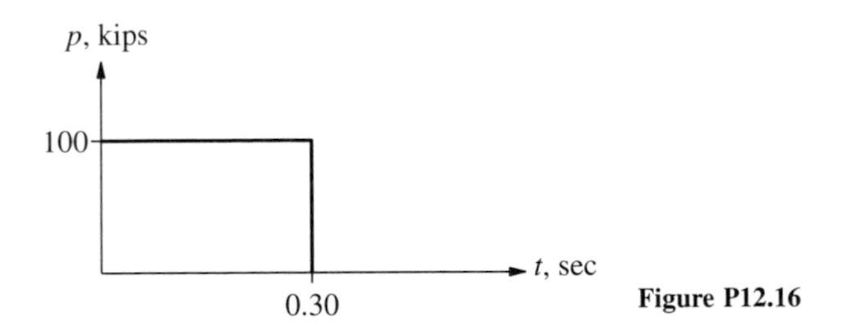

Figure P12.16

12.17 Repeat part (c) of Example 12.6 without using equivalent static forces. In other words, determine the shears and bending moments directly from the displacements and rotations.

12.18 For the system and excitation of Problem 12.14 determine the shears and bending moments at sections a, b, c, d, e, and f (Fig. P12.18) at $t = 0.1$ sec by using equivalent static forces. Draw the shear force and bending moment diagrams due to each mode separately and combined.

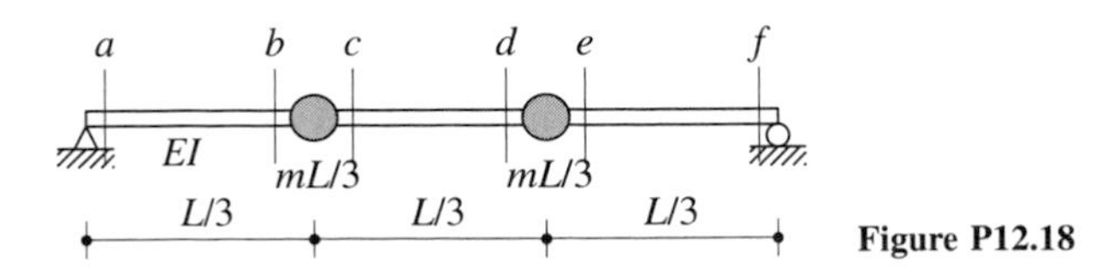

Figure P12.18

13.2.3 Total Response

Combining the response contributions of all the modes gives the earthquake response of the multistory building:

$$r(t) = \sum_{n=1}^{N} r_n(t) = \sum_{n=1}^{N} r_n^{\text{st}} A_n(t) \tag{13.2.10}$$

wherein Eq. (13.2.8) has been substituted for $r_n(t)$, the nth-mode response.

The modal analysis procedure can also provide floor accelerations, although these are not necessary to compute earthquake-induced forces in the structure. The floor accelerations can be computed from

$$\ddot{u}_j^t(t) = \ddot{u}_g(t) + \sum_{n=1}^{N} \Gamma_n \phi_{jn} \ddot{D}_n(t) \tag{13.2.11}$$

using the values of $\ddot{D}_n$ available at each time step from the numerical time-stepping procedure used to solve Eq. (13.1.8) for $D_n(t)$.

13.2.4 Summary

The response of an N-story building with plan symmetric about two orthogonal axes to earthquake ground motion along an axis of symmetry can be computed as a function of time by the procedure just developed, which is summarized next in step-by-step form:

1. Define the ground acceleration $\ddot{u}_g(t)$ numerically at every time step Δt.
2. Define the structural properties.
   - **a.** Determine the mass matrix $\mathbf{m}$ and lateral stiffness matrix $\mathbf{k}$ (Section 9.4).
   - **b.** Estimate the modal damping ratios ζ_n (Chapter 11).
3. Determine the natural frequencies ω_n (natural periods $T_n = 2\pi/\omega_n$) and natural modes ϕ_n of vibration (Chapter 10).
4. Determine the modal components $\mathbf{s}_n$ [Eq. (13.2.4)] of the effective earthquake force distribution.
5. Compute the response contribution of the nth mode by the following steps, which are repeated for all modes, $n = 1, 2, \ldots, N$:
   - **a.** Perform static analysis of the building subjected to lateral forces $\mathbf{s}_n$ to determine r_n^{st}, the modal static response for each desired response quantity r (Table 13.2.1).
   - **b.** Determine the pseudo-acceleration response $A_n(t)$ of the nth-mode SDF system to $\ddot{u}_g(t)$, using numerical time-stepping methods (Chapter 5).
   - **c.** Determine $r_n(t)$ from Eq. (13.2.8).
6. Combine the modal contributions $r_n(t)$ to determine the total response using Eq. (13.2.10).

As will be shown later, usually only the lower few modes contribute significantly to the response. Therefore, steps 3, 4, and 5 need to be implemented only for these modes, and the modal summation of Eq. (13.2.10) truncated accordingly.

Example 13.3

Derive equations for **(a)** the floor displacements and **(b)** the story shears for the shear frame of Example 13.2 subjected to ground motion $\ddot{u}_g(t)$.

Solution Steps 1 to 4 of the procedure summary have already been implemented in Example 13.2.

(a) *Floor displacements.* Substituting Γ_n and ϕ_{jn} from Example 13.2 in Eq. (13.2.5) gives the floor displacements due to the each mode:

$$\begin{Bmatrix} u_1(t) \\ u_2(t) \end{Bmatrix}_1 = \frac{4}{3}\begin{Bmatrix} \frac{1}{2} \\ 1 \end{Bmatrix} D_1(t) \qquad \begin{Bmatrix} u_1(t) \\ u_2(t) \end{Bmatrix}_2 = -\frac{1}{3}\begin{Bmatrix} -1 \\ 1 \end{Bmatrix} D_2(t) \tag{a}$$

Combining the contributions of the two modes gives the floor displacements:

$$u_1(t) = u_{11}(t) + u_{12}(t) = \tfrac{2}{3}D_1(t) + \tfrac{1}{3}D_2(t) \tag{b}$$

$$u_2(t) = u_{21}(t) + u_{22}(t) = \tfrac{4}{3}D_1(t) - \tfrac{1}{3}D_2(t) \tag{c}$$

(b) *Story shears.* Static analysis of the frame for external floor forces $\mathbf{s}_n$ gives V_{in}^{st}, $i = 1$ and 2, shown in Fig. E13.3. Substituting these results in Eq. (13.2.8) gives

$$V_{11}(t) = \tfrac{8}{3}mA_1(t) \qquad V_{21}(t) = \tfrac{4}{3}mA_1(t) \tag{d}$$

$$V_{12}(t) = \tfrac{1}{3}mA_2(t) \qquad V_{22}(t) = -\tfrac{1}{3}mA_2(t) \tag{e}$$

Combining the contributions of two modes gives the story shears

$$V_1(t) = V_{11}(t) + V_{12}(t) = \tfrac{8}{3}mA_1(t) + \tfrac{1}{3}mA_2(t) \tag{f}$$

$$V_2(t) = V_{21}(t) + V_{22}(t) = \tfrac{4}{3}mA_1(t) - \tfrac{1}{3}mA_2(t) \tag{g}$$

The floor displacements and story shears have been expressed in terms of $D_n(t)$ and $A_n(t)$. These responses of the nth-mode SDF system to prescribed $\ddot{u}_g(t)$ can be determined by numerical time-stepping methods (Chapter 5).

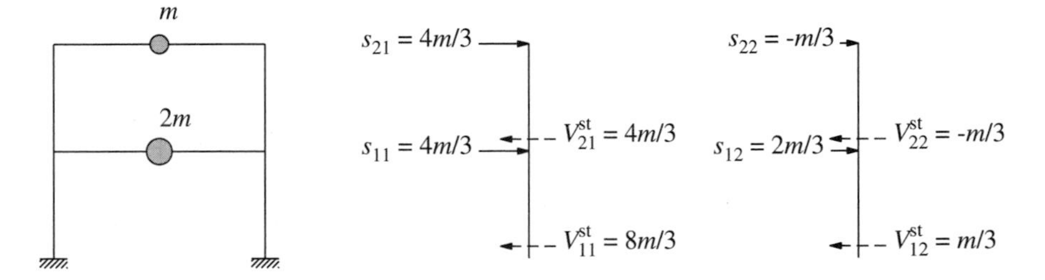

Figure E13.3

Example 13.4

Derive equations for **(a)** the floor displacements and **(b)** the element forces for the two-story frame of Fig. E13.4a due to horizontal ground motion $\ddot{u}_g(t)$.

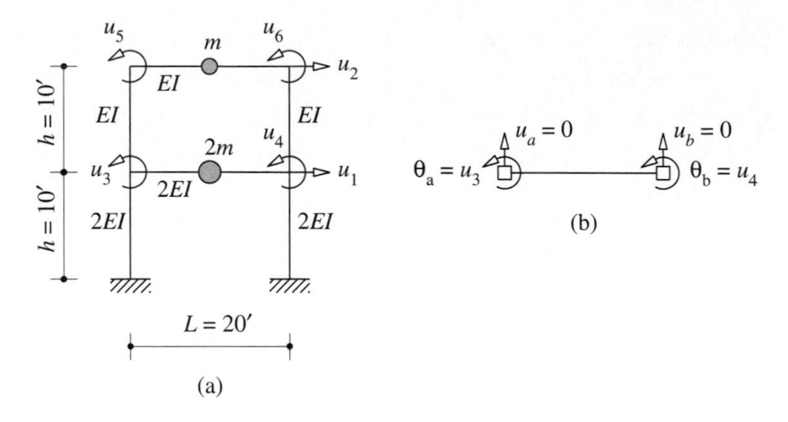

Figure E13.4

Solution Equation (9.3.4) with $\mathbf{p}_t(t) = -\mathbf{m}_{tt}\mathbf{1}\ddot{u}_g(t)$ governs the displacement vector $\mathbf{u}_t = \langle u_1 \quad u_2 \rangle$; where $\mathbf{m}_{tt}$ and $\mathbf{k}_{tt}$, determined in Example 9.9, are

$$\mathbf{m}_{tt} = m\begin{bmatrix} 2 & \\ & 1 \end{bmatrix} \qquad \hat{\mathbf{k}}_{tt} = \frac{EI}{h^3}\begin{bmatrix} 54.88 & -17.51 \\ -17.51 & 11.61 \end{bmatrix} \tag{a}$$

where $h = 10$ ft. The natural frequencies and modes of the system, determined in Example 10.5, are

$$\omega_1 = 2.198\sqrt{\frac{EI}{mh^3}} \qquad \omega_2 = 5.850\sqrt{\frac{EI}{mh^3}} \tag{b}$$

$$\phi_1 = \begin{Bmatrix} 0.3871 \\ 1 \end{Bmatrix} \qquad \phi_2 = \begin{Bmatrix} -1.292 \\ 1 \end{Bmatrix} \tag{c}$$

Thus steps 1 to 3 of Section 13.2.4 have already been implemented.

(**a**) *Floor displacements and joint rotations.* The floor displacements are given by Eq. (13.2.5), where Γ_n are computed from Eq. (13.2.3): $M_1 = 2m(0.3871)^2 + m(1)^2 = 1.300m$, $L_1^h = 2m(0.3871) + m(1) = 1.774m$, and $\Gamma_1 = 1.774m/1.300m = 1.365$. Similarly, $M_2 = 4.337m$, $L_2^h = -1.583m$, and $\Gamma_2 = -0.365$. Substituting these in Eq. (13.2.5) with $n = 1$ gives the floor displacements due to the first mode:

$$\mathrm{u}_1(t) = \begin{Bmatrix} u_1(t) \\ u_2(t) \end{Bmatrix}_1 = 1.365\begin{Bmatrix} 0.3871 \\ 1 \end{Bmatrix} D_1(t) = \begin{Bmatrix} 0.5284 \\ 1.365 \end{Bmatrix} D_1(t) \tag{d}$$

The joint rotations associated with these floor displacements are determined from Eq. (d) of Example 9.9 by substituting $\mathbf{u}_1$ from Eq. (d) for $\mathbf{u}_t$:

$$\mathbf{u}_{01}(t) = \begin{Bmatrix} u_3(t) \\ u_4(t) \\ u_5(t) \\ u_6(t) \end{Bmatrix}_1 = \frac{1}{h}\begin{bmatrix} -0.4426 & -0.2459 \\ -0.4426 & -0.2459 \\ 0.9836 & -0.7869 \\ 0.9836 & -0.7869 \end{bmatrix}\begin{Bmatrix} 0.5284 \\ 1.365 \end{Bmatrix} D_1(t)$$

$$= \frac{1}{h}\begin{Bmatrix} -0.5696 \\ -0.5696 \\ -0.5544 \\ -0.5544 \end{Bmatrix} D_1(t) \tag{e}$$

Similarly, the floor displacements $\mathbf{u}_2(t)$ and joint rotations $\mathbf{u}_{02}(t)$ due to the second mode are determined:

$$\mathbf{u}_2(t) = \begin{Bmatrix} u_1(t) \\ u_2(t) \end{Bmatrix}_2 = \begin{Bmatrix} 0.4716 \\ -0.3651 \end{Bmatrix} D_2(t)$$

$$\mathbf{u}_{02}(t) = \begin{Bmatrix} u_3(t) \\ u_4(t) \\ u_5(t) \\ u_6(t) \end{Bmatrix}_2 = \frac{1}{h}\begin{Bmatrix} -0.1189 \\ -0.1189 \\ 0.7511 \\ 0.7511 \end{Bmatrix} D_2(t) \tag{f}$$

Combining the contributions of the two modes gives the floor displacements and joint rotations:

$$\mathbf{u}(t) = \mathbf{u}_1(t) + \mathbf{u}_2(t) \qquad \mathbf{u}_0(t) = \mathbf{u}_{01}(t) + \mathbf{u}_{02}(t) \tag{g}$$

(b) *Element forces.* Instead of implementing step 5 of the procedure (Section 13.2.4), we will illustrate the computation of element forces from the floor displacements and joint rotations by using the beam stiffness coefficients (Appendix 1). For example, the bending moment at the left end of the first floor beam (Fig. E13.4b) is

$$M_a = \frac{4EI}{L}\theta_a + \frac{2EI}{L}\theta_b + \frac{6EI}{L^2}u_a - \frac{6EI}{L^2}u_b \tag{h}$$

The vertical displacements u_a and u_b are zero because the columns are assumed as axially rigid; joint rotations $\theta_a = u_3$ and $\theta_b = u_4$, where u_3 and u_4 are known from Eqs. (e), (f), and (g); thus

$$\theta_a(t) = -\frac{0.5696}{h}D_1(t) - \frac{0.1189}{h}D_2(t) \qquad \theta_b(t) = \theta_a(t) \tag{i}$$

Substituting for u_a, u_b, θ_a, and θ_b in Eq. (h), replacing EI by $2EI$, and using $D_n(t) = A_n(t)/\omega_n^2$ gives

$$M_a(t) = mh[-0.7077A_1(t) - 0.0209A_2(t)] \qquad M_b(t) = M_a(t) \tag{j}$$

Equations for forces in all beams and columns can be obtained similarly.

Comparing the two terms in Eq. (j) for $M_a(t)$ with Eq. (13.2.8) indicates that $M_{a1}^{\text{st}} = -0.7077mh$ and $M_{a2}^{\text{st}} = -0.0209mh$. These modal static responses could have been obtained by static analysis of the structure due to $\mathbf{s}_n$ determined from Eq. (13.2.4).

The various response quantities have been expressed in terms of $D_n(t)$ and $A_n(t)$; these responses of the nth-mode SDF system to given $\ddot{u}_g(t)$ can be determined by numerical time-stepping methods (Chapter 5).

13.2.5 Effective Modal Mass and Modal Height

In this section physically motivated interpretations of M_n^* and h_n^*, introduced in Eq. (13.2.9a), are presented. The base shear due to the nth mode is obtained by specializing Eq. (13.2.8) for V_b:

$$V_{bn}(t) = V_{bn}^{\text{st}} A_n(t) \tag{13.2.12a}$$

which after substituting for V_{bn}^{st} from Table 13.2.1 becomes

$$V_{bn}(t) = M_n^* A_n(t) \tag{13.2.12b}$$

Comparing Eqs. (13.2.12a) and (13.2.12b) indicates that M_n^* is equal to V_{bn}^{st} and may therefore be interpreted as the resultant of the forces $\mathbf{s}_n$ (Fig. 13.2.2a); similarly, $V_{bn}(t)$ is the resultant of equivalent static forces $f_{jn}(t)$ associated with the nth-mode response of the building (Fig. 13.2.3a).

In contrast to Eq. (13.2.12b), the base shear in a one-story system with mass m, natural frequency ω_n, and damping ratio ζ_n is [from Eq. (6.7.3)]

$$V_b(t) = mA_n(t) \tag{13.2.13}$$

Comparing Eqs. (13.2.12b) and (13.2.13) indicates that the base shear in a one-story system with lumped mass M_n^* (Fig. 13.2.3b) is the same as the nth-mode base shear in a multistory system with its mass distributed among the various floor levels. Thus M_n^* is called the *base shear effective modal mass* or, for brevity, *effective modal mass*.

Equation (13.2.13) implies that the total mass m of a one-story system is effective in producing the base shear. This is so because the mass and hence the equivalent static force are concentrated at one location, the roof. In contrast, only a portion of the mass of a multistory building is effective in producing the base shear due to the nth mode because the building mass is distributed among the various floor levels and the equivalent static forces vary over the height as $m_j\phi_{jn}$. This portion depends on the distribution of the mass of the building over its height and on the shape of the mode, as indicated by Eqs. (13.2.9a) and (13.2.3). The sum of the effective modal masses M_n^* over all the modes is equal to the total mass of the building (see Deriva-

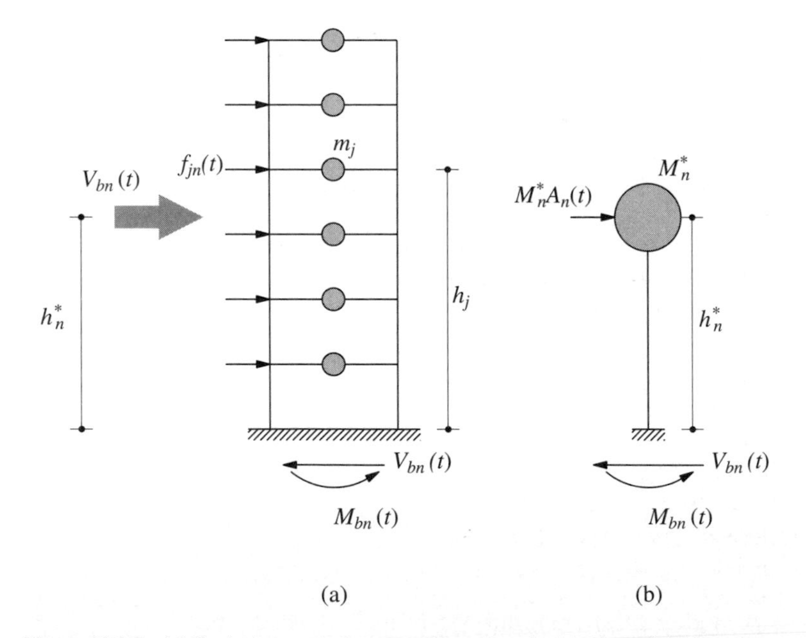

Figure 13.2.3 (a) Equivalent static forces and base shear in the nth mode; (b) one-story system with effective modal mass and effective modal height.

tion 13.1):

$$\sum_{n=1}^{N} M_n^* = \sum_{j=1}^{N} m_j \tag{13.2.14}$$

Now we compare the base overturning moment equations for multistory and one-story systems. The base overturning moment in a multistory building due to its nth mode is obtained by specializing Eq. (13.2.8) for M_b:

$$M_{bn}(t) = M_{bn}^{\text{st}} A_n(t)$$

which after substituting for M_{bn}^{st} from Table 13.2.1 becomes

$$M_{bn}(t) = h_n^* V_{bn}(t) \tag{13.2.15}$$

In contrast, the base overturning moment in a one-story system with mass m lumped at height h above the base is given by Eq. (6.7.3), repeated here for convenience:

$$M_b(t) = h V_b(t) \tag{13.2.16}$$

Comparing Eqs. (13.2.15) and (13.2.16) indicates that the base overturning moment for a one-story system with mass M_n^* lumped at height h_n^* (Fig. 13.2.3b) is the same as M_{bn}, the nth-mode base overturning moment, in a multistory building with its mass distributed among the various floor levels. Thus h_n^* is called the *base-moment effective modal height* or, for brevity, *effective modal height*. It may also be interpreted as the height of the resultant of the forces $\mathbf{s}_n$ (Fig. 13.2.2a) or of the forces $f_{jn}(t)$ (Fig. 13.2.3a).

Equation (13.2.16) implies that the total height h of the one-story system is effective in producing the base overturning moment. This is so because the mass of the structure and hence the equivalent static force is concentrated at height h above the base. In contrast, the *effective modal height* h_n^* is less than the total height of the building because the lumped masses and hence the equivalent static forces are located at the various floor levels; h_n^* depends on the distribution of the mass over the height of the building and on the shape of the mode [Eqs. (13.2.9) and (13.2.3)]. The sum of the first moments about the base of the effective modal masses M_n^* located at effective heights h_n^* is equal to the first moment of the floor masses about the base (see Derivation 13.2):

$$\sum_{n=1}^{N} h_n^* M_n^* = \sum_{j=1}^{N} h_j m_j \tag{13.2.17}$$

For some of the modes higher than the fundamental mode, the effective modal height may be negative. A negative value of h_n^* implies that at any instant of time, the modal static base shear V_{bn}^{st} and the modal static base overturning moment M_{bn}^{st} for the nth mode have opposite algebraic signs; the M_{b1}^{st} and V_{b1}^{st} for the first mode are both positive, by definition.

Derivation 13.1

Premultiplying both sides of Eq. (13.2.2) by $\mathbf{1}^T$ gives

$$\mathbf{1}^T \mathbf{m} \mathbf{1} = \sum_{n=1}^{N} \Gamma_n (\mathbf{1}^T \mathbf{m} \boldsymbol{\phi}_n)$$

Noting that $\mathbf{m}$ is a diagonal matrix with $m_{jj} = m_j$, this can be rewritten as

$$\sum_{j=1}^{N} m_j = \sum_{n=1}^{N} \Gamma_n L_n^h$$

This provides a proof for Eq. (13.2.14) because the nth term on the right side is M_n^*.

Derivation 13.2

A modal expansion of the force vector $\mathbf{mh}$ where $\mathbf{h} = \langle h_1 \quad h_2 \quad \cdots \quad h_N \rangle^T$ is obtained by substituting $\mathbf{s} = \mathbf{mh}$ in Eqs. (12.8.2) and (12.8.3):

$$\mathbf{mh} = \sum_{n=1}^{N} \frac{L_n^\theta}{M_n} \mathbf{m}\boldsymbol{\phi}_n$$

Premultiplying both sides by $\mathbf{1}^T$ gives

$$\mathbf{1}^T\mathbf{mh} = \sum_{n=1}^{N} \frac{L_n^\theta}{M_n} \mathbf{1}^T\mathbf{m}\boldsymbol{\phi}_n$$

Noting that $\mathbf{m}$ is a diagonal matrix with $m_{jj} = m_j$, this can be rewritten as

$$\sum_{j=1}^{N} m_j h_j = \sum_{n=1}^{N} \frac{L_n^\theta}{M_n} L_n^h = \sum_{n=1}^{N} h_n^* M_n^*$$

wherein Eq. (13.2.9) has been used. This provides a proof for Eq. (13.2.17).

Example 13.5

Determine the effective modal masses and effective modal heights for the two-story shear frame of Example 13.2. The height of each story is h.

Solution In Example 13.2 the $\mathbf{m}$, $\mathbf{k}$, ω_n, and $\boldsymbol{\phi}_n$ for this system were presented, and L_n^h and M_n for each of the two modes computed. These are listed next, together with the new computations for M_n^* and h_n^*. For the first mode: $L_1^h = 2m$, $M_1 = 3m/2$, $M_1^* = (L_1^h)^2/M_1 = \frac{8}{3}m$, $L_1^\theta = h(2m)\frac{1}{2} + 2h(m)1 = 3hm$, and $h_1^* = L_1^\theta/L_1^h = 3hm/2m = 1.5h$. Similarly, for the second mode: $L_2^h = -m$, $M_2 = 3m$, $M_2^* = (L_2^h)^2/M_2 = \frac{1}{3}m$, $L_2^\theta = h(2m)(-1) + 2h(m)1 = 0$, and $h_2^* = L_2^\theta/L_2^h = 0$.

Observe that $M_1^* + M_2^* = 3m$, the total mass of the frame, confirming that Eq. (13.2.14) is satisfied; also note that the effective height for the second mode is zero, implying that the base overturning moment $M_{b2}(t)$ due to that mode will be zero at all t. This is an illustration of a more general result developed in Example 13.6.

Example 13.6

Show that the base overturning moment in a multistory building due to the second and higher modes is zero if the first mode shape is linear (i.e., the floor displacements are proportional to floor heights above the base).

Solution Equation (13.2.15) gives the nth-mode contribution to the base overturning moment. A linear first mode implies that $\phi_{j1} = h_j/h_N$, where h_j is the height of the jth floor above the base and h_N is the total height of the building. Substituting $h_j = h_N \phi_{j1}$ in (13.2.9b) gives

$$L_n^\theta = \sum_{j=1}^{N} h_j m_j \phi_{jn} = h_N \boldsymbol{\phi}_1^T \mathbf{m} \boldsymbol{\phi}_n$$

and this is zero for all $n \neq 1$ because of the orthogonality property of modes. Therefore, for all $n \neq 1$, $h_n^* = 0$ from Eq. (13.2.9a) and $M_{bn}(t) = 0$ from Eq. (13.2.15).

13.2.6 Example: Five-Story Shear Frame

In this section the earthquake analysis procedure summarized in Section 13.2.4 is implemented for the five-story shear frame of Fig. 12.8.1, subjected to the El Centro ground motion shown in Fig. 6.1.4. The results presented are accompanied by interpretive comments that should assist us in developing an understanding of the response behavior of multistory buildings.

System properties. The lumped mass $m_j = m = 100$ kips/g at each floor, the lateral stiffness of each story is $k_j = k = 31.54$ kips/in., and the height of each story is 12 ft. The damping ratio for all natural modes is $\zeta_n = 5\%$. The mass matrix $\mathbf{m}$, stiffness matrix $\mathbf{k}$, natural frequencies, and natural modes of this system were presented in Section 12.8. For the given k and m, the natural periods are $T_n = 2.0$, 0.6852, 0.4346, 0.3383, and 0.2966 sec. (These natural periods, which are much longer than for typical five-story buildings, were chosen to accentuate the contributions of the second through fifth modes to the structural response.) Thus steps 1 to 3 of the analysis procedure (Section 13.2.4) have already been completed.

Modal expansion of m1. With the modes ϕ_n known, the modal properties M_n, L_n^h, and L_n^θ are computed from Eqs. (13.2.3) and (13.2.9b) (Table 13.2.2). The Γ_n are computed from Eq. (13.2.3) and substituted in Eq. (13.2.4), together with values for m_j and ϕ_{jn}, to obtain the $\mathbf{s}_n$ vectors shown in Fig. 13.2.4. The contribution of the fundamental mode to the force distribution $\mathbf{s} = \mathbf{m1}$ of the effective earthquake forces is the largest, and the modal contributions to these forces decrease progressively for higher modes.

TABLE 13.2.2 MODAL PROPERTIES

Mode	M_n	L_n^h	L_n^θ / h
1	1.000	1.067	3.750
2	1.000	−0.336	0.404
3	1.000	0.177	0.135
4	1.000	−0.099	0.059
5	1.000	0.045	0.023

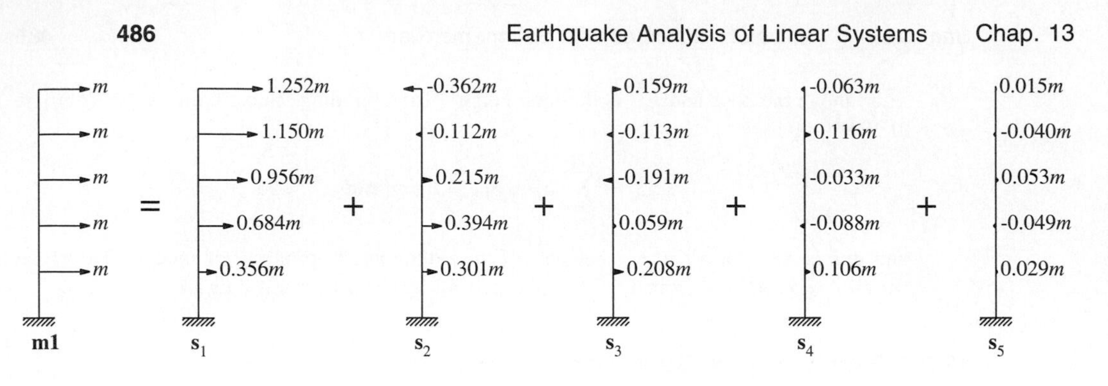

Figure 13.2.4 Modal expansion of **m1**.

Modal static responses. Table 13.2.3 gives the results for four response quantities—base shear V_b, fifth-story shear V_5, base overturning moment M_{bn}, and roof displacement u_5—obtained using the equations in Table 13.2.1 and the known s_{jn}, ϕ_{5n}, and ω_n^2. The effective modal masses $M_n^* = V_{bn}^{\text{st}}$ and effective modal heights $h_n^* = M_{bn}^{\text{st}}/V_{bn}^{\text{st}}$ are shown schematically in Fig. 13.2.5; note that the algebraic sign of h_n^* is ignored in the values shown. Observe that $\sum M_n^* = 5m$, confirming that Eq. (13.2.14) is satisfied. Also note that $\sum h_n^* M_n^* = 15mh$; this is the same as $\sum h_j m_j = 15mh$, confirming that Eq. (13.2.17) is satisfied.

TABLE 13.2.3 MODAL STATIC RESPONSES

Mode	V_{bn}^{st}/m	V_{5n}^{st}/m	M_{bn}^{st}/mh	u_{5n}^{st}
1	4.398	1.252	15.45	0.127
2	0.436	−0.362	−0.525	−0.004
3	0.121	0.159	0.092	0.0008
4	0.037	−0.063	−0.022	−0.0002
5	0.008	0.015	0.004	0.00003

Earthquake excitation. The ground acceleration $\ddot{u}_g(t)$ is defined by its numerical values at time instants equally spaced at every Δt. This time step $\Delta t = 0.01$ sec is chosen to be small enough to define $\ddot{u}_g(t)$ accurately and to determine accurately the response of SDF systems with natural periods T_n, the shortest of which is 0.2966 sec.

Response of SDF systems. The deformation response $D_n(t)$ of the nth-mode SDF system with natural period T_n and damping ratio ζ_n to the ground motion is determined (step 5b of Section 13.2.4). The time-stepping linear acceleration method (Chapter 5) was implemented to obtain discrete values of D_n at every Δt. For convenience, however, we continue to denote these discrete values as $D_n(t)$. At each time instant the pseudo-acceleration is calculated from $A_n(t) = \omega_n^2 D_n(t)$. These computations are implemented for the SDF systems corresponding to each of the five modes of the structure, and the results are presented in Fig. 13.2.6.

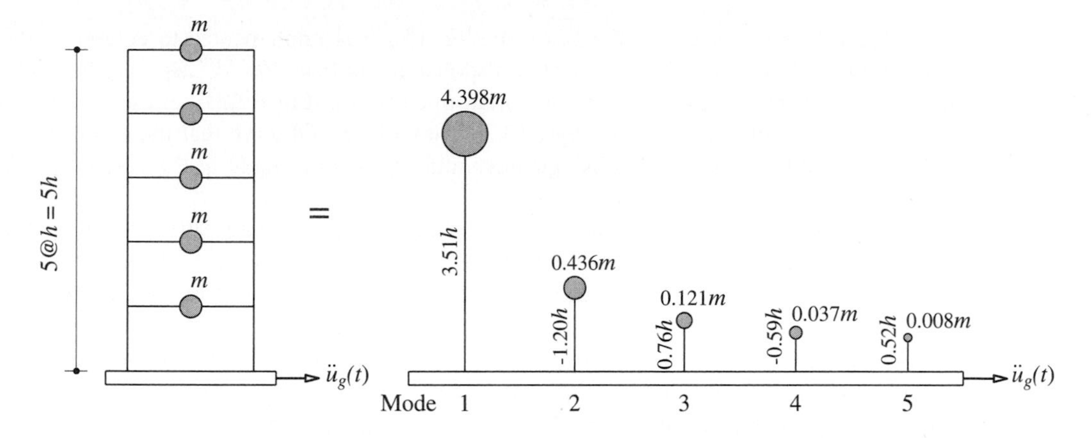

Figure 13.2.5 Effective modal masses and effective modal heights.

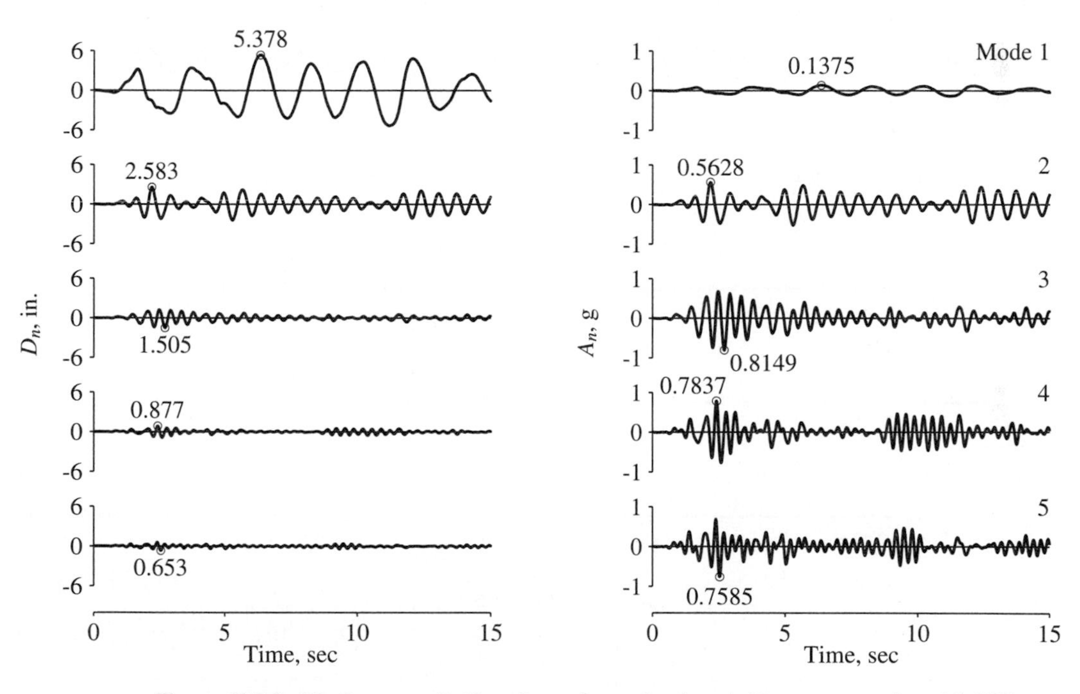

Figure 13.2.6 Displacement $D_n(t)$ and pseudo-acceleration $A_n(t)$ responses of modal SDF systems.

As seen in Chapter 6, the peak values of $D_n(t)$ and $A_n(t)$, noted in Fig. 13.2.6, can be determined directly from the response spectrum for the ground motion. This fact will enable us to determine the peak value of the nth-mode contribution to any response quantity directly from the response spectrum. Part B of this chapter is devoted to this development.

Modal responses. Step 5c of Section 13.2.4 is implemented to determine the contribution of the nth mode to selected response quantities: V_b, V_5, M_{bn}, and u_5. The modal static responses (Table 13.2.3) are multiplied by A_n (Fig. 13.2.6) at each time step to obtain the results presented in Figs. 13.2.7 and 13.2.8. Observe that the contribution of the nth mode to every response quantity attains its peak value at the same time as $A_n(t)$ does.

Figures 13.2.7 and 13.2.8 give us a first impression of the relative values of the response contributions of the different modes. The modal expansion of the excitation vector (Fig. 13.2.4) had suggested that the response will be largest in the fundamental mode and will tend to decrease in the higher modes. Such is the case in this example for roof displacement, base shear, and base overturning moment but not for the fifth-story shear. How the relative modal responses depend on the response quantity and on the building properties is discussed in Chapter 18.

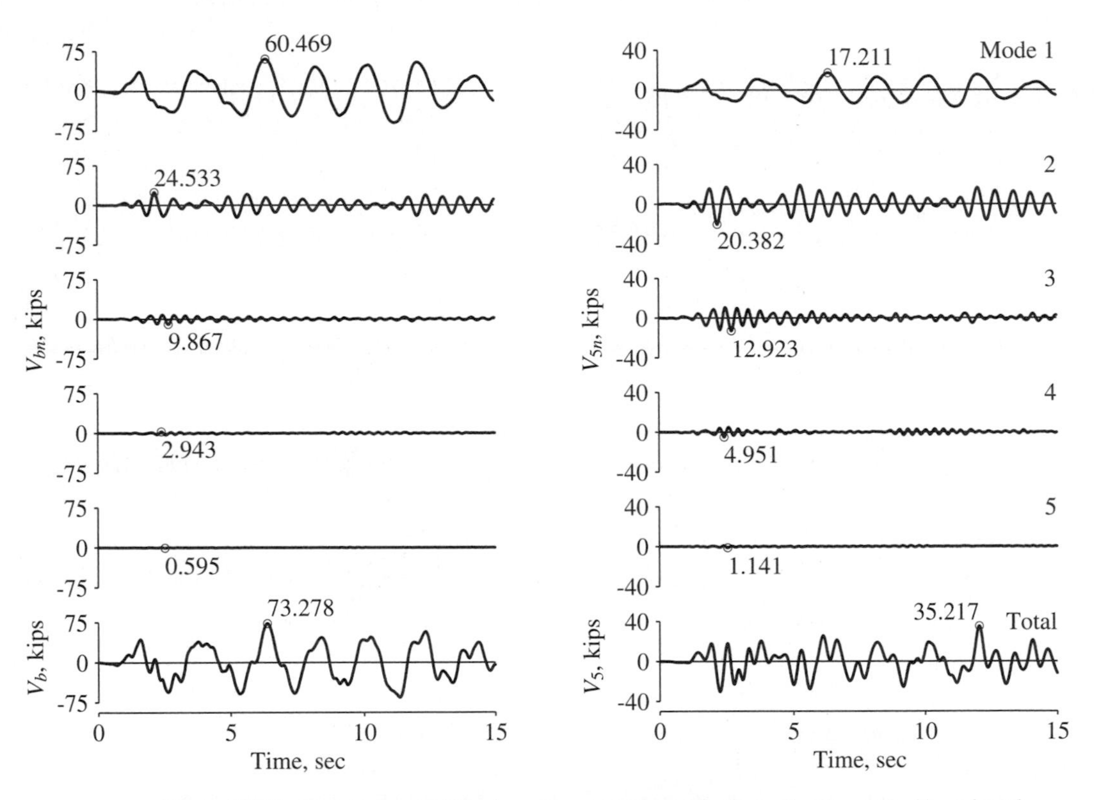

Figure 13.2.7 Base shear and fifth-story shear: modal contributions, $V_{bn}(t)$ and $V_{5n}(t)$, and total responses, $V_b(t)$ and $V_5(t)$.

Total responses. The total responses, determined by combining the modal contributions $r_n(t)$, according to Eq. (13.2.10), are shown in Figs. 13.2.7 and 13.2.8. The peak value of the total response occurs at a time instant different from when the indi-

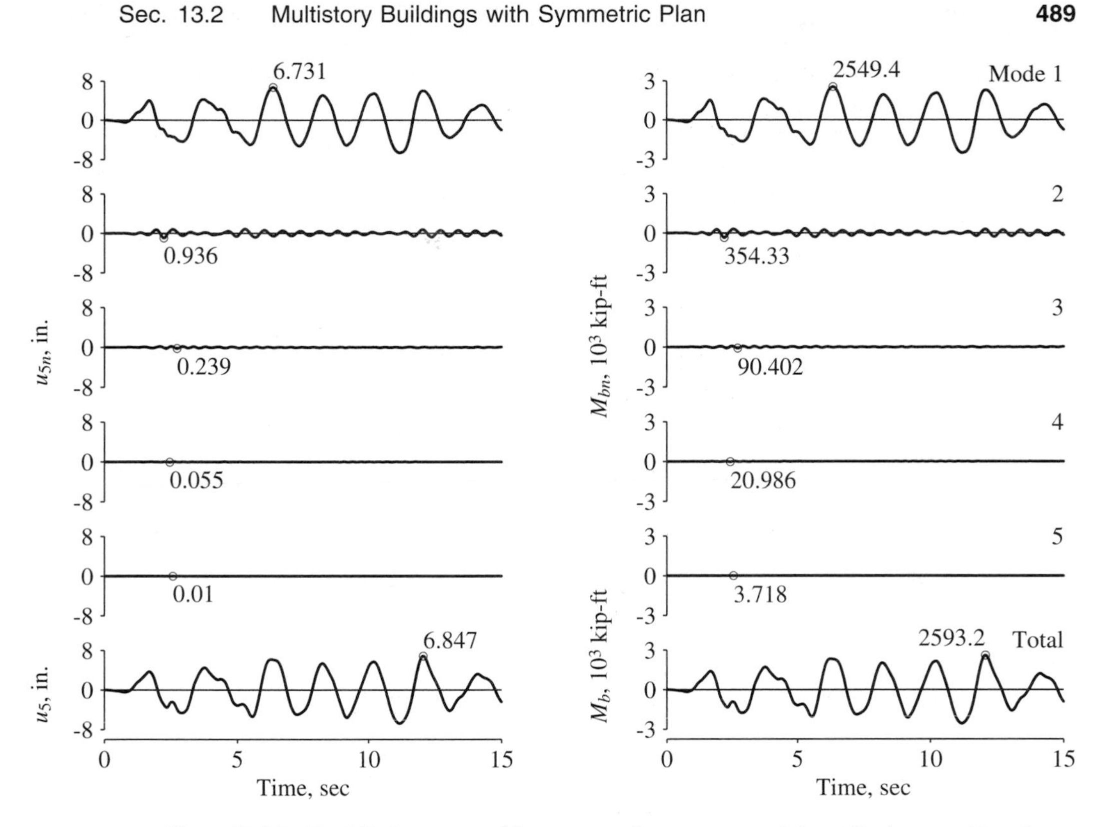

Figure 13.2.8 Roof displacement and base overturning moment: modal contributions, $u_{5n}(t)$ and $M_{bn}(t)$, and total responses, $u_5(t)$ and $M_b(t)$.

vidual modal peaks are attained. The peak values of the total responses for the four response quantities occur at different time instants because the relative values of the modal contributions vary with the response quantity.

The results presented indicate that it is not necessary to include the contributions of all the modes in computing the response of a multistory building; the lower few modes may suffice and the modal summations can be truncated accordingly. In this particular example, the contribution of the fourth and fifth modes could be neglected; the results would still be accurate enough for use in structural design. How many modes should be included depends on the earthquake ground motion and building properties. This issue is addressed in Chapter 18.

13.2.7 Example: Four-Story Frame with an Appendage

This section is concerned with the earthquake analysis and response of a five-story shear frame that has a very light and flexible fifth story. This may represent an idealization of a four-story building with a light appendage—a small housing for mechanical equipment, advertising billboard, or the like. This example is presented because it brings out certain special response features representative of a system with two natural frequencies that are close.

System properties. The lumped masses at the first four floors are $m_j = m$, at the fifth floor $m_5 = 0.01m$, and $m = 100$ kips/g. The lateral stiffness of each of the first four stories is $k_j = k$, the fifth-story stiffness $k_5 = 0.0012k$, and $k = 22.599$ kips/in. The height of each story is 12 ft. The damping ratio for all natural modes is $\zeta_n = 5\%$. The response of this system to the El Centro ground motion is determined. The analysis procedure and its implementation are identical to Section 13.2.6; therefore, only a summary of the results is presented.

Summary of results. The natural periods T_n and modes ϕ_n of this system are presented in Fig. 13.2.9. Observe that T_1 and T_2 are close and the corresponding modes show large deformations in the appendage. Table 13.2.4 gives the modal static responses for the base shear V_b and appendage shear V_5. Observe that V_{bn}^{st} for the first two modes are similar in magnitude and of the same algebraic sign; V_{5n}^{st} for the first two modes are also of similar magnitude but of opposite signs.

The responses $D_n(t)$ and $A_n(t)$ of the SDF systems corresponding to the five modes of the system are shown in Fig. 13.2.10. Note that $D_n(t)$—also $A_n(t)$—for the first two modes are essentially in phase because the two natural periods are close; the peak values are similar because of similar periods and identical damping in the two modes.

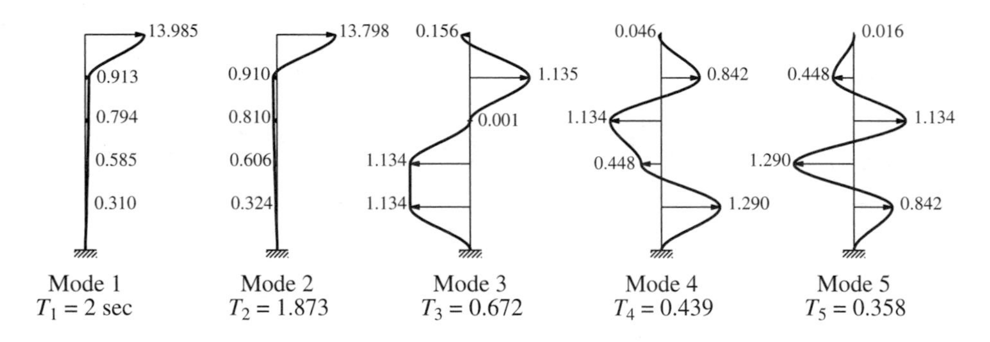

Figure 13.2.9 Natural periods and modes of vibration of building with appendage.

TABLE 13.2.4 MODAL STATIC RESPONSES

	Mode				
	1	2	3	4	5
V_{bn}^{st}/m	1.951	1.633	0.333	0.078	0.015
V_{5n}^{st}/m_5	9.938	−8.979	0.046	−0.007	0.0001

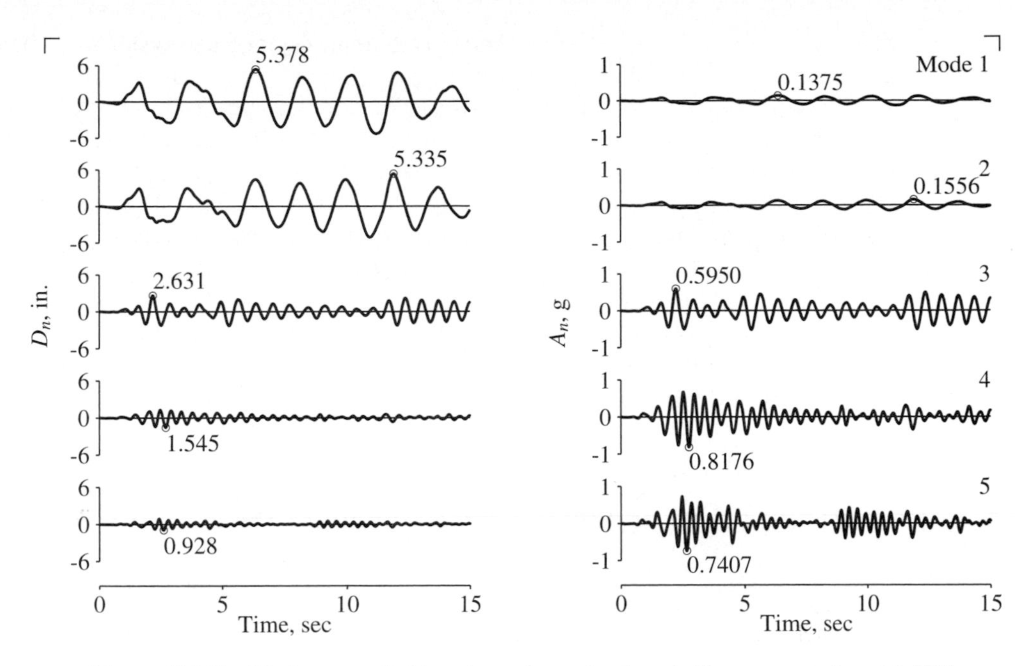

Figure 13.2.10 Displacement $D_n(t)$ and pseudo-acceleration $A_n(t)$ responses of modal SDF systems.

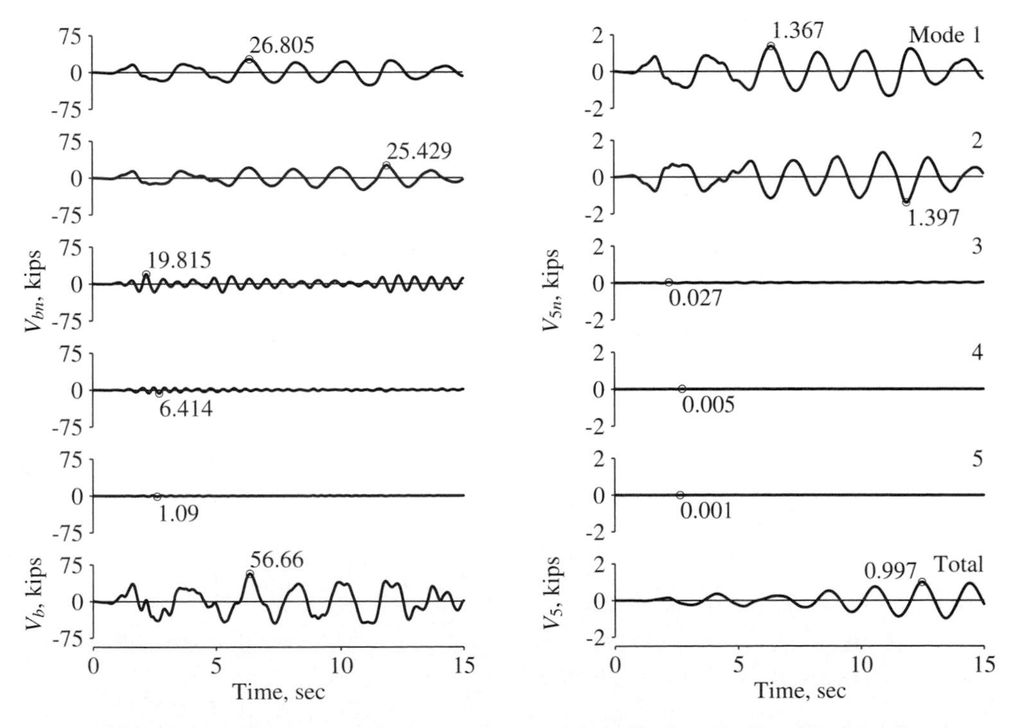

Figure 13.2.11 Base shear and appendage shear: modal contributions, $V_{bn}(t)$ and $V_{5n}(t)$, and total responses, $V_b(t)$ and $V_5(t)$.

The modal contributions to the base shear and to the appendage shear together with the total response are presented in Fig. 13.2.11. Observe that the response contributions of the first two modes are similar in magnitude because the modal static responses are about the same and the $A_n(t)$ are similar. In the case of base shear the two modal contributions are roughly in phase and hence additive because the two modal static responses are of the same algebraic sign. The combined response is therefore much larger than the individual modal responses. In contrast, these modal contributions to the appendage shear have opposite phase and they tend to cancel each other because the two modal static responses are of opposite sign. The combined response is therefore much smaller than the individual modal responses.

13.3 MULTISTORY BUILDINGS WITH UNSYMMETRIC PLAN

In this section the modal analysis of Section 13.1 is specialized for multistory buildings with their plans symmetric about the x-axis but unsymmetric about the y-axis subjected to ground motion $\ddot{u}_{gy}(t)$ in the y-direction. Equation (9.5.28) governs the motion of the system of Fig. 9.5.3, which has floor diaphragms with centers of mass along the same vertical axis and the same radius of gyration r; this equation is repeated for convenience:

$$\begin{bmatrix} \mathbf{m} & \mathbf{0} \\ \mathbf{0} & r^2\mathbf{m} \end{bmatrix} \begin{Bmatrix} \ddot{\mathbf{u}}_y \\ \ddot{\mathbf{u}}_\theta \end{Bmatrix} + \begin{bmatrix} \mathbf{k}_{yy} & \mathbf{k}_{y\theta} \\ \mathbf{k}_{\theta y} & \mathbf{k}_{\theta\theta} \end{bmatrix} \begin{Bmatrix} \mathbf{u}_y \\ \mathbf{u}_\theta \end{Bmatrix} = - \begin{bmatrix} \mathbf{m} & \mathbf{0} \\ \mathbf{0} & r^2\mathbf{m} \end{bmatrix} \begin{Bmatrix} \mathbf{1} \\ \mathbf{0} \end{Bmatrix} \ddot{u}_{gy}(t) \qquad (13.3.1)$$

The general analysis procedure developed in Section 13.1 is applicable to unsymmetric–plan buildings because Eq. (13.3.1) is of the same form as Eq. (13.1.1).

13.3.1 Modal Expansion of Effective Earthquake Forces

The effective earthquake forces $\mathbf{p}_{\text{eff}}(t)$ are defined by the right side of Eq. (13.3.1):

$$\mathbf{p}_{\text{eff}}(t) = - \begin{Bmatrix} \mathbf{m1} \\ \mathbf{0} \end{Bmatrix} \ddot{u}_{gy}(t) \qquad (13.3.2)$$

The modal expansion of the spatial distribution of these effective earthquake forces can be derived (following Section 12.8) to be

$$\begin{Bmatrix} \mathbf{m1} \\ \mathbf{0} \end{Bmatrix} = \sum_{n=1}^{2N} \mathbf{s}_n = \sum_{n=1}^{2N} \Gamma_n \begin{Bmatrix} \mathbf{m}\boldsymbol{\phi}_{yn} \\ r^2\mathbf{m}\boldsymbol{\phi}_{\theta n} \end{Bmatrix} \qquad (13.3.3)$$

In this equation $\boldsymbol{\phi}_n = \langle \boldsymbol{\phi}_{yn}^T \quad \boldsymbol{\phi}_{\theta n}^T \rangle$, where $\boldsymbol{\phi}_{yn}$ includes the translations and $\boldsymbol{\phi}_{\theta n}$ the rotations of the N floors about a vertical axis;

$$\Gamma_n = \frac{L_n^h}{M_n} \qquad (13.3.4)$$

where

$$L_n^h = \langle \phi_{yn}^T \quad \phi_{\theta n}^T \rangle \begin{Bmatrix} \mathbf{m1} \\ \mathbf{0} \end{Bmatrix} = \phi_{yn}^T \mathbf{m1} = \sum_{j=1}^{N} m_j \phi_{jyn} \tag{13.3.5}$$

and

$$M_n = \langle \phi_{yn}^T \quad \phi_{\theta n}^T \rangle \begin{bmatrix} \mathbf{m} & \\ & r^2\mathbf{m} \end{bmatrix} \begin{Bmatrix} \phi_{yn} \\ \phi_{\theta n} \end{Bmatrix}$$

or

$$M_n = \phi_{yn}^T \mathbf{m} \phi_{yn} + r^2 \phi_{\theta n}^T \mathbf{m} \phi_{\theta n} = \sum_{j=1}^{N} m_j \phi_{jyn}^2 + r^2 \sum_{j=1}^{N} m_j \phi_{j\theta n}^2 \tag{13.3.6}$$

where j denotes the floor number and m_j the floor mass. Equation (13.3.5) differs from Eq. (13.2.3b) for symmetric-plan systems because ϕ_{yn} is not necessarily the same as ϕ_n. In Eq. (13.3.3) the nth-mode contribution to the spatial distribution of effective earthquake forces is

$$\mathbf{s}_n = \begin{Bmatrix} \mathbf{s}_{yn} \\ \mathbf{s}_{\theta n} \end{Bmatrix} = \Gamma_n \begin{Bmatrix} \mathbf{m}\phi_{yn} \\ r^2\mathbf{m}\phi_{\theta n} \end{Bmatrix} \tag{13.3.7}$$

The jth element of these subvectors gives the lateral force s_{jyn} and torque $s_{j\theta n}$ at the jth floor level:

$$s_{jyn} = \Gamma_n m_j \phi_{jyn} \qquad s_{j\theta n} = \Gamma_n r^2 m_j \phi_{j\theta n} \tag{13.3.8}$$

Premultiplying each submatrix equation in Eq. (13.3.3) by $\mathbf{1}^T$, two interesting results can be derived:

$$\sum_{n=1}^{2N} M_n^* = \sum_{j=1}^{N} m_j \qquad \sum_{n=1}^{2N} I_{On}^* = 0 \tag{13.3.9}$$

where

$$M_n^* = \frac{(L_n^h)^2}{M_n} \qquad I_{On}^* = \sum_{j=1}^{N} r^2 \Gamma_n m_j \phi_{j\theta n} \tag{13.3.10}$$

Although this equation for M_n^* for unsymmetric-plan systems is the same as Eq. (13.2.9a) for symmetric-plan systems, the two may not be identical because the mode shapes ϕ_{yn} and ϕ_n in the two cases are not necessarily the same. We shall see later that M_n^* is the base shear effective modal mass for the nth mode, and also the modal static response for base shear. As for symmetric-plan buildings, Eq. (13.3.9a) implies that the sum of the effective modal masses over all modes is equal to the total mass of the building. As we shall see later, rI_{On}^* is the modal static response for base torque; their sum of over all modes is zero according to Eq. (13.3.9b).

Example 13.7

Determine the modal expansion for the distribution of the effective earthquake forces for the system of Example 10.6. Also compute the modal static responses for base shear and base torque, and verify Eq. (13.3.9)

Solution The DOFs are the lateral displacement u_y and rotation u_θ of the roof. With reference to these DOFs, the natural frequencies and modes were determined in Example 10.6:

$$\omega_1 = 5.878 \qquad \omega_2 = 6.794 \text{ rad/sec}$$

$$\phi_1 = \begin{Bmatrix} -0.5228 \\ 0.0493 \end{Bmatrix} \qquad \phi_2 = \begin{Bmatrix} -0.5131 \\ -0.0502 \end{Bmatrix}$$

Specializing Eqs. (13.3.6), (13.3.5), and (13.3.4) to a one-story system with roof mass m and radius of gyration r yields

$$M_n = m(\phi_{yn}^2 + r^2\phi_{\theta n}^2) \qquad L_n^h = m\phi_{yn} \qquad \Gamma_n = \frac{L_n^h}{M_n} \tag{a}$$

For this system, $m = 1.863$ kip-sec^2/ft and $r^2 = (b^2+d^2)/12 = (30^2+20^2)/12 = 108.\overline{3}$ ft^2 (see Example 10.6). In Eq. (a) with $n = 1$, substituting known values of m, r, ϕ_{y1}, and $\phi_{\theta 1}$, we obtain $M_1 = 1.863\left[(-0.5228)^2 + 108.\overline{3}(0.0493)^2\right] = 1.0$, $L_1^h = 1.863(-0.5228) = -0.974$, and $\Gamma_1 = -0.974$. Similarly, in Eq. (a) with $n = 2$, substituting for m, r, ϕ_{y2}, and $\phi_{\theta 2}$, we obtain $M_2 = 1.0$, $L_2^h = -0.956$, and $\Gamma_2 = -0.956$. Specializing Eq. (13.3.10) to a one-story system gives

$$M_n^* = \frac{(L_n^h)^2}{M_n} \qquad I_{On}^* = \Gamma_n r^2 m\phi_{\theta n} \tag{b}$$

With $n = 1$, substitute numerical values for Γ_n, L_n^h, M_n, r, m, and $\phi_{\theta n}$ to obtain

$$\frac{M_1^*}{m} = \frac{(-0.974)^2}{1.863} = 0.509 \qquad \frac{I_{O1}^*}{m} = (-0.974)108.\overline{3}(0.0493) = -5.203$$

Similarly, for the second mode,

$$\frac{M_2^*}{m} = 0.491 \qquad \frac{I_{O2}^*}{m} = 5.203$$

Observe that these data show that $M_1^* + M_2^* = m$ and $I_{O1}^* + I_{O2}^* = 0$, which provides a numerical confirmation of Eq. (13.3.9) for this one-story ($N = 1$) system.

The modal expansion of the effective earthquake force distribution is obtained from Eq. (13.3.3) by specializing it to a one-story system:

$$\begin{Bmatrix} m \\ 0 \end{Bmatrix} = \begin{Bmatrix} M_1^* + M_2^* \\ I_{O1}^* + I_{O2}^* \end{Bmatrix} \tag{c}$$

Substituting numerical values for M_n^* and I_{On}^* gives

$$m\begin{Bmatrix} 1 \\ 0 \end{Bmatrix} = m\begin{Bmatrix} 0.509 \\ -5.203 \end{Bmatrix} + m\begin{Bmatrix} 0.491 \\ 5.203 \end{Bmatrix}$$

This modal expansion is shown on the structural plan in Fig. E13.7.

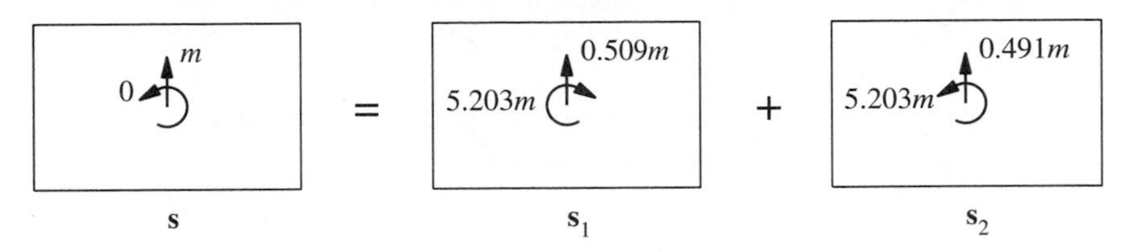

Figure E13.7 Modal expansion of effective force vector shown on plan view of the building.

13.3.2 Modal Responses

Displacements. The differential equation governing the nth modal coordinate is Eq. (13.1.7) with $\ddot{u}_g(t)$ replaced by $\ddot{u}_{gy}(t)$ and Γ_n defined by Eqs. (13.3.4) to (13.3.6). Using this Γ_n, Eq. (13.1.10) gives the contribution $\mathbf{u}_n(t)$ of the nth mode to the displacement $\mathbf{u}(t)$. The lateral displacement $\mathbf{u}_{yn}$ and torsional displacements $\mathbf{u}_{\theta n}$ are

$$\mathbf{u}_{yn}(t) = \Gamma_n \boldsymbol{\phi}_{yn} D_n(t) \qquad \mathbf{u}_{\theta n}(t) = \Gamma_n \boldsymbol{\phi}_{\theta n} D_n(t) \tag{13.3.11}$$

In particular, the lateral and torsional displacements of the jth floor are

$$u_{jyn}(t) = \Gamma_n \phi_{jyn} D_n(t) \qquad u_{j\theta n}(t) = \Gamma_n \phi_{j\theta n} D_n(t) \tag{13.3.12}$$

Building forces. The equivalent static forces $\mathbf{f}_n(t)$ associated with displacements $\mathbf{u}_n(t)$ include lateral forces $\mathbf{f}_{yn}(t)$ and torques $\mathbf{f}_{\theta n}(t)$. These forces are given by a generalization of Eq. (13.1.11):

$$\begin{Bmatrix} \mathbf{f}_{yn}(t) \\ \mathbf{f}_{\theta n}(t) \end{Bmatrix} = \begin{Bmatrix} \mathbf{s}_{yn} \\ \mathbf{s}_{\theta n} \end{Bmatrix} A_n(t) \tag{13.3.13}$$

The lateral force and torque at the jth floor level are

$$f_{jyn}(t) = s_{jyn} A_n(t) \qquad f_{j\theta n}(t) = s_{j\theta n} A_n(t) \tag{13.3.14}$$

Then the response r due to the nth mode is given by Eq. (13.1.13), repeated here for convenience:

$$r_n(t) = r_n^{\text{st}} A_n(t) \tag{13.3.15}$$

The modal static response r_n^{st} is determined by static analysis of the building due to external forces $\mathbf{s}_{yn}$ and $\mathbf{s}_{\theta n}$. In applying these forces to the structure, the direction of forces is controlled by the algebraic sign of ϕ_{jyn} and $\phi_{j\theta n}$. In particular for the fundamental mode, the lateral forces all act in the same direction, and the torques all act in the same direction (Fig. 13.3.1). However, for the second and higher modes, the lateral forces or torques, or both, will change direction as one moves up the structure.

The modal static responses are presented in Table 13.3.1 for eight response quantities: the shear V_i and torque T_i in the ith story, the overturning moment M_i at the ith floor, the base shear V_b, the base torque T_b, and the base overturning moment M_b, floor translations u_{jy}, and floor rotations $u_{j\theta}$. The equations for forces are determined by

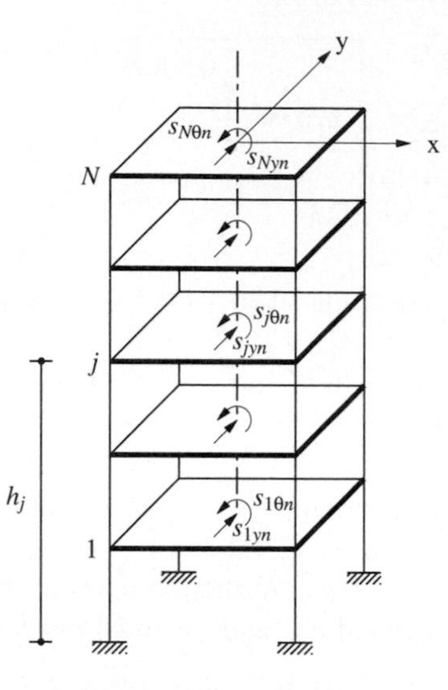

Figure 13.3.1 External forces s_{jyn} and $s_{j\theta n}$ for nth mode.

static analysis of the building subjected to lateral forces $\mathbf{s}_{yn}$ and torques $\mathbf{s}_{\theta n}$ (Fig. 13.3.1); and the results for u_{jy} and $u_{j\theta}$ are obtained by transforming Eqs. (13.3.12) into a form similar to Eq. (13.3.15). Parts of the equations for V_{bn}^{st}, M_{bn}^{st}, and T_{bn}^{st} are obtained by substituting Eq. (13.3.8) for s_{jyn} and $s_{j\theta n}$, Eq. (13.3.5) for L_n^h, Eq. (13.3.10) for M_n^* and I_{On}^*, and Eq. (13.2.9a) for h_n^* with ϕ_{jn} replaced by ϕ_{jyn} in Eq. (13.2.9b).

TABLE 13.3.1 MODAL STATIC RESPONSES

Response, r	Modal Static Response, r_n^{st}
V_i	$V_{in}^{\text{st}} = \sum_{j=i}^{N} s_{jyn}$
M_i	$M_{in}^{\text{st}} = \sum_{j=i}^{N} (h_j - h_i) s_{jyn}$
T_i	$T_{in}^{\text{st}} = \sum_{j=i}^{N} s_{j\theta n}$
V_b	$V_{bn}^{\text{st}} = \sum_{j=1}^{N} s_{jyn} = \Gamma_n L_n^h = M_n^*$
M_b	$M_{bn}^{\text{st}} = \sum_{j=1}^{N} h_j s_{jyn} = \Gamma_n L_n^\theta = h_n^* M_n^*$
T_b	$T_{bn}^{\text{st}} = \sum_{j=1}^{N} s_{j\theta n} = I_{On}^*$
u_{jy}	$u_{jyn}^{\text{st}} = (\Gamma_n/\omega_n^2)\phi_{jyn}$
$u_{j\theta}$	$u_{j\theta n}^{\text{st}} = (\Gamma_n/\omega_n^2)\phi_{j\theta n}$

Specializing Eq. (13.3.15) for V_b, M_b, and T_b and substituting for V_{bn}^{st}, M_{bn}^{st}, and T_{bn}^{st} from Table 13.3.1 gives

$$V_{bn}(t) = M_n^* A_n(t) \qquad T_{bn}(t) = I_{On}^* A_n(t) \qquad M_{bn}(t) = h_n^* M_n^* A_n(t) \tag{13.3.16}$$

For reasons mentioned in Section 13.2.5, M_n^* is called the *effective modal mass* and h_n^* the *effective modal height*.

Frame forces. In addition to the overall story forces for the building, it is desired to determine the element forces—bending moments, shears, etc.—in structural elements—beams, columns, walls, etc.—of each frame of the building. For this purpose, the lateral displacements $\mathbf{u}_{in}$ of the ith frame associated with displacements $\mathbf{u}_n$ in the floor DOFs of the building are determined from Eq. (9.5.21). Substituting Eq. (9.5.22) for $\mathbf{a}_{xi}$ and $\mathbf{a}_{yi}$, $\mathbf{u}_n^T = \langle \mathbf{u}_{yn}^T \quad \mathbf{u}_{\theta n}^T \rangle$, and Eq. (13.3.11) for $\mathbf{u}_{yn}$ and $\mathbf{u}_{\theta n}$ leads to

$$\mathbf{u}_{in}(t) = \Gamma_n(-y_i \boldsymbol{\phi}_{\theta n}) D_n(t) \qquad \mathbf{u}_{in}(t) = \Gamma_n(\boldsymbol{\phi}_{yn} + x_i \boldsymbol{\phi}_{\theta n}) D_n(t) \tag{13.3.17}$$

The first equation is for frames oriented in the x-direction and the second for frames in the y-direction. At each time instant, the internal forces in elements of frame i can be determined from these displacements and joint rotations (see Example 13.4) using the element stiffness properties (Appendix 1).

Alternatively, equivalent static forces $\mathbf{f}_{in}$ can be defined for the ith frame with lateral stiffness matrix $\mathbf{k}_{xi}$ if the frame is oriented in the x-direction or $\mathbf{k}_{yi}$ for a frame along the y-direction. Thus

$$\mathbf{f}_{in}(t) = \mathbf{k}_{xi}\mathbf{u}_{in}(t) = \Gamma_n \mathbf{k}_{xi}(-y_i \boldsymbol{\phi}_{\theta n}) D_n(t) \tag{13.3.18a}$$

$$\mathbf{f}_{in}(t) = \mathbf{k}_{yi}\mathbf{u}_{in}(t) = \Gamma_n \mathbf{k}_{yi}(\boldsymbol{\phi}_{yn} + x_i \boldsymbol{\phi}_{\theta n}) D_n(t) \tag{13.3.18b}$$

where Eq. (13.3.17) is used to obtain the second part of these equations. At each instant of time the element forces are determined by static analysis of the ith frame subjected to the lateral forces $\mathbf{f}_{in}(t)$ shown in Fig. 13.3.2.

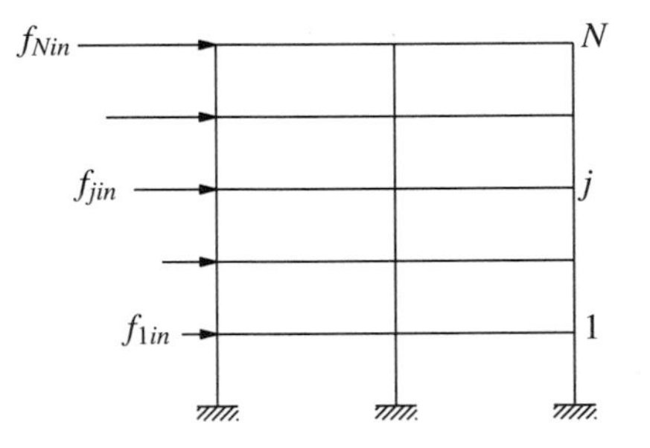

Figure 13.3.2 Equivalent static forces for ith frame associated with response of the building in its nth natural mode.

13.3.3 Total Response

Combining the response contributions of all the modes gives the total response of the unsymmetric-plan building to earthquake excitation:

$$r(t) = \sum_{n=1}^{2N} r_n(t) = \sum_{n=1}^{2N} r_n^{\text{st}} A_n(t) \tag{13.3.19}$$

wherein Eq. (13.3.15) has been substituted for $r_n(t)$, the nth-mode response.

13.3.4 Summary

The response history of an N-story building with plan unsymmetric about the y-axis to earthquake ground motion in the y-direction can be computed by the procedure just developed, which is summarized next in step-by-step form:

1. Define the ground acceleration $\ddot{u}_{gy}(t)$ numerically at every time step Δt.
2. Define the structural properties.
   a. Determine the mass and stiffness matrices from Eqs. (13.3.1) and (9.5.26).
   b. Estimate the modal damping ratios ζ_n (Chapter 11).
3. Determine the natural frequencies ω_n (natural periods $T_n = 2\pi/\omega_n$) and natural modes of vibration.
4. Determine the modal components $\mathbf{s}_n^T = \langle \mathbf{s}_{yn}^T \quad \mathbf{s}_{\theta n}^T \rangle$—defined by Eqs. (13.3.7) and (13.3.8)—of the effective force distribution.
5. Compute the response contribution of the nth mode by the following steps, which are repeated for all modes, $n = 1, 2, \ldots, 2N$:
   a. Perform static analysis of the building subjected to lateral forces $\mathbf{s}_{yn}$ and torques $\mathbf{s}_{\theta n}$ to determine r_n^{st}, the modal static response for each desired response quantity r (Table 13.3.1).
   b. Determine the deformation response $D_n(t)$ and pseudo-acceleration response $A_n(t)$ of the nth-mode SDF system to $\ddot{u}_{gy}(t)$, using numerical time-stepping methods (Chapter 5).
   c. Determine $r_n(t)$ from Eq. (13.3.15). This equation may also be used to determine the element forces in the ith frame provided that the modal static responses are derived for these response quantities. Alternatively, these internal forces can be determined by static analysis of the frame subjected to the lateral forces of Eq. (13.3.18).
6. Combine the modal contributions $r_n(t)$ to determine the total response using Eq. (13.3.19).

Only the modes with significant response contributions need to be included in modal analysis. The system considered has coupled lateral-torsional motion in $2N$ modes or N pairs of modes. For many buildings both modes in a pair have similar natural frequencies and responses of similar magnitude (see Example 13.8). Usually, only the lower few pairs of modes contribute significantly to the response. Therefore, steps 3, 4, and 5 need to be implemented only for these modal pairs, and the modal summation in step 6 truncated accordingly.

Extension to arbitrary-plan buildings. The procedure just summarized is for earthquake analysis of multistory buildings with plan unsymmetric about one axis, say the y-axis, but symmetric about the other axis, the x-axis, subjected to ground motion in the y-direction. This procedure can be extended to multistory buildings with arbitrary plan with no axis of symmetry. In this case the system has $3N$ dynamic degrees of freedom; each of the $3N$ modes generally contains coupled x-lateral, y-lateral, and torsional motions and is excited by ground motion in the x or y directions.

Example 13.8

Determine the response of the system of Examples 13.7 and 10.6 with modal damping ratios $\zeta_n = 5\%$ to the El Centro ground motion acting along the y-axis. The response quantities of interest are floor displacements, base shear, and base torque in the building, and base shear in frames A and B.

Solution Steps 1 to 4 of the analysis procedure (Section 13.3.4) have already been implemented in Examples 10.6 and 13.7.

Step 5a: Static analysis of the structure subjected to forces $\mathbf{s}_n$ (Fig. E13.7) gives the modal static responses: $V_{b1}^{\text{st}} = 0.509m$ and $V_{b2}^{\text{st}} = 0.491m$; $T_{b1}^{\text{st}} = -5.203m$ and $T_{b2}^{\text{st}} = 5.203m$. The modal static responses for the lateral displacement u_y and rotation u_θ of the roof are obtained by specializing the equations in Table 13.3.1 for this one-story system:

$$u_{yn}^{\text{st}} = \frac{\Gamma_n}{\omega_n^2}\phi_{yn} \qquad u_{\theta n}^{\text{st}} = \frac{\Gamma_n}{\omega_n^2}\phi_{\theta n} \tag{a}$$

Substituting numerical values for Γ_n, ϕ_{yn}, and $\phi_{\theta n}$ (from Example 13.7) for the first mode in Eq. (a) gives $u_{y1}^{\text{st}} = 0.509/\omega_1^2$ and $u_{\theta 1}^{\text{st}} = -0.0480/\omega_1^2$. Similarly, for the second mode, $u_{y2}^{\text{st}} = 0.491/\omega_2^2$ and $u_{\theta 2}^{\text{st}} = 0.0480/\omega_2^2$. Observe that the modal static responses u_{yn}^{st} and V_{bn}^{st} for the two modes are similar in magnitude and of the same algebraic sign; $u_{\theta n}^{\text{st}}$ and T_{bn}^{st} for the two modes are also of similar (identical for a one-story system) magnitude but of opposite signs.

Step 5b: Response analysis of the first-mode SDF system ($T_1 = 2\pi/\omega_1 = 2\pi/5.878 = 1.069$ sec and $\zeta_1 = 5\%$) and the second-mode SDF system ($T_2 = 2\pi/\omega_2 = 2\pi/6.794 = 0.9248$ sec and $\zeta_2 = 5\%$) to the El Centro ground motion gives the $D_n(t)$ and $A_n(t)$ shown in Fig. E13.8a. Observe that $D_n(t)$—also $A_n(t)$—for the two modes are essentially in phase because their natural periods are similar.

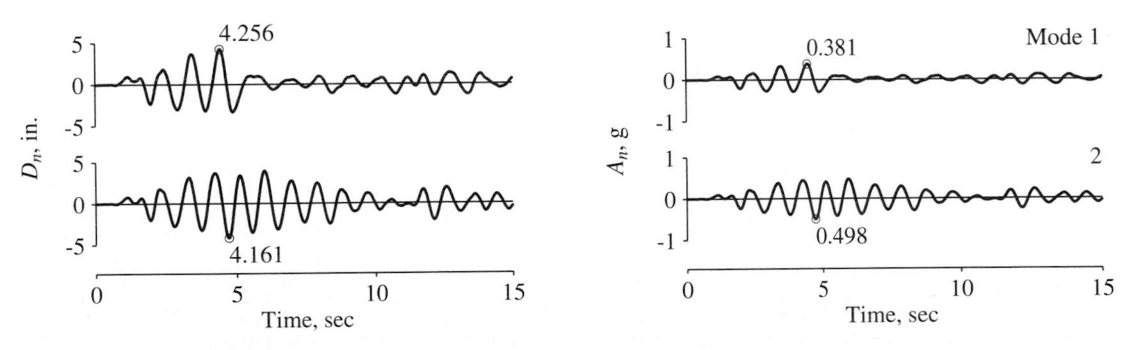

Figure E13.8a Displacement $D_n(t)$ and pseudo-acceleration $A_n(t)$ responses of modal SDF systems.

Step 5c: Substituting V_{bn}^{st} and T_{bn}^{st} from step 5a in Eq. (13.3.15) gives the contributions of the nth mode to the base shear and base torque:

$$V_{b1}(t) = 0.509mA_1(t) \qquad T_{b1}(t) = -5.203mA_1(t) \tag{b}$$

$$V_{b2}(t) = 0.491mA_2(t) \qquad T_{b2}(t) = 5.203mA_2(t) \tag{c}$$

Figure E13.8b shows $V_{bn}(t)$ and $T_{bn}(t)$ computed from Eqs. (b) and (c) using $m = 1.863$ kip–sec²/ft and the $A_n(t)$ in Fig. E13.8a.

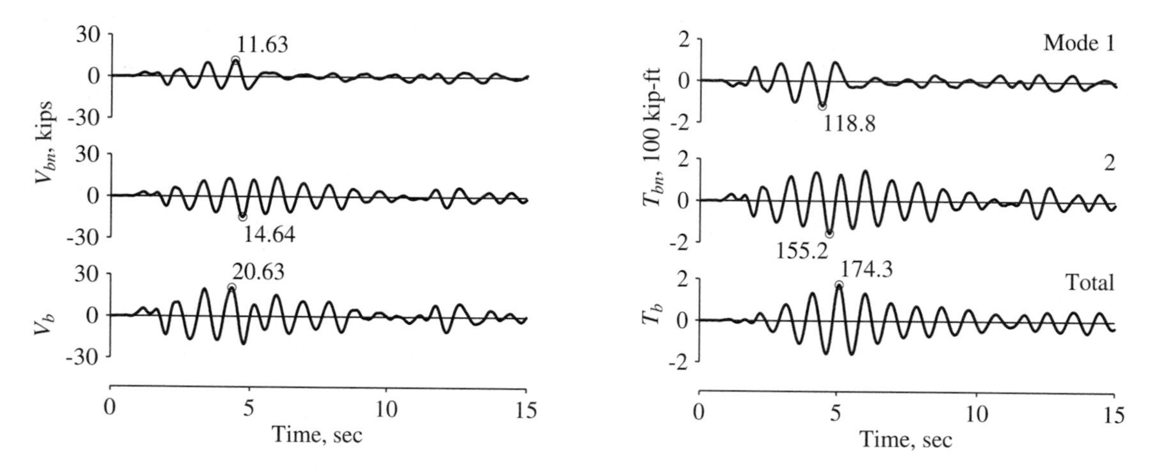

Figure E13.8b Base shear and base torque: modal contributions, $V_{bn}(t)$ and $T_{bn}(t)$, and total responses, $V_b(t)$ and $T_b(t)$.

Substituting u_{yn}^{st} and $u_{\theta n}^{\text{st}}$ from step 5a in Eq. (13.3.15) give the contributions of the nth mode to roof displacements:

$$u_{y1}(t) = 0.509D_1(t) \qquad u_{\theta 1}(t) = -0.0480D_1(t) \tag{d}$$

$$u_{y2}(t) = 0.491D_2(t) \qquad u_{\theta 2}(t) = 0.0480D_2(t) \tag{e}$$

where $D_n(t)$ is in units of feet. Figure E13.8c shows $u_{yn}(t)$ and $(b/2)u_{\theta n}(t)$ computed from Eqs. (d) and (e) using the $D_n(t)$ in Fig. E13.8a.

Observe that the modal contributions to lateral displacement (and to base shear) are similar in magnitude because the modal static responses are about the same and the $D_n(t)$ and $A_n(t)$ are similar for the two modes. The modal contributions are roughly in phase because the two periods are similar and the two modal static responses are of the same algebraic sign. The combined response is therefore much larger than the individual modal responses. In contrast, the modal contributions to roof rotation (and to base torque), although similar in magnitude, have different phase and their peaks are not directly additive. The combined response is therefore only slightly larger than the modal responses.

Consider another one-story unsymmetric-plan system which has the two natural periods much closer than for the structure analyzed in this example. For such a system, the modal contributions to roof rotation (and to base torque) will have opposite phase and tend to cancel each other because the two modal static responses are of opposite sign. The combined values of roof rotation and base torque will then be much smaller than the individual modal responses.

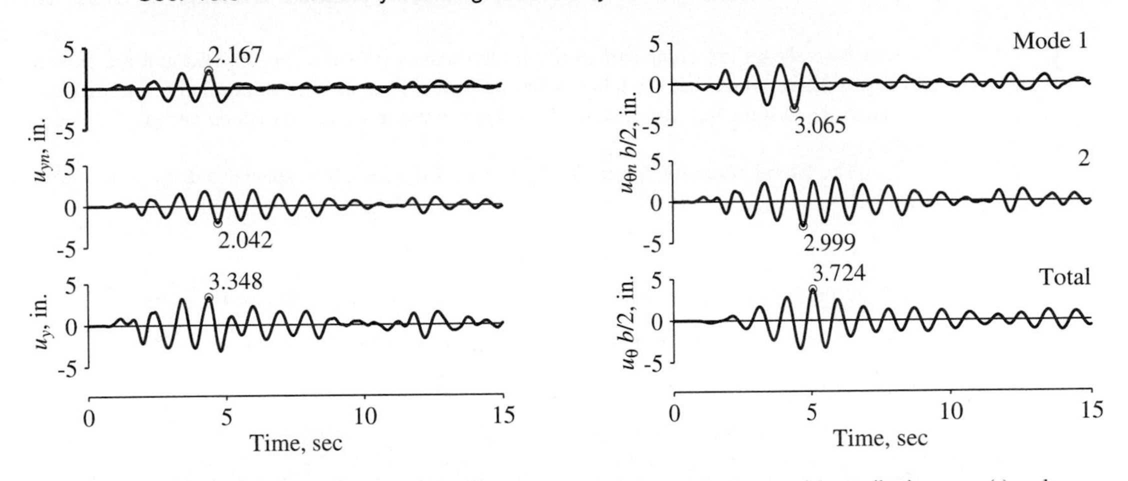

Figure E13.8c Lateral displacement and $b/2$ times roof rotation: modal contributions, $u_{yn}(t)$ and $(b/2)u_{\theta_n}(t)$, and total responses, $u_y(t)$ and $(b/2)u_\theta(t)$.

The lateral force for frame A is given by Eq. (13.3.18b) specialized for a one-story frame:

$$f_{An}(t) = k_A\left[u_{yn}(t) + x_A u_{\theta n}(t)\right] \tag{f}$$

Substituting for $k_A = 75$ kips/ft, $x_A = 1.5$ ft, $u_{yn}(t)$ and $u_{\theta n}(t)$ from Eqs. (d) and (e) gives

$$f_{A1}(t) = 2.733D_1(t)\text{kips} \qquad f_{A2}(t) = 3.516D_2(t)\text{kips}$$

where $D_n(t)$ are in feet. The base shear in a one-story frame is equal to the lateral force; thus the base shear due to the two modes is

$$V_{bA1}(t) = 2.733D_1(t) \qquad V_{bA2}(t) = 3.516D_2(t) \tag{g}$$

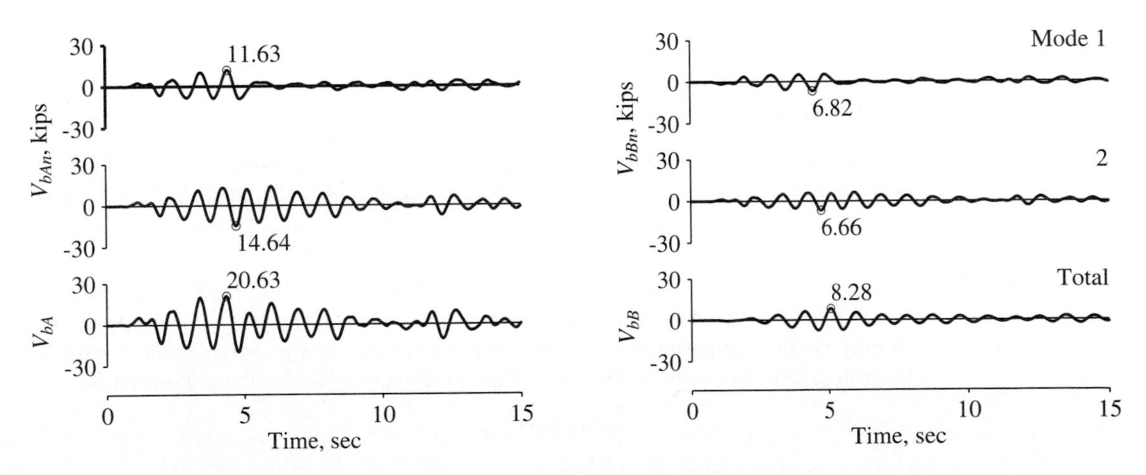

Figure E13.8d Base shears in frames A and B: modal contributions, V_{bAn} and V_{bBn}, and total responses, V_{bA} and V_{bB}.

These base shears are computed using the known $D_n(t)$ from Fig. E13.8a and are shown in Fig. E13.8d. Note that the base shears for frame A alone and the building are identical because the system has only frame A in the y direction and this frame carries the entire force.

The lateral force for frame B is given by Eq. (13.3.18a) specialized for a one-story frame:

$$f_{Bn}(t) = k_B\,[-y_B u_{\theta n}(t)] \tag{h}$$

Substituting for $k_B = 40$ kips/ft, $y_B = 10$ ft, $u_{\theta n}(t)$ from Eqs. (d) and (e) gives

$$f_{B1}(t) = 19.2D_1(t) \qquad f_{B2}(t) = -19.2D_2(t)$$

where $D_n(t)$ are in feet. The base shear due to the two modes is

$$V_{bB1}(t) = 19.2D_1(t) \qquad V_{bB2}(t) = -19.2D_2(t)$$

These base shears in frame B are computed using the known $D_n(t)$ from Fig. E13.8a and are shown in Fig. E13.8d.

Figure E13.8b–d show that the two modes contribute similarly to the response of this one-story system. This is typical of unsymmetric-plan systems where pairs of modes in a structure with one axis of symmetry (or triplets of modes if the system has no axis of symmetry) may have similar response contributions.

Step 6: Combining the modal contributions gives the total response for this two-DOF system:

$$r(t) = r_1(t) + r_2(t)$$

The combined values of lateral displacement, rotation, base shear, and base torque for the building, and base shear for frames A and B are shown in Fig. E13.8b–d. Observe that the combined response attains its peak value at a time instant different from when the modal peaks are attained.

Example 13.9

Identify the effects of plan asymmetry on the earthquake response of the one-story system of Example 13.8 by comparing its response with that of the symmetric-plan one-story system defined in Section 9.5.3.

Solution The response of the symmetric-plan system to ground motion in the y-direction is governed by the second of the three differential equations (9.5.20). Dividing this equation by m and introducing damping gives the familiar equation for an SDF system:

$$\ddot{u}_y + 2\zeta_y\omega_y\dot{u}_y + \omega_y^2 u_y = -\ddot{u}_{gy}(t) \tag{a}$$

where $\omega_y = \sqrt{k_y/m}$. As mentioned in Section 9.5.3, the y-component of ground motion will only produce lateral response in the y-direction without any torsion about a vertical axis or displacements in the x-direction. The lateral displacement in the y-direction is

$$u_y(t) = D(t, \omega_y, \zeta_y) \tag{b}$$

and the associated base shear in frame A is

$$V_{bA} = mA(t, \omega_y, \zeta_y) \tag{c}$$

where $D(t, \omega_y, \zeta)$ and $A(t, \omega_y, \zeta)$ denote the deformation and pseudo-acceleration responses, respectively, of an SDF system with natural frequency ω_y and damping ratio ζ to ground acceleration $\ddot{u}_{gy}(t)$. Frames B and C would experience no forces.

For the symmetric-plan system associated with Example 13.8, $\omega_y = 6.344$ (see Example 10.7) and the damping ratio is the same, $\zeta = 5\%$. The response of this SDF system is computed from Eqs. (a) to (c) and shown in Fig. E13.9, where it is also compared with the response of the unsymmetric-plan system (Example 13.8). It is clear that plan asymmetry has the effect of (1) modifying the lateral displacement and base shear in frame A, and (2) causing torsion in the system and forces in frames B and C that do not exist if the building plan is symmetric. In this particular case, the base shear in frame A is reduced because of plan asymmetry, but such is not always the case, depending on the natural period of the structure, ground motion characteristics, and the location of the frame in the building plan.

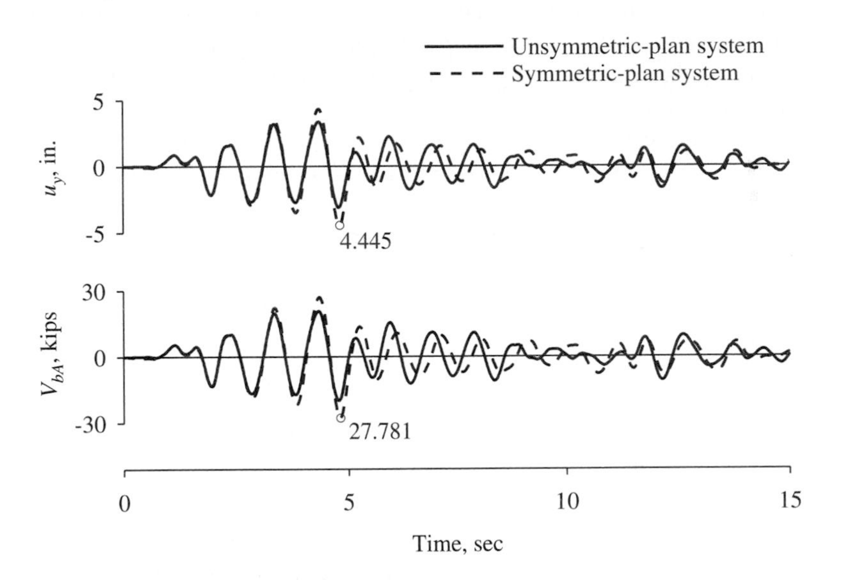

Figure E13.9 Responses of unsymmetric-plan system and symmetric-plan system.

13.4 TORSIONAL RESPONSE OF SYMMETRIC-PLAN BUILDINGS

In this section the torsional response of multistory buildings with their plans nominally symmetric about two orthogonal axis is discussed briefly. Such structures may undergo "accidental" torsional motions for several reasons, including the two principal factors: the building is usually not *perfectly* symmetric, and the spatial variations in ground motion may cause rotation (about the vertical axis) of the building's base, which will induce torsional motion of the building even if its plan is perfectly symmetric.

Consider first the analysis of torsional response of a building with a perfectly symmetric plan due to rotation of its base. For a given rotational excitation $\ddot{u}_{g\theta}(t)$, the governing equations (9.6.1) can be solved by the modal analysis procedure, considering the purely torsional vibration modes of the building. This procedure could be developed along the lines of Section 13.3. It is not presented, however, for two reasons: (1) it is straightforward; and (2) in structural engineering practice, buildings are not analyzed for rotational excitation. Therefore, we devote this brief section to a discussion of the results of such analysis.

Consider the building shown in Fig. 13.4.1, located in Pomona, California. This reinforced-concrete frame building has two stories, a partial basement and a light penthouse structure. For all practical and code design purposes, the building has a nominally symmetric floor plan, as indicated by its framing plan in Fig. 13.4.2. The lateral force-resisting system in the building consists of peripheral columns interconnected by longitudinal and transverse beams, but the L-shaped exterior corner columns as well as the interior columns in the building are not designed especially for earthquake resistance. The floor decking system is formed by a 6-in.-thick concrete slab. The building also includes walls in the stairwell system—concrete walls in the basement and masonry walls in upper stories. Foundations of columns and interior walls are supported on piles.

The accelerograph channels located as shown in Fig. 13.4.3 recorded the motion of the building during the Upland (February 28, 1990) earthquake, including three channels of horizontal motion at each of three levels: roof, second floor, and basement. The peak accelerations of the basement were 0.12g and 0.13g in the x and y directions, respectively. These motions were amplified to 0.24g in the x-direction and 0.39g in the y-direction at the roof. The building experienced no structural damage during this earthquake.

Figure 13.4.1 First Federal Savings building, a two-story R/C building (with a partial basement) in Pomona, California. (Courtesy of California Strong Motion Instrumentation Program.)

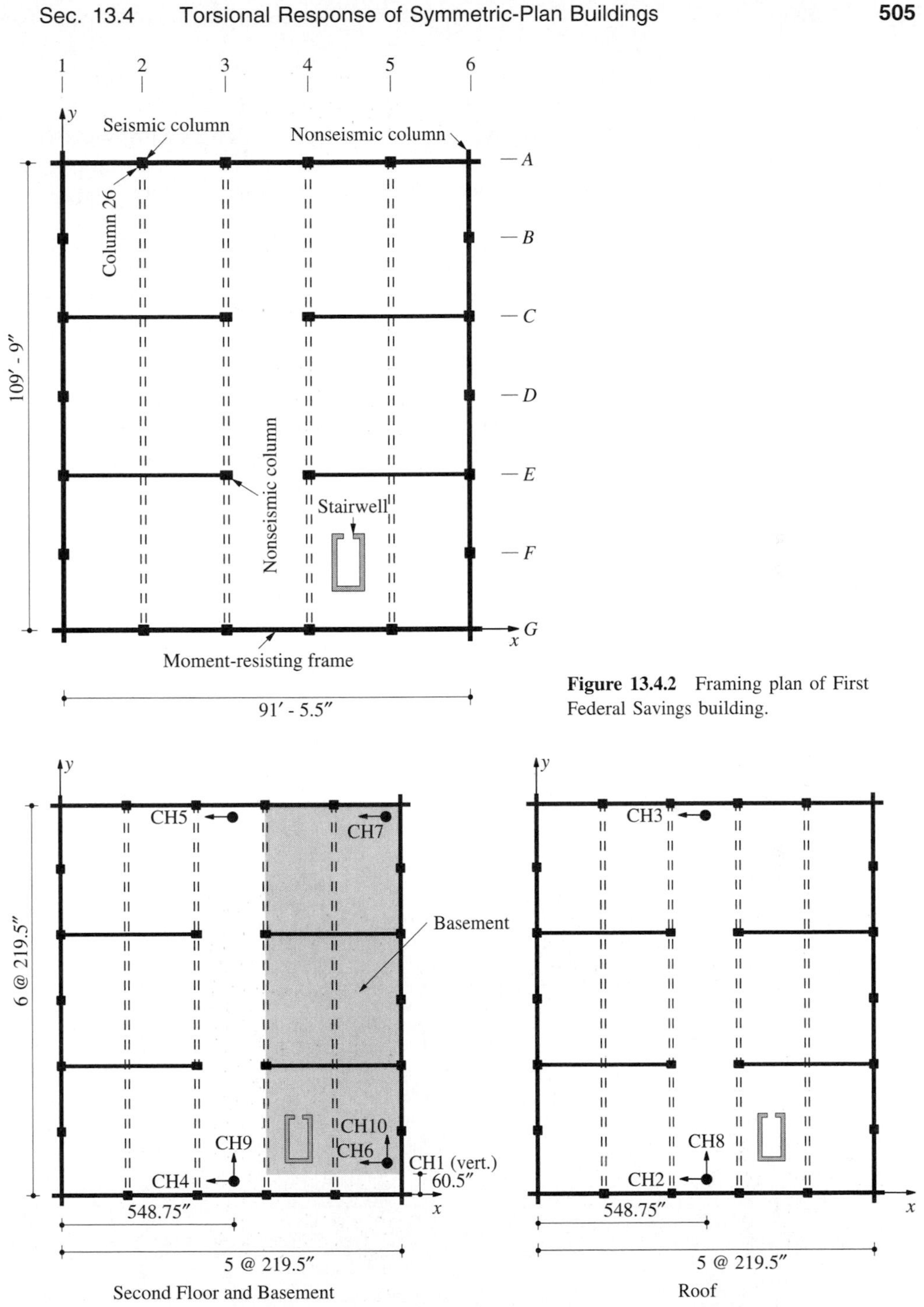

Figure 13.4.2 Framing plan of First Federal Savings building.

Figure 13.4.3 Accelerograph channels in First Federal Savings building.

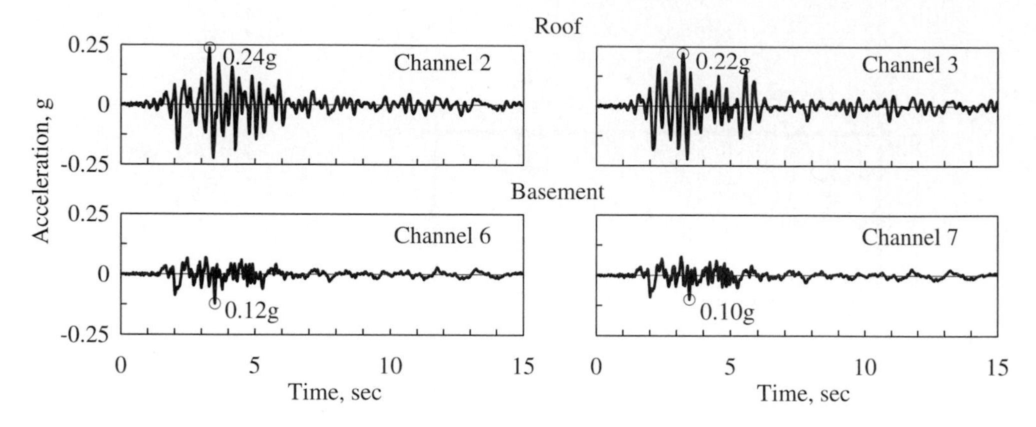

Figure 13.4.4 Motions recorded at First Federal Savings building during the Upland earthquake of February 28, 1990.

Some of the recorded motions are shown in Fig. 13.4.4. These include the x-translational accelerations at two locations at the basement of the building and at two locations at the roof level. By superimposing the motions at two locations on the roof in Fig. 13.4.5 it is clear that this building experienced some torsion; otherwise, these two motions would have been identical. Assuming rigid base, its rotational acceleration is computed as the difference between the two x-translational records at the basement of the building divided by the distance between the two locations. This rotational base acceleration is multiplied by $b/2$, where the building-plan dimension $b = 109.75$ ft, and plotted in Fig. 13.4.6. The peak value of $(b/2)\ddot{u}_{g\theta}(t)$ is 0.029g compared with the peak acceleration of 0.12g in the x-direction.

The torsional response of the building to the rotational motion of the basement, Fig. 13.4.6, is determined by modal solution of Eq. (9.6.1) with modal damping ratios of 5%. These damping ratios were estimated from the recorded motions at the roof and basement using some of the procedures mentioned in Chapter 11, Part A. The response

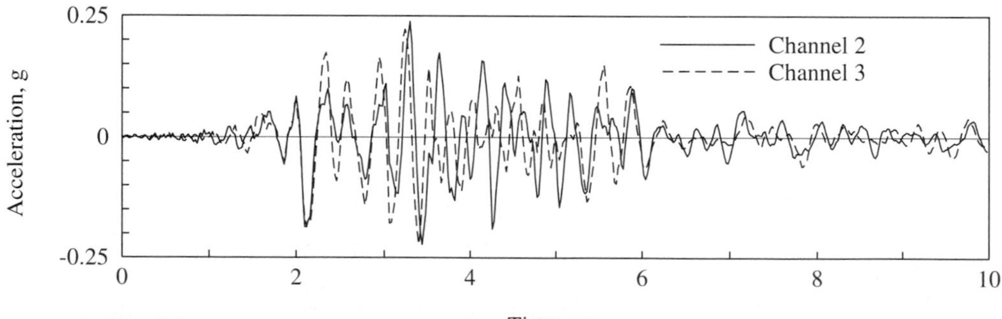

Figure 13.4.5 Motions recorded at two locations on the roof of First Federal Savings building during the Upland earthquake of February 28, 1990.

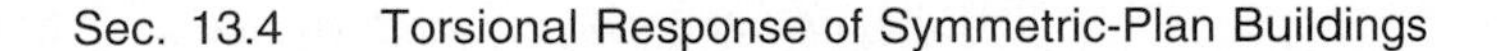

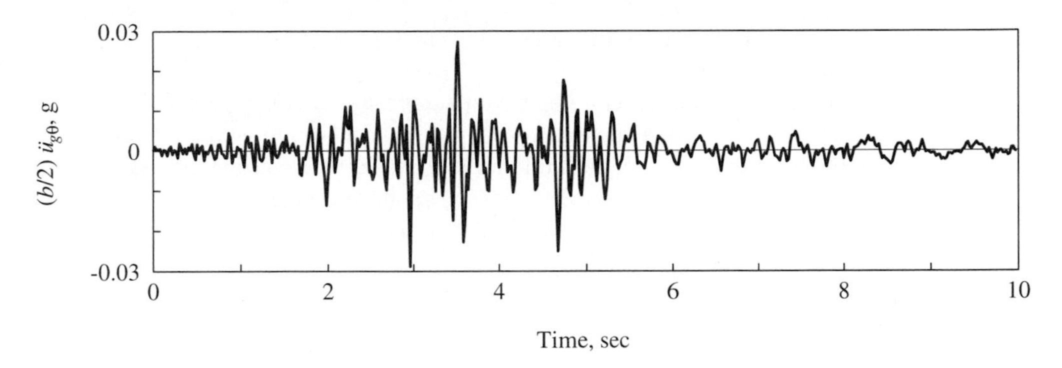

Figure 13.4.6 Rotational acceleration of basement multiplied by $b/2$. [From De la Llera and Chopra (1994).]

history of the shear force in a selected column of the building is presented in Fig. 13.4.7. This is only a part of the element force due to the actual torsional motion of the building during the earthquake, as will be demonstrated next.

Approximate values of the element forces due to recorded torsion can be determined at each instant of time by static analysis of the building subjected to floor inertia torques $I_{Oj}\ddot{u}_{j\theta}(t)$ at all floors ($j = 1, 2, \ldots, N$), where I_{Oj} is the moment of inertia of the jth floor mass about the vertical axis through O, the center of mass (CM) of the floor, and $\ddot{u}_{j\theta}$ is the torsional acceleration of the jth floor diaphragm. By using these inertia forces as equivalent static forces, we have included the damping forces and this is a source of approximation. The results of these static analyses for the shear force in the same column are also presented in Fig. 13.4.7.

The peak force due to rotational basement motion is about 45% of the peak force due to the actual torsional motions of the building. The remaining effects—55% for this building and earthquake—of torsional motion of a building arise, in part, because a building is usually not *perfectly* symmetric. This building is not perfectly symmetric because of several factors, the most obvious of them being the stairwell system shown

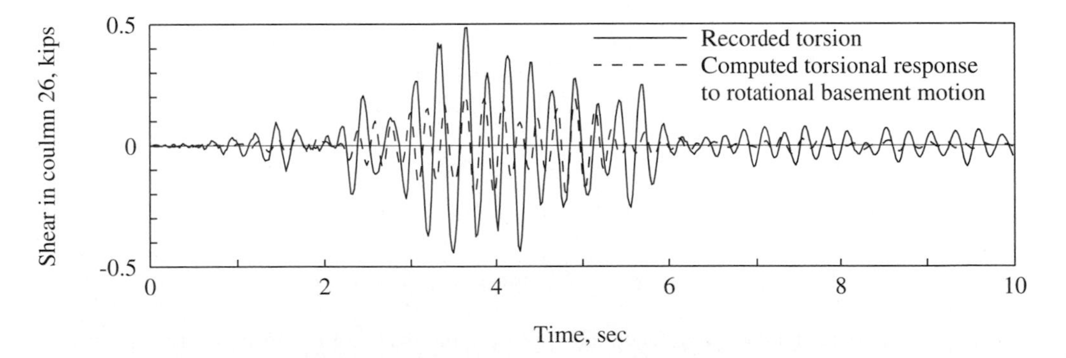

Figure 13.4.7 Comparison of shear force (x-component) in column 26 due to recorded torsion of the building and computed torsional response of the building to rotational basement motion. [From De la Llera and Chopra (1994).]

in Fig. 13.4.2, and because the basement, which is under one-half of the floor plan, is not symmetrically located.

Torsional motion of buildings with nominally symmetric plan is usually called *accidental torsion*. Such motion contributes a small fraction of the total earthquake forces in the structure. For the building and earthquake considered, accidental torsion contributed about 4% of the total force (results not presented here), but larger contributions have been identified in the earthquake response of other buildings. The structural response associated with accidental torsion is not amenable to calculation in structural design for two reasons: (1) the rotational base motion is not defined, and (2) it is not practical to identify and analyze the effect of each source of asymmetry in a nominally symmetric plan. Therefore, building codes include a simple design provision to account for accidental torsion in symmetric and unsymmetric buildings; in the latter case it is considered in addition to torsion arising from plan asymmetry (Section 13.3). Research has demonstrated deficiencies in this code provision.

13.5 RESPONSE ANALYSIS FOR MULTIPLE SUPPORT EXCITATION

In this section the modal analysis procedure of Section 13.1 is extended to MDF systems excited by prescribed motions $\ddot{u}_{gl}(t)$ at the various supports ($l = 1, 2, \ldots, N_g$) of the structure. In Section 9.7 the governing equations were shown to be the same as Eq. (13.1.1), with the effective earthquake forces

$$\mathbf{p}_{\text{eff}}(t) = -\sum_{l=1}^{N_g} \mathbf{m}\boldsymbol{\iota}_l \ddot{u}_{gl}(t) \tag{13.5.1}$$

instead of Eq. (13.1.2). The modal Eq. (13.1.7) now becomes

$$\ddot{q}_n + 2\zeta_n\omega_n\dot{q}_n + \omega_n^2 q_n = -\sum_{l=1}^{N_g} \Gamma_{nl}\ddot{u}_{gl}(t) \tag{13.5.2}$$

where

$$\Gamma_{nl} = \frac{L_{nl}}{M_n} \qquad L_{nl} = \boldsymbol{\phi}_n^T \mathbf{m}\boldsymbol{\iota}_l \qquad M_n = \boldsymbol{\phi}_n^T \mathbf{m}\boldsymbol{\phi}_n \tag{13.5.3}$$

The solution of Eq. (13.5.2) can be written as a generalization of Eq. (13.1.9):

$$q_n(t) = \sum_{l=1}^{N_g} \Gamma_{nl} D_{nl}(t) \tag{13.5.4}$$

where $D_{nl}(t)$ is the deformation response of the nth-mode SDF system to support acceleration $\ddot{u}_{gl}(t)$.

The displacement response of the structure, Eq. (9.7.2), contains two parts:

1. The dynamic displacements are obtained by combining Eqs. (13.1.3) and (13.5.4):

$$\mathbf{u}(t) = \sum_{l=1}^{N_g}\sum_{n=1}^{N} \Gamma_{nl}\boldsymbol{\phi}_n D_{nl}(t) \tag{13.5.5}$$

2. The quasi-static displacements $\mathbf{u}^s$ are given by Eq. (9.7.9). Combining the two parts gives the total displacements in the structural DOFs:

$$\mathbf{u}^t(t) = \sum_{l=1}^{N_g} \boldsymbol{\iota}_l u_{gl}(t) + \sum_{l=1}^{N_g} \sum_{n=1}^{N} \Gamma_{nl} \boldsymbol{\phi}_n D_{nl}(t) \tag{13.5.6}$$

The forces in structural elements can be obtained from the structural displacements $\mathbf{u}^t(t)$ and prescribed support displacements $\mathbf{u}_g(t)$ without additional dynamic analyses by using either of the two procedures mentioned in Section 9.10. In the first method, the element forces are calculated from the known nodal displacements using the element stiffness properties. This method is usually preferred in computer implementation of force calculations. It is instructive, however, to generalize the second method based on equivalent static forces for multiple support excitation. The rest of this section is devoted to this development.

The equivalent static forces in the structural DOF are given by the last term on the left side of Eq. (9.7.1):

$$\mathbf{f}_S = \mathbf{k}\mathbf{u}^t + \mathbf{k}_g \mathbf{u}_g \tag{13.5.7}$$

Substituting Eq. (9.7.2) for $\mathbf{u}^t$ and using Eq. (9.7.7) gives

$$\mathbf{f}_S(t) = \mathbf{k}\mathbf{u}(t) \tag{13.5.8}$$

These forces depend only on the dynamic displacements, Eq. (13.5.5). Therefore,

$$\mathbf{f}_S(t) = \sum_{l=1}^{N_g} \sum_{n=1}^{N} \Gamma_{nl} \mathbf{k} \boldsymbol{\phi}_n D_{nl}(t) \tag{13.5.9}$$

which can be written in terms of the mass matrix by utilizing Eq. (10.2.4):

$$\mathbf{f}_S(t) = \sum_{l=1}^{N_g} \sum_{n=1}^{N} \Gamma_{nl} \mathbf{m} \boldsymbol{\phi}_n A_{nl}(t) \tag{13.5.10}$$

where

$$A_{nl}(t) = \omega_n^2 D_{nl}(t) \tag{13.5.11}$$

is the pseudo-acceleration response of the nth-mode SDF system to support acceleration $\ddot{u}_{gl}(t)$.

The equivalent static forces along the support DOF are also given by the last term on the left side of Eq. (9.7.1):

$$\mathbf{f}_{Sg} = \mathbf{k}_g^T \mathbf{u}^t + \mathbf{k}_{gg} \mathbf{u}_g \tag{13.5.12}$$

Substituting Eq. (9.7.2) for $\mathbf{u}^t$ and using Eq. (9.7.3) for the quasi-static support forces $\mathbf{p}_g^s(t)$ gives

$$\mathbf{f}_{Sg}(t) = \mathbf{k}_g^T \mathbf{u}(t) + \mathbf{p}_g^s(t) \tag{13.5.13}$$

Observe that the support forces $\mathbf{f}_{Sg}$ depend on the displacements in the structural DOFs as well as on support displacements, and can no longer be obtained by statics from the force vector $\mathbf{f}_S$. This is different from Section 13.1, where for a structure excited at its only support, or excited by identical motion at all supports, the base shear could be determined from $\mathbf{f}_S$. By utilizing Eqs. (9.7.9) and (13.5.5), the support forces can be expressed as

$$\mathbf{f}_{Sg}(t) = \sum_{l=1}^{N_g} \left(\mathbf{k}_g^T \iota_l + \mathbf{k}_{gg}^l\right) u_{gl}(t) + \sum_{l=1}^{N_g} \sum_{n=1}^{N} \Gamma_{nl} \mathbf{k}_g^T \phi_n D_{nl}(t) \tag{13.5.14}$$

where $\mathbf{k}_{gg}^l$ is the lth column of $\mathbf{k}_{gg}$.

The element forces at each time instant are evaluated by static analysis of the structure subjected to the forces $\mathbf{f}_S(t)$ and $\mathbf{f}_{Sg}(t)$, given by Eqs. (13.5.10) and (13.5.14), respectively. While this procedure was presented to show that the equivalent static force concept can be generalized to structures excited by multiple support excitation, as mentioned earlier, in computer analysis it is usually preferable to evaluate the element forces directly from the nodal displacements using the element stiffness properties.

Example 13.10

In the two-span continuous bridge of Example 9.10, support A undergoes vertical motion $u_g(t)$, support B describes the same motion as A, but it does so t' seconds later, and support C undergoes the same motion $2t'$ after support A. Determine the following responses as a function of time: displacement of the two masses; bending moments at the midpoint of each span; and bending moment at the center support. Express all results in terms of $D_n(t)$ and $A_n(t)$, the displacement and pseudo-acceleration responses of the nth-mode SDF system to $\ddot{u}_g(t)$. Compare the preceding results with the response of the bridge if all supports undergo identical motion $u_g(t)$.

1. *Evaluate the natural frequencies and modes.* The eigenvalue problem to be solved is

$$\mathbf{k}\phi = \omega^2 \mathbf{m}\phi$$

where Eqs. (c) and (e) of Example 9.10 give

$$\mathbf{k} = \frac{EI}{L^3}\begin{bmatrix} 78.86 & 30.86 \\ 30.86 & 78.86 \end{bmatrix} \qquad \mathbf{m} = m\begin{bmatrix} 1 & \\ & 1 \end{bmatrix} \tag{a}$$

Solution of the eigenvalue problem gives

$$\omega_1 = 6.928\sqrt{\frac{EI}{mL^3}} \qquad \omega_2 = 10.47\sqrt{\frac{EI}{mL^3}} \tag{b}$$

$$\phi_1 = \begin{Bmatrix} -1 \\ 1 \end{Bmatrix} \qquad \phi_2 = \begin{Bmatrix} 1 \\ 1 \end{Bmatrix} \tag{c}$$

2. *Determine* $\Gamma_{nl} = L_{nl}/M_n$.

$$L_{nl} = \phi_n^T \mathbf{m} \iota_l \qquad l = 1, 2, 3, \quad n = 1, 2$$

Substituting for ϕ_n and $\mathbf{m}$ from Eqs. (c) and (a), respectively, and for ι_l from Eq. (g) of Example 9.10 gives

$$\mathbf{L} = [L_{nl}] = \begin{bmatrix} -0.5000m & 0 & 0.5000m \\ 0.3125m & 1.375m & 0.3125m \end{bmatrix} \begin{matrix} \leftarrow \text{mode 1} \\ \leftarrow \text{mode 2} \end{matrix} \tag{d}$$
$$\begin{matrix} \uparrow & \uparrow & \uparrow \\ \ddot{u}_{g1} & \ddot{u}_{g2} & \ddot{u}_{g3} \end{matrix}$$

$$M_n = \phi_n^T \mathbf{m} \phi_n \qquad n = 1, 2$$

Substituting for ϕ_n and $\mathbf{m}$ gives $M_n = 2m$, $n = 1, 2$. Then $\Gamma_{nl} = L_{nl}/M_n$ gives

$$\mathbf{\Gamma} = [\Gamma_{nl}] = \begin{bmatrix} -0.25000 & 0 & 0.25000 \\ 0.15625 & 0.6875 & 0.15625 \end{bmatrix} \begin{matrix} \leftarrow \text{mode 1} \\ \leftarrow \text{mode 2} \end{matrix} \tag{e}$$
$$\begin{matrix} \uparrow & \uparrow & \uparrow \\ \ddot{u}_{g1} & \ddot{u}_{g2} & \ddot{u}_{g3} \end{matrix}$$

3. *Determine the response of the nth-mode SDF system to $\ddot{u}_{gl}(t)$.* Given $\ddot{u}_{g1}(t) = \ddot{u}_g(t)$, $\ddot{u}_{g2}(t) = \ddot{u}_g(t - t')$, $\ddot{u}_{g3}(t) = \ddot{u}_g(t - 2t')$. Then

$$D_{n1}(t) = D_n(t) \qquad D_{n2}(t) = D_n(t - t') \qquad D_{n3}(t) = D_n(t - 2t') \tag{f}$$

$$A_{n1}(t) = A_n(t) \qquad A_{n2}(t) = A_n(t - t') \qquad A_{n3}(t) = A_n(t - 2t') \tag{g}$$

4. *Determine the displacement response.* In Eq. (13.5.6) with $N = 2$ and $N_g = 3$, substituting for Γ_{nl}, ϕ_n, and D_{nl} from Eqs. (e), (c), and (f), respectively, and for ι_l from Eq. (g) of Example 9.10 gives

$$\begin{aligned} \begin{Bmatrix} u_1^t(t) \\ u_2^t(t) \end{Bmatrix} = {} & \begin{Bmatrix} 0.40625 \\ -0.09375 \end{Bmatrix} u_g(t) + \begin{Bmatrix} 0.6875 \\ 0.6875 \end{Bmatrix} u_g(t - t') + \begin{Bmatrix} -0.09375 \\ 0.40625 \end{Bmatrix} u_g(t - 2t') \\ & - 0.25 \begin{Bmatrix} -1 \\ 1 \end{Bmatrix} D_1(t) + 0 \begin{Bmatrix} -1 \\ 1 \end{Bmatrix} D_1(t - t') + 0.25 \begin{Bmatrix} -1 \\ 1 \end{Bmatrix} D_1(t - 2t') \\ & + 0.15625 \begin{Bmatrix} 1 \\ 1 \end{Bmatrix} D_2(t) + 0.6875 \begin{Bmatrix} 1 \\ 1 \end{Bmatrix} D_2(t - t') + 0.15625 \begin{Bmatrix} 1 \\ 1 \end{Bmatrix} D_2(t - 2t') \end{aligned} \tag{h}$$

5. *Compute the equivalent static forces.* In Eq. (13.5.10) with $N = 2$ and $N_g = 3$, substituting for $\mathbf{m}$, ϕ_n, Γ_{nl}, and $A_{nl}(t)$ from Eqs. (a), (c), (e), and (g), respectively, gives

$$\begin{aligned} \mathbf{f}_S(t) = {} & - 0.25 \begin{Bmatrix} -1 \\ 1 \end{Bmatrix} mA_1(t) + 0 \begin{Bmatrix} -1 \\ 1 \end{Bmatrix} mA_1(t - t') + 0.25 \begin{Bmatrix} -1 \\ 1 \end{Bmatrix} mA_1(t - 2t') \\ & + 0.15625 \begin{Bmatrix} 1 \\ 1 \end{Bmatrix} mA_2(t) + 0.6875 \begin{Bmatrix} 1 \\ 1 \end{Bmatrix} mA_2(t - t') + 0.15625 \begin{Bmatrix} 1 \\ 1 \end{Bmatrix} mA_2(t - 2t') \end{aligned} \tag{i}$$

6. *Compute the equivalent static support forces.* In Eq. (13.5.12) substituting for $\mathbf{k}_g$ and $\mathbf{k}_{gg}$ from Eq. (d) of Example 9.10 and $\mathbf{u}^t(t)$ from Eq. (h) gives

$$\mathbf{f}_{Sg}(t) = \begin{Bmatrix} -0.125 \\ 0 \\ 0.125 \end{Bmatrix} mA_1(t) + \begin{Bmatrix} 0 \\ 0 \\ 0 \end{Bmatrix} mA_1(t-t') + \begin{Bmatrix} 0.125 \\ 0 \\ -0.125 \end{Bmatrix} mA_1(t-2t')$$
$$+ \begin{Bmatrix} -0.0488 \\ -0.2148 \\ -0.0488 \end{Bmatrix} mA_2(t) + \begin{Bmatrix} -0.2149 \\ -0.9454 \\ -0.2149 \end{Bmatrix} mA_2(t-t') + \begin{Bmatrix} -0.0488 \\ -0.2148 \\ -0.0488 \end{Bmatrix} mA_2(t-2t')$$
$$+ \begin{Bmatrix} 1.5 \\ -3.0 \\ 1.5 \end{Bmatrix} \frac{EI}{L^3} u_g(t) + \begin{Bmatrix} -3 \\ 6 \\ -3 \end{Bmatrix} \frac{EI}{L^3} u_g(t-t') + \begin{Bmatrix} 1.5 \\ -3 \\ 1.5 \end{Bmatrix} u_g(t-2t') \qquad \text{(j)}$$

where Eq. (b) was used to express EI/L^3 in terms of ω_n and Eq. (13.5.11) to express D_{nl} in terms of A_{nl}. The equivalent static forces given by Eqs. (i) and (j) are shown in Fig. E13.10. Observe that at each time instant, these forces defined by Eqs. (i) and (j) are in equilibrium.

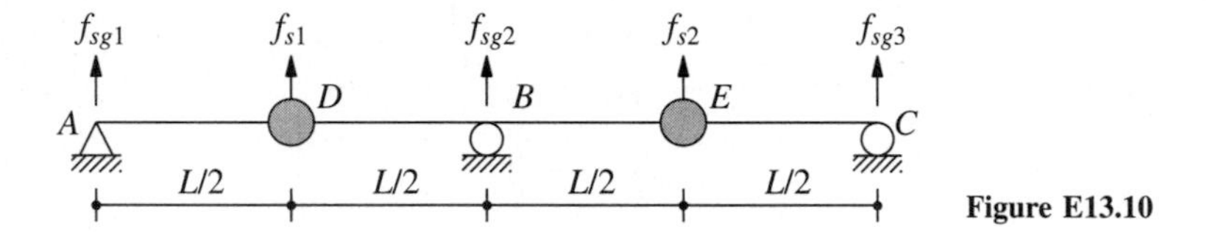

Figure E13.10

7. *Compute the bending moments.* The bending moments M_D, M_E, and M_B at the locations of the left mass, right mass, and support B, respectively, are obtained by static analysis of the system subjected to the forces shown in Fig. E13.10:

$$M_D = mL\Big(-0.0625A_1(t) + 0A_1(t-t') + 0.0625A_1(t-2t')\Big)$$
$$+ mL\Big(-0.0244A_2(t) - 0.1074A_2(t-t') - 0.0244A_2(t-2t')\Big)$$
$$+ \frac{EI}{L^2}\Big(0.75u_g(t) - 1.50u_g(t-t') + 0.75u_g(t-2t')\Big) \qquad \text{(k)}$$

$$M_E = mL\Big(0.0625A_1(t) + 0A_1(t-t') - 0.0625A_1(t-2t')\Big)$$
$$+ mL\Big(-0.0244A_2(t) - 0.1074A_2(t-t') - 0.0244A_2(t-2t')\Big)$$
$$+ \frac{EI}{L^2}\Big(0.75u_g(t) - 1.50u_g(t-t') + 0.75u_g(t-2t')\Big) \qquad \text{(l)}$$

$$M_B = mL\Big(0.0293A_2(t) + 0.1289A_2(t-t') + 0.0293A_2(t-2t')\Big)$$
$$+ \frac{EI}{L^2}\Big(1.5u_g(t) - 3.0u_g(t-t') + 1.5u_g(t-2t')\Big) \qquad \text{(m)}$$

Observe that the first mode does not contribute to M_B because B is a point of inflection for this mode.

8. *Identical support motions.* If all the supports undergo identical motion $u_g(t)$, the motion of the structure is given by Eq. (13.1.15), where Γ_n is defined by Eq. (13.1.5) with $\boldsymbol{\iota} = \mathbf{1}$. For this system

$$\Gamma_1 = 0 \qquad \Gamma_2 = 1$$

When we substitute these data, Eq. (13.1.15) gives

$$\mathbf{u}(t) = \Gamma_2 \phi_2 D_2(t) = \begin{Bmatrix} 1 \\ 1 \end{Bmatrix} D_2(t) \tag{n}$$

Observe that the first mode, which is antisymmetric, is not excited by the symmetric excitation, and all the response is due to the second mode.

The equivalent static forces are given by Eq. (13.1.11):

$$\mathbf{f}_S(t) = \Gamma_2 \mathbf{m} \phi_2 A_2(t) = m \begin{Bmatrix} 1 \\ 1 \end{Bmatrix} A_2(t) \tag{o}$$

The support forces can be obtained by static analysis of the bridge subjected to the external forces of Eq. (o). Alternatively, the support forces are given by Eq. (j), specialized by substituting $A_2(t) = A_2(t-t') = A_2(t-2t')$ and $u_g(t) = u_g(t-t') = u_g(t-2t')$. Either method gives

$$\mathbf{f}_{Sg}(t) = \begin{Bmatrix} -0.3125 \\ -1.3750 \\ -0.3125 \end{Bmatrix} m A_2(t) \tag{p}$$

Determined by static analysis of the structure due to the forces of Eq. (o), the bending moments are:

$$M_D = -0.15625 mLA_2(t) \qquad M_E = -0.15625 mLA_2(t) \qquad M_B = 0.1875 mLA_2(t) \tag{q}$$

9. *Comparison.* If the support motions are identical, the quasi-static forces $\mathbf{p}_g^s(t)$ in Eq. (13.5.13) are zero, there is no quasi-static component in the bending moments, and all support forces and internal forces can be computed directly (by statics) from the equivalent static forces in the structural DOFs. In contrast, if the support motions are different, the calculation of forces is more involved. In particular, the quasi-static forces associated with the different displacements of the supports must be included, and support forces cannot be obtained from only the equivalent static forces in the structural DOF.

13.6 STRUCTURAL IDEALIZATION AND EARTHQUAKE RESPONSE

With the development of earthquake analysis procedures presented in this chapter and the availability of modern computers, it is now possible to determine the linearly elastic response of an idealization (mathematical model) of any structure to prescribed ground motion. How well the computed response agrees with the actual response of a structure during an earthquake depends primarily on the quality of the structural idealization.

To illustrate this concept, we return to the natural periods and damping ratios for the Millikan Library building. Presented in Chapter 11 were these data from low-amplitude forced vibration tests, and from the Lytle Creek and San Fernando earthquakes, which caused roof accelerations of approximately 0.05g and 0.31g, respectively. These results demonstrated that with increasing levels of motion the natural periods lengthen and the damping ratios increase. The loss of stiffness indicated by this period change is believed to be primarily the result of cracking and other types of degradation of the so-called nonstructural elements during the higher-level earthquake responses, especially from the San Fernando earthquake. A nonlinear structural idealization having stiffness and damping properties varying with deformation level would be necessary to reproduce this period change and to describe the behavior of a structure through the complete range of deformation amplitudes.

However, if the structure experiences no structural damage, good estimates of the response during the earthquake can usually be computed from an equivalent linear model with viscous damping. If the computed natural periods and modes and the estimated damping ratios represent the properties of the structure during the earthquake, modal analysis with viscously damped SDF systems describing the response of natural vibration modes is an accurate approximation for analysis of "linear" response. This has been demonstrated by numerous analyses of recorded motions of structures during earthquakes; one such example is the response of the Millikan Library building during the San Fernando earthquake (Figs. 11.1.3 and 11.1.4). Using the natural periods and damping ratios of this building determined from these recorded motions and system identification procedures (Table 11.1.1), the displacement response of this building to the basement motion calculated by modal analysis was shown to agree almost perfectly with the displacements (relative to the ground) shown in Fig. 11.1.5, which were determined from the accelerations recorded at the roof and at the basement.

The usual situation, however, is different in the sense that the natural periods and modes are computed from an idealization of the structure. It is the quality of this idealization that determines the accuracy of response. Therefore, only those structural and nonstructural elements that contribute to the mass and stiffness of the structure at the amplitudes of motion expected during the earthquake should be included in the structural idealization; and their stiffness properties should be determined using realistic assumptions. Similarly, as discussed in Chapter 11, selection of damping values for analysis of a structure should be based on available data from recorded earthquake responses of similar structures.

PART B: RESPONSE SPECTRUM ANALYSIS

13.7 PEAK RESPONSE FROM EARTHQUAKE RESPONSE SPECTRUM

The response history analysis (RHA) procedure presented in Part A provides structural response $r(t)$ as a function of time, but structural design is usually based on the peak values of forces and deformations over the duration of the earthquake-induced response. Can the peak response be determined directly from the response spectrum for the ground motion without carrying out a response history analysis? For SDF systems the answer to this question is yes (Chapter 6). However, for MDF systems the answer is a qualified yes. The peak response of MDF systems can be calculated from the response spectrum, but the result is not exact—in the sense that it is not identical to the RHA result; the estimate obtained is accurate enough for structural design applications, however. In Part B we present such response spectrum analysis (RSA) procedures for structures excited by a single component of ground motion; thus simultaneous action of the other two components is excluded and multiple support excitation is not considered. However, these more general cases have been solved by researchers and the interested reader should consult the published literature.

13.7.1 Peak Modal Responses

The *exact* value of the peak response of an MDF system in its nth natural mode can be obtained from the earthquake response spectrum. This becomes evident from Eq. (13.1.13) for the response history by recalling that the peak value of $A_n(t)$ is available from the pseudo-acceleration response spectrum as its ordinate $A(T_n, \zeta_n)$ corresponding to natural period T_n and damping ratio ζ_n; for brevity, $A(T_n, \zeta_n)$ will be denoted as A_n. Therefore, the peak modal response is†

$$r_{no} = r_n^{\mathrm{st}} A_n \tag{13.7.1}$$

The algebraic sign of r_{no} is the same as that of r_n^{st} because A_n is positive by definition. This algebraic sign must be retained because it can be important, as will be seen in Section 13.7.2. All response quantities $r_n(t)$ associated with a particular mode, say the nth mode, reach their peak values at the same time instant as $A_n(t)$ reaches its peak (see Figs. 13.2.6 to 13.2.8).

13.7.2 Modal Combination Rules

How do we combine the peak modal responses r_{no} $(n = 1, 2, \ldots, N)$ to determine the peak value $r_o \equiv \max_t |r(t)|$ of the total response? It will not be possible to determine the exact value of r_o from r_{no} because, in general, the modal responses $r_n(t)$ attain their peaks at different time instants and the combined response $r(t)$ attains its peak at yet a different instant. This phenomenon can be observed in Fig. 13.2.7b, where results for the shear in the top story of a five-story frame are presented. The individual modal responses $V_{5n}(t), n = 1, 2, \ldots, 5$, are shown together with the total response $V_5(t)$.

Approximations must be introduced in combining the peak modal responses r_{no} determined from the earthquake response spectrum because no information is available when these peak modal values occur. The assumption that all modal peaks occur at the same time and their algebraic sign is ignored provides an upper bound to the peak value of the total response:

$$r_o \leq \sum_{n=1}^{N} |r_{no}| \tag{13.7.2}$$

†This notation r_{no} should not be confused with the use of a subscript o in Chapter 6 to denote the maximum (over time) of the absolute value of the response quantity, which is positive by definition.

This upper-bound value is usually too conservative, as we shall see in example computations to be presented later. Therefore, this *absolute sum* (ABSSUM) modal combination rule, is not popular in structural design applications.

The *square-root-of-sum-of-squares* (SRSS) rule for modal combination, developed in E. Rosenblueth's Ph.D. thesis (1951), is

$$r_o \simeq \left(\sum_{n=1}^{N} r_{no}^2\right)^{1/2} \tag{13.7.3}$$

The peak response in each mode is squared, the squared modal peaks are summed, and the square root of the sum provides an estimate of the peak total response. As will be seen later, this modal combination rule provides excellent response estimates for structures with well-separated natural frequencies. This limitation has not always been recognized in applying this rule to practical problems, and at times it has been misapplied to systems with closely spaced natural frequencies, such as piping systems in nuclear power plants and multistory buildings with unsymmetrical plan.

The *complete quadratic combination* (CQC) rule for modal combination is applicable to a wider class of structures as it overcomes the limitations of the SRSS rule. According to the CQC rule,

$$r_o \simeq \left(\sum_{i=1}^{N}\sum_{n=1}^{N} \rho_{in} r_{io} r_{no}\right)^{1/2} \tag{13.7.4}$$

Each of the N^2 terms on the right side of this equation is the product of the peak responses in the ith and nth modes and the correlation coefficient ρ_{in} for these two modes; ρ_{in} varies between 0 and 1 and $\rho_{in} = 1$ for $i = n$. Thus Eq. (13.7.4) can be rewritten as

$$r_o \simeq \left(\sum_{n=1}^{N} r_{no}^2 + \underbrace{\sum_{i=1}^{N}\sum_{n=1}^{N}}_{i \neq n} \rho_{in} r_{io} r_{no}\right)^{1/2} \tag{13.7.5}$$

to show that the first summation on the right side is identical to the SRSS combination rule of Eq. (13.7.3); each term in this summation is obviously positive. The double summation includes all the cross ($i \neq n$) terms; each of these terms may be positive or negative. A cross term is negative when the modal static responses r_i^{st} and r_n^{st} assume opposite signs—for the algebraic sign of r_{no} is the same as that of r_n^{st} because A_n is positive by definition. Thus the estimate for r_o obtained by the CQC rule may be larger or smaller than the estimate provided by the SRSS rule. [It can be shown that the double summation inside the parentheses of Eq. (13.7.4) is always positive.]

Starting in the late 1960s and continuing through the 1970s and early 1980s, several formulations for the peak response to earthquake excitation were published. Some of these are identical or similar to Eq. (13.7.4) but differ in the mathematical expressions given for the correlation coefficient. Here we include two: one due to E. Rosenblueth and J. Elorduy for historical reasons because it was apparently the earliest (1969) result; and a second (1981) due to A. Der Kiureghian because it is now widely used.

The 1971 textbook *Fundamentals of Earthquake Engineering* by N. M. Newmark and E. Rosenblueth gives the Rosenblueth–Elorduy equations for the correlation coefficient:

$$\rho_{in} = \frac{1}{1 + \epsilon_{in}^2} \tag{13.7.6}$$

where

$$\epsilon_{in} = \frac{\omega_i\sqrt{1-\zeta_i^2} - \omega_n\sqrt{1-\zeta_n^2}}{\zeta_i'\omega_i + \zeta_n'\omega_n} \qquad \zeta_n' = \zeta_n + \frac{2}{\omega_n s} \tag{13.7.7}$$

and s is the duration of the strong phase of the earthquake excitation. Equations (13.7.6) and (13.7.7) show that $\rho_{in} = \rho_{ni}$; $0 \le \rho_{in} \le 1$; and $\rho_{in} = 1$ for $i = n$ or for two modes with equal frequencies and equal damping ratios. It is instructive to specialize Eq. (13.7.6) for systems with the same damping ratio in all modes subjected to earthquake excitation with duration s long enough to replace Eq. (13.7.7b) by $\zeta_n' = \zeta_n$. We substitute $\zeta_i = \zeta_n = \zeta$ in Eq. (13.7.7a), introduce $\beta_{in} = \omega_i/\omega_n$, and insert Eq. (13.7.7a) in Eq. (13.7.6) to obtain

$$\rho_{in} = \frac{\zeta^2(1+\beta_{in})^2}{(1-\beta_{in})^2 + 4\zeta^2\beta_{in}} \tag{13.7.8}$$

The equation for the correlation coefficient due to Der Kiureghian is

$$\rho_{in} = \frac{8\sqrt{\zeta_i\zeta_n}(\zeta_i + \beta_{in}\zeta_n)\beta_{in}^{3/2}}{(1-\beta_{in}^2)^2 + 4\zeta_i\zeta_n\beta_{in}(1+\beta_{in}^2) + 4(\zeta_i^2+\zeta_n^2)\beta_{in}^2} \tag{13.7.9}$$

This equation also implies that $\rho_{in} = \rho_{ni}$, $\rho_{in} = 1$ for $i = n$ or for two modes with equal frequencies and equal damping ratios. For equal modal damping $\zeta_i = \zeta_n = \zeta$ this equation simplifies to

$$\rho_{in} = \frac{8\zeta^2(1+\beta_{in})\beta_{in}^{3/2}}{(1-\beta_{in}^2)^2 + 4\zeta^2\beta_{in}(1+\beta_{in})^2} \tag{13.7.10}$$

Figure 13.7.1 shows Eqs. (13.7.8) and (13.7.10) for the correlation coefficient ρ_{in} plotted as a function of $\beta_{in} = \omega_i/\omega_n$ for four damping values: $\zeta = 0.02$, 0.05, 0.10, and 0.20. Observe that the two expressions give essentially identical values for ρ_{in}.

This figure also provides an understanding of the correlation coefficient. Observe that this coefficient diminishes rapidly as the two natural frequencies ω_i and ω_n move farther apart. This is especially the case at small damping values that are typical of structures. In other words, it is only in a narrow range of β_{in} around $\beta_{in} = 1$ that ρ_{in} has significant values; and this range depends on damping. For example, $\rho_{in} < 0.1$ for systems with 5% damping over the frequency ratio range $1/1.5 \le \beta_{in} \le 1.5$. If the damping is 2%, this range is reduced to $1/1.2 \le \beta_{in} \le 1.2$. For structures with well-separated natural frequencies the coefficients ρ_{in} vanish; as a result all cross ($i \ne n$) terms in the CQC rule, Eq. (13.7.5), can be neglected and it reduces to the SRSS rule, Eq. (13.7.3). It is now clear that the SRSS rule applies to structures with well-separated natural frequencies.

The SRSS and CQC rules for combination of peak modal responses have been presented without the underlying derivations based on random vibration theory, a subject

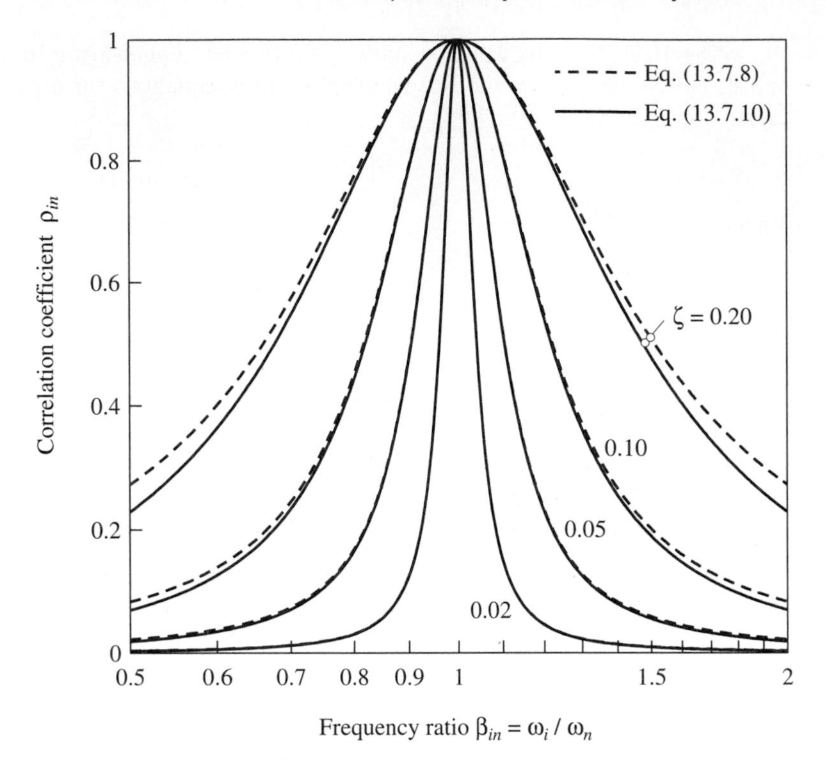

Figure 13.7.1 Variation of correlation coefficient ρ_{in} with modal frequency ratio, $\beta_{in} = \omega_i/\omega_n$, as given by two different equations for four damping values; abcissa scale is logarithmic.

beyond the scope of this book. It is important, however, to recognize the implications of the assumptions behind the derivations. These assumptions indicate that the modal combination rules would be most accurate for earthquake excitations that contain a wide band of frequencies with long phases of strong shaking, which are several times longer than the fundamental periods of the structures, which are not too lightly damped ($\zeta_n > 0.005$). In particular, these modal combination rules will become less accurate for short-duration impulsive ground motions and are not recommended for ground motions that contain many cycles of essentially harmonic excitation.

Considering that the SRSS and CQC modal combination rules are based on random vibration theory, r_o should be interpreted as the mean of the peak values of response to an ensemble of earthquake excitations. Thus the modal combination rules are intended for use when the excitation is characterized by a smooth response (or design) spectrum, based on the response spectra for many earthquake excitations. The smooth spectrum may be the mean or median of the individual response spectra or it may be a more conservative spectrum, such as the mean-plus-one-standard-deviation spectrum (Section 6.9). The CQC or SRSS modal combination rule (as appropriate depending on the closeness of natural frequencies) when used in conjunction with, say, the mean spectrum provides an estimate of the peak response that is reasonably close to the mean of the peak values of response to individual excitations. The error in the estimate of the peak may be on either side, conservative or unconservative, and is usually no more than several percent for typical structures and earthquakes; see examples later.

It has been found that Eq. (13.7.3) or (13.7.4) also approximates the peak response to a single ground motion characterized by a jagged response spectrum. The errors are larger, however, in this case: perhaps in the range of 10 to 30%, depending on the fundamental period of the structure.

13.7.3 Interpretation of Response Spectrum Analysis

The response spectrum analysis (RSA) described in the preceding section is a procedure for dynamic analysis of a structure subjected to earthquake excitation, but it reduces to a series of static analyses. For each mode considered, static analysis of the structure subjected to forces $\mathbf{s}_n = \Gamma_n \mathbf{m} \phi_n$ provides the modal static response r_n^{st}, which is multiplied by the spectral ordinate A_n to obtain the peak modal response r_{no} [Eq. (13.7.1)]. Thus the RSA procedure avoids the dynamic analysis of SDF systems necessary for response history analysis (Fig. 13.1.1). However, the RSA is still a dynamic analysis procedure, because it uses the vibration properties—natural frequencies, natural modes, and modal damping ratios—of the structure and the dynamic characteristics of the ground motion through its response spectrum. It is just that the user does not have to carry out any response history calculations; somebody has already done these in developing the earthquake response spectrum or the earthquake excitation has been characterized by a smooth design spectrum.

13.8 MULTISTORY BUILDINGS WITH SYMMETRIC PLAN

13.8.1 Response Spectrum Analysis Procedure

In this section the response spectrum analysis procedure of Section 13.7 is specialized for multistory buildings with their plans having two axes of symmetry subjected to horizontal ground motion along one of these axes. The peak value† of the nth-mode contribution $r_n(t)$ to a response quantity is given by Eq. (13.7.1). The modal static response r_n^{st} is calculated by static analysis of the building subjected to lateral forces $\mathbf{s}_n$ of Eq. (13.2.4). Equations for r_n^{st} for several response quantities are available in Table 13.2.1. Substituting these formulas for floor displacement u_j, story drift Δ_j, base shear V_b, and base overturning moment M_b in Eq. (13.7.1) gives

$$u_{jn} = \Gamma_n \phi_{jn} D_n \qquad \Delta_{jn} = \Gamma_n (\phi_{jn} - \phi_{j-1,n}) D_n \tag{13.8.1a}$$

$$V_{bn} = M_n^* A_n \qquad M_{bn} = h_n^* M_n^* A_n \tag{13.8.1b}$$

†From now on, the subscript o is dropped from r_o for brevity [i.e., r will denote the peak value of $r(t)$].

where $D_n \equiv D(T_n, \zeta_n)$, the deformation spectrum ordinate corresponding to natural period T_n and damping ratio ζ_n; $D_n = A_n/\omega_n^2$.

Equations (13.8.1) for the peak modal responses are identical to static analysis of the building subjected to the equivalent static forces associated with the nth-mode peak response:

$$\mathbf{f}_n = \mathbf{s}_n A_n \qquad f_{jn} = \Gamma_n m_j \phi_{jn} A_n \tag{13.8.2}$$

where $\mathbf{f}_n$ is the vector of forces f_{jn} at the various floor levels, $j = 1, 2, \ldots, N$; $\mathbf{s}_n$ is defined by Eq. (13.2.4). The force vector $\mathbf{f}_n$ is the peak value of $\mathbf{f}_n(t)$, obtained by replacing $A_n(t)$ in Eq. (13.2.7) by the spectral ordinate A_n. Because only one static analysis is required for each mode, it is more direct to do so for the forces $\mathbf{f}_n$ instead of $\mathbf{s}_n$. In contrast, the use of the modal static response r_n^{st} was emphasized in response history analysis because it highlighted the fact that the static analysis for forces $\mathbf{s}_n$ was needed only once even though the response was computed at many time instants.

Thus the peak value r_n of the nth-mode contribution to a response quantity r is determined by static analysis of the building due to lateral forces $\mathbf{f}_n$; the direction of forces f_{jn} is controlled by the algebraic sign of ϕ_{jn}. Hence these forces for the fundamental mode will act in the same direction, Fig. 13.8.1, but for the second and higher modes they will change direction as one moves up the building. Observe that this static analysis is not necessary to determine floor displacements or story drifts; Eq. (13.8.1a) provides the more convenient alternative. The peak value of the total response is estimated using the modal combination rules of Eq. (13.7.3) or (13.7.4), as appropriate, including all modes that contribute significantly to the response.

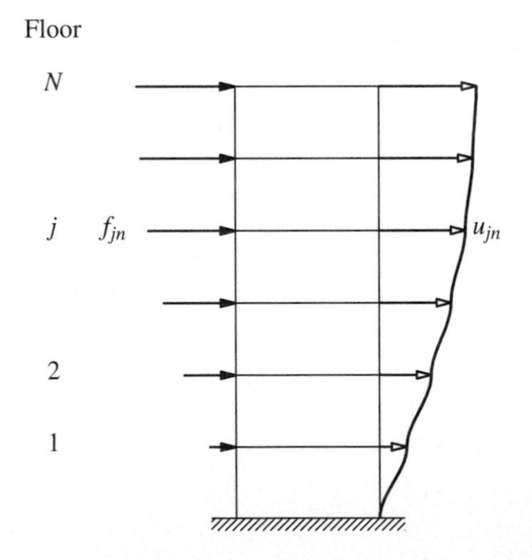

Figure 13.8.1 Peak values of lateral displacements and equivalent static lateral forces associated with the nth mode.

Summary. The procedure to compute the peak response of an N-story building with plan symmetric about two orthogonal axes to earthquake ground motion along an axis of symmetry, characterized by a response spectrum or design spectrum, is summarized in step-by-step form:

1. Define the structural properties.
   a. Determine the mass matrix **m** and lateral stiffness matrix **k** (Section 9.4).
   b. Estimate the modal damping ratios ζ_n (Chapter 11).
2. Determine the natural frequencies ω_n (natural periods $T_n = 2\pi/\omega_n$) and natural modes ϕ_n of vibration (Chapter 10).
3. Compute the peak response in the nth mode by the following steps to be repeated for all modes, $n = 1, 2, \ldots, N$:
   a. Corresponding to natural period T_n and damping ratio ζ_n, read D_n and A_n, the deformation and pseudo-acceleration, from the earthquake response spectrum or the design spectrum.
   b. Compute the floor displacements and story drifts from Eq. (13.8.1a).
   c. Compute the equivalent static lateral forces $\mathbf{f}_n$ from Eq. (13.8.2).
   d. Compute the story forces—shear and overturning moment—and element forces—bending moments and shears—by static analysis of the structure subjected to lateral forces $\mathbf{f}_n$.
4. Determine an estimate for the peak value r of any response quantity by combining the peak modal values r_n according to the SRSS rule, Eq. (13.7.3), if the natural frequencies are well separated. The CQC rule, Eq. (13.7.4), should be used if the natural frequencies are closely spaced.

Usually, only the lower modes contribute significantly to the response. Therefore, steps 2 and 3 need to be implemented for only these modes and the modal combinations of Eqs. (13.7.3) and (13.7.4) truncated accordingly.

Example 13.11

The peak response of the two-story frame of Example 13.4, shown in Fig. E13.11a to ground motion characterized by the design spectrum of Fig. 6.9.5 scaled to 0.5g peak ground acceleration is to be determined. This reinforced-concrete frame has the following properties: $E = 3 \times 10^3$ ksi, $I = 1000$ in^4, $h = 10$ ft, $L = 20$ ft. Determine the lateral displacements of the frame and bending moments at both ends of each beam and column.

Solution Steps 1 and 2 of the summary have already been implemented and the results are available in Examples 10.5 and 13.4. Substituting for E, I, and h in Eq. (b) of Example 13.4 gives ω_n and $T_n = 2\pi/\omega_n$:

$$\omega_1 = 4.023 \qquad \omega_2 = 10.71 \text{ rad/sec}$$

$$T_1 = 1.562 \qquad T_2 = 0.5868 \text{ sec}$$

Step 3a: Corresponding to these periods, the spectral ordinates are $D_1 = 13.72$ in. and $D_2 = 4.578$ in.

1. *Determine the floor displacements.*

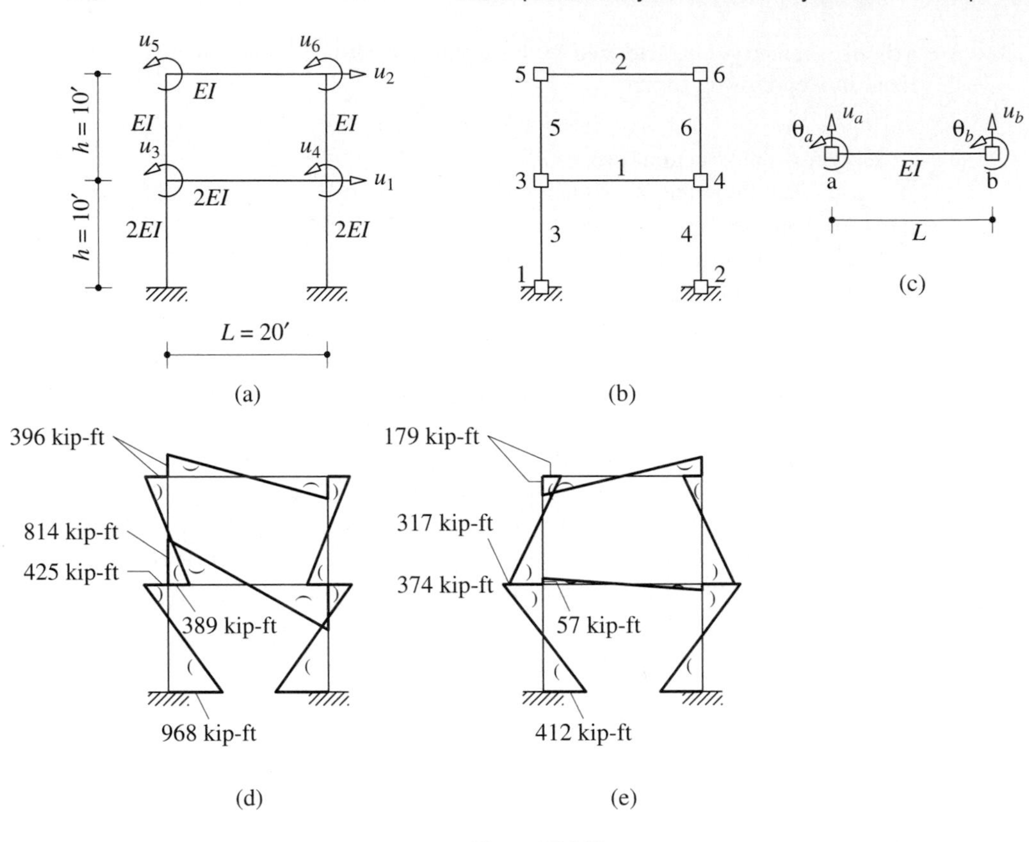

Figure E13.11

Step 3b: Using Eq. (13.8.1a) with numerical values for Γ_n and ϕ_{jn} from Example 13.4 and D_n from step 3(a) gives the peak displacements $\mathbf{u}_n$ due to the two modes:

$$\mathbf{u}_1 = \begin{Bmatrix} u_1 \\ u_2 \end{Bmatrix}_1 = 1.365 \begin{Bmatrix} 0.3871 \\ 1 \end{Bmatrix} 13.72 = \begin{Bmatrix} 7.252 \\ 18.73 \end{Bmatrix} \text{ in.}$$

$$\mathbf{u}_2 = \begin{Bmatrix} u_1 \\ u_2 \end{Bmatrix}_2 = -0.365 \begin{Bmatrix} -1.292 \\ 1 \end{Bmatrix} 4.578 = \begin{Bmatrix} 2.159 \\ -1.672 \end{Bmatrix} \text{ in.}$$

Step 4: Using the SRSS rule for modal combination, estimates for the peak values of the floor displacements are

$$u_1 \approx \sqrt{(7.252)^2 + (2.159)^2} = 7.566 \text{ in.}$$

$$u_2 \approx \sqrt{(18.73)^2 + (-1.672)^2} = 18.81 \text{ in.}$$

2. *Determine the element forces.* Instead of implementing steps 3c and 3d as described in the summary, here we will illustrate the computation of element forces from the floor displacements and joint rotations. The elements and nodes are numbered as shown in Fig. E13.11b.

First mode. Joint rotations are obtained from Eq. (d) of Example 9.9 with $\mathbf{u}_t$ replaced by $\mathbf{u}_1$:

$$\mathbf{u}_{01} = \begin{Bmatrix} u_3 \\ u_4 \\ u_5 \\ u_6 \end{Bmatrix}_1 = \frac{1}{120} \begin{bmatrix} -0.4426 & -0.2459 \\ -0.4426 & -0.2459 \\ 0.9836 & -0.7869 \\ 0.9836 & -0.7869 \end{bmatrix} \begin{bmatrix} 7.252 \\ 18.73 \end{bmatrix} = \begin{bmatrix} -6.514 \\ -6.514 \\ -6.340 \\ -6.340 \end{bmatrix} \times 10^{-2}$$

From $\mathbf{u}_1$ and $\mathbf{u}_{01}$ all element forces can be calculated. For example, the bending moment at the left end of the first floor beam (Fig. E13.11c) is

$$M_a = \frac{4EI}{L}\theta_a + \frac{2EI}{L}\theta_b + \frac{6EI}{L^2}u_a - \frac{6EI}{L^2}u_b$$

Substituting $E = 3 \times 10^3$ ksi, $I = 2000$ in^4, $L = 240$ in., $\theta_a = u_3$, $\theta_b = u_4$, $u_a = u_b = 0$ gives $M_a = -9770$ kip-in. $= -814$ kip-ft. Bending moments in all elements can be calculated similarly. The results are summarized in Table E13.11 and in Fig. E13.11d.

TABLE E13.11 PEAK BENDING MOMENTS (KIP-FT)

Element	Node	Mode 1	Mode 2	Total
Beam 1	3	−814	− 57	816
	4	−814	− 57	816
Beam 2	5	−396	179	435
	6	−396	179	435
Column 3	3	425	374	566
	1	968	412	1052
Column 5	5	396	−179	435
	3	389	−317	502

Second mode. Joint rotations $\mathbf{u}_{02}$ are obtained from Eq. (d) of Example 9.9 with $\mathbf{u}_t$ replaced by $\mathbf{u}_2$. Computations for the element forces parallel those shown for the first mode, but using $\mathbf{u}_2$ and $\mathbf{u}_{02}$, leading to the results in Table E13.11 and in Fig. E13.11e.

Step 4: The peak value of each element force is estimated by combining its peak modal values by the SRSS rule. The results are shown in Table E13.11. Note that the algebraic signs of the bending moments are lost in the total values; therefore, it is not meaningful to draw the bending moment diagram and the total moments do not satisfy equilibrium at joints.

13.8.2 Example: Five-Story Shear Frame

In this section the RSA procedure is implemented for the five-story shear frame of Fig. 12.8.1. The complete history of this structure's response to the El Centro ground motion was determined in Section 13.2.6. We now estimate its peak response directly from the response spectrum for this excitation (i.e., without computing its response history).

Presented in Sections 12.8 and 13.2.6 were the mass and stiffness matrices and the natural vibration periods and modes of this structure. From these data, the modal properties M_n and L_n^h were computed (Table 13.2.2). The damping ratios are estimated as $\zeta_n = 5\%$.

Response spectrum ordinates. The response spectrum for the El Centro ground motion for 5% damping gives the values of D_n and A_n noted in Fig. 13.8.2 corresponding to the natural periods T_n. These are the precise values for the spectral ordinates, the peak values of $D_n(t)$ and $A_n(t)$ in Fig. 13.2.6, thus eliminating any errors in reading spectral ordinates. Such errors are inherent in practical implementation of the RSA procedure with a jagged response spectrum, but are eliminated if a smooth design spectrum, such as Fig. 6.9.5, is used.

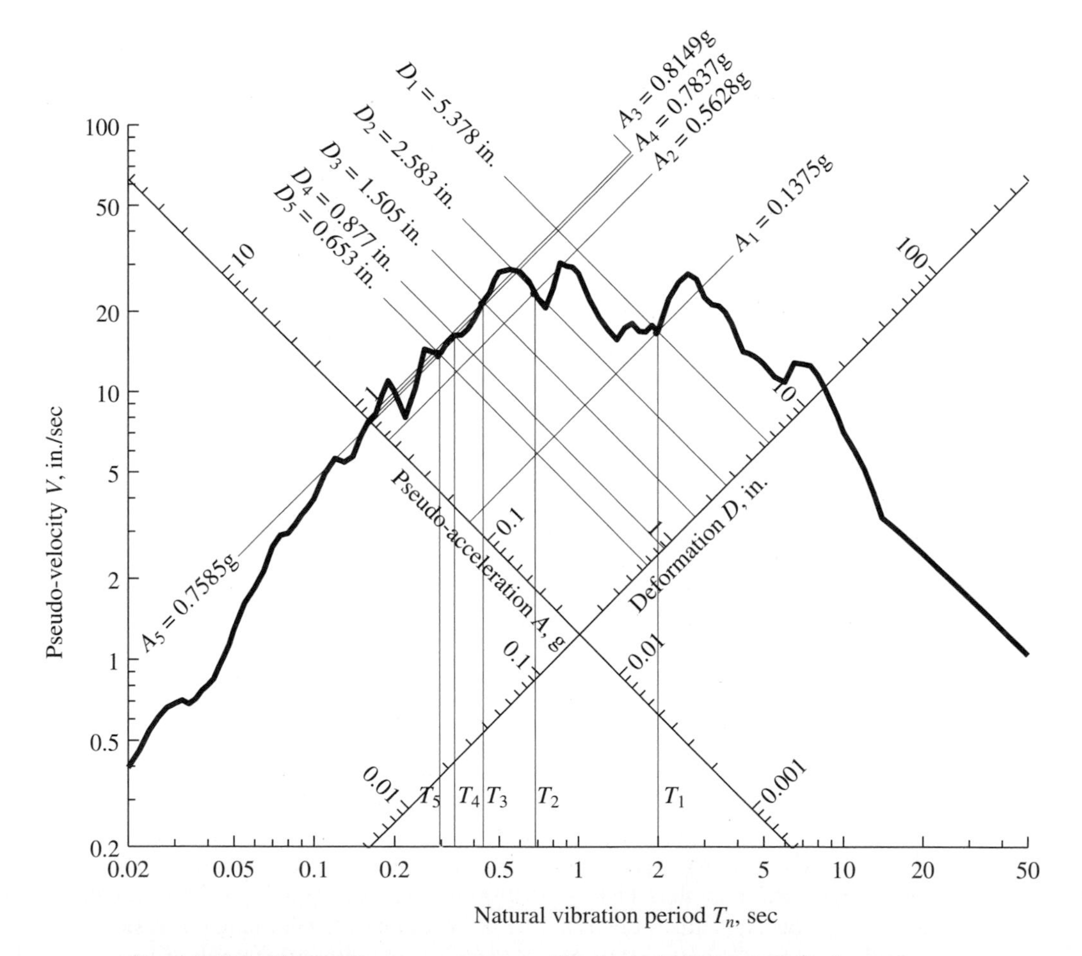

Figure 13.8.2 Earthquake response spectrum with natural vibration periods T_n of example structure shown together with spectral values D_n and A_n.

Peak modal responses. The floor displacements are determined from Eq. (13.8.1a) using known values of ϕ_{jn} (Section 12.8), of L_n^h (Table 13.2.2) and $\Gamma_n = L_n^h$ (because $M_n = 1$), and of D_n (Fig. 13.8.2). For example, the floor displacements due to the first mode are computed as follows:

$$\mathbf{u}_1 = \Gamma_1 \phi_1 D_1 = 1.067 \begin{Bmatrix} 0.334 \\ 0.641 \\ 0.895 \\ 1.078 \\ 1.173 \end{Bmatrix} 5.378 = \begin{Bmatrix} 1.916 \\ 3.677 \\ 5.139 \\ 6.188 \\ 6.731 \end{Bmatrix} \text{ in.}$$

These displacements are shown in Fig. 13.8.3a. The equivalent static forces for the first mode are computed from Eq. (13.8.2) using known values of Γ_n, ϕ_{jn}, $m_j = m = 100$ kips/g, and A_n (Fig. 13.8.2). For example, the forces associated with the first mode

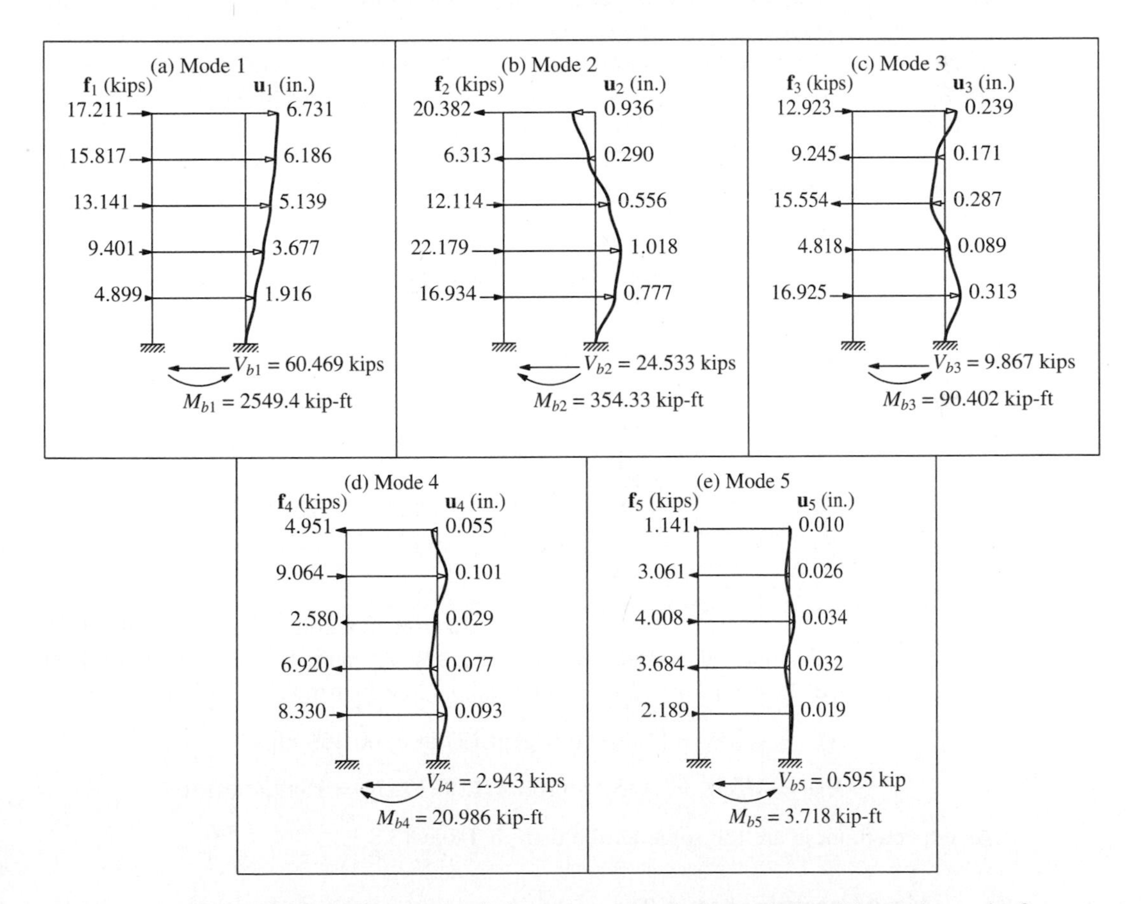

Figure 13.8.3 Peak values of displacements and equivalent static lateral forces due to the five natural vibration modes.

are computed as follows:

$$\mathbf{f}_1 = \Gamma_1 \begin{Bmatrix} m_1\phi_{11} \\ m_2\phi_{21} \\ m_3\phi_{31} \\ m_4\phi_{41} \\ m_5\phi_{51} \end{Bmatrix} A_1 = 1.067\frac{100}{\text{g}} \begin{Bmatrix} 0.334 \\ 0.641 \\ 0.895 \\ 1.078 \\ 1.173 \end{Bmatrix} 0.1375\text{g} = \begin{Bmatrix} 4.899 \\ 9.401 \\ 13.141 \\ 15.817 \\ 17.211 \end{Bmatrix} \text{kips}$$

These forces are also shown in Fig. 13.8.3a. Repeating these computations for modes $n = 2$, 3, 4, and 5 leads to the remaining results of Fig. 13.8.3. Observe that the equivalent static forces for the first mode all act in the same direction, but for the second and higher modes they change direction as one moves up the building; the direction of forces is controlled by the algebraic sign of ϕ_{jn}.

For each mode the peak value of any story force or element force is computed by static analysis of the structure subjected to the equivalent static lateral forces $\mathbf{f}_n$. Table 13.8.1 summarizes these peak values for the base shear V_b, top-story shear V_5, and base overturning moment M_b. The earlier data for roof displacement u_5 are also included. These peak modal values are exact because the errors in reading spectral ordinates had been eliminated in this example. This is apparent by comparing the data in Table 13.8.1 and the peak modal values from response history analysis in Figs. 13.2.7 and 13.2.8. The two sets of data agree except possibly for their algebraic signs because the peak values D_n and A_n are positive by definition.

TABLE 13.8.1 PEAK MODAL RESPONSES

Mode	V_b (kips)	V_5 (kips)	M_b (kip-ft)	u_5 (in.)
1	60.469	17.211	2549.4	6.731
2	24.533	−20.382	−354.33	−0.936
3	9.867	12.923	90.402	0.239
4	2.943	−4.951	−20.986	−0.055
5	0.595	1.141	3.718	0.010

Alternatively, Eq. (13.7.1) could have been used for computing the peak modal response. For example, the modal static responses V_{bn}^{st} and M_{bn}^{st} are available from Table 13.2.3 and A_n from Fig. 13.8.2. For example, the first-mode calculations are

$$V_{b1} = V_{b1}^{\text{st}} A_1 = [4.398(100/\text{g})]0.1375\text{g} = 60.469 \text{ kips}$$

$$M_{b1} = M_{b1}^{\text{st}} A_1 = [(15.45)(100/\text{g})12]0.1375\text{g} = 2549.4 \text{ kip-ft}$$

As expected, these are the same as the data in Table 13.8.1.

Modal combination. The peak value r of the total response $r(t)$ is estimated by combining the peak modal responses according to the ABSSUM, SRSS, and CQC rules of Eqs. (13.7.2) to (13.7.4). Their use is illustrated for one response quantity, the base shear.

The ABSSUM rule of Eq. (13.7.2) is specialized for the base shear:

$$V_b \leq \sum_{n=1}^{5} |V_{bn}| \tag{13.8.3}$$

Substituting for the known values of V_{bn} from Table 13.8.1 gives

$$V_b \leq 60.469 + 24.533 + 9.867 + 2.943 + 0.595 \quad \text{or} \quad V_b \leq 98.407 \text{ kips}$$

As expected, the ABSSUM estimate of 98.407 kips is much larger than the exact value of 73.278 kips (Fig. 13.2.7).

The SRSS rule of Eq. (13.7.3) is specialized for the base shear:

$$V_b \simeq \left(\sum_{n=1}^{5} V_{bn}^2 \right)^{1/2} \tag{13.8.4}$$

Substituting for the known values of V_{bn} from Table 13.8.1 gives

$$V_b \simeq \sqrt{(60.469)^2 + (24.533)^2 + (9.867)^2 + (2.943)^2 + (0.595)^2} = 66.066 \text{ kips}$$

Observe that the contributions of modes higher than the second are small.

The CQC rule of Eq. (13.7.4) is specialized for the base shear:

$$V_b \simeq \left(\sum_{i=1}^{5} \sum_{n=1}^{5} \rho_{in} V_{bi} V_{bn} \right)^{1/2} \tag{13.8.5}$$

Needed in this equation are the correlation coefficients ρ_{in}, which depend on the frequency ratios $\beta_{in} = \omega_i/\omega_n$, computed from the known natural frequencies (Section 13.2.6) and repeated in Table 13.8.2 for convenience.

TABLE 13.8.2 NATURAL FREQUENCY RATIOS β_{in}

Mode, i	$n = 1$	$n = 2$	$n = 3$	$n = 4$	$n = 5$	ω_i (rad/sec)
1	1.000	0.343	0.217	0.169	0.148	3.1416
2	2.919	1.000	0.634	0.494	0.433	9.1703
3	4.602	1.576	1.000	0.778	0.683	14.4561
4	5.911	2.025	1.285	1.000	0.877	18.5708
5	6.742	2.310	1.465	1.141	1.000	21.1810

For each β_{in} value in Table 13.8.2, ρ_{in} is determined from Eq. (13.7.10) for $\zeta = 0.05$ and presented in Table 13.8.3. Observe that the cross-correlation coefficients ρ_{in} ($i \neq n$) are small because the natural frequencies of the five-story shear frame are well separated.

These results bring out several response features of systems with two modes having close natural frequencies and contributing significantly to the response (e.g., the first two modes of the four-story building with an appendage). The cross-correlation coefficient for these two modes is 0.698, which is significant relative to its largest possible value of unity. As a result, the 1–2 cross terms for V_5 and V_b are comparable in magnitude to the individual modal (1–1 or 2–2) terms (Tables 13.8.9 and 13.8.10). Therefore, the SRSS and CQC modal combination rules provide very different estimates of peak responses (Table 13.8.11). The CQC rule gives a base shear that is larger than its value from the SRSS rule because all the cross ($i \neq n$) terms are positive (Table 13.8.9). For the appendage shear, however, the significant cross-term associated with the first two modes is negative (Table 13.8.10). Therefore, the CQC rule gives an appendage shear that is smaller than that obtained from the SRSS rule. Table 13.8.11 shows that only the CQC modal combination rule provides estimates of peak response that are close to the RHA results of Fig. 13.2.11. The errors in the SRSS estimates are unacceptably large; and they are even larger in the ABSSUM results.

An examination of the RHA results reveals the reasons for these large errors in the SRSS combination rule. Observe that the SDF system responses $A_n(t)$ for the first two modes are highly correlated, as they are essentially in phase because the two natural periods are close (Fig. 13.2.10); the peak values of the two modal responses are similar because their natural periods are close and their damping ratios are identical. As a result, the response contributions of the first two modes are similar in magnitude (Fig. 13.2.11). These modal contributions to the base shear are almost directly additive because they are essentially in phase (Fig. 13.2.11a). This feature of the response is not represented by the SRSS rule, whereas it is recognized in the CQC rule by the significant cross term (between modes 1 and 2) with positive value (Table 13.8.9). In contrast, the two modal contributions to the appendage shear tend to cancel each other because they have essentially opposite phase (Fig. 13.2.11b). This feature of the response is again not represented by the SRSS rule, whereas it is recognized in the CQC rule by the significant cross term (between modes 1 and 2) with negative value (Table 13.8.10). It is clear from this discussion that the SRSS modal combination rule should not be used for systems with closely spaced natural frequencies.

13.9 MULTISTORY BUILDINGS WITH UNSYMMETRIC PLAN

In this section the response spectrum analysis procedure of Section 13.7 is specialized for multistory buildings with their plans symmetric about the x-axis but unsymmetric about the y-axis subjected to ground motion in the y-direction. The peak value of the nth-mode contribution $r_n(t)$ to a response quantity is given by Eq. (13.7.1). The modal static response r_n^{st} is calculated by static analysis of the building subjected to lateral forces $\mathbf{s}_{yn}$ and torques $\mathbf{s}_{\theta n}$ of Eq. (13.3.7). Equations for the modal static response r_n^{st} for several response quantities are available in Table 13.3.1. Substituting these formulas for floor translation u_{jy}, floor rotation $u_{j\theta}$, base shear V_b, base overturning moment

M_b, and base torque T_b in Eq. (13.7.1) gives

$$u_{jyn} = \Gamma_n \phi_{jyn} D_n \qquad u_{j\theta n} = \Gamma_n \phi_{j\theta n} D_n \tag{13.9.1a}$$

$$V_{bn} = M_n^* A_n \qquad M_{bn} = h_n^* M_n^* A_n \qquad T_{bn} = I_{On}^* A_n \tag{13.9.1b}$$

Equations (13.9.1) for the peak modal responses are identical to static analysis of the building subjected to the equivalent static forces associated with the nth-mode peak response; the lateral forces $\mathbf{f}_{yn}$ and torques $\mathbf{f}_{\theta n}$ are

$$\begin{Bmatrix} \mathbf{f}_{yn} \\ \mathbf{f}_{\theta n} \end{Bmatrix} = \begin{Bmatrix} \mathbf{s}_{yn} \\ \mathbf{s}_{\theta n} \end{Bmatrix} A_n \tag{13.9.2}$$

The lateral force and torque at the jth floor level (Fig. 13.9.1) are

$$f_{jyn} = \Gamma_n m_j \phi_{jyn} A_n \qquad f_{j\theta n} = \Gamma_n r^2 m_j \phi_{j\theta n} A_n \tag{13.9.3}$$

For reasons mentioned in Section 13.8, it is more direct to do the static analysis for the forces $\mathbf{f}_{yn}$ and $\mathbf{f}_{\theta n}$ instead of $\mathbf{s}_{yn}$ and $\mathbf{s}_{\theta n}$.

For any response quantity, therefore, the peak value of the nth-mode response is determined by static analysis of the building subjected to lateral forces $\mathbf{f}_{yn}$ and torques $\mathbf{f}_{\theta n}$; the direction of forces f_{jyn} and $f_{j\theta n}$ is controlled by the algebraic signs of ϕ_{jyn} and $\phi_{j\theta n}$. Observe that this static analysis is not necessary to determine floor displacements or rotations; Eq. (13.9.1a) provides the more convenient alternative. Although such a three-dimensional static analysis of an unsymmetric-plan building provides the element forces in all frames of the building, it may be useful to recognize that the element forces in an individual (ith) frame can also be determined by planar analysis of the frame subjected to lateral forces:

$$\mathbf{f}_{in} = \Gamma_n \mathbf{k}_{xi} (-y_i \phi_{\theta n}) D_n \qquad \mathbf{f}_{in} = \Gamma_n \mathbf{k}_{yi} (\phi_{yn} + x_i \phi_{\theta n}) D_n \tag{13.9.4}$$

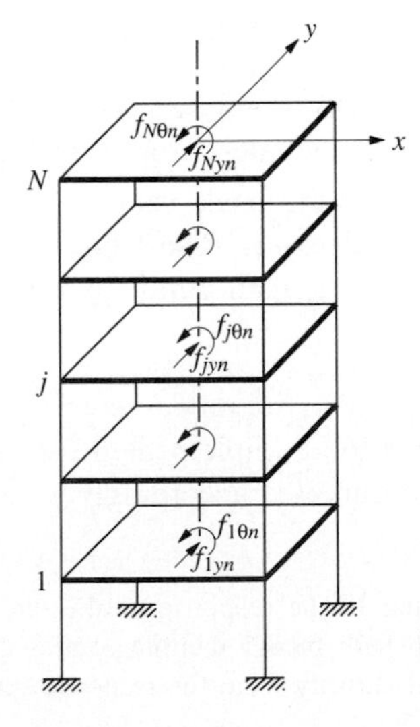

Figure 13.9.1 Peak values of equivalent static forces: lateral forces and torques.

The first of these equations applies to frames in the x-direction and the second to frames in the y-direction. They are obtained from Eq. (13.3.18) with $D_n(t)$ replaced by the corresponding spectral value D_n.

Once these peak modal responses have been determined for all the modes that contribute significantly to the total response, they can be combined using the CQC rule, Eq. (13.7.4), with N replaced by $2N$ in both summations, to obtain an estimate of the peak total response. The SRSS rule for modal combination should not be used because many unsymmetric-plan buildings have pairs (or triplets) of closely spaced natural frequencies.

Summary. The procedure to compute the peak response of an N-story building with its plan symmetric about the x-axis but unsymmetric about the y-axis subjected to the y-component of ground motion, characterized by a response spectrum or design spectrum, is summarized in step-by-step form:

1. Define the structural properties.
   a. Determine the mass and stiffness matrices from Eqs. (13.3.1) and (9.5.26).
   b. Estimate the modal damping ratios (Chapter 11).
2. Determine the natural frequencies ω_n (natural periods $T_n = 2\pi/\omega_n$) and natural modes ϕ_n of vibration (Chapter 10).
3. Compute the peak response in the nth mode by the following steps, to be repeated for all modes, $n = 1, 2, \ldots, 2N$:
   a. Corresponding to the natural period T_n and damping ratio ζ_n, read D_n and A_n, the deformation and pseudo-acceleration, from the earthquake response spectrum or design spectrum.
   b. Compute the lateral displacements and rotations of the floors from Eq. (13.9.1a).
   c. Compute the equivalent static forces: lateral forces $\mathbf{f}_{yn}$ and torques $\mathbf{f}_{\theta n}$ from Eq. (13.9.2) or (13.9.3).
   d. Compute the story forces—shear, torque, and overturning moment—and element forces—bending moments and shears—by three-dimensional static analysis of the structure subjected to external forces $\mathbf{f}_{yn}$ and $\mathbf{f}_{\theta n}$. Alternatively, the element forces in the ith frame can be calculated by planar static analysis of this frame subjected to the lateral forces of Eq. (13.9.4).
4. Determine an estimate for the peak value r of any response quantity by combining the peak modal values r_n. The CQC rule for modal combination should be used because unsymmetric-plan buildings usually have pairs of closely spaced frequencies.

Usually, only the lower pairs of modes contribute significantly to the response. Therefore, steps 2 and 3 need to be implemented for only these modes and the double summations in the CQC rule truncated accordingly.

Example 13.12

Determine the peak values of the response of the one-story unsymmetric-plan system of Examples 13.7 and 10.6 with modal damping ratios $\zeta_n = 5\%$ to the El Centro ground motion in the y-direction, directly from the response spectrum for this ground motion.

Solution Steps 1 and 2 of the procedure summary just presented have already been implemented in Example 10.6.

Step 3a: Corresponding to the known T_n and $\zeta_n = 5\%$, Fig. 6.6.4 gives the ordinates D_n and A_n. For $T_1 = 1.069$ sec: $D_1 = 4.256$ in. $= 0.3547$ ft and $A_1/g = 0.381$. For $T_2 = 0.9248$ sec: $D_2 = 4.161$ in. $= 0.3468$ ft and $A_2/g = 0.497$. (Obviously, numbers cannot be read to four significant figures from the response spectrum; they were obtained from the numerical data used in plotting Fig. 6.6.4; see also Fig. E13.8a.)

Step 3b: The peak values of roof displacement and rotation are obtained by specializing Eq. (13.9.1a) for the one-story system:

$$u_{yn} = \Gamma_n \phi_{yn} D_n \qquad u_{\theta n} = \Gamma_n \phi_{\theta n} D_n \tag{a}$$

Substituting numerical values for Γ_n, ϕ_{yn}, and $\phi_{\theta n}$ (from Example 13.7) in Eq. (a) with $n = 1$ gives the first-mode peak responses:

$$\begin{aligned} u_{y1} &= (-0.974)(-0.5228)(0.3547) = 0.1806 \text{ ft} = 2.168 \text{ in.} \\ \frac{b}{2}u_{\theta 1} &= \frac{30}{2}(-0.974)(0.0493)(0.3547) = -0.2555 \text{ ft} = -3.065 \text{ in.} \end{aligned} \tag{b}$$

where $(b/2)u_{\theta 1}$ represents the lateral displacement at the edge of the plan due to floor rotation. Similarly, the second-mode peak responses are

$$\begin{aligned} u_{y2} &= (-0.956)(-0.5131)(0.3468) = 0.1701 \text{ ft} = 2.042 \text{ in.} \\ \frac{b}{2}u_{\theta 2} &= \frac{30}{2}(-0.956)(-0.0502)(0.3468) = 0.2497 \text{ ft} = 2.999 \text{ in.} \end{aligned} \tag{c}$$

Step 3c: The peak values of f_{yn} and $f_{\theta n}$, the lateral force and torque, are obtained by specializing Eq. (13.9.3) for this one-story frame:

$$f_{yn} = \Gamma_n m \phi_{yn} A_n \qquad f_{\theta n} = \Gamma_n r^2 m \phi_{\theta n} A_n \tag{d}$$

By statics, the base shear and base torque are

$$V_{bn} = f_{yn} \qquad T_{bn} = f_{\theta n} \tag{e}$$

Alternatively, Eq. (13.7.1) can be used for computing the peak modal response. For example, the modal static responses V_{bn}^{st} and M_{bn}^{st} are available from Example 13.8. Substituting $V_{b1}^{\text{st}} = 0.509m$, $T_{b1}^{\text{st}} = -5.203m$, $m = 60$ kips/g, and $A_1 = 0.381$g in Eq. (13.7.1) gives

$$\begin{aligned} V_{b1} &= [0.509(60/\text{g})]0.381\text{g} = 11.63 \text{ kips} \\ T_{b1} &= [-5.203(60/\text{g})](0.381\text{g}) = -118.8 \text{ kip-ft} \end{aligned} \tag{f}$$

Substituting $V_{b2}^{\text{st}} = 0.491m$, $T_{b2}^{\text{st}} = 5.203m$, and $A_2 = 0.497$g in Eq. (13.7.1) gives

$$\begin{aligned} V_{b2} &= [0.491(60/\text{g})]0.497\text{g} = 14.64 \text{ kips} \\ T_{b2} &= [5.203(60/g)](0.497\text{g}) = 155.2 \text{ kip-ft} \end{aligned} \tag{g}$$

Step 3d: The peak lateral force for frame A is given by Eq. (13.9.4b) specialized for a one-story frame:

$$f_{An} = k_A \Gamma_n (\phi_{yn} + x_A \phi_{\theta n}) D_n \tag{h}$$

Substituting $k_A = 75$ kips/ft, $x_a = 1.5$ ft, and numerical values for Γ_n, ϕ_{yn}, $\phi_{\theta n}$, and D_n gives

$$f_{A1} = 75(-0.974)[-0.5228 + 1.5(0.0493)]0.3547 = 11.63 \text{ kips}$$

$$f_{A2} = 75(-0.956)[-0.5131 + 1.5(-0.0502)]0.3468 = 14.64 \text{ kips}$$

The base shear in a one-story frame is equal to the lateral force; thus

$$V_{bA1} = 11.63 \text{ kips} \qquad V_{bA2} = 14.64 \text{ kips} \tag{i}$$

The lateral force for frame B is given by Eq. (13.9.4a) specialized for a one-story frame:

$$f_{Bn} = k_B \Gamma_n (-y_B \phi_{\theta n}) D_n \tag{j}$$

Substituting $k_B = 40$ kips/ft, $y_B = 10$ ft, and numerical values for Γ_n, $\phi_{\theta n}$, and D_n gives

$$f_{B1} = 40(-0.974)[-10(0.0493)]0.3547 = -6.814 \text{ kips}$$

$$f_{B2} = 40(-0.956)[-10(-0.0502)]0.3468 = 6.662 \text{ kips}$$

The corresponding base shears are

$$V_{bB1} = -6.814 \text{ kips} \qquad V_{bB2} = 6.662 \text{ kips} \tag{k}$$

The results for peak modal responses are presented in Table E13.12a.

TABLE E13.12a PEAK MODAL RESPONSES

Mode	u_y (in.)	$b/2u_\theta$ (in.)	V_b (kips)	T_b (kip-ft)	V_{bA} (kips)	V_{bB} (kips)
1	2.168	−3.065	11.63	−118.8	11.63	−6.814
2	2.042	2.999	14.64	155.2	14.64	6.662

Step 4: For this system with two modes, the ABSSUM, SRSS, and CQC rules, Eqs. (13.7.2)–(13.7.4), specialize to

$$r \leq |r_1| + |r_2| \qquad r \simeq (r_1^2 + r_2^2)^{1/2} \qquad r \simeq (r_1^2 + r_2^2 + 2\rho_{12} r_1 r_2)^{1/2} \tag{l}$$

For this system, $\beta_{12} = \omega_1/\omega_2 = 5.878/6.794 = 0.865$. For this value of β_{12} and $\zeta = 0.05$, Eq. (13.7.10) gives $\rho_{12} = 0.322$. The results from Eq. (l) are summarized in Table E13.12b, wherein the peak values of total responses determined by RHA are also included. These

TABLE 13.12b RSA AND RHA VALUES OF PEAK RESPONSE

	u_y (in.)	$(b/2)u_\theta$ (in.)	V_b (kips)	T_b (kip-ft)	V_{bA} (kips)	V_{bB} (kips)
ABSSUM	4.210	6.064	26.27	274.0	26.27	13.48
SRSS	2.978	4.289	18.70	195.5	18.70	9.530
CQC	3.423	3.532	21.43	162.3	21.43	7.848
RHA	3.349	3.724	20.63	174.3	20.63	8.275

were computed using the results of Example 13.8, where $D_n(t)$ and $A_n(t)$ were computed by dynamic analysis of the nth-mode SDF system.

As expected, the ABSSUM estimate is always larger than the RHA value. The SRSS estimate is better, but the CQC estimate is the best because it accounts for the cross-correlation term in the modal combination, which is significant in this example because the natural frequencies are close, a situation common for unsymmetric-plan systems.

Example 13.13

Figure E13.13a–c shows a two-story building consisting of rigid diaphragms supported by three frames, A, B, and C. The lumped weights at the first and second floor levels are 120 and 60 kips, respectively. The lateral stiffness matrices of these frames, each idealized as a shear frame, are

$$\mathbf{k}_{yA} = \mathbf{k}_y = \begin{bmatrix} 225 & -75 \\ -75 & 75 \end{bmatrix} \qquad \mathbf{k}_{xB} = \mathbf{k}_{xC} = \mathbf{k}_x = \begin{bmatrix} 120 & -40 \\ -40 & 40 \end{bmatrix}$$

The design spectrum for $\zeta_n = 5\%$ is given by Fig. 6.9.5 scaled to 0.5g peak ground acceleration. Determine the peak value of the base shear in frame A.

Solution This system has four DOFs: u_{yj} and $u_{\theta j}$ (Fig. E13.13a); $j = 1$ and 2. The stiffness matrix of Eqs. (9.5.25) and (9.5.26) is specialized for this system with three frames:

$$\mathbf{k} = \begin{bmatrix} \mathbf{k}_y & e\mathbf{k}_y \\ e\mathbf{k}_y & e^2\mathbf{k}_y + (d^2/2)\mathbf{k}_x \end{bmatrix}$$

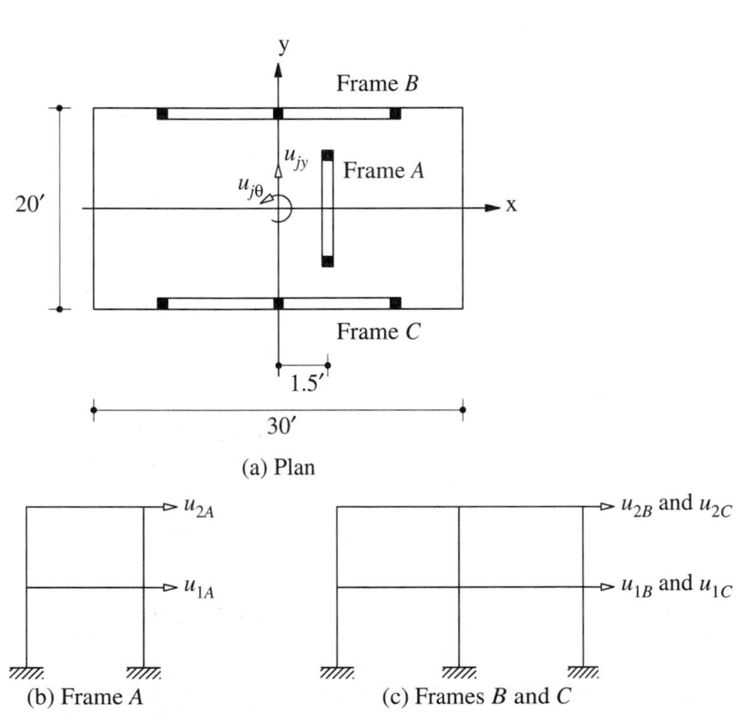

Figure E13.13a–c

Substituting for $\mathbf{k}_x$, $\mathbf{k}_y$, $e = 1.5$ ft, $d = 20$ ft, gives

$$\mathbf{k} = \begin{bmatrix} 225.0 & -75.00 & 337.5 & -112.5 \\ & 75.00 & -112.5 & 112.5 \\ & (\text{sym}) & 24,506 & -8,169 \\ & & & 8,169 \end{bmatrix}$$

The floor masses are $m_1 = 120/\text{g} = 3.727$ kip-sec^2/ft and $m_2 = 60/\text{g} = 1.863$ kip-sec^2/ft, and the floor moments of inertia are $I_{Oj} = m_j(b^2 + d^2)/12 = m_j(30^2 + 20^2)/12 = 1300m_j/12$. Substituting these data in the mass matrix of Eq. (9.5.27) gives

$$\mathbf{m} = \begin{bmatrix} 3.727 & & & \\ & 1.863 & & \\ & & 403.7 & \\ & & & 201.9 \end{bmatrix}$$

The eigenvalue problem is solved to determine the natural periods T_n and modes ϕ_n shown in Fig. E13.13d. Observe that each mode includes lateral and torsional motion. In the first mode the two floors displace in the same lateral direction and the two floors rotate in the same direction. In the second mode the two floors rotate in the same direction, which is opposite to the first mode. In the third and fourth modes the lateral displacements at the two floors are in opposite directions; the same is true for the rotations of the two floors.

The Γ_n are computed from Eqs. (13.3.4) to (13.3.6): $\Gamma_1 = 1.591$, $\Gamma_2 = 1.561$, $\Gamma_3 = -0.562$, and $\Gamma_4 = 0.552$.

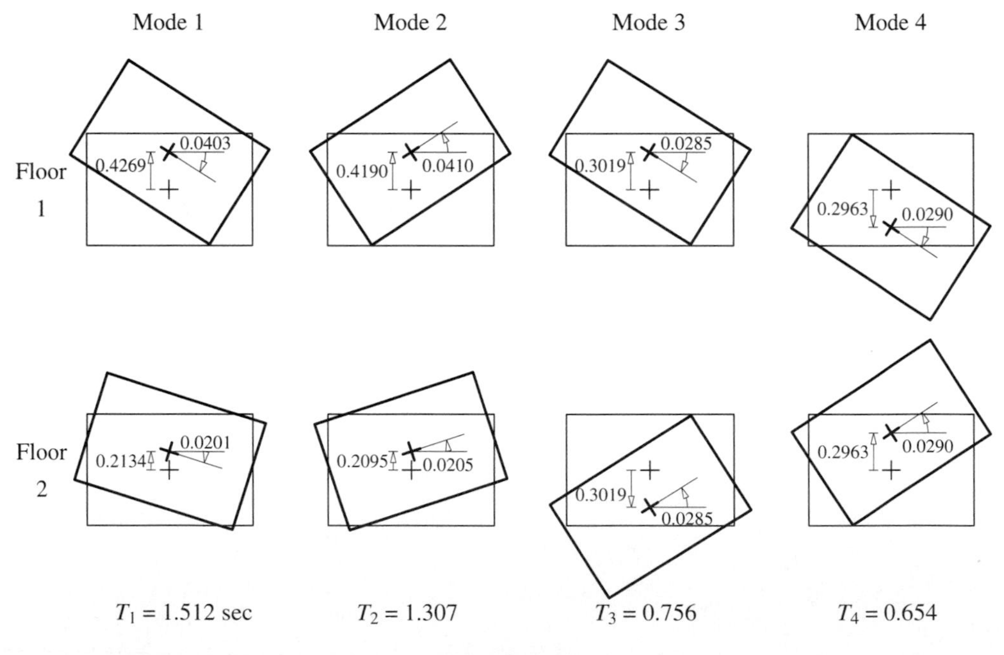

Figure E13.13d

For $T_n = 1.512$, 1.307, 0.756, and 0.654 sec, the design spectrum gives $A_1/g = 0.595$, $A_2/g = 0.688$, $A_3/g = 1.191$, and $A_4/g = 1.355$. The corresponding values of $D_n = A_n/\omega_n^2$ are $D_1 = 13.29$, $D_2 = 11.50$, $D_3 = 6.643$, and $D_4 = 5.671$ in.

The peak values of the equivalent static lateral forces for frame A are [from Eq. (13.9.4b)]

$$\mathbf{f}_{An} = \Gamma_n \mathbf{k}_y (\boldsymbol{\phi}_{yn} + e\boldsymbol{\phi}_{\theta n}) D_n$$

Substituting for Γ_1, $\mathbf{k}_y$, D_1, $\boldsymbol{\phi}_{y1}$, and $\boldsymbol{\phi}_{\theta 1}$ gives the lateral forces associated with the first mode:

$$\begin{Bmatrix} f_{A1} \\ f_{A2} \end{Bmatrix}_1 = 1.591 \frac{13.29}{12} \begin{bmatrix} 225 & -75 \\ -75 & -75 \end{bmatrix} \left(\begin{Bmatrix} 0.2134 \\ 0.4269 \end{Bmatrix} + 1.5 \begin{Bmatrix} -0.0201 \\ -0.0403 \end{Bmatrix} \right) = \begin{Bmatrix} 24.2 \\ 24.2 \end{Bmatrix}$$

Static analysis of the frame subjected to these lateral forces (Fig. E13.13e) gives the internal forces. In particular, the base shear is $V_{bA1} = f_{11} + f_{21} = 48.4$ kips. Similar computations lead to the peak base shear due to the second, third, and fourth modes: $V_{bA2} = 53.9$, $V_{bA3} = 12.1$, and $V_{bA4} = 13.3$ kips.

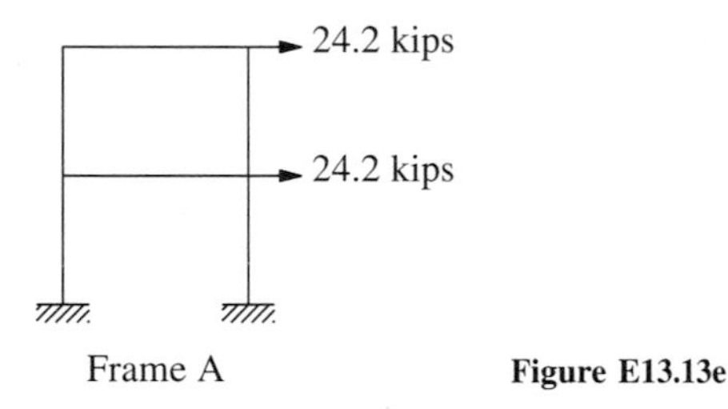

Figure E13.13e

The peak value r of the total response $r(t)$ will be estimated by combining the peak modal responses according to the CQC rule, Eq. (13.7.4). For this purpose it is necessary to determine the frequency ratios $\beta_{in} = \omega_i/\omega_n$; these are given in Table E13.13a. For each of the β_{in} values the correlation coefficient ρ_{in} is computed from Eq. (13.7.10) with $\zeta = 0.05$ and presented in Table E13.3b.

TABLE E13.13a NATURAL FREQUENCY RATIOS β_{in}

Mode, i	$n = 1$	$n = 2$	$n = 3$	$n = 4$	ω_i (rad/sec)
1	1.000	0.865	0.500	0.433	4.157
2	1.156	1.000	0.578	0.500	4.804
3	2.000	1.730	1.000	0.865	8.313
4	2.312	2.000	1.156	1.000	9.608

Substituting the peak modal values V_{bAn} and the correlation coefficients ρ_{in} in the CQC rule, we obtain the 16 terms in the double summation of Eq. (13.7.4) (Table E13.13c). Adding the 16 terms and taking the square root gives $V_{bA} = 86.4$ kips. Table E13.13c shows that the terms with significant values are the $i = n$ terms, and the cross terms between modes 1 and 2 and between modes 3 and 4. The cross terms between modes 1 and 3, 1 and 4, 2 and 3, or 2 and 4 are small because those frequencies are well separated. The

TABLE E13.3b CORRELATION COEFFICIENTS ρ_{in}

Mode, i	$n = 1$	$n = 2$	$n = 3$	$n = 4$
1	1.000	0.322	0.018	0.012
2	0.322	1.000	0.030	0.018
3	0.018	0.030	1.000	0.322
4	0.012	0.018	0.322	1.000

TABLE E13.13c INDIVIDUAL TERMS IN CQC RULE: BASE SHEAR V_{bA} IN FRAME A

Mode, i	$n = 1$	$n = 2$	$n = 3$	$n = 4$
1	2344.039	839.912	10.833	7.839
2	839.913	2905.669	19.748	13.250
3	10.833	19.748	146.502	51.797
4	7.839	13.250	51.797	176.807

square root of the sum of the four $i = n$ terms in Table E13.3c gives the SRSS estimate: $V_{bA} = 74.7$ kips. This is less accurate.

FURTHER READING

De la Llera, J. C., and Chopra, A. K., "Evaluation of Code Accidental Torsional Provisions from Building Records," *Journal of Structural Engineering, ASCE*, **120**, 1994, pp. 597–616.

Der Kiureghian, A., "A Response Spectrum Method for Random Vibration Analysis of MDF Systems," *Earthquake Engineering and Structural Dynamics*, **9**, 1981, pp. 419–435.

Newmark, N. M., and Rosenblueth E., *Fundamentals of Earthquake Engineering*, Prentice Hall, Englewood Cliffs, N.J., 1971, pp. 308–312.

Rosenblueth, E., "A Basis for Aseismic Design," Ph.D. thesis, University of Illinois, Urbana, Ill., 1951.

Rosenblueth, E., and Elorduy, J., "Responses of Linear Systems to Certain Transient Disturbances," *Proceedings of the 4th World Conference on Earthquake Engineering*, Santiago, Chile, Vol. I, 1969, pp. 185–196.

PROBLEMS

Part A

13.1 For the inverted L-shaped frame of Fig. E9.6a excited by vertical ground motion $\ddot{u}_g(t)$, determine (**a**) the modal expansion of effective earthquake forces, (**b**) the displacement response in terms of $D_n(t)$, and (**c**) the bending moment at the base of the column in terms of $A_n(t)$.

13.2 For the umbrella structure of Fig. P13.2 (also of Problem 9.5) excited by horizontal ground motion $\ddot{u}_g(t)$, determine (**a**) the modal expansion of effective earthquake forces, (**b**) the displacement response in terms of $D_n(t)$, and (**c**) the bending moments at the base of the column and at location a of the beam in terms of $A_n(t)$.

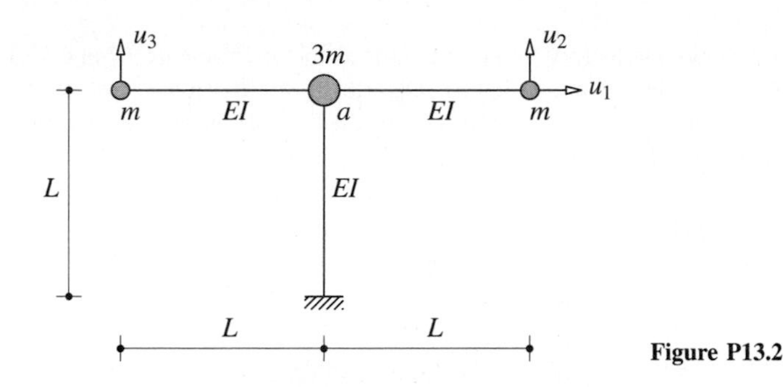

Figure P13.2

13.3 Repeat Problem 13.2 if the excitation is vertical ground motion.

13.4 For the two-story shear frame of Fig. P13.4 (also of Problems 9.6 and 10.7) excited by horizontal ground motion $\ddot{u}_g(t)$, determine (**a**) the modal expansion of effective earthquake forces, (**b**) the floor displacement response in terms of $D_n(t)$, (**c**) the story shear response in terms of $A_n(t)$, and (**d**) the first-floor and base overturning moments in terms of $A_n(t)$.

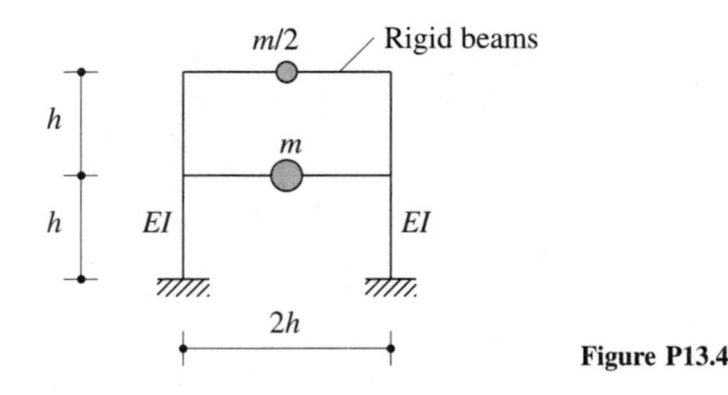

Figure P13.4

***13.5** The response of the two-story shear frame of Fig. P13.4 (also of Problems 9.6 and 10.7) to El Centro ground motion is to be computed as a function of time. The properties of the frame are $h = 12$ ft, $m = 100$ kips/g, $I = 727$ in^4, $E = 29{,}000$ ksi, and $\zeta_n = 5\%$. The ground acceleration data are available in Appendix 6 at every $\Delta t = 0.02$ sec.

(**a**) Determine the SDF system responses $D_n(t)$ and $A_n(t)$ using a numerical time-stepping method of your choice with an appropriate Δt; plot $D_n(t)$ and $A_n(t)$.

(**b**) For each natural mode calculate as a function of time the following response quantities: (i) the displacements at each floor, (ii) the story shears, and (iii) the floor and base overturning moments.

(**c**) At each instant of time combine the modal contributions to each of the response quantities to obtain the total response; determine the peak value of the total responses. For selected response quantities plot as a function of time the modal responses and total response.

*Denotes that a computer is necessary to solve this problem.

13.6 Determine the effective modal masses and effective modal heights for the two-story shear frame of Fig. P13.4 (also of Problems 9.6 and 10.7); the height of each story is h. Display this information on the SDF systems for the modes. Verify that Eqs. (13.2.14) and (13.2.17) are satisfied.

***13.7** Figure P13.7 shows a two-story frame (the same as that in Problems 9.7 and 10.11) with flexural rigidity EI for beams and columns. Determine the dynamic response of this structure to horizontal ground motion $\ddot{u}_g(t)$. Express **(a)** the floor displacements and joint rotations in terms of $D_n(t)$, and **(b)** the bending moments in a first-story column and in the second-floor beam in terms of $A_n(t)$.

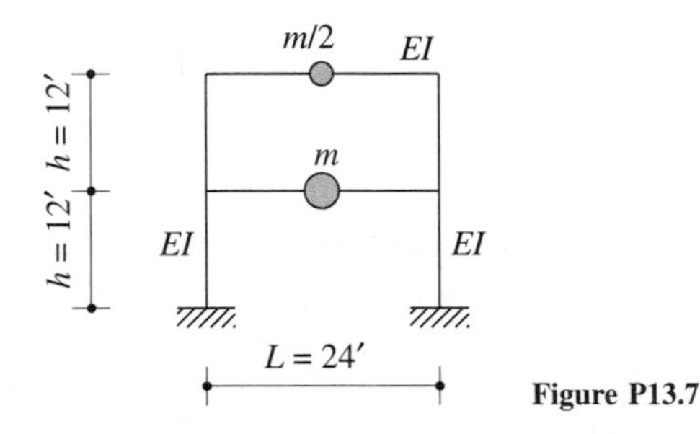

Figure P13.7

13.8 For the three-story shear frame of Fig. P13.8 (also of Problems 9.8 and 10.12) excited by horizontal ground motion $\ddot{u}_g(t)$, determine **(a)** the modal expansion of effective earthquake forces, **(b)** the floor displacement response in terms of $D_n(t)$, and **(c)** the story shear response in terms of $A_n(t)$, and **(d)** the base overturning moment in terms of $A_n(t)$.

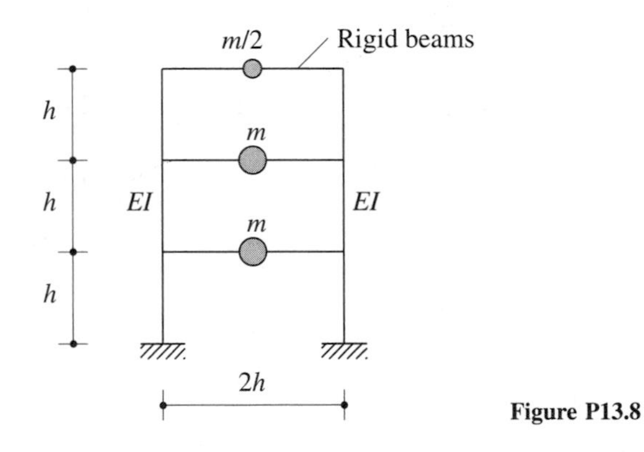

Figure P13.8

***13.9** The response of the three-story shear frame of Fig. P13.8 (also of Problems 9.8 and 10.12) to El Centro ground motion is to be computed as a function of time. The properties of the frame are $h = 12$ ft, $m = 100$ kips/g, $I = 1400$ in^4, $E = 29{,}000$ ksi, and $\zeta_n = 5\%$. The ground acceleration data are available in Appendix 6 at every $\Delta t = 0.02$ sec.
(a) Determine the SDF system responses $D_n(t)$ and $A_n(t)$ using a numerical time-stepping method of your choice with an appropriate Δt; plot $D_n(t)$ and $A_n(t)$.

*Denotes that a computer is necessary to solve this problem.

(b) For each natural mode calculate as a function of time the following response quantities: (i) the roof displacement, (ii) the story shears, and (iii) the base overturning moment.
(c) At each instant of time combine the modal contributions to each of the response quantities to obtain the total response; determine the peak value of the total responses. For selected response quantities plot as a function of time the modal responses and total response.

13.10 Determine the effective modal masses and effective modal heights for the three-story shear frame of Fig. P13.8; the height of each story is h. Display this information on the SDF systems for the modes. Verify that Eqs. (13.2.14) and (13.2.17) are satisfied.

***13.11** Figure P13.11 shows a three-story frame (the same as that in Problems 9.9 and 10.16) with flexural rigidity EI for beams and columns. Determine the dynamic response of this three-story frame to horizontal ground motion $\ddot{u}_g(t)$. Express **(a)** the floor displacements and joint rotations in terms of $D_n(t)$, and **(b)** the bending moments in a first-story column and in the second-floor beam in terms of $A_n(t)$.

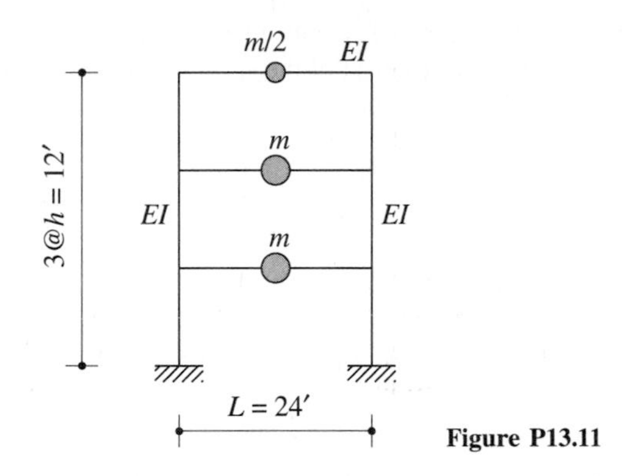

Figure P13.11

***13.12** A cantilever tower is shown in Fig. P13.12 with three lumped masses and its flexural stiffness properties; $m = 0.486$ kip-sec^2/in., $EI/L^3 = 56.26$ kips/in., and $EI'/L^3 = 0.0064$ kip/in. Note that the top mass and its supporting element are an appendage to the main tower. Damping is defined by modal damping ratios, with $\zeta_n = 5\%$ for all modes.

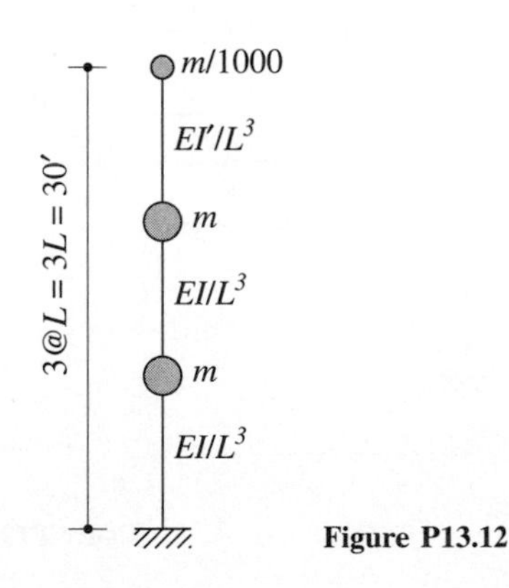

Figure P13.12

*Denotes that a computer is necessary to solve this problem.

(**a**) Determine the natural vibration periods and modes; sketch the modes.

(**b**) Expand the effective earthquake forces into their modal components and show this expansion graphically.

(**c**) Compute the modal static responses for three quantities: (i) the displacement of the appendage mass, (ii) the shear force at the base of the appendage, and (iii) the shear force at the base of the tower.

(**d**) What can you predict about the relative values of modal contributions to each of the response quantity from the results of parts (a) and (c)?

***13.13** The response of the tower with appendage of Fig. P13.12 to El Centro ground motion is to be computed as a function of time. The ground acceleration is available in Appendix 6 at every $\Delta t = 0.02$ sec. The damping of the structure is defined by the modal damping ratios $\zeta_n = 5\%$ for all modes.

(**a**) Determine the SDF system responses $D_n(t)$ and $A_n(t)$ using a numerical time-stepping method of your choice with an appropriate Δt.

(**b**) For each vibration mode calculate and plot as a function of time the following response quantities: (i) the displacement of the appendage mass, (ii) the shear force in the appendage, and (iii) the shear force at the base of the tower. Determine the peak value of each modal response.

(**c**) Calculate and plot as a function of time the total values of the three response quantities determined in part (b); determine the peak values of the total responses.

(**d**) Compute the seismic coefficients (defined as the shear force normalized by the weight) for the appendage and the tower. Why is the seismic coefficient for the appendage much larger than for the tower?

13.14 For the one-story, unsymmetric-plan system of Fig. P13.14 (the same as that defined in Problem 9.10 for which the natural vibration frequencies and modes were to be determined in Problem 10.17) which is excited by ground motion $\ddot{u}_{gy}(t)$ in the y-direction:

(**a**) Expand the effective earthquake forces in terms of their modal components and show this expansion graphically.

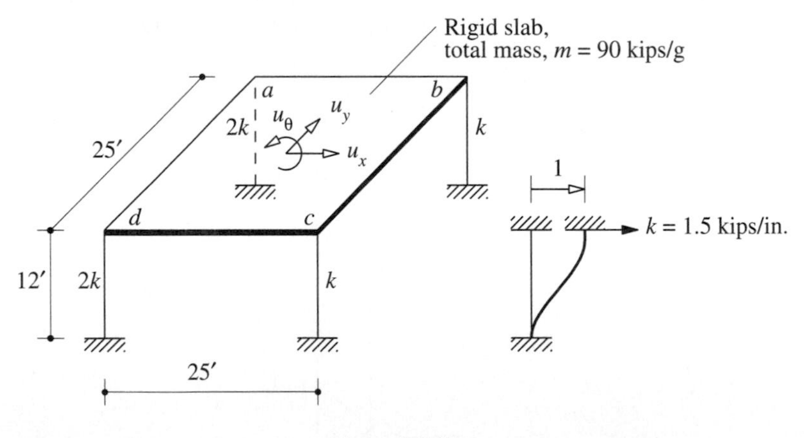

Figure P13.14

*Denotes that a computer is necessary to solve this problem.

(**b**) Verify that Eq. (13.3.9) is satisfied.
(**c**) Determine the displacement u_y and rotation u_θ of the slab in terms of $D_n(t)$.
(**d**) Determine the base shear and base torque in terms of $A_n(t)$.

13.15 For the one-story, unsymmetric-plan system of Fig. P13.14 (the same as that defined in Problem 9.10 for which the natural vibration frequencies and modes were to be determined in Problem 10.17) which is excited by ground motion $\ddot{u}_g(t)$ along the diagonal *d–b*:
(**a**) Expand the effective earthquake forces in terms of their modal components and show this expansion graphically.
(**b**) Verify that Eq. (13.3.9) is satisfied.
(**c**) Determine the displacement u_y and rotation u_θ of the slab in terms of $D_n(t)$.
(**d**) Determine the x and y components of the base shear and base torque in terms of $A_n(t)$.

***13.16** The response history of the system of Problem 13.14 (the same as that in Problem 9.10 for which the natural vibration frequencies and modes were to be determined in Problem 10.17) to El Centro ground motion along the y-direction is to be determined. In addition to the system properties given in Fig. P13.14, $\zeta_n = 5\%$ for all natural vibration modes. The ground acceleration is available in Appendix 6 at every $\Delta t = 0.02$ sec.
(**a**) Determine the SDF system responses $D_n(t)$ and $A_n(t)$ using a numerical time-stepping method of your choice with an appropriate Δt; plot $D_n(t)$ and $A_n(t)$.
(**b**) For each vibration mode calculate and plot as a function of time the following response quantities: u_y, $b/2u_\theta$, base shear V_b, and base torque T_b.
(**c**) Calculate and plot as a function of time the total responses; determine the peak values of the total responses.

13.17 The system of Fig. P13.17 (and of Problem 9.12) is subjected to support motions $u_{g1}(t)$ and $u_{g2}(t)$. Determine the motion of the two masses as a function of time for two excitations: (**a**) $u_{g1}(t) = -u_{g2}(t) = u_g(t)$, and (**b**) $u_{g2}(t) = u_{g1}(t) = u_g(t)$; express all results in terms of $D_n(t)$, the deformation response of the nth-mode SDF system to $\ddot{u}_g(t)$. Comment on how the response to the two excitations differs and why.

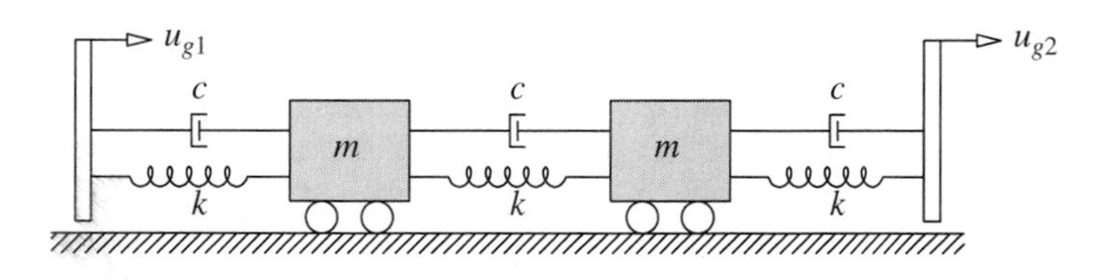

Figure P13.17

13.18 The undamped system of Fig. P13.18 (and of Problem 9.13), with $L = 50$ ft, $m = 0.2$ kip-sec^2/in., and $EI = 5 \times 10^8$ kip-in^2 is subjected to support motions $u_{g1}(t)$ and $u_{g2}(t)$. Determine the steady-state motion of the lumped mass and the steady-state value of the bending moment at the midspan due to two harmonic excitations: (i) $u_{g1}(t) = u_{go} \sin \omega t$, $u_{g2}(t) = 0$; and (ii) $u_{g1}(t) = u_{g2}(t) = u_{go} \sin \omega t$. The excitation frequency ω is $0.8\omega_n$, where ω_n is the natural vibration frequency of the ω is $0.8\omega_n$, where ω_n is the natural vibration frequency of the system. Express your results in terms of u_{go}. Comment on (**a**) the relative

*Denotes that a computer is necessary to solve this problem.

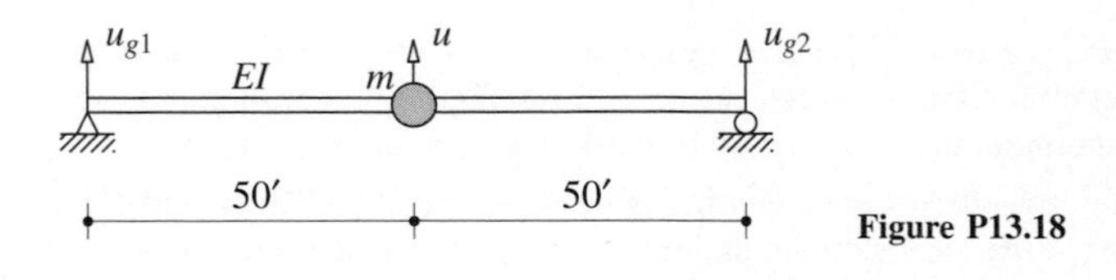

Figure P13.18

contributions of the quasi-static and dynamic components in each response quantity due to each excitation case, and (**b**) how the responses to the two excitations differ and why.

13.19 (**a**) In the intake tower of Problem 9.14 the base of the tower undergoes horizontal motion $u_g(t)$, and the right end of the bridge undergoes the same motion as the base A, but it does so t' seconds later. Determine the following responses as a function of time: (i) the displacement at the top of the tower, (ii) the shear and bending moment at the tower base, and (iii) the axial force in the bridge. Express the displacements in terms of $D_n(t)$ and forces in terms of $A_n(t)$, where $D_n(t)$ and $A_n(t)$ are the deformation and pseudo-acceleration response of the nth-mode SDF system to $\ddot{u}_g(t)$.
(**b**) Compare the preceding results with the response of the tower if both supports undergo identical motion $u_g(t)$. Comment on how the responses in the two cases differ and why.

Part B

***13.20** Figure P13.7 shows a two-story frame (the same as that in Problems 9.7 and 10.11) with $m = 100$ kips/g, $I = 727$ in^4 for beams and columns, and $E = 29{,}000$ ksi. Determine the response of this frame to ground motion characterized by the design spectrum of Fig. 6.9.5 (for 5% damping) scaled to $\frac{1}{3}$g peak ground acceleration. Compute (**a**) the floor displacements, and (**b**) the bending moments in a first-story column and in the second-floor beam.

13.21 The two-story shear frame of Fig. P13.4 (also of Problems 9.6 and 10.7) has the following properties: $h = 12$ ft, $m = 100$ kips/g, $I = 727$ in^4 for columns, $E = 29{,}000$ ksi, and $\zeta_n = 5\%$. The peak response of this structure to El Centro ground motion is to be estimated by response spectrum analysis (RSA) and compared with the results of Problem 13.5 from response history analysis (RHA). For the purposes of this comparison the RSA is to be implemented as follows.
(**a**) Determine the spectral ordinates D_n and A_n for the nth-mode SDF system as the peak values of $D_n(t)$ and $A_n(t)$, respectively, determined in part (a) of Problem 13.5. [We are doing so to avoid errors inherent in reading D_n and A_n from the response spectrum. However, in the standard application of RSA, $D_n(t)$ or $A_n(t)$ would not be available and D_n or A_n will be read from the response or design spectrum.]
(**b**) For each mode calculate the peak values of the following response quantities: (i) the floor displacements, (ii) the story shears, and (iii) the floor and base overturning moments.
(**c**) Combine the peak modal responses using an appropriate modal combination rule to obtain the peak value of the total response for each response quantity in part (b).
(**d**) Comment on the accuracy of the modal combination rule by comparing the RSA results from part (c) with the RHA results of Problem 13.5.

*Denotes that a computer is necessary to solve this problem.

***13.22** Figure P13.11 shows a three-story frame (the same as that in Problems 9.9 and 10.16) with $m = 100$ kips/g, $I = 1400$ in^4 for beams and columns, and $E = 29{,}000$ ksi. Determine the response of this frame to ground motion characterized by the design spectrum of Fig. 6.9.5 (for 5% damping) scaled to $\frac{1}{3}$g peak ground acceleration. Compute (**a**) the floor displacements, and (**b**) the bending moments in a first-story column and in the second-floor beam.

13.23 The three-story shear frame of Fig. P13.8 (also of Problems 9.8 and 10.12) has the following properties: $h = 12$ ft, $m = 100$ kips/g, $I = 1400$ in^4 for columns, $E = 29{,}000$ ksi, and $\zeta_n = 5\%$. The peak response of this structure to El Centro ground motion is to be estimated by response spectrum analysis (RSA) and compared with the results of Problem 13.9 from response history analysis (RHA). For the purposes of this comparison the RSA is to be implemented as follows.

(**a**) Determine the spectral ordinates D_n and A_n for the nth-mode SDF system as the peak values of $D_n(t)$ and $A_n(t)$, respectively, determined in part (a) of Problem 13.9. [We are doing so to avoid errors inherent in reading D_n and A_n from the response spectrum. However, in the standard application of RSA, $D_n(t)$ or $A_n(t)$ would not be available, and D_n or A_n will be read from the response or design spectrum.]

(**b**) For each mode calculate the peak values of the following response quantities: (i) the floor displacements, (ii) the story shears, and (iii) the floor and base overturning moments.

(**c**) Combine the peak modal responses using an appropriate modal combination rule to obtain the peak value of the total response for each response quantity in part (b).

(**d**) Comment on the accuracy of the modal combination rule by comparing the RSA results from part (c) with the RHA results of Problem 13.9.

13.24 The peak earthquake response of the tower with the appendage of Fig. P13.12 is to be determined. The ground motion is characterized by the design spectrum of Fig. 6.9.5 (for 5% damping), scaled to $\frac{1}{3}$g peak ground acceleration.

(**a**) Using the SRSS and CQC modal combination rules, calculate the peak values of the following response quantities: (i) the displacement of the appendage mass, (ii) the shear force at the base of the appendage, and (iii) the shear force at the base of the tower.

(**b**) Comment on the differences between the results from the two modal combination rules and the reasons for these differences. Which of the two methods is accurate?

13.25 The peak response of the tower with the appendage of Fig. P13.12 to El Centro ground motion is to be estimated by response spectrum analysis (RSA) and compared with the results of Problem 13.13 from response history analysis (RHA). For the purposes of this comparison the RSA is to be implemented as follows.

(**a**) Determine the spectral ordinates D_n and A_n for the nth-mode SDF system as the peak values of $D_n(t)$ and $A_n(t)$, respectively, determined in part (a) of Problem 13.13. [We are doing so to avoid errors inherent in reading D_n and A_n from the response spectrum. However, in the standard application of RSA, $D_n(t)$ or $A_n(t)$ would not be available and D_n or A_n will be read from the response or design spectrum.]

(**b**) For each mode calculate the peak values of the following response quantities: (i) the displacement of the appendage mass, (ii) the shear force in the appendage, and (iii) the shear force at the base of the tower.

(**c**) Using the CQC method, combine the modal peak to determine the peak value of each of the response quantities of part (b). Which of the modal correlation terms must be retained and which could be dropped from CQC calculations, and why?

*Denotes that a computer is necessary to solve this problem.

(**d**) Repeat part (c) using the SRSS method.

(**e**) Comment on the accuracy of the CQC and SRSS modal combination rules by comparing the RSA results from parts (c) and (d) with the RHA results by solving Problem 13.13.

13.26 The peak response of the one-story, unsymmetric-plan system of Fig. P13.14 with $\zeta_n = 5\%$ is to be estimated by response spectrum analysis (RSA) and compared with the results of Problem 13.16 from response history analysis (RHA). For purposes of this comparison the RSA is implemented as follows.

(**a**) Determine the spectral ordinates D_n and A_n for the nth-mode SDF system as the peak values of $D_n(t)$ and $A_n(t)$, respectively, determined in part (a) of Problem 13.16. [We are doing so to avoid errors inherent in reading D_n or A_n from the response spectrum. However, in the standard application of RSA, $D_n(t)$ or $A_n(t)$ would not be available, and D_n or A_n will be read from the response or design spectrum.]

(**b**) For each mode calculate the peak values of the following response quantities: u_y, $(b/2)u_\theta$, the base shear V_b, and the base torque T_b.

(**c**) Using the SRSS and CQC modal combination rules, compute the peak value for each response quantity.

(**d**) Comment on the accuracy of the SRSS and CQC methods by comparing the RSA results from part (c) with the RHA results of Problem 13.16.

13.27 Determine the peak response of the one-story, unsymmetric-plan system of Fig. P13.14 to ground motion along the y-direction. The excitation is characterized by the design spectrum of Fig. 6.9.4 (for 5% damping), scaled to 0.5g peak ground acceleration:

(**a**) Using the SRSS and CQC modal combination rules, calculate the peak values of the following response quantities: u_x, u_y, $b/2u\theta$, the base shears in the x and y directions and the base torque, and the bending moments about the x and y axes at the base of each column.

(**b**) Comment on the differences between the results from the two modal combination rules and the reasons for these differences. Which of the two methods is accurate?

13.28 Determine the peak response of the one-story, unsymmetric-plan system of Fig. P13.14 to ground motion along the diagonal d–b. The excitation is characterized by the design spectrum of Fig. 6.9.4 (for 5% damping), scaled to 0.5g peak ground acceleration.

(**a**) Using the SRSS and CQC modal combination rules, calculate the peak values of the following response quantities: (i) u_x, (ii) u_y, (iii) $b/2u\theta$, (iv) the base shears in the x and y directions and the base torque, and (v) the bending moments about the x and y axes at the base of each column.

(**b**) Comment on the differences between the results from the two modal combination rules and the reasons for these differences. Which of the two methods is accurate?

14

Reduction of Degrees of Freedom

PREVIEW

Although our objective in this book is the analysis of structures for dynamic excitation, we recognize that in practice a dynamic analysis is usually preceded by static analysis for dead and live loads. The structural idealization for the static analysis is dictated by the complexity of the structure, and several hundred to a few thousand DOFs may be necessary for accurate evaluation of the internal element forces and stresses in a complex structure.

The same refined idealization may be used for dynamic analysis of the structure, but this may be unnecessarily refined and drastically fewer DOFs could suffice. Such is the case because the dynamic response of many structures can be represented well by the first few natural vibration modes, and these modes can be determined accurately from a structural idealization with drastically fewer DOFs than required for static analysis. Thus we are interested in reducing the number of DOFs as much as reasonably possible before proceeding with computation of natural frequencies and modes, which is perhaps the most demanding phase of dynamic analysis.

Presented in this chapter are two approaches to reduce the number of DOFs: mass lumping in selected DOFs and the Rayleigh–Ritz method. Before presenting these procedures we mention how kinematic constraints based on structural properties can be used to reduce the number of DOFs in the structural idealization for static analysis; this idealization is the starting point for dynamic analysis.

14.1 KINEMATIC CONSTRAINTS

The configuration and properties of a structure may suggest kinematic constraints which express the displacements of many DOFs in terms of a smaller set of displacements. For example, the floor diaphragms (or slabs) of a multistory building, although flexible in the vertical direction, are usually very stiff in their own plane and can be assumed as rigid without introducing significant error. With this assumption the horizontal displacements of all the joints at one floor level are related to the three rigid-body DOFs of the diaphragm in its own plane: the two horizontal components of displacement and rotation about a vertical axis.

As a result of the kinematic constraint, the number of DOFs that would be considered in a static analysis can be reduced almost by half for dynamic analysis. Consider, for example, the 20-story building shown in Fig. 14.1.1, consisting of eight frames in the x-direction and four in the y-direction. With 640 joints and six DOFs (three translations and three rotations) per joint, the system has 3840 DOFs. Assuming the floor diaphragms to be rigid in their own planes, the system has only 1980 DOFs. These include the vertical displacement and two rotations (in xz and yz planes) of each joint and three rigid-body DOFs per floor.

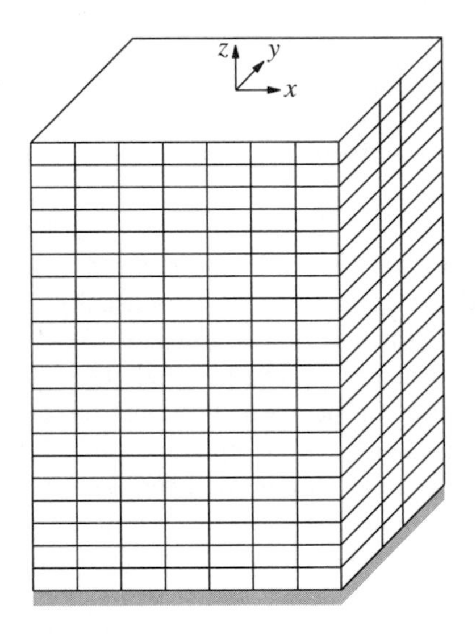

Figure 14.1.1 Twenty-story building.

Another kinematic constraint is sometimes assumed in building analysis is that the columns are axially rigid. This assumption should be used with discretion because it may be reasonable only in special circumstances: for example, buildings that are *not* slender. If justifiable, the assumption leads to further reduction in the number of DOFs; for static analysis of the multistory building of Fig. 14.1.1 this number reduces to 1340.

Once the structural idealization has been established for static analysis after considering kinematic constraints appropriate to the structure, the number of DOFs can be reduced for dynamic analysis by the procedures presented next.

14.2 STATIC CONDENSATION

In the static condensation method, developed in Section 9.3, the number of DOFs in the structural idealization established for static analysis is reduced by static equilibrium constraints. These DOFs are subdivided into two parts: $\mathbf{u}_t$, which have mass, and the remaining $\mathbf{u}_0$, which have zero mass and no external dynamic force, but are necessary for accurate representation of the stiffness properties of the structure. The DOFs $\mathbf{u}_0$ are related to $\mathbf{u}_t$ by Eq. (9.3.3) and, as indicated by Eq. (9.3.4), the equations of motion can be formulated in terms of only $\mathbf{u}_t$, the dynamic DOFs. The static condensation method is especially effective in earthquake analysis of multistory buildings; drastic reduction in the number of DOFs is possible because the inertial effects associated with rotations and vertical displacements of the joints are usually not significant. Assigning zero mass to these DOFs leaves only the three rigid-body DOFs of each floor diaphragm to be included in dynamic analysis. For the 20-story building of Fig. 14.1.1, this method reduces the number of degrees of freedom from 1980 to 60. The reduction in actual computational effort may be much less significant, however, than the reduction in the number of DOFs. This is because the efficiency of computation permitted by the narrow banding of the stiffness matrix $\mathbf{k}$ in Eq. (9.2.12) is in part lost in using the fully populated condensed stiffness matrix $\hat{\mathbf{k}}_{tt}$ in Eq. (9.3.4).

The relationship between $\mathbf{u}_0$ and $\mathbf{u}_t$, Eq. (9.3.3), although exact only if the $\mathbf{u}_0$ DOFs have zero mass, can also be used if this condition is not satisfied. In such cases Eq. (9.3.3) provides a basis to select displacement shapes for use in the Rayleigh–Ritz method described in the next section.

14.3 RAYLEIGH–RITZ METHOD

A most general technique for reducing the number of DOFs and finding approximations to the lower natural frequencies and modes is the Rayleigh–Ritz method. It is an extension of Rayleigh's method suggested by W. Ritz in 1909. Originally developed for systems with distributed mass and elasticity (see Chapter 16), the method is presented next for discretized systems.

14.3.1 Reduced Equations of Motion

The equations of motion for a system with N DOFs subjected to forces $\mathbf{p}(t) = \mathbf{s}p(t)$ are

$$\mathbf{m}\ddot{\mathbf{u}} + \mathbf{c}\dot{\mathbf{u}} + \mathbf{k}\mathbf{u} = \mathbf{s}p(t) \tag{14.3.1}$$

In Rayleigh's method we expressed the structural displacements as $\mathbf{u}(t) = z(t)\psi$, where ψ was an assumed shape vector; this led to an approximate value for the fundamental natural frequency and reduced the system to one with a single degree of freedom. In the Rayleigh–Ritz method, the displacements are expressed as a linear combination of several shape vectors ψ_j:

$$\mathbf{u}(t) = \sum_{j=1}^{J} z_j(t)\psi_j = \mathbf{\Psi z}(t) \tag{14.3.2}$$

where $z_j(t)$ are called the generalized coordinates, and the Ritz vectors ψ_j—$j = 1, 2, \ldots, J$—must be linearly independent vectors satisfying the geometric boundary conditions. They are selected appropriate for the system to be analyzed, as discussed in Section 14.4. The vectors ψ_j make up the columns of the $N \times J$ matrix $\mathbf{\Psi}$ in Eq. (14.3.2) and $\mathbf{z}$ is the vector of the J generalized coordinates.

Substituting the Ritz transformation of Eq. (14.3.2) in Eq. (14.3.1) gives

$$\mathbf{m\Psi\ddot{z}} + \mathbf{c\Psi\dot{z}} + \mathbf{k\Psi z} = \mathbf{s}p(t)$$

Each term is premultiplied by $\mathbf{\Psi}^T$ to obtain

$$\tilde{\mathbf{m}}\ddot{\mathbf{z}} + \tilde{\mathbf{c}}\dot{\mathbf{z}} + \tilde{\mathbf{k}}\mathbf{z} = \tilde{\mathbf{L}}p(t) \tag{14.3.3}$$

where

$$\tilde{\mathbf{m}} = \mathbf{\Psi}^T\mathbf{m\Psi} \qquad \tilde{\mathbf{c}} = \mathbf{\Psi}^T\mathbf{c\Psi} \qquad \tilde{\mathbf{k}} = \mathbf{\Psi}^T\mathbf{k\Psi} \qquad \tilde{\mathbf{L}} = \mathbf{\Psi}^T\mathbf{s} \tag{14.3.4}$$

Equation (14.3.3) is a system of J differential equations in the J generalized coordinates $\mathbf{z}(t)$. Observe that Eq. (14.3.4) defining $\tilde{\mathbf{m}}$, $\tilde{\mathbf{c}}$, and $\tilde{\mathbf{k}}$ is of the same form as Eqs. (12.3.4) and (12.4.3) for $\mathbf{M}$, $\mathbf{C}$, and $\mathbf{K}$ in the modal equations. The two differ, however, in an important sense that the Ritz vectors are used in one case, whereas the natural vibration modes in the other. Because the Ritz vectors are generally different from the natural modes, $\tilde{\mathbf{m}}$ and $\tilde{\mathbf{k}}$ are not diagonal matrices, whereas $\mathbf{M}$ and $\mathbf{K}$ are diagonal; see Eq. (12.3.6).

In summary, the Ritz transformation of Eq. (14.3.2) has made it possible to reduce the original set of N equations (14.3.1) in the nodal displacements $\mathbf{u}$ to a smaller set of J equations (14.3.3) in the generalized coordinates $\mathbf{z}$. This could be very advantageous if J is much smaller than N (i.e., if the displacements of the structure could be represented satisfactorily by a relatively few Ritz vectors). This is a powerful procedure because the approximations to the natural modes of the system determined by solving the eigenvalue problem associated with Eq. (14.3.3) represent the "best" solution among all possible solutions that are linear combinations of the selected Ritz vectors. In the next subsection we determine in what sense the solution "best" approximates the natural modes sought.

14.3.2 "Best" Approximation

We first determine Rayleigh's quotient, Eq. (10.12.1), for a vector $\tilde{\phi}$ defined consistent with Eq. (14.3.2) as a linear combination of the Ritz vectors:

$$\tilde{\phi} = \mathbf{\Psi z} \tag{14.3.5}$$

Substituting Eq. (14.3.5) in Eq. (10.12.1) gives

$$\rho(\mathbf{z}) = \frac{\sum_{i=1}^{J}\sum_{j=1}^{J} z_i z_j \tilde{k}_{ij}}{\sum_{i=1}^{J}\sum_{j=1}^{J} z_i z_j \tilde{m}_{ij}} \equiv \frac{\tilde{k}(\mathbf{z})}{\tilde{m}(\mathbf{z})} \tag{14.3.6}$$

where $\tilde{k}(\mathbf{z})$ and $\tilde{m}(\mathbf{z})$ are scalar quantities and

$$\tilde{k}_{ij} = \psi_i^T \mathbf{k} \psi_j \qquad \tilde{m}_{ij} = \psi_i^T \mathbf{m} \psi_j \tag{14.3.7}$$

Rayleigh's quotient cannot be determined from Eq. (14.3.6) because the generalized coordinates z_n are unknown. From Section 10.12 it is known, however, that

$$\omega_1^2 \le \rho(z) \le \omega_N^2 \tag{14.3.8}$$

where ω_1^2 and ω_N^2 are the smallest and largest eigenvalues.

In Rayleigh–Ritz analysis we aim to determine the specific vectors $\tilde{\phi}_n$, $n = 1, 2, \ldots, J$, that best approximate the natural modes. For this purpose we invoke *Rayleigh's stationarity condition*, the property that Rayleigh's quotient is a minimum in the neighborhood of the true modes (or true values of $\mathbf{z}$). Because z_i are the only variables, the necessary condition for a minimum of $\rho(\mathbf{z})$ is

$$\frac{\partial \rho}{\partial z_i} = 0 \qquad i = 1, 2, \ldots, J \tag{14.3.9}$$

For the ρ given by Eq. (14.3.6),

$$\frac{\partial \rho}{\partial z_i} = \frac{2\tilde{m}\sum_{j=1}^{J} z_j \tilde{k}_{ij} - 2\tilde{k}\sum_{j=1}^{J} z_j \tilde{m}_{ij}}{\tilde{m}^2}$$

This condition for a minimum of $\rho(\mathbf{z})$ can be rewritten by substituting $\rho = \tilde{k}/\tilde{m}$ from Eq. (14.3.6):

$$\sum_{j=1}^{J} (\tilde{k}_{ij} - \rho \tilde{m}_{ij}) z_j = 0 \qquad i = 1, 2, \ldots, J \tag{14.3.10}$$

Writing these J equations in matrix form gives the reduced eigenvalue problem

$$\tilde{\mathbf{k}}\mathbf{z} = \rho \tilde{\mathbf{m}} \mathbf{z} \tag{14.3.11}$$

where $\tilde{\mathbf{k}}$ and $\tilde{\mathbf{m}}$ are the $J \times J$ matrices defined by Eq. (14.3.4) with their typical elements given by Eq. (14.3.7), and $\mathbf{z}$ is the vector of generalized coordinates that remain to be determined.

The solution to Eq. (14.3.11) by methods of Chapter 10 yields J eigenvalues ρ_n—$n = 1, 2, \ldots, J$—and the corresponding eigenvectors

$$\mathbf{z}_n = \langle z_{1n}, z_{2n}, \ldots, z_{Jn} \rangle^T \qquad n = 1, 2, \ldots, J \tag{14.3.12}$$

The eigenvalues provide $\tilde{\omega}_n = \sqrt{\rho_n}$, which are approximations to the true natural frequencies ω_n. The eigenvectors $\mathbf{z}_n$ substituted in Eq. (14.3.5) provide the vectors

$$\tilde{\phi}_n = \mathbf{\Psi} \mathbf{z}_n \qquad n = 1, 2, \ldots, J \tag{14.3.13}$$

which are approximations to the true natural modes ϕ_n. The accuracy of these approximate results is generally better for the lower modes than for higher modes.

An important property of the approximate natural frequencies is that they are not smaller than the true frequencies, that is,

$$\omega_1 \le \tilde{\omega}_1 \qquad \omega_2 \le \tilde{\omega}_2 \qquad \cdots \qquad \omega_J \le \tilde{\omega}_J \tag{14.3.14}$$

Thus, by minimizing Rayleigh's quotient, we have obtained approximate values $\tilde{\omega}_n$ that are "best" in the sense that they are closest to the true values ω_n. These best results are the solution of Eq. (14.3.11), which is the eigenvalue problem associated with Eq. (14.3.3), the reduced system of equations in terms of the Ritz coordinates. This proves the assertion at the end of Section 14.3.1.

14.3.3 Orthogonality of Approximate Modes

In this section we demonstrate that the vectors $\tilde{\phi}_n$ satisfy the orthogonality conditions

$$\tilde{\phi}_n^T \mathbf{k} \tilde{\phi}_r = 0 \qquad \tilde{\phi}_n^T \mathbf{m} \tilde{\phi}_r = 0 \qquad n \ne r \tag{14.3.15}$$

This result is by no means obvious because the vectors $\tilde{\phi}_n$ are only approximations of the natural modes ϕ_n, which are known to satisfy Eq. (10.4.1).

The eigenvectors $\mathbf{z}_n$ of Eq. (14.3.11) satisfy the orthogonality conditions:

$$\mathbf{z}_n^T \tilde{\mathbf{k}} \mathbf{z}_r = 0 \qquad \mathbf{z}_n^T \tilde{\mathbf{m}} \mathbf{z}_r = 0 \qquad n \ne r \tag{14.3.16}$$

Using this property and Eq. (14.3.13), the first orthogonality condition in Eq. (14.3.15) can be proven as follows:

$$\tilde{\phi}_n^T \mathbf{k} \tilde{\phi}_r = \mathbf{z}_n^T \mathbf{\Psi}^T \mathbf{k} \mathbf{\Psi} \mathbf{z}_r = \mathbf{z}_n^T \tilde{\mathbf{k}} \mathbf{z}_r = 0 \qquad n \ne r$$

The second orthogonality condition in Eq. (14.3.15) can be demonstrated similarly.

If the eigenvectors $\mathbf{z}_n$ were made to be mass orthonormal (Section 10.6), then

$$\mathbf{z}_n^T \tilde{\mathbf{m}} \mathbf{z}_n = 1 \qquad \mathbf{z}_n^T \tilde{\mathbf{k}} \mathbf{z}_n = \tilde{\omega}_n^2 \tag{14.3.17}$$

This implies, as can be easily demonstrated, that the approximate modes $\tilde{\phi}_n$ are also mass orthonormal:

$$\tilde{\phi}_n \mathbf{m} \tilde{\phi}_n = 1 \qquad \tilde{\phi}_n^T \mathbf{k} \tilde{\phi}_n = \tilde{\omega}_n^2 \tag{14.3.18}$$

Because the approximate modes $\tilde{\phi}_n$ satisfy the orthogonality conditions of Eq. (14.3.15), they can be used in classical modal solution of Eq. (14.3.1). Therefore, in the rest of this chapter we do not distinguish between the approximate values ($\tilde{\omega}_n$, $\tilde{\phi}_n$) and the exact values (ω_n, ϕ_n).

14.4 SELECTION OF RITZ VECTORS

The success of the Rayleigh–Ritz method depends on how well linear combinations of Ritz vectors can approximate the natural modes of vibration. Therefore, it is important that the Ritz vectors be selected judiciously. In this section we present two very different approaches; the first is based on physical insight into shapes of natural modes, and the second is a formal computational procedure.

14.4.1 Physical Insight into Natural Mode Shapes

If we can visualize the shapes of the first few natural vibration modes of a structure, the Ritz vectors can be selected as approximations to these modes. In particular, the nth Ritz vector ψ_n is selected to approximate the nth natural mode ϕ_n of the structure. For example, based on the examples solved in Chapters 10 and 12, we can visualize the first two natural modes in planar vibration of a multistory frame. Thus the two Ritz vectors shown in Fig. 14.4.1 could be used in the Rayleigh–Ritz method to determine approximations to the first two natural frequencies and modes of this structure.

This approach may not be possible for complex systems because it may be difficult to visualize their mode shapes if we have never determined the natural modes of similar structures. Such visualization can be especially difficult if the natural mode includes two- or three-dimensional motions. A general procedure to select Ritz vectors that does not depend on physical visualization of the natural modes is therefore developed in the next section. This systematic procedure is suitable for implementation on a computer.

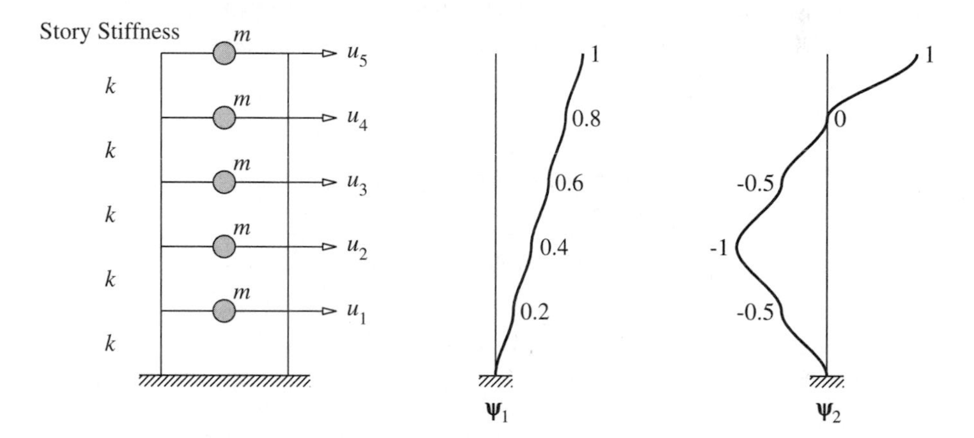

Figure 14.4.1 Ritz vectors for a five-story frame.

Example 14.1

By the Rayleigh–Ritz method, determine the first two natural frequencies and modes of a uniform five-story shear frame with story stiffnesses k and lumped floor masses m. Use the two Ritz vectors shown in Fig. 14.4.1.

Solution

1. *Formulate the stiffness and mass matrices.*

$$\mathbf{k} = k \begin{bmatrix} 2 & -1 & 0 & 0 & 0 \\ -1 & 2 & -1 & 0 & 0 \\ 0 & -1 & 2 & -1 & 0 \\ 0 & 0 & -1 & 2 & -1 \\ 0 & 0 & 0 & -1 & 1 \end{bmatrix} \qquad \mathbf{m} = m \begin{bmatrix} 1 & & & & \\ & 1 & & & \\ & & 1 & & \\ & & & 1 & \\ & & & & 1 \end{bmatrix}$$

2. *Compute* $\tilde{\mathbf{k}}$ *and* $\tilde{\mathbf{m}}$.

$$\boldsymbol{\Psi} = [\psi_1 \quad \psi_2] = \begin{bmatrix} 0.2 & -0.5 \\ 0.4 & -1.0 \\ 0.6 & -0.5 \\ 0.8 & 0 \\ 1.0 & 1.0 \end{bmatrix}$$

$$\tilde{\mathbf{k}} = \boldsymbol{\Psi}^T\mathbf{k}\boldsymbol{\Psi} = k\begin{bmatrix} 0.2 & 0.2 \\ 0.2 & 2.0 \end{bmatrix} \qquad \tilde{\mathbf{m}} = \boldsymbol{\Psi}^T\mathbf{m}\boldsymbol{\Psi} = m\begin{bmatrix} 2.2 & 0.2 \\ 0.2 & 2.5 \end{bmatrix}$$

3. *Solve the reduced eigenvalue problem, Eq.* (14.3.11). Substituting $\tilde{\mathbf{m}}$ and $\tilde{\mathbf{k}}$, Eq. (14.3.11) gives

$$\begin{bmatrix} 0.2 & 0.2 \\ 0.2 & 2.0 \end{bmatrix}\begin{bmatrix} z_1 \\ z_2 \end{bmatrix} = \left(\rho\frac{m}{k}\right)\begin{bmatrix} 2.2 & 0.2 \\ 0.2 & 2.5 \end{bmatrix}\begin{bmatrix} z_1 \\ z_2 \end{bmatrix} \tag{a}$$

The eigenvalue problem of Eq. (a) is solved to obtain

$$\rho_1 = 0.08238(k/m) \qquad \rho_2 = 0.8004(k/m)$$

$$\mathbf{z}_1 = \begin{Bmatrix} 1.329 \\ -0.1360 \end{Bmatrix} \qquad \mathbf{z}_2 = \begin{Bmatrix} 0.03170 \\ 1.240 \end{Bmatrix}$$

4. *Determine the approximate frequencies and modes.*

$$\tilde{\omega}_n = \sqrt{\rho_n} \qquad \tilde{\phi}_n = \boldsymbol{\Psi}\mathbf{z}_n$$

The results are presented in Table E14.1.

TABLE E14.1 COMPARISON OF APPROXIMATE AND EXACT RESULTS

Approximate	Exact
$\omega_1 = 0.2870\sqrt{k/m}$	$\omega_1 = 0.2846\sqrt{k/m}$
$\omega_2 = 0.8947\sqrt{k/m}$	$\omega_2 = 0.8308\sqrt{k/m}$
$\tilde{\boldsymbol{\Phi}} = \begin{bmatrix} 0.3338 & -0.6135 \\ 0.6676 & -1.227 \\ 0.8654 & -0.6008 \\ 1.063 & 0.02536 \\ 1.193 & 1.271 \end{bmatrix}$	$\boldsymbol{\Phi} = \begin{bmatrix} 0.3338 & -0.8954 \\ 0.6405 & -1.173 \\ 0.8954 & -0.6411 \\ 1.078 & 0.3338 \\ 1.173 & 1.078 \end{bmatrix}$

5. *Compare with the exact results.* The approximate values of the natural frequencies and modes are compared in Table E14.1 with their exact values obtained in Section 12.8. The errors in the approximate frequencies and modes are less than 1% in the first frequency, 8% in the second frequency, and 4% in the first mode. However, the second mode is so much in error that it may be useless.

14.4.2 Force-Dependent Ritz Vectors

It is desired to determine Ritz vectors appropriate for analysis of a structure subjected to external dynamic forces:

$$\mathbf{p}(t) = \mathbf{s}p(t) \tag{14.4.1}$$

The spatial distribution of forces defined by the vector $\mathbf{s}$ does not vary with time, and the time dependence of all forces is given by the same scalar function $p(t)$. Using the vector $\mathbf{s}$, a procedure is presented next to generate a sequence of orthonormal Ritz vectors.

The first Ritz vector ψ_1 is defined as the static displacements due to applied forces $\mathbf{s}$. It is determined by solving

$$\mathbf{k}\mathbf{y}_1 = \mathbf{s} \tag{14.4.2}$$

The vector $\mathbf{y}_1$ is normalized to be mass orthonormal; thus

$$\psi_1 = \frac{\mathbf{y}_1}{\left(\mathbf{y}_1^T\mathbf{m}\mathbf{y}_1\right)^{1/2}} \tag{14.4.3}$$

The second Ritz vector ψ_2 is determined from the vector $\mathbf{y}_2$ of static displacements due to applied forces given by the inertia force distribution associated with the first Ritz vector ψ_1. The vector $\mathbf{y}_2$ is obtained by solving

$$\mathbf{k}\mathbf{y}_2 = \mathbf{m}\psi_1 \tag{14.4.4}$$

The vector $\mathbf{y}_2$ will in general contain a component of the previous vector, ψ_1. It can therefore be expressed as

$$\mathbf{y}_2 = \hat{\psi}_2 + a_{12}\psi_1 \tag{14.4.5}$$

where $\hat{\psi}_2$ is a pure vector that does not contain the previous vector and $a_{12}\psi_1$ is the component of the previous vector present in $\mathbf{y}_2$. The vector $\hat{\psi}_2$ will by definition be orthogonal to, and hence linearly independent of, ψ_1. The coefficient a_{12} is determined by premultiplying both sides of Eq. (14.4.5) by $\psi_1^T\mathbf{m}$ to obtain

$$\psi_1^T\mathbf{m}\mathbf{y}_2 = \psi_1^T\mathbf{m}\hat{\psi}_2 + a_{12}\left(\psi_1^T\mathbf{m}\psi_1\right)$$

Note that $\psi_1^T\mathbf{m}\hat{\psi}_2 = 0$ by definition of $\hat{\psi}_2$, and $\psi_1^T\mathbf{m}\psi_1 = 1$ from Eq. (14.4.3). Thus

$$a_{12} = \psi_1^T\mathbf{m}\mathbf{y}_2 \tag{14.4.6}$$

The pure vector $\hat{\psi}_2$ is given by

$$\hat{\psi}_2 = \mathbf{y}_2 - a_{12}\psi_1 \tag{14.4.7}$$

Finally, the vector $\hat{\psi}_2$ is normalized so that it is mass orthonormal to obtain the second Ritz vector:

$$\psi_2 = \frac{\hat{\psi}_2}{\left(\hat{\psi}_2^T\mathbf{m}\hat{\psi}_2\right)^{1/2}} \tag{14.4.8}$$

Generalizing this procedure, the nth Ritz vector ψ_n is determined from the vector $\mathbf{y}_n$ of the static displacements due to applied forces given by the inertia force distribution associated with the $(n-1)$th Ritz vector ψ_{n-1}. The vector $\mathbf{y}_n$ is determined by solving

$$\mathbf{k}\mathbf{y}_n = \mathbf{m}\psi_{n-1} \tag{14.4.9}$$

The vector $\mathbf{y}_n$ will in general contain components of previous Ritz vectors ψ_j and can therefore be expressed as

$$\mathbf{y}_n = \hat{\psi}_n + \sum_{j=1}^{n-1} a_{jn}\psi_j \tag{14.4.10}$$

where $\hat{\psi}_n$ is a pure vector that does not contain the previous vectors, and $a_{jn}\psi_j$ are the components of the previous vectors present in $\mathbf{y}_n$. The vector $\hat{\psi}_n$ will by definition be orthogonal to, and hence linearly independent of, all the previous vectors. The coefficient a_{jn} is determined by premultiplying both sides of Eq. (14.4.10) by $\psi_i^T\mathbf{m}$:

$$\psi_i^T\mathbf{m}\mathbf{y}_n = \psi_i^T\mathbf{m}\hat{\psi}_n + \sum_{j=1}^{n-1} a_{jn}\left(\psi_i^T\mathbf{m}\psi_j\right)$$

Observe that $\psi_i^T\mathbf{m}\hat{\psi}_n = 0$ by definition of $\hat{\psi}_n$, $\psi_i^T\mathbf{m}\psi_j = 0$ for $i \neq j$ because all previous vectors are mutually orthogonal with respect to the mass matrix, and $\psi_i^T\mathbf{m}\psi_i = 1$ because all previous vectors are mass orthonormal [Eqs. (14.4.3) and (14.4.8)]. Thus

$$a_{in} = \psi_i^T\mathbf{m}\mathbf{y}_n \qquad i = 1, 2, \ldots, n-1 \tag{14.4.11}$$

The pure vector $\hat{\psi}_n$ is given by

$$\hat{\psi}_n = \mathbf{y}_n - \sum_{i=1}^{n-1} a_{in}\psi_i \tag{14.4.12}$$

where a_{in} are known from Eq. (14.4.11). Finally, the vector ψ_n is normalized so that it is mass orthonormal to obtain the nth Ritz vector:

$$\psi_n = \frac{\hat{\psi}_n}{\left(\hat{\psi}_n^T\mathbf{m}\hat{\psi}_n\right)^{1/2}} \tag{14.4.13}$$

The sequence of vectors $\psi_1, \psi_2, \ldots, \psi_J$ are mutually mass-orthogonal and hence they satisfy the linear independence requirement of the Rayleigh–Ritz method.

While the Gram–Schmidt orthogonalization procedure of Eqs. (14.4.11) and (14.4.12) should theoretically mass-orthogonalize the new vector with respect to all previous vectors, the actual computer implementation may be fraught with loss-of-orthogonality problems due to numerical round-off errors. To overcome these difficulties the Gram–Schmidt procedure is modified as follows. After computation of *each* a_{in} from Eq. (14.4.11), an improved vector $\hat{\psi}_n$ is calculated from Eq. (14.4.12), which is used instead of $\mathbf{y}_n$ in Eq. (14.4.11) to calculate the next a_{in}. Including this modification, the procedure to generate the force-dependent Ritz vectors is summarized in Table 14.4.1 as it might be implemented on the computer.

The procedure to generate these vectors is reminiscent of the vector sequence $\mathbf{x}_1, \mathbf{k}^{-1}\mathbf{m}\mathbf{x}_1, (\mathbf{k}^{-1}\mathbf{m})^2\,\mathbf{x}_1, \ldots$ generated in the inverse iteration procedure (Section 10.13). When obtained without making the vectors orthogonal, this vector sequence converges

TABLE 14.4.1 GENERATION OF FORCE-DEPENDENT RITZ VECTORS

1. Determine the first vector, ψ_1.

a. Determine $\mathbf{y}_1$ by solving: $\mathbf{k}\mathbf{y}_1 = \mathbf{s}$.

b. Normalize $\mathbf{y}_1$: $\psi_1 = \mathbf{y}_1 \div (\mathbf{y}_1^T \mathbf{m} \mathbf{y}_1)^{1/2}$.

2. Determine additional vectors, ψ_n, $n = 2, 3, \ldots, J$.

a. Determine $\mathbf{y}_n$ by solving: $\mathbf{k}\mathbf{y}_n = \mathbf{m}\psi_{n-1}$.

b. Orthogonalize $\mathbf{y}_n$ with respect to previous $\psi_1, \psi_2, \ldots, \psi_{n-1}$ by repeating the following steps for $i = 1, 2, \ldots, n-1$:

- $a_{in} = \psi_i^T \mathbf{m} \mathbf{y}_n$.
- $\hat{\psi}_n = \mathbf{y}_n - a_{in}\psi_i$.
- $\mathbf{y}_n = \hat{\psi}_n$.

c. Normalize $\hat{\psi}_n$: $\psi_n = \hat{\psi}_n \div (\hat{\psi}_n^T \mathbf{m} \hat{\psi}_n)^{1/2}$.

to the lowest natural mode. With Gram–Schmidt orthogonalization, as in Table 14.4.1, this sequence provides the force-dependent Ritz vectors.

Example 14.2

The vibration properties of the uniform five-story shear frame of Example 14.1 with $m =$ 100 kips/g $= 0.2591$ kip-sec^2/in. and $k = 31.56$ kips/in. are to be determined by the Rayleigh–Ritz method using Ritz vectors determined from a force distribution $\mathbf{s} = \langle m \quad m \quad m \quad m \quad m \rangle^T$. Using two force-dependent vectors, determine the first two natural frequencies and modes of vibration.

Solution

1. The stiffness and mass matrices, **k** and **m**, are given in Example 14.1 with $k = 31.56$ kips/in. and $m = 0.2591$ kip-sec^2/in.

2. *Determine the first Ritz vector,* ψ_1.

- Solve $\mathbf{k}\mathbf{y}_1 = m\mathbf{1}$ to obtain $\mathbf{y}_1 = \langle 0.0410 \quad 0.0739 \quad 0.0985 \quad 0.1149 \quad 0.1231 \rangle^T$.
- Divide $\mathbf{y}_1$ by $(\mathbf{y}_1^T \mathbf{m} \mathbf{y}_1)^{1/2} = 0.1082$ to obtain the normalized vector:

$$\psi_1 = \langle 0.3792 \quad 0.6826 \quad 0.9102 \quad 1.062 \quad 1.138 \rangle^T$$

3. *Determine the second Ritz vector* ψ_2.

- Solve $\mathbf{k}\mathbf{y}_2 = \mathbf{m}\psi_1$ to obtain $\mathbf{y}_2 = \langle 0.0342 \quad 0.0654 \quad 0.0909 \quad 0.1090 \quad 0.1183 \rangle^T$.
- Orthogonalize $\mathbf{y}_2$ with respect to ψ_1:

$$a_{12} = \psi_1^T \mathbf{m} \mathbf{y}_2 = 0.1012$$

$$\hat{\psi}_2 = \mathbf{y}_2 - 0.1012\psi_1$$

$$= 10^{-2}\langle -0.4134 \quad -0.3705 \quad -0.1204 \quad 0.1500 \quad 0.3164 \rangle^T$$

- Divide $\hat{\psi}_2$ by $(\hat{\psi}_2 \mathbf{m} \hat{\psi}_2)^{1/2} = 0.3396 \times 10^{-2}$ to get the normalized vector:

$$\psi_2 = \langle -1.217 \quad -1.091 \quad -0.3546 \quad 0.4418 \quad 0.9316 \rangle^T$$

4. *Compute* $\tilde{\mathbf{k}}$ *and* $\tilde{\mathbf{m}}$.

$$\boldsymbol{\Psi} = [\psi_1 \ \psi_2]$$

$$\tilde{\mathbf{k}} = \boldsymbol{\Psi}^T \mathbf{k} \boldsymbol{\Psi} = \begin{bmatrix} 9.986 & -3.086 \\ -3.086 & 91.95 \end{bmatrix} \qquad \tilde{\mathbf{m}} = \boldsymbol{\Psi}^T \mathbf{m} \boldsymbol{\Psi} = \begin{bmatrix} 1.0 & \\ & 1.0 \end{bmatrix}$$

5. *Solve the reduced eigenvalue problem, Eq.* (14.3.11).

$$\tilde{\omega}_1 = 3.142 \qquad \tilde{\omega}_2 = 9.595$$

$$\mathbf{z}_1 = \begin{Bmatrix} 0.9993 \\ 0.0376 \end{Bmatrix} \qquad \mathbf{z}_2 = \begin{Bmatrix} -0.0376 \\ 0.9993 \end{Bmatrix}$$

6. *Determine the natural modes.* Substituting $\boldsymbol{\Psi}_n$ and $\mathbf{z}_n$ in Eq. (14.3.13) gives

$$\tilde{\phi}_1 = \langle 0.3332 \quad 0.6412 \quad 0.8962 \quad 1.078 \quad 1.172 \rangle^T$$

$$\tilde{\phi}_2 = \langle -1.230 \quad -1.116 \quad -0.3886 \quad 0.4016 \quad 0.8882 \rangle^T$$

7. *Compare with the exact results.* Table E14.1 gives the exact modes and frequencies; the latter, after substituting for k and m, are:

$$\omega_1 = 3.142 \qquad \omega_2 = 9.170 \text{ rad/sec}$$

The approximate frequencies $\tilde{\omega}_n$ and modes from this example, using force-dependent Ritz vectors, are better than those determined in Example 14.1 from assumed vectors.

14.5 DYNAMIC ANALYSIS USING RITZ VECTORS

Now that we have developed procedures to generate Ritz vectors, we return to the solution of Eq. (14.3.3), the reduced system of equations. With J Ritz vectors included, these J equations are coupled because in general the matrices $\tilde{\mathbf{m}}$, $\tilde{\mathbf{c}}$, and $\tilde{\mathbf{k}}$ in Eq. (14.3.3) are not diagonal. However, if the force-dependent Ritz vectors of Section 14.4.2 are used, $\tilde{\mathbf{m}} = \mathbf{I}$, the identity matrix. The set of J coupled equations can be solved for the unknowns $z_j(t)$—$j = 1, 2, \ldots, J$—by numerical time-stepping methods (Chapter 15). Then at each time instant, the nodal displacement vector $\mathbf{u}$ is determined from Eq. (14.3.2) and the element forces by the methods of Section 9.10. This method is quite general in the sense that it applies to classically damped systems as well as nonclassically damped systems. For systems with classical damping, an alternative procedure is presented at the end of this section.

The number of force-dependent Ritz vectors included in the dynamic analysis should be sufficient to represent accurately the vector $\mathbf{s}$ that defines the spatial distribution of forces. Because these Ritz vectors are mass-orthonormal, following the concepts of Section 12.8, the vector $\mathbf{s}$ can be expanded as follows:

$$\mathbf{s} = \sum_{n=1}^{N} \tilde{\Gamma}_n \mathbf{m} \psi_n \qquad \text{where} \qquad \tilde{\Gamma}_n = \psi_n^T \mathbf{s} \tag{14.5.1}$$

The J Ritz vectors included in dynamic analysis provide an approximation to $\mathbf{s}$, and an error vector can be defined as

$$\mathbf{e}_J = \mathbf{s} - \sum_{n=1}^{J} \tilde{\Gamma}_n \mathbf{m} \psi_n \tag{14.5.2}$$

Considering that a logical norm for the vector $\mathbf{s}$ is its length $(\mathbf{s}^T\mathbf{s})^{1/2}$, an error norm e_J is defined as

$$e_J = \frac{\mathbf{s}^T \mathbf{e}_J}{\mathbf{s}^T \mathbf{s}} \tag{14.5.3}$$

This error e_J will be zero when all N Ritz vectors are included ($J = N$) because of Eq. (14.5.1), and e_J will equal unity when no Ritz vectors are included ($J = 0$). Thus, enough Ritz vectors should be included so that e_J is sufficiently small.

To illustrate these concepts, the error is computed for the five-story uniform shear frame of Example 14.1. The results presented in Fig. 14.5.1 are for three different force distributions: $\mathbf{s}_a = \langle 0 \quad 0 \quad 0 \quad 0 \quad 1 \rangle^T$, $\mathbf{s}_b = \langle 0 \quad 0 \quad 0 \quad -2 \quad 1 \rangle^T$, and $\mathbf{s}_c = \langle 1 \quad 1 \quad 1 \quad 1 \quad 1 \rangle^T$. For a given force distribution, the error decreases as more Ritz vectors are included, and is zero when all five Ritz vectors are included. For a fixed number of Ritz vectors, the error is smallest for the force distribution $\mathbf{s}_c$, largest for $\mathbf{s}_b$, and has an intermediate value for $\mathbf{s}_a$.

Figure 14.5.1 also provides a comparison of the error e_J if J Ritz vectors are included in the analysis versus the error e_J if J natural vibration modes of the system are considered. The latter was calculated from formulas similar to Eqs. (14.5.2) and (14.5.3), with the Ritz vectors ψ_n replaced by the natural modes ϕ_n. The error is smaller when Ritz vectors are used because they are derived from the force distribution. While this property would indicate that Ritz vectors are preferable to natural modes, the latter lead to uncoupled modal equations, which have several advantages. In particular, they permit estimation of the peak value of the earthquake response of a structure by response spectrum analysis (Chapter 13, Part B).

Do we need a static correction term (see Section 12.12) to supplement the response obtained by dynamic analysis using a truncated set of Ritz vectors? This is not necessary because the static correction effect is contained in the first Ritz vector because it is obtained from the static displacements due to the applied forces.

For dynamic analysis of systems with classical damping, classical modal analysis of Eq. (14.3.1) may be preferable over solution of the coupled equations (14.3.3) in Ritz coordinates, especially if the natural frequencies and modes of the system are desired. The Rayleigh–Ritz concept is still useful, however, because these vibration properties are obtained by solving Eq. (14.3.11), a smaller eigenvalue problem of order J, instead of the original eigenvalue problem of size N. As mentioned earlier, the resulting frequencies $\tilde{\omega}_n$ and eigenvectors $\tilde{\phi}_n$—$n = 1, 2, \ldots, J$—are approximations to the first J natural vibration frequencies and modes of the system; the approximate results are generally more accurate for the lower modes and gradually deteriorate for the higher modes. Usually, the second half of the set of J frequencies and modes is not accurate enough to be useful. Thus the number of Ritz vectors included should be sufficient to determine accurately the desired number of natural modes. Because these approximate modes $\tilde{\phi}_n$

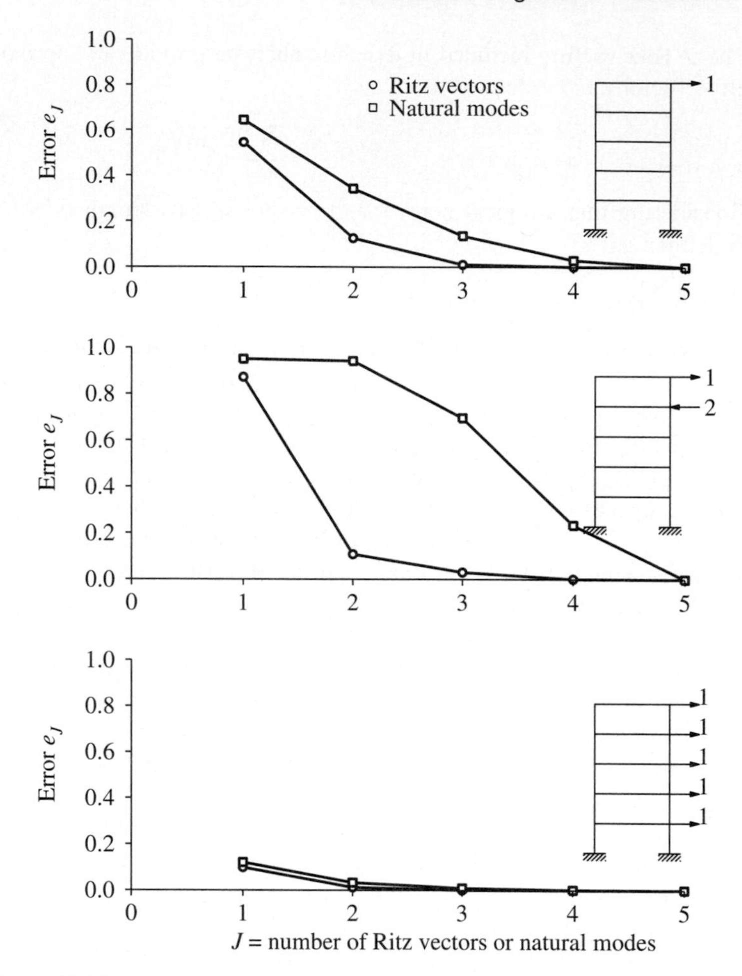

Figure 14.5.1 Variation of error e_J with the number J of Ritz vectors and of natural modes for three distributions of lateral forces.

are orthogonal with respect to the mass and stiffness matrices **m** and **k** (Section 14.3.3), they can be used just like the exact modes in classical modal analysis of the system.

FURTHER READING

Clough, R. W., and Penzien, J., *Dynamics of Structures*, McGraw-Hill, New York, 1993, pp. 314–323.

Humar, J. L., *Dynamics of Structures*, Prentice Hall, Englewood Cliffs, N.J., 1990, pp. 548–586.

Leger, P., Wilson, E. L., and Clough, R. W., "The Use of Load Dependent Vectors for Dynamic and Earthquake Analysis," *Report No. UCB/EERC 86–04*, Earthquake Engineering Research Center, University of California, Berkeley, Calif., 1986.

PROBLEMS

14.1 By the Rayleigh–Ritz method, determine the first two natural vibration frequencies and modes of the system in Fig. 14.4.1 using the following two Ritz vectors:

$$\psi_1 = \langle 0.3 \quad 0.6 \quad 0.8 \quad 0.9 \quad 1 \rangle^T$$

$$\psi_2 = \langle -1 \quad -1 \quad -0.5 \quad 0.5 \quad 1 \rangle^T$$

Compare these results with those obtained in Example 14.1 and the exact values presented in Section 12.8 and comment on how the selected Ritz vectors influence the accuracy of the results.

***14.2** Resolve Example 14.2 using Ritz vectors determined from the force distribution $\mathbf{s} = \langle 0 \quad 0 \quad 0 \quad 0 \quad 1 \rangle^T$. Comment on the accuracy of the results and how the force distribution used in generating Ritz vectors influences the accuracy.

14.3 By the Rayleigh–Ritz method, determine the first two natural vibration frequencies and modes of the five-story shear frame in Fig. P14.3 using the following two Ritz vectors:

$$\psi_1 = \langle 0.2 \quad 0.4 \quad 0.6 \quad 0.8 \quad 1 \rangle^T$$

$$\psi_2 = \langle -0.5 \quad -1 \quad -0.5 \quad 0 \quad 1 \rangle^T$$

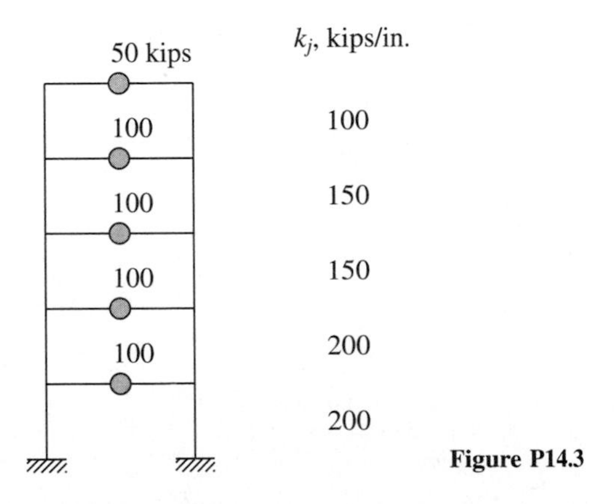

Figure P14.3

***14.4** Solve Problem 14.3 using force-dependent Ritz vectors determined from the force distribution $\mathbf{s} = \langle 1 \quad 1 \quad 1 \quad 1 \quad 0.5 \rangle^T$. Comment on the relative accuracy of the results of Problems 14.3 and 14.4.

*Denotes that a computer is necessary to solve this problem.

Numerical Evaluation of Dynamic Response

PREVIEW

So far, we have been concerned primarily with modal analysis of MDF systems with classical damping responding within their linearly elastic range. The uncoupled modal equations could be solved in closed form if the excitation were a simple function (Chapter 12), but the numerical methods of Chapter 5 were necessary for complex excitations such as earthquake ground motion (Chapter 13). Uncoupling of modal equations is not possible if the system has nonclassical damping or it responds into the nonlinear range. For such systems coupled equations of motion in nodal, modal, or Ritz coordinates—Eqs. (9.8.2), (12.4.4), or (14.3.3), respectively—need to be solved by numerical methods. A vast body of literature, including major chapters of several textbooks, exists about these methods. This chapter includes only a few methods, however, that build upon the procedures presented in Chapter 5 for SDF systems. It provides the basic concepts underlying these methods and the computational algorithms needed to implement the methods.

15.1 TIME-STEPPING METHODS

The objective is to solve numerically the system of differential equations governing the response of MDF systems:

$$\mathbf{m}\ddot{\mathbf{u}} + \mathbf{c}\dot{\mathbf{u}} + \mathbf{f}_S(\mathbf{u}, \dot{\mathbf{u}}) = \mathbf{p}(t) \tag{15.1.1}$$

with the initial conditions

$$\mathbf{u} = \mathbf{u}(0) \quad \text{and} \quad \dot{\mathbf{u}} = \dot{\mathbf{u}}(0) \tag{15.1.2}$$

at $t = 0$. The solution will provide the displacement vector $\mathbf{u}(t)$ as a function of time.

As in Chapter 5, the time scale is divided into a series of time steps, usually of constant duration Δt. The excitation is defined at discrete time instants $t_i = i\,\Delta t$; at this time, denoted as time i, the excitation vector is $\mathbf{p}_i \equiv \mathbf{p}(t_i)$. The response will be determined at the same time instants and is denoted by $\mathbf{u}_i \equiv \mathbf{u}(t_i)$, $\dot{\mathbf{u}}_i \equiv \dot{\mathbf{u}}(t_i)$, and $\ddot{\mathbf{u}}_i \equiv \ddot{\mathbf{u}}(t_i)$.

Starting with the known response of the system at time i that satisfies Eq. (15.1.1) at time i:

$$\mathbf{m}\ddot{\mathbf{u}}_i + \mathbf{c}\dot{\mathbf{u}}_i + (\mathbf{f}_S)_i = \mathbf{p}_i \tag{15.1.3}$$

time-stepping methods enable us to step ahead to determine the response $\mathbf{u}_{i+1}$, $\dot{\mathbf{u}}_{i+1}$, and $\ddot{\mathbf{u}}_{i+1}$ of the system at time $i+1$ that satisfies Eq. (15.1.1) at time $i+1$:

$$\mathbf{m}\ddot{\mathbf{u}}_{i+1} + \mathbf{c}\dot{\mathbf{u}}_{i+1} + (\mathbf{f}_S)_{i+1} = \mathbf{p}_{i+1} \tag{15.1.4}$$

When applied successively with $i = 0, 1, 2, 3, \ldots$, the time-stepping procedure gives the desired response at all time instants $i = 1, 2, 3, \ldots$. The known initial conditions at time $i = 0$, Eq. (15.1.2), provide the information necessary to start the procedure.

The numerical procedure requires three matrix equations to determine the three unknown vectors $\mathbf{u}_{i+1}$, $\dot{\mathbf{u}}_{i+1}$, and $\ddot{\mathbf{u}}_{i+1}$. Two of these equations are derived from either finite difference equations for the velocity and acceleration vectors or from an assumption on how the response varies during a time step. The third is Eq. (15.1.1) at a selected time instant. If it is the current time i, the method of integration is said to be an *explicit method.* If the time $i+1$ at the end of the time step is used, the method is known as an *implicit method;* see Chapter 5.

As mentioned in Chapter 5, for a numerical procedure to be useful, it should (1) converge to the exact solution as Δt decreases, (2) be stable in the presence of numerical round-off errors, and (3) be accurate (i.e., the computational errors should be within an acceptable limit). The stability criteria were shown not to be restrictive in the response analysis of SDF systems because Δt must be considerably smaller than the stability limit to ensure adequate accuracy in the numerical results. Stability of the numerical method is a critical consideration, however, in the analysis of MDF systems, as we shall see in this chapter. In particular, conditionally stable procedures can be used effectively for analysis of linear response of large MDF systems, but unconditionally stable procedures are generally necessary for nonlinear response analysis of such systems.

In the following sections we present some of the numerical methods for each type of response analysis.

15.2 ANALYSIS OF LINEAR SYSTEMS WITH NONCLASSICAL DAMPING

The N differential equations (15.1.1) to be solved for the nodal displacements $\mathbf{u}$, when specialized for linear systems, are

$$\mathbf{m}\ddot{\mathbf{u}} + \mathbf{c}\dot{\mathbf{u}} + \mathbf{k}\mathbf{u} = \mathbf{p}(t) \tag{15.2.1}$$

If the system has a few DOFs, it may be appropriate to solve these equations in their present form. For large systems it is usually advantageous to transform Eq. (15.2.1) to a smaller set of equations by expressing the displacements in terms of the first few natural vibration modes ϕ_n of the undamped system (Chapter 12) or an appropriate set of Ritz vectors (Chapter 14). In this section we use the modal transformation; extension of the concepts to use Ritz vector transformation is straightforward.

A major reduction in the number of equations is possible if the nodal displacements of the system can be approximated by a linear combination of a few natural modes:

$$\mathbf{u}(t) \simeq \sum_{n=1}^{J} \phi_n \mathbf{q}_n(t) = \mathbf{\Phi}\mathbf{q}(t) \tag{15.2.2}$$

Using this transformation, as shown in Section 12.4, Eq. (15.2.1) becomes

$$\mathbf{M}\ddot{\mathbf{q}} + \mathbf{C}\dot{\mathbf{q}} + \mathbf{K}\mathbf{q} = \mathbf{P}(t) \tag{15.2.3}$$

where

$$\mathbf{M} = \mathbf{\Phi}^T\mathbf{m}\mathbf{\Phi} \qquad \mathbf{C} = \mathbf{\Phi}^T\mathbf{c}\mathbf{\Phi} \qquad \mathbf{K} = \mathbf{\Phi}^T\mathbf{k}\mathbf{\Phi} \qquad \mathbf{P}(t) = \mathbf{\Phi}^T\mathbf{p}(t) \tag{15.2.4}$$

Equation (15.2.3) is a system of J equations in the unknowns $q_n(t)$, and if J is much smaller than N, it may be advantageous to solve them numerically instead of Eq. (15.2.1). The resulting computational savings can more than compensate for the additional computational effort necessary to determine the first J modes.

The J equations (15.2.3) may be coupled or uncoupled depending on the form of the damping matrix. They are uncoupled for systems with classical damping, and each modal equation can be solved numerically by the methods of Chapter 5. For systems with nonclassical damping, $\mathbf{C}$ is not a diagonal matrix and the equations are coupled. In this section numerical methods are presented for solving such coupled equations for linear systems. Although these methods are presented with reference to Eq. (15.2.3), they can be extended to the reduced set of equations (14.3.3) using Ritz vectors.

Conditionally stable numerical methods can be used to solve Eq. (15.2.3); that is, we need not insist on an unconditionally stable procedure (see Section 5.5.1). The time step Δt should be chosen so that $\Delta t/T_n$ is small enough to ensure an accurate solution for each of the modes included, $n = 1, 2, \ldots, J$; T_n is the natural period of the nth mode of the undamped system. The choice for Δt would be dictated by the period of the Jth mode because it has the shortest period; thus $\Delta t/T_J$ should be small, say less than 0.1. This choice implies that $\Delta t/T_n$ for all the lower modes is even smaller, ensuring an accurate solution for all the modes included. The Δt chosen to satisfy the accuracy requirement, say $\Delta t < 0.1T_J$, would obviously satisfy the stability requirement. For example, $\Delta t = 0.1T_J$ is much smaller than the stability limits of T_J/π and $0.551T_J$ for the central difference method and the linear acceleration method, respectively (Chapter 5).

Direct solution of Eq. (15.2.1)—without transforming to modal coordinates—may be preferable for systems with few DOFs or for systems and excitations where most of the modes contribute significantly to the response, because in these situations there is little to be gained by modal transformation. The numerical methods presented next are readily adaptable to such direct solution as long as the time step Δt is chosen to satisfy the stability requirement relative to the shortest natural period T_N of the undamped system.

Two conditionally stable procedures are presented next for linear response analysis of MDF systems. These are the central difference method and Newmark's method.

15.2.1 Central Difference Method

Developed in Section 5.3 for SDF systems, this is an explicit method that is especially easy to implement on the computer. The scalar equations (5.3.1) that relate the response quantities at time $i + 1$ to those at time i, and the scalar equation of equilibrium at time i, all now become matrix equations. The other new feature arises from the need to transform the initial conditions on nodal displacements, Eq. (15.1.2), to modal coordinates, and to transform back the solution of Eq. (15.2.3) in modal coordinates to nodal displacements. Putting all these ideas together leads to Table 15.2.1, where the central difference method is presented as it might be implemented on the computer.

Two observations regarding the central difference method may be useful. First, the algebraic equations to be solved in step 1.3 to determine $\ddot{\mathbf{q}}_0$ are uncoupled because $\mathbf{M}$ is a diagonal matrix when modal coordinates or force-dependent Ritz vectors are used. Second, step 2.3 is based on equilibrium at time i, and the stiffness matrix $\mathbf{K}$ does not enter into the system of algebraic equations solved to determine $\mathbf{q}_{i+1}$ at time $i + 1$, implying that the central difference method is an explicit numerical method.

The central difference method can also be used for direct solution of the original equations in nodal displacements, Eq. (15.2.1), without transforming them to modal coordinates, by modifying Table 15.2.1 as follows: Delete steps 1.1, 1.2, 2.1, and 2.5. Replace (1) $\mathbf{q}$, $\dot{\mathbf{q}}$, and $\ddot{\mathbf{q}}$ by $\mathbf{u}$, $\dot{\mathbf{u}}$, and $\ddot{\mathbf{u}}$; (2) $\mathbf{M}$, $\mathbf{C}$, and $\mathbf{K}$ by $\mathbf{m}$, $\mathbf{c}$, and $\mathbf{k}$; (3) $\mathbf{P}$ by $\mathbf{p}$; and (4) $\hat{\mathbf{K}}$ and $\hat{\mathbf{P}}$ by $\hat{\mathbf{k}}$ and $\hat{\mathbf{p}}$.

TABLE 15.2.1 CENTRAL DIFFERENCE METHOD: LINEAR SYSTEMS

1.0 *Initial calculations*

1.1 $$(q_n)_0 = \frac{\boldsymbol{\phi}_n^T \mathbf{m} \mathbf{u}_0}{\boldsymbol{\phi}_n^T \mathbf{m} \boldsymbol{\phi}_n}, \qquad (\dot{q}_n)_0 = \frac{\boldsymbol{\phi}_n^T \mathbf{m} \dot{\mathbf{u}}_0}{\boldsymbol{\phi}_n^T \mathbf{m} \boldsymbol{\phi}_n}.$$

$$\mathbf{q}_0^T = \langle (q_1)_0, \ldots, (q_J)_0 \rangle \qquad \dot{\mathbf{q}}_0^T = \langle (\dot{q}_1)_0, \ldots, (\dot{q}_J)_0 \rangle$$

1.2 $\mathbf{P}_0 = \mathbf{\Phi}^T \mathbf{p}_0$.

1.3 Solve: $\mathbf{M}\ddot{\mathbf{q}}_0 = \mathbf{P}_0 - \mathbf{C}\dot{\mathbf{q}}_0 - \mathbf{K}\mathbf{q}_0 \Rightarrow \ddot{\mathbf{q}}_0$.

1.4 Select Δt.

1.5 $\mathbf{q}_{-1} = \mathbf{q}_0 - \Delta t\, \dot{\mathbf{q}}_0 + \dfrac{(\Delta t)^2}{2} \ddot{\mathbf{q}}_0$.

1.6 $\hat{\mathbf{K}} = \dfrac{1}{(\Delta t)^2}\mathbf{M} + \dfrac{1}{2\Delta t}\mathbf{C}$.

1.7 $\mathbf{a} = \dfrac{1}{(\Delta t)^2}\mathbf{M} - \dfrac{1}{2\Delta t}\mathbf{C}, \quad \mathbf{b} = \mathbf{K} - \dfrac{2}{(\Delta t)^2}\mathbf{M}$.

2.0 *Calculations for each time step i*

2.1 $\mathbf{P}_i = \mathbf{\Phi}^T \mathbf{p}_i$.

2.2 $\hat{\mathbf{P}}_i = \mathbf{P}_i - \mathbf{a}\mathbf{q}_{i-1} - \mathbf{b}\mathbf{q}_i$.

2.3 Solve: $\hat{\mathbf{K}}\mathbf{q}_{i+1} = \hat{\mathbf{P}}_i \Rightarrow \mathbf{q}_{i+1}$.

2.4 If required:

$$\dot{\mathbf{q}}_i = \frac{1}{2\Delta t}(\mathbf{q}_{i+1} - \mathbf{q}_i) \qquad \ddot{\mathbf{q}}_i = \frac{1}{(\Delta t)^2}(\mathbf{q}_{i+1} - 2\mathbf{q}_i + \mathbf{q}_{i-1})$$

2.5 $\mathbf{u}_{i+1} = \mathbf{\Phi}\mathbf{q}_{i+1}$.

3.0 *Repetition for the next time step.* Replace i by $i + 1$ and repeat steps 2.1 to 2.5 for the next time step.

15.2.2 Newmark's Method

Developed in Section 5.4 for SDF systems, this implicit method can readily be extended to MDF systems. The scalar equations (5.4.9) that relate the response—displacement, velocity, and acceleration—increments over the time step i to $i + 1$ to each other and the response values at time i, and the scalar equation (5.4.12) of incremental equilibrium, all now become matrix equations. In Table 15.2.2 the time-stepping solution using Newmark's method is summarized as it might be implemented on the computer.

The two special cases of Newmark's method that are commonly used are: (1) $\gamma = \frac{1}{2}$ and $\beta = \frac{1}{4}$, which gives the average acceleration method, and (2) $\gamma = \frac{1}{2}$ and $\beta = \frac{1}{6}$, corresponding to the linear acceleration method. The average acceleration method is unconditionally stable, whereas the linear acceleration method is conditionally stable for $\Delta t \leq 0.551 T_J$. For a given time step that does not approach this stability limit, the linear acceleration method is more accurate than the average acceleration method. Therefore, it is especially useful for linear systems because the Δt chosen to obtain accurate response in the highest mode included would satisfy the stability requirements. Observe that the stiffness matrix **K** enters into the system of algebraic equations solved in step 2.3 to determine q_{i+1} at time $i+1$, implying that Newmark's method is an implicit method.

Newmark's method can also be used for direct solution of the original equations in nodal displacements, Eq. (15.2.1), without transforming them to modal coordinates, by modifying Table 15.2.2 appropriately.

TABLE 15.2.2 NEWMARK'S METHOD: LINEAR SYSTEMS

1.0 *Initial calculations*

1.1 $$(q_n)_0 = \frac{\phi_n^T \mathbf{m}\mathbf{u}_0}{\phi_n^T \mathbf{m}\phi_n}, \qquad (\dot{q}_n)_0 = \frac{\phi_n^T \mathbf{m}\dot{\mathbf{u}}_0}{\phi_n^T \mathbf{m}\phi_n}.$$

$$\mathbf{q}_0^T = \langle (q_1)_0, \ldots, (q_J)_0 \rangle \qquad \dot{\mathbf{q}}_0^T = \langle (\dot{q}_1)_0, \ldots, (\dot{q}_J)_0 \rangle$$

1.2 $\mathbf{P}_0 = \mathbf{\Phi}^T \mathbf{p}_0$.

1.3 Solve: $\mathbf{M}\ddot{\mathbf{q}}_0 = \mathbf{P}_0 - \mathbf{C}\dot{\mathbf{q}}_0 - \mathbf{K}\mathbf{q}_0 \Rightarrow \ddot{\mathbf{q}}_0$.

1.4 Select Δt.

1.5 $$\hat{\mathbf{K}} = \mathbf{K} + \frac{\gamma}{\beta \Delta t}\mathbf{C} + \frac{1}{\beta (\Delta t)^2}\mathbf{M}.$$

1.6 $$\mathbf{a} = \frac{1}{\beta \Delta t}\mathbf{M} + \frac{\gamma}{\beta}\mathbf{C}, \quad \mathbf{b} = \frac{1}{2\beta}\mathbf{M} + \Delta t \left(\frac{\gamma}{2\beta} - 1 \right) \mathbf{C}.$$

2.0 *Calculations for each time step i*

2.1 $\mathbf{P}_i = \mathbf{\Phi}^T \mathbf{p}_i$.

2.2 $\Delta \hat{\mathbf{P}}_i = \Delta \mathbf{P}_i + \mathbf{a}\dot{\mathbf{q}}_i + \mathbf{b}\ddot{\mathbf{q}}_i$.

2.3 Solve: $\hat{\mathbf{K}} \Delta \mathbf{q}_i = \Delta \hat{\mathbf{P}}_i \Rightarrow \Delta \mathbf{q}_i$.

2.4 $$\Delta \dot{\mathbf{q}}_i = \frac{\gamma}{\beta \Delta t}\Delta \mathbf{q}_i - \frac{\gamma}{\beta}\dot{\mathbf{q}}_i + \Delta t \left(1 - \frac{\gamma}{2\beta} \right) \ddot{\mathbf{q}}_i.$$

2.5 $$\Delta \ddot{\mathbf{q}}_i = \frac{1}{\beta (\Delta t)^2}\Delta \mathbf{q}_i - \frac{1}{\beta \Delta t}\dot{\mathbf{q}}_i - \frac{1}{2\beta}\ddot{\mathbf{q}}_i.$$

2.6 $\mathbf{q}_{i+1} = \mathbf{q}_i + \Delta \mathbf{q}_i, \quad \dot{\mathbf{q}}_{i+1} = \dot{\mathbf{q}}_i + \Delta \dot{\mathbf{q}}_i, \quad \ddot{\mathbf{q}}_{i+1} = \ddot{\mathbf{q}}_i + \Delta \ddot{\mathbf{q}}_i$.

2.7 $\mathbf{u}_{i+1} = \mathbf{\Phi}\mathbf{q}_{i+1}$.

3.0 *Repetition for the next time step.* Replace i by $i + 1$ and implement steps 2.1 to 2.7 for the next time step.

Example 15.1

A reinforced-concrete chimney idealized as the lumped mass cantilever (Fig. E15.1a) is subjected at the top to a step force $p(t)$ of 1000 kips (Fig. E15.1b); $m = 208.6$ kip-sec^2/ft and $EI = 5.469 \times 10^{10}$ kip-ft^2. Solve the equations of motion after transforming them to the first two modes by the linear acceleration method with $\Delta t = 0.1$ sec.

Solution First, we set up modal equations. The stiffness matrix is determined by the procedure illustrated in Example 9.8:

$$\mathbf{k} = \frac{EI}{h^3}\begin{bmatrix} 18.83 & -11.90 & 4.773 & -1.193 & 0.1989 \\ & 14.65 & -10.71 & 4.177 & -0.6961 \\ & & 14.06 & -9.514 & 2.586 \\ (\text{sym}) & & & 9.878 & -3.646 \\ & & & & 1.608 \end{bmatrix} \tag{a}$$

where $EI = 5.469 \times 10^{10}$ kip-ft^2 and $h = 120$ ft. $\mathbf{k}$ is in units of kips/ft. The mass matrix and the applied force vector are

$$\mathbf{m} = m\begin{bmatrix} 1 & & & & \\ & 1 & & & \\ & & 1 & & \\ & & & 1 & \\ & & & & 0.5 \end{bmatrix} \qquad \mathbf{p} = 1000\begin{bmatrix} 0 \\ 0 \\ 0 \\ 0 \\ 1 \end{bmatrix} \tag{b}$$

where $m = 208.6$ kip-sec^2/ft.

Solving the eigenvalue problem gives the first two natural frequencies and modes:

$$\omega = \begin{bmatrix} 1.701 & \\ & 10.22 \end{bmatrix} \qquad \mathbf{\Phi} = 10^{-1}\begin{bmatrix} 0.0386 & 0.1728 \\ 0.1391 & 0.4020 \\ 0.2796 & 0.3697 \\ 0.4411 & 0.0041 \\ 0.6098 & -0.5502 \end{bmatrix} \tag{c}$$

Substituting $\mathbf{m}$, $\mathbf{k}$, $\mathbf{p}$, and $\mathbf{\Phi}$ in Eq. (15.2.4) gives

$$\mathbf{M} = \begin{bmatrix} 1 & \\ & 1 \end{bmatrix} \qquad \mathbf{K} = \begin{bmatrix} 2.895 & 0 \\ 0 & 104.4 \end{bmatrix} \qquad \mathbf{P} = \begin{bmatrix} 60.98 \\ -55.02 \end{bmatrix} \tag{d}$$

The two equations (15.2.3) in modal coordinates are uncoupled for the undamped system. For generality, this uncoupling property is not used in solving this example, however.

The step increase in the applied force from $p(0^-) = 0$ to $p(0^+) = 1000$ kips can be handled in one of two ways: (1) Define $p(0) = 1000$ at $t = 0$; or (2) define $p(0) = 0$ and $p(\Delta t) = 1000$. The two solutions can be made as close as desired by selecting a small enough time step Δt. In the solution that follows we have chosen the second approach in implementing the procedure of Table 15.2.2.

1.0 *Initial calculations*

1.1 Since the system starts from rest, $\mathbf{u}_0 = \dot{\mathbf{u}}_0 = \mathbf{0}$; therefore, $\mathbf{q}_0 = \dot{\mathbf{q}}_0 = \mathbf{0}$.

1.2 $\mathbf{p}_0 = \mathbf{0}$; therefore, $\mathbf{P}_0 = \mathbf{0}$.

1.3 $\ddot{\mathbf{q}}_0 = \mathbf{0}$.

1.4 $\Delta t = 0.1$ sec..

1.5 Substituting $\mathbf{K}$, $\mathbf{M}$, Δt, $\mathbf{C} = \mathbf{0}$, and $\beta = \frac{1}{6}$ in step 1.5 gives

$$\hat{\mathbf{K}} = \begin{bmatrix} 602.9 & 0 \\ 0 & 704.4 \end{bmatrix}$$

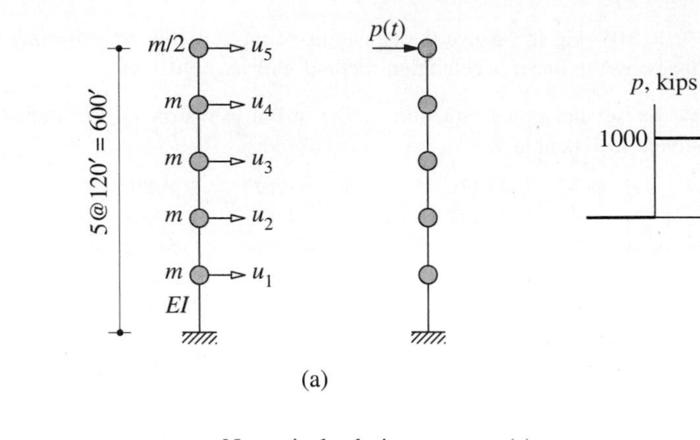

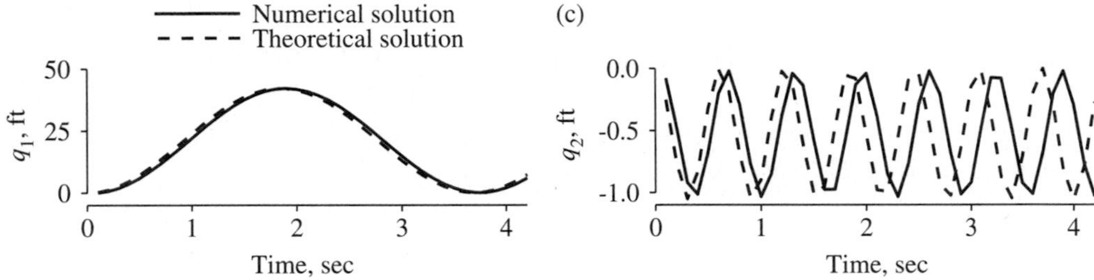

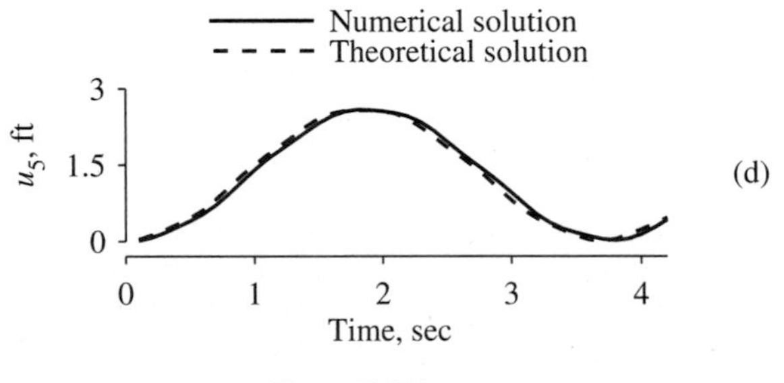

Figure E15.1

1.6 Substituting $\mathbf{M}$, Δt, $\mathbf{C} = \mathbf{0}$, $\beta = \frac{1}{6}$, and $\gamma = \frac{1}{2}$ in step 1.6 gives

$$\mathbf{a} = \begin{bmatrix} 60 & 0 \\ 0 & 60 \end{bmatrix} \qquad \mathbf{b} = \begin{bmatrix} 3 & 0 \\ 0 & 3 \end{bmatrix}$$

2.0 *Calculations for each time step i*. For the parameters of this example, computational steps 2.1 through 2.5 are specialized and implemented for each time step i as follows:

2.1 $\mathbf{P}_i = \begin{bmatrix} 60.98 \\ -55.02 \end{bmatrix}$.

2.2 $\Delta\hat{\mathbf{P}}_i = \Delta\mathbf{P}_i + \mathbf{a}\dot{\mathbf{q}}_i + \mathbf{b}\ddot{\mathbf{q}}_i$; $\Delta\mathbf{P}_1 = \begin{bmatrix} 60.98 \\ -55.02 \end{bmatrix}$ and $\Delta\mathbf{P}_i = \begin{bmatrix} 0 \\ 0 \end{bmatrix}$, $i > 1$ and

$$\begin{bmatrix} \Delta\hat{P}_1 \\ \Delta\hat{P}_2 \end{bmatrix}_i = \begin{bmatrix} \Delta P_1 & +60\dot{q}_1 + 3\ddot{q}_1 \\ \Delta P_2 & +60\dot{q}_2 + 3\ddot{q}_2 \end{bmatrix}_i.$$

2.3 Solve: $\begin{bmatrix} 602.9 & 0 \\ 0 & 704.4 \end{bmatrix} \begin{bmatrix} \Delta q_1 \\ \Delta q_2 \end{bmatrix}_i = \begin{bmatrix} \Delta\hat{P}_1 \\ \Delta\hat{P}_2 \end{bmatrix}_i \Longrightarrow \Delta\mathbf{q}_i.$

2.4 $\begin{bmatrix} \Delta\dot{q}_1 \\ \Delta\dot{q}_2 \end{bmatrix}_i = 30 \begin{bmatrix} \Delta q_1 \\ \Delta q_2 \end{bmatrix}_i - 3 \begin{bmatrix} \dot{q}_1 \\ \dot{q}_2 \end{bmatrix}_i - 0.05 \begin{bmatrix} \ddot{q}_1 \\ \ddot{q}_2 \end{bmatrix}_i.$

2.5 $\begin{bmatrix} \Delta\ddot{q}_1 \\ \Delta\ddot{q}_2 \end{bmatrix}_i = 600 \begin{bmatrix} \Delta q_1 \\ \Delta q_2 \end{bmatrix}_i - 60 \begin{bmatrix} \dot{q}_1 \\ \dot{q}_2 \end{bmatrix}_i - 3 \begin{bmatrix} \ddot{q}_1 \\ \ddot{q}_2 \end{bmatrix}_i.$

2.6 With $\Delta\mathbf{q}_i$, $\Delta\dot{\mathbf{q}}_i$, and $\Delta\ddot{\mathbf{q}}_i$ known from steps 2.3, 2.4, and 2.5, respectively, $\mathbf{q}_i$, $\dot{\mathbf{q}}_i$, and $\ddot{\mathbf{q}}_i$ are updated to determine the responses $\mathbf{q}_{i+1}$, $\dot{\mathbf{q}}_{i+1}$, and $\ddot{\mathbf{q}}_{i+1}$ at the end of the time step. The modal displacements $\mathbf{q}_i$ for the first 20 time steps are shown in Table E15.1 and Fig. E15.1c.

2.7 $\begin{bmatrix} u_1 \\ u_2 \\ u_3 \\ u_4 \\ u_5 \end{bmatrix}_{i+1} = 10^{-1} \begin{bmatrix} 0.0386 & 0.1727 \\ 0.1391 & 0.4020 \\ 0.2796 & 0.3697 \\ 0.4411 & 0.0041 \\ 0.6098 & -0.5502 \end{bmatrix} \begin{bmatrix} q_1 \\ q_2 \end{bmatrix}_{i+1}.$

These displacements are also presented in Table E15.1, and u_5 is plotted in Fig. E15.1d as a function of time.

TABLE E15.1 NUMERICAL SOLUTION OF MODAL EQUATIONS BY THE LINEAR ACCELERATION METHOD

Time	q_1	q_2	u_1	u_2	u_3	u_4	u_5
0.1	0.1011	−0.0781	−0.0010	−0.0017	−0.0001	0.0044	0.0105
0.2	0.7051	−0.4773	−0.0055	−0.0094	0.0021	0.0309	0.0693
0.3	1.8956	−0.9207	−0.0086	−0.0107	0.0190	0.0832	0.1663
0.4	3.6384	−1.0141	−0.0035	0.0098	0.0642	0.1601	0.2777
0.5	5.8832	−0.6744	0.0110	0.0547	0.1396	0.2592	0.3959
0.6	8.5654	−0.2036	0.0295	0.1109	0.2320	0.3777	0.5335
0.7	11.6080	−0.0205	0.0444	0.1606	0.3238	0.5120	0.7090
0.8	14.9230	−0.2878	0.0526	0.1960	0.4066	0.6581	0.9258
0.9	18.4140	−0.7678	0.0577	0.2252	0.4865	0.8119	1.1651
1.0	21.9820	−1.0338	0.0669	0.2641	0.5764	0.9692	1.3974
1.1	25.5240	−0.8491	0.0838	0.3208	0.6823	1.1255	1.6032
1.2	28.9370	−0.3781	0.1050	0.3872	0.7951	1.2763	1.7854
1.3	32.1240	−0.0395	0.1232	0.4451	0.8968	1.4170	1.9611
1.4	34.9920	−0.1344	0.1326	0.4812	0.9735	1.5434	2.1412
1.5	37.4580	−0.5785	0.1345	0.4976	1.0260	1.6520	2.3160
1.6	39.4530	−0.9768	0.1353	0.5094	1.0670	1.7398	2.4596
1.7	40.9170	−0.9752	0.1409	0.5298	1.1081	1.8044	2.5488
1.8	41.8100	−0.5751	0.1513	0.5583	1.1478	1.8440	2.5812
1.9	42.1050	−0.1323	0.1601	0.5802	1.1724	1.8572	2.5748
2.0	41.7940	−0.0405	0.1605	0.5796	1.1671	1.8435	2.5508

Comparison with theoretical solution. The modal equations with a step force can also be solved analytically, following the procedure of Example 12.6. Considering the first two modes of the system, such theoretical results were derived. They were computed at every 0.1 sec and are presented as the dashed lines in Fig. E15.1c and d. The dashed line in Fig. E15.1d also represents the theoretical solution including all five modes, indicating that the response contributions of the third, fourth, and fifth modes are negligible.

The numerical results for q_1 are accurate because the chosen time step $\Delta t = 0.1$ sec and natural period $T_1 = 2\pi/1.701 = 3.69$ sec, implying a very small $\Delta t/T_1 = 0.027$. However, the same Δt implies that $\Delta t/T_2 = 0.16$, which is not small enough to provide good accuracy for q_2. The numerical solution for u_5 is quite accurate, however, because the contribution of the second mode is small.

15.3 ANALYSIS OF NONLINEAR SYSTEMS

Numerical evaluation of the dynamic response of systems responding beyond their linearly elastic range is computationally demanding for systems with a large number of DOF. The N equations for an N-DOF system are usually solved in their original form, Eq. (15.1.1), because classical modal analysis is not applicable to nonlinear systems. However, even the displacements of a nonlinear system can be expressed as a combination of the natural modes of the undamped system vibrating within the range of its linear behavior:

$$\mathbf{u}(t) = \sum_{n=1}^{N} \phi_n q_n(t) \tag{15.3.1}$$

This transformation will serve to uncouple the equations of motion of a classically damped system only as long as the structure remains linear. After yielding, the modal equations would become coupled, precluding classical modal analysis. Despite this complication, it may seem attractive to truncate the modal transformation of Eq. (15.3.1) to include only the first J (typically, $J \ll N$) modes that contribute significantly to the response, and then solve the J coupled equations in modal coordinates instead of the N equations in physical coordinates. However, this approach is usually not effective for general nonlinear systems but can be used with advantage for structures composed of linear subsystems connected through nonlinear elements. Although the equations (15.1.1) being solved are not uncoupled equations, it is convenient for the discussion to follow to think of the response in terms of its modal decomposition, Eq. (15.3.1).

Direct solution of Eq. (15.1.1) is equivalent to including all the N modes in the analysis, although only the first J terms in Eq. (15.3.1) may be sufficient to represent accurately the structural response. It would seem that the choice of Δt should be based on the accuracy requirements for the Jth mode, say $\Delta t = T_J/10$, where T_J is the period of the Jth mode of undamped linear vibration. This choice of Δt implies that the higher-mode ($J+1$ to N) terms in Eq. (15.3.1) would be inaccurate, but this should not be of concern because we had concluded that these higher-mode contributions to the response were negligible. Although this choice of Δt would seem to provide accurate results, it may not be sufficiently small to ensure stability of the numerical procedure.

Accuracy is required only for the first J modes, but stability must be ensured for all modes because even if the response in the higher modes is insignificant, it will "blow up" if the stability requirements are not satisfied relative to these modes. This problem is illustrated in Fig. 15.3.1, where the response of the cantilever tower of Example 15.1 to a step force is presented as obtained by two numerical methods. The dashed curve shows the results from Table E15.1 determined by solving the first two modal equations by the linear acceleration method with $\Delta t = 0.1$ sec. When the original equations (15.1.1) are solved by the same method using the same time step, this direct solution "blows up" around $t = 1$ sec.

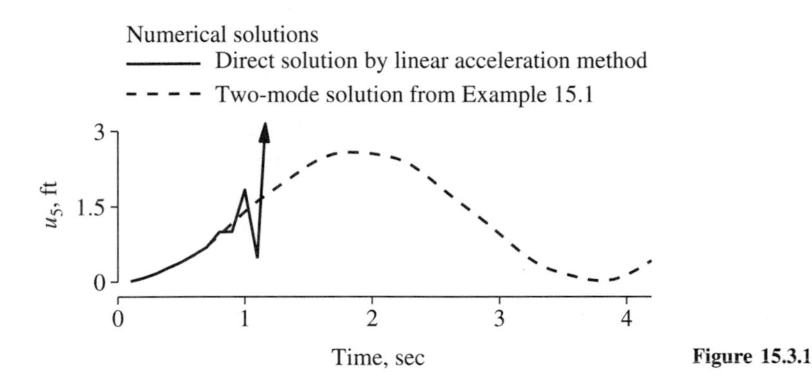

Figure 15.3.1

Requiring stability for all modes imposes very severe restrictions on Δt, as illustrated by the following example. Consider a system in which the highest mode with significant response contribution has a period $T_J = 0.10$ sec, whereas the period of the highest mode is $T_N = 0.001$ sec. If the linear acceleration method is used, the numerical solution would be reasonably accurate if Δt is chosen as $T_J/10$ (i.e., $\Delta t = 0.01$ sec). To ensure stability of the procedure, however, Δt should be less than $0.551T_N$ (i.e., $\Delta t < 0.00055$ sec). This choice of Δt implies that about 2000 time steps are necessary to compute the response of the system for 1 sec of the excitation. In light of this excessive computational requirement for a conditionally stable method, it is obvious that the numerical procedure used should be unconditionally stable. Then, in the example above, a time step of 0.01 sec could be used without the solution blowing up. The same conclusion applies to linear systems if their equations of motion are not transformed to a truncated set of J modal coordinates.

Only two unconditionally stable methods are presented in this section: the average acceleration method and Wilson's method. They are intended for the solution of Eq. (15.1.1) for linear or nonlinear systems. The average acceleration method has the drawback that it provides no numerical damping (Fig. 5.5.2). This is a disadvantage because it is desirable to filter out the response contributions of modes higher than the J significant modes because these higher modes and their frequencies, which have been calculated from an idealization of the structure, are usually not accurate relative to the actual properties of the structure. One approach for achieving this goal is to define the damping matrix consistent with increasing damping ratio for modes higher than the Jth mode (see Section 11.4). Researchers have also been interested in formulating numerical time-stepping algorithms which, in some sense, have optimal numerical damping. Wilson's method, which is developed in Section 15.3.2, provides for numerical damping in modes with period T_n such that $\Delta t/T_n \geq 1.0$; other methods are also available.

15.3.1 Average Acceleration Method

The average acceleration method has already been presented for nonlinear response analysis of SDF systems; it is the procedure summarized in Table 5.7.2, specialized for $\gamma = \frac{1}{2}$ and $\beta = \frac{1}{4}$. This procedure carries over directly to MDF systems with each scalar equation in the procedure for SDF systems now becoming a matrix equation for MDF systems. Table 15.3.1 summarizes the procedure as it might be implemented on the computer.

Steps 2.2 and 2.4 in Table 15.3.1 are the most time-consuming steps. The calculation of the tangent stiffness matrix and of the resisting forces from the displacements $\mathbf{u}_i$ at each time i can be quite complicated for nonlinear MDF systems. Such procedures are available in textbooks on static structural analysis and are not included here.

As in the case of SDF systems, the time step Δt should be shortened appropriately to detect accurately the transitions from unloading to loading branches, or around sharp corners, of the force–deformation curves. Keeping track of these branches for every structural element makes the procedure computationally demanding for systems with many elements.

TABLE 15.3.1 AVERAGE ACCELERATION METHOD: NONLINEAR SYSTEMS

1.0 *Initial calculations*

1.1 Solve: $\mathbf{m}\ddot{\mathbf{u}}_0 = \mathbf{p}_0 - \mathbf{c}\dot{\mathbf{u}}_0 - (\mathbf{f}_S)_0 \Longrightarrow \ddot{\mathbf{u}}_0$.

1.2 Select Δt.

1.3 $\mathbf{a} = \dfrac{4}{\Delta t}\mathbf{m} + 2\mathbf{c}$; and $\mathbf{b} = 2\mathbf{m}$.

2.0 *Calculations for each time step i*

2.1 $\Delta\hat{\mathbf{p}}_i = \Delta\mathbf{p}_i + \mathbf{a}\dot{\mathbf{u}}_i + \mathbf{b}\ddot{\mathbf{u}}_i$.

2.2 Determine the tangent stiffness matrix $\mathbf{k}_i$.

2.3 $\hat{\mathbf{k}}_i = \mathbf{k}_i + \dfrac{2}{\Delta t}\mathbf{c} + \dfrac{4}{(\Delta t)^2}\mathbf{m}$.

2.4 Solve for $\Delta\mathbf{u}_i$ from $\hat{\mathbf{k}}_i$ and $\Delta\hat{\mathbf{p}}_i$ using the iterative procedure of Table 15.3.2.

2.5 $\Delta\dot{\mathbf{u}}_i = \dfrac{2}{\Delta t}\Delta\mathbf{u}_i - 2\dot{\mathbf{u}}_i$.

2.6 $\Delta\ddot{\mathbf{u}}_i = \dfrac{4}{(\Delta t)^2}\Delta\mathbf{u}_i - \dfrac{4}{\Delta t}\dot{\mathbf{u}}_i - 2\ddot{\mathbf{u}}_i$.

2.7 $\mathbf{u}_{i+1} = \mathbf{u}_i + \Delta\mathbf{u}_i$, $\dot{\mathbf{u}}_{i+1} = \dot{\mathbf{u}}_i + \Delta\dot{\mathbf{u}}_i$, and $\ddot{\mathbf{u}}_{i+1} = \ddot{\mathbf{u}}_i + \Delta\ddot{\mathbf{u}}_i$.

3.0 *Repetition for the next time step.* Replace i by $i + 1$ and implement steps 2.1 to 2.6 for the next time step.

Example 15.2

Determine the response of the system of Example 15.1 to the given excitation by direct solution of Eq. (15.1.1) using the average acceleration method and a time step of $\Delta t = 0.1$ sec.

Solution The 5×5 mass and stiffness matrices of the system were defined in Example 15.1; the damping matrix is excluded because the system is undamped. For the system starting from rest, $\mathbf{u}_0 = \mathbf{0}$ and $\dot{\mathbf{u}}_0 = \mathbf{0}$. Because the system is linear, the computational steps of Table 15.3.1 are simplified in three ways: (1) Step 2.2 is eliminated because $\mathbf{k}_i = \mathbf{k}$; and (2) step 2.3 needs to be computed only once:

$$\hat{\mathbf{k}} = \mathbf{k} + \frac{2}{\Delta t}\mathbf{c} + \frac{4}{(\Delta t)^2}\mathbf{m} = \mathbf{k} + \frac{4}{(0.1)^2}\mathbf{m} = \mathbf{k} + 400\mathbf{m}$$

(3) In step 2.4 no iteration is necessary and $\Delta\mathbf{u}_i$ is determined by solving the algebraic equations $\hat{\mathbf{k}}\Delta\mathbf{u}_i = \Delta\hat{\mathbf{p}}_i$.

We now implement the procedure of Table 15.3.1 as follows:

1.0 *Initial calculations*

1.1 Solve $\mathbf{m}\ddot{\mathbf{u}}_0 = \mathbf{p}_0$, where $\mathbf{p}_0 = \mathbf{0}$ to obtain $\ddot{\mathbf{u}}_0 = \mathbf{0}$.

1.2 $\Delta t = 0.1$ sec.

1.3 $\mathbf{a} = \dfrac{4}{\Delta t}\mathbf{m} + 2\mathbf{c} = \dfrac{4}{0.1}\mathbf{m} = 40\mathbf{m}$; and $\mathbf{b} = 2\mathbf{m}$.

TABLE E15.2 NUMERICAL SOLUTION BY THE AVERAGE ACCELERATION METHOD

Time	u_1	u_2	u_3	u_4	u_5
0.1	−0.0003	−0.0009	−0.0003	0.0051	0.0172
0.2	−0.0024	−0.0046	0.0028	0.0297	0.0753
0.3	−0.0057	−0.0056	0.0214	0.0830	0.1682
0.4	−0.0025	0.0122	0.0660	0.1602	0.2799
0.5	0.0123	0.0553	0.1367	0.2574	0.4044
0.6	0.0301	0.1095	0.2270	0.3755	0.5436
0.7	0.0440	0.1598	0.3222	0.5105	0.7127
0.8	0.0557	0.2013	0.4085	0.6553	0.9228
0.9	0.0631	0.2343	0.4900	0.8075	1.1588
1.0	0.0690	0.2683	0.5780	0.9657	1.3921
1.1	0.0831	0.3177	0.6769	1.1216	1.6068
1.2	0.1037	0.3811	0.7841	1.2700	1.7974
1.3	0.1220	0.4403	0.8885	1.4115	1.9676
1.4	0.1341	0.4838	0.9736	1.5389	2.1341
1.5	0.1406	0.5095	1.0319	1.6457	2.3012
1.6	0.1412	0.5199	1.0718	1.7339	2.4462
1.7	0.1408	0.5290	1.1053	1.8000	2.5469
1.8	0.1479	0.5486	1.1352	1.8388	2.5953
1.9	0.1575	0.5705	1.1582	1.8518	2.5926
2.0	0.1600	0.5781	1.1636	1.8402	2.5540

2.0 *Calculations for each time step, i.* Computational steps 2.1, 2.4 (modified as above), 2.5, and 2.6 are implemented for $i = 1, 2, 3, \ldots$ to obtain the displacements u_1, u_2, u_3, u_4, and u_5 presented in Table E15.2. The displacement u_5 is plotted as a function of time in Fig. E15.2 and compared with the theoretical solution first presented in Fig. E15.1d.

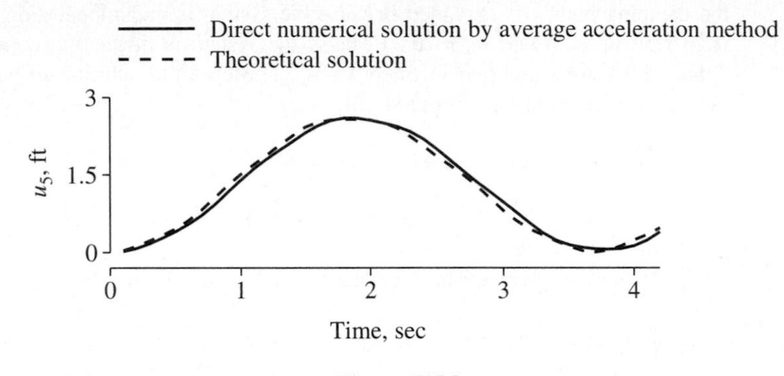

Figure E15.2

Modified Newton–Raphson iteration. As in the case of an SDF system, iteration within a time step (step 2.4) is necessary to reduce the error introduced by the use of the tangent stiffness matrix instead of the unknown secant stiffness matrix. Such iteration can be carried out by the modified Newton–Raphson method described in Table 5.7.1 for SDF systems. When generalized for MDF systems, this iterative process for the time step i to $i + 1$ may be summarized as in Table 15.3.2.

TABLE 15.3.2 MODIFIED NEWTON–RAPHSON ITERATION

1.0 *Initialize data.*

$$\mathbf{u}_{i+1}^{(0)} = \mathbf{u}_i \qquad \mathbf{f}_S^{(0)} = (\mathbf{f}_S)_i \qquad \Delta\mathbf{R}^{(1)} = \Delta\hat{\mathbf{p}}_i \qquad \hat{\mathbf{k}}_T = \hat{\mathbf{k}}_i$$

2.0 *Calculations for each iteration,* $j = 1, 2, 3, \ldots$

2.1 Solve: $\hat{\mathbf{k}}_T \Delta\mathbf{u}^{(j)} = \Delta\mathbf{R}^{(j)} \Longrightarrow \Delta\mathbf{u}^{(j)}$

2.2 $\mathbf{u}_{i+1}^{(j)} = \mathbf{u}_{i+1}^{(j-1)} + \Delta\mathbf{u}^{(j)}$

2.3 $\Delta\mathbf{f}^{(j)} = \mathbf{f}_S^{(j)} - \mathbf{f}_S^{(j-1)} + (\hat{\mathbf{k}}_T - \mathbf{k}_i)\Delta\mathbf{u}^{(j)}$

2.4 $\Delta\mathbf{R}^{(j+1)} = \Delta\mathbf{R}^{(j)} - \Delta\mathbf{f}^{(j)}$

3.0 *Repetition for the next iteration.* Replace j by $j + 1$ and repeat calculation steps 2.1 to 2.4.

In the modified Newton–Raphson method the tangent stiffness matrix is calculated at time i, the beginning of the time step, and is used through all iterations within that time step. Thus the matrix $\hat{\mathbf{k}}_T$ needs to be factorized only once and used repeatedly in solving the algebraic equations in step 2.1. This method is usually preferred over the full Newton–Raphson method, where, as mentioned in Section 5.7, the tangent stiffness matrix and $\hat{\mathbf{k}}_T$ is updated for each iteration. This improves the convergence, but additional computational effort is required in forming a new tangent stiffness matrix $\hat{\mathbf{k}}_T$ and factorizing it at each iteration cycle.

The iterative process is terminated after ℓ iterations when the incremental displacement vector $\mathbf{\Delta u}^{(\ell)}$ becomes small enough, say smaller than ϵ. The smallness of this vector is judged by requiring its Euclidean norm to be small compared to that of the current estimate, $\mathbf{\Delta u} = \sum_{j=1}^{\ell} \mathbf{\Delta u}^{(j)}$ of the incremental displacements. In other words,

$$\frac{\mathbf{\Delta u}^{(\ell)}{}_{2}}{\mathbf{\Delta u}_{2}} < \epsilon \tag{15.3.2}$$

This intuitively appealing criterion may not be satisfactory if different elements of the displacement vector are measured in different units and their numerical values differ greatly. Such is the case in analysis of buildings where floor displacements and joint rotations have different units and the joint rotations are an order of magnitude smaller than floor displacements. In this situation Eq. (15.3.2) may indicate convergence because the floor displacements that dominate the norm have converged; however, the joint rotations may still be in significant error.

To avoid such difficulties and to ensure that both the displacements and forces are near their final values, a convergence criterion that has been found useful is

$$\frac{\left[\mathbf{\Delta R}^{(j)}\right]^T \mathbf{\Delta u}^{(j)}}{\left[\mathbf{\Delta \hat{p}}_i\right]^T \mathbf{\Delta u}} < \epsilon \tag{15.3.3}$$

In this criterion the work done by the residual forces $\mathbf{\Delta R}^{(j)}$ in the displacement increments $\mathbf{\Delta u}^{(j)}$ is compared to the work associated with the total incremental (over the time step) force $\mathbf{\Delta \hat{p}}_i$ and the current estimate of the total incremental displacement $\mathbf{\Delta u}$.

15.3.2 Wilson's Method

A method developed by E. L. Wilson is a modification of the conditionally stable linear acceleration method that makes it unconditionally stable. This modification is based on the assumption that the acceleration varies linearly over an extended time step $\delta t = \theta\,\Delta t$, as shown in Fig. 15.3.2. The accuracy and stability properties of the method depend on the value of the parameter θ, which is always greater than 1.

The numerical procedure can be derived merely by rewriting the basic relationships of the linear acceleration method. For SDF systems these are given by Eq. (5.4.9), specialized for $\gamma = \frac{1}{2}$ and $\beta = \frac{1}{6}$. The corresponding matrix equations that apply to MDF systems are

$$\mathbf{\Delta \dot{u}}_i = (\Delta t)\mathbf{\ddot{u}}_i + \frac{\Delta t}{2}\mathbf{\Delta \ddot{u}}_i \qquad \mathbf{\Delta u}_i = (\Delta t)\mathbf{\dot{u}}_i + \frac{(\Delta t)^2}{2}\mathbf{\ddot{u}}_i + \frac{(\Delta t)^2}{6}\mathbf{\Delta \ddot{u}}_i \tag{15.3.4}$$

Replacing Δt by $\theta\,\delta t$ and the incremental responses by $\boldsymbol{\delta}\mathbf{u}_i$, $\boldsymbol{\delta}\mathbf{\dot{u}}_i$, and $\boldsymbol{\delta}\mathbf{\ddot{u}}_i$ (see Fig. 15.3.2 for an SDF system) gives the corresponding equations for the extended time step:

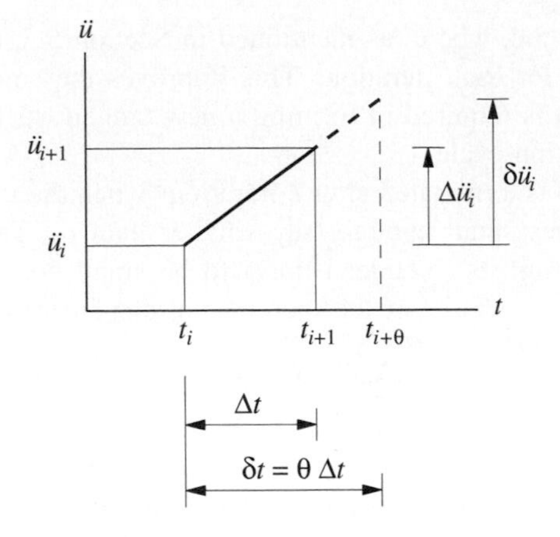

Figure 15.3.2 Linear variation of acceleration over normal and extended time steps shown for an SDF system.

$$\boldsymbol{\delta}\dot{\mathbf{u}}_i = (\delta t)\ddot{\mathbf{u}}_i + \frac{\delta t}{2}\boldsymbol{\delta}\ddot{\mathbf{u}}_i \qquad \boldsymbol{\delta}\mathbf{u}_i = (\delta t)\dot{\mathbf{u}}_i + \frac{(\delta t)^2}{2}\ddot{\mathbf{u}}_i + \frac{(\delta t)^2}{6}\boldsymbol{\delta}\ddot{\mathbf{u}}_i \tag{15.3.5}$$

Equation (15.3.5b) can be solved for

$$\boldsymbol{\delta}\ddot{\mathbf{u}}_i = \frac{6}{(\delta t)^2}\boldsymbol{\delta}\mathbf{u}_i - \frac{6}{\delta t}\dot{\mathbf{u}}_i - 3\ddot{\mathbf{u}}_i \tag{15.3.6}$$

Substituting Eq. (15.3.6) in Eq. (15.3.5a) gives

$$\boldsymbol{\delta}\dot{\mathbf{u}}_i = \frac{3}{\delta t}\boldsymbol{\delta}\mathbf{u}_i - 3\dot{\mathbf{u}}_i - \frac{\delta t}{2}\ddot{\mathbf{u}}_i \tag{15.3.7}$$

Next, Eqs. (15.3.6) and (15.3.7) are substituted into the incremental (over the extended time step) equation of motion:

$$\mathbf{m}\,\boldsymbol{\delta}\ddot{\mathbf{u}}_i + \mathbf{c}\,\boldsymbol{\delta}\dot{\mathbf{u}}_i + \mathbf{k}_i\,\boldsymbol{\delta}\mathbf{u}_i = \boldsymbol{\delta}\mathbf{p}_i \tag{15.3.8}$$

where, based on the assumption that the exciting force vector also varies linearly over the extended time step,

$$\boldsymbol{\delta}\mathbf{p}_i = \theta(\boldsymbol{\Delta}\mathbf{p}_i) \tag{15.3.9}$$

This substitution leads to

$$\hat{\mathbf{k}}_i\,\boldsymbol{\delta}\mathbf{u}_i = \boldsymbol{\delta}\hat{\mathbf{p}}_i \tag{15.3.10}$$

where

$$\hat{\mathbf{k}}_i = \mathbf{k}_i + \frac{3}{\theta\,\Delta t}\mathbf{c} + \frac{6}{(\theta\,\Delta t)^2}\mathbf{m} \tag{15.3.11}$$

$$\boldsymbol{\delta}\hat{\mathbf{p}}_i = \theta(\boldsymbol{\Delta}\mathbf{p}_i) + \left(\frac{6}{\theta\,\Delta t}\mathbf{m} + 3\mathbf{c}\right)\dot{\mathbf{u}}_i + \left(3\mathbf{m} + \frac{\theta\,\Delta t}{2}\mathbf{c}\right)\ddot{\mathbf{u}}_i \tag{15.3.12}$$

Equation (15.3.10) is solved for $\boldsymbol{\delta}\mathbf{u}_i$, and $\boldsymbol{\delta}\ddot{\mathbf{u}}_i$ is computed from Eq. (15.3.6). The incremental acceleration over the normal time step is then given by (see Fig. 15.3.2)

$$\boldsymbol{\Delta}\ddot{\mathbf{u}}_i = \frac{1}{\theta}\boldsymbol{\delta}\ddot{\mathbf{u}}_i \tag{15.3.13}$$

and the incremental velocity and displacement are determined from Eq. (15.3.4). The procedure is summarized in Table 15.3.3.

TABLE 15.3.3 WILSON'S METHOD: NONLINEAR SYSTEMS

1.0 *Initial calculations*

1.1 Solve $\mathbf{m}\ddot{\mathbf{u}}_0 = \mathbf{p}_0 - \mathbf{c}\dot{\mathbf{u}}_0 - (\mathbf{f}_S)_0 \Longrightarrow \ddot{\mathbf{u}}_0$.

1.2 Select Δt and θ.

1.3 $\mathbf{a} = \dfrac{6}{\theta\,\Delta t}\mathbf{m} + 3\mathbf{c}$; and $\mathbf{b} = 3\mathbf{m} + \dfrac{\theta\,\Delta t}{2}\mathbf{c}$.

2.0 *Calculations for each time step, i*

2.1 $\delta\hat{\mathbf{p}}_i = \theta(\Delta\mathbf{p}_i) + \mathbf{a}\dot{\mathbf{u}}_i + \mathbf{b}\ddot{\mathbf{u}}_i$.

2.2 Determine the tangent stiffness matrix $\mathbf{k}_i$.

2.3 $\hat{\mathbf{k}}_i = \mathbf{k}_i + \dfrac{3}{\theta\,\Delta t}\mathbf{c} + \dfrac{6}{(\theta\,\Delta t)^2}\mathbf{m}$.

2.4 Solve for $\delta\mathbf{u}_i$ from $\hat{\mathbf{k}}_i$ and $\Delta\hat{\mathbf{p}}_i$ using the iterative procedure of Table 15.3.2.

2.5 $\delta\ddot{\mathbf{u}}_i = \dfrac{6}{(\theta\,\Delta t)^2}\delta\mathbf{u}_i - \dfrac{6}{\theta\,\Delta t}\dot{\mathbf{u}}_i - 3\ddot{\mathbf{u}}_i$; and $\Delta\ddot{\mathbf{u}}_i = \dfrac{1}{\theta}\delta\ddot{\mathbf{u}}_i$.

2.6 $\Delta\dot{\mathbf{u}}_i = (\Delta t)\ddot{\mathbf{u}}_i + \dfrac{\Delta t}{2}\Delta\ddot{\mathbf{u}}_i$; and $\Delta\mathbf{u}_i = (\Delta t)\dot{\mathbf{u}}_i + \dfrac{(\Delta t)^2}{2}\ddot{\mathbf{u}}_i + \dfrac{(\Delta t)^2}{6}\Delta\ddot{\mathbf{u}}_i$.

2.7 $\mathbf{u}_{i+1} = \mathbf{u}_i + \Delta\mathbf{u}_i$, $\dot{\mathbf{u}}_{i+1} = \dot{\mathbf{u}}_i + \Delta\dot{\mathbf{u}}_i$, and $\ddot{\mathbf{u}}_{i+1} = \ddot{\mathbf{u}}_i + \Delta\ddot{\mathbf{u}}_i$.

3.0 *Repetition for the next time step.* Replace i by $i+1$ and implement steps 2.1 to 2.7 for the next time step.

As mentioned earlier, the value of θ governs the stability characteristics of Wilson's method. If $\theta = 1$, this method reverts to the linear acceleration method, which is stable if $\Delta t < 0.551 T_N$, where T_N is the shortest natural period of the system. If $\theta \geq 1.37$, Wilson's method is unconditionally stable, making it suitable for direct solution of Eq. (15.1.1); $\theta = 1.42$ gives optimal accuracy.

Example 15.3

Find the response of the system of Example 15.1 by direct solution of Eq. (15.1.1) using Wilson's method, with $\theta = 1.42$ and a time step of $\Delta t = 0.1$ sec.

Solution The mass and stiffness matrices of the system were defined in Example 15.1; the damping matrix is excluded because the system is undamped. For the system starting from rest, $\mathbf{u}_0 = 0$ and $\dot{\mathbf{u}}_0 = 0$. Because the system is linear, the computational steps of Table 15.3.3 are simplified in three ways: (1) Step 2.2 is eliminated because $\mathbf{k}_i = \mathbf{k}$; (2) step 2.3 needs to be computed only once:

$$\hat{\mathbf{k}} = \mathbf{k} + \frac{3}{\theta\Delta t}\mathbf{c} + \frac{6}{(\theta\Delta t)^2}\mathbf{m} = \mathbf{k} + \frac{6}{(1.42\times 0.1)^2}\mathbf{m} = \mathbf{k} + 297.6\mathbf{m}$$

and (3) in step 2.4 no iteration is necessary and $\delta\mathbf{u}_i$ is determined by solving the algebraic equations (15.3.9).

We now implement the procedure of Table 15.3.3 as follows:

1.0 *Initial calculations*

1.1 Solve $\mathbf{m}\ddot{\mathbf{u}}_0 = \mathbf{p}_0$ where $\mathbf{p}_0 = \mathbf{0}$ to obtain $\ddot{\mathbf{u}}_0 = \mathbf{0}$.

1.2 $\Delta t = 0.1$ sec.

1.3 $\mathbf{a} = \dfrac{6}{\theta\,\Delta t}\mathbf{m} + 3\mathbf{c} = \dfrac{6}{1.42(0.1)}\mathbf{m} = 42.25\mathbf{m}$; and $\mathbf{b} = 3\mathbf{m} + \dfrac{\theta\,\Delta t}{2}\mathbf{c} = 3\mathbf{m}$.

2.0 *Calculations for each time step, i.* Computational steps 2.1, 2.4 (modified as above), 2.5, 2.6, and 2.7 are implemented for $i = 1, 2, 3, \ldots, 20$ to obtain the displacements u_1, u_2, u_3, u_4, and u_5 presented in Table E15.3. The displacement u_5 is plotted as a function of time in Fig. E15.3 and compared with the theoretical solution first presented in Fig. E15.1d.

TABLE E15.3 NUMERICAL SOLUTION BY WILSON'S METHOD

Time	u_1	u_2	u_3	u_4	u_5
0.1	−0.0002	−0.0004	0.0000	0.0025	0.0077
0.2	−0.0014	−0.0027	0.0020	0.0198	0.0520
0.3	−0.0037	−0.0039	0.0151	0.0622	0.1311
0.4	−0.0029	0.0068	0.0494	0.1295	0.2339
0.5	0.0060	0.0375	0.1080	0.2182	0.3537
0.6	0.0211	0.0839	0.1872	0.3268	0.4895
0.7	0.0369	0.1349	0.2779	0.4532	0.6479
0.8	0.0501	0.1819	0.3697	0.5932	0.8360
0.9	0.0605	0.2224	0.4579	0.7421	1.0516
1.0	0.0694	0.2601	0.5454	0.8963	1.2805
1.1	0.0797	0.3016	0.6373	1.0516	1.5049
1.2	0.0935	0.3509	0.7348	1.2033	1.7123
1.3	0.1099	0.4050	0.8338	1.3470	1.8994
1.4	0.1253	0.4555	0.9256	1.4788	2.0691
1.5	0.1366	0.4949	1.0018	1.5948	2.2253
1.6	0.1430	0.5205	1.0586	1.6916	2.3659
1.7	0.1459	0.5356	1.0982	1.7664	2.4813
1.8	0.1478	0.5458	1.1252	1.8173	2.5598
1.9	0.1504	0.5547	1.1422	1.8428	2.5938
2.0	0.1532	0.5610	1.1477	1.8422	2.5840

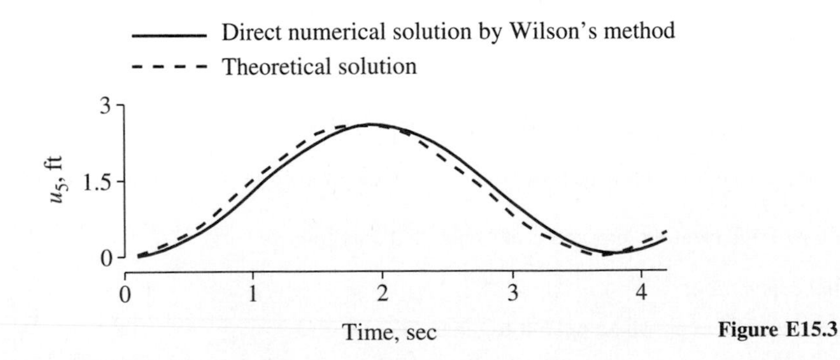

Figure E15.3

FURTHER READING

Bathe, K.-J., *Finite Element Procedures in Engineering Analysis,* Prentice Hall, Englewood Cliffs, N.J., 1982, Chapter 9.

Hughes, T. J. R., *The Finite Element Method,* Prentice Hall, Englewood Cliffs, N.J., 1987, Chapter 9.

Humar, J. L., *Dynamics of Structures,* Prentice Hall, Englewood Cliffs, N.J., 1990, Chapter 13.

Newmark, N. M., "A Method of Computation for Structural Dynamics," *Journal of the Engineering Mechanics Division, ASCE,* **85**, 1959, pp. 67–94.

PROBLEMS

***15.1** Solve the problem in Example 15.1 by the central difference method, implemented by a computer program in a language of your choice using $\Delta t = 0.1$ sec.

***15.2** Repeat Problem 15.1 using $\Delta t = 0.05$ sec. How does the time step affect the accuracy of the solution?

***15.3** Solve the problem in Example 15.1 by the average acceleration method, implemented by a computer program in a language of your choice using $\Delta t = 0.1$ sec. Based on these results and those from Problem 15.1, comment on the relative accuracy of the average acceleration and central difference methods.

***15.4** Repeat Problem 15.3 using $\Delta t = 0.05$ sec. How does the time step affect the accuracy of the solution?

***15.5** Solve the problem in Example 15.1 by the linear acceleration method, implemented by a computer program in a language of your choice using $\Delta t = 0.1$ sec. Based on these results and those from Problem 15.3, comment on the relative accuracy of the linear acceleration and average acceleration methods. Note that this problem was solved as Example 15.1 and the results were presented in Table E15.1.

***15.6** Repeat Problem 15.5 using $\Delta t = 0.05$ sec. How does the time step affect the accuracy of the solution?

***15.7** Find the response of the system of Example 15.1 to the given excitation by direct solution of Eq. (15.1.1) using the average acceleration method and a time step of $\Delta t = 0.1$ sec. Implement this method by a computer program in a language of your choice. Note that this problem was solved as Example 15.2 and that the results are presented in Table E15.2.

***15.8** Repeat Problem 15.7 using $\Delta t = 0.05$ sec. How does the time step affect the accuracy of the solution?

***15.9** Find the response of the system of Example 15.1 to the given excitation by direct solution of Eq. (15.1.1) using Wilson's method, with $\theta = 1.42$ and a time step of $\Delta t = 0.1$ sec. Implement this method by a computer program in a language of your choice. Note that this problem was solved as Example 15.3 and that the results are presented in Table E15.3.

***15.10** Repeat Problem 15.9 using $\Delta t = 0.05$ sec. How does the time step affect the accuracy of the solution?

*Denotes that a computer is necessary to solve this problem.

16

Systems with Distributed Mass and Elasticity

PREVIEW

So far in this book we have focused on discretized systems, typically with lumped masses; such a system is an assemblage of rigid elements having mass (e.g., the floor diaphragms of a multistory building) and massless elements that are flexible (e.g., the beams and columns of a building). A major part of this book is devoted to lumped-mass discretized systems, for two reasons. First, such systems can effectively idealize many classes of structures, especially multistory buildings. Second, effective methods that are ideal for computer implementation are available to solve the system of ordinary differential equations governing the motion of such systems. However, a lumped-mass idealization, although applicable, is not a natural approach for certain types of structures, such as a chimney (Fig. 2.1.2f), an arch dam (Fig. 1.10.2), or a nuclear containment structure (Fig. 1.10.1).

In this chapter we formulate the structural dynamics problem for one-dimensional systems with distributed mass, such as a beam or a tower, and solutions are presented for simple systems (e.g., a uniform beam and a uniform tower). The solutions presented for these simple cases provide insight into the dynamics of distributed-mass systems which have an infinite number of DOFs and how they differ from lumped-mass systems with a finite number of DOFs. The chapter ends with a discussion of why this infinite-DOF approach is not feasible for practical systems, pointing to the need for discretized methods for distributed-mass systems. The results presented for the simple systems provide the exact solution against which results from discretized methods can be compared (Chapter 17).

16.1 EQUATION OF UNDAMPED MOTION: APPLIED FORCES

In this section we develop the equation governing the transverse vibration of a straight beam without damping subjected to external force. Figure 16.1.1a shows such a beam with flexural rigidity $EI(x)$ and mass $m(x)$ per unit length, both of which may vary with position x. The external forces $p(x, t)$, which may vary with position and time, cause motion of the beam described by the transverse displacement $u(x, t)$ (Fig. 16.1.1b). The equation of motion to be developed will be valid for support conditions other than the simple supports shown and for beams with intermediate supports.

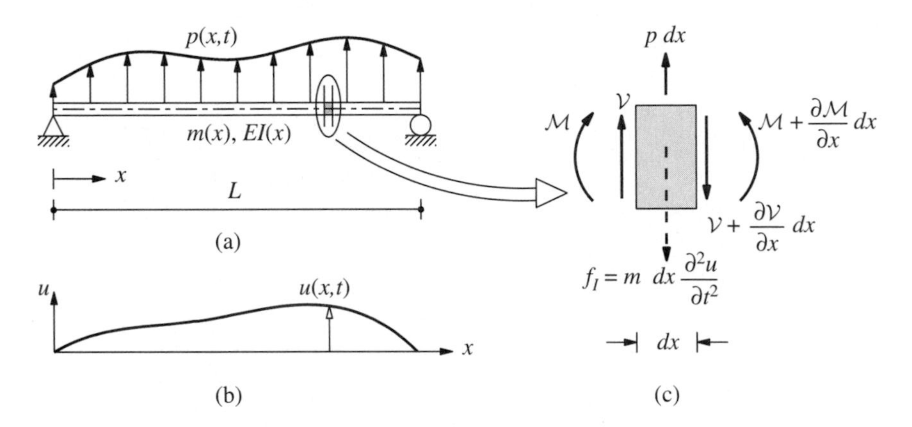

Figure 16.1.1 System with distributed mass and elasticity: (a) beam and applied force; (b) displacement; (c) forces on element.

The system has an infinite number of DOFs because its mass is distributed. Therefore, we consider a differential element of the beam, isolated by two adjoining sections. The forces on the element are shown in Fig. 16.1.1c, where an inertia force has been included following D'Alembert's principle (Section 1.5.2); $\mathcal{V}(x, t)$ is the transverse shear force and $\mathcal{M}(x, t)$ is the bending moment. Equilibrium of forces in the y-direction gives

$$\frac{\partial \mathcal{V}}{\partial x} = p - m \frac{\partial^2 u}{\partial t^2} \tag{16.1.1}$$

Without the inertia force this equation is the familiar relation between the shear force in a beam and external transverse force. The inertia force modifies the external force in recognition of the dynamics of the problem. If the inertial moment associated with angular acceleration of the element is neglected, rotational equilibrium of the element gives the standard relation

$$\mathcal{V} = \frac{\partial \mathcal{M}}{\partial x} \tag{16.1.2}$$

We now use Eqs. (16.1.1) and (16.1.2) to write the equation governing the transverse displacement $u(x, t)$. With shear deformation neglected, the moment–curvature relation is

$$\mathcal{M} = EI\frac{\partial^2 u}{\partial x^2} \tag{16.1.3}$$

Substituting Eqs. (16.1.3) and (16.1.2) into Eq. (16.1.1) gives

$$m(x)\frac{\partial^2 u}{\partial t^2} + \frac{\partial^2}{\partial x^2}\left[EI(x)\frac{\partial^2 u}{\partial x^2}\right] = p(x,t) \tag{16.1.4}$$

This is the partial differential equation governing the motion $u(x,t)$ of the beam subjected to external dynamic forces $p(x,t)$. To obtain a unique solution to this equation, we must specify two boundary conditions at each end of the beam and the initial displacement $u(x,0)$ and initial velocity $\dot{u}(x,0)$.

16.2 EQUATION OF UNDAMPED MOTION: SUPPORT EXCITATION

Consider two simple cases: a cantilever beam subjected to horizontal base motion (Fig. 16.2.1a) or a beam with multiple supports subjected to identical motion in the vertical direction (Fig. 16.2.1b). The total displacement of the beam is

$$u^t(x,t) = u(x,t) + u_g(t) \tag{16.2.1}$$

where the beam displacement $u(x,t)$, measured relative to the support motion $u_g(t)$, results from the deformations of the beam.

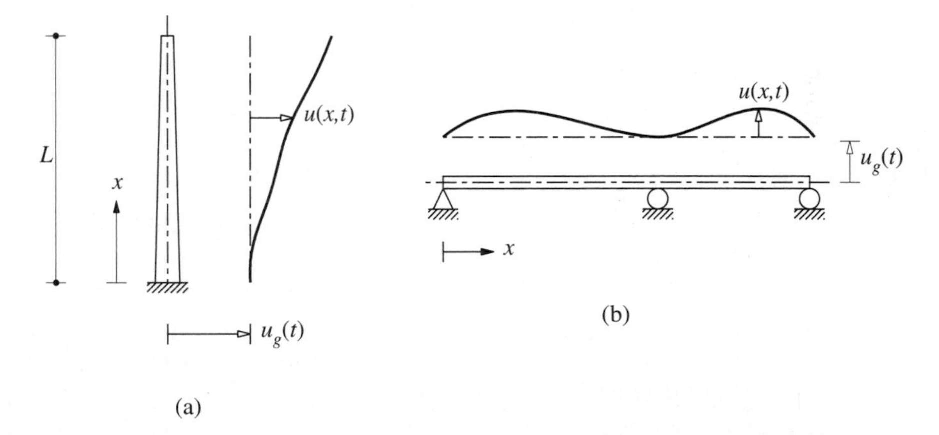

Figure 16.2.1 (a) Cantilever beam subjected to base excitation; (b) continuous beam subjected to identical motion at all supports.

For these simple cases of a beam excited by support motion the derivation of the equation of motion is only slightly different than that for applied forces. Recognizing that the inertia forces are now related to the total accelerations and that external forces

$p(x, t)$ do not exist, Eq. (16.1.1) becomes

$$\frac{\partial \mathcal{V}}{\partial x} = -m\frac{\partial^2 u^t}{\partial t^2} = -m\frac{\partial^2 u}{\partial t^2} - m\frac{d^2 u_g}{dt^2} \tag{16.2.2}$$

wherein Eq. (16.2.1) has been used to obtain the second half of the equation. Substituting Eqs. (16.1.3) and (16.1.2) into Eq. (16.2.2) gives

$$m(x)\frac{\partial^2 u}{\partial t^2} + \frac{\partial^2}{\partial x^2}\left[EI(x)\frac{\partial^2 u}{\partial x^2}\right] = -m(x)\ddot{u}_g(t) \tag{16.2.3}$$

By comparing Eqs. (16.2.3) and (16.1.4), it is clear that the deformation response $u(x, t)$ of the beam to support acceleration $\ddot{u}_g(t)$ will be identical to the response of the system with stationary supports due to external forces $= -m(x)\ddot{u}_g(t)$. The support excitation can therefore be replaced by effective forces (Fig. 16.2.2):

$$p_{\text{eff}}(x, t) = -m(x)\ddot{u}_g(t) \tag{16.2.4}$$

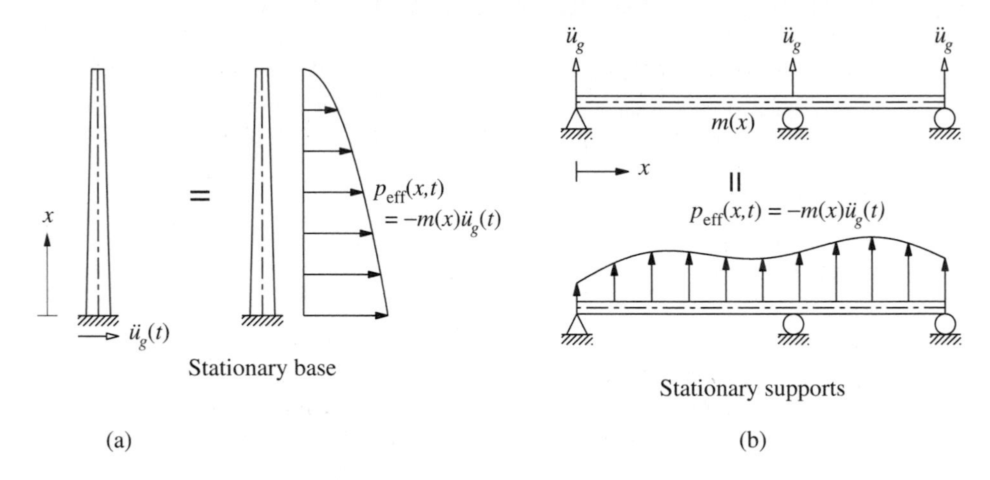

Figure 16.2.2 Effective forces $p_{\text{eff}}(x, t)$.

This formulation can be generalized to include the possibility of different motions of the various supports of a structure. Such multiple support excitation may exist in several practical situations (Section 9.7), but is not included in this chapter because it is usually not possible to analyze such practical problems as infinite-DOF systems. They are usually discretized by the finite element method (Chapter 17) and analyzed by extensions of the procedures of Section 13.5.

16.3 NATURAL VIBRATION FREQUENCIES AND MODES

For the case of free vibration, Eqs. (16.1.4) and (16.2.3) become

$$m(x)\frac{\partial^2 u}{\partial t^2} + \frac{\partial^2}{\partial x^2}\left[EI(x)\frac{\partial^2 u}{\partial x^2}\right] = 0 \tag{16.3.1}$$

We attempt a solution of the form

$$u(x, t) = \phi(x)q(t) \tag{16.3.2}$$

Then

$$\frac{\partial^2 u}{\partial t^2} = \phi(x)\ddot{q}(t) \qquad \frac{\partial^2 u}{\partial x^2} = \phi''(x)q(t) \tag{16.3.3}$$

where overdots denote a time derivative and primes denote an x derivative; thus $\dot{q}(t) \equiv \partial q/\partial t$, $\ddot{q}(t) = \partial^2 q/\partial t^2$, and $\phi''(x) = \partial^2\phi/\partial x^2$. Substituting Eq. (16.3.3) in Eq. (16.3.1) leads to

$$m(x)\phi(x)\ddot{q}(t) + q(t)\left[EI(x)\phi''(x)\right]'' = 0$$

which, when divided by $m(x)\phi(x)q(t)$, becomes

$$\frac{-\ddot{q}(t)}{q(t)} = \frac{[EI(x)\phi''(x)]''}{m(x)\phi(x)} \tag{16.3.4}$$

The expression on the left is a function of t only and the one on the right depends only on x. For Eq. (16.3.4) to be valid for all values of x and t, the two expressions must therefore be constant, say ω^2. Thus the partial differential equation (16.3.1) becomes two ordinary differential equations, one governing the time function $q(t)$ and the other governing the spatial function $\phi(x)$:

$$\ddot{q} + \omega^2 q = 0 \tag{16.3.5}$$

$$\left[EI(x)\phi''(x)\right]'' - \omega^2 m(x)\phi(x) = 0 \tag{16.3.6}$$

Equation (16.3.5) has the same form as the equation governing free vibration of an SDF system with natural frequency ω. For any given stiffness and mass functions, $EI(x)$ and $m(x)$, respectively, there is an infinite set of frequencies ω and associated modes $\phi(x)$ that satisfy the eigenvalue problem defined by Eq. (16.3.6) and the support (boundary) conditions for the beam.

For the special case of a uniform beam, $EI(x) = EI$ and $m(x) = m$, and Eq. (16.3.6) becomes

$$EI\phi^{iv}(x) - \omega^2 m\phi(x) = 0 \quad \text{or} \quad \phi^{iv}(x) - \beta^4\phi(x) = 0 \tag{16.3.7}$$

where

$$\beta^4 = \frac{\omega^2 m}{EI} \tag{16.3.8}$$

The general solution of Eq. (16.3.7) is (see Derivation 16.1)

$$\phi(x) = C_1 \sin\beta x + C_2 \cos\beta x + C_3 \sinh\beta x + C_4 \cosh\beta x \tag{16.3.9}$$

This solution contains four unknown constants, C_1, C_2, C_3, and C_4, and the eigenvalue parameter β. Application of the four boundary conditions for a single-span beam, two at each end of the beam, will provide a solution for β and hence for the natural frequency ω [from Eq. (16.3.8)] and for three constants in terms of the fourth, resulting in the natural mode of Eq. (16.3.9). This procedure is illustrated next by two examples: a simply supported beam and a cantilever beam. Results are also available for other boundary conditions but are not included in this book.

16.3.1 Uniform Simply Supported Beam

The natural frequencies and modes of vibration of a uniform beam simply supported at both ends are determined next. At $x = 0$ and $x = L$, the displacement and bending moment are zero. Thus, using Eqs. (16.3.2), (16.1.3), and (16.3.9) at $x = 0$ gives

$$u(0) = 0 \Rightarrow \phi(0) = 0 \Rightarrow C_2 + C_4 = 0 \tag{16.3.10a}$$

$$\mathcal{M}(0) = 0 \Rightarrow EI\phi''(0) = 0 \Rightarrow \beta^2(-C_2 + C_4) = 0 \tag{16.3.10b}$$

These two equations give $C_2 = C_4 = 0$ and the general solution reduces to

$$\phi(x) = C_1 \sin \beta x + C_3 \sinh \beta x \tag{16.3.11}$$

Then at $x = L$,

$$u(L) = 0 \Rightarrow \phi(L) = 0 \Rightarrow C_1 \sin \beta L + C_3 \sinh \beta L = 0 \tag{16.3.12a}$$

$$\mathcal{M}(L) = 0 \Rightarrow EI\phi''(L) = 0 \Rightarrow \beta^2(-C_1 \sin \beta L + C_3 \sinh \beta L) = 0 \tag{16.3.12b}$$

Adding these two equations gives

$$C_3 \sinh \beta L = 0$$

Since $\sinh \beta L$ cannot be zero (otherwise, ω will be zero, a trivial solution implying no vibration at all), so C_3 must be zero. This leads to the frequency equation:

$$C_1 \sin \beta L = 0 \tag{16.3.13}$$

This equation can be satisfied by selecting $C_1 = 0$, which gives $\phi(x) = 0$, a trivial solution. Therefore, $\sin \beta L$ must be zero, from which

$$\beta L = n\pi \qquad n = 1, 2, 3, \ldots \tag{16.3.14}$$

Equation (16.3.8) then gives the natural vibration frequencies:

$$\omega_n = \frac{n^2\pi^2}{L^2}\sqrt{\frac{EI}{m}} \qquad n = 1, 2, 3, \ldots \tag{16.3.15}$$

The natural vibration mode corresponding to ω_n is obtained by substituting Eq. (16.3.14) in Eq. (16.3.11) with $C_3 = 0$ as determined earlier:

$$\phi_n(x) = C_1 \sin \frac{n\pi x}{L} \tag{16.3.16}$$

The value of C_1 is arbitrary; $C_1 = 1$ will make the maximum value of $\phi_n(x)$ equal to unity. These natural modes are shown in Fig. 16.3.1.

For a simply supported uniform beam, we have determined an infinite series of modes each with its vibration frequency. Equations (16.3.15) and (16.3.16) and Fig. 16.3.1 tell us that the first mode is a half sine wave and that its frequency $\omega_1 =$

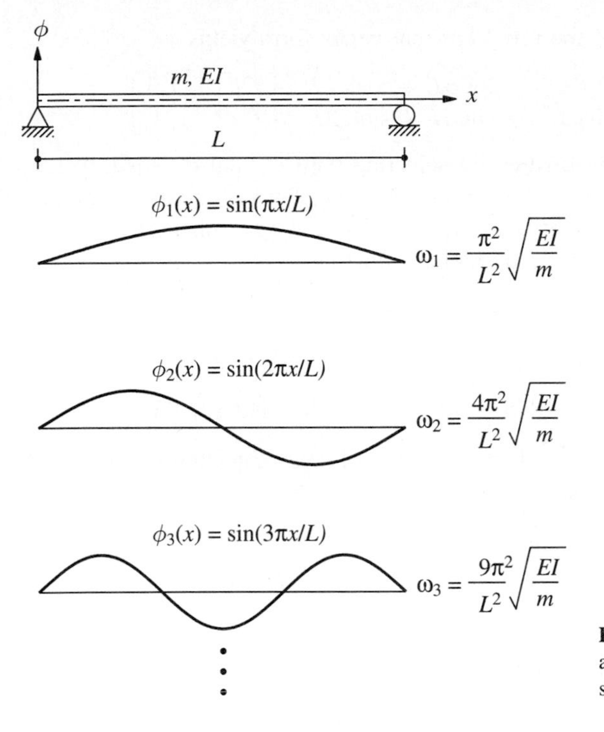

Figure 16.3.1 Natural vibration modes and frequencies of uniform simply supported beams.

$\pi^2(EI/mL^4)^{1/2}$. The second mode is a complete sine wave with frequency $\omega_2 = 4\omega_1$; the third is one and a half sine waves with frequency $\omega_3 = 9w_1$; and so on.

16.3.2 Uniform Cantilever Beam

In this section the natural vibration frequencies and modes of a uniform cantilever beam are determined. At the clamped end, $x = 0$, the displacement and slope are zero. Thus Eq. (16.3.9) gives

$$u(0) = 0 \Rightarrow \phi(0) = 0 \Rightarrow C_2 + C_4 = 0 \Rightarrow C_4 = -C_2 \tag{16.3.17a}$$

$$u'(0) = 0 \Rightarrow \phi'(0) = 0 \Rightarrow \beta(C_1 + C_3) = 0 \Rightarrow C_3 = -C_1 \tag{16.3.17b}$$

At the free end, $x = L$, of the cantilever the bending moment and shear are both zero. Thus, from Eqs. (16.3.9) and after using Eq. (16.3.17), we obtain

$$\mathcal{M}(L) = 0 \Rightarrow EI\phi''(L) = 0$$

$$\Rightarrow C_1(\sin\beta L + \sinh\beta L) + C_2(\cos\beta L + \cosh\beta L) = 0 \tag{16.3.18a}$$

$$\mathcal{V}(L) = 0 \Rightarrow EI\phi'''(L) = 0$$

$$\Rightarrow C_1(\cos\beta L + \cosh\beta L) + C_2(-\sin\beta L + \sinh\beta L) = 0 \tag{16.3.18b}$$

Rewriting Eqs. (16.3.18a) and (16.3.18b) in matrix form yields

$$\begin{bmatrix} \sin\beta L + \sinh\beta L & \cos\beta L + \cosh\beta L \\ \cos\beta L + \cosh\beta L & -\sin\beta L + \sinh\beta L \end{bmatrix} \begin{bmatrix} C_1 \\ C_2 \end{bmatrix} = \begin{bmatrix} 0 \\ 0 \end{bmatrix} \tag{16.3.19}$$

Equation (16.3.19) can be satisfied by selecting both C_1 and C_2 equal to zero, but this would give a trivial solution of no vibration at all. For either or both of C_1 and C_2 to be nonzero, the coefficient matrix in Eq. (16.3.19) must be singular (i.e., its determinant must be zero). This leads to the frequency equation:

$$1 + \cos\beta L \cosh\beta L = 0 \tag{16.3.20}$$

No simple solution is available for βL, so Eq. (16.3.20) is solved numerically to obtain

$$\beta_n L = 1.8751,\ 4.6941,\ 7.8548,\ \text{and } 10.996 \tag{16.3.21}$$

for $n = 1, 2, 3$, and 4. For $n > 4$, $\beta_n L \simeq (2n-1)\pi/2$. Equation (16.3.8) then gives the first four natural frequencies:

$$\omega_1 = \frac{3.516}{L^2}\sqrt{\frac{EI}{m}} \qquad \omega_2 = \frac{22.03}{L^2}\sqrt{\frac{EI}{m}} \qquad \omega_3 = \frac{61.70}{L^2}\sqrt{\frac{EI}{m}} \qquad \omega_4 = \frac{120.9}{L^2}\sqrt{\frac{EI}{m}} \tag{16.3.22}$$

Corresponding to each value of $\beta_n L$, the natural vibration mode is

$$\phi_n(x) = C_1\left[\cosh\beta_n x - \cos\beta_n x - \frac{\cosh\beta_n L + \cos\beta_n L}{\sinh\beta_n L + \sin\beta_n L}(\sinh\beta_n x - \sin\beta_n x)\right] \tag{16.3.23}$$

where C_1 is an arbitrary constant. To arrive at Eq. (16.3.23), C_2 is expressed in terms of C_1 from Eq. (16.3.18a) and substituted in the general solution, Eq. (16.3.9), and Eq. (16.3.17) is used. The first four natural vibration modes are shown in Fig. 16.3.2.

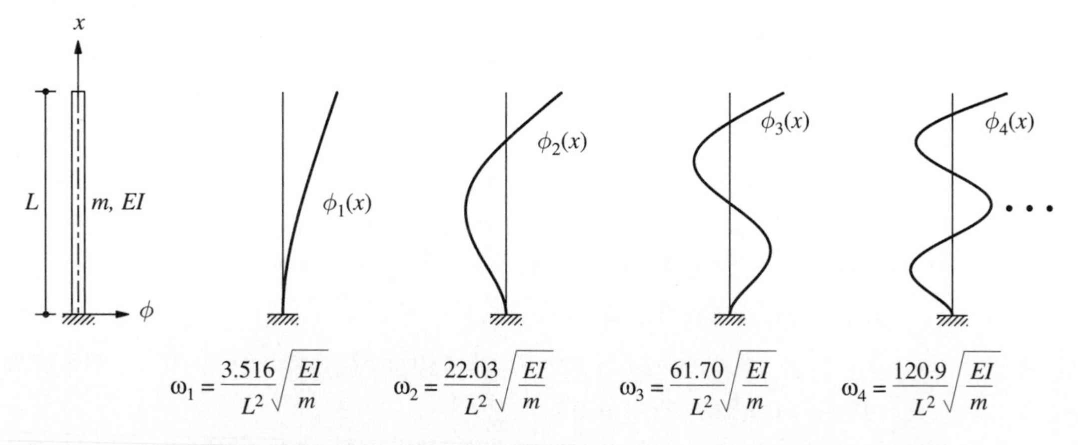

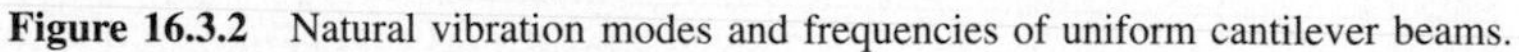

Figure 16.3.2 Natural vibration modes and frequencies of uniform cantilever beams.

Derivation 16.1

The solution of the fourth-order ordinary differential equation (16.3.7) is of the form

$$\phi(x) = Ae^{ax} \tag{a}$$

where A is an arbitrary constant. Substituting for $\phi(x)$ and its fourth derivative in Eq. (a) yields the characteristic equation

$$a^4 - \beta^4 = 0 \quad \text{or} \quad (a^2 - \beta^2)(a^2 + \beta^2) = 0 \tag{b}$$

which gives $a = \pm\beta$ and $a = \pm i\beta$. Thus the general solution of Eq. (16.3.7) is

$$\phi(x) = A_1 e^{i\beta x} + A_2 e^{-i\beta x} + A_3 e^{\beta x} + A_4 e^{-\beta x} \tag{c}$$

Equation (c) can be rewritten as Eq. (16.3.9) because

$$e^{\pm\beta x} = \cosh\beta x \pm \sinh\beta x \qquad e^{\pm i\beta x} = \cos\beta x \pm i\sin\beta x \tag{d}$$

16.3.3 Shear Deformation and Rotational Inertia

In the preceding derivation of the equation of motion for the transverse vibration of a beam, the inertial moment associated with rotation of the beam sections was ignored in Eq. (16.1.2), and only the deflection associated with bending stress in the beam is included in Eq. (16.1.3), thus ignoring the deflection due to shear stress in the beam. The analysis of beam vibration including both the effects of rotational inertia and shear deformation is called the *Timoshenko beam theory*.

The following equation governs such free vibration of a uniform beam with $m(x) = m$ and $EI(x) = EI$:

$$m\frac{\partial^2 u}{\partial t^2} + EI\frac{\partial^4 u}{\partial x^4} - mr^2\left(1 + \frac{E}{\kappa G}\right)\frac{\partial^4 u}{\partial x^2 t^2} + \frac{m^2 r^2}{\kappa GA}\frac{\partial^4 u}{\partial t^4} = 0 \tag{16.3.24}$$

where G is the modulus of rigidity; $r = \sqrt{I/A}$, the radius of gyration of the beam cross section; A the area of cross section; and κ is a constant that depends on the cross-sectional shape and accounts for the nonuniform distribution of shear stress across the section. The constant κ is derived for various cross-sectional shapes in textbooks on solid mechanics (e.g., κ is $\frac{5}{6}$ for rectangular cross section and $\frac{9}{10}$ for circular cross section).

Assuming a solution of the form $u(x, t) = C\sin(n\pi x/L)\sin\omega_n' t$, which satisfies the necessary end conditions, the frequency equation is obtained for both ends simply supported. Denoting a natural frequency of the beam by ω_n' if shear and rotational inertia effects are included, by ω_n if these effects are neglected [Eq. (16.3.15)], and defining $\Omega_n = \omega_n'/\omega_n$, this frequency equation can be written as

$$(1 - \Omega_n^2) - \Omega_n^2\left(\frac{n\pi r}{L}\right)^2\left(1 + \frac{E}{\kappa G}\right) + \Omega_n^4\left(\frac{n\pi r}{L}\right)^4\frac{E}{\kappa G} = 0 \tag{16.3.25}$$

If it is assumed that $nr/L << 1$, Eq. (16.3.25) reduces to

$$\omega_n' = \omega_n\frac{1}{\sqrt{1 + (n\pi r/L)^2(1 + E/\kappa G)}} \tag{16.3.26}$$

The correction due to rotational inertia is represented by the term $(n\pi r/L)^2$ in the denominator, whereas the shear deformation correction appears as $(n\pi r/L)^2(E/\kappa G)$. Thus the correction term for shear deformation is $E/\kappa G$ times larger than the rotational inertia correction term. For steel beams of rectangular cross section, $E/\kappa G$ is approximately 3.2. Values of $\Omega_n = \omega_n'/\omega_n$ are plotted in Fig. 16.3.3 using the solution of Eq. (16.3.25), a quadratic equation in Ω_n^2, for three values of $E/\kappa G$; these results are valid for all natural frequencies because n is included in the abscissa scale. Similar results are presented in Fig. 16.3.4 for the first five natural frequencies of a beam with $E/\kappa G = 3.2$. Also included is the approximate value of the frequency given by Eq. (16.3.26). When $nr/L < 0.2$ the error in the approximate equation is less than 5%.

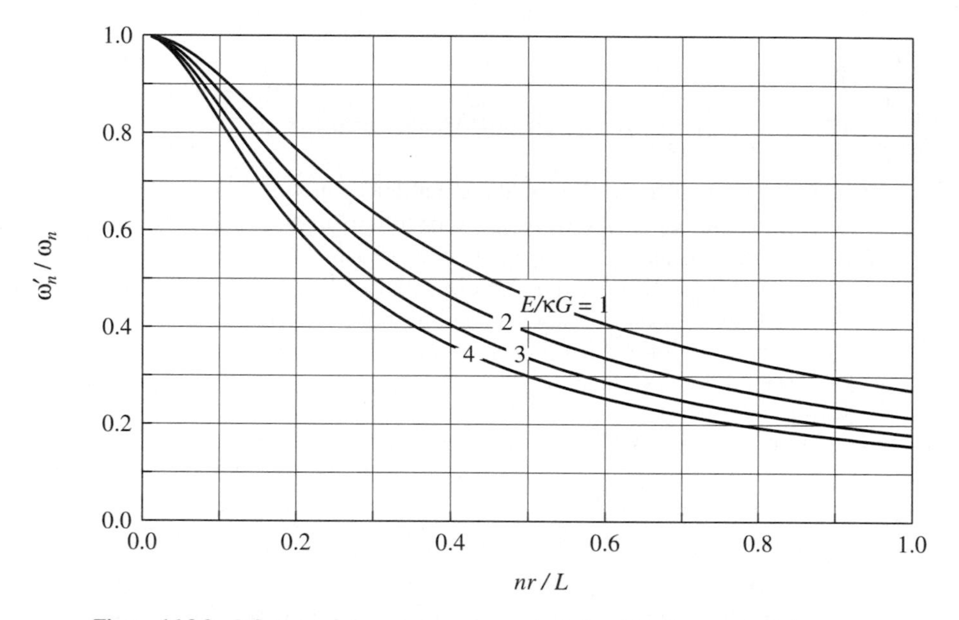

Figure 16.3.3 Influence of shear deformation and rotational inertia on natural frequencies of simply supported beams.

Shear deformation and rotational inertia have the effect of lowering the natural frequencies, as shown in Figs. 16.3.3 and 16.3.4. For a fixed value of the slenderness ratio L/r of the beam, the frequency reduction due to shear deformation and rotational inertia increases with mode number. This implies that while the corrections due to shear deformation and rotational inertia may be unimportant for the fundamental natural frequency, they could be significant for the higher frequencies. From the results presented one can estimate whether these corrections need to be included in a particular problem. For earthquake response analysis of many practical structures, these corrections are not significant, but it is important to realize that these corrections do exist. If significant, they can be included in the finite element formulation for practical structures which are not amenable to solution as infinite-DOF systems.

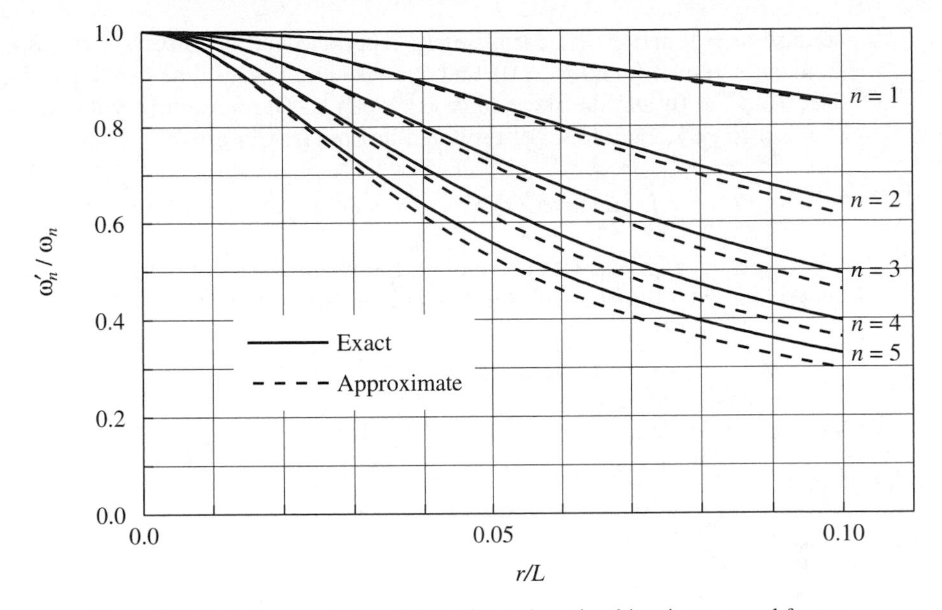

Figure 16.3.4 Influence of shear deformation and rotational inertia on natural frequencies of simply supported beams.

16.4 MODAL ORTHOGONALITY

In this section we derive the orthogonality properties of natural vibration modes of systems with distributed mass and elasticity. For convenience, the derivation is restricted to single-span beams with hinged, clamped, or free ends and without any lumped mass at the ends, although the final result applies in general.

The starting point for this derivation is Eq. (16.3.6), which governs the natural frequencies and modes; for mode r,

$$\left[EI(x)\phi_r''(x)\right]'' = \omega_r^2 m(x)\phi_r(x) \tag{16.4.1}$$

Multiplying both sides by $\phi_n(x)$ and integrating from 0 to L gives

$$\int_0^L \phi_n(x)\left[EI(x)\phi_r''(x)\right]'' dx = \omega_r^2 \int_0^L m(x)\phi_n(x)\phi_r(x)\, dx \tag{16.4.2}$$

The left side of this equation is integrated by parts; applying this procedure twice leads to

$$\int_0^L \phi_n(x)\left[EI(x)\phi_r''(x)\right]'' dx = \left\{\phi_n(x)[EI(x)\phi_r''(x)]'\right\}_0^L - \left\{\phi_n'(x)[EI(x)\phi_r''(x)]\right\}_0^L + \int_0^L EI(x)\phi_n''(x)\phi_r''(x)\, dx \tag{16.4.3}$$

It is easy to see that the quantities enclosed in $\{\cdots\}$ are zero at $x = 0$ and L if the ends of the beam are free, simply supported, or clamped. This is true at a clamped end because $\phi = 0$ and $\phi' = 0$, at a simply supported end because $\phi = 0$ and the bending moment is zero (i.e., $EI\phi'' = 0$), and at a free end because the bending moment is zero (i.e., $EI\phi'' = 0$) and the shear force is zero [i.e., $(EI\phi'') = 0$]. With the quantities in $\{\cdots\}$ set to zero, Eq. (16.4.3) substituted in Eq. (16.4.2) gives

$$\int_0^L EI(x)\phi_n''(x)\phi_r''(x)\,dx = \omega_r^2 \int_0^L m(x)\phi_n(x)\phi_r(x)\,dx \tag{16.4.4}$$

Starting with Eq. (16.3.6) written for mode n, multiplying both sides by $\phi_r(x)$, integrating from 0 to L, and using integration by parts twice leads to

$$\int_0^L EI(x)\phi_n''(x)\phi_r''(x)\,dx = \omega_n^2 \int_0^L m(x)\phi_n(x)\phi_r(x)\,dx \tag{16.4.5}$$

Subtracting Eq. (16.4.4) from Eq. (16.4.5) gives

$$(\omega_n^2 - \omega_r^2)\int_0^L m(x)\phi_n(x)\phi_r(x)\,dx = 0$$

Therefore, if $\omega_n \neq \omega_r$,

$$\int_0^L m(x)\phi_n(x)\phi_r(x)\;dx = 0 \tag{16.4.6a}$$

and this substituted in Eq. (16.4.2) leads to

$$\int_0^L \phi_n(x)\left[EI(x)\phi_r''(x)\right]''\,dx = 0 \tag{16.4.6b}$$

Equations (16.4.6a) and (16.4.6b) are the orthogonality relations for the natural vibration modes. If a system has repeated frequencies, modes $\phi_n(x)$ still exist such that any two modes, $n \neq r$, satisfy the orthogonality relations even if $\omega_n = \omega_r$.

16.5 MODAL ANALYSIS OF FORCED DYNAMIC RESPONSE

We now return to the partial differential equation (16.1.4), which is to be solved for a given applied force $p(x, t)$. Assuming that the associated eigenvalue problem of Eq. (16.3.6) has been solved for the natural frequencies and modes, the displacement is given by a linear combination of the modes:

$$u(x, t) = \sum_{r=1}^{\infty} \phi_r(x)q_r(t) \tag{16.5.1}$$

Thus the response $u(x, t)$ has been expressed as the superposition of the contributions of the individual modes; the rth term in the series of Eq. (16.5.1) is the contribution of the rth mode to the response.

We will see next that Eq. (16.1.4) can be transformed to an infinite set of ordinary differential equations, each of which has one modal coordinate $q_n(t)$ as the unknown.

Substituting Eq. (16.5.1) in Eq. (16.1.4) gives

$$\sum_{r=1}^{\infty} m(x)\phi_r(x)\ddot{q}_r(t) + \sum_{r=1}^{\infty} \left[EI(x)\phi_r''(x)\right]'' q_r(t) = p(x,t)$$

Now we multiply each term by $\phi_n(x)$, integrate it over the length of the beam, and interchange the order of integration and summation to get

$$\sum_{r=1}^{\infty} \ddot{q}_r(t) \int_0^L m(x)\phi_n(x)\phi_r(x)\,dx + \sum_{r=1}^{\infty} q_r(t) \int_0^L \phi_n(x)\left[EI(x)\phi_r''(x)\right]''\,dx$$

$$= \int_0^L p(x,t)\phi_n(x)\,dx$$

By virtue of the orthogonality properties of modes given by Eq. (16.4.6), all terms in each of the summations on the left side vanish except the one term for which $r = n$, leaving

$$\ddot{q}_n(t) \int_0^L m(x)\left[\phi_n(x)\right]^2 dx + q_n(t) \int_0^L \phi_n(x)\left[EI(x)\phi_n''(x)\right]''\,dx = \int_0^L p(x,t)\phi_n(x)\,dx$$

This equation can be rewritten as

$$M_n\ddot{q}_n(t) + K_n q_n(t) = P_n(t) \tag{16.5.2}$$

where

$$M_n = \int_0^L m(x)\left[\phi_n(x)\right]^2 dx \qquad K_n = \int_0^L \phi_n(x)\left[EI(x)\phi_n''(x)\right]''\,dx$$

$$P_n(t) = \int_0^L p(x,t)\phi_n(x)\,dx \tag{16.5.3}$$

If each end of the beam is free, hinged, or clamped, adapting the results of Section 16.4 gives an alternative equation for K_n:

$$K_n = \int_0^L EI(x)\left[\phi_n''(x)\right]^2 dx \tag{16.5.4}$$

The *generalized mass* M_n and *generalized stiffness* K_n for the nth mode are related:

$$K_n = \omega_n^2 M_n \tag{16.5.5}$$

This relation can be derived by writing Eq. (16.4.1) for the nth mode, multiplying both sides by $\phi_n(x)$, integrating over 0 to L, and utilizing the definitions of M_n and K_n. The term $P_n(t)$ in Eq. (16.5.2) is called the *generalized force* for the nth mode. Equation (16.5.2) governs the nth modal coordinate $q_n(t)$, and the generalized properties M_n, K_n, and $P_n(t)$ depend only on the nth mode $\phi_n(x)$. Thus we have an infinite number of equations like Eq. (16.5.2), one for each mode. The partial differential equation (16.1.4) in the unknown function $u(x,t)$ has been transformed to an infinite set of ordinary differential equations (16.5.2) in unknowns $q_n(t)$.

For applied dynamic forces defined by $p(x,t)$, the motion $u(x,t)$ of the system can be determined by solving the modal equations for $q_n(t)$. The equation for each mode is independent of the equations for all other modes and can therefore be solved separately. Furthermore, each modal equation is of the same form as the equation of motion for an SDF system. Thus the results obtained in Chapters 3 and 4 for the response of SDF systems to various dynamic forces—harmonic force, impulsive force, etc.—can be adapted to obtain solutions $q_n(t)$ for the modal equations.

Once the $q_n(t)$ have been determined, the contribution of the nth mode to the displacement $u(x,t)$ is given by

$$u_n(x,t) = \phi_n(x)q_n(t) \tag{16.5.6}$$

The total displacement is the combination of the contributions of all the modes:

$$u(x,t) = \sum_{n=1}^{\infty} u_n(x,t) = \sum_{n=1}^{\infty} \phi_n(x)q_n(t) \tag{16.5.7}$$

The bending moment and shear force at any section along the length of the beam are related to the displacements $u(x)$ as follows:

$$\mathcal{M}(x) = EI(x)u''(x) \qquad \mathcal{V}(x) = \left[EI(x)u''(x)\right]' \tag{16.5.8}$$

These static relationships apply at each instant of time with $u(x)$ replaced by $u(x,t)$, which is given by Eq. (16.5.7). Thus

$$\mathcal{M}(x,t) = \sum_{n=1}^{\infty} EI(x)\phi_n''(x)q_n(t) \qquad \mathcal{V}(x,t) = \sum_{n=1}^{\infty} \left[EI(x)\phi_n''(x)\right]' q_n(t) \tag{16.5.9}$$

Example 16.1

Derive mathematical expressions for the dynamic response—displacement and bending moments—of a uniform simply supported beam to a step-function force p_o at distance ξ from the left end (Fig. E16.1). Specialize the results for the force applied at midspan.

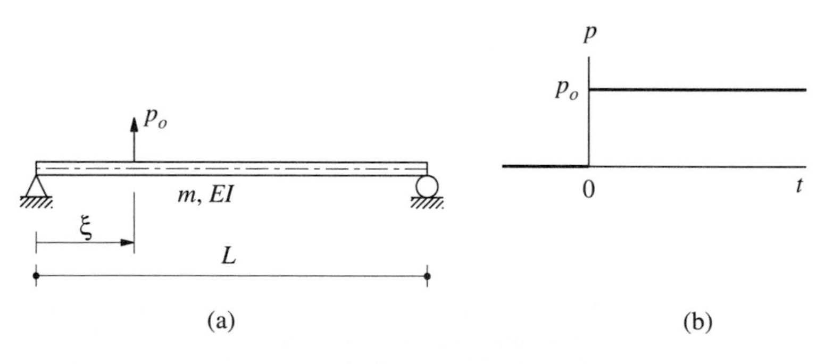

Figure E16.1

Solution

1. *Determine the natural vibration frequencies and modes.*

$$\omega_n = \frac{n^2\pi^2}{L^2}\sqrt{\frac{EI}{m}} \qquad \phi_n(x) = \sin\frac{n\pi x}{L} \tag{a}$$

2. *Set up the modal equations.* Substituting $\phi_n(x)$ in Eq. (16.5.3a) gives M_n, which is substituted in Eq. (16.5.5) together with ω_n^2 to get K_n:

$$M_n = \frac{mL}{2} \qquad K_n = \frac{n^4\pi^4 EI}{2L^3} \tag{b}$$

Substituting $p(x,t) = p_o\delta(x-\xi)$, where $\delta(x-\xi)$ is the Dirac delta function centered at ξ, in Eq. (16.5.3c) gives

$$P_n(t) = p_o\phi_n(\xi) \tag{c}$$

Then the nth modal equation is

$$M_n\ddot{q}_n(t) + K_n q_n(t) = p_o\phi_n(\xi) \tag{d}$$

3. *Solve the modal equations.* Equation (4.3.2) describes the response of an SDF system to a step force. We will adapt this solution to Eq. (d) by changing the notation $u(t)$ to $q_n(t)$ and noting that $(u_{st})_o = p_o\phi_n(\xi)/K_n$. Thus

$$q_n(t) = \frac{p_o\phi_n(\xi)}{K_n}(1-\cos\omega_n t) = \frac{2p_oL^3}{\pi^4 EI}\frac{\phi_n(\xi)}{n^4}(1-\cos\omega_n t) \tag{e}$$

Substituting Eq. (e) in Eq. (16.5.7) and noting that $\phi_n(x)$ is known from Eq. (a), we obtain the displacement response $u(x,t)$.

4. *Specialize for $\xi = L/2$.* Substituting $\xi = L/2$ in Eq. (e) and the latter in Eq. (16.5.7) gives

$$u(x,t) = \frac{2p_oL^3}{\pi^4 EI}\sum_{n=1}^{\infty}\frac{\phi_n(L/2)}{n^4}(1-\cos\omega_n t)\sin\frac{n\pi x}{L} \tag{f}$$

where

$$\phi_n\left(\frac{L}{2}\right) = \begin{cases} 0 & n = 2,4,6,\ldots \\ 1 & n = 1,5,9,\ldots \\ -1 & n = 3,7,11,\ldots \end{cases} \tag{g}$$

Substituting Eq. (g) in Eq. (f) gives

$$u(x,t) = \frac{2p_oL^3}{\pi^4 EI}\left(\frac{1-\cos\omega_1 t}{1}\sin\frac{\pi x}{L} - \frac{1\cos\omega_3 t}{81}\sin\frac{3\pi x}{L}\right.$$
$$\left. + \frac{1-\cos\omega_5 t}{625}\sin\frac{5\pi x}{L} - \frac{1-\cos\omega_7 t}{2401}\sin\frac{7\pi x}{L} + \cdots\right) \tag{h}$$

The displacement at midspan is

$$u\left(\frac{L}{2},t\right) = \frac{2p_oL^3}{\pi^4 EI}\left(\frac{1-\cos\omega_1 t}{1} + \frac{1-\cos\omega_3 t}{81} + \frac{1-\cos\omega_5 t}{625} + \frac{1-\cos\omega_7 t}{2401} + \cdots\right) \tag{i}$$

The coefficients 1, 81, 625, 2401, and so on, in the denominator suggest that the first-mode contribution is dominant and that the series converges rapidly.

The bending moments are obtained by substituting Eq. (h) in Eq. (16.5.9a):

$$\mathcal{M}(x,t) = -\frac{2p_oL}{\pi^2}\left(\frac{1-\cos\omega_1 t}{1}\sin\frac{\pi x}{L} - \frac{1-\cos\omega_3 t}{9}\sin\frac{3\pi x}{L}\right.$$
$$\left. + \frac{1-\cos\omega_5 t}{25}\sin\frac{5\pi x}{L} - \frac{1-\cos\omega_7 t}{49}\sin\frac{7\pi x}{L} + \cdots\right) \tag{j}$$

The bending moment at midspan is

$$\mathcal{M}\left(\frac{L}{2}, t\right) = -\frac{2p_o L}{\pi^2}\left(\frac{1-\cos\omega_1 t}{1} + \frac{1-\cos\omega_3 t}{9} + \frac{1-\cos\omega_5 t}{25} + \frac{1-\cos\omega_7 t}{49} + \cdots\right) \tag{k}$$

This series with n^2 in the denominator converges slowly compared to Eq. (i) with n^4 in the denominator. This difference implies that higher modes contribute more significantly to forces than to displacements, a result consistent with the conclusions of Chapters 12 and 13 for discretized systems.

16.6 EARTHQUAKE RESPONSE HISTORY ANALYSIS

As shown earlier, when the excitation is acceleration $\ddot{u}_g(t)$ of the supports, the equation of motion for a beam is the same as for applied force $p(x, t)$ except that this force is replaced by $p_{\text{eff}}(x, t)$ given by Eq. (16.2.4). Thus the modal analysis procedure of Section 16.5 can readily be extended to the earthquake problem.

Before developing this extension, it is instructive and will later be useful to expand the effective earthquake forces, $p_{\text{eff}}(x, t)$, as a summation of modal inertia forces:

$$p_{\text{eff}}(x, t) = -m(x)\ddot{u}_g(t) = -\ddot{u}_g(t)\sum_{r=1}^{\infty}\Gamma_r m(x)\phi_r(x) \tag{16.6.1}$$

Premultiplying both sides by $\phi_n(x)$, integrating over the length of the beam, and utilizing modal orthogonality with respect to the mass distribution, Eq. (16.4.6a), leads to

$$\Gamma_n = \frac{L_n^h}{M_n} \qquad \text{where} \qquad L_n^h = \int_0^L m(x)\phi_n(x)\,dx \tag{16.6.2}$$

Noting that $\ddot{u}_g(t)$ appears on both sides of Eq. (16.6.1), it is more meaningful to rewrite it as

$$m(x) = \sum_{n=1}^{\infty}\Gamma_n m(x)\phi_n(x) \tag{16.6.3}$$

This is an expansion of $m(x)$, which defines the spatial distribution of the effective earthquake forces, as a summation of the modal inertia force distributions:

$$s_n(x) = \Gamma_n m(x)\phi_n(x) \tag{16.6.4}$$

Observe that these modal expansion equations for a distributed-mass system are similar to the corresponding equations (13.2.2) and (13.2.4) for lumped-mass systems. For uniform cantilever towers with mass m per unit length the modal expansion of Eq. (16.6.3) is as shown in Fig. 16.6.1. The functions $s_n(x)$ were evaluated from Eq. (16.6.4) using Eqs. (16.6.2) and (16.5.3a) and the $\phi_n(x)$ given by Eqs. (16.3.23) and (16.3.21).

Returning now to the modal analysis procedure of Section 16.5, $p(x, t)$ in Eq. (16.5.3c) is replaced by $p_{\text{eff}}(x, t)$ given by Eq. (16.2.4), to obtain

$$P_n(t) = -L_n^h\ddot{u}_g(t) \tag{16.6.5}$$

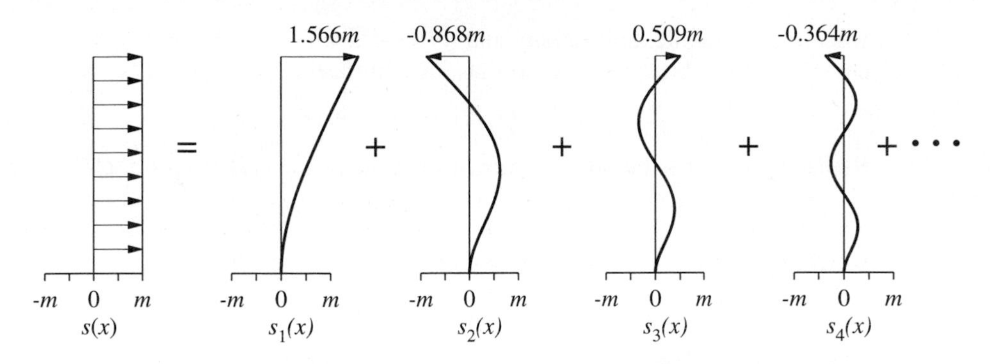

Figure 16.6.1 Modal expansion of effective earthquake forces for uniform towers.

Substituting Eq. (16.6.5) in Eq. (16.5.2), dividing by M_n, and using Eqs. (16.5.5) and (16.6.2a), gives the modal equations of an undamped tower subjected to earthquake excitation:

$$\ddot{q}_n + \omega_n^2 q_n = -\Gamma_n \ddot{u}_g(t) \tag{16.6.6}$$

For classically damped systems, Eq. (16.6.6) becomes

$$\ddot{q}_n + 2\zeta_n \omega_n \dot{q}_n + \omega_n^2 q_n = -\Gamma_n \ddot{u}_g(t) \tag{16.6.7}$$

where ζ_n is the damping ratio for the nth mode. This is the same as Eq. (13.1.7) derived earlier for N-DOF systems, except that L_n^h and M_n that enter into Γ_n are now given by Eqs. (16.6.2) and (16.5.3), respectively. As shown in Section 13.1.3, the solution of Eq. (16.6.7) is

$$q_n(t) = \Gamma_n D_n(t) \tag{16.6.8}$$

where $D_n(t)$ is the deformation response of the nth-*mode SDF system.* This is an SDF system with vibration properties—natural frequency ω_n and damping ratio ζ_n—of the nth mode of the distributed mass system. Thus $q_n(t)$ is readily available once the SDF system response has been determined by the methods of Chapter 6. The contribution of the nth mode to the earthquake response of the tower can be expressed in terms of $D_n(t)$. Substituting Eq. (16.6.8) in Eqs. (16.5.6) and (16.5.9) gives the displacements, bending moments, and shear forces due to the nth mode:

$$u_n(x,t) = \Gamma_n \phi_n(x) D_n(t) \tag{16.6.9}$$

$$\mathcal{M}_n(x,t) = \Gamma_n EI(x)\phi_n''(x) D_n(t) \qquad \mathcal{V}_n(x,t) = \Gamma_n \left[EI(x)\phi_n''(x)\right]' D_n(t) \tag{16.6.10}$$

Alternatively, as in Chapter 13 for an N-DOF system, the internal forces can be determined from the equivalent static forces associated with displacements $u_n(x,t)$ computed from dynamic analysis. To derive an equation for these forces, we introduce a familiar equation from elementary beam theory relating deflections $u(x)$ to applied forces $f(x)$. For a uniform beam

$$EIu^{iv}(x) = f(x) \tag{16.6.11}$$

where EI is the flexural rigidity and $u^{iv} = d^4u/dx^4$. The more general version of this equation applicable to nonuniform beams with flexural rigidity $EI(x)$ is

$$\left[EI(x)u''(x)\right]'' = f(x) \tag{16.6.12}$$

Replacing $u(x)$ by the time-varying displacements $u_n(x,t)$ from Eq. (16.6.9) gives

$$f_n(x,t) = \Gamma_n \left[EI(x)\phi_n''(x)\right]'' D_n(t) \tag{16.6.13}$$

which, by using Eq. (16.4.1) rewritten for the nth mode, becomes

$$f_n(x,t) = s_n(x)A_n(t) \tag{16.6.14}$$

where $s_n(x)$ is given by Eq. (16.6.4), and $A_n(t)$, the pseudo-acceleration response of the nth-mode SDF system, is given by Eq. (13.1.12), which is repeated:

$$A_n(t) = \omega_n^2 D_n(t) \tag{16.6.15}$$

Observe the similarity between Eqs. (16.6.14) and (13.2.7) for a lumped-mass system. At any time instant the contribution $r_n(t)$ of the nth mode to any response quantity $r(t)$—deflection, shear force, or bending moment at any location of the beam—is determined by static analysis of the beam subjected to external forces $f_n(x,t)$, and can be expressed as

$$r_n(t) = r_n^{\text{st}} A_n(t) \tag{16.6.16}$$

The modal static response r_n^{st} is determined by static analysis of the tower due to external forces $s_n(\xi)$, Fig. 16.6.2. As shown in Fig. 16.6.1, these forces due to the fundamental mode all act in the same direction, but for the second and higher modes they will change direction as one moves up the tower.

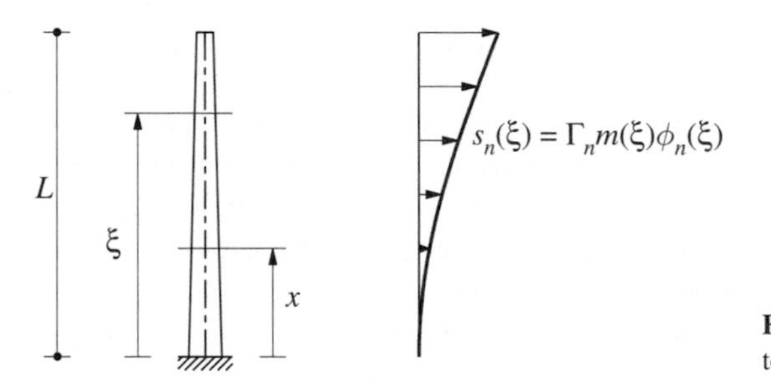

Figure 16.6.2 Static problem to be solved to determine modal static responses.

The modal static responses are presented in Table 16.6.1 for five response quantities: the shear $\mathcal{V}(x)$ at location x, the bending moment $\mathcal{M}(x)$ at location x, the base shear $\mathcal{V}_b = \mathcal{V}(0)$, the base moment $\mathcal{M}_b = \mathcal{M}(0)$, and deflection $u(x)$. The first four equations come from static analysis of the system in Fig. 16.6.2. The result for $u(x)$ is obtained by comparing Eqs. (16.6.9) and (16.6.16) and using Eq. (16.6.15). Parts of the equations for $\mathcal{V}_{bn}^{\text{st}}$ and $\mathcal{M}_{bn}^{\text{st}}$ are obtained by substituting Eq. (16.6.4) for $s_n(\xi)$, using Eq. (16.6.2) for L_n^h, and defining

$$M_n^* = \Gamma_n L_n^h \qquad h_n^* = \frac{L_n^\theta}{L_n^h} \qquad L_n^\theta = \int_0^L x m(x)\phi_n(x) \tag{16.6.17}$$

TABLE 16.6.1 MODAL STATIC RESPONSES

Response, r	Modal Static Response, r_n^{st}
$\mathcal{V}(x)$	$\mathcal{V}_n^{st}(x) = \int_x^L s_n(\xi)\,d\xi$
$\mathcal{M}(x)$	$\mathcal{M}_n^{st}(x) = \int_x^L (\xi - x)s_n(\xi)\,d\xi$
$\mathcal{V}_b$	$\mathcal{V}_{bn}^{st} = \int_0^L s_n(\xi)\,d\xi = \Gamma_n L_n^h = M_n^*$
$\mathcal{M}_b$	$\mathcal{M}_{bn}^{st} = \int_0^L \xi s_n(\xi)\,d\xi = \Gamma_n L_n^\theta = h_n^* M_n^*$
$u(x)$	$u_n^{st}(x) = (\Gamma_n/\omega_n^2)\phi_n(x)$

Observe the similarity between the equations in Table 16.6.1 and those for a lumped-mass system in Table 13.2.1. The approach symbolized by Eq. (16.6.16) to determine shear and moment is preferable over Eq. (16.6.10) because it avoids computation of the second and third derivatives of the mode shapes; obviously, both methods will give identical results.

The base shear $\mathcal{V}_{bn}(t)$ and base moment $\mathcal{M}_{bn}(t)$ due to the nth mode are obtained by specializing Eq. (16.6.16) for $\mathcal{V}_b$ and $\mathcal{M}_b$ and substituting for $\mathcal{V}_{bn}^{st}$ and $\mathcal{M}_{bn}^{st}$ from Table 16.6.1:

$$\mathcal{V}_{bn}(t) = M_n^* A_n(t) \qquad \mathcal{M}_{bn}(t) = h_n^* \mathcal{V}_{bn}(t) \tag{16.6.18}$$

Because Eq. (16.6.18) is identical to Eqs. (13.2.12b) and (13.2.15) for lumped-mass systems, following Section 13.2.5, M_n^* and h_n^* may be interpreted as the *effective modal mass* and *effective modal height* for the nth mode. Observe that Eq. (16.6.17a and b) is identical to Eq. (13.2.9a) for lumped-mass systems; the definitions of M_n, L_n^h, and L_n^θ differ, however, between distributed-mass and lumped-mass systems.

The sum of the effective modal masses over all modes is equal to the total mass of the tower:

$$\sum_{n=1}^{\infty} M_n^* = \int_0^L m(x)\,dx \tag{16.6.19}$$

and the sum of the first moments about the base of the effective modal masses M_n^* located at heights h_n^* is equal to the first moment of the distributed mass about the base:

$$\sum_{n=1}^{\infty} h_n^* M_n^* = \int_0^L x m(x)\,dx \tag{16.6.20}$$

These relations can be proven in the same manner as the analogous equations (13.2.14) and (13.2.17) for a lumped-mass system. In particular, Eq. (16.6.19) can be proven by integrating Eq. (16.6.3) over the height of the tower and using Eq. (16.6.2b). Similarly, Eq. (16.6.20) can be derived from the modal expansion of forces $xm(x)$. The effective modal masses M_n^* and effective modal heights h_n^* for a uniform cantilever tower are shown in Fig. 16.6.3; these were determined from Eq. (16.6.17) using the known modes (Fig. 16.3.2). Observe that the sum of M_n^* for the first four modes gives 90% of the total mass of the tower.

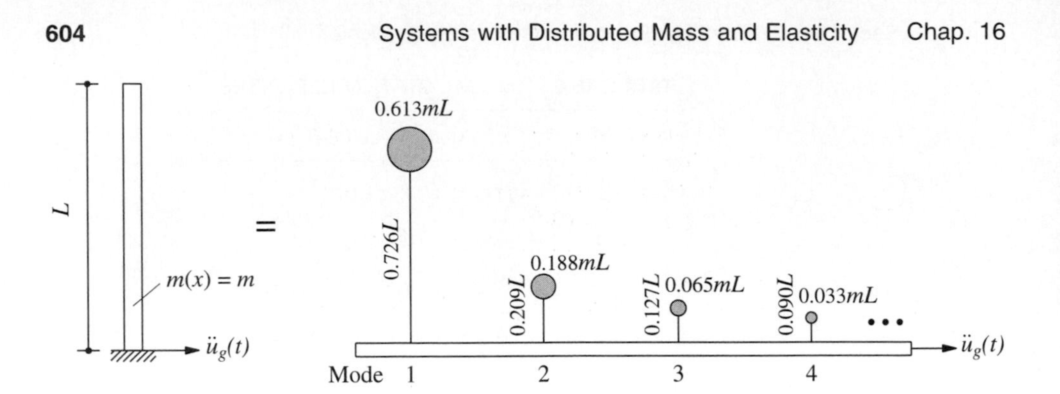

Figure 16.6.3 Effective modal masses and effective modal heights.

Combining the response contributions of all the modes gives the earthquake response of the system:

$$r(t) = \sum_{n=1}^{\infty} r_n(t) = \sum_{n=1}^{\infty} r_n^{\text{st}} A_n(t) \tag{16.6.21}$$

where Eq. (16.6.16) has been used for $r_n(t)$. This nth-mode contribution to the response can be determined from the modal static response (Table 16.6.1) and $A_n(t)$, the pseudo-acceleration response of the nth-mode SDF system.

16.7 EARTHQUAKE RESPONSE SPECTRUM ANALYSIS

The peak response of a distributed mass system, such as a cantilever tower, can be estimated from the earthquake response (or design) spectrum by procedures analogous to those developed in Chapter 13, Part B for lumped-mass systems.

The exact peak value of the nth-mode response $r_n(t)$ is

$$r_{no} = r_n^{\text{st}} A_n \tag{16.7.1}$$

where $A_n \equiv A(T_n, \zeta_n)$ is the ordinate of the pseudo-acceleration spectrum corresponding to natural period T_n and damping ratio ζ_n. Alternatively, r_{no} may be viewed as the result of static analysis of the tower subjected to external forces

$$f_{no}(x) = s_n(x) A_n \tag{16.7.2}$$

which are the peak values of the equivalent static forces $f_n(x, t)$ defined in Eq. (16.6.14).

The peak value r_o of the total response $r(t)$ can be estimated by combining the modal peaks r_{no} according to one of the modal combination rules presented in Section 13.7.2. Because the natural frequencies of transverse vibration of a beam are well separated, the SRSS combination rule is satisfactory. Thus

$$r_o \simeq \left(\sum_{n=1}^{\infty} r_{no}^2 \right)^{1/2} \tag{16.7.3}$$

Example 16.2

A reinforced-concrete chimney, 600 ft high, has a uniform hollow circular cross section with outside diameter 50 ft and wall thickness 2 ft 6 in. (Fig. E16.2a). For purposes of earthquake analysis, the chimney is assumed clamped at the base and its mass and flexural stiffness are computed from the gross area of the concrete (neglecting the reinforcing steel). The elastic modulus for concrete $E_c = 3600$ ksi, and its unit weight is 150 lb/ft^3. Modal damping ratios are estimated as 5%. Determine the displacements, shear forces, and bending moments due to an earthquake characterized by the design spectrum of Fig. 6.9.5 scaled to a peak ground acceleration of 0.25g. Neglect shear deformations and rotational inertia.

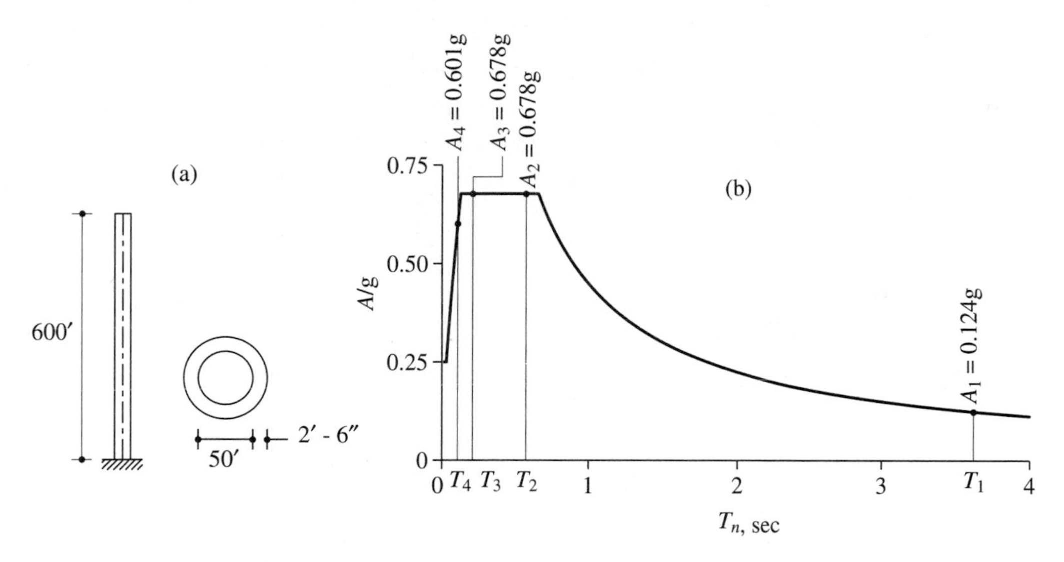

Figure E16.2a, b

Solution

1. *Determine the chimney properties.*

$$m = \pi \frac{[25^2 - (22.5)^2]0.15}{32.2} = 1.738 \text{ kip-sec}^2/\text{ft}^2$$

$$EI = (3600 \times 144)\frac{\pi}{4}\left[25^4 - (22.5)^4\right] = 5.469 \times 10^{10} \text{ kip-ft}^2$$

2. *Determine the natural vibration periods and modes.* Equation (16.3.22) gives the natural frequencies of vibration, and the corresponding periods, in seconds, are $T_1 = 3.626$, $T_2 = 0.5787$, $T_3 = 0.2067$, $T_4 = 0.1055$, and so on. The natural modes, given by Eq. (16.3.23) with $\beta_n L$ defined by Eq. (16.3.21), were evaluated numerically for many values of x and are shown in Fig. 16.3.2, normalized to unit value at the top.

3. *Compute the modal properties.* With the mode shapes known, the properties M_n, L_n^h, L_n^θ, M_n^*, and h_n^* were obtained by numerically evaluating their respective integrals, and are presented in Table E16.2.

4. *Read the design spectrum ordinates.* The design spectrum of Fig. 6.9.5 scaled to a peak acceleration of 0.25g is shown in Fig. E16.2b, wherein the pseudo-acceleration

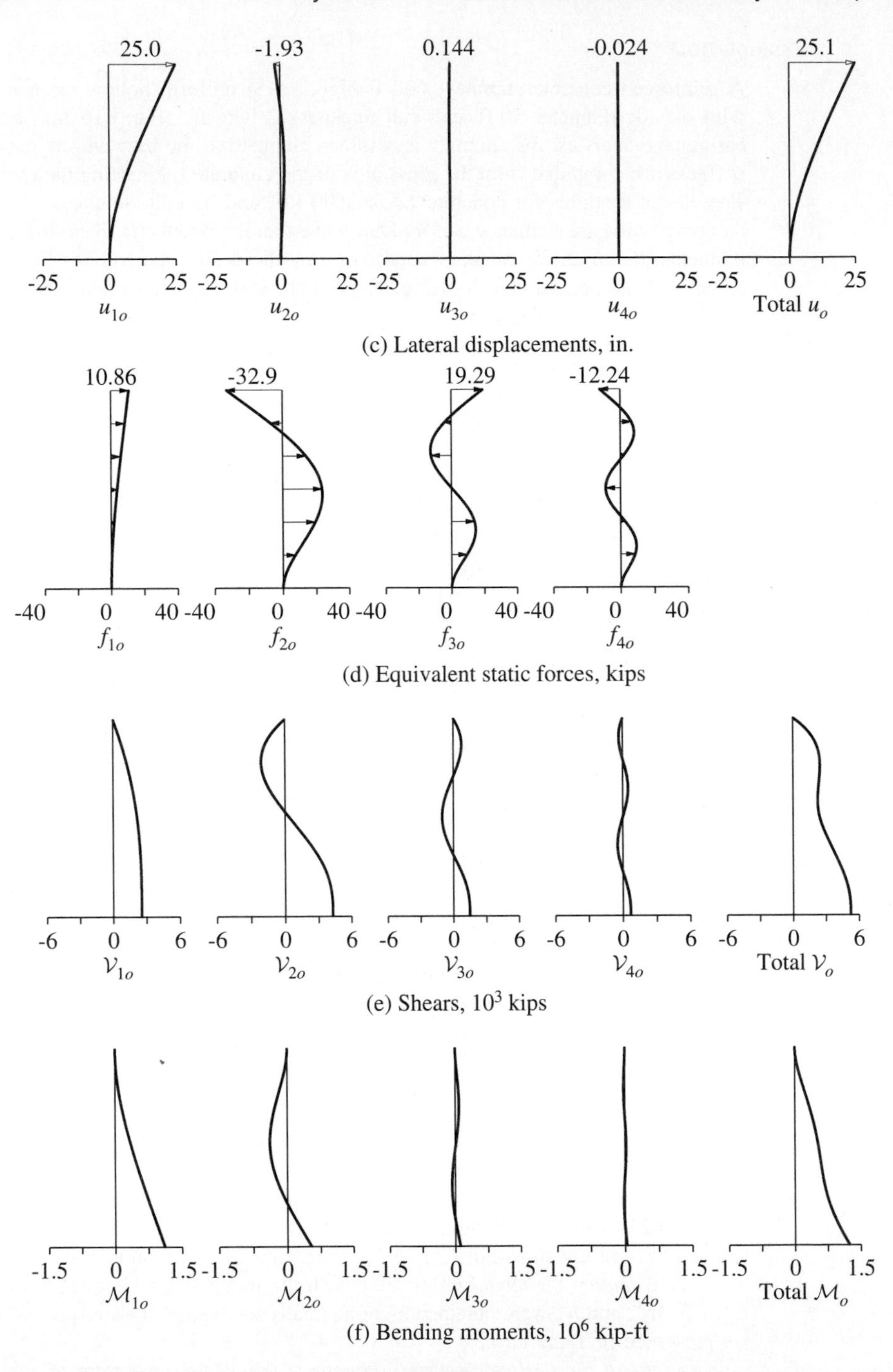

Figure E16.2c–f

TABLE E16.2 MODAL PROPERTIES

Mode	M_n/mL	L_n^h/mL	Γ_n	L_n^θ/mL^2
1	0.2500	0.3915	1.5660	0.2844
2	0.2500	−0.2170	−0.8679	−0.0454
3	0.2500	0.1272	0.5089	0.0162
4	0.2498	−0.0909	−0.3637	−0.0082

ordinates corresponding to the first four periods are noted: $A_n/g = 0.124$, 0.678, 0.678, and 0.601.

5. *Compute the displacements.* The peak displacements $u_{no}(x)$ due to the nth mode are given by Eq. (16.7.1), where the modal static response r_n^{st} becomes $u_n^{st}(x)$ given in Table 16.6.1. Substituting known values of Γ_n, $\phi_n(x)$, ω_n^2, and A_n leads to $u_{no}(x)$, $n = 1$, 2, 3, and 4, shown in Fig. E16.2c. At each location x these peak modal displacements are combined according to the SRSS rule, Eq. (16.7.3), to obtain an estimate of the total displacements $u_o(x)$, which are also shown. Observe that the total displacements are due primarily to the first mode.

6. *Determine the modal expansion of* $m(x)$, *Eq.* (16.6.3). With $\phi_n(x)$ known from Fig. 16.3.2 and Γ_n from Table E16.2, the functions $s_n(x)$ are determined from Eq. (16.6.4). Actually, these were presented in Fig. 16.6.1.

7. *Compute the equivalent static forces for the nth mode.* These forces $f_{no}(x)$ are determined from Eq. (16.7.2) using the $s_n(x)$ of Fig. 16.6.1 and the A_n values of Fig. E16.2b. The results for the first four modes are shown in Fig. E16.2d.

8. *Compute the shears and bending moments.* For each mode the peak values of shears and bending moments at location x are computed by static analysis of the chimney subjected to forces $f_{no}(x)$. The resulting shears and bending moments due to the first four modes are shown in Fig. E16.2e and f. At each section x, these modal responses are combined by the SRSS rule [Eq. (16.7.3)] to obtain an estimate of the total forces, which are also shown. Observe that the first two modes contribute significantly to the total response, with the second-mode contribution more significant for the shears than for moments.

9. *Compare with Rayleigh's method.* It is of interest to compare the results above considering response in four modes with the approximate solution using Rayleigh's method (Example 8.3). The approximate analysis predicts the displacements reasonably well but not the bending moments or shears. There are two reasons for the larger errors in forces: (a) The approximate results differ from the exact response due to the first mode because the assumed shape function in Rayleigh's method is an approximation to this mode; this discrepancy introduces larger errors in forces than in displacements. (b) The second and higher modes, whose response contributions to forces are more significant than they are for displacements, are neglected in Rayleigh's method.

16.8 DIFFICULTY IN ANALYZING PRACTICAL SYSTEMS

It is evident that the dynamic response of systems with distributed mass and elasticity can be determined by the modal analysis procedure once the natural vibration frequencies and modes of the system have been determined. Both examples solved in Section 16.3

involved uniform beams and we found the natural frequencies and modes analytically although the frequency equation for the cantilever had to be solved numerically. This classical approach is rarely feasible if the flexural rigidity EI or mass m vary along the length of the beam, several intermediate supports are involved, or the system is an assemblage of several members with distributed mass. In this section we identify some of the difficulties in obtaining analytical solutions for the above-mentioned systems.

Consider a single-span beam with mass $m(x)$ and flexural stiffness $EI(x)$. To determine the natural frequencies and modes, we need to solve Eq. (16.3.6), which can be rewritten as

$$EI(x)\phi^{iv}(x) + 2EI'(x)\phi'''(x) + EI''(x)\phi''(x) - \omega^2 m(x)\phi(x) = 0 \tag{16.8.1}$$

Because the coefficients $EI(x)$, $EI'(x)$, $EI''(x)$, and $m(x)$ of this fourth-order differential equation vary with x, an analytical solution is rarely feasible for ω^2 and $\phi(x)$. Therefore, it is not practical to use the classical approach for practical problems in which $EI(x)$ and $m(x)$ may be complicated functions.

In finding the natural frequencies and modes of a beam on multiple supports, the uniform segment between each pair of supports is considered as a separate beam with its origin at the left end of the segment. Equation (16.3.9) applies to each segment, there is one such equation for each segment, and the necessary boundary conditions are:

1. At each end of the beam the usual boundary conditions are applicable, depending on the type of support.
2. At each intermediate support the deflection is zero, and since the beam is continuous, the slope and the moment just to the left and to the right of the support are the same.

This process quickly becomes unmanageable because of the four constants in Eq. (16.3.9), which must be evaluated in each segment. An analytical solution is rarely feasible for ω^2 and $\phi(x)$, especially if the span lengths vary and $m(x)$ and $EI(x)$ vary within each segment, as would often be the case for a multispan bridge.

Consider the two-member frame shown in Fig. 16.8.1. Each member is axially rigid and has uniform properties—flexural rigidity and mass—as indicated; however,

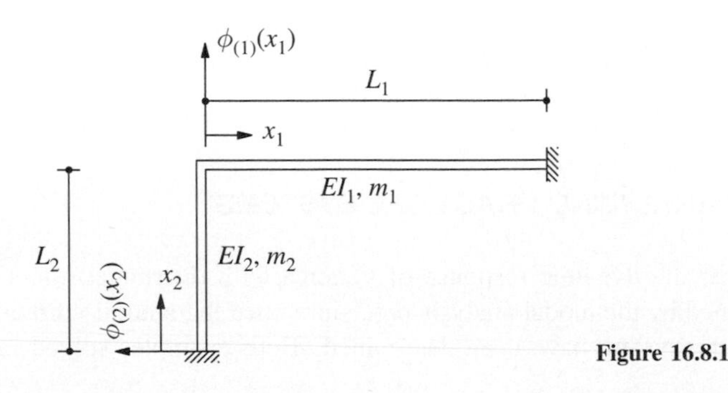

Figure 16.8.1

they may differ from one member to the other. Each member is considered as a separate beam with its origin at one end. Equation (16.3.9) applies to each uniform member, there is one such equation for each member, and the necessary end and joint conditions are:

1. At the supports of the frame the usual boundary conditions are applicable, depending on the type of support, resulting in four equations for the frame of Fig. 16.8.1:

$$\phi_{(1)}(L_1) = 0 \qquad \phi'_{(1)}(L_1) = 0 \qquad \phi_{(2)}(0) = 0 \qquad \phi'_{(2)}(0) = 0$$

2. At the joint the end displacements of the joining members should be compatible; this condition for axially rigid members gives

$$\phi_{(1)}(0) = 0 \qquad \phi_{(2)}(L_2) = 0$$

3. At the joint the end slopes of the joining members should be compatible; thus

$$\phi'_{(1)}(0) = \phi'_{(2)}(L_2)$$

4. At the joint the bending moments should be in equilibrium; thus

$$EI_1\phi''_1(0) + EI_2\phi''_2(L_2) = 0$$

A simple two-member frame requires setting up these eight conditions and the evaluation of eight constants. The process becomes unmanageable for a frame with many members.

It should now be evident that the classical procedure to determine the natural frequencies and modes of a distributed mass system with infinite number of DOF, is not feasible for practical structures. Such problems can be analyzed by discretizing them as systems with a finite number of DOFs, as discussed in the next chapter.

FURTHER READING

Clough, R. W., and Penzien, J., *Dynamics of Structures*, McGraw-Hill, New York, 1993, Chapters 17–19,

Humar, J. L., *Dynamics of Structures*, Prentice Hall, Englewood Cliffs, N.J., 1990, Chapters 14–17.

Stokey, W. F., "Vibration of Systems Having Distributed Mass and Elasticity," Chapter 8 in *Shock and Vibration Handbook* (ed. C. M. Harris), McGraw-Hill, New York, 1988.

Timoshenko, S., Young, D. H., and Weaver, W., Jr., *Vibration Problems in Engineering*, Wiley, New York, 1974.

PROBLEMS

16.1 Find the first three natural vibration frequencies and modes of a uniform beam clamped at both ends. Sketch the modes. Comment on how these frequencies compare with those of a simply supported beam.

16.2 Find the first three natural vibration frequencies and modes of a uniform beam clamped at one end and simply supported at the other. Sketch the modes.

16.3 Find the first five natural vibration frequencies and modes of a uniform beam free at both ends. Sketch the modes. Comment on how these frequencies compare with those for a beam clamped at both ends. (*Hint*: The first two modes are rigid-body modes.)

16.4 A weight W is suspended from the midspan of a simply supported beam as shown in Fig. P16.4. If the wire by which the weight is suspended suddenly snaps, describe the subsequent vibration of the beam. Specialize the general result to obtain the deflection at midspan. Neglect damping.

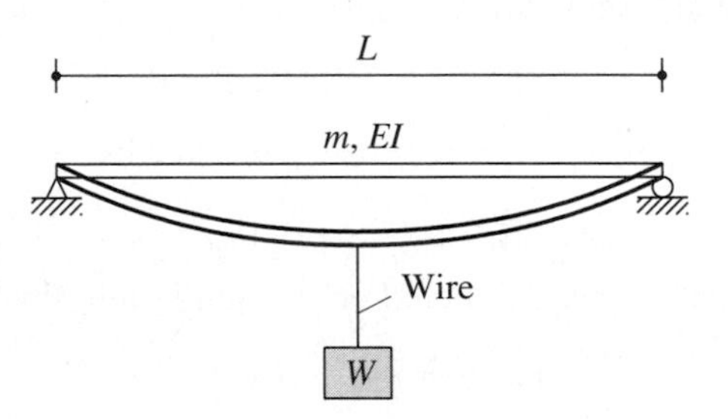

Figure P16.4

16.5 Derive mathematical expressions for the displacement response of a simply supported uniform beam to the force distribution shown in Fig. P16.5; the time variation of the force is a step function. Express the displacements $u(x, t)$ in terms of the natural vibration modes of the beam. Identify the modes that do not contribute to the response. Specialize the general result to obtain the deflection at midspan. Neglect damping.

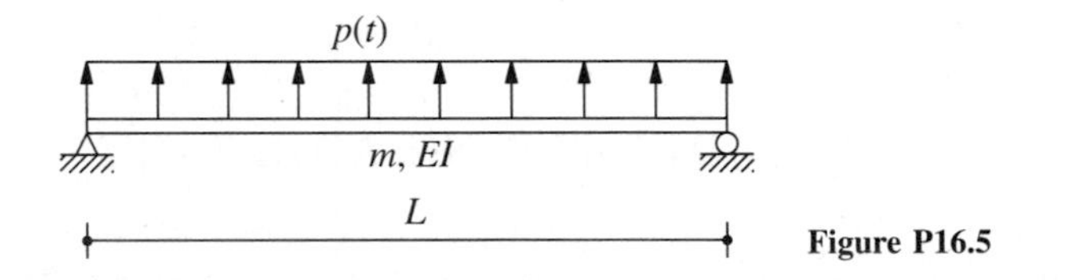

Figure P16.5

16.6 Derive mathematical expressions for the displacement response of a simply supported uniform beam to the force distribution shown in Fig. P16.6; the time variation of the force is a step function. Express the displacements $u(x, t)$ in terms of the natural vibration modes of the beam. Identify the modes that do not contribute to the response. Specialize the general result to obtain the deflection of at quarter span. Neglect damping.

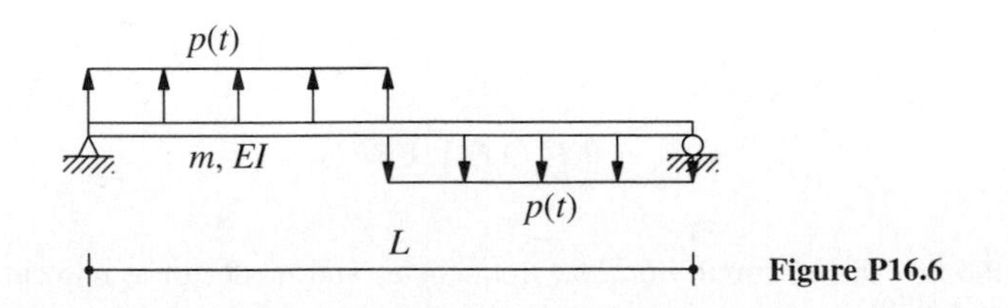

Figure P16.6

16.7 Prove that Eq. (16.6.19) is valid for a vertical cantilever beam.

16.8 Prove that Eq. (16.6.20) is valid for a vertical cantilever beam.

16.9 A free-standing intake–outlet tower 200 ft high has a uniform hollow circular cross section with outside diameter 25 ft and wall thickness 1 ft 3 in. Assume that the tower is clamped at the base and that its mass and flexural stiffness are computed from the gross area of the concrete (neglecting reinforcing steel). The elastic modulus for concrete is 3600 ksi, and its unit weight is 150 lb/ft^3. Modal damping ratios are estimated as 5%. Determine the top displacement, base shear, and base overturning moment due to an earthquake characterized by the design spectrum of Fig. 6.9.5 scaled to the peak ground acceleration of $\frac{1}{3}$g. Neglect shear deformations and rotational inertia.

Introduction to the Finite Element Method

PREVIEW

The classical analysis of distributed mass systems with infinite number of DOFs is not feasible for practical structures, for reasons mentioned in Chapter 16. In this chapter, two methods are presented for discretizing one-dimensional distributed-mass systems: the Rayleigh–Ritz method and the finite element method. As a result, the governing partial differential equation is replaced by a system of ordinary differential equations, as many as the DOFs in the discretized system, which can be solved by the methods presented in the Chapters 10 to 15. The consistent mass matrix concept is introduced and the accuracy and convergence of the approximate natural frequencies of a cantilever beam, determined by the finite element method using consistent or lumped-mass matrices, is demonstrated. The chapter ends with a short discussion on application of the finite element method to the dynamic analysis of structural continua.

PART A: RAYLEIGH–RITZ METHOD

17.1 FORMULATION USING CONSERVATION OF ENERGY

Developed in Chapter 14 for lumped-mass systems, the Rayleigh–Ritz method is also applicable to systems with distributed mass and elasticity. It was for the latter class of systems that the method was originally developed by W. Ritz in 1909. The method, applicable to any one-, two-, or three-dimensional system with distributed mass and elasticity, reduces the system with an infinite number of DOFs to one with a finite number of DOFs. In this section we present the Rayleigh–Ritz method for the transverse vibration of a straight beam.

Consider such a beam with flexural rigidity $EI(x)$ and mass $m(x)$ per unit length, both of which may vary arbitrarily with position x. The deflections $u(x)$ of the system are expressed as a linear combination of several trial functions $\psi_j(x)$:

$$u(x) = \sum_{j=1}^{N} z_j \psi_j(x) = \mathbf{\Psi}(x)\mathbf{z} \tag{17.1.1}$$

where z_j are the generalized coordinates, which vary with time in a dynamic problem, $\mathbf{z}$ is the $N \times 1$ vector of generalized coordinates, and the $1 \times N$ matrix of trial functions is

$$\mathbf{\Psi}(x) = [\psi_1(x) \quad \psi_2(x) \quad \cdots \quad \psi_N(x)]$$

Each trial function $\psi_j(x)$—also known as Ritz function or shape function—must be admissible: that is, continuous and have a continuous first derivative, and satisfy the displacement boundary conditions on the system. All the trial functions must be linearly independent and are selected appropriate for the system to be analyzed.

The starting point for formulation of the Rayleigh–Ritz method is the Rayleigh quotient [Eq. (8.5.11)], derived from the principle of energy conservation, which is rewritten as

$$\rho = \frac{\int_0^L EI(x)\left[u''(x)\right]^2 dx}{\int_0^L m(x)\left[u(x)\right]^2 dx} \tag{17.1.2}$$

After substituting Eq. (17.1.1), Eq. (17.1.2) can be expressed in the form

$$\rho(\mathbf{z}) = \frac{\sum_{i=1}^N \sum_{j=1}^N z_i z_j \tilde{k}_{ij}}{\sum_{i=1}^N \sum_{j=1}^N z_i z_j \tilde{m}_{ij}} \tag{17.1.3}$$

where

$$\tilde{k}_{ij} = \int_0^L EI(x)\psi_i''(x)\psi_j''(x)\,dx \qquad \tilde{m}_{ij} = \int_0^L m(x)\psi_i(x)\psi_j(x)\,dx \tag{17.1.4}$$

Rayleigh's quotient cannot be determined from Eq. (17.1.3) because the N generalized coordinates z_i are unknown. Our objective is to find their values that provide the "best" approximate solution for the natural vibration frequencies and modes of the system.

For this purpose we invoke the property that Rayleigh's quotient is a minimum in the neighborhood of the true modes (or true values of $\mathbf{z}$), which implies that $\partial\rho/\partial z_i = 0$—$i = 1, 2, \ldots, N$. We do not need to go through the details of the derivation because the $\rho(z)$ for distributed mass systems, Eq. (17.1.3), is of the same form as that for discretized systems, Eq. (14.3.6). Thus the minimum condition on Eq. (17.1.3) leads to the eigenvalue problem of Eq. (14.3.11), which is repeated here for convenience:

$$\tilde{\mathbf{k}}\mathbf{z} = \rho\tilde{\mathbf{m}}\mathbf{z} \tag{17.1.5}$$

where $\tilde{\mathbf{k}}$ and $\tilde{\mathbf{m}}$ are square matrices of order N, the number of Ritz functions used to represent the deflections $u(x)$ in Eq. (17.1.1), with their elements given by Eq. (17.1.4). The original eigenvalue problem for distributed mass systems, Eq. (16.3.6), has been reduced to a matrix eigenvalue problem of order N by using Rayleigh's stationarity condition.

The solution of Eq. (17.1.5), obtained by the methods of Chapter 10, yields N eigenvalues $\rho_1, \rho_2, \ldots, \rho_N$ and the corresponding eigenvectors

$$\mathbf{z}_n = \langle z_{1n} \quad z_{2n} \quad \cdots \quad z_{Nn} \rangle^T \qquad n = 1, 2, \ldots, N \tag{17.1.6}$$

The eigenvalues provide $\tilde{\omega}_n = \sqrt{\rho_n}$, which are approximations to the true natural frequencies ω_n of the system; the approximate values are not smaller than the true frequencies, that is,

$$\omega_n \le \tilde{\omega}_n \qquad n = 1,\ 2, \ldots N \tag{17.1.7}$$

The eigenvectors $\mathbf{z}_n$ substituted in Eq. (17.1.1) provide the functions:

$$\tilde{\phi}_n(x) = \sum_{j=1}^{N} z_{jn}\psi_j(x) = \mathbf{\Psi}(x)\mathbf{z}_n \qquad n = 1,\ 2, \ldots, N \tag{17.1.8}$$

which are approximations to the true natural modes $\phi_n(x)$ of the system. The quality of these approximate results is generally better for the lower modes than for higher modes. Therefore, more Ritz functions should be included than the number of modes desired for dynamic response analysis of the system. Although the natural modes $\tilde{\phi}_n(x)$ are approximate, they satisfy the orthogonality properties of Eq. (16.4.6)—as demonstrated for discretized systems (Section 14.3.3)—and can therefore be used in modal analysis of the system as described in Sections 16.5 to 16.7.

Example 17.1

Find approximations for the first two natural frequencies and modes of lateral vibration of a uniform cantilever beam of Fig. E17.1a by the Rayleigh–Ritz method using the shape functions shown in Fig. E17.1b:

$$\psi_1(x) = 1 - \cos\frac{\pi x}{2L} \qquad \psi_2(x) = 1 - \cos\frac{3\pi x}{2L} \tag{a}$$

They satisfy the admissibility requirements, the two geometric boundary conditions at the clamped end, and the zero moment condition at the free end, but not the zero shear condition at the free end.

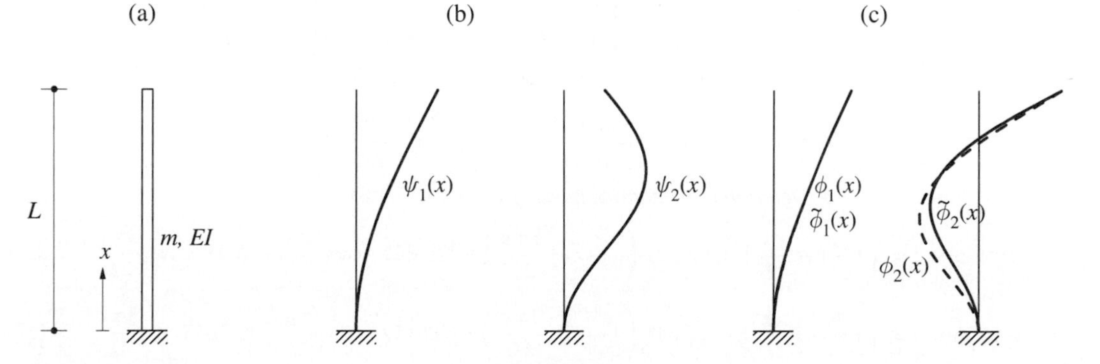

Figure E17.1

Solution

1. Set up $\tilde{\mathbf{k}}$ and $\tilde{\mathbf{m}}$. For the selected Ritz functions of Eq. (a), the stiffness coefficients are computed from Eq. (17.1.4a):

$$\tilde{k}_{11} = \frac{1}{16}\frac{\pi^4 EI}{L^4}\int_0^L \left(\cos\frac{\pi x}{2L}\right)^2 dx = \frac{1}{32}\frac{\pi^4 EI}{L^3} \tag{b1}$$

$$\tilde{k}_{12} = \frac{9}{16}\frac{\pi^4 EI}{L^4}\int_0^L \cos\frac{\pi x}{2L}\cos\frac{3\pi x}{2L}dx = 0 \tag{b2}$$

$$\tilde{k}_{21} = \tilde{k}_{12} = 0 \tag{b3}$$

$$\tilde{k}_{22} = \frac{81}{16}\frac{\pi^4 EI}{L^4}\int_0^L \left(\cos\frac{3\pi x}{2L}\right)^2 dx = \frac{81}{32}\frac{\pi^4 EI}{L^3} \tag{b4}$$

Similarly, the mass coefficients are determined from Eq. (17.1.4b):

$$\tilde{m}_{11} = m\int_0^L \left(1-\cos\frac{\pi x}{2L}\right)^2 dx = 0.2268mL \tag{c1}$$

$$\tilde{m}_{12} = m\int_0^L \left(1-\cos\frac{\pi x}{2L}\right)\left(1-\cos\frac{3\pi x}{2L}\right) dx = 0.5756mL \tag{c2}$$

$$\tilde{m}_{21} = \tilde{m}_{12} = 0.5756mL \tag{c3}$$

$$\tilde{m}_{22} = m\int_0^L \left(1-\cos\frac{3\pi x}{2L}\right)^2 dx = 1.9244mL \tag{c4}$$

Substituting Eqs. (b) and (c) in Eq. (17.1.5) gives

$$\frac{\pi^4 EI}{32L^3}\begin{bmatrix}1 & 0\\ 0 & 81\end{bmatrix}\begin{Bmatrix}z_1\\ z_2\end{Bmatrix} = \overline{\rho}\begin{bmatrix}0.2268 & 0.5756\\ 0.5756 & 1.9244\end{bmatrix}\begin{Bmatrix}z_1\\ z_2\end{Bmatrix} \tag{d}$$

where

$$\overline{\rho} = \frac{32mL^4}{\pi^4 EI}\rho \tag{e}$$

2. *Solve the reduced eigenvalue problem.*

$$\overline{\rho}_1 = 4.0775 \qquad \overline{\rho}_2 = 188.87$$

$$\mathbf{z}_1 = \begin{bmatrix}z_{11}\\ z_{21}\end{bmatrix} = \begin{bmatrix}1\\ 0.0321\end{bmatrix} \qquad \mathbf{z}_2 = \begin{bmatrix}z_{12}\\ z_{22}\end{bmatrix} = \begin{bmatrix}1\\ -0.3848\end{bmatrix}$$

3. *Determine the natural frequencies from Eq. (e).*

$$\tilde{\omega}_1 = \frac{3.523}{L^2}\sqrt{\frac{EI}{m}} \qquad \tilde{\omega}_2 = \frac{23.978}{L^2}\sqrt{\frac{EI}{m}} \tag{f}$$

4. *Determine the natural modes from Eq.* (17.1.8).

$$\tilde{\phi}_1(x) = \left(1-\cos\frac{\pi x}{2L}\right) + 0.0321\left(1-\cos\frac{3\pi x}{2L}\right) = 1.0321 - \cos\frac{\pi x}{2L} - 0.0321\cos\frac{3\pi x}{2L} \tag{g1}$$

$$\tilde{\phi}_2(x) = \left(1-\cos\frac{\pi x}{2L}\right) - 0.3848\left(1-\cos\frac{3\pi x}{2L}\right) = 0.6152 - \cos\frac{\pi x}{2L} + 0.3848\cos\frac{3\pi x}{2L} \tag{g2}$$

These approximate modes are plotted in Fig. E17.1c.

5. *Compare with the exact solution.* The exact values for natural frequencies and modes of a cantilever beam were determined in Section 16.3.2. The exact frequencies are

$$\omega_1 = \frac{3.516}{L^2}\sqrt{\frac{EI}{m}} \qquad \omega_2 = \frac{22.03}{L^2}\sqrt{\frac{EI}{m}} \tag{h}$$

As expected, both approximate frequencies are higher than the corresponding exact values, and the error is larger in the second frequency. The exact modes are also plotted in Fig. E17.1c. It is clear that the approximate solution is excellent for the first mode but not as good for the second mode.

The approximate value for the fundamental frequency in Eq. (h) is lower and hence better than the result obtained in Example 8.2 by Rayleigh's method with $\psi_1(x)$ as the only trial function.

17.2 FORMULATION USING VIRTUAL WORK

In this section the equation governing the transverse vibration of a straight beam due to external forces will be formulated using the principle of virtual displacements. At each time instant the system is in equilibrium under the action of the external forces $p(x, t)$, internal resisting bending moments $\mathcal{M}(x, t)$, and the fictitious inertia forces, which by D'Alembert's principle are

$$f_I(x, t) = -m(x)\ddot{u}(x, t) \tag{17.2.1}$$

If the system in equilibrium is subjected to virtual displacements $\delta u(x)$, the external virtual work δW_E is equal to the internal virtual work δW_I:

$$\delta W_I = \delta W_E \tag{17.2.2}$$

Based on the development of Section 8.3.2, these work quantities are

$$\delta W_E = -\int_0^L m(x)\ddot{u}(x, t)\delta u(x)\, dx + \int_0^L p(x, t)\delta u(x)\, dx \tag{17.2.3a}$$

$$\delta W_I = \int_0^L EI(x)u''(x, t)\delta[u''(x)]\, dx \tag{17.2.3b}$$

The displacements $u(x, t)$ are given by Eq. (17.1.1) and $\delta u(x)$ is any admissible virtual displacement:

$$\delta u(x) = \psi_i(x)\,\delta z_i \qquad i = 1, 2, \ldots, N \tag{17.2.4}$$

Substituting Eqs. (17.1.1) and (17.2.4) in Eq. (17.2.3) leads to

$$\delta W_E = -\delta z_i \sum_{j=1}^{N} \ddot{z}_j \tilde{m}_{ij} + \delta z_i \tilde{p}_i(t) \tag{17.2.5a}$$

$$\delta W_I = \delta z_i \sum_{j=1}^{N} z_j \tilde{k}_{1j} \tag{17.2.5b}$$

where

$$\tilde{m}_{ij} = \int_0^L m(x)\psi_i(x)\psi_j(x)\,dx$$

$$\tilde{k}_{ij} = \int_0^L EI(x)\psi_i''(x)\psi_j''(x)\,dx \tag{17.2.6}$$

$$\tilde{p}_i(t) = \int_0^L p(x,t)\psi_i(x)\,dx$$

Substituting Eq. (17.2.5) in Eq. (17.2.2) gives

$$\delta z_i \left(\sum_{j=1}^{N} \ddot{z}_j \tilde{m}_{ij} + \sum_{j=1}^{N} z_j \tilde{k}_{ij} \right) = \delta z_i \tilde{p}_i(t) \tag{17.2.7}$$

and δz_i can be dropped from both sides because this equation is valid for any δz_i.

Corresponding to the N independent virtual displacements of Eq. (17.2.4), there are N equations like Eq. (17.2.7). Together they can be expressed in matrix notation:

$$\tilde{\mathbf{m}}\ddot{\mathbf{z}} + \tilde{\mathbf{k}}\mathbf{z} = \tilde{\mathbf{p}}(t) \tag{17.2.8}$$

where $\mathbf{z}$ is the vector of N generalized coordinates, $\tilde{\mathbf{m}}$ the generalized mass matrix with its elements defined by Eq. (17.2.6a), $\tilde{\mathbf{k}}$ the generalized stiffness matrix whose elements are given by Eq. (17.2.6b), and $\tilde{\mathbf{p}}(t)$ the generalized applied force vector with its elements defined by Eq. (17.2.6c). It is obvious from Eq. (17.2.6) that $\tilde{\mathbf{m}}$ and $\tilde{\mathbf{k}}$ are symmetric matrices. A damping matrix can also be included in the virtual work formulation if the damping mechanisms can be defined. The system of coupled differential equations (17.2.8) can be solved for the unknowns $z_j(t)$ using the numerical procedures presented in Chapter 15. Then at each time instant the displacement $u(x)$ is determined from Eq. (17.1.1). This is an alternative to the classical modal analysis of the system mentioned at the end of Section 17.1.

The mass and stiffness matrices obtained using the principle of virtual displacements are identical to those derived in Section 17.1 from the principle of energy conservation, the concept underlying the original development of the Rayleigh–Ritz method. We will draw upon the virtual work approach when we introduce the finite element method in Part B of this chapter.

17.3 DISADVANTAGES OF RAYLEIGH–RITZ METHOD

The Rayleigh–Ritz method leads to natural frequencies $\tilde{\omega}_n$ and natural modes $\tilde{\phi}_n(x)$ that approximate the true values ω_n and $\phi_n(x)$, respectively, best among the admissible class of functions described by Eq. (17.1.1). However, the method is not practical for general application and automated computer implementation to analyze complex structures for several reasons: (1) It is difficult to select the Ritz trial functions because they should be suitable for the particular system and its boundary conditions. (2) It is not clear how to select additional functions to improve the accuracy of an approximate solution obtained using fewer functions. (3) It may be difficult to evaluate the integrals of Eq. (17.2.6) over the entire structure, especially with higher-order trial functions. (4) It is computationally expensive to work with the matrices $\tilde{\mathbf{m}}$ and $\tilde{\mathbf{k}}$ because they are full matrices. (5) It is difficult to interpret the generalized coordinates, as they do not necessarily represent displacements at physical locations on the structure. These difficulties are overcome by the finite element method introduced next.

PART B: FINITE ELEMENT METHOD

The finite element method is one of the most important developments in applied mechanics. Although the method is applicable to a wide range of problems, only an introduction is included here with reference to systems that can be idealized as an assemblage of one-dimensional finite elements. The presentation, although self-contained, is based on the presumption that the reader is familiar with the finite element method for analysis of static problems.

17.4 FINITE ELEMENT APPROXIMATION

In the finite element method the trial functions are selected in a special way to overcome the aforementioned difficulties of the Rayleigh–Ritz method. To illustrate this concept, consider the cantilever beam shown in Fig. 17.4.1, which is subdivided into a number of segments, called *finite elements*. Their size is arbitrary; they may be all of the same size or all different. The elements are interconnected only at *nodes* or *nodal points*. In this simple case the nodal points are the ends of the element, and each node has two DOFs, transverse displacement and rotation. In the finite element method nodal displacements are selected as the generalized coordinates, and the equations of motion are formulated in terms of these physically meaningful displacements.

The deflection of the beam is expressed in terms of the nodal displacements through trial functions $\hat{\psi}_i(x)$ shown in Fig. 17.4.1. Corresponding to each DOF, a trial function is selected with the following properties: unit value at the DOF; zero value at all other DOFs; continuous function with continuous first derivative. No trial functions are shown for the node at the clamped end because the displacement and slope are both zero. These trial functions satisfy the requirements of admissibility because they are linearly independent, continuous with continuous first derivative, and consistent with the geometric boundary conditions. The deflection of the beam is expressed as

$$u(x) = \sum_i u_i \hat{\psi}_i(x) \tag{17.4.1}$$

where u_i is the nodal displacement in the ith DOF and $\hat{\psi}_i(x)$ is the associated trial function. Because these functions define the displacements between nodal points (in contrast to global displacements of the structure in the Rayleigh–Ritz method), they are called *interpolation functions*.

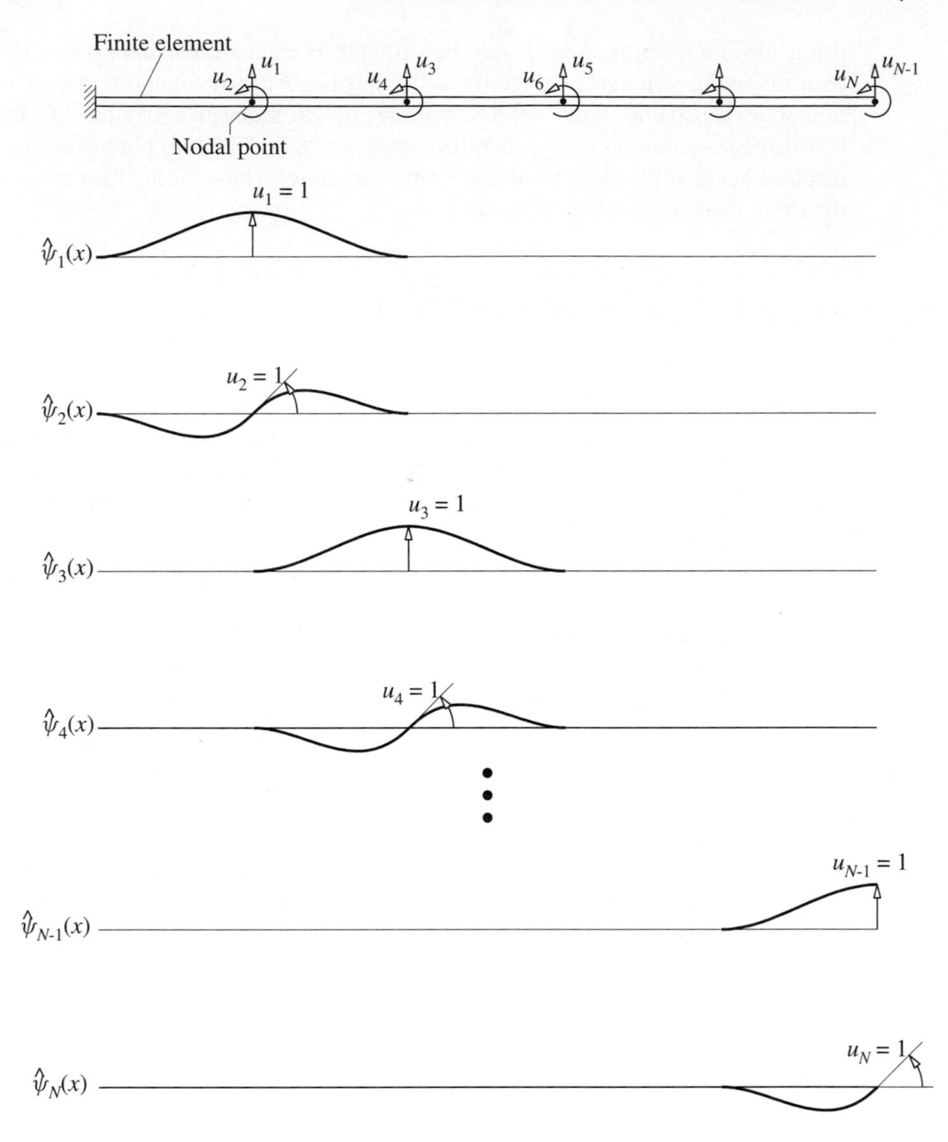

Figure 17.4.1

The finite element method offers several important advantages over the Rayleigh–Ritz method. Stated in the same sequence as the disadvantages of the Rayleigh–Ritz method mentioned at the end of Section 17.3, the advantages of the finite element method are: (1) Simple interpolation functions can be chosen for each finite element. (2) Accuracy of the solution can be improved by increasing the number of finite elements in the structural idealization. (3) Computation of the integrals of Eq. (17.2.6) is much easier because the interpolation functions are simple, and the same functions may be chosen for each finite element. (4) The structural stiffness and mass matrices developed by the finite element method are narrowly banded, a property that reduces the computational effort necessary to solve the equations of motion. (5) The generalized displacements are physically meaningful, as they give the nodal displacements directly.

17.5 ANALYSIS PROCEDURE

The formulation of the equations of motion for a structure by the finite element method may be summarized as a sequence of the following steps:

1. Idealize the structure as an assemblage of finite elements interconnected only at nodes (Fig. 17.5.1a); define the DOF $\mathbf{u}$ at these nodes (Fig. 17.5.1b).
2. For each finite element form the element stiffness matrix $\mathbf{k}_e$, the element mass matrix $\mathbf{m}_e$, and the element (applied) force vector $\mathbf{p}_e(t)$ with reference to the DOF for the element (Fig. 17.5.1c). For each element the force–displacement relation and the inertia force–acceleration relation are

$$(\mathbf{f}_S)_e = \mathbf{k}_e\mathbf{u}_e \qquad (\mathbf{f}_I)_e = \mathbf{m}_e\ddot{\mathbf{u}}_e \tag{17.5.1}$$

   In the finite element formulation these relations are obtained by assuming the displacement field over the element, expressed in terms of nodal displacements.
3. Form the transformation matrix $\mathbf{a}_e$ that relates the displacements $\mathbf{u}_e$ and forces $\mathbf{p}_e$ for the element to the displacements $\mathbf{u}$ and forces $\mathbf{p}$ for the finite element assemblage:

$$\mathbf{u}_e = \mathbf{a}_e\mathbf{u} \qquad \mathbf{p}(t) = \mathbf{a}_e^T\mathbf{p}_e(t) \tag{17.5.2}$$

   where $\mathbf{a}_e$ is a Boolean matrix consisting of zeros and ones. It simply locates the elements of $\mathbf{k}_e$, $\mathbf{m}_e$, and $\mathbf{p}_e$ at the proper locations in the mass matrix, stiffness matrix, and (applied) force vector for the finite element assemblage. Therefore, it is not necessary to carry out the transformations: $\hat{\mathbf{k}}_e = \mathbf{a}_e^T\mathbf{k}_e\mathbf{a}_e$, $\hat{\mathbf{m}}_e = \mathbf{a}_e^T\mathbf{m}_e\mathbf{a}_e$, or

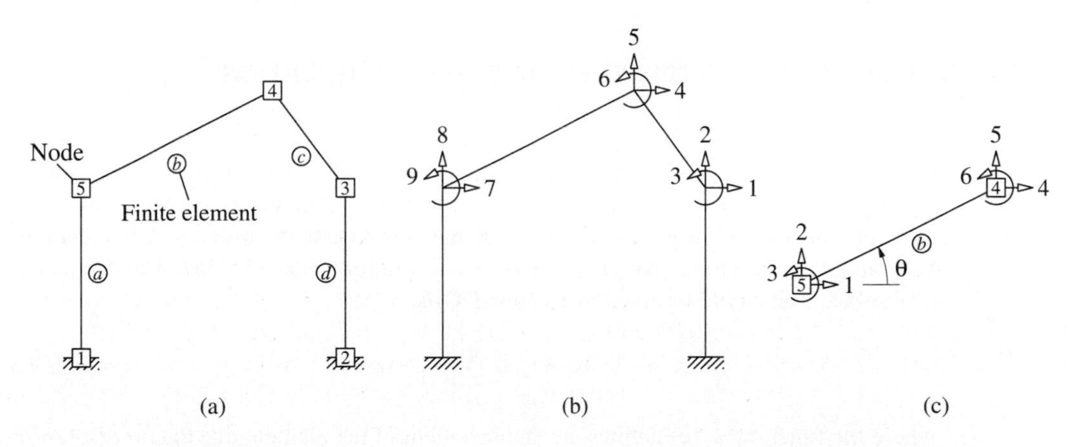

Figure 17.5.1 (a) Finite elements and nodes; (b) assemblage DOF $\mathbf{u}$; (c) element DOF $\mathbf{u}_e$.

$\hat{\mathbf{p}}_e(t) = \mathbf{a}_e^T \mathbf{p}_e(t)$ to transform the element stiffness and mass matrices and applied force vector to the nodal displacements for the assemblage.

4. Assemble the element matrices to determine the stiffness and mass matrices and the applied force vector for the assemblage of finite elements:

$$\mathbf{k} = \mathcal{A}_{e=1}^{N_e} \mathbf{k}_e \qquad \mathbf{m} = \mathcal{A}_{e=1}^{N_e} \mathbf{m}_e \qquad \mathbf{p}(t) = \mathcal{A}_{e=1}^{N_e} \mathbf{p}_e(t) \tag{17.5.3}$$

The operator $\mathcal{A}$ denotes the *direct assembly procedure* for assembling according to the matrix $\mathbf{a}_e$, the element stiffness matrix, element mass matrix, and the element force vector—for each element $e = 1$ to N_e, where N_e is the number of elements—into the assemblage stiffness matrix, assemblage mass matrix, and assemblage force vector, respectively.

5. Formulate the equations of motion for the finite element assemblage:

$$\mathbf{m}\ddot{\mathbf{u}} + \mathbf{c}\dot{\mathbf{u}} + \mathbf{k}\mathbf{u} = \mathbf{p}(t) \tag{17.5.4}$$

where the damping matrix $\mathbf{c}$ is established by the methods of Chapter 11.

The governing equations (17.5.4) for a finite element system are of the same form as formulated in Chapter 9 for frame structures. It should be clear from the outline above that the only difference between the displacement method for analysis of frame structures and the finite element method is in the formulation of the element mass and stiffness matrices. Therefore, Eq. (17.5.4) can be solved for $\mathbf{u}(t)$ by the methods developed in preceding chapters. The classical modal analysis procedure of Chapters 12 and 13 is applicable if the system has classical damping and the direct methods of Chapter 15 enable analysis of nonclassically damped systems.

In the subsequent sections, which are restricted to assemblages of one-dimensional finite elements, we define the interpolation functions and develop the element stiffness matrix $\mathbf{k}_e$, the element mass matrix $\mathbf{m}_e$, and the element (applied) force vector $\mathbf{p}_e(t)$. Assembly of these element matrices to construct the corresponding matrices for the finite element assemblage is illustrated by an example.

17.6 ELEMENT DEGREES OF FREEDOM AND INTERPOLATION FUNCTIONS

Consider a straight-beam element of length L, mass per unit length $m(x)$, and flexural rigidity $EI(x)$. The two nodes by which the finite element can be assembled into a structure are located at its ends. If only planar displacements are considered, each node has two DOFs: the transverse displacement and rotation (Fig. 17.6.1a). The displacement of the beam element is related to its four DOFs:

$$u(x,t) = \sum_{i=1}^{4} u_i(t)\psi_i(x) \tag{17.6.1}$$

where the function $\psi_i(x)$ defines the displacement of the element due to unit displacement u_i while constraining other DOFs to zero. Thus $\psi_i(x)$ satisfies the following boundary

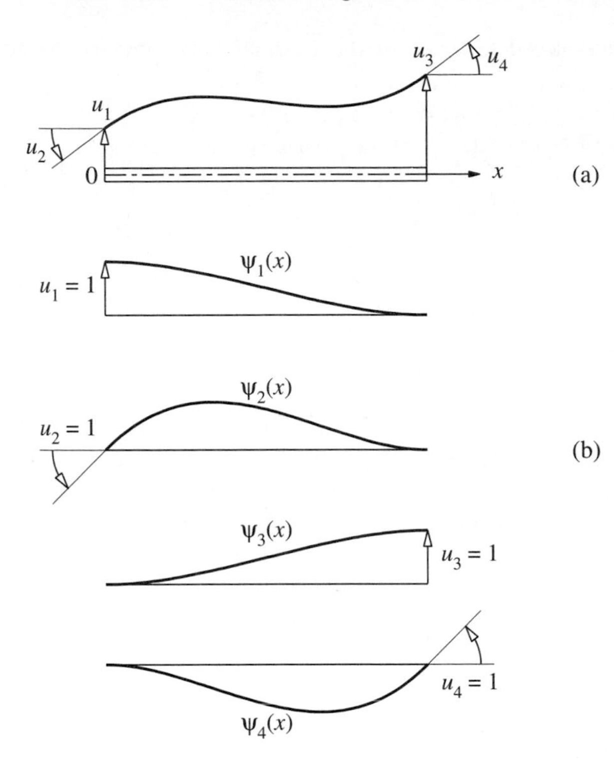

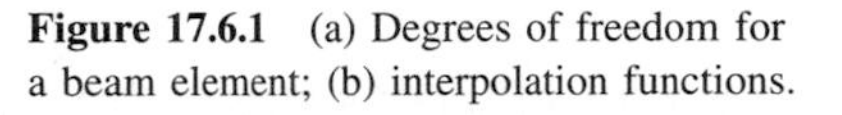

Figure 17.6.1 (a) Degrees of freedom for a beam element; (b) interpolation functions.

conditions:

$$i = 1: \quad \psi_1(0) = 1, \;\; \psi_1'(0) = \psi_1(L) = \psi_1'(L) = 0 \tag{17.6.2a}$$

$$i = 2: \quad \psi_2'(0) = 1, \;\; \psi_2(0) = \psi_2(L) = \psi_2'(L) = 0 \tag{17.6.2b}$$

$$i = 3: \quad \psi_3(L) = 1, \;\; \psi_3(0) = \psi_3'(0) = \psi_3'(L) = 0 \tag{17.6.2c}$$

$$i = 4: \quad \psi_4'(L) = 1, \;\; \psi_4(0) = \psi_4'(0) = \psi_4(L) = 0 \tag{17.6.2d}$$

These interpolation functions could be any arbitrary shapes satisfying the boundary conditions. One possibility is the exact deflected shapes of the beam element due to the imposed boundary conditions, but these are difficult to determine if the flexural rigidity varies over the length of the element. However, they can conveniently be obtained for a uniform beam as illustrated next. Neglecting shear deformations, the equilibrium equation for a beam loaded only at its ends is

$$EI\frac{d^4u}{dx^4} = 0 \tag{17.6.3}$$

The general solution of Eq. (17.6.3) for a uniform beam is a cubic polynomial

$$u(x) = a_1 + a_2\left(\frac{x}{L}\right) + a_3\left(\frac{x}{L}\right)^2 + a_4\left(\frac{x}{L}\right)^3 \tag{17.6.4}$$

The constants a_i can be determined for each of the four sets of boundary conditions, Eq. (17.6.2), to obtain

$$\psi_1(x) = 1 - 3\left(\frac{x}{L}\right)^2 + 2\left(\frac{x}{L}\right)^3 \tag{17.6.5a}$$

$$\psi_2(x) = L\left(\frac{x}{L}\right) - 2L\left(\frac{x}{L}\right)^2 + L\left(\frac{x}{L}\right)^3 \tag{17.6.5b}$$

$$\psi_3(x) = 3\left(\frac{x}{L}\right)^2 - 2\left(\frac{x}{L}\right)^3 \tag{17.6.5c}$$

$$\psi_4(x) = -L\left(\frac{x}{L}\right)^2 + L\left(\frac{x}{L}\right)^3 \tag{17.6.5d}$$

These interpolation functions, illustrated in Fig. 17.6.1b, can be used in formulating the element matrices for nonuniform elements.

The foregoing approach is possible for a beam finite element—because the governing differential equation (17.6.3) could be solved for a uniform beam—but not for two- or three-dimensional finite elements. Therefore, the finite element method is based on assumed relationships between the displacements at interior points of the element and the displacements at the nodes. Proceeding in this manner makes the problem tractable but introduces approximations in the solution.

17.7 ELEMENT STIFFNESS MATRIX

Consider a beam element of length L with flexural rigidity $EI(x)$. By definition, the stiffness influence coefficient k_{ij} of the beam element is the force in DOF i due to unit displacement in DOF j. Using the principle of virtual displacement in a manner similar to Section 17.2, we can derive a general equation for k_{ij}:

$$k_{ij} = \int_0^L EI(x)\psi_i''(x)\psi_j''(x)\,dx \tag{17.7.1}$$

The symmetric form of this equation shows that the stiffness matrix is symmetric; $k_{ij} = k_{ji}$. Observe that this result for an element has the same form as Eq. (17.2.6) for the structure. Equation (17.7.1) is a general result in the sense that it is applicable to elements with arbitrary variation of flexural rigidity $EI(x)$, although the interpolation functions of Eq. (17.6.5) are exact only for uniform elements. The associated errors can be reduced to any desired degree by reducing the element size and increasing the number of finite elements in the structural idealization.

For a uniform finite element with $EI(x) = EI$, the integral of Eq. (17.7.1) can be evaluated analytically, resulting in the element stiffness matrix:

$$\bar{\mathbf{k}}_e = \frac{EI}{L^3}\begin{bmatrix} 12 & 6L & -12 & 6L \\ 6L & 4L^2 & -6L & 2L^2 \\ -12 & -6L & 12 & -6L \\ 6L & 2L^2 & -6L & 4L^2 \end{bmatrix} \tag{17.7.2}$$

These stiffness coefficients are the exact values for a uniform beam, neglecting shear deformation, because the interpolation functions of Eq. (17.6.5) are the true deflection shapes for this case. Observe that the stiffness matrix of Eq. (17.7.2) is equivalent to the force–displacement relations for a uniform beam that are familiar from classical structural analysis (see Chapter 1, Appendix 1). For nonuniform elements, such as a haunched beam, approximate values for the stiffness coefficients can be determined by numerically evaluating Eq. (17.7.1).

The 4×4 element stiffness matrix $\overline{\mathbf{k}}_e$ of Eq. (17.7.2) in local element coordinates (Fig. 17.6.1a) is transformed to the 6×6 $\mathbf{k}_e$ of Eq. (17.5.1) in global element coordinates (Fig. 17.5.1c). Before carrying out this transformation, $\overline{\mathbf{k}}_e$ is expanded to a 6×6 matrix that includes the stiffness coefficients associated with the axial DOF at each node. The transformation matrix that depends on the orientation θ of the member (Fig. 17.5.1c) should be familiar to the reader.

17.8 ELEMENT MASS MATRIX

As defined in Section 9.2.4, the mass influence coefficient m_{ij} for a structure is the force in the ith DOF due to unit acceleration in the jth DOF. Applying this definition to a beam element with distributed mass $m(x)$ and using the principle of virtual displacement along the lines of Section 17.2, a general equation for m_{ij} can be derived:

$$m_{ij} = \int_0^L m(x)\psi_i(x)\psi_j(x)\,dx \tag{17.8.1}$$

The symmetric form of this equation shows that the mass matrix is symmetric; $m_{ij} = m_{ji}$. Observe that the result for an element has the same form as Eq. (17.2.6) for the structure.

If we use the same interpolation functions in Eq. (17.8.1) as were used to derive the element stiffness matrix, the result obtained is known as the *consistent mass matrix*. The integrals of Eq. (17.8.1) are evaluated numerically or analytically depending on the function $m(x)$. For an element with uniform mass [i.e., $m(x) = m$], the integrals can be evaluated analytically to obtain the element (consistent) mass matrix:

$$\overline{\mathbf{m}}_e = \frac{mL}{420}\begin{bmatrix} 156 & 22L & 54 & -13L \\ 22L & 4L^2 & 13L & -3L^2 \\ 54 & 13L & 156 & -22L \\ -13L & -3L^2 & -22L & 4L^2 \end{bmatrix} \tag{17.8.2}$$

Observe that the consistent-mass matrix is not diagonal, whereas the lumped-mass approximation leads to a diagonal matrix, as we shall see next.

The mass matrix of a finite element can be simplified by assuming that the distributed mass of the element can be lumped as point masses along the translational DOF u_1 and u_3 at the ends (Fig. 17.6.1), with the two masses being determined by static analysis of the beam under its own mass. For example, if the mass of a uniform element is m per unit length, a point mass of $mL/2$ will be assigned to each end, leading to

$$\overline{\mathbf{m}}_e = mL \begin{bmatrix} \frac{1}{2} & 0 & 0 & 0 \\ 0 & 0 & 0 & 0 \\ 0 & 0 & \frac{1}{2} & 0 \\ 0 & 0 & 0 & 0 \end{bmatrix} \tag{17.8.3}$$

Observe that for a lumped-mass idealization of the finite element, the mass matrix is diagonal. The off-diagonal terms m_{ij} of this matrix are zero because an acceleration of any point mass produces an inertia force only in the same DOF. The diagonal terms m_{ii} associated with the rotational degrees of freedom are zero because of the idealization that the mass is lumped in points that have no rotational inertia.

The 4×4 element mass matrix $\overline{\mathbf{m}}_e$ given by Eq. (17.8.2) or (17.8.3) in local element coordinates (Fig. 17.6.1a) is transformed to the 6×6 $\mathbf{m}_e$ of Eq. (17.5.1) in global element coordinates (Fig. 17.5.1c). The procedure and transformation matrix are the same as described earlier for the stiffness matrix.

The dynamic analysis of a consistent-mass system requires considerably more computational effort than does a lumped-mass idealization, for two reasons: (1) The lumped-mass matrix is diagonal, whereas the consistent-mass matrix has off-diagonal terms; and (2) the rotational DOF can be eliminated by static condensation (see Section 9.3) from the equations of motion for a lumped-mass system, whereas all DOFs must be retained in a consistent-mass system.

However, the consistent-mass formulation has two advantages. First, it leads to greater accuracy in the results and rapid convergence to the exact results with an increasing number of finite elements, as we shall see later by an example, but in practice the improvement is often only slight because the inertia forces associated with node rotations are generally not significant in many structural earthquake engineering problems. Second, with a consistent-mass approach, the potential energy and kinetic energy quantities are evaluated in a consistent manner, and therefore we know how the computed values of the natural frequencies relate to the exact values (see Part A of this chapter). The second advantage seldom outweighs the additional computational effort required to achieve a slight increase in accuracy, and therefore the lumped-mass idealization is widely used.

17.9 ELEMENT (APPLIED) FORCE VECTOR

If the external forces $p_i(t)$, $i = 1, 2, 3$ and 4, are applied along the four DOFs at the two nodes of the finite element, the element force vector can be written directly. On the other hand, if the external forces include distributed force $p(x, t)$ and concentrated forces $p'_j(t)$ at locations x_j, the nodal force in the ith DOF is

$$p_i(t) = \int_0^L p(x,t)\psi_i(x)\,dx + \sum_j p'_j \psi_i(x_j) \tag{17.9.1}$$

This equation can be obtained by the principle of virtual displacements, following the derivation of the similar equation (17.2.6c) for the complete structure. If we use the same interpolation functions in Eq. (17.9.1) as were used to derive the element stiffness matrix, the results obtained are called *consistent nodal forces.*

A simpler, less accurate approach is to use linear interpolation functions:

$$\psi_1(x) = 1 - \frac{x}{L} \qquad \psi_3(x) = \frac{x}{L} \tag{17.9.2}$$

Then Eq. (17.9.1) gives the nodal forces $p_1(t)$ and $p_3(t)$ in the translational DOF; the forces $p_2(t)$ and $p_4(t)$ in the rotational DOF are zero unless external moments are applied directly to the nodes.

The 4×1 element force vector $\overline{\mathbf{p}}_e$ given by Eq. (17.9.1) in local element coordinates is transformed to the 6×1 $\mathbf{p}_e$ of Eq. (17.5.2) in global element coordinates. The transformation matrix is the transpose of the one described earlier for the stiffness matrix.

Example 17.2

Determine the natural frequencies and modes of vibration of a uniform cantilever beam, idealized as an assemblage of two finite elements (Fig. E17.2a) using the consistent mass matrix. The flexural rigidity is EI and mass per unit length is m.

Solution

1. *Identify the assemblage and element DOFs.* The six DOFs for the finite element assemblage are shown in Fig. E17.2b, and the two finite elements and their local DOFs in Fig. E17.2c.

2. *Form the element stiffness matrices.* Replacing L in Eq. (17.7.2) by $L/2$, the length of each finite element in Fig. E17.2a, gives the stiffness matrices $\overline{\mathbf{k}}_1$ and $\overline{\mathbf{k}}_2$ for the two finite elements in their local DOFs. Since both local element DOFs and assemblage DOFs are defined along the same set of Cartesian coordinates, no transformation of coordinates is

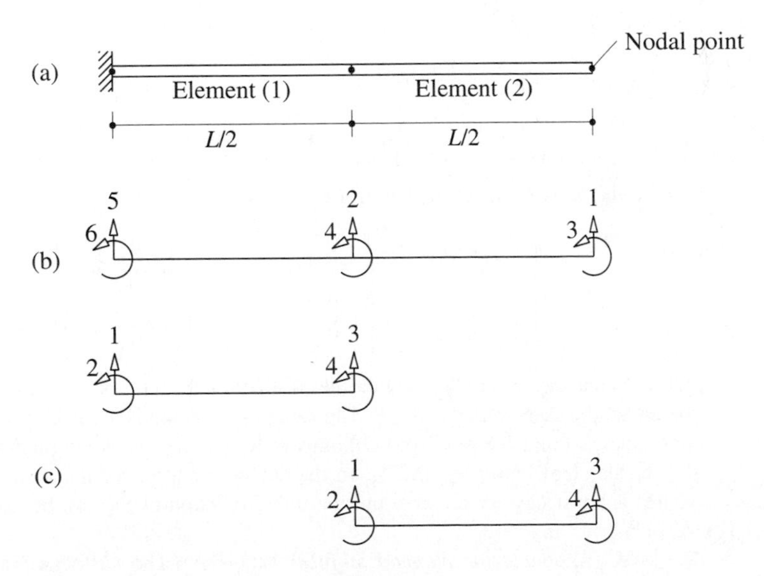

Figure E17.2

required. Therefore, $\mathbf{k}_1 = \overline{\mathbf{k}}_1$ and $\mathbf{k}_2 = \overline{\mathbf{k}}_2$; thus

$$\mathbf{k}_1 = \frac{8EI}{L^3} \begin{array}{c} \begin{matrix} (5) & (6) & (2) & (4) \end{matrix} \\ \begin{bmatrix} 12 & 3L & -12 & 3L \\ 3L & L^2 & -3L & L^2/2 \\ -12 & -3L & 12 & -3L \\ 3L & L^2/2 & -3L & L^2 \end{bmatrix} \end{array} \begin{matrix} (5) \\ (6) \\ (2) \\ (4) \end{matrix}$$

$$\mathbf{k}_2 = \frac{8EI}{L^3} \begin{array}{c} \begin{matrix} (2) & (4) & (1) & (3) \end{matrix} \\ \begin{bmatrix} 12 & 3L & -12 & 3L \\ 3L & L^2 & -3L & L^2/2 \\ -12 & -3L & 12 & -3L \\ 3L & L^2/2 & -3L & L^2 \end{bmatrix} \end{array} \begin{matrix} (2) \\ (4) \\ (1) \\ (3) \end{matrix}$$

The numbers in parentheses alongside rows and columns of $\mathbf{k}_e$ refer to the assemblage DOFs that correspond to the element DOFs. This information enables $\mathbf{k}_e$ to be assembled.

3. *Transform the element stiffness matrices to assemblage DOFs.* This step is not required for solving this problem but is included to assist in better understanding of the procedure. The element stiffness matrix with reference to the nodal displacements of the finite element assemblage is given by

$$\hat{\mathbf{k}}_e = \mathbf{a}_e^T \mathbf{k}_e \mathbf{a}_e \tag{a}$$

The nodal displacements of elements (1) and (2) are related to the assemblage displacements by

$$(u_1)_1 = u_5 \quad (u_2)_1 = u_6 \quad (u_3)_1 = u_2 \quad (u_4)_1 = u_4$$

$$(u_1)_2 = u_2 \quad (u_2)_2 = u_4, \quad (u_3)_2 = u_1 \quad (u_4)_2 = u_3$$

These relationships for the two elements can be expressed as

$$\mathbf{u}_1 = \mathbf{a}_1\mathbf{u} \qquad \mathbf{u}_2 = \mathbf{a}_2\mathbf{u} \tag{b}$$

where the transformation matrices are

$$\mathbf{a}_1 = \begin{bmatrix} 0 & 0 & 0 & 0 & 1 & 0 \\ 0 & 0 & 0 & 0 & 0 & 1 \\ 0 & 1 & 0 & 0 & 0 & 0 \\ 0 & 0 & 0 & 1 & 0 & 0 \end{bmatrix} \qquad \mathbf{a}_2 = \begin{bmatrix} 0 & 1 & 0 & 0 & 0 & 0 \\ 0 & 0 & 0 & 1 & 0 & 0 \\ 1 & 0 & 0 & 0 & 0 & 0 \\ 0 & 0 & 1 & 0 & 0 & 0 \end{bmatrix}$$

Observe that $a_{ij} = 1$ indicates that the element DOF i corresponds to (and becomes renumbered as) the assemblage DOF j. The same information is available through the numbers in parentheses alongside rows and columns of $\mathbf{k}_1$ and $\mathbf{k}_2$. Thus the matrices $\mathbf{a}_1$ and $\mathbf{a}_2$ simply locate the elements of $\mathbf{k}_1$ and $\mathbf{k}_2$ in the stiffness matrix for the assemblage. This implies that it is not necessary to carry out the transformation of Eq. (a), because $\mathbf{a}_1$ and $\mathbf{a}_2$ consist of only ones and zeros.

4. *Assemble the element stiffness matrices.* The stiffness matrix $\mathbf{k}$ for the finite element system is determined by assembling $\mathbf{k}_1$ and $\mathbf{k}_2$ by locating the elements of $\mathbf{k}_1$ and

$\mathbf{k}_2$ in $\mathbf{k}$ according to $\mathbf{a}_1$ and $\mathbf{a}_2$, respectively.

$$\mathbf{k} = \mathcal{A}_{i=1}^{2}\mathbf{k}_i$$

$$= \frac{8EI}{L^3}\left[\begin{array}{cccc|cc} 12 & -12 & -3L & -3L & 0 & 0 \\ -12 & 24 & 3L & 0 & -12 & -3L \\ -3L & 3L & L^2 & L^2/2 & 0 & 0 \\ -3L & 0 & L^2/2 & 2L^2 & 3L & L^2/2 \\ \hline 0 & -12 & 0 & 3L & 12 & 3L \\ 0 & -3L & 0 & L^2/2 & 3L & L^2 \end{array}\right] \Big\} \text{ support DOFs}$$

support DOFs

5. *Form the element mass matrices.* Replacing L in Eq. (17.8.2) by $L/2$ gives the element mass matrices $\overline{\mathbf{m}}_1$ and $\overline{\mathbf{m}}_2$ in their local DOF; and, as for the stiffness matrices, $\mathbf{m}_1 = \overline{\mathbf{m}}_1$ and $\mathbf{m}_2 = \overline{\mathbf{m}}_2$. Thus

$$\mathbf{m}_1 = \frac{mL}{840}\begin{bmatrix} 156 & 11L & 54 & -6.5L \\ 11L & L^2 & 6.5L & -0.75L^2 \\ 54 & 6.5L & 156 & -11L \\ -6.5L & -0.75L^2 & -11L & L^2 \end{bmatrix} \begin{matrix} (5) \\ (6) \\ (2) \\ (4) \end{matrix} \qquad \text{columns: } (5)\ (6)\ (2)\ (4)$$

$$\mathbf{m}_2 = \frac{mL}{840}\begin{bmatrix} 156 & 11L & 54 & -6.5L \\ 11L & L^2 & 6.5L & -0.75L^2 \\ 54 & 6.5L & 156 & -11L \\ -6.5L & -0.75L^2 & -11L & L^2 \end{bmatrix} \begin{matrix} (2) \\ (4) \\ (1) \\ (3) \end{matrix} \qquad \text{columns: } (2)\ (4)\ (1)\ (3)$$

6. *Assemble the element mass matrices.* The mass matrix for the finite element system is determined by assembling $\mathbf{m}_1$ and $\mathbf{m}_2$ in a manner analogous to the stiffness matrix assembly:

$$\mathbf{m} = \mathcal{A}_{i=1}^{2}\mathbf{m}_i$$

$$= \frac{mL}{840}\left[\begin{array}{cccc|cc} 156 & 54 & -11L & 6.5L & 0 & 0 \\ 54 & 312 & -6.5L & 0 & 54 & 6.5L \\ -11L & -6.5L & L^2 & -0.75L^2 & 0 & 0 \\ 6.5L & 0 & -0.75L^2 & 2L^2 & -6.5L & -0.75L^2 \\ \hline 0 & 54 & 0 & -6.5L & 156 & 11L \\ 0 & 6.5L & 0 & -0.75L^2 & 11L & L^2 \end{array}\right] \Big\} \text{ support DOFs}$$

support DOFs

7. *Formulate the equations of motion.* Before writing the equation of motion, the support conditions must be imposed. For the cantilever beam of Fig. E17.2a, $u_5 = u_6 = 0$. Thus the fifth and sixth rows and columns are deleted from the matrices $\mathbf{m}$ and $\mathbf{k}$ to obtain

$$\mathbf{m}\ddot{\mathbf{u}} + \mathbf{k}\mathbf{u} = \mathbf{0} \tag{c}$$

or

$$\frac{mL}{840}\begin{bmatrix} 156 & 54 & -11L & 6.5L \\ & 312 & -6.5L & 0 \\ \text{(sym)} & & L^2 & -0.75L^2 \\ & & & 2L^2 \end{bmatrix}\begin{bmatrix} \ddot{u}_1 \\ \ddot{u}_2 \\ \ddot{u}_3 \\ \ddot{u}_4 \end{bmatrix} + \frac{8EI}{L^3}\begin{bmatrix} 12 & -12 & -3L & -3L \\ & 24 & 3L & 0 \\ \text{(sym)} & & L^2 & 0.5L^2 \\ & & & 2L^2 \end{bmatrix}\begin{bmatrix} u_1 \\ u_2 \\ u_3 \\ u_4 \end{bmatrix} = \begin{bmatrix} 0 \\ 0 \\ 0 \\ 0 \end{bmatrix} \tag{d}$$

Note that the stiffness matrix in Eq. (d) is the same as the one obtained by classical methods in Example 9.4.

8. *Solve the eigenvalue problem.* The natural frequencies are determined by solving $\mathbf{k}\phi = \omega^2\mathbf{m}\phi$:

$$\omega_1 = 3.51772\sqrt{\frac{EI}{mL^4}} \qquad \omega_2 = 22.2215\sqrt{\frac{EI}{mL^4}}$$
$$\omega_3 = 75.1571\sqrt{\frac{EI}{mL^4}} \qquad \omega_4 = 218.138\sqrt{\frac{EI}{mL^4}} \tag{e}$$

Example 17.3

Repeat Example 17.2 using the lumped-mass approximation.

Solution The only change is in formulation of the mass matrix. Using the lumped-mass matrix of Eq. (17.8.3) for each element and proceeding as in steps 5, 6, and 7 of Example 17.2 leads to

$$\mathbf{m}\ddot{\mathbf{u}} + \mathbf{k}\mathbf{u} = \mathbf{0} \tag{a}$$

where **u** and **k** are the same as in Eq. (d) of Example 17.2 but **m** is different:

$$\mathbf{m} = \begin{bmatrix} mL/4 & 0 & 0 & 0 \\ & mL/2 & 0 & 0 \\ \text{(sym)} & & 0 & 0 \\ & & & 0 \end{bmatrix} \tag{b}$$

Because the mass associated with the rotational DOFs u_3 and u_4 is zero, they can be eliminated from the stiffness matrix by static condensation. The resulting 2 × 2 stiffness matrix in terms of the translational DOF was presented in Example 9.5. The eigenvalue problem was solved in Example 10.2 to obtain the natural frequencies:

$$\omega_1 = 3.15623\sqrt{\frac{EI}{mL^4}} \qquad \omega_2 = 16.2580\sqrt{\frac{EI}{mL^4}}$$

17.10 COMPARISON OF FINITE ELEMENT AND EXACT SOLUTIONS

In this section the approximate values for the natural frequencies of a uniform cantilever beam determined by the finite element method are compared with the exact solutions presented in Chapter 16. The approximate results are obtained by discretizing the beam into

N finite elements of equal length and analyzing it by the finite element method. Such results for the coefficient α_n in $\omega_n = \alpha_n\sqrt{EI/mL^4}$ obtained for $N_e = 1, 2, 3, 4,$ and 5 finite elements and using the consistent-mass matrix are presented in Table 17.10.1. Example 17.2 provides the results for $N_e = 2$ and those for other N_e were obtained similarly.

TABLE 17.10.1 NATURAL FREQUENCIES OF A UNIFORM CANTILEVER BEAM: CONSISTENT-MASS FINITE ELEMENT AND EXACT SOLUTIONS

	Number of Finite Elements, N_e					
Mode	1	2	3	4	5	Exact
1	3.53273	3.51772	3.51637	3.51613	3.51606	3.51602
2	34.8069	22.2215	22.1069	22.0602	22.0455	22.0345
3		75.1571	62.4659	62.1749	61.9188	61.6972
4		218.138	140.671	122.657	122.320	120.902
5			264.743	228.137	203.020	199.860
6			527.796	366.390	337.273	298.556
7				580.849	493.264	416.991
8				953.051	715.341	555.165
9					1016.20	713.079
10					1494.88	890.732

Source: R. R. Craig, Jr., *Structural Dynamics*, Wiley, New York, 1981.

Observe from these results that the accuracy of the natural frequencies deteriorates for the higher modes of a particular N_e-element system, but the accuracy is improved by increasing N_e and hence the number of DOFs. The accuracy is quite good for a number of modes equal to the number of elements, but the frequencies of the higher modes are poor. As expected from Rayleigh–Ritz or consistent finite element formulations, the frequencies converge from above to the exact solution.

The natural frequencies obtained by a lumped-mass idealization of the finite element system and using the condensed stiffness matrix with $N_e = 1, 2, 3, 4,$ and 5 finite elements are presented in Table 17.10.2. Example 17.3 provides the results for $N_e = 2$ and those for other N_e were obtained similarly.

TABLE 17.10.2 NATURAL FREQUENCIES OF A UNIFORM CANTILEVER BEAM: LUMPED-MASS FINITE ELEMENT AND EXACT SOLUTIONS

	Number of Finite Elements, N_e					
Mode	1	2	3	4	5	Exact
1	2.44949	3.15623	3.34568	3.41804	3.45266	3.51602
2		16.2580	18.8859	20.0904	20.7335	22.0345
3			47.0284	53.2017	55.9529	61.6972
4				92.7302	104.436	120.902
5					153.017	199.860

Source: R. R. Craig, Jr., *Structural Dynamics*, Wiley, New York, 1981.

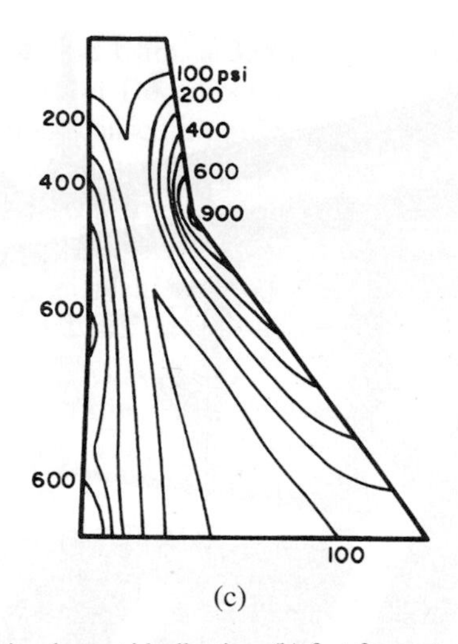

Figure 17.11.5 (a) Finite element idealization, (b) first four natural vibration modes and periods, and (c) stresses in the dam due to the Koyna earthquake, computed by linear analysis. [From A. K. Chopra, "Earthquake Response Analysis of Concrete Dams," in *Advanced Dam Engineering for Design, Construction and Rehabilitation* (ed. R. B. Jansen), Van Nostrand Reinhold, New York, 1988.]

downstream face changes. These stresses, which exceed 600 psi on the upstream face and 900 psi on the downstream face, are approximately two to three times the tensile strength (350 psi) of the concrete used in the upper parts of the dam. Hence, based on the finite element analysis results and concrete strength data, it was possible to identify the locations in the dam where significant cracking of concrete can be expected. These are consistent with the locations where the dam was damaged by the Koyna earthquake.

FURTHER READING

Bathe, K. J., *Finite Element Procedures in Engineering Analysis*, Prentice Hall, Englewood Cliffs, N.J., 1982.

Chopra, A. K., "Earthquake Analysis of Complex Structures," in *Applied Mechanics in Earthquake Engineering*, AMD Vol. 8 (ed. by W. D. Iwan), ASME, New York, 1974, pp. 163–204.

Chopra, A. K., "Earthquake Response Analysis of Concrete Dams," Chapter 15 in *Advanced Dam Engineering for Design, Construction and Rehabilitation* (ed. R. B. Jansen), Van Nostrand Reinhold, New York, 1988, pp. 416–465.

Clough, R. W., "The Finite Element Method in Plane Stress Analysis," *Proceedings of the 2nd ASCE Conference on Electronic Computation*, Pittsburgh, Pa., September 1960.

Cook, D., *Concepts and Applications of Finite Element Analysis*, Wiley, New York, 1981.

Hughes, T. J. R., *The Finite Element Method: Linear Static and Dynamic Finite Element Analysis*, Prentice Hall, Englewood Cliffs, N.J., 1987.

Zienkiewicz, O. C., and Taylor, R. L., *The Finite Element Method*, Vols. 1 and 2, 4th ed., McGraw-Hill, London, 1989.

PROBLEMS

Part A

17.1 A chimney of height L has been idealized as a cantilever beam with mass per unit length varying linearly from m at the base to $m/2$ at the top, and with second moment of cross-sectional area varying linearly from I at the base to $I/2$ at the top. Estimate the first two natural frequencies and modes of lateral vibration of the chimney using the shape functions $\psi_1(x) = 1 - \cos(\pi x/2L)$ and $\psi_2(x) = 1 - \cos(3\pi x/2L)$, where x is measured from the base.

17.2 A simply supported beam of length L has constant flexural rigidity EI and a mass distribution

$$m(x) = \begin{cases} 2m_o(x/L) & 0 \leq x \leq L/2 \\ 2m_o(1 - x/L) & L/2 \leq x \leq L \end{cases}$$

Determine the first two natural frequencies and modes of symmetric vibration of this nonuniform beam by the Rayleigh–Ritz method using two modes of the uniform beam as the shape functions.

Part B

*__17.3__ Determine the natural vibration frequencies and modes of a simply supported uniform beam, idealized as an assemblage of two finite elements (Fig. P17.3), using **(a)** the consistent-mass matrix, and **(b)** the lumped-mass matrix. Compare these results with the exact solutions obtained in Example 16.1.

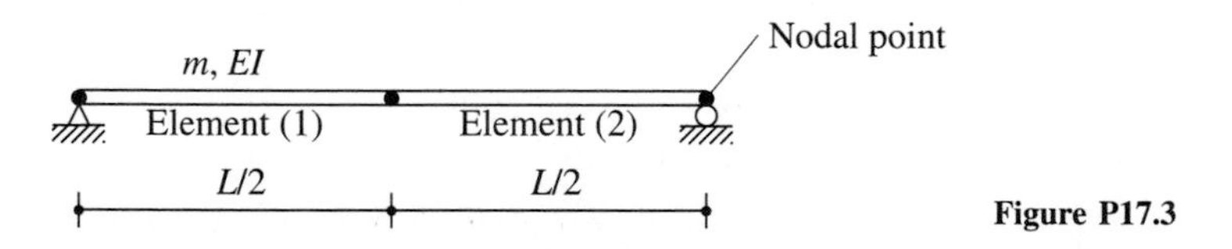

Figure P17.3

17.4 Determine the natural vibration frequencies and modes of a uniform beam clamped at both ends, idealized as an assemblage of two finite elements (Fig. P17.4), using **(a)** the consistent-mass matrix, and **(b)** the lumped-mass matrix. Compare these results with the exact solutions obtained by solving Problem 16.1.

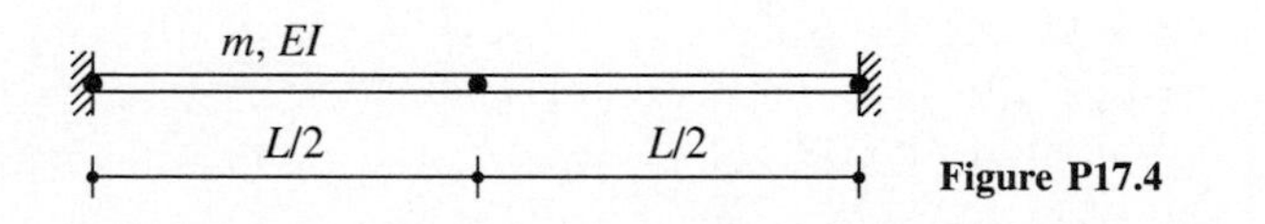

Figure P17.4

*Denotes that a computer is necessary to solve this problem.

17.5 Determine the natural vibration frequencies and modes of a uniform beam clamped at one end and simply supported at the other, idealized as an assemblage of two finite elements (Fig. P17.5), using **(a)** the consistent-mass matrix, and **(b)** the lumped-mass matrix. Compare these results with the exact solutions obtained by solving Problem 16.2.

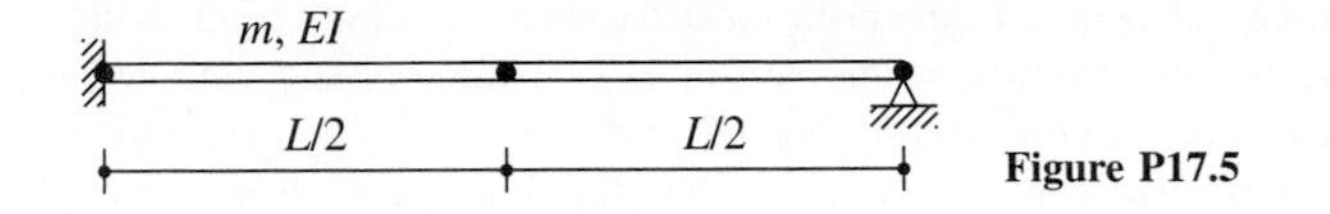

Figure P17.5

*__17.6__ Figure P17.6 shows a one-story, one-bay frame with mass per unit length and second moment of cross-sectional area given for each member. The frame, idealized as an assemblage of three finite elements, has the three DOFs shown if axial deformations are neglected in all elements.

(a) Starting with the definition of influence coefficients, formulate the stiffness matrix and the consistent-mass matrix. Express these matrices in terms of m, EI, and h.

(b) Determine the natural vibration frequencies and modes of the frame; express rotations in terms of h. Sketch the modes showing translations and rotations of the nodes.

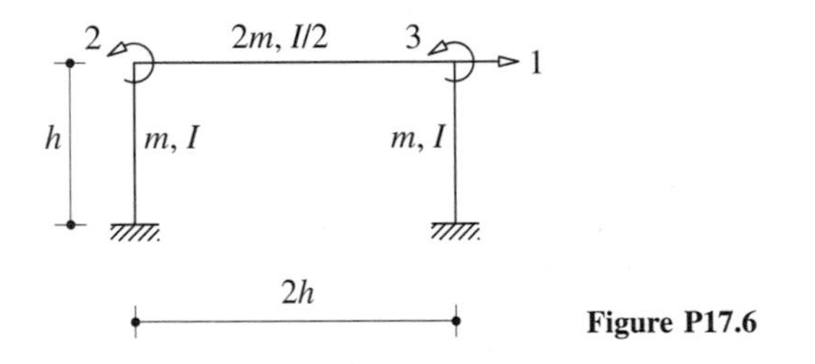

Figure P17.6

17.7 Repeat Problem 17.6 using the lumped-mass approximation. Comment on the effects of mass lumping on the vibration properties.

17.8 Repeat part (a) of Problem 17.6 starting with the stiffness and mass matrices for each element and using the direct assembly procedure of Section 17.5.

*Denotes that a computer is necessary to solve this problem.

PART III

Earthquake Response and Design of Multistory Buildings

Earthquake Response of Linearly Elastic Buildings

PREVIEW

Developed in Chapter 13 were two procedures—response history analysis (RHA) and response spectrum analysis (RSA)—for calculating the earthquake response of any structure described as a linearly elastic system with a finite number of degrees of freedom. Determined by RSA, the earthquake response of multistory buildings to excitations characterized by a design spectrum is presented in this chapter for a wide range of the two key parameters: fundamental natural vibration period and beam-to-column stiffness ratio. Based on these results, we develop an understanding of how these parameters affect the earthquake response of buildings and how they affect the relative response contributions of the different natural vibration modes. These results also enable us to identify the conditions under which the first mode or the first two modes are sufficient to provide a useful approximation to the total response. The understanding we develop of the significance of the higher modes in building response will be useful in Chapter 21, where we evaluate the equivalent static force procedure in seismic building codes in light of the results of dynamic analyses.

18.1 SYSTEMS ANALYZED, DESIGN SPECTRUM, AND RESPONSE QUANTITIES

18.1.1 Systems Analyzed

The systems analyzed are single-bay, five-story frames with constant story height $= h$ and bay width $= 2h$ (Fig. 18.1.1). All the beams have the same flexural rigidity, EI_b, and the column rigidity, EI_c, does not vary with height. The building is idealized as a

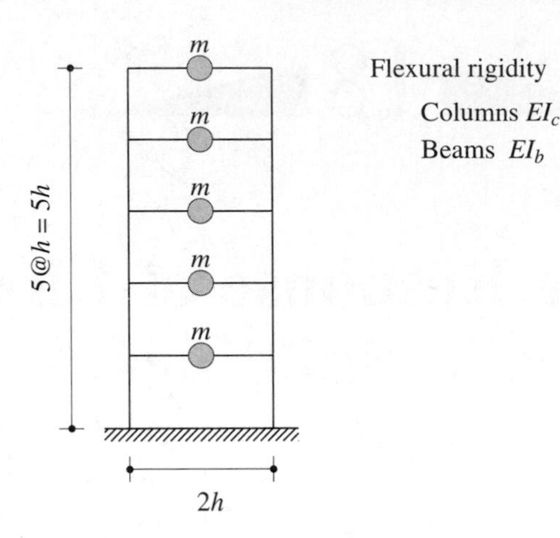

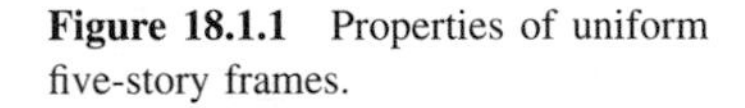

Figure 18.1.1 Properties of uniform five-story frames.

lumped-mass system with the same mass m at all the floor levels. The damping ratio for all five natural vibration modes is assumed to be 5%.

Only two additional parameters are needed to define the system completely: the fundamental natural vibration period T_1 and the *beam-to-column stiffness ratio* ρ. The latter parameter is based on the properties of the beams and columns in the story closest to the midheight of the frame:

$$\rho = \frac{\sum_{\text{beams}} EI_b/L_b}{\sum_{\text{columns}} EI_c/L_c} \tag{18.1.1}$$

where L_b and L_c are the lengths of the beams and columns and the summations include all the beams and columns in the midheight story. For the uniform, one-bay frame defined in the preceding paragraph, Eq. (18.1.1) reduces to

$$\rho = \frac{I_b}{4I_c} \tag{18.1.2}$$

which was introduced in Section 1.3 for a one-story frame. This parameter is a measure of the relative beam-to-column stiffness and indicates how much the system may be expected to behave as a frame. For $\rho = 0$ the beams impose no restraint on joint rotations, and the frame behaves as a flexural beam (Fig. 18.1.2a). For $\rho = \infty$ the beams restrain completely the joint rotations, and the structure behaves as a shear beam with double-curvature bending of the columns in each story (Fig. 18.1.2c). An intermediate value of ρ represents a frame in which beams and columns undergo bending deformation with joint rotation (Fig. 18.1.2b). As an example for the frame of Fig. 18.1.1, $\rho = \frac{1}{8}$ represents $I_b = I_c/2$, which implies a frame with columns stiffer than the beams, typical of earthquake-resistant construction.

The parameter ρ controls several properties of the frame: the fundamental natural period, the relative closeness or separation of the natural periods, and the shapes of the natural modes. These vibration properties of the frame of Fig. 18.1.1 are calculated by the procedures of Chapters 9 and 10. The variation of the fundamental period with ρ is shown in Fig. 18.1.3, which indicates that for a fixed column stiffness EI_c and floor mass m,

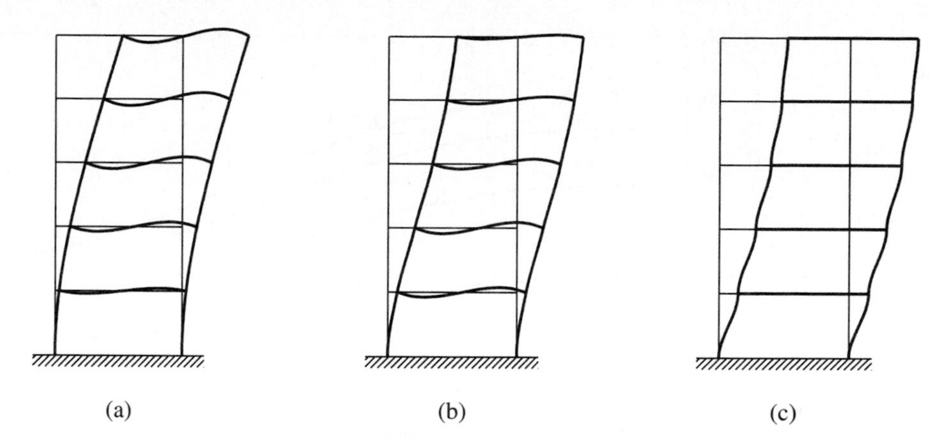

Figure 18.1.2 Deflected shapes: (a) $\rho = 0$; (b) $\rho = \frac{1}{8}$; (c) $\rho = \infty$.

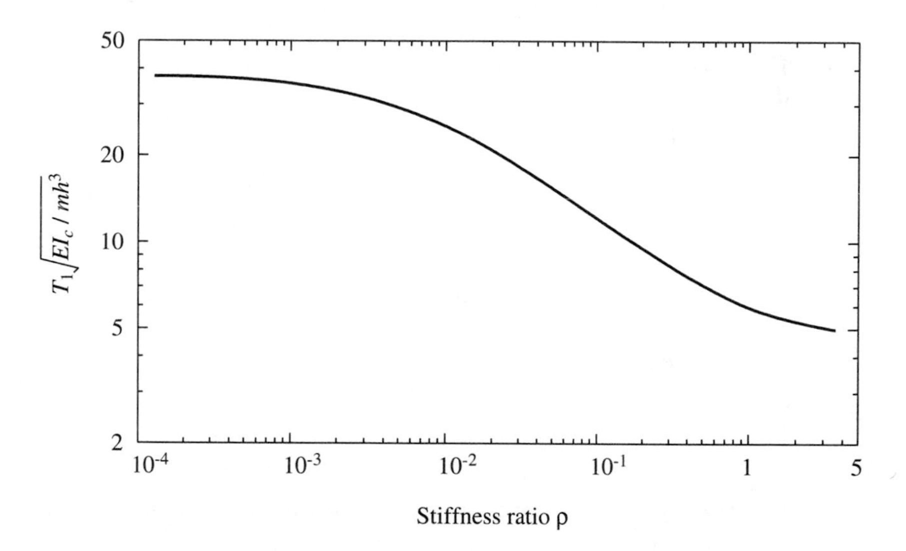

Figure 18.1.3 Fundamental natural vibration period of uniform five-story frames.

the fundamental period is reduced by a factor of over 8 as ρ increases from 0 to ∞. The ratios of the natural periods are independent of T_1 but strongly depend on ρ, especially the higher-mode periods, as shown in Fig. 18.1.4. As a result, the natural periods of a frame with small ρ are more separated from each other than if ρ is large. The shapes of the natural modes depend significantly on ρ, as shown in Fig. 18.1.5. It is clear from these results that the stiffness ratio ρ must have great importance in determining the dynamic (and static) behavior of the frame. We demonstrate this in the sections that follow.

The fundamental period T_1 will be varied over a range that is much wider than is reasonable for five-story frames. However, this is appropriate for the objectives of this chapter, where we are studying the influence of T_1 on the response of buildings.

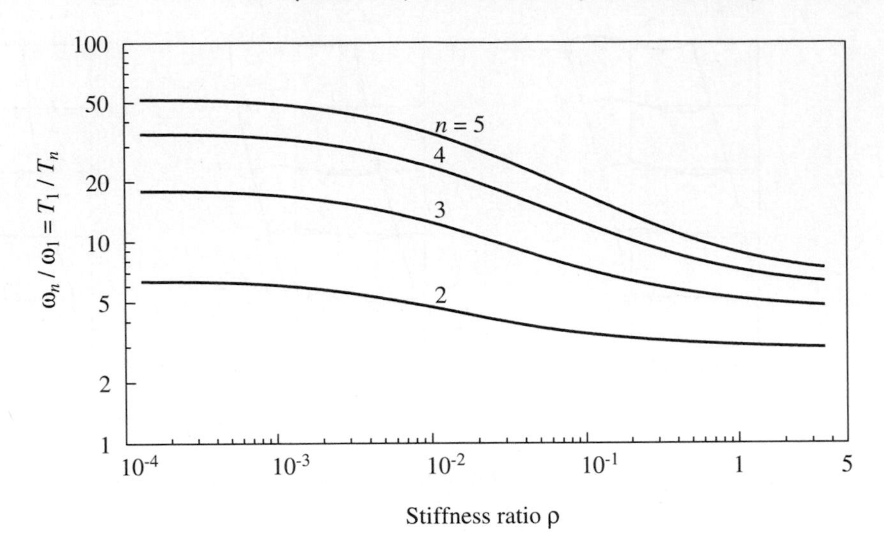

Figure 18.1.4 Natural vibration period ratios for uniform five-story frames. (After Roehl, 1971.)

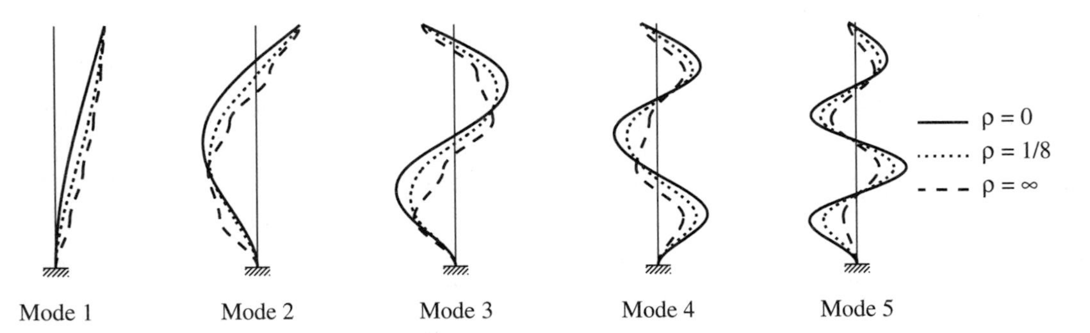

Figure 18.1.5 Natural vibration modes of uniform five-story frame for three values of ρ.

The response behavior is controlled by T_1 primarily and affected only secondarily by the number of stories. Hence the observations we glean from the results presented are not restricted to five-story buildings.

18.1.2 Design Spectrum

The earthquake excitation is characterized by the design spectrum of Fig. 6.9.5, multiplied by 0.5, so that it applies to ground motions with a peak ground acceleration $\ddot{u}_{go} = 0.5$g, velocity $\dot{u}_{go} = 24$ in./sec, and displacement $u_{go} = 18$ in. This design spectrum for 5% damping is shown in Fig. 18.1.6, where the acceleration-sensitive, velocity-sensitive, and displacement-sensitive regions (defined in Chapter 6) are identified.

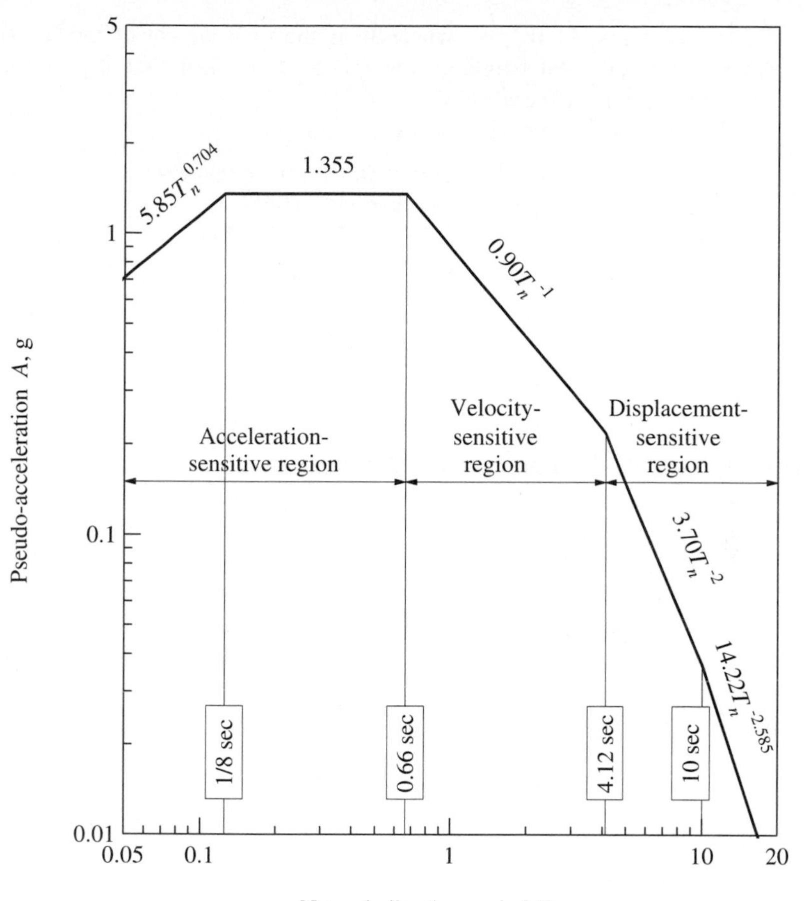

Figure 18.1.6 Design spectrum for ground motions with $\ddot{u}_{go} = 0.5\text{g}$, $\dot{u}_{go} = 24$ in./sec, and $u_{go} = 18$ in.; $\zeta = 5\%$.

18.1.3 Response Quantities

The peak values of the response of a frame described in Section 18.1.1 with specified T_1 and ρ to ground motion characterized by the design spectrum of Fig. 18.1.6 are determined by the RSA procedure. Such analyses were repeated for three values of ρ—0, $\frac{1}{8}$, and ∞—and many values of T_1. Among the many response quantities, we will examine four of them: top-floor displacement u_5 relative to the ground, base shear V_b, base overturning moment M_b, and top-story shear V_5. The first three will be normalized as follows: (1) u_5/u_{go}, where u_{go} is the peak ground displacement; (2) V_b/W_1^*, where $W_1^* = M_1^* g$ and M_1^* is the effective modal mass for the first mode; and (3) $M_b/W_1^* h_1^*$, where h_1^* is the effective modal height for the first mode. The values of W_1^* and h_1^* are computed using Eq. (13.2.9) and the shape of the first mode (Fig. 18.1.5). Presented in

Table 18.1.1 are (1) W_1^*/W, where W is the total weight of the frame, and (2) $h_1^*/5h$, where $5h$ is the total height of the frame. It is clear that W_1^* and h_1^* depend on the beam-to-column stiffness ratio ρ.

TABLE 18.1.1 FUNDAMENTAL MODE PROPERTIES

	$\rho = 0$	$\rho = \frac{1}{8}$	$\rho = \infty$
W_1^*/W	0.679	0.796	0.880
$h_1^*/5h$	0.794	0.742	0.703

18.2 INFLUENCE OF T_1 AND ρ ON RESPONSE

Shown in Fig. 18.2.1 are three normalized responses of the frame plotted against its fundamental period T_1 for three values of ρ. Over a wide range of T_1 values, the top-floor displacement varies very little with ρ (i.e., it is not sensitive to variations in the beam-to-column stiffness ratio). For very-long-period systems the top-floor displacement approaches the ground displacement because the floor masses of such a system remain stationary while the ground beneath moves; such behavior of a one-story frame is shown in Fig. 6.8.5.

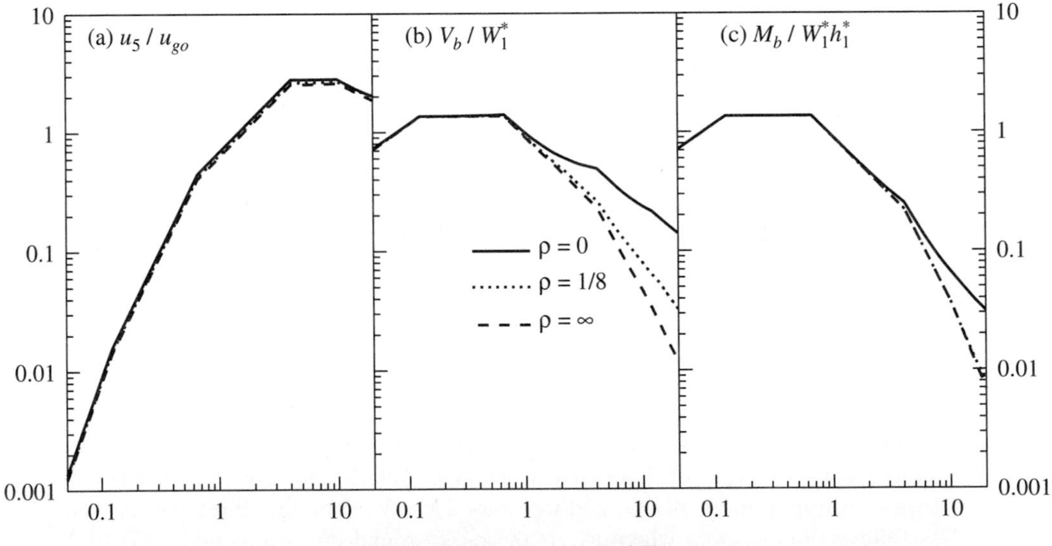

Figure 18.2.1 Normalized values of top-floor displacement u_5, base shear V_b, and base overturning moment M_b in uniform five-story frames for three values of ρ.

The shear and overturning moment at the base of the frame are of special interest because their design values are specified in building codes; they are also the forces needed in the design of the foundation system. When these normalized forces are plotted against T_1, as in Fig. 18.2.1, the curves have the general appearance of the pseudo-acceleration spectrum of Fig. 18.1.6. Thus the individual curves tend to the peak ground acceleration, 0.5g, for short T_1, and to zero for long T_1, as does the pseudo-acceleration spectrum.

The normalized base shear and base overturning moment vary significantly with ρ for buildings with T_1 in the velocity- or displacement-sensitive regions of the spectrum, with the variation in M_b not as great as in V_b. However, these normalized responses do not vary appreciably with ρ in the acceleration-sensitive region of the spectrum. Observe that W_1^* and h_1^* themselves vary with ρ, as shown in Table 18.1.1, and hence they influence the values of the actual (in contrast to normalized) responses V_b and M_b and how they depend on ρ.

The variation in the normalized responses with ρ is closely related to the significance of the response contributions of the second and higher modes, which generally increase with decreasing ρ (Section 18.5) and—for the design spectrum selected—generally increase with increasing T_1 (Section 18.4). To study individual modal responses, we use the modal contribution factors introduced in Chapter 12, Part C.

18.3 MODAL CONTRIBUTION FACTORS

The peak value of the nth-mode contribution to a response quantity r is given by Eq. (13.7.1), repeated here for convenience:

$$r_n = r_n^{\text{st}} A_n \tag{18.3.1}$$

where A_n is the ordinate of the pseudo-acceleration response (or design) spectrum corresponding to natural period T_n and damping ratio ζ_n of the nth mode; and r_n^{st} is the modal static response. As defined in Section 13.2.2, r_n^{st} is the static value of response quantity r due to external forces $\mathbf{s}_n$, given by Eq. (13.2.4). These modal static responses are presented in Table 13.2.1 and repeated here for base shear V_b, top-story shear V_5, base overturning moment M_b, and top-floor displacement u_5:

$$V_{bn}^{\text{st}} = M_n^* \qquad V_{5n}^{\text{st}} = \Gamma_n m \phi_{5n} \qquad M_{bn}^{\text{st}} = h_n^* M_n^* \qquad u_{5n}^{\text{st}} = \frac{\Gamma_n}{\omega_n^2}\phi_{5n} \tag{18.3.2}$$

Alternatively, Eq. (18.3.1) can be expressed as

$$r_n = r^{\text{st}} \bar{r}_n A_n \tag{18.3.3}$$

where

$$r^{\text{st}} = \sum_{n=1}^{N} r_n^{\text{st}} \quad \text{and} \quad \bar{r}_n = \frac{r_n^{\text{st}}}{r^{\text{st}}} \tag{18.3.4}$$

As demonstrated in Section 12.10, r^{st} is also the static value of r due to external forces $\mathbf{s} = \mathbf{m1}$. The modal contribution factors $\bar{r}_n$ for a response quantity r are dimensionless, are independent of how the modes are normalized, and add up to unity when summed over all modes:

$$\sum_{n=1}^{N} \bar{r}_n = 1 \tag{18.3.5}$$

The modal contribution factors $\bar{r}_n$ are calculated for four response quantities of the five-story frame from Eqs. (18.3.2) and (18.3.4) using the known system properties and computed natural frequencies and modes. The results presented in Tables 18.3.1a and 18.3.1b for $\rho = 0$, $\frac{1}{8}$, and ∞ are independent of T_1. They permit three useful observations that have a bearing on relative values of the modal responses. First, for a fixed value of ρ and each of the response quantities, the modal contribution factor $\bar{r}_1$ for the first mode is larger than the factors $\bar{r}_n$ for the higher modes, suggesting that the fundamental mode should have the largest contribution to each of these responses. However, this tentative conclusion may or may not hold true, depending on the shape of the design spectrum as it influences the relative values of A_n for the various modes. Second, for a fixed value of ρ, the absolute values of $\bar{r}_n$ for the second and higher modes are larger for V_5 than for V_b, and the values for V_b in turn are larger than those for M_b and u_5. This observation suggests that the second- and higher-mode response contributions are expected to be more significant for base shear V_b than for the base overturning moment M_b or top-floor displacement u_5. Among the story shears the higher-mode responses should be more significant for the fifth-story shear than for the base shear. Third, for V_5, V_b, and M_b, the absolute values of $\bar{r}_n$ for the second and higher modes

TABLE 18.3.1a MODAL CONTRIBUTION FACTORS FOR V_b AND V_5

	Base Shear V_b			Top-Story Shear V_5		
Mode	$\rho = 0$	$\rho = \frac{1}{8}$	$\rho = \infty$	$\rho = 0$	$\rho = \frac{1}{8}$	$\rho = \infty$
1	0.679	0.796	0.879	1.38	1.30	1.25
2	0.206	0.117	0.087	−0.528	−0.441	−0.362
3	0.070	0.051	0.024	0.204	0.211	0.159
4	0.033	0.026	0.007	−0.080	−0.089	−0.063
5	0.012	0.009	0.002	0.020	0.023	0.015

TABLE 18.3.1b MODAL CONTRIBUTION FACTORS FOR M_b AND u_5

	Base Overturning Moment M_b			Top-Story Displacement u_5		
Mode	$\rho = 0$	$\rho = \frac{1}{8}$	$\rho = \infty$	$\rho = 0$	$\rho = \frac{1}{8}$	$\rho = \infty$
1	0.898	0.985	1.030	1.009	1.027	1.030
2	0.078	−0.003	−0.035	−0.009	−0.030	−0.035
3	0.016	0.014	0.006	0.0005	0.003	0.006
4	0.006	0.003	−0.001	−0.00005	−0.0005	−0.001
5	0.002	0.001	0.0003	0.000005	0.00007	0.0003

increase (but for minor exceptions) as ρ decreases; this increase is especially significant for the second mode. This observation suggests that the higher-mode contributions to any of these forces should become a larger fraction of the total response as ρ decreases and should be largest for a flexural beam with $\rho = 0$. However, this tentative conclusion may or may not be valid, depending on the shape of the design spectrum.

18.4 INFLUENCE OF T_1 ON HIGHER-MODE RESPONSE

In this section we analyze how and why the fundamental natural period T_1 influences the combined responses due to all modes higher than the first mode. For this purpose we compare the total base shear due to all five modes with the base shear due to the first mode only. Both values of the normalized base shear are plotted against T_1 for three values of ρ in Fig. 18.4.1. The one-mode curves are independent of ρ and identical to

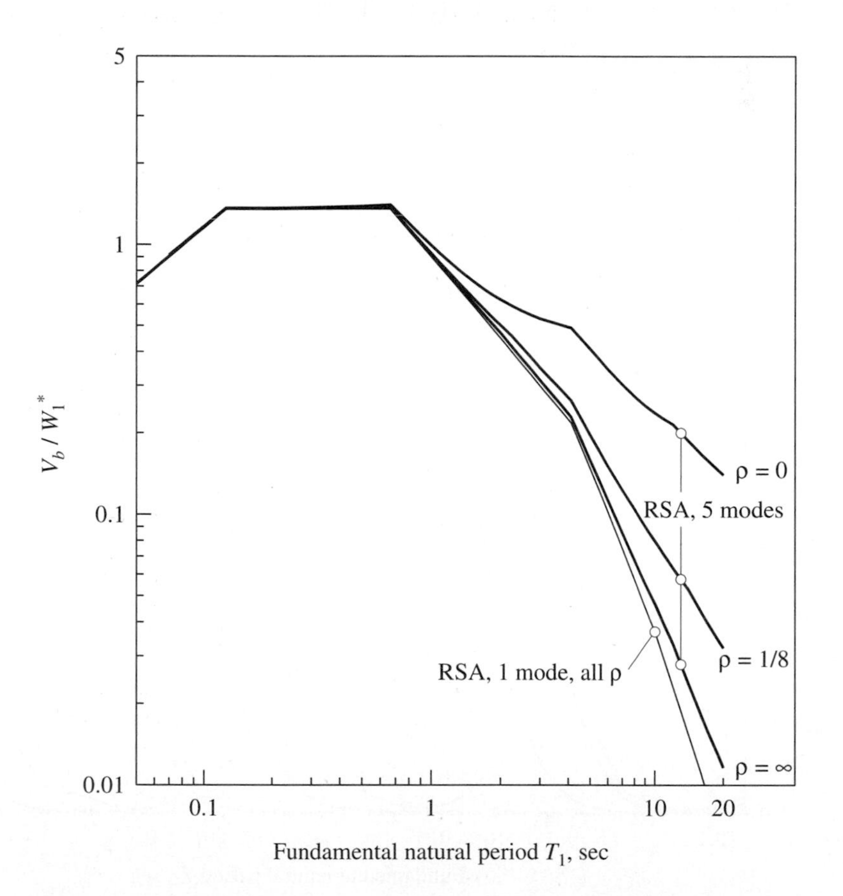

Figure 18.4.1 Normalized base shear in uniform five-story frames for three values of ρ. Results were obtained by response spectrum analysis (RSA), including one or five modes.

the design spectrum of Fig. 18.1.6 because Eqs. (18.3.1) and (18.3.2a) give

$$V_{b1} = \frac{A_1}{g} W_1^* \quad \text{or} \quad \frac{V_{b1}}{W_1^*} = \frac{A_1}{g} \tag{18.4.1}$$

The higher-mode response, expressed as a percentage of the total response, is presented in Fig. 18.4.2 for the four response quantities. The data for base shear are obtained from the results of Fig. 18.4.1, and those for other response quantities are obtained similarly. The higher-mode response of buildings is negligible for T_1 in the acceleration-sensitive region of the spectrum and increases with increasing T_1 in the velocity- and displacement-sensitive regions.

These trends can be explained by examining the three factors that enter into Eq. (18.3.3) for the peak modal response: (1) The static value r^{st} of r is a common factor in all modal responses and therefore does not influence the relative values of the modal responses. (2) As mentioned in Section 18.3, for a fixed ρ the modal contribution factors $\bar{r}_n$ are independent of T_1. (3) The pseudo-acceleration spectrum ordinate A_n is the only factor in Eq. (18.3.3) that depends on the fundamental period T_1 and period ratios T_n/T_1, which, for a fixed ρ, do not depend on T_1 (see Section 18.1). Thus the increase in higher-mode response with increasing T_1 must be related to the shape of the design spectrum. This is illustrated for the selected design spectrum in parts (a) and (b) of Fig. 18.4.3, wherein the natural periods T_n of two shear frames ($\rho = \infty$) with fundamental natural periods $T_1 = 0.5$ and 3.0 sec, respectively, are identified. For the building with $T_1 = 3$ sec, the A_n values for the higher modes are larger than A_1 for the fundamental mode, whereas for the building with $T_1 = 0.5$ sec, the A_n ($n \geq 2$) values are either equal to or smaller than A_1. Thus the higher-mode response, expressed as a

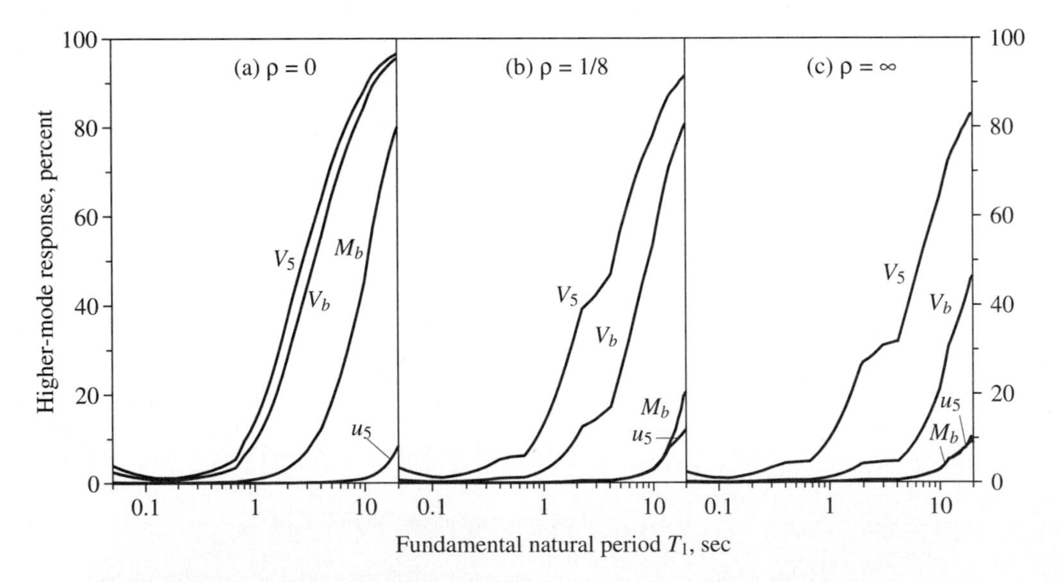

Figure 18.4.2 Higher-mode response in V_b, V_5, M_b, and u_5 for uniform five-story frames for three values of ρ.

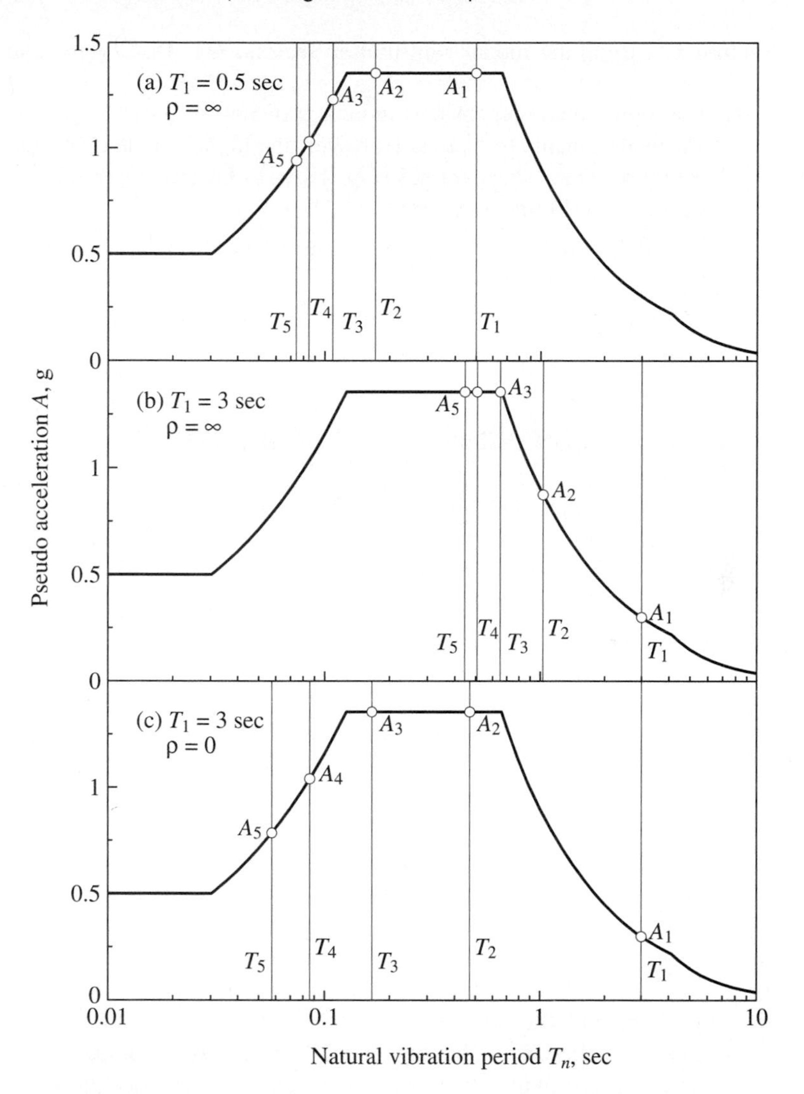

Figure 18.4.3 Natural vibration periods and spectral ordinates for three cases: (a) $T_1 = 0.5$ sec, $\rho = \infty$; (b) $T_1 = 3$ sec, $\rho = \infty$; and (c) $T_1 = 3$ sec, $\rho = 0$.

percentage of the total response, should be larger for the building with $T_1 = 3$ sec than for the $T_1 = 0.5$ sec building. This is indeed what is observed in Figs. 18.4.1 and 18.4.2 from the results of dynamic analyses. It is clear from this discussion that for the spectrum selected, as T_1 increases within the velocity- and displacement-sensitive regions of the spectrum, the higher-mode response will become an increasing percentage of the total response; this explains the trends observed in Figs. 18.4.1 and 18.4.2.

For a fixed T_1 and ρ the significance of the response contributions of the higher modes varies with the response quantity. As suggested earlier by the observations in

Section 18.3 from the modal contribution factors, Fig. 18.4.2 demonstrates:

1. The higher-mode response is more significant for forces (e.g., V_5, V_b, and M_b) than for displacements (e.g., u_5). However, the higher-mode contributions to u_5 and M_b are identical for shear frames (Fig. 18.4.2c) because the modal contribution factors are identical (Table 18.3.1b).
2. The higher-mode response is more significant for base shear than for base overturning moment.
3. The higher-mode response is more significant for top-story shear than for base shear.

18.5 INFLUENCE OF ρ ON HIGHER-MODE RESPONSE

We next examine how and why the beam-to-column stiffness ratio ρ influences the combined responses due to all modes higher than the first mode. Figures 18.4.1 and 18.4.2 demonstrate that for each response quantity, the higher-mode response is least significant for frames behaving like shear beams ($\rho = \infty$), becomes increasingly significant as ρ decreases, and is largest for buildings deforming like flexural beams ($\rho = 0$). There are two reasons for this trend. First, the absolute values of the modal contribution factors suggest this trend. In particular, as ρ decreases, the higher-mode contribution factors for the base shear and top-story shear increase especially in the second mode (Table 18.3.1a). Second, as shown in Fig. 18.1.4, the ratios T_1/T_n become larger as ρ decreases; hence the T_n values are spread out over a wider period range of the design spectrum. This is illustrated in parts (b) and (c) of Fig. 18.4.3. Both frames have the same fundamental period, $T_1 = 3$ sec, but they differ in ρ—one is a shear beam ($\rho = \infty$), and the other a flexural beam ($\rho = 0$). As a result, the ratio A_2 for the second mode—generally the most significant of the higher modes—to A_1 for the first mode is larger for buildings with $\rho = 0$ than for the $\rho = \infty$ case. Thus, putting these two reasons together, both the modal contribution factor $\bar{r}_n$ and the spectral ordinate A_n for the second mode are larger for the $\rho = 0$ frame; therefore, the higher-mode response is more significant in this case than for the frame with $\rho = \infty$. In general, for the design spectrum selected and for T_1 within the velocity- and displacement-sensitive regions of the spectrum, the ratio A_n/A_1 increases with decreasing ρ (or more precisely, does not decrease with decreasing ρ), resulting in increased higher-mode response. Consequently, the higher-mode response, expressed as a percent of the total response, becomes larger as ρ decreases.

The effect of ρ on the higher-mode response varies with the response quantity, as shown in Fig. 18.4.2. For a fixed T_1 these trends can again be explained based on the influence of ρ on modal contribution factors for the different response quantities (Table 18.3.1) and on the period ratios T_n/T_1 (Fig. 18.1.4). The top-floor displacement displays trends opposite to the forces in the sense that higher-mode contributions decrease with decreasing ρ, but this reverse trend is supported by the modal contribution factors.

The displacements due to higher modes are so small, however, that they are of little consequence.

18.6 HEIGHTWISE VARIATION OF HIGHER-MODE RESPONSE

In this section we examine how the response contributions of modes higher than the first mode to the story shears and floor overturning moments vary over building height. Compared in Figs. 18.6.1 and 18.6.2 are these forces due to the first mode only and the total forces considering all five modes. The difference between the two sets of forces is the combined response contribution of all modes higher than the first mode to each force; this higher-mode response is expressed as a percentage of the total force and presented in Figs. 18.6.3 and 18.6.4. These results indicate, as before, that (1) for a fixed ρ, the higher-mode response is more significant for longer-period buildings, and (2) for a fixed T_1, the higher-mode response is more significant for frames with smaller ρ.

The results presented in Figs. 18.6.3 and 18.6.4 provide information on how the higher modes affect the forces in different stories of a building. For a particular building with fixed values of T_1 and ρ, the percentage contribution of the higher-mode response tends to increase as one moves up the building, although the trend is not perfect in all cases. This trend is essentially consistent for story overturning moments. Observe that the influence of the higher modes is small on the base overturning moment but considerable for the overturning moments in the upper stories, especially for frames with

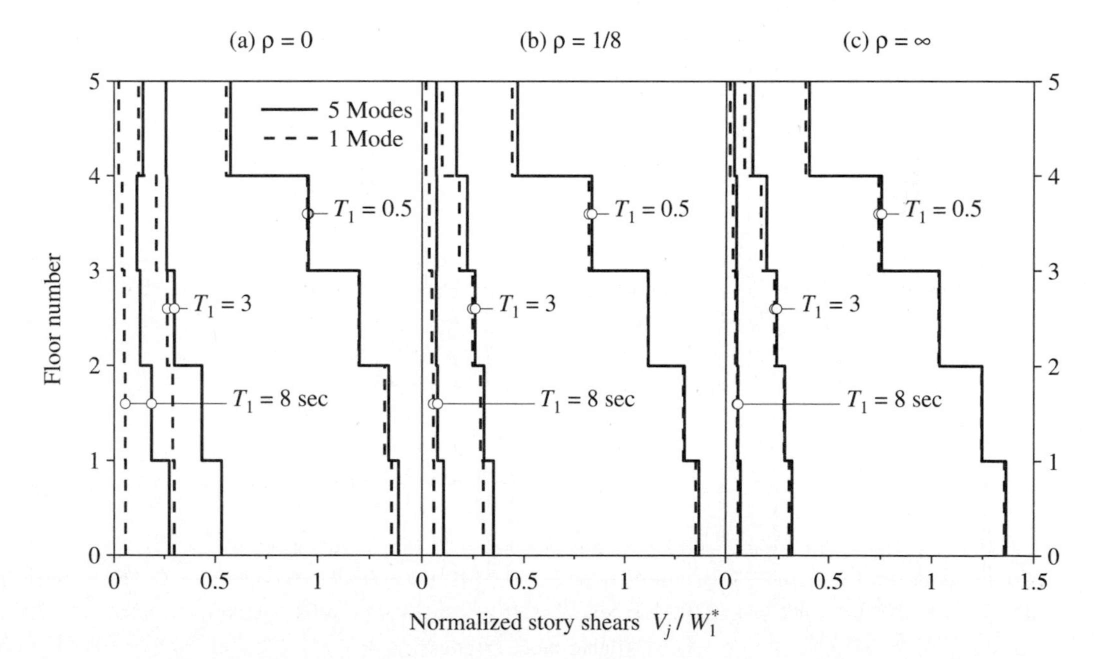

Figure 18.6.1 Normalized story shears in uniform five-story frames for three values of ρ and three values of T_1. Results were obtained by RSA, including one or five modes.

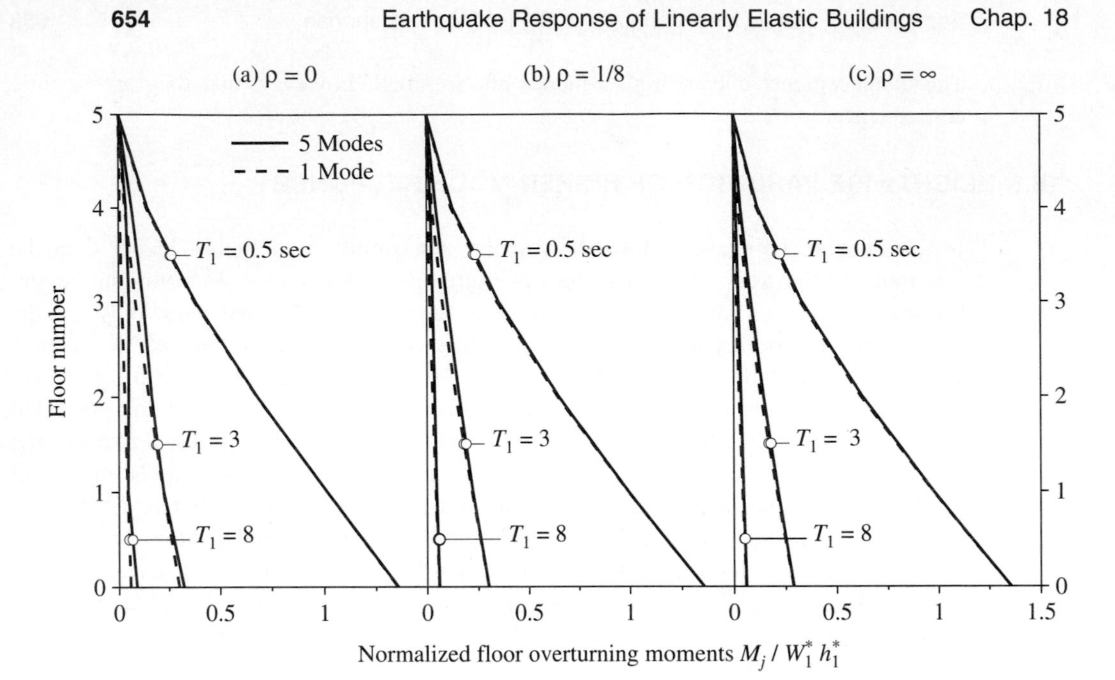

Figure 18.6.2 Normalized floor overturning moments in uniform five-story frames for three values of ρ and three values of T_1. Results were obtained by RSA including one or five modes.

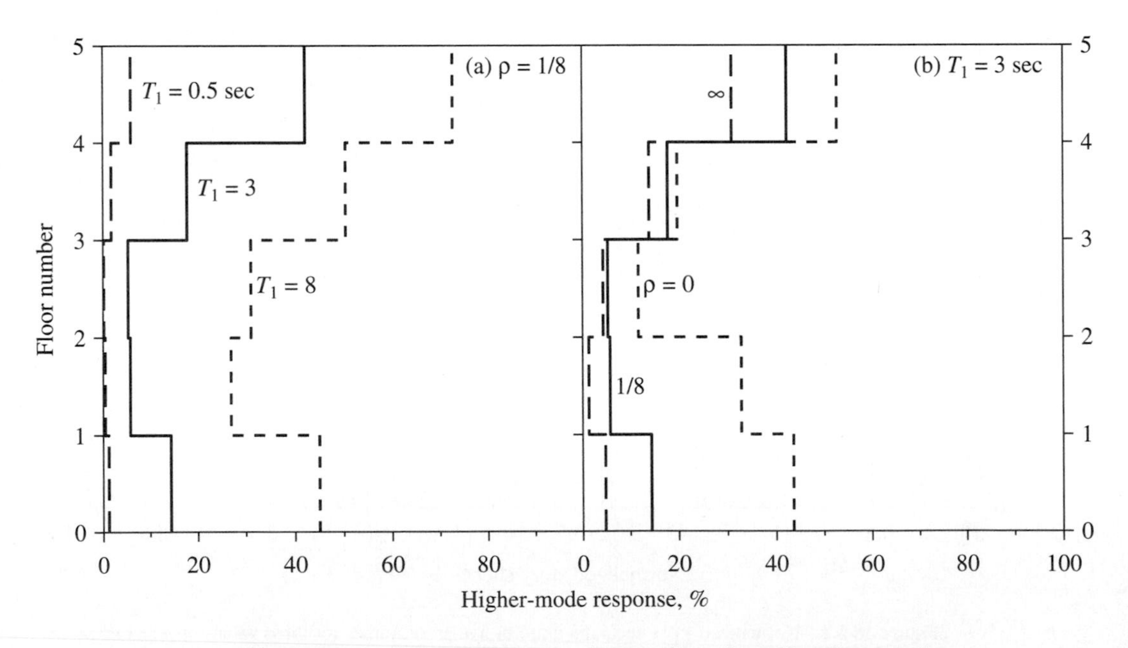

Figure 18.6.3 Higher-mode response in story shears for uniform five-story frames.

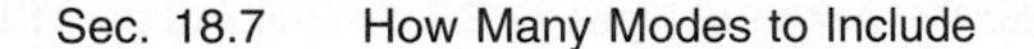

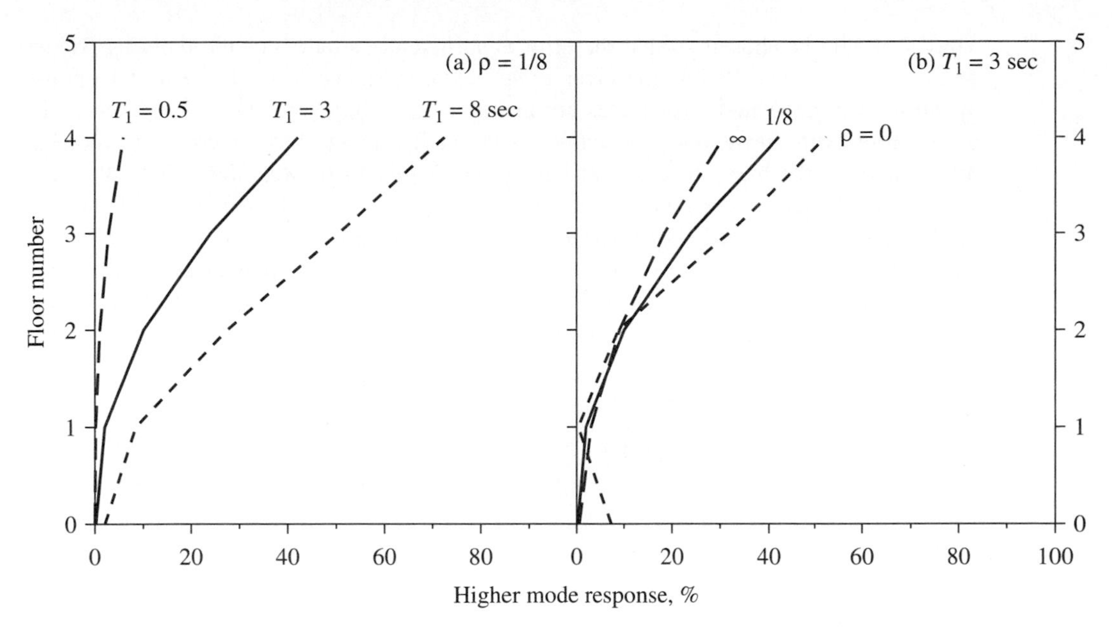

Figure 18.6.4 Higher-mode response in floor overturning moments for uniform five-story frames.

longer-period T_1 and smaller ρ. The above-noted trend is not always consistent for story shears, however, because the higher-mode response tends to increase for the forces in the stories near the bottom of the building in addition to the stories near the top of the building. This lack of perfect consistency in the above-noted trend is an indication of the complex dependence of the earthquake response of buildings on the system parameters and on the earthquake excitation.

18.7 HOW MANY MODES TO INCLUDE

The response contributions of all the natural modes must be included if the "exact" value of the structural response to earthquake excitation is desired, but the first few modes can usually provide sufficiently accurate results. In this section we bring together several ideas from the preceding sections to discuss the number of modes that should be included in earthquake response analysis. This discussion must consider two factors, modal contribution factor $\bar{r}_n$ and spectral ordinate A_n, that enter into the modal response equation (18.3.3).

We utilize the important result that the sum of the modal contribution factors over all the modes is unity, Eq. (18.3.5). If only the first J modes are included, the error in the static solution is

$$e_J = 1 - \sum_{n=1}^{J} \bar{r}_n \qquad (18.7.1)$$

Thus J should be chosen so that the error is sufficiently small. For the building frames described in Section 18.1.1, the error e_2 is below 0.15 or 15% for the four response quantities when the first two modes are included (see Table 18.7.1). For a fixed ρ the error varies with the response quantity. It is smaller in the base overturning moment M_b relative to the base shear V_b, and in V_b compared to the top-story shear V_5. The error is much smaller for the top-floor displacement, and it is less than 3% if the first mode alone is considered. For a particular response quantity the error e_2 (Table 18.7.1) varies with ρ, being smallest for $\rho = \infty$ (i.e., shear beams) and largest for $\rho = 0$ (i.e., flexural beams). These data suggest that the first one or two modes may provide a good approximation to the total response, with the accuracy depending on the response quantity and on ρ.

TABLE 18.7.1 $e_2 = 1 - \sum_{n=1}^{2} \bar{r}_n$

Response	$\rho = 0$	$\rho = \frac{1}{8}$	$\rho = \infty$
V_5	0.144	0.144	0.110
V_b	0.115	0.086	0.033
M_b	0.024	0.018	0.005
u_5	0.0004	0.003	0.005

The spectral ordinates A_n also influence the relative values of the modal responses due to various modes and hence the number of modes that should be included in the analysis. For a fixed ρ and T_1 in the velocity- or displacement-sensitive regions of the spectrum, the ratio A_n/A_1 is larger for frames with longer fundamental period T_1 (Fig. 18.4.3a and b). Thus for the same desired accuracy, more modes should be included in the analysis of buildings with longer T_1 than the number of modes necessary for shorter-period buildings. For a fixed T_1 in the velocity-sensitive or displacement-sensitive regions of the spectrum, the ratio A_n/A_1 is larger for frames with smaller ρ (Fig. 18.4.3b and c). Thus, for the same desired accuracy, more modes should be included in the analysis of buildings with smaller ρ compared to the number of modes necessary for buildings with larger ρ. In particular, accurate analysis of flexural frames ($\rho = 0$) requires more modes than shear frames ($\rho = \infty$).

These expectations regarding how T_1 and ρ influence the number of modes that should be included in earthquake response analysis are confirmed by the results of Fig. 18.7.1, where, for each ρ value, five response curves for base shear are identified by indicating the number of modes included in the analysis. From these results several observations can be made. It is clear that the first two modes provide a reasonably accurate value for the base shear in frames with T_1 in the velocity-sensitive region of the spectrum, and one mode is sufficient in the acceleration-sensitive region. This conclusion is also valid for shears in all the stories and overturning moments at all floors. The first mode alone provides accurate results for u_5 over the entire range of T_1, and for all ρ values, as indicated in Fig. 18.4.2.

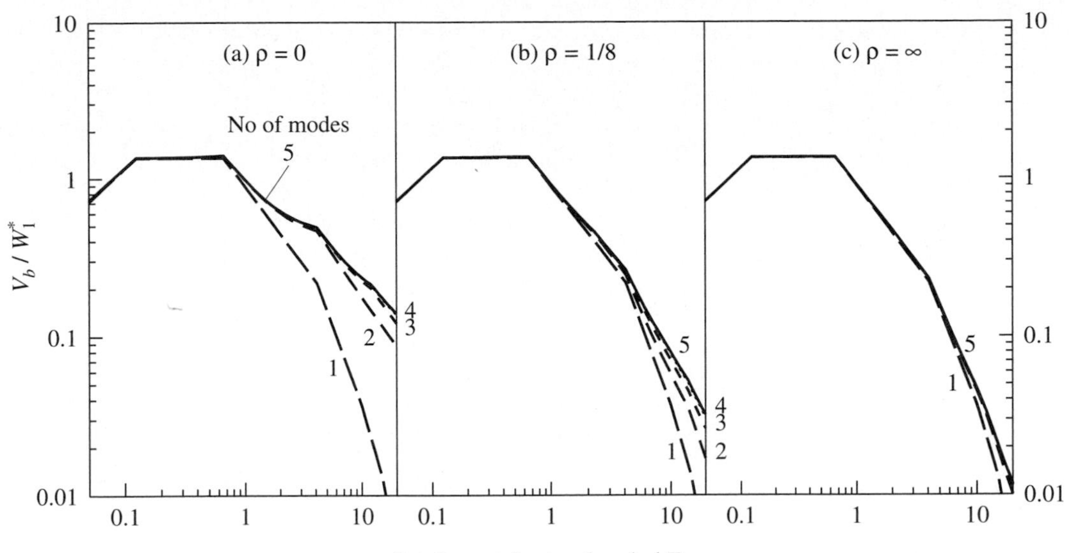

Figure 18.7.1 Normalized base shear in uniform five-story frames for three values of ρ. Results were determined by RSA considering one, two, three, four, or five modes.

FURTHER READING

Blume, J. A., "Dynamic Characteristics of Multistory Buildings," *Journal of the Structural Division, ASCE,* **94**, (ST2), February 1968, pp. 337–402.

Cruz, E. F., and Chopra, A. K., "Elastic Earthquake Response of Building Frames," *Journal of Structural Engineering, ASCE,* **112**, 1986, pp. 443–459.

Cruz, E. F., and Chopra, A. K., "Simplified Procedures for Earthquake Analysis of Buildings," *Journal of Structural Engineering, ASCE,* **112**, 1986, pp. 461-480.

Roehl, J. L., "Dynamic Response of Ground-Excited Building Frames," Ph.D. thesis, Rice University, Houston, Texas, October 1971.

Earthquake Response of Inelastic Buildings

PREVIEW

As mentioned in Chapter 7, most buildings are expected to deform beyond the limit of linearly elastic behavior when subjected to strong ground shaking. Thus the earthquake response of buildings deforming into their inelastic range is of central importance in earthquake engineering. This chapter is concerned with this important subject. First, we identify the differences between the ductility demands imposed by earthquake excitation on multistory buildings and on SDF systems. Then we demonstrate that the heightwise variation of floor displacements and of story ductility demands in a building depends on the relative yield strengths of its various stories. In particular, yielding is shown to be confined to the first story if it is "weak" or "soft" relative to the upper stories, situations that impose large ductility demand on this story, while the upper stories remain elastic. This is followed by a study of the ductility demands on buildings designed for the lateral force distribution specified by the 1994 *Uniform Building Code*, and of how these demands compare with the allowable ductility. Thereafter we present an approach to determine the base shear strength necessary, in conjunction with heightwise distribution of code forces, to ensure that the story ductility demands do not exceed the allowable ductility. The chapter closes with a recognition of some of the many aspects of the complicated problem of inelastic earthquake response of buildings that are not addressed in this introductory presentation.

19.1 ALLOWABLE DUCTILITY AND DUCTILITY DEMAND

19.1.1 Single-Degree-of-Freedom Systems

The response spectrum for elastoplastic systems developed in Chapter 7 provides a convenient basis to determine the design yield strength corresponding to an allowable ductility:

$$f_y = \frac{A_y}{g} w$$

where w is the weight of the system and A_y is the pseudo-acceleration corresponding to the allowable ductility and known values of the vibration properties—natural period T_n and damping ratio ζ—of the system in its linear range of vibration. Consider two different designs of an SDF system with $T_n = 0.8$ sec and $\zeta = 5\%$ to be designed for the El Centro ground motion. In one case the allowable ductility $\mu = 1$, which implies elastic design, and in the second case $\mu = 4$. The yield deformation and yield strength corresponding to each μ value is obtained from the response spectrum for the El Centro ground motion (Fig. 7.5.2 or 7.5.3): For $\mu = 1$, $u_y = 3.105$ in., $A_y/g = 0.4962$, and $f_y = 0.4962w$. Corresponding to $\mu = 4$, $u_y = 1.081$ in., $A_y/g = 0.1727$, and $f_y = 0.1727w$. (Obviously, we could not read u_y or A_y/g to four significant figures from a spectrum plot; these values were obtained from the computer data that led to the spectrum.)

The ductility demand imposed by the El Centro ground motion on the systems we have just "designed" will be exactly equal to the allowable ductility. This is demonstrated in Fig. 19.1.1, where the deformation history for each system, determined by the numerical time-stepping methods of Section 5.7, is presented. The peak deformation of the system designed for allowable ductility $\mu = 1$ is $u_m = 3.105$ in. and the corresponding ductility demand is $\mu = u_m/u_y = 1$ (i.e., the system deforms exactly up to its linear elastic limit). Similarly, the peak deformation of the system designed for an allowable ductility $\mu = 4$ is $u_m = 4.322$ in. and the corresponding ductility demand is $\mu = u_m/u_y = 4$. Such exact correspondence between ductility demand and allowable ductility always exists for SDF systems when the yield strength f_y is determined from the earthquake response spectrum corresponding to the allowable ductility. This is not true for MDF systems, however, as is demonstrated next.

19.1.2 Multistory Buildings: System Properties

Consider a five-story building idealized as a shear frame (Fig. 19.1.2a) with elastoplastic relation (defined in Section 7.1.2) between the shear force V_j and story drift, $\Delta_j = u_j - u_{j-1}$, in each story (Fig. 19.1.2c). The initial, linearly elastic values of the story stiffnesses k_j are given in Fig. 19.1.2a. These were determined by satisfying two requirements: (1) the fundamental natural period is $T_1 = 0.8$ sec for this system with equal floor masses $m_j = m = 100$ kips/g, and (2) static application of code forces, assuming the structure to be linearly elastic, should cause equal story drift in all five stories, resulting in floor deflections increasing linearly with height (because all stories have the same height). For a building with $T_1 = 0.8$ sec and $m = 100$ kips/g, the 1994 *Uniform Building Code*

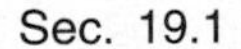

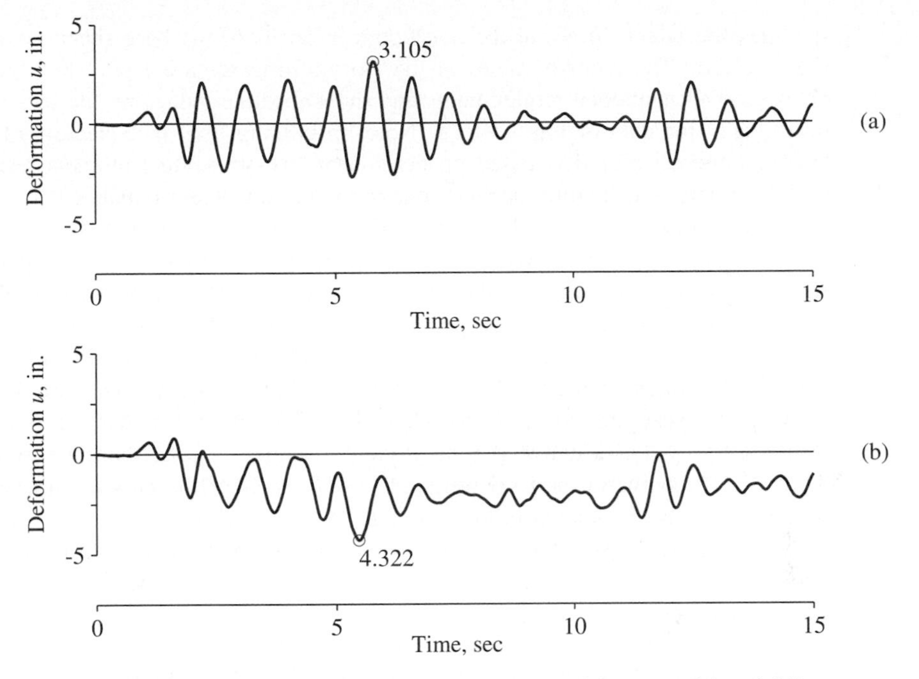

Figure 19.1.1 Response of two SDF systems with $T_n = 0.8$ sec, $\zeta = 5\%$, and different yield strengths to El Centro ground motion: (a) $f_y = 0.4962w$ (corresponding to $\mu = 1$); (b) $f_y = 0.1727w$ (for $\mu = 4$).

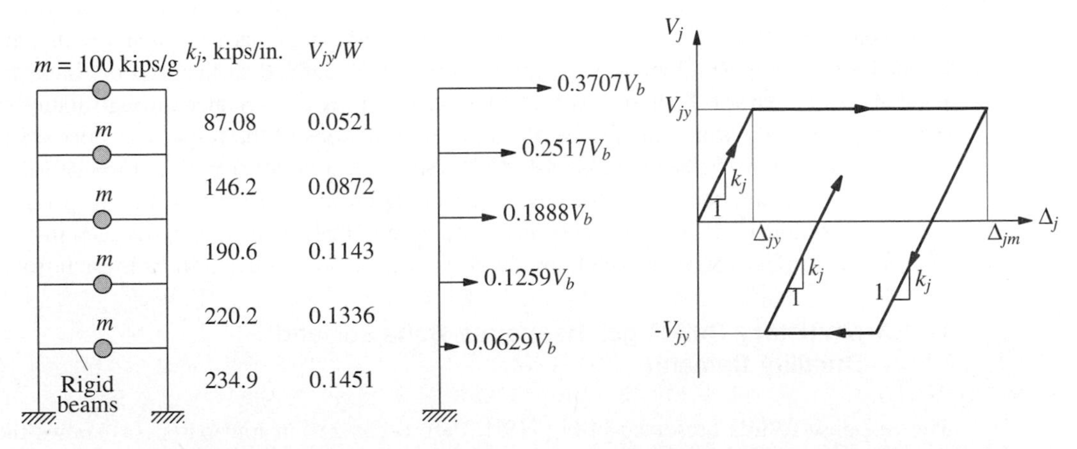

Figure 19.1.2 Five-story building: (a) system properties; (b) code forces; (c) elastoplastic relation between story shear and story drift.

specifies the lateral forces at the five floors in terms of the base shear V_b, as shown in Fig. 19.1.2b. The relative values of the story stiffnesses are established by the second of the above-mentioned requirements and their absolute values by the first requirement, resulting in the data of Fig. 19.1.2a. Note, in passing, that the selection of 0.8 sec for the fundamental period is based on periods of five-story steel moment-resisting frame buildings determined from their motions recorded during earthquakes.

The story yield strengths V_{jy} (see Fig. 19.1.2c) are determined from the response spectrum for elastoplastic SDF systems corresponding to an allowable ductility. If the structure is to be designed for the El Centro ground motion with an allowable ductility of $\mu = 4$, the relevant response spectrum is the curve identified by $\mu = 4$ in Fig. 7.5.2 or 7.5.3. Using this spectrum and assuming the structure to be linearly elastic with mass and stiffness properties defined in Fig. 19.1.2a, the story shears are determined by the response spectrum analysis procedure, including all five modes (Section 13.8). The resulting story shears define the story yield strengths V_{jy} of the elastoplastic system. Normalized relative to the total weight $W = 500$ kips of the building, these story yield strengths V_{jy}/W are shown in Fig. 19.1.2a.

Assuming the structure to be linearly elastic with mass and stiffness properties defined in Fig. 19.1.2a, the viscous damping in the structure is defined by Rayleigh damping. The coefficients a_0 and a_1 in Eq. (11.4.7) are determined by the procedures of Section 11.4 to be consistent with 5% damping in the first two natural vibration modes. The resulting damping matrix gives damping ratios of 6.5, 8.2, and 10.1% in the third, fourth, and fifth modes, respectively. Observe that there is a discrepancy between these damping values and the 5% damping implied in the response spectrum of Fig. 7.5.2 used to calculate story yield strengths. As a result, the contributions of the third, fourth, and fifth modes were overestimated in computing the story yield strengths. This inconsistency is not significant, however, because we know from Section 18.4 that these higher modes should contribute little to the response of shear frames with a fundamental vibration period of 0.8 sec.

Next we determine the response to the El Centro ground motion of the five-story shear frame described in the preceding paragraphs. The differential equations to be solved are given by Eq. (9.8.2), where the mass matrix $\mathbf{m}$ is a diagonal matrix with $m_{jj} = m_j$, the lumped mass at the jth floor, the damping matrix $\mathbf{c}$ was defined earlier, and the inelastic resisting force vector is $\mathbf{f}_S(\mathbf{u}, \dot{\mathbf{u}})$. At each time instant, $\mathbf{f}_S(\mathbf{u}, \dot{\mathbf{u}})$ is related through statics to the story shears V_j, which in turn are elastoplastic functions of the respective story drifts Δ_j and depend on whether the associated velocity is positive or negative (Fig. 19.1.2c). These equations are solved by the average acceleration method using a time step $\Delta t = 0.02$ sec (Section 15.3.1). The results of this computer analysis included, for each instant of time, the displacements of each floor, the drift in each story, and the shear in each story.

19.1.3 Multistory Buildings: Response Behavior and Ductility Demand

The response results presented in Fig. 19.1.3 are organized in four parts: (a) shows the displacement $u_5(t)$ of the top floor; (b) shows the fifth-story shear $V_5(t)$; (c) identifies the time intervals during which the fifth story is yielding; and (d) shows the relation

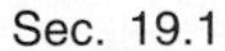

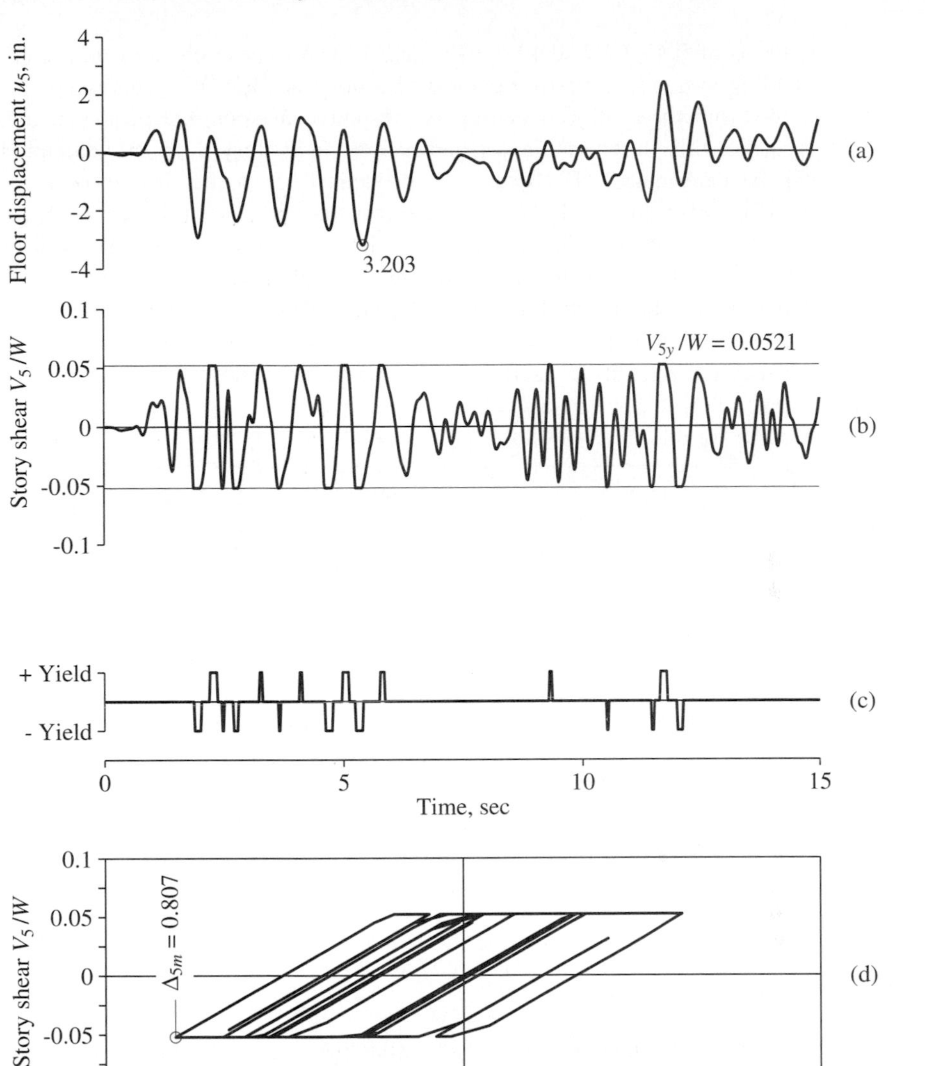

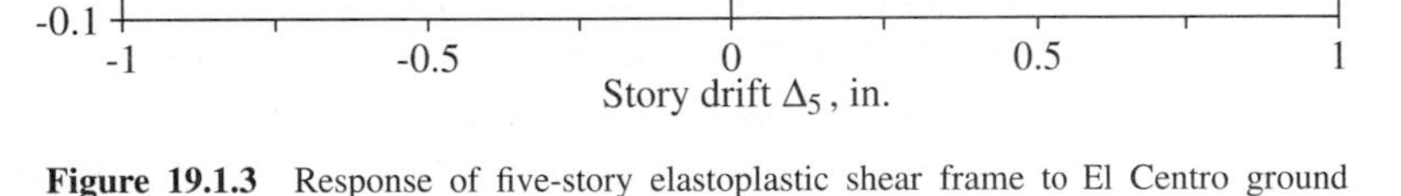

Figure 19.1.3 Response of five-story elastoplastic shear frame to El Centro ground motion.

between story shear V_5 and story drift Δ_5 as the structure goes through several cycles of motion. The peak displacement u_{5m} is 3.203 in. and the system is seen to oscillate, but not always about its initial undeformed position. Because of yielding, the system drifts from its initial position, then oscillates around this new deformed position until it gets shifted by another episode of yielding. After the end of the full 30 sec of ground shaking, the structure is permanently deformed with floor displacements (from top to bottom) of 0.43, 0.77, 0.40, 0.013, and 0.13 in. These phenomena are characteristic of yielding systems, as first observed in Section 7.4 for SDF systems.

From the results of earthquake response analysis, the peak values of story drifts Δ_{jm}, floor displacements u_{jm} (Fig. 19.1.4a), and story ductility demands (Fig. 19.1.4b) can be determined. For example, the peak drift in the fifth story is $\Delta_{5m} = 0.8070$ in. (Fig. 19.1.3d), and the yield deformation for the fifth story, $\Delta_{5y} = V_{5y}/k_5 = 0.0521W/87.08 = 0.2992$ in. The ductility demand $\mu_5 = \Delta_{5m}/\Delta_{5y} = 0.8070/0.2992 = 2.697$; ductility demands are determined similarly for other stories. These ductility demands and the peak story drifts Δ_{jm} represent design requirements imposed by the ground motion on the system. The structure should be designed and detailed to possess the required ductility capacity and deformation capacity.

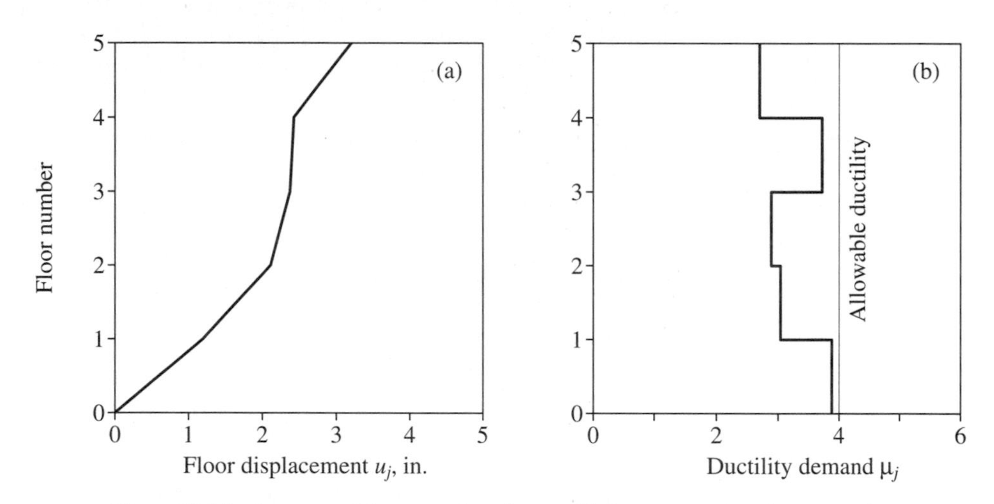

Figure 19.1.4 Response of five-story shear frame to El Centro ground motion: (a) peak floor displacements; (b) peak story-ductility demands.

The story ductility demands for a multistory building vary over its height and differ from the allowable ductility used in defining the design spectrum and computing the story yield strengths. This is clear from Fig. 19.1.4 for the five-story shear building with an allowable ductility of $\mu = 4$; observe the variability of the ductility demands over the height and how they differ from the allowable ductility. Also observe that for this system and excitation, the variations in ductility demand are not especially large, and in none of the stories does the ductility demand exceed the allowable ductility. However, we shall see later that ductility demands may vary considerably over the building height, depending on the building height and on the relative values of story strengths, and these demands on some stories may exceed the allowable ductility.

These are limitations of the design procedure wherein story yield strengths are determined by linear analysis of the system using the inelastic response (or design) spectrum corresponding to the allowable ductility. Thus a different approach to inelastic structural design is necessary so that the ductility demands remain within the allowable ductility. We present one such approach in Section 19.3.

19.2 BUILDINGS WITH "WEAK" OR "SOFT" FIRST STORY

The heightwise variation in ductility demands on multistory buildings depends, in part, on the relative yield strengths of the various stories. To demonstrate this important concept, we determine ductility demands for two extreme cases: (1) a building with a first story that is very "weak" (smaller yield strength) compared to the upper stories, and (2) a building with a first story that is very "soft" (smaller stiffness and smaller yield strength) compared to the other stories.

Consider first a five-story building with a weak first story. Its linearly elastic properties—and hence its natural periods, natural modes, and damping ratios—are identical to the "standard" shear frame described in Section 19.1.2; however, it is made stronger by increasing the yield strengths of the second through fifth stories by a factor of 4 while keeping the yield strength of the first story the same; the story yield strengths are given in Fig. 19.2.1. Thus the first story is no weaker than before; it is weak only relative to the second and higher stories, implying a sharp discontinuity of strength across the first floor. Using the numerical time-stepping procedures described in Section 19.1.2, the response of this weak-first-story system to the El Centro ground motion is determined, resulting in the peak floor displacements and peak story ductility demands shown in Fig. 19.2.2. A ductility factor less than or equal to 1 implies linearly elastic behavior.

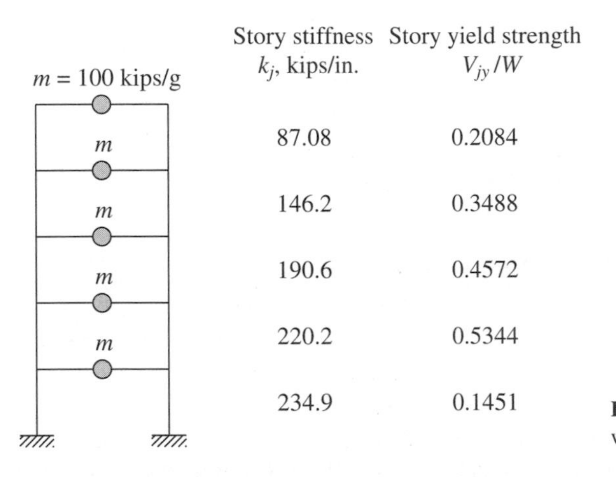

Figure 19.2.1 Five-story shear frame with weak first story.

It is apparent by comparing Fig. 19.2.2 with Fig. 19.1.4 that an increase in the yield strength of upper stories has a great influence on the heightwise variation of story drifts and yielding. The drifts were roughly similar in three of the five stories of the standard building, but the deformations are concentrated in the first story of the weak-first-story building. In particular, the top-floor displacement of 4.24 in. results from a 3.15-in. deformation in the first story, whereas the drifts in all the upper stories combined are

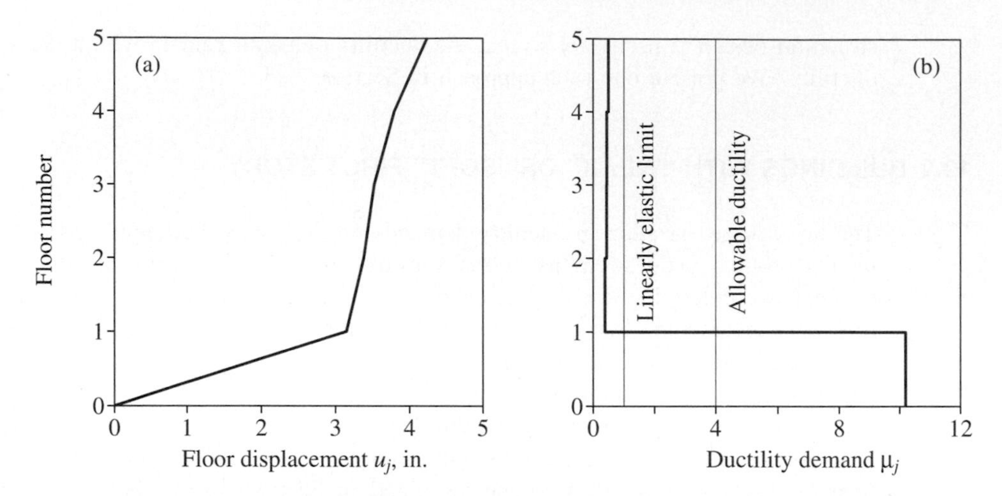

Figure 19.2.2 Response of five-story shear frame with weak first story to El Centro ground motion: (a) peak floor displacements; (b) peak story-ductility demands.

only 1.09 in. All stories of the standard building yield, but the upper stories of the weak-first-story building remain elastic because of their increased yield strength, and yielding is confined to the first story. Thus all the energy that was dissipated through yielding of the upper stories in the standard building must be dissipated by the weak first story, resulting in a ductility demand of about 10. Before discussing the implications of this large ductility demand on the first story, we consider the response of a more realistic soft-story building.

In actual buildings, if the first story is relatively weak, it is usually also relatively flexible because stiffness and strength are interrelated. Consider next such a soft-first-story building which is a variation on the "standard" shear frame described in Section 19.1.2. The stiffness and yield strength of the second through fifth stories are increased by a factor of 4, while keeping the first story unchanged, resulting in the properties shown in Fig. 19.2.3. Thus the first story is no weaker or more flexible than before;

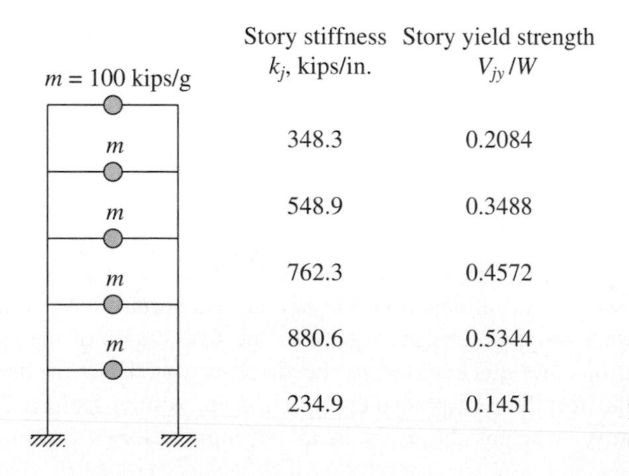

Figure 19.2.3 Five-story shear frame with soft first story.

it is weak and flexible only relative to the upper stories, implying sharp discontinuities in strength and stiffness across the first floor. Because the upper stories are stiffer, the natural periods and modes of the structure are modified. In particular, the fundamental period of the structure is reduced from 0.8 sec for the standard building to 0.56 sec for the soft-first-story building. The Rayleigh damping matrix is redefined to be consistent with 5% damping in the first two modes of this stiffer system.

Figure 19.2.4 shows the peak values of the floor displacements and of the story ductility demands for the soft-first-story building. As in the weak-first-story building, the upper stories remain elastic, and yielding is confined to the first story, resulting in a ductility demand of over 6. A comparison of Figs. 19.2.4 and 19.2.2 indicates that the increased stiffness of the upper stories has the effect of significantly reducing the drifts in these stories; the drift in the first story and the associated ductility demand are also reduced for this particular structure and excitation. These are still large, however, because yielding is confined to the first story.

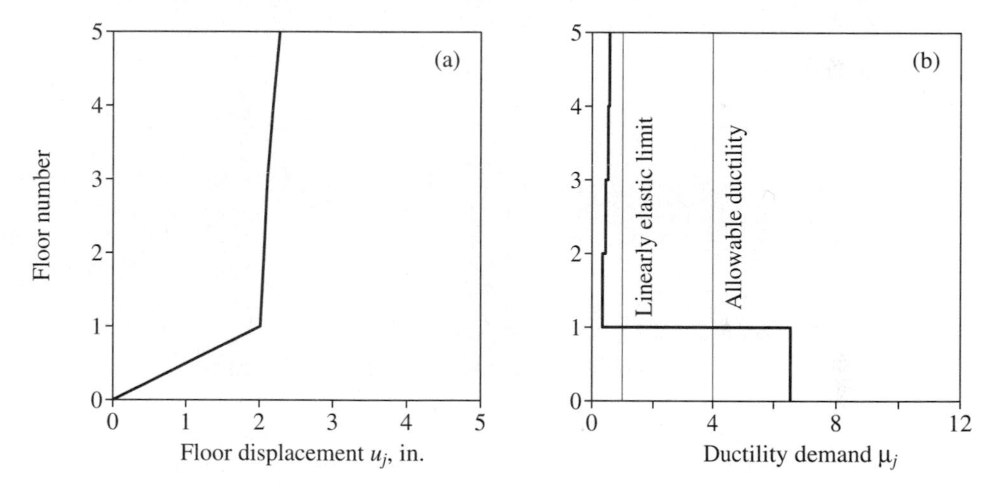

Figure 19.2.4 Response of five-story shear frame with soft first story to El Centro ground motion: (a) peak floor displacements; (b) peak story-ductility demands.

A well-known example of a building with a soft first story is the Olive View Hospital building. This was a six-story reinforced-concrete building with its first story partially under ground. The lateral force-resisting system included large walls in the upper four stories which did not extend down to the lower two stories (Fig. 19.2.5). These discontinuous shear walls created a large discontinuity in strength and stiffness at the second-floor level. During the February 9, 1971 San Fernando earthquake this structure behaved as suggested by the preceding results from dynamic analysis of a hypothetical building. The upper four stories of this building escaped with minor damage, with the damage decreasing toward the top. Most of the damage was concentrated in the partially underground story and the first aboveground story, with permanent drift in the latter

Figure 19.2.5 Olive View Hospital building. The shear walls in the upper four stories did not extend to the lower two stories. (From K. V. Steinbrugge Collection, courtesy of the Earthquake Engineering Reserch Center, University of California at Berkeley.)

story exceeding 30 in. (Fig. 19.2.6). This large drift imposed very severe deformation and ductility demands on the first-story columns. As a result, the tied columns failed in a brittle manner (Fig. 19.2.7); however, ductile behavior of the spirally reinforced columns prevented the collapse of the building (Fig. 19.2.8). This building, completed only a few months prior to the earthquake, was damaged so severely that it had to be demolished. There are many such examples of severe damage to buildings with a soft first story during past earthquakes.

Although soft-first-story buildings are obviously not appropriate for earthquake-prone regions, their response during past earthquakes suggests the possibility of reducing the damage to a building by a base isolation system that acts like a soft first story. This topic is discussed in Chapter 20.

Figure 19.2.6 Large deformations in the first aboveground story of the Olive View Hospital building due to the San Fernando earthquake of February 9, 1971. (Courtesy of G. W. Housner.)

Figure 19.2.7 Brittle failure of a tied corner column of the Olive View Hospital building. (From K. V. Steinbrugge Collection, courtesy of the Earthquake Engineering Research Center, University of California at Berkeley.)

Figure 19.2.8 Large permanent deformation of a spirally reinforced column of the Olive View Hospital building. (From K. V. Steinbrugge Collection, courtesy of the Earthquake Engineering Research Center, University of California at Berkeley.)

19.3 BUILDINGS DESIGNED FOR CODE FORCE DISTRIBUTION

The preliminary design of most buildings is based on the equivalent static forces specified by the governing building code (Chapter 21). In this section, therefore, we evaluate the ductility demands on multistory buildings designed for lateral forces distributed over their height in accordance with a building code, and compare these demands with the allowable ductility. This comparison will provide some insight into how well the heightwise distribution of code forces ensures that the story ductility demands are uniform over the building height and close to the allowable ductility.

19.3.1 Systems Considered

Buildings with 2, 5, 10, 20, 30, or 40 stories, idealized as elastoplastic shear frames, are analyzed. The properties of the five-story frame are defined first to emphasize similarities and differences relative to the system defined in Section 19.1.2. The mass, damping, and linear elastic stiffness properties are unchanged. However, now the story yield strengths are not determined by a response spectrum analysis using an inelastic response spectrum; instead, their heightwise distribution is based on the 1994 *Uniform Building Code*. For a five-story building with equal floor masses, equal story heights, and fundamental period $T_1 = 0.8$ sec, the code-specified forces are shown in Fig. 19.3.1b in terms of the base shear V_b. The resulting story shears shown in Fig. 19.3.1c define the story yield strengths. The base shear, which is also the yield strength for the first story, is not taken as the code value but is selected as

$$V_{by} = \frac{A_y}{g} W \tag{19.3.1}$$

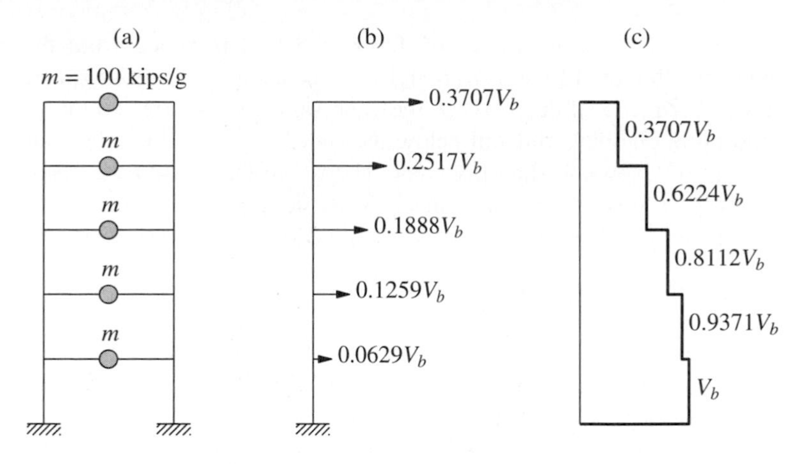

Figure 19.3.1 (a) Five-story shear frame; (b) UBC forces; (c) story yield strengths.

where W is the total weight of the building ($W = 5w = 500$ kips for the five-story building); and A_y is the pseudo-acceleration corresponding to $T_1 = 0.8$ sec, $\zeta = 5\%$, and the allowable ductility, determined from the inelastic response spectrum for the El Centro ground motion (Fig. 7.5.2 or 7.5.3). The base shear values for the five-story building, corresponding to four values of the allowable ductility factor—$\mu = 1, 2, 4$, and 8—are presented in Table 19.3.1.

TABLE 19.3.1 BASE SHEAR FOR MULTISTORY BUILDINGS: $V_{by}/W = A_y/\mathrm{g}$

Allowable Ductility	$T_1 = 0.5$ sec (2-Story)	$T_1 = 0.8$ sec (5-Story)	$T_1 = 1.6$ sec (10-Story)	$T_1 = 3$ sec (20-Story)	$T_1 = 4$ sec (30-Story)	$T_1 = 5$ sec (40-Story)
1	0.7585	0.4962	0.1757	0.1227	0.0646	0.0415
2	0.4465	0.2350	0.0876	0.0512	0.0347	0.0244
4	0.2177	0.1727	0.0455	0.0213	0.0117	0.0086
8	0.1588	0.0598	0.0276	0.0141	0.0072	0.0036

In a similar manner, the base shear is determined for the 2-, 10-, 20-, 30-, and 40-story buildings. The fundamental period T_1 of these buildings is selected as 0.5, 1.6, 3, 4, and 5 sec, respectively, based as before on periods of steel moment-resisting frame buildings determined from their motions recorded during earthquakes. The resulting base shear values are presented in Table 19.3.1. For these buildings the story stiffnesses are determined by the procedure of Section 19.1.2 and the heightwise distribution of story yield strengths from the code-specified forces (as in Fig. 19.3.1 for the five-story building). These properties are listed in Appendix 19. In each case the viscous damping is represented by Rayleigh damping, with the coefficients a_o and a_1 in Eq. (11.4.7) selected to give 5% damping in the first two natural vibration modes. For each building, the lumped masses at all the floors are the same, 100 kips.

The base shear values of Table 19.3.1, determined from the inelastic response spectrum for the El Centro ground motion, differ from those specified by the building code (Chapter 21); they exceed the code values for shorter buildings or lower values of allowable ductility, and fall below the code values for the taller buildings or the larger values of allowable ductility. Our choice to determine base shear from the inelastic response spectrum was motivated by the desire to compare the ductility demand in a multistory building with a corresponding SDF system described next.

19.3.2 Corresponding SDF System

In the linearly elastic range, the natural period and damping ratio of the corresponding SDF system are the same as the fundamental mode properties of the multistory building. The weight of the corresponding SDF system is the same as the total weight of the multistory building. Both systems are elastoplastic with identical values for the yield base shear, defined in Table 19.3.1. By defining the yield base shear of the multistory building as the same as that of the corresponding SDF system, instead of the code value, the computed ductility demands will permit direct comparison between the two systems and with the allowable ductility.

19.3.3 Story-Ductility Demands

Consider first the buildings "designed" in Section 19.3.1 for allowable ductility $\mu = 1$. The dynamic response of these elastoplastic systems to the El Centro ground motion was determined (by methods described in Section 19.1.2), leading to the story ductility demands presented in Fig. 19.3.2. Before interpreting these results, we recall from Fig. 19.1.1a that the peak deformation in the corresponding SDF systems would just reach, but not exceed, the yield value (i.e., the ductility demand will be exactly 1). In contrast, the ductility demand exceeds 1 in a few stories of some of these multistory buildings, implying that these stories yield; and in other stories the ductility demand remains below 1, implying that these stories remain elastic. For the ground motion selected, there is no consistent pattern for the dependence of ductility demand on the number of stories in a building. However, research results for ductility demands averaged over many ground motions have shown consistent trends: as the fundamental period increases, the building has a greater tendency to yield in the lower and upper stories because of the increasing significance of the response due to higher vibration modes, demonstrated in Section 18.4. This trend can be especially pronounced for buildings with smaller values of the beam-to-column stiffness ratio ρ [Eq. (18.1.2)] because, as seen in Sections 18.4 and 18.5, the higher-mode responses are especially significant for such buildings.

Consider next the buildings "designed" in Section 19.3.1 for allowable ductility of $\mu = 4$. The dynamic response of these elastoplastic systems to the El Centro ground motion was determined leading to the story ductility demands presented in Fig. 19.3.3. Before interpreting these results, we recall from Fig. 19.1.1b that the ductility demand in the corresponding SDF systems would exactly equal the allowable ductility. In contrast, the ductility demand exceeds the allowable ductility in some stories of a multistory building. The results of Fig. 19.3.3 for one ground motion suggest trends that become

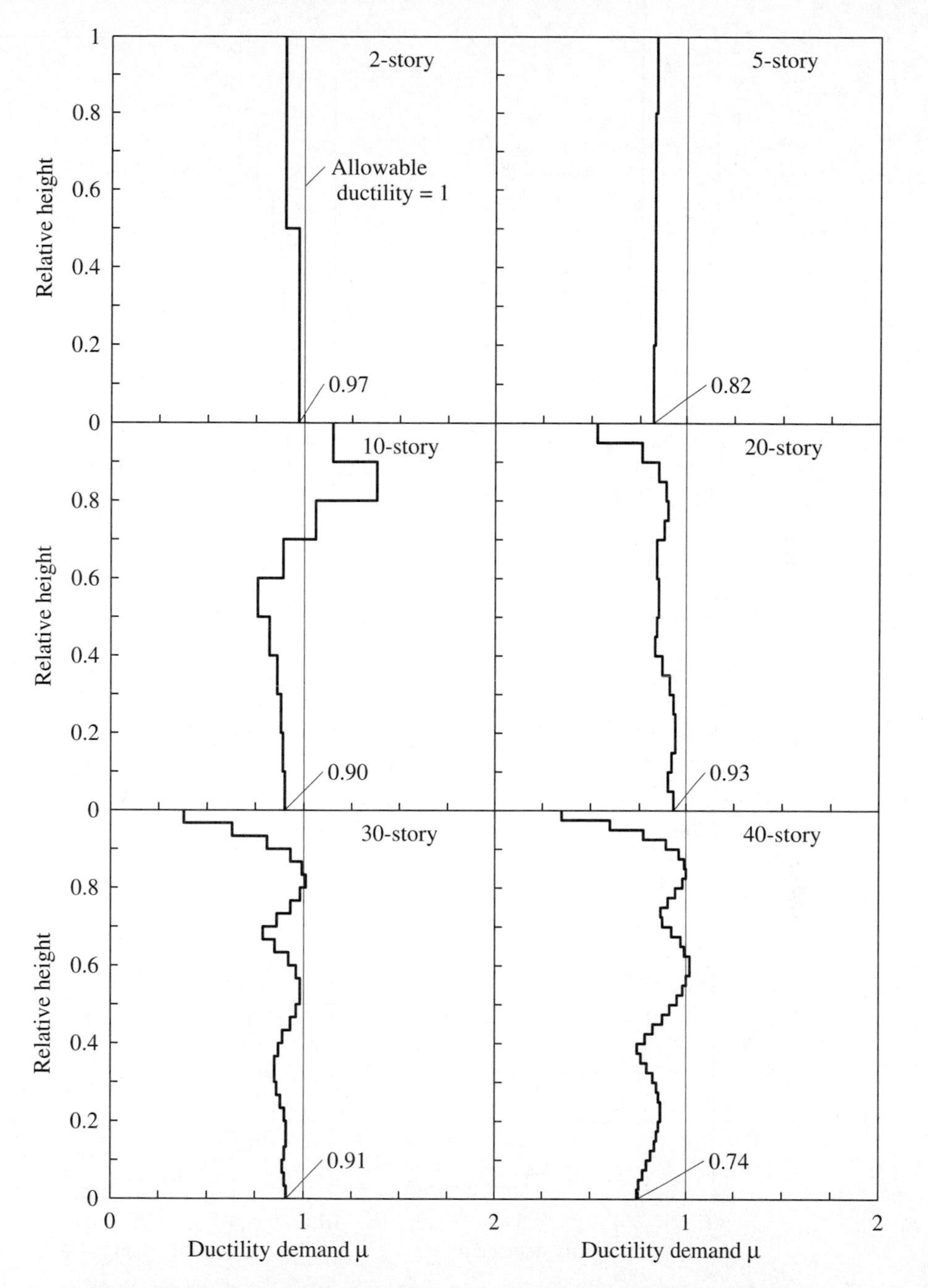

Figure 19.3.2 Story ductility demands in 2-, 5-, 10-, 20-, 30-, and 40-story shear frames due to El Centro ground motion, compared with allowable ductility $\mu = 1$.

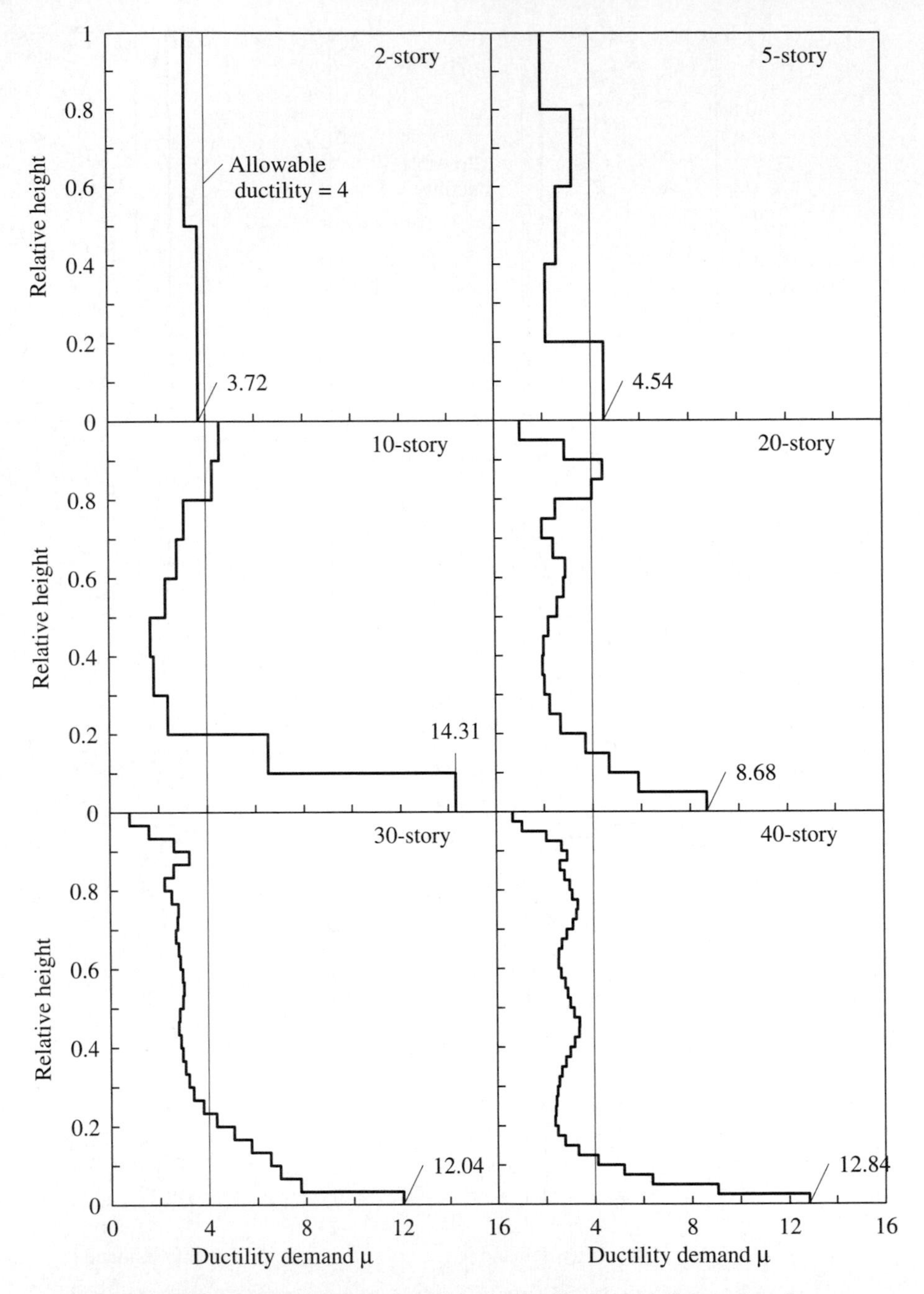

Figure 19.3.3 Story ductility demands in 2-, 5-, 10-, 20-, 30-, and 40-story shear frames due to El Centro ground motion, compared with allowable ductility $\mu = 4$.

evident from the story ductility demands averaged over many ground motions (not presented in this book):

1. The variation of ductility demands is roughly uniform over the height of two- and five-story buildings. For the taller (longer period) buildings the ductility demands are larger in the upper and lower stories and decrease in the middle stories. This pattern is related to the higher-mode response contributions that, as shown in Section 18.4, are more significant for longer-period buildings.
2. The deviation of story ductility demands from the allowable ductility increases for taller buildings.
3. The relative story yield strengths, which were chosen in conformance with the heightwise distribution of the earthquake forces specified in the 1994 *Uniform Building Code*, do not lead to equal ductility demand in all stories, but this is a desirable design objective.
4. In most cases the ductility demand in the first story is largest among all stories.

The ductility demand in the first story of buildings designed for an allowable ductility of $\mu = 4$ is plotted against the fundamental period T_1 in Fig. 19.3.4. The

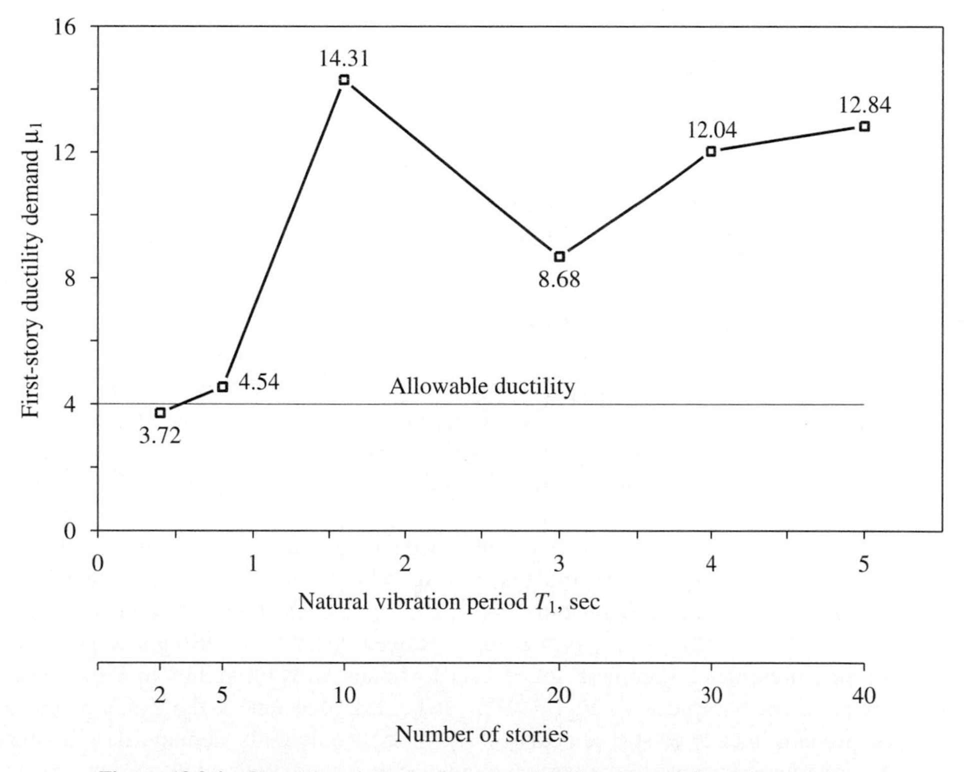

Figure 19.3.4 Comparison between first-story ductility demand for multistory shear frames and allowable ductility; El Centro ground motion.

ground motion selected imposes on some of the buildings a ductility demand over three times the allowable ductility, but there is no consistent pattern of variation with T_1. If results similar to Fig. 19.3.4 were generated for many ground motions and averaged, they would indicate that the first-story ductility demand increases with T_1 (i.e., it is small for short buildings and increases for taller structures).

Results similar to Figs. 19.3.2 and 19.3.3 were also developed for two additional "designs" of each of the six buildings: 2-, 5-, 10-, 20-, 30-, and 40-story systems. These additional designs correspond to allowable ductility of $\mu = 2$ and 8, with the corresponding yield values of base shear noted in Table 19.3.1. The entire set of results for four values of allowable ductility enable us to study further how the ductility demand compares with the allowable ductility.

For this purpose, consider a 20-story building and a corresponding SDF system, both with natural period of 3.0 sec. The base shear yield strength V_{by} of the two systems is the same and is given in Table 19.3.1; these data corresponding to allowable ductility $\mu = 1$, 2, 4, and 8 are repeated in Table 19.3.2. Dividing these V_{by} values by the value corresponding to $\mu = 1$ gives the data for normalized strength $\overline{V}_{by}$.

TABLE 19.3.2 DUCTILITY DEMANDS IN 20-STORY BUILDING AND CORRESPONDING SDF SYSTEM

Allowable Ductility, μ	Base Shear, V_{by}/W	Normalized Strength, $\overline{V}_{by}$	Ductility Demands	
			20-Story System	SDF System
1	0.1227	1.0	0.93	1
2	0.0512	0.417	2.55	2
4	0.0213	0.173	8.68	4
8	0.0141	0.115	15.8	8

Response analysis of the 20-story building with $V_{by}/W = 0.1227$, corresponding to allowable ductility $\mu = 1$, led to a first-story ductility demand of 0.93 (Fig. 19.3.2). Similar to $V_{by}/W = 0.0213$, corresponding to allowable ductility $\mu = 4$, the ductility demand is 8.68 (Fig. 19.3.3). The first-story ductility demands in Table 19.3.2 for allowable ductility $\mu = 2$ and 8 were determined by additional, similar analyses. As mentioned before, the ductility demand on the corresponding SDF system is identical to the allowable ductility; this is also noted in Table 19.3.2. As the base shear or normalized strength decreases, the ductility demand increases and these relationships for the 20-story system and the corresponding SDF system are shown in Fig. 19.3.5. Observe that for each system Table 19.3.2 gives only four data points, corresponding to the four normalized strength values. However, for SDF systems with $T_n = 3$ sec, we had presented a complete set of data for many strength values in Fig. 7.5.1; these data have been included in Fig. 19.3.5. It is clear that unless the systems are designed to remain linearly elastic or nearly so ($\mu \approx 1$), the ductility demand in a 20-story building is larger compared to the corresponding SDF system, both with the same base shear yield strength, and the discrepancy between the two increases with the allowable ductility.

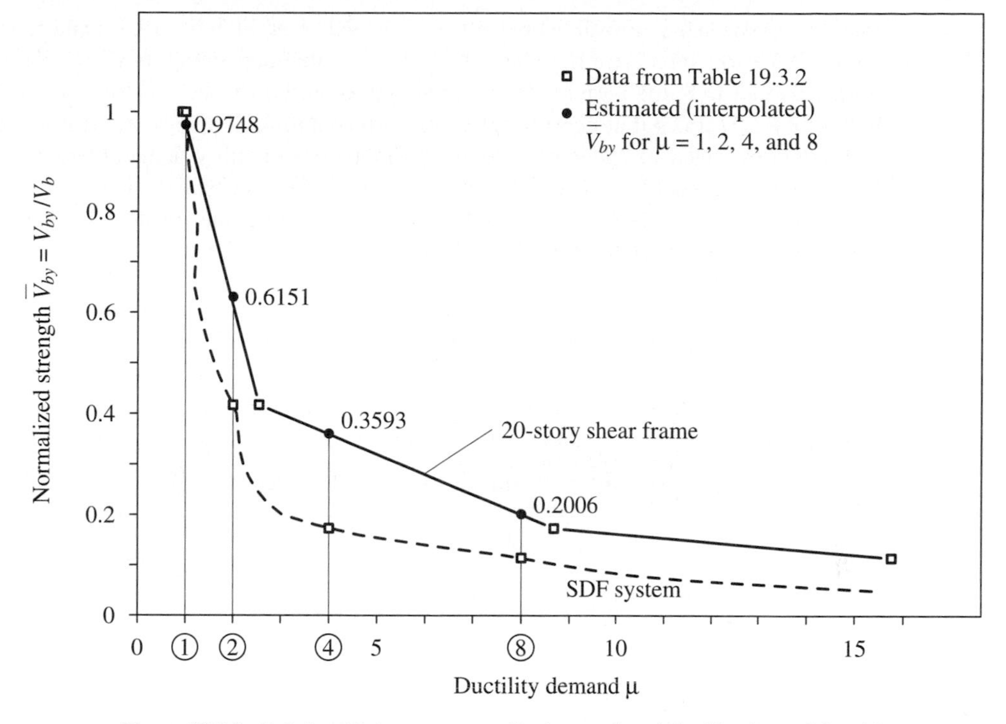

Figure 19.3.5 Relationship between normalized strength and ductility demand for: (1) 20-story shear frame with $T_1 = 3$ sec, and (2) corresponding SDF system; El Centro ground motion.

19.3.4 Base Shear Yield Strength

How should the previously designed multistory buildings be modified to ensure that the ductility demands do not exceed the allowable ductility? This question may be approached in several ways. For example, we may try to optimize the heightwise distribution of story stiffness and strength with the objective that the ductility demands are similar in all stories and less than the allowable ductility. Alternatively, we may determine the base shear yield strength necessary, in conjunction with heightwise distribution of forces according to the building code, to ensure that the first-story ductility demand (usually, the largest among all stories) does not exceed the allowable ductility.

We adopt the latter approach, which is much simpler and appeals to intuition. Corresponding to allowable ductility of $\mu = 1$, 2, 4, and 8, the normalized strengths $\overline{V}_{by}$ for 20-story buildings are noted in Fig. 19.3.5, and these are multiplied by the normalizing value, $V_b/W = 0.1227$ (from Table 19.3.2), to obtain the V_{by} values given in Table 19.3.3 for 20-story systems (e.g., for $\mu = 4$, $\overline{V}_{by} = 0.3593$ from Fig. 19.3.5 and $V_{by}/W = 0.3593 \times 0.1227 = 0.0441$). This result and similar calculations for the other values of μ lead to the data of Table 19.3.3. These base shear yield strengths for 20-story buildings are compared with the base shear yield strengths for the corresponding SDF systems for the same allowable ductility values, presented in Table 19.3.2 and repeated in Table 19.3.3 for convenient reference. It is apparent that depending on allowable ductility in the range 2 to 8, the base shear yield strength required for the 20-story system should be 1.48 to 2.07 times that necessary for the corresponding SDF system. If this increased strength is provided in 20-story systems, the first-story ductility demand should be close to the allowable ductility. The only reason the two may not be exactly equal is because of the coarseness of the data in Fig. 19.3.5, requiring linear interpolation between widely spaced data points to determine the normalized strength of 20-story systems.

TABLE 19.3.3 BASE SHEAR YIELD STRENGTH OF 20-STORY BUILDINGS AND CORRESPONDING SDF SYSTEMS

Allowable Ductility	V_{by}/W		$\frac{(V_{by})_{MDF}}{(V_{by})_{SDF}}$
	20-Story System	SDF System	
1	0.1196	0.1227	0.975
2	0.0755	0.0512	1.48
4	0.0441	0.0213	2.07
8	0.0246	0.0141	1.74

The required base shear yield strength is plotted in Fig. 19.3.6 as a function of the natural vibration period for multistory buildings. In this figure for an allowable ductility $\mu = 4$, the data point for the 20-story building ($T_1 = 3$ sec) comes from Table 19.3.3. Similar calculations (not included here) for 2-, 5-, 10-, 30-, and 40-story buildings described earlier with fundamental periods $T_1 = 0.5$, 0.8, 1.6, 4.0, and 5.0 sec, respectively, led to the remaining data points. Also shown in Fig. 19.3.6 is the inelastic response spectrum for a constant ductility of 4 (first presented in Fig. 7.5.2), which gives the base shear for SDF systems. It is clear that for the same allowable ductility of 4, a multistory building should be designed for a larger base shear relative to an SDF system having the same weight and natural period. How much larger should be the base shear yield strength of an MDF system depends on the fundamental natural period, among other factors. In particular, for a 20-story building with $T_1 = 3$ sec, the multiplying factor is 2.07 (Table 19.3.3).

Finally, we address how the results presented can be incorporated in the inelastic design of multistory buildings. The response (or design) spectrum for the base shear yield strength V_{by} for SDF systems needs to be modified to account for two factors that exist in MDF systems (but not in SDF systems): multimode effects and heightwise variation of ductility demand. Presented in Fig. 19.3.7 is the modification factor, $(V_{by})_{MDF}/(V_{by})_{SDF}$, where $(V_{by})_{MDF}$ and $(V_{by})_{SDF}$ are the base shear yield strengths of MDF and SDF systems, respectively. For an allowable ductility of 4, these strength values are available from Table 19.3.3 and their ratio gives the $\mu = 4$ curve in Fig. 19.3.7; similar analyses led to the curves for $\mu = 1$ and 8. These results suggest trends that would become evident

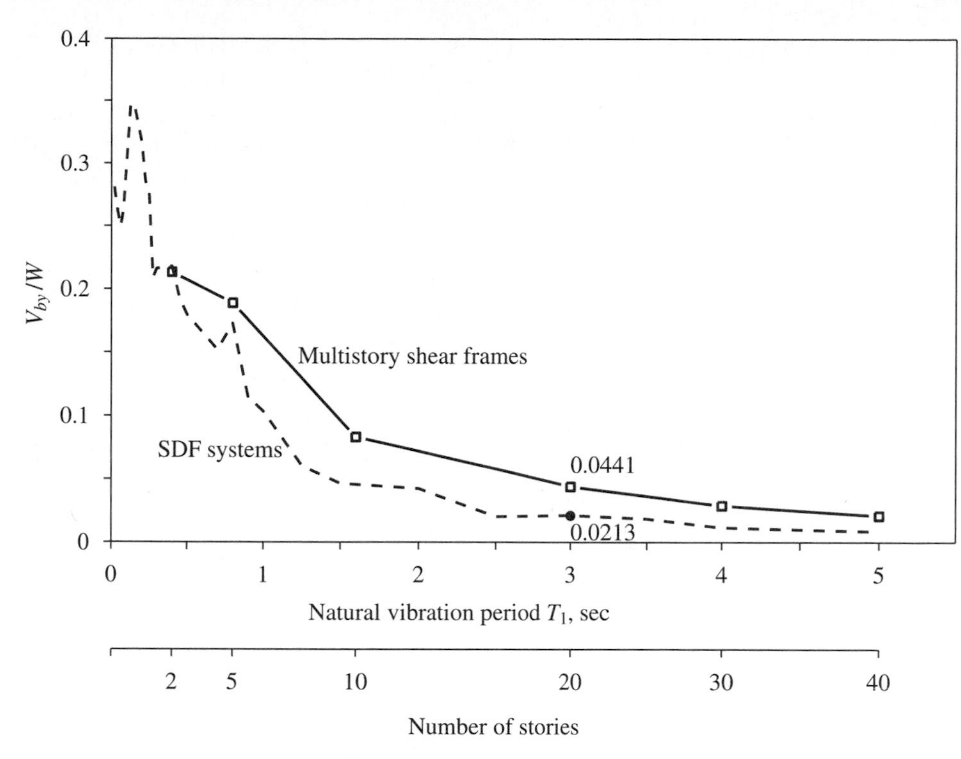

Figure 19.3.6 Required base shear yield strength for multistory shear frames and SDF systems associated with allowable ductility of 4; El Centro ground motion.

from the modification factor averaged over many ground motions (not presented in this book): (1) The required strength modification factor for MDF effects varies between 1 and 4, depending significantly on the allowable ductility and on the fundamental vibration period T_1; (2) the modification factor increases with T_1 and hence the number of stories; (3) the modification factor increases with the allowable ductility; and (4) elastic MDF systems attract lower base shears than those predicted by corresponding SDF systems. This is consistent with the results of Fig. 18.4.1; for five-story shear buildings ($\rho = \infty$), $W_1^* = 0.880W$ and their base shear is smaller than for corresponding SDF systems if $T_1 < 6$ sec.

Several factors should be considered in estimating the base shear yield strength of multistory buildings in order to keep the ductility demands below the allowable ductility. These factors include the fundamental period and the allowable ductility, as demonstrated by Fig. 19.3.7. In addition, the criteria should consider different plastic failure mechanisms: story mechanisms (as in the shear buildings considered here and in strong-beam/weak-column systems), beam mechanisms (as would develop for strong-column/weak-beam systems), and plastic mechanisms only in the first story (as in a soft-first-story building). Generally, ductility demands depend strongly on the failure mechanism.

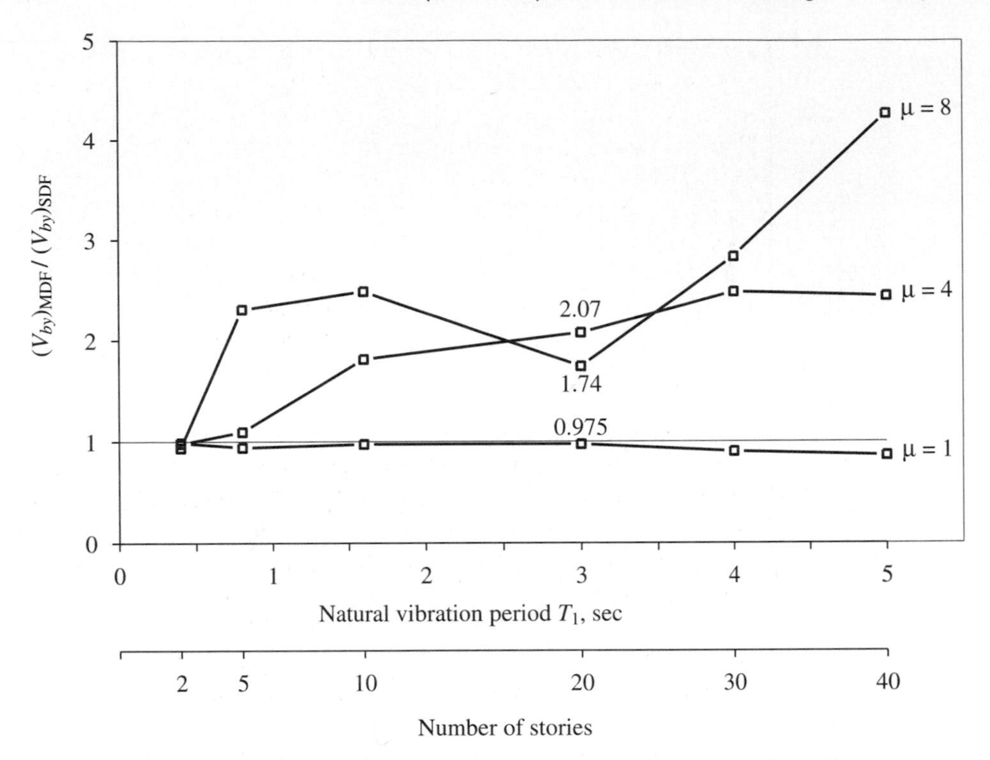

Figure 19.3.7 Modification factor to obtain the base shear yield strength of a multistory shear frame from the base shear yield strength of the corresponding SDF system for three values of allowable ductility μ.

19.4 LIMITED SCOPE

This chapter has been limited to identifying how the inelastic response behavior of multistory buildings differs from SDF systems and to developing an approach for calculating the base shear yield strength required for multistory buildings from the inelastic response (or design) spectra (for SDF systems). The results presented were restricted to shear buildings, elastoplastic force–deformation relations, and a single ground motion. In this section, the implications of each of these restrictions is discussed briefly.

While the shear building assumption was convenient to develop basic concepts, for practical results it is necessary to work with realistic idealizations of multistory buildings considering relative flexibility and relative strength of beams, columns, and other structural elements. For such systems the ductility demand and allowable ductility introduced here should be associated with the overall force–deformation relation of the structure rather than with the local behavior of individual structural elements and joints. However, local ductility is an important consideration and can be expressed either as the ratio of total to yield-limit curvatures at a given section of a structural element or as the ratio of total to yield-limit rotations at the end of an element. These ductility demands on elements and joints are influenced by the relative strengths of beams, columns, and other structural elements, in addition to the relative story strengths. The local ductility demand on elements and connections is generally much larger than the overall ductility demand, with this difference depending on the structural system and whether inelastic deformations are distributed throughout a building or concentrated in one story or a few members of the building. The structural elements and joints should be designed and detailed in such a way that the structure is capable of mobilizing the ductility demands and deformation demands at the overall and local levels. Detailed discussion of these issues is beyond the scope of this book.

While the elastoplastic force–deformation relation was convenient for our limited objective, the results may not apply directly to practical structures exhibiting strain hardening, stiffness degradation, or strength degradation. The earthquake response of such systems has been the subject of many research studies.

The response of inelastic systems is especially sensitive to the detailed time variation of the ground motions. Therefore, results for one excitation do not provide systematic trends, as seen in Fig. 19.3.7. As mentioned earlier, researchers have repeated computations for many ground motions and studied the statistics of response to obtain consistent trends for base shear strength with increasing fundamental period and allowable ductility.

Starting in the 1960s, the inelastic response of multistory buildings to earthquakes has attracted the attention of many researchers and a large body of literature exists; therein many facets of the subject are discussed that could not be included here.

FURTHER READING

Clough, R. W., and Benuska, K. L., "Nonlinear Earthquake Behavior of Tall Buildings," *Journal of the Engineering Mechanics Division, ASCE*, **93**, 1967, pp. 129–146.

Mahin, S. A., Bertero, V. V., Chopra, A. K., and Collins, R. G., "Response of the Olive View Hospital Main Building During the San Fernando Earthquake," *Report No. EERC 76-22*, Earthquake Engineering Research Center, University of California, Berkeley, Calif., October 1976.

Nassar, A. A., and Krawinkler, H., "Seismic Demands for SDOF and MDOF Systems," *Report No. 95*, The John A. Blume Earthquake Engineering Center, Stanford University, Stanford, Calif., June 1991.

Veletsos, A. S., and Vann, W. P., "Response of Ground Excited Elastoplastic Frames," *Journal of the Structural Division, ASCE*, **97**, 1971, pp. 1257–1281.

APPENDIX 19: PROPERTIES OF MULTISTORY BUILDINGS

The story stiffnesses k_j (kips/in.) and story yield strengths V_{jy}, normalized by base shear V_b, of the multistory buildings described in Section 19.3.1 are listed below.

Two-story building

k_j = 191.6, 127.7
V_{jy}/V_b = 1, 0.6667

Five-story building

k_j = 234.9, 220.1, 190.6, 146.2, 87.08
V_{jy}/V_b = 1, 0.9371, 0.8112, 0.6224, 0.3707

Ten-story building

k_j = 210.8, 207.4, 200.6, 190.4, 176.8, 159.8, 139.4, 115.5, 88.29, 57.65
V_{jy}/V_b = 1, 0.9839, 0.9516, 0.9031, 0.8385, 0.7578, 0.6609, 0.5479, 0.4188, 0.2735

Twenty-story building

k_j = 222.4, 221.5, 219.9, 217.3, 214.0, 209.8, 204.8, 198.9, 192.3, 184.7, 176.4, 167.2, 157.1, 146.2, 134.5, 122.0, 108.6, 94.38, 79.32, 63.43

V_{jy}/V_b = 1, 0.9962, 0.9887, 0.9774, 0.9624, 0.9436, 0.9210, 0.8947, 0.8646, 0.8307, 0.7931, 0.7517, 0.7066, 0.6577, 0.6050, 0.5486, 0.4884, 0.4244, 0.3567, 0.2852

Thirty-story building

k_j = 274.1, 273.6, 272.8, 271.4, 269.7, 267.5, 264.8, 261.7, 258.2, 254.2, 249.8, 244.9, 239.6, 233.9, 227.7, 221.0, 214.0, 206.5, 198.5, 190.1, 181.3, 172.0, 162.2, 152.1, 141.5, 130.4, 118.9, 107.0, 94.61, 81.78

V_{jy}/V_b = 1, 0.9984, 0.9952, 0.9903, 0.9839, 0.9758, 0.9661, 0.9548, 0.9419, 0.9274, 0.9113, 0.8935, 0.8742, 0.8532, 0.8306, 0.8065, 0.7806, 0.7532, 0.7242, 0.6935, 0.6613, 0.6274, 0.5919, 0.5548, 0.5161, 0.4758, 0.4339, 0.3903, 0.3452, 0.2984

Forty-story building

k_j = 309.3, 309.0, 308.4, 307.6, 306.5, 305.1, 303.4, 301.4, 299.1, 296.6, 293.7, 290.6, 287.2, 283.6, 279.6, 275.3, 270.8, 266.0, 260.9, 255.5, 249.9, 243.9, 237.7, 231.2, 224.4, 217.4, 210.0, 202.4, 194.4, 186.2, 177.8, 169.0, 159.9, 150.6, 141.0, 131.1, 120.9, 110.4, 99.67, 88.64

V_{jy}/V_b = 1, 0.9991, 0.9973, 0.9945, 0.9909, 0.9863, 0.9808, 0.9744, 0.9671, 0.9588, 0.9497, 0.9396, 0.9287, 0.9168, 0.9040, 0.8902, 0.8756, 0.8601, 0.8436, 0.8262, 0.8079, 0.7887, 0.7686, 0.7476, 0.7256, 0.7027, 0.6790, 0.6543, 0.6287, 0.6021, 0.5747, 0.5463, 0.5171, 0.4869, 0.4558, 0.4238, 0.3909, 0.3570, 0.3223, 0.2866

20

Earthquake Dynamics of Base-Isolated Buildings

PREVIEW

The concept of protecting a building from the damaging effects of an earthquake by introducing some type of support that isolates it from the shaking ground is an attractive one, and many mechanisms to achieve this result have been proposed. Although the early proposals go back 100 years, it is only in recent years that *base isolation* has become a practical strategy for earthquake-resistant design. In this chapter we study the dynamic behavior of buildings supported on base isolation systems with the limited objective of understanding why and under what conditions isolation is effective in reducing the earthquake-induced forces in a structure. Base isolation is currently an active and expanding subject, however, and a large body of literature exists on various aspects of base isolation: testing and mechanics of hardware in isolation systems, nonlinear dynamic analysis, shaking table tests, design projects, field installation, and field performance.

20.1 ISOLATION SYSTEMS

Despite wide variation in detail, base isolation techniques follow two basic approaches with certain common features. In the first approach the isolation system introduces a layer of low lateral stiffness between the structure and the foundation. With this isolation layer the structure has a natural period that is much longer than its fixed-base natural period. As shown by the elastic design spectrum of Fig. 20.1.1, this lengthening of period can reduce the pseudo-acceleration and hence the earthquake-induced force in the structure,

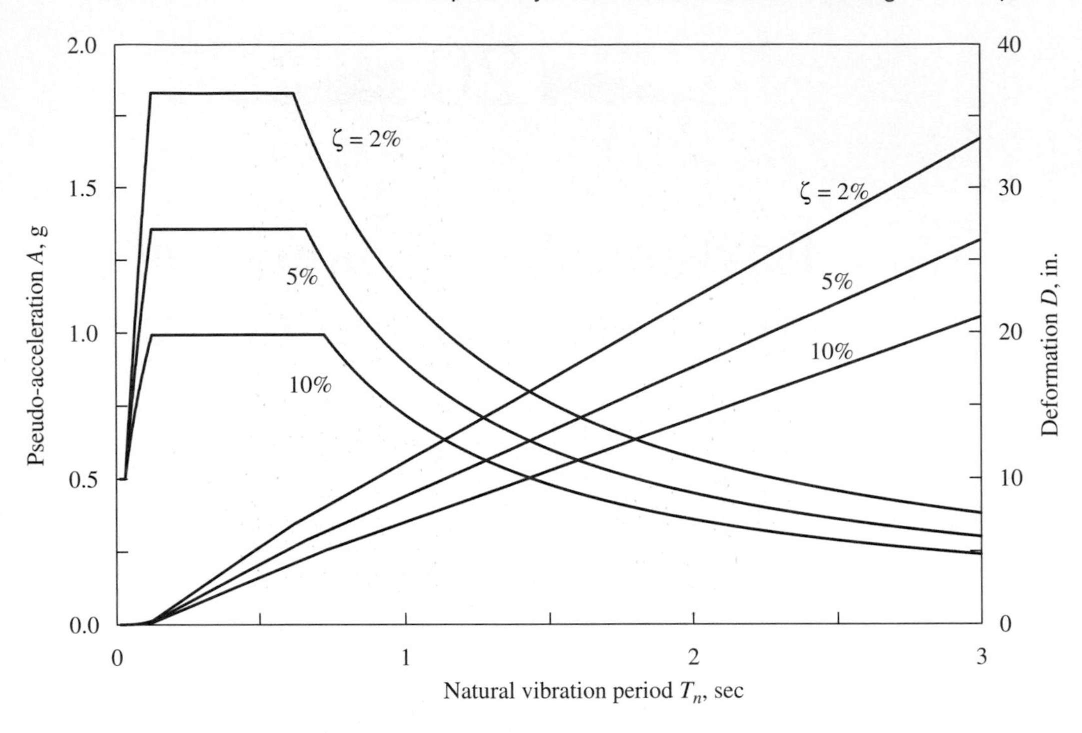

Figure 20.1.1 Elastic design spectrum.

but the deformation is increased; this is the deformation across the isolation system. This type of isolation system is effective even if the system is linear and undamped. Damping is beneficial, however, in further reducing the forces in the structure and the deformation in the isolation system.

The most common system of this type uses short, cylindrical bearings with alternating layers of steel plates and hard rubber (Fig. 20.1.2). Interposed between the base of the structure and the foundation, these laminated bearings are strong and stiff under vertical loads, yet very flexible under lateral forces (Fig. 20.1.3). Because the natural damping of the rubber is low, additional damping is usually provided by some form of mechanical damper. These have included lead plugs within the bearings, hydraulic dampers, or steel coils. Metallic dampers provide energy dissipation through yielding, thus introducing nonlinearity in the system.

The second type of isolation system uses rollers or sliders between the foundation and the base of the structure. The shear force transmitted to the structure across the isolation interface is limited by keeping the coefficient of friction as low as practical. However, the friction must be sufficiently high to sustain strong winds and small earthquakes without sliding, a requirement that reduces the isolation effect. In this type of isolation system, the sliding displacements are controlled by high-tension springs or laminated rubber bearings, or by using a concave dish for the rollers; these mechanisms provide a restoring force, otherwise unavailable in this type of system, to re-

Figure 20.1.2 Cross section of a laminated rubber bearing. (Courtesy of I. D. Aiken.)

Figure 20.1.3 Deformed laminated rubber bearing. (Courtesy of I. D. Aiken.)

turn the structure to its equilibrium position. The dynamics of structures on roller or slider type of isolation systems is complicated because the slip process is intrinsically nonlinear.

This introductory presentation is limited to understanding the dynamic behavior of structures using the isolation system with laminated rubber bearings.

20.2 BASE-ISOLATED ONE-STORY BUILDINGS

In this section we identify why base isolation is effective in reducing the earthquake-induced forces in buildings. For this purpose we consider a one-story building with an isolation system between the base of the building and the ground. Most isolation systems are nonlinear in their force–deformation relationships, but it is not necessary to consider these nonlinear effects in this introductory treatment of the subject. A linear analysis of the system would serve our purpose of gaining insight into the dynamics of base-isolated buildings. Nonlinearity in the force–deformation relation should be considered for final design, however.

System considered and parameters. The one-story building to be isolated is shown idealized in Fig. 20.2.1a together with its parameters: lumped mass m, lateral stiffness k, and lateral damping c. This is the familiar SDF system with natural frequency ω_n, natural period T_n, and damping ratio ζ. Here we use the subscript f instead of n to emphasize that these are properties of the structure on a fixed base (i.e., without any isolation system); thus

$$\omega_f = \sqrt{\frac{k}{m}} \qquad T_f = \frac{2\pi}{\omega_f} \qquad \zeta_f = \frac{c}{2m\omega_f} \tag{20.2.1}$$

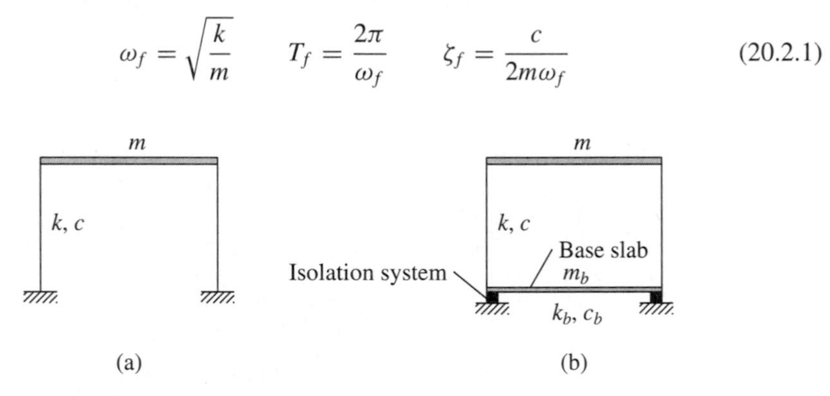

Figure 20.2.1 (a) Fixed-base structure; (b) isolated structure.

As shown in Fig. 20.2.1b, this one-story building is mounted on a base slab of mass m_b that in turn is supported on a base isolation system with lateral stiffness k_b and linear viscous damping c_b. Two parameters, T_b and ζ_b, are introduced to characterize the isolation system:

$$T_b = \frac{2\pi}{\omega_b} \qquad \text{where} \qquad \omega_b = \sqrt{\frac{k_b}{m + m_b}} \tag{20.2.2a}$$

$$\zeta_b = \frac{c_b}{2(m + m_b)\omega_b} \tag{20.2.2b}$$

We may interpret T_b as the natural vibration period, and ζ_b as the damping ratio, of (1) the isolated building with the building assumed to be rigid, or (2) the isolation system with the building assumed to be rigid. For base isolation to be effective in reducing the forces in the building, T_b must be much longer than T_f. The one-story building on a base isolation system (Fig. 20.2.1b) is a two-DOF system with mass, stiffness, and damping matrices denoted by **m**, **k**, and **c**, respectively. The disparity between the high damping in rubber bearings and the low damping of the building means that damping in the combined system is nonclassical.

Analysis procedure. As mentioned in Section 12.14, analysis of such systems with nonclassical damping requires numerical solution of the coupled equations of motion. While this can be implemented using the methods of Chapter 15, it is not convenient for our objective to understand the dynamics of base-isolated buildings. Although, strictly speaking, modal analysis is not applicable to nonclassically damped systems, it can provide approximate results that suffice for our limited objective. Therefore, we use classical modal analysis to determine structural response to ground motion characterized by a design spectrum.

The two-DOF system that defines the one-story building on an isolation system is analyzed by the methods presented earlier in this book. With **m**, **k**, and **c** appropriately defined, Eq. (9.4.4) gives the equations of motion for the system; the natural vibration periods and modes of the system are determined following Example 10.4, and the earthquake response of the system is estimated by response spectrum analysis following Section 13.8. The results of this analysis are presented next for an example system.

Effects of base isolation. To understand the dynamics of base isolation, let us consider a specific system: $m_b = 2m/3$, $T_f = 0.4$ sec, $T_b = 2.0$ sec, $\zeta_f = 2\%$, and $\zeta_b = 10\%$. The base shear V_b in the building and the base displacement u_b are to be estimated using the elastic design spectrum of Fig. 20.1.1, shown for damping ratios 2, 5, and 10%. For 5% damping this design spectrum is the same as in Fig. 6.9.5 scaled to peak ground acceleration of 0.5g. The other two spectra were constructed similarly using appropriate amplification factors from Table 6.9.1.

Observe that we have chosen damping in the structure as 2% of critical damping, lower than the 5% typically assumed in earthquake analysis and design of structures. As mentioned in Chapter 11, the higher damping value accounts for the additional energy dissipation through nonstructural damage expected in conventional structures at the larger motion during earthquakes. The aim of base isolation is to reduce the forces imparted to the structure to such a level that no damage to the structure or nonstructural elements occur and thus a lower value of damping is appropriate.

The natural vibration periods T_n and modes ϕ_n of the one-story building on an isolation system are shown in Fig. 20.2.2. In the first mode the isolator undergoes deformation but the structure behaves as essentially rigid; this mode is therefore called the *isolation mode*. The natural period of this mode, $T_1 = 2.024$ sec, indicates that the isolation system period, $T_b = 2.0$ sec, is changed only slightly by flexibility of the structure. The second mode involves deformation of the structure as well as in the isolation system, and the structural deformation is larger. Therefore, this is called the *structural mode*, although as we shall see later, this mode contributes little to the earthquake-induced forces in the structure. The natural period of this mode, $T_2 = 0.25$ sec,

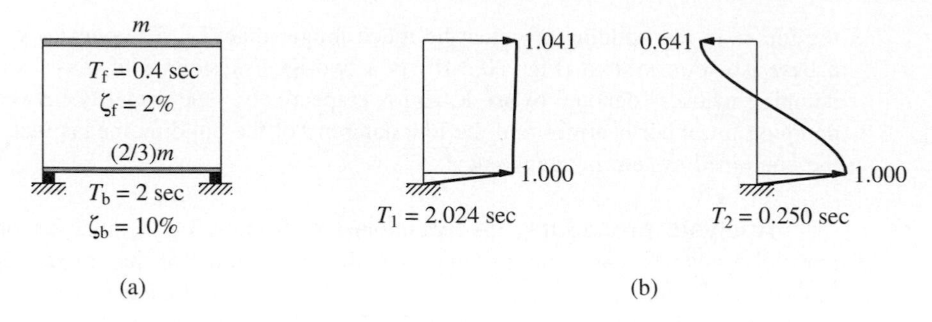

Figure 20.2.2 (a) One-story building on isolation system; (b) natural vibration modes and periods.

is significantly shorter than the fixed-base period, $T_f = 0.4$ sec, of the structure. The natural periods of the combined system are more separate than the isolation system period T_b and fixed-base period T_f of the structure.

Introduced in Section 13.2.1, the modal expansion of the effective earthquake force distribution, $\mathbf{s} = \mathbf{m1}$, for the system of Fig. 20.2.1b is shown in Fig. 20.2.3. These striking results indicate that the first-mode forces $\mathbf{s}_1$ are essentially the same as the total forces $\mathbf{s}$, and the second-mode forces $\mathbf{s}_2$ are very small. Static analysis of the system for forces $\mathbf{s}_n$ gives the modal static responses r_n^{st} for response quantity $r(t)$. In particular, for the base shear $V_b(t)$ in the structure and displacement $u_b(t)$ at the base, which is also the deformation of the isolation system, the modal static responses in the two modes are (see Fig. 20.2.3)

$$V_{b1}^{st} = 1.015m \qquad V_{b2}^{st} = -0.015m \tag{20.2.3a}$$

$$\omega_1^2 u_{b1}^{st} = 0.976 \qquad \omega_2^2 u_{b2}^{st} = 0.024 \tag{20.2.3b}$$

It is clear that these quantities for the second mode are negligible compared to the first mode. This result and the fact that the natural vibration period of the first mode, providing most of the response, is much longer than the fixed-base period of the structure reveal the underlying reasons for the effectiveness of a base isolation system. Note that we have identified the fundamental reasons for the effectiveness of base isolation without consideration of damping in the isolation system; the associated dissipation of energy is not a primary reason but is only a secondary factor in reducing structural response. This we shall see next.

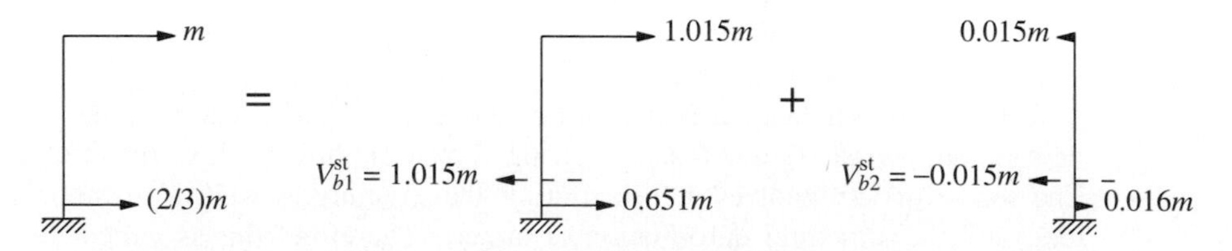

Figure 20.2.3 Modal expansion of effective earthquake forces and modal static responses for base shear.

The modal damping ratios are determined by Eq. (10.9.11), repeated here for convenience:

$$\zeta_n = \frac{C_n}{2M_n\omega_n} \tag{20.2.4a}$$

where

$$M_n = \phi_n^T \mathbf{m} \phi_n \quad \text{and} \quad C_n = \phi_n^T \mathbf{c} \phi_n \tag{20.2.4b}$$

For the system chosen, these equations give

$$\zeta_1 = 9.65\% \qquad \zeta_2 = 5.06\% \tag{20.2.5}$$

Observe that the 9.65% damping in the first mode is very similar to the isolation-system damping, $\zeta_b = 10\%$; damping in the structure has little influence on modal damping because the structure behaves as a rigid body for this mode. In contrast, high damping of the isolation system has increased the damping in the structural mode from 2% to 5.06%. Because damping in the system is nonclassical, the coupling terms $C_{21} = C_{12} = \phi_1^T \mathbf{c} \phi_2$ are nonzero and the modal equations are coupled (see Section 12.4). It is this coupling we are neglecting in using classical modal analysis for this system.

The peak value of the nth-mode contribution $r_n(t)$ to response $r(t)$ is given by Eq. (13.7.1), repeated here for convenience:

$$r_n = r_n^{\text{st}} A_n$$

where $A_n \equiv A(T_n, \zeta_n)$ is the ordinate of the pseudo-acceleration response (or design) spectrum at period T_n for damping ratio ζ_n. Specializing this equation for the two response quantities of interest, base shear V_b in the structure and isolator deformation u_b, gives

$$V_{bn} = V_{bn}^{\text{st}} A_n \qquad u_{bn} = (\omega_n^2 u_{bn}^{\text{st}}) D_n \tag{20.2.6}$$

where $D_n = A_n/\omega_n^2$ is the deformation spectrum ordinate. These calculations are summarized in Table 20.2.1 using the A_n values noted in Fig. 20.2.4; these were obtained for the actual damping values, Eq. (20.2.5), by using Table 6.9.2 instead of interpolating between the spectrum curves for 2, 5, and 10% damping. Observe that the response due to the second mode, the structural mode, is negligible. Obtained by combining modal responses by the SRSS combination rule, the deformation in the isolator is 14.042 in. and the base shear is 36.5% of the building weight excluding the base slab.

TABLE 20.2.1 CALCULATION OF BASE SHEAR AND ISOLATOR DEFORMATION

	Base Shear			Isolator Deformation		
Mode	A_n/g	V_{bn}^{st}/m	V_{bn}/w	D_n (in.)	$\omega_n^2 u_{bn}^{\text{st}}$	u_{bn} (in.)
1	0.359	1.015	0.365	14.390	0.976	14.042
2	1.347	−0.015	−0.021	0.823	0.024	0.020
SRSS			0.365			14.042

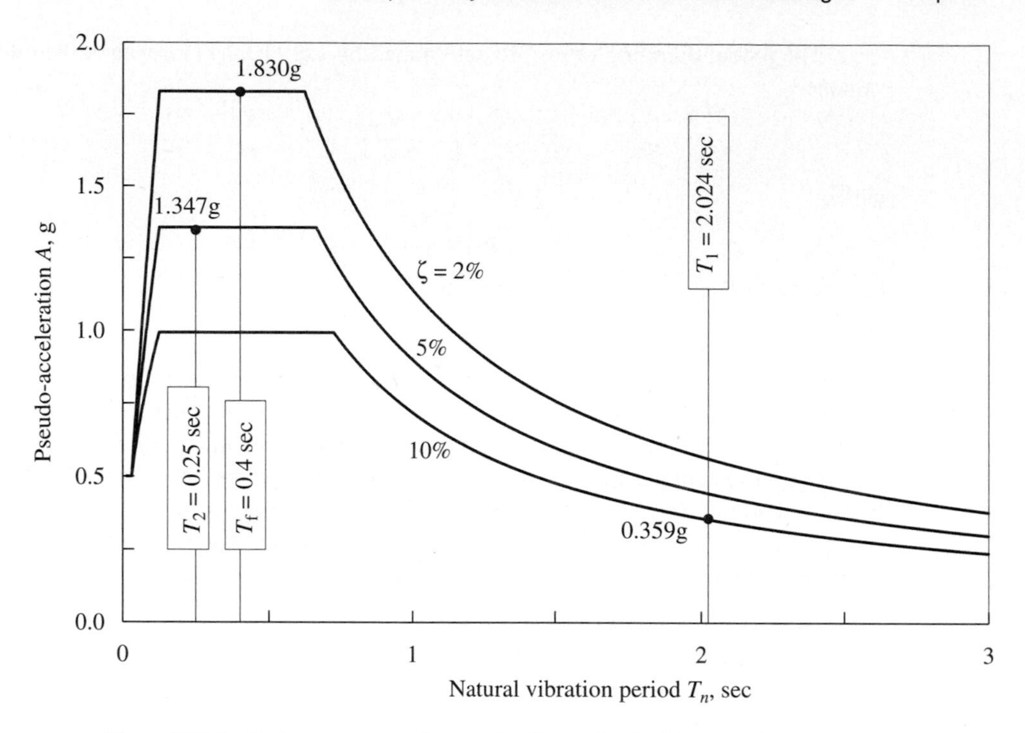

Figure 20.2.4 Design spectrum and spectral ordinates for fixed-base and isolated buildings.

The base shear is much larger if the structure is not isolated. This fixed-base structure has a natural period $T_f = 0.4$ sec and damping ratio $\zeta_f = 2\%$. For these parameters the design spectrum of Fig. 20.2.4 gives $A(T_f, \zeta_f) = 1.830\text{g}$. Thus the base shear in the fixed-base structure is

$$V_b = mA(T_f, \zeta_f) = m(1.830\text{g}) \quad \text{or} \quad \frac{V_b}{w} = 1.830 \tag{20.2.7}$$

that is, 183% of the weight w of the building excluding the base slab, about five times the base shear in the isolated building.

The isolation system reduces the base shear primarily because the natural period of the first mode, the isolation mode, providing most of the response, is much longer than the fixed-base period of the structure, leading to a smaller spectral ordinate, as seen in Fig. 20.2.4. This becomes clear by reexamining the terms entering into the base shear due to the first mode:

$$V_{b1} = V_{b1}^{\text{st}} A_1 = (1.015m)(0.359\text{g}) \tag{20.2.8}$$

Comparing Eq. (20.2.8) with (20.2.7a), it is apparent that because of the isolation system, the pseudo-acceleration is reduced from 1.830g to 0.359g, whereas the effective modal mass is essentially unaffected.

Rigid-structure approximation. The base shear in the building and the deformation of the isolation system can be estimated by a simpler analysis treating the building as rigid. With this assumption the combined system has only one DOF. For this SDF system with natural period T_b and damping ratio ζ_b, the design spectrum gives the pseudo-acceleration $A(T_b, \zeta_b)$ and deformation $D(T_b, \zeta_b)$. Thus the isolator deformation is

$$u_b = D(T_b, \zeta_b) \tag{20.2.9}$$

and the base shear in the structure is

$$V_b = mA(T_b, \zeta_b) \tag{20.2.10}$$

The approximate results of Eqs. (20.2.9) and (20.2.10) are accurate for base-isolated systems if the period T_b of the isolation system (assuming rigid structure) is much longer than the fixed-base period T_f of the structure. This is illustrated using the system of Fig. 20.2.2: For $T_b = 2$ sec and $\zeta_b = 10\%$, the design spectrum gives $A(T_b, \zeta_b) = 0.359\text{g}$ and $D(T_b, \zeta_b) = 14.036$ in., and Eq. (20.2.10) gives

$$V_b = m(0.359\text{g}) \qquad \text{or} \qquad \frac{V_b}{w} = 0.359 \tag{20.2.11}$$

Comparing Eq. (20.2.11) with Eq. (20.2.8), it is clear why this approximate analysis assuming a rigid structure gives almost "exact" results. Because the vibration properties with the rigid-structure assumption, $T_b = 2$ sec and $\zeta_b = 10\%$, are very close to the first-mode values, $T_1 = 2.024$ sec and $\zeta_1 = 9.65\%$, the spectral accelerations $A(T_b, \zeta_b)$ and $A(T_1, \zeta_1)$ are identical to three digits. Furthermore, the effective masses that enter into Eqs. (20.2.8) and (20.2.11) are essentially identical. Similarly, the isolator deformation from Eq. (20.2.9),

$$u_b = 14.036 \text{ in.} \tag{20.2.12}$$

is essentially identical to the first-mode response in Table 20.2.1.

Because of its accuracy, the rigid-structure approximation provides an expedient means to estimate the effectiveness of a base isolation system and to estimate the isolator deformation. First, the ratio $A(T_b, \zeta_b)/A(T_f, \zeta_f)$ of two spectral ordinates gives the base shear in the isolated system as a fraction of the base shear in the fixed-base structure. Second, the deformation spectrum ordinate $D(T_b, \zeta_b)$ is the isolator deformation.

20.3 EFFECTIVENESS OF BASE ISOLATION

It is clear that the effectiveness of base isolation in reducing structural forces is closely tied to the lengthening of the natural period of the structure, and for this purpose the period ratio T_b/T_f should be as large as practical. In the example of the preceding section, the natural period of the fixed-base structure located the structure at the peak of the selected design spectrum. With base isolation, the natural period (of the isolation mode contributing almost all of the response) was shifted to the velocity-sensitive region of the spectrum with much smaller pseudo-acceleration. As a result, the base shear is reduced from 183% of the structural weight (excluding the base slab) to 36.5%. Whether the forces in the structure are reduced because of this period shift depends on the natural period of the fixed-base structure and on the shape of the earthquake design spectrum, among other factors. We illustrate these concepts next.

First, consider the same one-story building and base isolation system as in the preceding section to be located in Mexico City at the site where ground motions recorded during the 1985 earthquake produced the response spectrum shown in Fig. 20.3.1. Noted on this spectrum are the pseudo-acceleration values $A(T_f, \zeta_f) = 0.25\text{g}$ for the fixed-base structure and $A(T_b, \zeta_b) = 0.63\text{g}$ for the isolated structure (with the building assumed to be rigid). These spectral values correspond to previously defined properties of the fixed-base structure: $T_f = 0.4$ sec and $\zeta_f = 2\%$, and of the isolation system: $T_b = 2.0$ sec and $\zeta_b = 10\%$. The ratio $A(T_b, \zeta_b)/A(T_f, \zeta_f) = 0.63\text{g}/0.25\text{g} = 2.52$ implies that the base shear in the isolated structure is approximately 2.52 times the base shear in the fixed-base structure. In this case base isolation is not helpful; in fact, it is harmful.

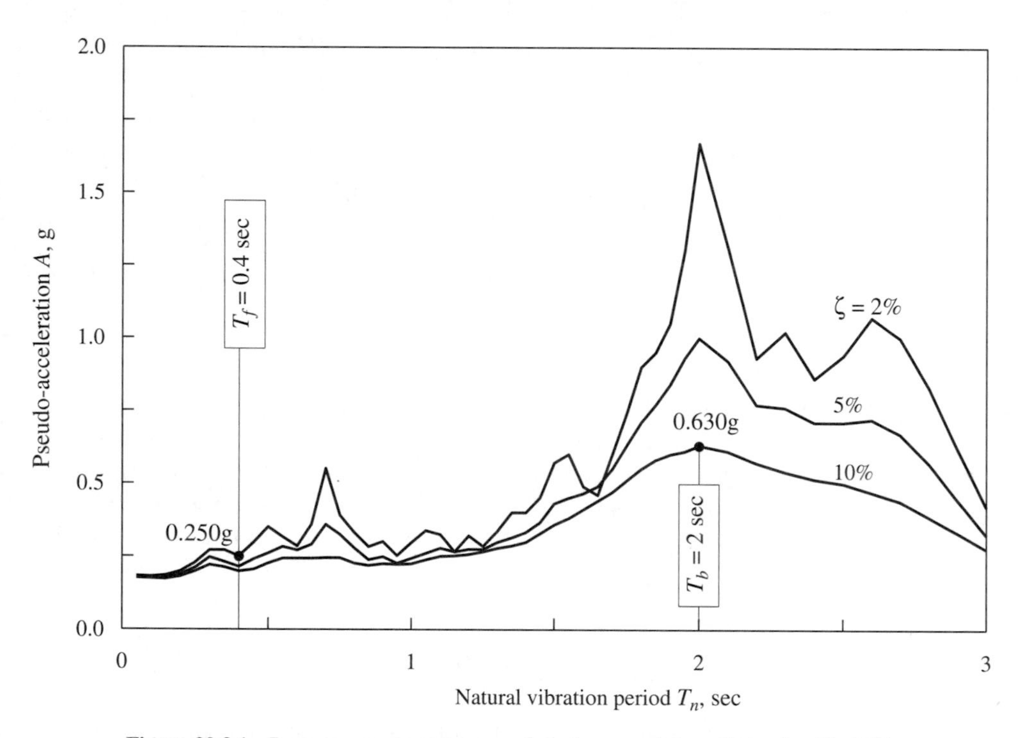

Figure 20.3.1 Response spectrum for ground motion recorded on September 19, 1985 at SCT site in Mexico City and spectral ordinates for fixed-base and isolated buildings.

Next, consider a structure with a relatively long fixed-base period and the ground motion characterized by the original design spectrum (Fig. 20.1.1). In this case we shall see that base isolation is only slightly beneficial—much less than when the fixed-base period was relatively short. To illustrate these results, consider a structure with a fixed-base period of 2 sec with other parameters for the structure and isolation system as before. Thus the system parameters are $T_f = 2$ sec, $\zeta_f = 2\%$, $m_b = \frac{2}{3}m$, $T_b = 2$ sec, and $\zeta_b = 10\%$. Analysis of this system with $T_b = T_f$ by the procedures used for the example of Section 20.2 gives the natural vibration periods and modes (Fig. 20.3.2), the modal expansion of the effective earthquake force distribution: $\mathbf{s} = \mathbf{m1}$ (Fig. 20.3.3), and the modal damping ratios: 4.50% and 12.64%. In contrast to the previous system with $T_b >> T_f$: (1) the structure does not behave as rigidly in the first mode, and this natural period is significantly affected by the flexibility of the structure; (2) the second-mode contribution to the effective earthquake forces is no longer negligible; and (3) the first-mode damping of 4.5% is no longer close to the isolation system damping of 10%.

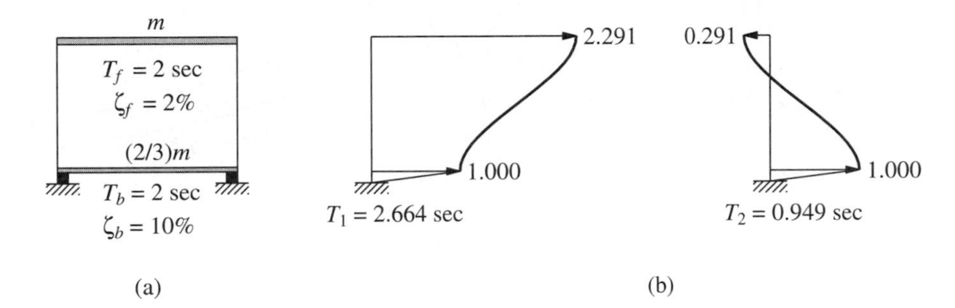

Figure 20.3.2 (a) One-story building on isolation system; (b) natural vibration modes and periods.

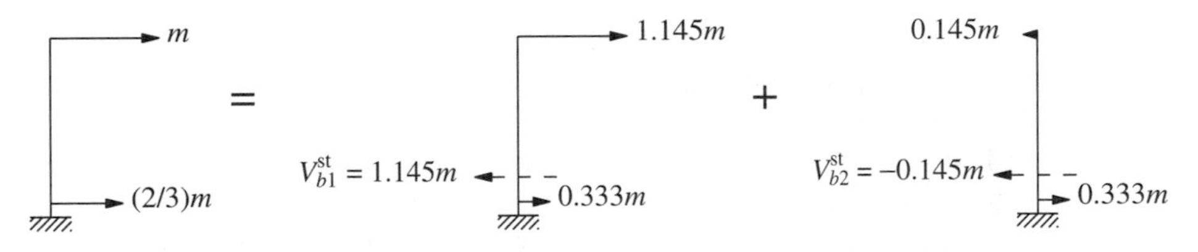

Figure 20.3.3 Modal expansion of effective earthquake forces and modal static responses for base shear.

A summary of the calculations to obtain the base shear and isolator deformation is presented in Table 20.3.1, with the spectral values identified in Fig. 20.3.4. In contrast

TABLE 20.3.1 CALCULATION OF BASE SHEAR AND ISOLATOR DEFORMATION

	Base Shear			Isolator Deformation		
Mode	A_n/g	V_{bn}^{st}/m	V_{bn}/w	D_n (in.)	$\omega_n^2 u_{bn}^{st}$	u_{bn} (in.)
1	0.348	1.145	0.398	24.136	0.500	12.068
2	0.691	−0.145	−0.101	6.095	0.500	3.047
SRSS			0.411			12.447

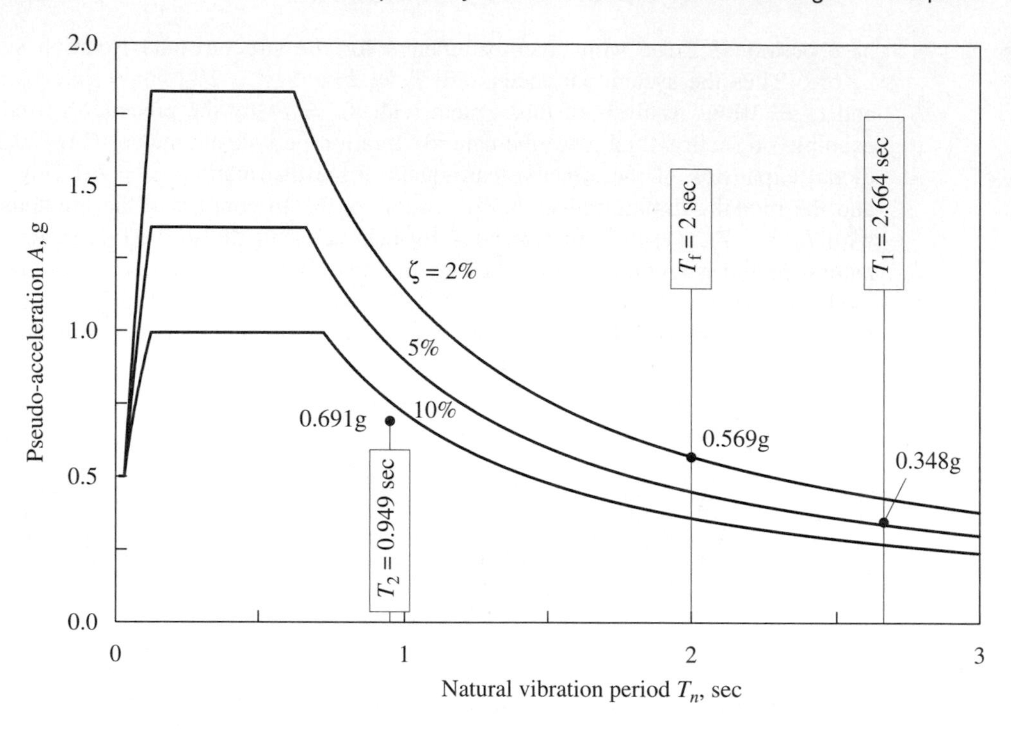

Figure 20.3.4 Design spectrum and spectral ordinates for fixed-base and isolated buildings.

to the previous system with $T_b >> T_f$, the response of the second mode is significant, although it does not contribute much when it is combined—using the SRSS rule—with the first-mode response. We find the isolator deformation to be 12.447 in., and the base shear is 41.1% of the structural weight excluding the base slab.

The fixed-base structure has a natural period $T_f = 2$ sec and damping ratio $\zeta_f = 2\%$, and the corresponding spectral ordinate is $A(T_f, \zeta_f) = 0.569\text{g}$ (Fig. 20.3.4). Thus if the structure were not isolated, the base shear is

$$V_b = mA(T_f, \zeta_f) \quad \text{or} \quad \frac{V_b}{w} = 0.569 \tag{20.3.1}$$

(i.e., 56.9% of the building weight). It is clear that some benefit is obtained by base isolation, although it is much less than if the vibration period of the structure had been shorter, as in the original example. It is for this reason that base isolation is rarely used for structures with natural period well into in the velocity-sensitive region of the spectrum.

In passing we also note that the approximate analysis based on the assumption of a rigid structure is not accurate for a structure with a relatively long natural period. The approximate analysis gives a base shear coefficient of 35.9%, Eq. (20.2.11), compared to 41.1% from the complete analysis. The isolator deformation from the approximate analysis is 14.036 in., Eq. (20.2.12), compared to 12.447 in. from the complete analysis.

20.4 BASE-ISOLATED MULTISTORY BUILDINGS

In the preceding sections we were able to identify the underlying reasons for the effectiveness of a base isolation system by studying the dynamics of a one-story building. In this section we investigate how the dynamics of a multistory building is modified by base isolation. As before, we assume the system to be linear. We will see that the key concepts underlying base isolation, identified by the dynamics of one-story systems, carry over to multistory systems.

System considered and parameters. The N-story building to be isolated is shown idealized in Fig. 20.4.1a. On a fixed base, this system is defined by mass matrix $\mathbf{m}_f$, damping matrix $\mathbf{c}_f$, and stiffness matrix $\mathbf{k}_f$, which can be constructed by the methods developed in Chapters 9 and 11; the subscript f denotes "fixed base." If the mass of the structure is idealized as lumped at the floor levels, as shown in Fig. 20.4.1a, $\mathbf{m}_f$ is a diagonal matrix with diagonal element $m_{jj} = m_j$, the mass lumped at the jth floor. The total mass of the building, $M = \sum m_j$. The natural periods and modes of vibration of the fixed-base system are denoted by T_{nf} and ϕ_{nf}, respectively, where $n = 1, 2, \ldots, N$. Damping in the structure is assumed to be of classical form and defined by modal damping ratios ζ_{nf}, $n = 1, 2, \ldots, N$.

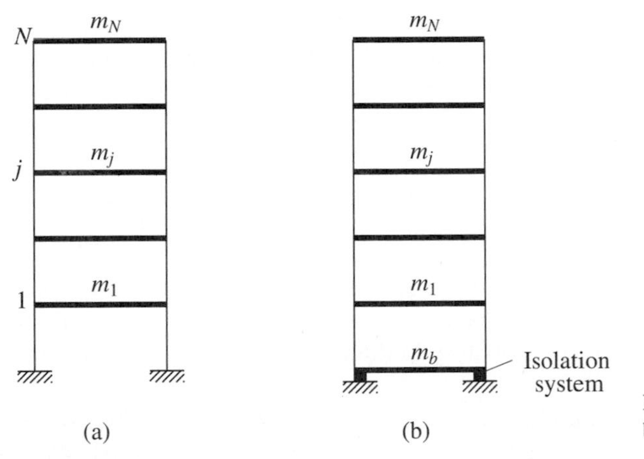

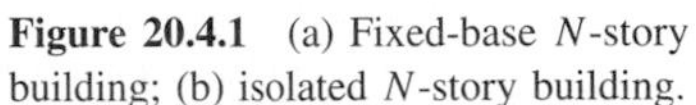
Figure 20.4.1 (a) Fixed-base N-story building; (b) isolated N-story building.

As shown in Fig. 20.4.1b, this N-story building is mounted on a base slab of mass m_b, supported in turn on a base isolation system with lateral stiffness k_b and linear viscous damping c_b. As in Section 20.2, two parameters, T_b and ζ_b, are introduced to characterize the isolation system:

$$T_b = \frac{2\pi}{\omega_b} \qquad \text{where} \quad \omega_b = \sqrt{\frac{k_b}{M + m_b}} \tag{20.4.1a}$$

$$\zeta_b = \frac{c_b}{2(M + m_b)\omega_b} \tag{20.4.1b}$$

As before, we may interpret T_b as the natural vibration period and ζ_b the damping ratio of (1) the isolated building with the building assumed to be rigid, or (2) the isolation system with the building assumed to be rigid. For base isolation to be effective in reducing the earthquake-induced forces in the building, T_b must be much longer than T_{1f}, the fundamental period of the fixed-base building.

The N-story building on a base isolation system is an $(N + 1)$-DOF system with nonclassical damping because damping in the isolation system is typically much more than in the building. The mass, stiffness, and damping matrices of order $N + 1$ for the combined system are denoted by $\mathbf{m}$, $\mathbf{k}$, and $\mathbf{c}$, respectively.

Analysis procedure. With ground motion characterized by a design spectrum, the RSA procedure of Chapter 13, Part B, will be used to analyze two systems: (1) a building on a fixed base, and (2) the same structure supported on an isolation system. In applying the RSA procedure to the isolated structure we are ignoring the coupling of modal equations due to nonclassical damping, typical of structures on isolation systems. The modal damping ratios of the isolated structure are given by Eq. (20.2.4). We focus on two response quantities: base shear in the building and the base displacement (or isolator deformation). The peak responses due to the nth mode of vibration are determined using Eq. (20.2.6), and these modal responses are combined by the SRSS rule.

Effects of base isolation. To understand how base isolation affects the dynamics of buildings, we consider a specific system. The fixed-base structure is a five-story shear frame (i.e., beam-to-column stiffness ratio $\rho = \infty$) with mass and stiffness properties uniform over its height: lumped mass $m = 100$ kips/g at each floor, and stiffnesses k for each story; k is chosen so that the fundamental natural vibration period $T_{1f} = 0.4$ sec. The classical damping matrix $\mathbf{c}_f = a_1 \mathbf{k}_f$ with a_1 chosen to obtain 2% damping in the fundamental mode. The base slab mass $m_b = m$ and the stiffness and damping of the isolation system are such that $T_b = 2.0$ sec and $\zeta_b = 10\%$ [Eq. (20.4.1)]. Next we examine the vibration properties—natural periods, natural modes, and modal damping ratios—and the earthquake response of two systems: (1) this five-story building on a fixed base, and (2) the same five-story building supported on the isolation system described above.

The natural periods, natural modes, and modal damping ratios of both systems are presented in Fig. 20.4.2 and Table 20.4.1. The fixed-base structure has the familiar mode shapes and ratios of natural periods; the modal damping ratios decrease linearly with natural period (i.e., increase linearly with natural frequency) because the damping is stiffness proportional (Section 11.4.1). The first natural mode of the isolated building is the isolation mode because the isolator undergoes deformation but the building behaves as essentially rigid. The natural period of this mode, $T_1 = 2.030$ sec, indicates that the isolation-system period, $T_b = 2.0$ sec, is only slightly changed by the flexibility of the structure. The other modes involve deformation in the structure as well as in the isolation system. We refer to these modes as structural modes, although as we shall see later, these modes contribute little to the earthquake-induced forces in the structure. It is clear that the isolation system has a large effect on the natural period of the first structural mode,

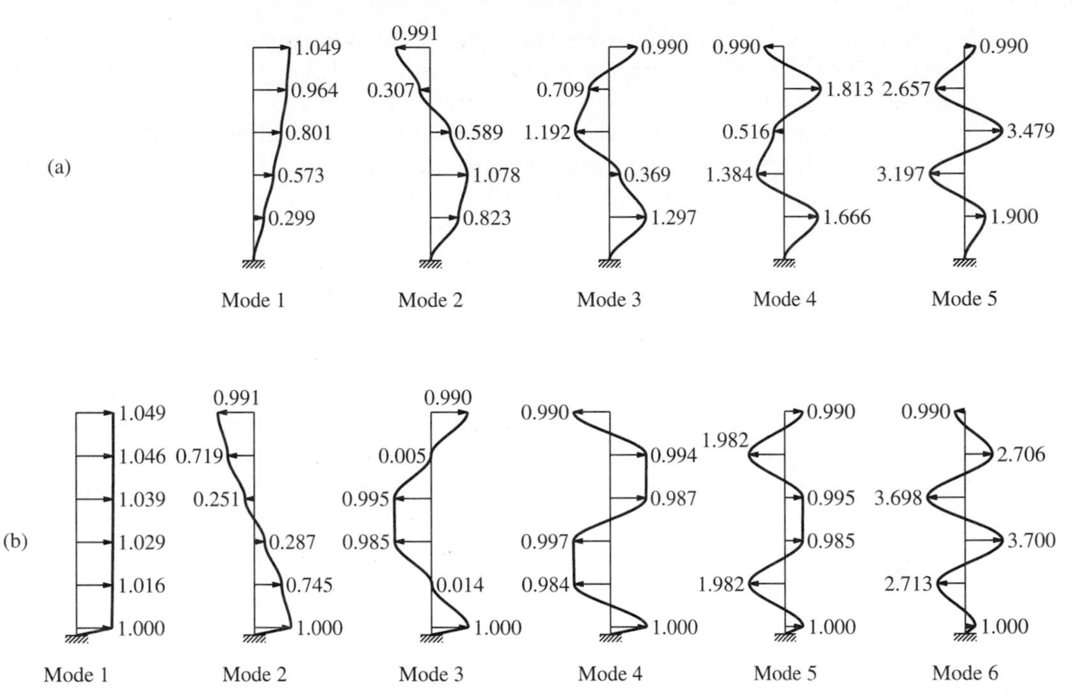

Figure 20.4.2 Natural vibration modes: (a) fixed-base building; (b) isolated building.

TABLE 20.4.1 NATURAL PERIODS AND MODAL DAMPING RATIOS

Fixed-Base Building			Isolated Building		
Model	T_{nf} (sec)	ζ_{nf} (%)	Mode	T_n (sec)	ζ_n (%)
			1	2.030	9.58
1	0.400	2.00	2	0.217	5.64
2	0.137	5.84	3	0.114	7.87
3	0.087	9.20	4	0.080	10.3
4	0.068	11.8	5	0.066	12.3
5	0.059	13.5	6	0.059	13.6

but a decreasing effect on the higher-mode periods. In these higher modes the motion of the base mass decreases relative to the structural motions, and the base mass is acting as a fixed base.

The damping ratios display a similar trend. The high damping of the isolation system has increased the damping in the first structural mode from 2.0% to 5.64%, whereas damping of the higher modes increases to a smaller degree. Observe that the damping of 9.58% in the isolation mode, the first mode of the combined system, is very similar to the isolation-system damping, $\zeta_b = 10\%$; damping in the structure has little influence on modal damping because the structure behaves as a rigid body for this mode.

We now compare the modal static response in the natural modes of both systems, the fixed-base and isolated buildings. The modal components of the effective earthquake force distribution, $\mathbf{s} = \mathbf{m1}$, are shown in Fig. 13.2.4 for the fixed-base structure and in Fig. 20.4.3 for the base-isolated structure. In the latter case, forces in the first mode, the isolation mode, are essentially the same as the total forces, and the forces associated with all the structural modes are very small. Static analysis of both systems for their respective modal forces gives the modal static shears V_{bn}^{st} at the base of the structure and modal static base displacements or isolator deformations u_{bn}^{st}. The results are given in Tables 20.4.2 and 20.4.3. It is clear that V_{bn}^{st} has significant values in the first two modes of the fixed-base structure. However, for the isolated building, V_{bn}^{st} is small in all the structural modes and the response in these modes will be negligible. The isolation mode provides the dominant value V_{b1}^{st} and therefore will provide most of the response.

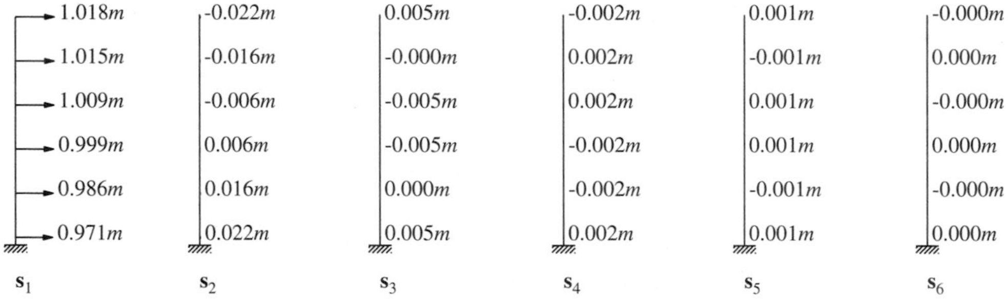

$\mathbf{s}_1$	$\mathbf{s}_2$	$\mathbf{s}_3$	$\mathbf{s}_4$	$\mathbf{s}_5$	$\mathbf{s}_6$
$1.018m$	$-0.022m$	$0.005m$	$-0.002m$	$0.001m$	$-0.000m$
$1.015m$	$-0.016m$	$-0.000m$	$0.002m$	$-0.001m$	$0.000m$
$1.009m$	$-0.006m$	$-0.005m$	$0.002m$	$0.001m$	$-0.000m$
$0.999m$	$0.006m$	$-0.005m$	$-0.002m$	$0.001m$	$0.000m$
$0.986m$	$0.016m$	$0.000m$	$-0.002m$	$-0.001m$	$-0.000m$
$0.971m$	$0.022m$	$0.005m$	$0.002m$	$0.001m$	$0.000m$

Figure 20.4.3 Modal components of effective earthquake forces for a five-story building on isolation system.

TABLE 20.4.2 CALCULATION OF BASE SHEAR IN FIXED-BASE AND ISOLATED BUILDINGS

Fixed-Base Building				Isolated Building			
Mode	A_n/g	V_{bn}^{st}/m	V_b/W	Mode	A_n/g	V_{bn}^{st}/m	V_b/W
				1	0.359	5.028	0.361
1	1.830	4.398	1.609	2	1.291	−0.021	−0.005
2	1.272	0.436	0.111	3	1.058	−0.005	−0.001
3	0.859	0.121	0.021	4	0.792	−0.002	−0.000
4	0.700	0.038	0.005	5	0.682	−0.0005	−0.000
5	0.638	0.008	0.001	6	0.635	−0.0001	−0.000
SRSS			1.613	SRSS			0.361

TABLE 20.4.3 CALCULATION OF ISOLATOR DEFORMATION

Mode	D_n	$\omega_n^2 u_{bn}^{st}$	u_{bn} (in.)
1	14.470	0.971	14.045
2	0.597	0.022	0.013
3	0.133	0.005	0.001
4	0.050	0.002	0.000
5	0.029	0.001	0.000
6	0.022	0.0001	0.000
SRSS			14.045

Next we compare the peak value of the earthquake response due to each natural mode of both systems (Tables 20.4.2 and 20.4.3). These peak responses are determined from Eq. (20.2.6), where the spectral ordinates A_n/g are shown in Figs. 20.4.4 and 20.4.5 and $D_n = A_n/\omega_n^2$. Observe that as predicted from the modal static responses, the dynamic response of the isolated building due to all its structural modes is negligible. The isolation mode alone produces essentially the entire response: isolation system deformation of 14.045 in. and base shear equal to 36.1% of W, the 500-kip weight of the building excluding the base slab.

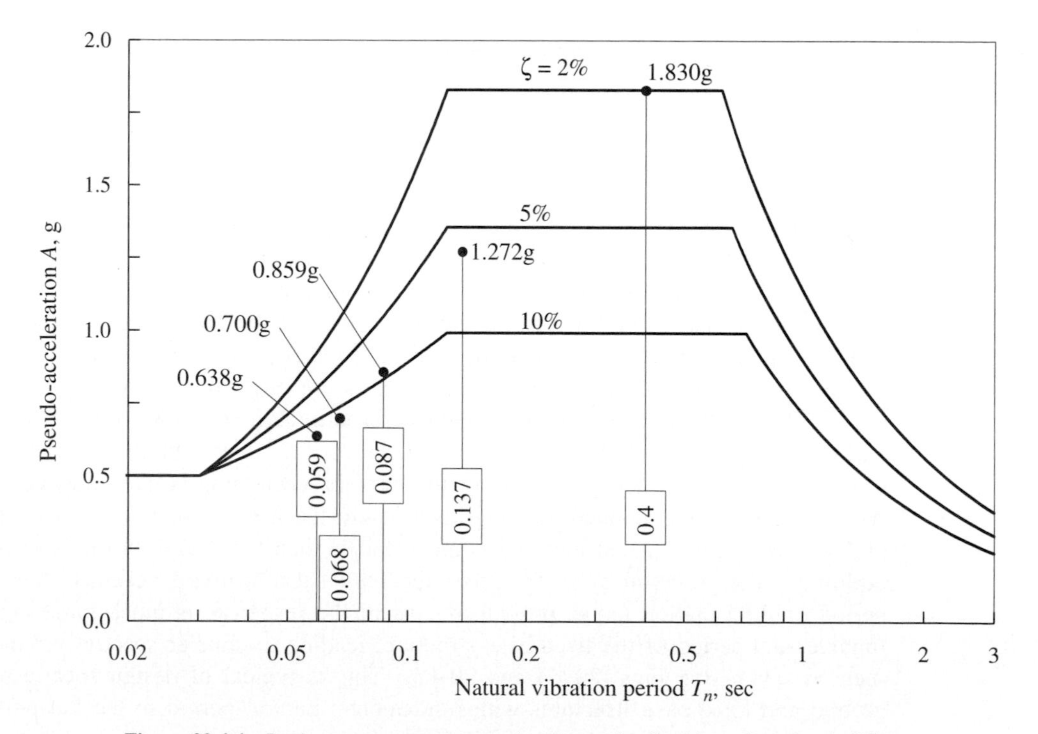

Figure 20.4.4 Design spectrum and spectral ordinates for fixed-base five-story building.

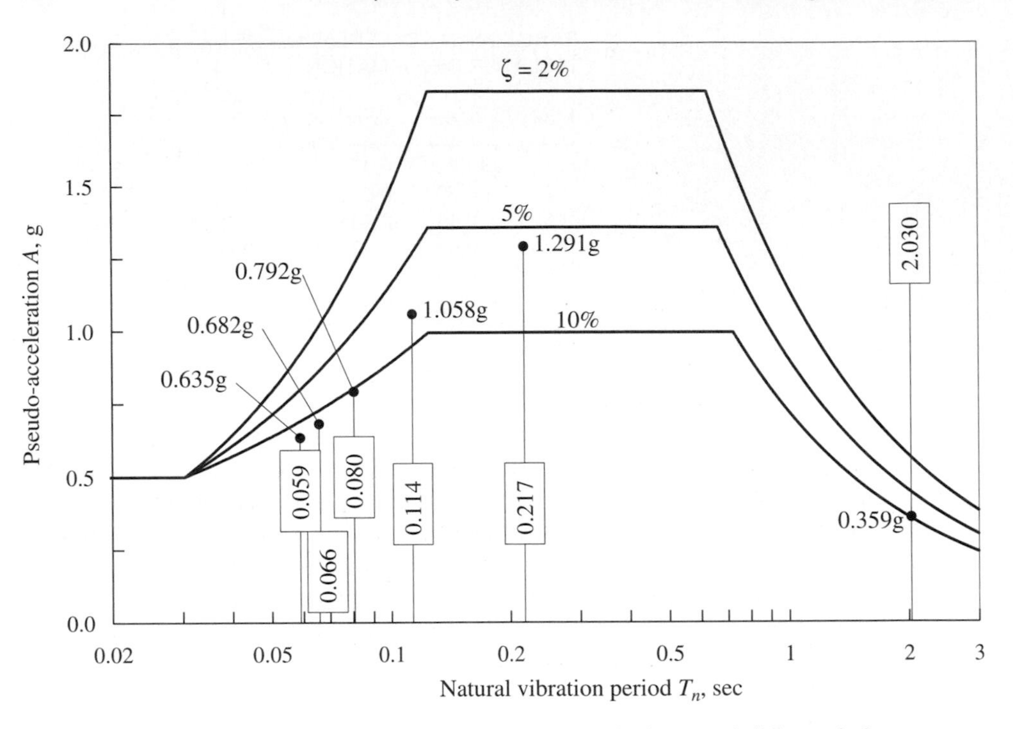

Figure 20.4.5 Design spectrum and spectral ordinates for five-story building on isolation system.

Table 20.4.2 shows that the base shear is much larger if the building is not isolated. To understand the underlying reasons, we examine the modal responses in both fixed-base and isolated systems. Each modal response is the product of two parts: the static response V_{bn}^{st} and the pseudo-acceleration A_n. Let us examine both parts for the isolation mode of the isolated building and for the first mode of the fixed-base building, modes that provide most of the response in the respective cases; observe that while the response in the second mode of the fixed-base structure is not negligible, it contributes little to the combined SRSS value. The modal static response $V_{b1}^{st} = 4.398m$ for the fixed-base building is somewhat smaller than $V_{b1}^{st} = 5.028m$ for the isolated building. However, the pseudo-acceleration of $A_1 = 1.830\text{g}$ for the fixed-base building is five times larger than $A_1 = 0.359\text{g}$ for the isolated building; as a result, the first-mode base shear coefficient of 160.9% for the isolated building is much larger than the 36.1% in the base-isolated building. The isolation system reduces the base shear primarily because the natural period of the isolation mode, providing most of the response, is much longer than the fundamental period of the fixed-base structure, leading to smaller spectral ordinates, as seen by comparing Figs. 20.4.4 and 20.4.5. This is typical of design spectra on firm ground and fixed-base structures with fundamental natural period in the flat portion of the acceleration-sensitive region of the spectrum.

Rigid-structure approximation. The base shear in the isolated building and the deformation of the isolation system can be estimated by a simpler analysis, treating the building as rigid. The natural period of the resulting SDF system is T_b and its damping ratio is ζ_b; the associated design spectrum ordinates are $A(T_b, \zeta_b)$ for the pseudo-acceleration and $D(T_b, \zeta_b)$ for the deformation. Thus the base shear in the structure and the isolator deformation are

$$V_b = MA(T_b, \zeta_b) \qquad u_b = D(T_b, \zeta_b) \tag{20.4.2}$$

This approximate procedure will provide excellent results if the isolation-system period T_b is much longer than the fundamental period T_{1f} of the fixed-base structure. This is illustrated using the system of Fig. 20.4.1b, analyzed earlier. For this system, $T_b =$ 2.0 sec and $\zeta_b = 10\%$ and the spectral values are $A(T_b, \zeta_b) = 0.359\text{g}$ and $D(T_b, \zeta_b) =$ 14.036, as noted in Section 20.2. Substituting these values in Eq. (20.4.2) gives

$$V_b = 0.359W \qquad u_b = 14.036 \text{ in.} \tag{20.4.3}$$

which are essentially identical to the responses due to the isolation mode (and to the total response) presented in Tables 20.4.2 and 20.4.3.

20.5 APPLICATIONS OF BASE ISOLATION

Base isolation provides an alternative to the conventional, fixed-base design of structures and may be cost-effective for some new buildings in locations where very strong ground shaking is likely. It is an attractive alternative for buildings that must remain functional after a major earthquake (e.g., hospitals, emergency communications centers, computer processing centers, etc).

Base isolation has also been used for retrofit of existing buildings that are brittle and weak: for example, unreinforced masonry buildings or reinforced-concrete buildings of early design, not including the type of detailing of the reinforcement necessary for ductile performance. Conventional seismic strengthening designs require adding new structural members, such as shear walls, frames, and bracing. Base isolation minimizes the need for such strengthening measures by reducing the earthquake forces imparted to the building. It is therefore an attractive retrofit approach for buildings of historical or architectural merit whose appearance and character must be preserved. However, it is difficult and expensive to construct a new foundation system for the isolators, to modify the base of the building so that it can be supported on isolators, and to shore up the building during construction of the isolation and foundation systems.

A good example of a retrofit application is the City and County Building in Salt Lake City, Utah. Constructed in 1894, this is an elegant five-story, unreinforced masonry and stone bearing-wall structure with a central 12-story clock tower exceeding 200 ft in height (Fig. 20.5.1). The building is nearly symmetrical along two major axes and is approximately 130 ft $\times$ 260 ft in plan. Its fixed-base fundamental natural period is approximately 0.5 sec, implying that large ductility demands can be imposed on the structure by the strong ground shaking expected at the building site from an earthquake

Figure 20.5.1 City & County Building, Salt Lake City, Utah. (Courtesy of Salt Lake City Corporation.)

on the nearby Wasatch fault. To improve the earthquake resistance of this brittle structure, base isolation was adopted because it preserved this monumental building at an acceptable cost. A total of 447 isolator bearings were used to support the exterior and interior walls of the building. The laminated rubber-steel bearings, with a lead plug to provide stiffness and energy absorption, were approximately 16 in. square by 15 in. high.

FURTHER READING

Kelly, J. M., *Earthquake Resistant Design with Rubber*, Springer-Verlag, New York, 1993, Chapters 1, 3, and 4.

Seismic Isolation: From Idea to Reality, theme issue of *Earthquake Spectra*, **6**, 1990, pp. 161–432.

Skinner, R. I., Robinson, W. H., and McVerry, G. H., *An Introduction to Seismic Isolation*, Wiley, Chichester, U.K., 1993.

Warburton, G. B., *Reduction of Vibrations*, The 3rd Mallet-Milne Lecture, Wiley, Chichester, U.K., 1992, pp. 21–46.

21

Structural Dynamics in Building Codes

PREVIEW

Most seismic building codes require that structures be designed to resist specified *static* lateral forces related to the properties of the structure and the seismicity of the region. Based on an estimate of the fundamental natural vibration period of the structure, formulas are specified for the base shear and the distribution of lateral forces over the height of the building. Static analysis of the building for these forces provides the design forces, including shears and overturning moments for the various stories, with some codes permitting reductions in the statically computed overturning moments. These seismic design provisions in three building codes—*Uniform Building Code* (United States), *National Building Code of Canada*, and *Mexico Federal District Code*—are presented in Part A of this chapter together with their relationship to the theory of structural dynamics developed in Chapters 6, 7, 8, and 13. The code provisions presented are not complete; those provisions that we are unprepared to evaluate based on this book have been excluded or only mentioned: effects of local soil conditions, torsional moments about a vertical axis, combination of earthquake forces due to the simultaneous action of ground motion components, and the requirements for detailing structures to ensure ductile behavior, among others. In Part B of the chapter the code provisions are evaluated in light of the results of dynamic analysis of buildings presented in Chapters 18 and 19.

Most codes also permit dynamic analysis procedures, both response spectrum analysis (RSA) and response history analysis (RHA). The code versions of these procedures are not presented because they are essentially equivalent to those that were developed in Chapter 13.

PART A: BUILDING CODES AND STRUCTURAL DYNAMICS

21.1 *UNIFORM BUILDING CODE* (UNITED STATES), 1994

21.1.1 Base Shear

The 1990 edition of the recommendations of the Structural Engineers Association of California, incorporated in the 1994 *Uniform Building Code* (UBC), specify the base shear as

$$V_b = C_s W \tag{21.1.1}$$

where W is the total dead load, and applicable portions of other loads, and the seismic coefficient,

$$C_s = \frac{C_e}{R_w} \tag{21.1.2}$$

This coefficient corresponding to $R_w = 1$ is called the elastic seismic coefficient:

$$C_e = ZIC \tag{21.1.3}$$

C_e is the product of three factors:

1. Seismic zone factor $Z = 0.4$ for zone 4, 0.3 for zone 3, 0.2 for zone 2A, 0.15 for zone 2B, 0.075 for zone 1, and zero for zone 0; the United States is divided into six zones (Fig. 21.1.1).
2. Importance factor $I = 1.0$ or 1.25. For most structures $I = 1$, but for "essential facilities" and "hazardous facilities" $I = 1.25$.
3. The numerical coefficient C is defined as

$$C = \frac{1.25S}{T_1^{2/3}} \le 2.75 \tag{21.1.4}$$

where T_1 is the fundamental natural vibration period of the structure in seconds; and the site coefficient $S = 1.0$, 1.2, 1.5, and 2.0 for soil profiles S_1, S_2, S_3, and S_4, respectively, defined in the code. These coefficients account for local soil effects on earthquake ground motion, a topic not covered in this book.

For the fundamental natural period of vibration the code provides the formula

$$T_1 = 2\pi \left[\frac{\sum_{i=1}^{N} w_i u_i^2}{g \sum_{i=1}^{N} F_i u_i} \right]^{1/2} \tag{21.1.5}$$

where w_i is the weight at the ith floor, and u_i are the floor displacements due to static application of a set of lateral forces F_i at floor levels $i = 1, 2, \ldots, N$ in an N-story

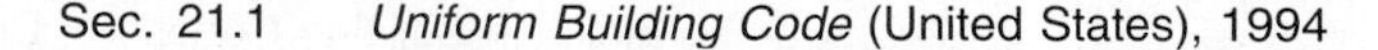

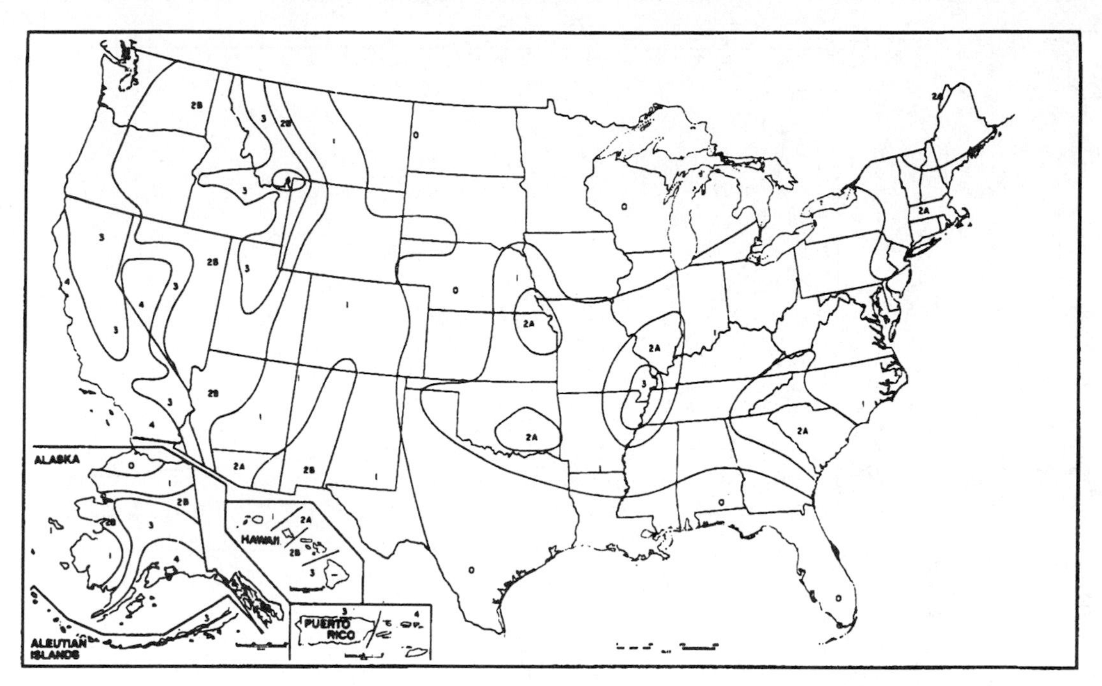

Figure 21.1.1 Seismic zone map of the United States. (Reproduced from the 1994 edition of the *Uniform Building Code*, copyright 1994, with the permission of the publisher, the International Conference of Building Officials.)

building. These forces F_i in Eq. (21.1.5) may be any reasonable distribution over the building height, and need not be exactly the design lateral forces specified in the code, Eqs. (21.1.7) and (21.1.8). Alternatively, the code gives empirical formulas for T_1 that depend on building material (steel, reinforced concrete, etc.), building type (frame, shear wall, etc.), and overall dimensions.

Figure 21.1.2 shows the elastic seismic coefficient C_e for $Z = 0.4$, applicable to zone 4; $I = 1$, valid for most structures; and $S = 1.0$ for soil profile S_1, which includes rock and stiff soils. The code also specifies (for use in dynamic analysis) the elastic pseudo-acceleration design spectrum that presumably is the basis for defining C_e; this spectrum for soil profile S_1 and $Z = 0.4$ is

$$A/g = \begin{cases} 0.4 + 4T_n & 0 \leq T_n \leq 0.15 \\ 1.0 & 0.15 < T_n \leq 0.39 \\ 0.39/T_n & T_n > 0.39 \end{cases} \tag{21.1.6}$$

where T_n is the natural vibration period (in seconds) of an SDF system.

The structural system coefficient R_w depends on several factors, including the ductility capacity and inelastic performance of structural materials and systems during earthquakes. Specified values of R_w vary between 4 (for certain bearing wall systems) and 12 (for a moment-resisting frame with special detailing for ductile connections).

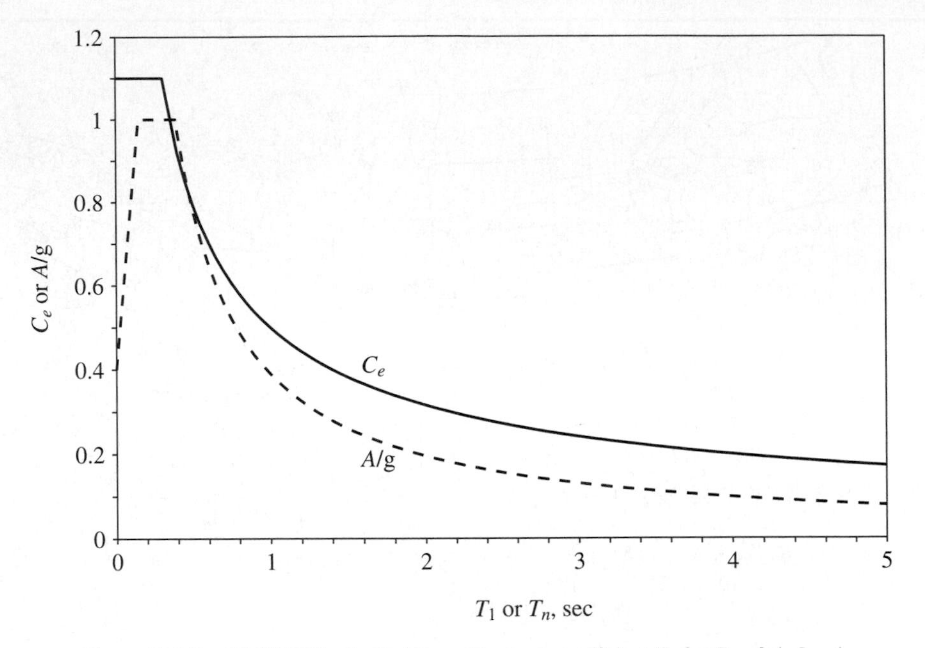

Figure 21.1.2 *UBC* (1994): elastic ($R_w = 1$) seismic coefficient C_e for $Z = 0.4$, $I = 1$, $S = 1$; and pseudo-acceleration A/g for $Z = 0.4$.

21.1.2 Lateral Forces

The distribution of lateral forces over the height of the building is to be determined from the base shear in accordance with the formula for the lateral (or horizontal) force at the jth floor:

$$F_j = (V_b - F_t)\frac{w_j h_j}{\sum_{i=1}^{N} w_i h_i} \tag{21.1.7}$$

with the exception that the force at the top floor (or roof) computed from Eq. (21.1.7) is increased by an additional force, the top force:

$$F_t = \begin{cases} 0 & T_1 \leq 0.7 \\ 0.07 T_1 V_b & 0.7 < T_1 < 3.6 \\ 0.25 V_b & T_1 \geq 3.6 \end{cases} \tag{21.1.8}$$

where h_j = height of the jth floor above the base. These lateral forces are shown in Fig. 21.1.3.

21.1.3 Story Forces

The design values of story shears, story overturning moments, and element forces are determined by static analysis of the building subjected to the lateral forces defined by the foregoing equations.

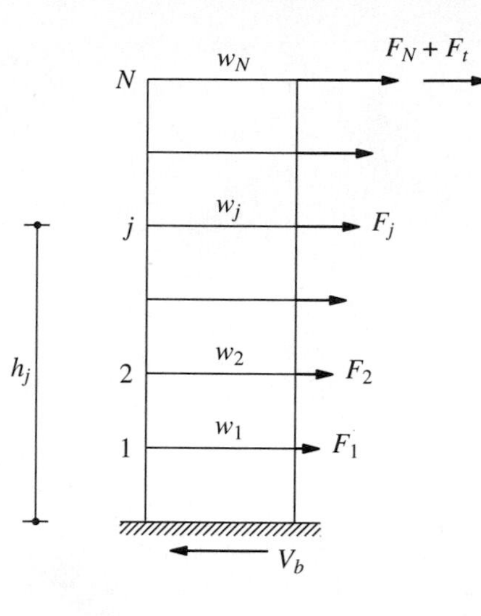

Figure 21.1.3 *UBC* lateral forces.

21.2 *NATIONAL BUILDING CODE OF CANADA*, 1995

21.2.1 Base Shear

The base shear formula in the 1995 edition of the *National Building Code of Canada* (NBCC) may also be expressed as Eq. (21.1.1), where the seismic coefficient is given by

$$C_s = \frac{C_e}{R} U \tag{21.2.1}$$

In this equation, $U = 0.6$ and $1/U = 1.67$ may be interpreted as the *overstrength factor*, by which the actual strength of the building is expected to be larger than its calculated strength. The elastic seismic coefficient

$$C_e = vSIF \tag{21.2.2}$$

is the product of four factors:

1. Zonal velocity ratio v varies between 0 for the least seismic zone and 0.4 for the most seismic zone (Fig. 21.2.1b).
2. Seismic importance factor $I = 1.5$ for "post-disaster buildings," 1.3 for schools, and 1.0 for all other buildings.
3. Foundation factor $F = 1.0$, 1.3, 1.5, or 2.0, depending on the soil category defined in the code.
4. Seismic response factor S varies with fundamental natural vibration period, as shown in Fig. 21.2.2. Observe that for $T_1 < 0.5$ sec, S depends on whether $Z_a > Z_v$, $Z_a = Z_v$ or $Z_a < Z_v$, where Z_a is the acceleration-related seismic zone and Z_v is the velocity-related seismic zone. Canada is divided into seven zones based on each of the two criteria (Fig. 21.2.1).

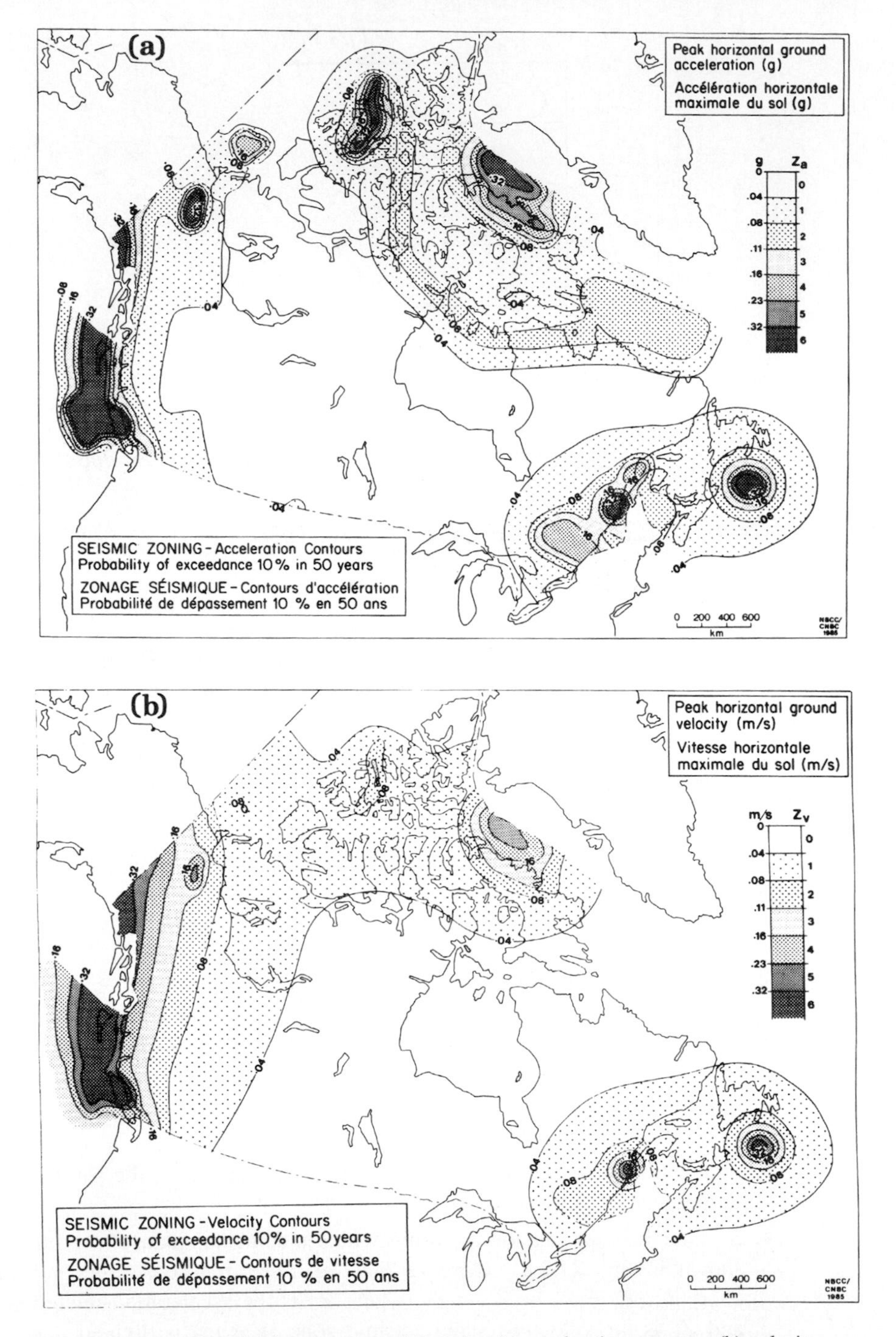

Figure 21.2.1 Seismic zone map of Canada: (a) acceleration contours; (b) velocity contours. (Courtesy of Geological Survey of Canada.)

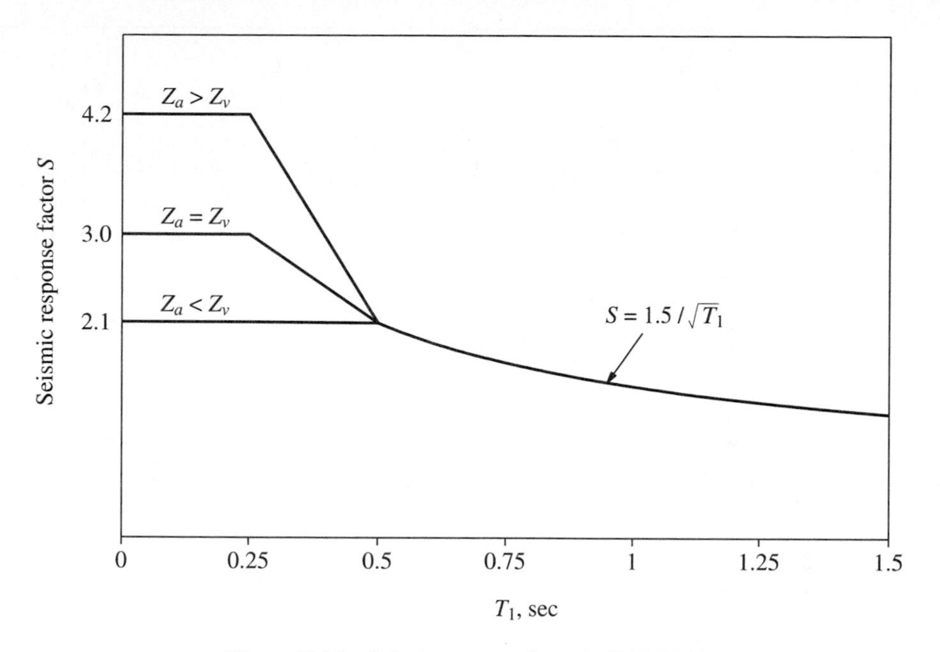

Figure 21.2.2 Seismic response factor in 1990 *NBCC*.

In addition to empirical formulas for estimating the fundamental period, the *NBCC* gives a formula similar to Eq. (21.1.5):

$$T_1 = 2\pi \left[\frac{\sum_{i=1}^{N} w_i u_i^2}{g \sum_{i=1}^{N} w_i u_i} \right]^{1/2} \tag{21.2.3}$$

where u_i are the floor displacements due to static application of a set of lateral forces w_i at floor levels $i = 1, 2, \ldots, N$.

Shown in Fig. 21.2.3 is the elastic seismic coefficient C_e for $v = 0.4$, $Z_a = Z_v$, $I = 1$, and $F = 1$. This definition of C_e is presumably based on the elastic pseudo-acceleration design spectrum, scaled to a peak ground velocity of 0.4 m/sec (which corresponds to $v = 0.4$):

$$\frac{A}{g} = \begin{cases} 1.2 & 0.03 \le T_n \le 0.427 \\ 0.512/T_n & T_n > 0.427 \end{cases} \tag{21.2.4}$$

The force modification factor R assigned to different types of structural systems reflects design and construction experience, as well as the performance of structures during earthquakes. It endeavors to account for the inelastic energy absorption capacity of the structure. Specified values of R range from 1.0 for brittle structures such as unreinforced masonry to 4.0 for ductile moment-resisting space frames.

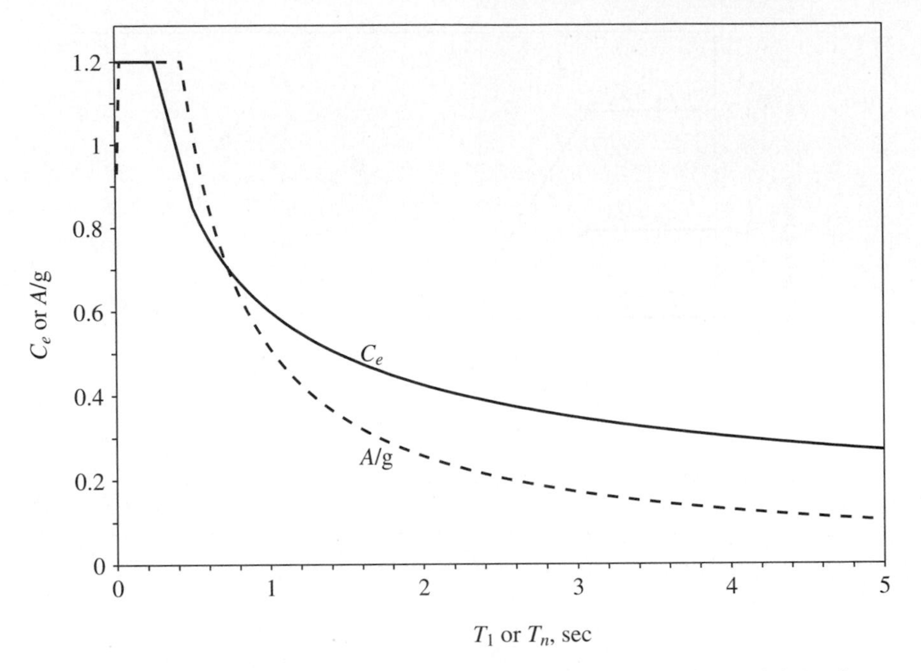

Figure 21.2.3 *NBCC* (1990): elastic ($R = 1$) seismic coefficient C_e for $v = 0.4$, $I = 1$, $F = 1$, $Z_a = Z_v$; and pseudo-acceleration A/g for $v = 0.4$.

21.2.2 Lateral Forces

The formulas in *NBCC* (1990) to determine the lateral forces at the various floor levels from the base shear are identical to Eqs. (21.1.7) and (21.1.8) in *UBC* (1994).

21.2.3 Story Forces

The design values of story shears are determined by static analysis of the structure subjected to these lateral forces. Similarly determined overturning moments are multiplied by reduction factors J and J_i at the base of the structure and at the ith floor level, respectively, where

$$J = \begin{cases} 1 & T_1 < 0.5 \\ 1.1 - 0.2T_1 & 0.5 \le T_1 \le 1.5 \\ 0.8 & T_1 > 1.5 \end{cases} \tag{21.2.5}$$

and

$$J_i = J + (1 - J)\left(\frac{h_i}{h_N}\right)^3 \tag{21.2.6}$$

21.3 *MEXICO FEDERAL DISTRICT CODE*, 1987

21.3.1 Base Shear

The base shear formula in the 1987 edition of the *Mexico Federal District Code* (MFDC), which applies to Mexico City, may also be expressed as Eq. (21.1.1) with the seismic coefficient

$$C_s = \frac{C_e}{Q'} \tag{21.3.1}$$

The elastic seismic coefficient

$$C_e = \begin{cases} A/\mathrm{g} & T_1 \leq T_c \\ A/\mathrm{g}\,\{1 + 0.5r\,[1 - (T_c/T_1)^r]\} & T_1 \geq T_c \end{cases} \tag{21.3.2}$$

where the pseudo-acceleration design spectrum A/g is given by

$$\frac{A}{\mathrm{g}} = \begin{cases} (1 + 3T_n/T_b)\,A_m/4 & T_n \leq T_b \\ A_m & T_b \leq T_n \leq T_c \\ A_m(T_c/T_n)^r & T_n \geq T_c \end{cases} \tag{21.3.3}$$

and T_1 is the fundamental period given by the same formula as in *UBC*, Eq. (21.1.5); T_b and T_c denote the periods at the beginning and end of the constant pseudo-acceleration region of the design spectrum (Fig. 21.3.1). The coefficients A_m and r and period values

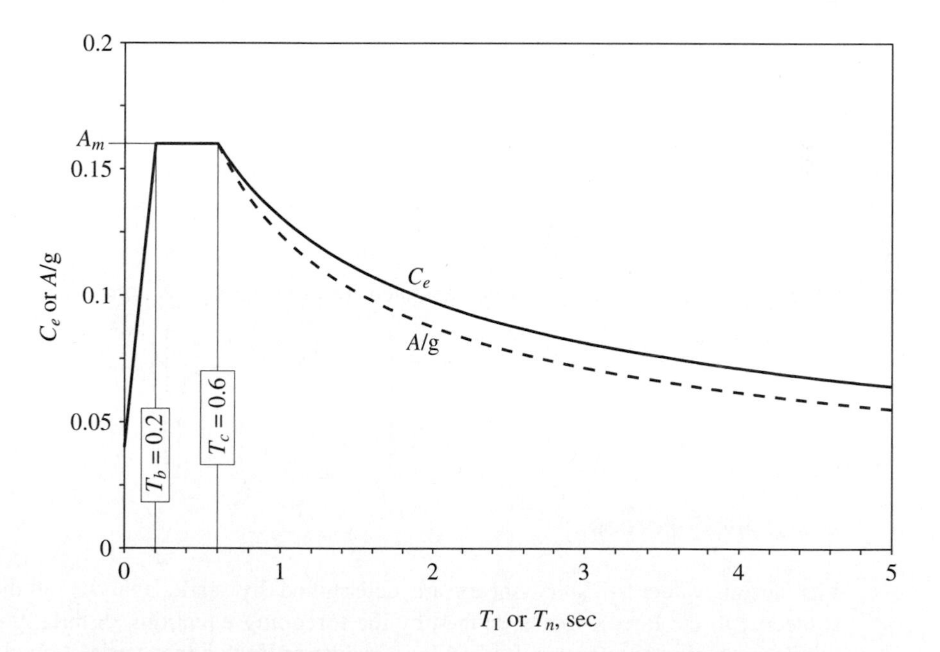

Figure 21.3.1 *MFDC* (1987): elastic ($Q' = 1$) seismic coefficient C_e for zone I; and pseudo-acceleration A/g.

T_b and T_c are given in Table 21.3.1 for the three zones of the Mexico Federal District, depending primarily on the local soil conditions. Figure 21.3.1 shows A/g and C_e for zone I.

The elastic seismic coefficient is divided by the seismic reduction factor:

$$Q' = \begin{cases} 1 + (T_1/T_b)(Q-1) & T_1 \leq T_b \\ Q & T_1 \geq T_b \end{cases} \tag{21.3.4}$$

The seismic behavior factor Q may be interpreted as the allowable ductility; it varies between 1 and 4, depending on several factors, including the structural system and structural materials.

TABLE 21.3.1 PSEUDO-ACCELERATION DESIGN SPECTRUM PARAMETERS

Zone	A_m	T_b (sec)	T_c (sec)	r
I: Hard ground	0.16	0.2	0.6	1/2
II: Transition	0.32	0.3	1.5	2/3
III: Soft soil	0.40	0.6	3.9	1

21.3.2 Lateral Forces

The formula for the lateral force F_j at the jth floor depends on whether $T_1 \leq T_c$ or $T_1 \geq T_c$:

$$F_j = V_b \frac{w_j h_j}{\sum_{i=1}^{N} w_i h_i} \qquad T_1 \leq T_c \tag{21.3.5a}$$

and

$$F_j = V_b^{(1)} \frac{w_j h_j}{\sum_{i=1}^{N} w_i h_i} + V_b^{(2)} \frac{w_j h_j^2}{\sum_{i=1}^{N} w_i h_i^2} \qquad T_1 \geq T_c \tag{21.3.5b}$$

where the base shear V_b has been separated into two parts, $V_b^{(1)}$ and $V_b^{(2)}$:

$$V_b^{(1)} = \frac{W(A/g)}{Q'} \left\{ 1 - r \left[1 - \left(\frac{T_c}{T_1} \right)^r \right] \right\} \tag{21.3.6a}$$

$$V_b^{(2)} = \frac{W(A/g)}{Q'} \left\{ 1.5r \left[1 - \left(\frac{T_c}{T_1} \right)^r \right] \right\} \tag{21.3.6b}$$

21.3.3 Story Forces

The design values of story shears are determined by static analysis of the structure subjected to the lateral forces defined by the foregoing equations. Similarly determined overturning moments are multiplied by a reduction factor that varies linearly from 1.0 at the top of the building to 0.8 at its base to obtain the design values. There is an additional requirement, however, that the reduced moments not be less than the product of the story shear at that elevation and the distance to the center of gravity of the portion of the building above the elevation being considered.

21.4 STRUCTURAL DYNAMICS IN BUILDING CODES

Most of the seismic provisions contained in building codes are either derived from or related to the theory of structural dynamics. In this section we discuss this interrelationship for several aspects of the code provisions.

21.4.1 Fundamental Vibration Period

The period formula in the *UBC* and *MFDC*, Eq. (21.1.5), is identical to the result obtained from Rayleigh's method using the shape function given by the static deflections u_i due to a set of lateral forces F_i at the floor levels. This becomes apparent by comparing Eq. (21.1.5) with Eq. (8.6.3b). The period formula in the *NBCC*, Eq. (21.2.3), has the same basis except that the lateral forces used to determine the static deflections are assumed equal to the lumped weights at the floor levels. This becomes apparent by comparing Eq. (21.2.3) with Eq. (8.6.3c).

21.4.2 Elastic Seismic Coefficient

The elastic seismic coefficient C_e is related to the pseudo-acceleration spectrum for linearly elastic systems. In a linear SDF (one-story) system of weight w, the peak base shear is (see Section 6.6.3)

$$V_b = \frac{A}{g} w \qquad (21.4.1)$$

and $C_e = A/g$. For buildings the three codes—*UBC*, *NBCC*, and *MFDC*—give the design base shear as

$$V_b = \frac{1}{R_w} C_e W \qquad V_b = \frac{U}{R} C_e W \quad \text{and} \quad V_b = \frac{1}{Q'} C_e W \qquad (21.4.2)$$

respectively. By taking $R_w = R = Q' = 1$ it is clear that C_e in building codes corresponds to A/g, the pseudo-acceleration for linearly elastic systems normalized with respect to gravitational acceleration. The two, C_e and A/g, as specified in codes, are not identical, however, as seen in Figs. 21.1.2, 21.2.3, and 21.3.1. The ratio $C_e \div A/g$ is plotted as a function of period in Fig. 21.4.1, where we note that it exceeds unity for almost all periods and increases with T_1.

The seismic coefficient C_e is specified larger than A/g to account for the more complex dynamics of multistory buildings responding in several natural modes of vibration and to recognize uncertainties in a calculated value of the fundamental vibration period. In Section 13.8 we demonstrated that the peak value of the base shear due to the

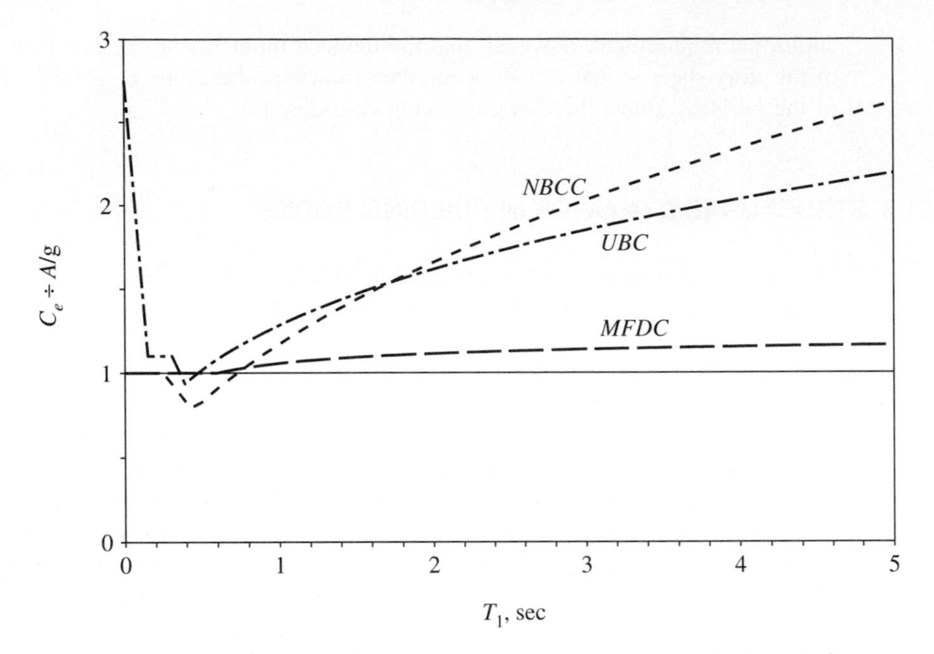

Figure 21.4.1 Ratio of elastic seismic coefficient C_e and pseudo-acceleration A/g for three building codes.

nth mode is

$$V_{bn} = \frac{A_n}{g} W_n^* \tag{21.4.3}$$

where W_n^* is the effective weight and A_n/g the normalized pseudo-acceleration, both for the nth mode. The peak value of the base shear considering several modes is generally estimated by the SRSS formula, Eq. (13.7.3). For the present purpose, however, we use the upper bound result of Eq. (13.7.2), specialized for the base shear:

$$V_b \le \sum_{n=1}^{N} | V_{bn} | = \sum_{n=1}^{N} \frac{A_n}{g} W_n^* \tag{21.4.4}$$

If all the A_n values were equal to A_1, which they are not, Eq. (21.4.4) reduces to

$$V_b \le \frac{A_1}{g} \sum_{n=1}^{N} W_n^* = \frac{A_1}{g} W \tag{21.4.5}$$

where the second half of this equation is obtained after using Eq. (13.2.14). Thus for an MDF system C_e and A_1/g have similar but by no means identical meaning. In Part B we discuss these conceptual differences between C_e and A_1/g further, as well as their numerical differences, shown in Fig. 21.4.1.

It is of interest to compare the pseudo-acceleration design spectrum specified in the three codes with two levels—median and median plus one standard deviation—of the design spectra for firm ground sites developed by the procedures of Fig. 6.9.3 (see also Figs. 6.9.4 and 6.9.5). All five of these spectra are presented in Fig. 21.4.2, where the

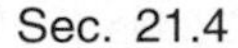

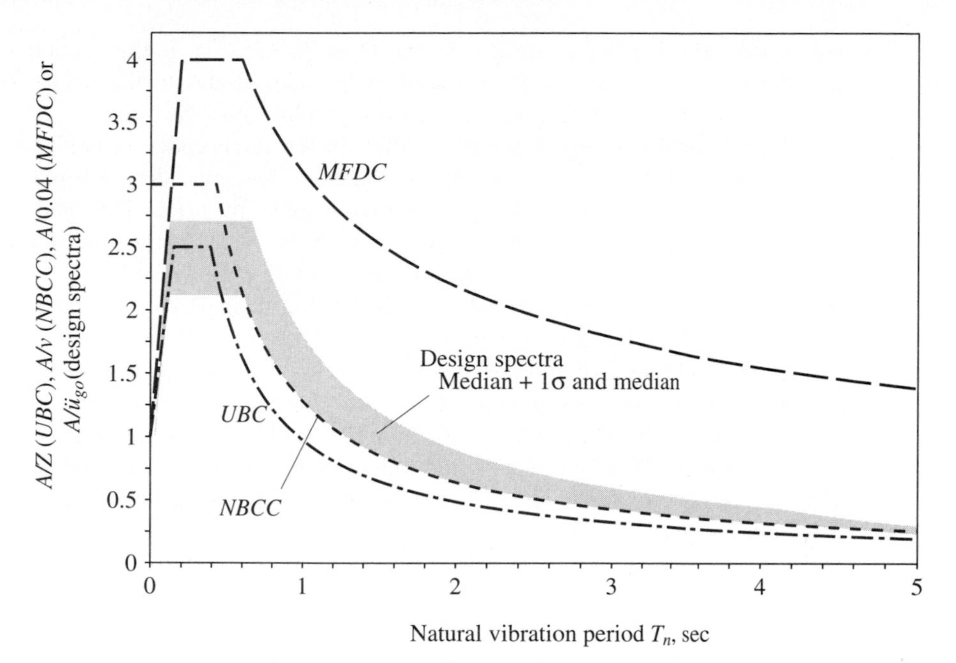

Figure 21.4.2 Comparison of pseudo-acceleration design spectrum in building codes compared with the design spectra developed in Chapter 6; the latter are median and median $+ 1\sigma$ spectra for 5% damping.

pseudo-acceleration is normalized relative to its value at zero period; such normalizing with respect to peak ground acceleration is widely used but is not the best option. This normalization removes any differences in the peak ground accelerations implied in the five spectra and provides a comparison of the spectral shapes. Relative to the median design spectrum, the *UBC* spectrum is lower in the velocity and displacement-sensitive period regions but higher in a part of the acceleration-sensitive region; in this period range the *UBC* spectrum is closer to the median $+1\sigma$ design spectrum. The *NBCC* spectrum is close to the median design spectrum in the velocity- and displacement-sensitive period regions but higher than even the median $+1\sigma$ spectrum in part of the acceleration-sensitive region. This period region for both code spectra is relatively narrow compared to the design spectrum. The *MFDC* spectrum is much higher than the median $+1\sigma$ spectrum at all periods, presumably to account for the different characteristics of ground motions expected on *hard* ground in Mexico City. This difference limits the meaning of some of the comparisons between *MFDC* and dynamic response analysis based on the selected design spectra, presented in Part B of this chapter.

21.4.3 Design Force Reduction

Most codes specify the design base shear to be smaller than the elastic base shear (determined using the elastic seismic coefficient C_e). For the three codes described earlier, the reduction factors are R_w, R, and Q' in Eq. (21.4.2). In this section we examine how the reduction in design force specified in codes relates to the results obtained in Chapter 7 from dynamic response analysis of yielding systems.

The normalized design force specified in the three codes is compared with the normalized yield strength of elastoplastic systems. The code design force for inelastic systems normalized by the elastic design force is given by $1/R_w$, $1/R$, and $1/Q'$ for the *UBC*, *NBCC*, and *MFDC*, respectively. These normalized design forces are plotted in Fig. 21.4.3 as a function of the fundamental vibration period T_1 for $R_w = R = Q = 4$. They are independent of T_1 in *UBC* and *NBCC* but their period dependence in *MFDC* is defined by Eq. (21.3.4). Determined from dynamic analysis of SDF systems, the normalized yield strength of elastoplastic systems corresponding to a ductility factor of 4 (Fig. 7.10.1) is also shown. It is clear from this comparison that the *MFDC* seismic reduction factor varies with vibration period in a manner consistent with structural dynamics theory. However, the period independence of factors R_w and R in *UBC* and *NBCC* contradicts dynamic response results for structures with fundamental period in the acceleration-sensitive region of the design spectrum. The resulting discrepancy in the design spectra is seen in Fig. 21.4.4 for two values of the ductility factor μ. The inelastic design spectra shown are from Fig. 7.10.4 scaled by 0.4 so that they correspond to peak ground acceleration $\ddot{u}_{go} = 0.4$g. The elastic design spectrum reduced by the period-independent factor μ is lower in the acceleration-sensitive period region, as shown in

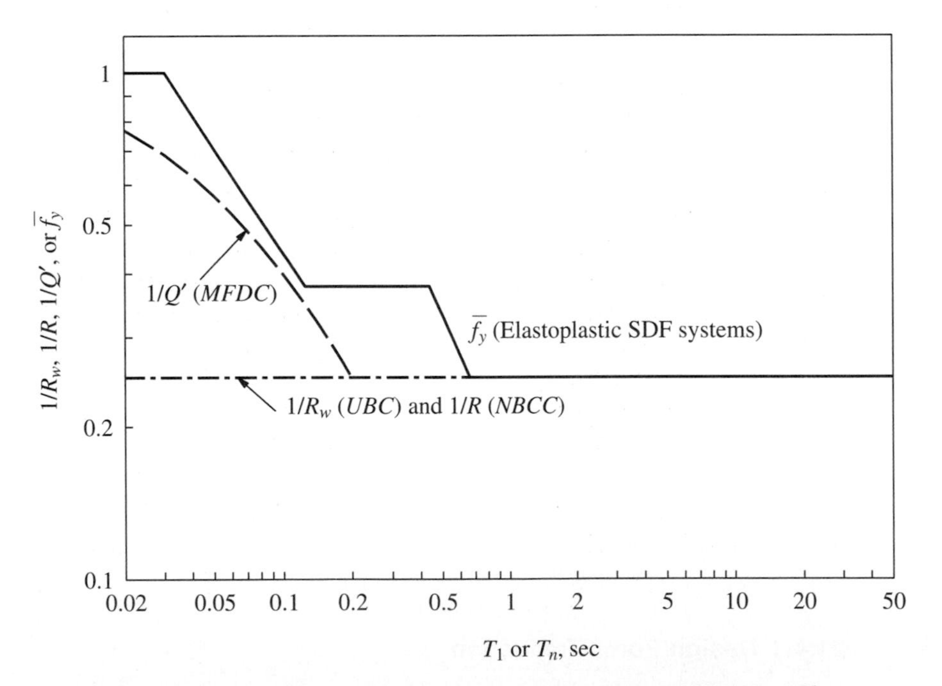

Figure 21.4.3 Comparison of normalized design force—$1/R_w$, $1/R$, and $1/Q'$—in building codes and the normalized yield strength $\bar{f}_y$ of elastoplastic SDF systems; $R_w = R = Q' = \mu = 4$.

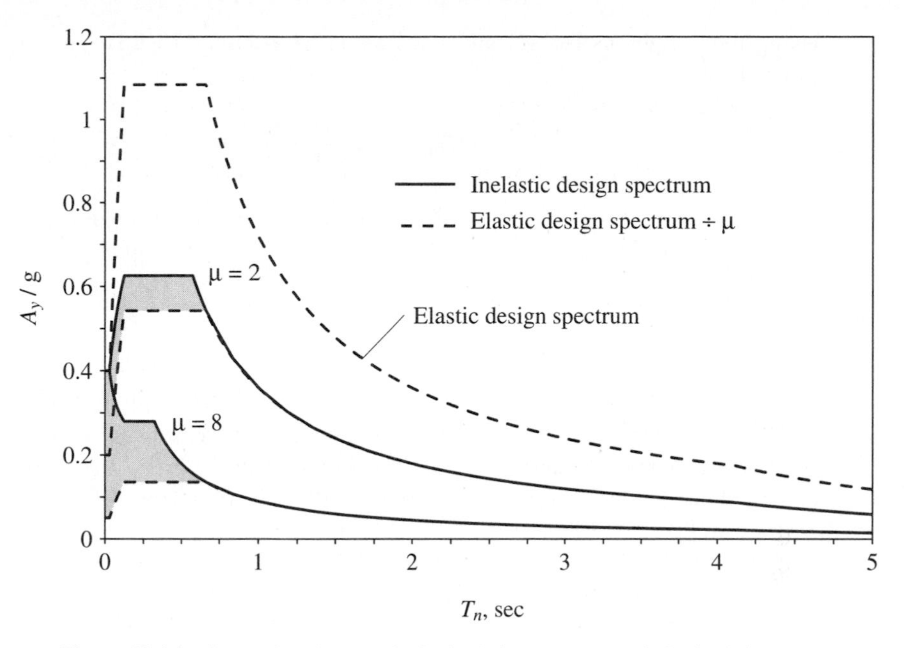

Figure 21.4.4 Comparison between inelastic design spectrum and elastic design spectrum reduced by the period-independent factor μ; results are presented for $\mu = 2$ and 8.

Fig. 21.4.4. Thus, by ignoring the period dependence of the normalized strength, the code may give excessively small design force for structures in this period region.

This may imply that the code provisions are unconservative in certain situations, but we will not get into this issue because of several practical considerations, which are beyond the scope of this book. One of these is worth mentioning, however. The actual strength of a building exceeds its design strength, especially for short-period systems. Overstrength can come from a variety of sources. Examples are the difference between design force and the theoretical strength of structural elements because of difference between allowable and yield stresses, the effects of gravity loads on element strengths, element overstrength due to discrete choices of member sizes, element overstrength due to stiffness (drift) requirements, and increase in structural strength due to redistribution of element forces in the inelastic range, and the contributions of all structural and non-structural elements that, in the design process, are not considered as part of the lateral force-resisting system. This overstrength of a building is recognized *explicitly* in *NBCC* but not in *UBC* or *MFDC*.

21.4.4 Lateral Force Distribution

Structural dynamics gives the base shear and equivalent static lateral force at floor level j for mode n of a multistory building (Section 13.8.1):

$$V_{bn} = M_n^* A_n \qquad f_{jn} = \Gamma_n m_j \phi_{jn} A_n$$

Using the definitions for M_n^* and Γ_n, Eqs. (13.2.9a) and (13.2.3), f_{jn} can be expressed in terms of V_b:

$$f_{jn} = V_{bn} \frac{w_j \phi_{jn}}{\sum_{i=1}^{N} w_i \phi_{in}} \tag{21.4.6}$$

We now compare this force distribution from structural dynamics with code specifications. If the top force F_t were zero, the *UBC* and *NBCC* give

$$F_j = V_b \frac{w_j h_j}{\sum_{i=1}^{N} w_i h_i} \tag{21.4.7}$$

This force distribution agrees with Eq. (21.4.6) if ϕ_{jn} were proportional to h_j, that is, if the mode shape were linear. The linear shape is a reasonable approximation to the fundamental mode of many buildings; as shown in Fig. 18.1.5, it is in between the fundamental mode shapes for the two extreme values, 0 and ∞, of the beam-to-column stiffness ratio ρ. Assignment of the additional force F_t at the top of the building is intended by the code to consider approximately and simplistically the influence of the higher vibration modes on the force distribution. The force F_t increases the shear force in the upper stories relative to the base shear. This is consistent with the predictions of structural dynamics that the higher modes affect the forces in the upper stories more than in lower stories (Section 18.6). Equation (21.1.8) gives F_t that ranges from zero for short-period buildings to $0.25V$ for long-period buildings for which structural dynamics theory demonstrates that higher-mode responses are more significant (Section 18.4).

In the *MFDC*, if $T_1 \le T_c$, the heightwise distribution of lateral forces is also given by Eq. (21.4.7), based on the response in only the fundamental vibration mode assumed to have a linear shape. For $T_1 \ge T_c$, the *MFDC* gives Eq. (21.3.5b), based on specifying floor displacements proportional to h_j at $T_1 = T_c$, to h_j^2 at T_1 much longer than T_c, and intermediate between the linear and parabolic shapes at intermediate values of T_1. This variation in deflected shape and hence force distribution is intended to recognize the changing shape of the fundamental vibration mode and increasing higher-mode responses with increasing fundamental period.

21.4.5 Overturning Moments

Some building codes, including *NBCC* and *MFDC*, allow reduction of overturning moments relative to the values computed from lateral forces F_j by statics, because the response contributions of higher modes are more significant for story shears than for overturning moments (Chapter 18). In particular, if the first mode were linear, the higher modes would provide no contribution to the overturning moment at the base (Example 13.6), although they would affect overturning moments at higher levels and story shears at all levels. Thus the overturning moments computed from the code forces, supposedly calibrated against dynamic response results to provide the correct story shears, would exceed the values predicted by dynamic analysis and could therefore be reduced.

In the *NBCC* the reduction factor at the building base ranges from 1.0 (no reduction) for buildings with $T_1 < 0.5$ sec to 0.8 (20% reduction) for buildings with $T_1 = 1.5$ sec or longer. This is generally consistent with the prediction of structural dynamics theory that higher-mode responses are more significant for longer-period buildings (Section 18.4).

Earlier editions of the *UBC* also permitted reduction of overturning moments relative to their values computed form lateral forces by statics. For example, the 1967 edition of *UBC* specified the base-overturning-moment reduction factor as

$$J = \frac{0.5}{T_1^{2/3}} \qquad 0.33 \le J \le 1$$

permitting up to 67% reduction for long-period buildings. Later the reduction factor was eliminated, implying that $J = 1$, after several tall buildings in Caracas were severely damaged due to overturning moments in the Caracas earthquake of July 29, 1967 (Fig. 21.4.5).

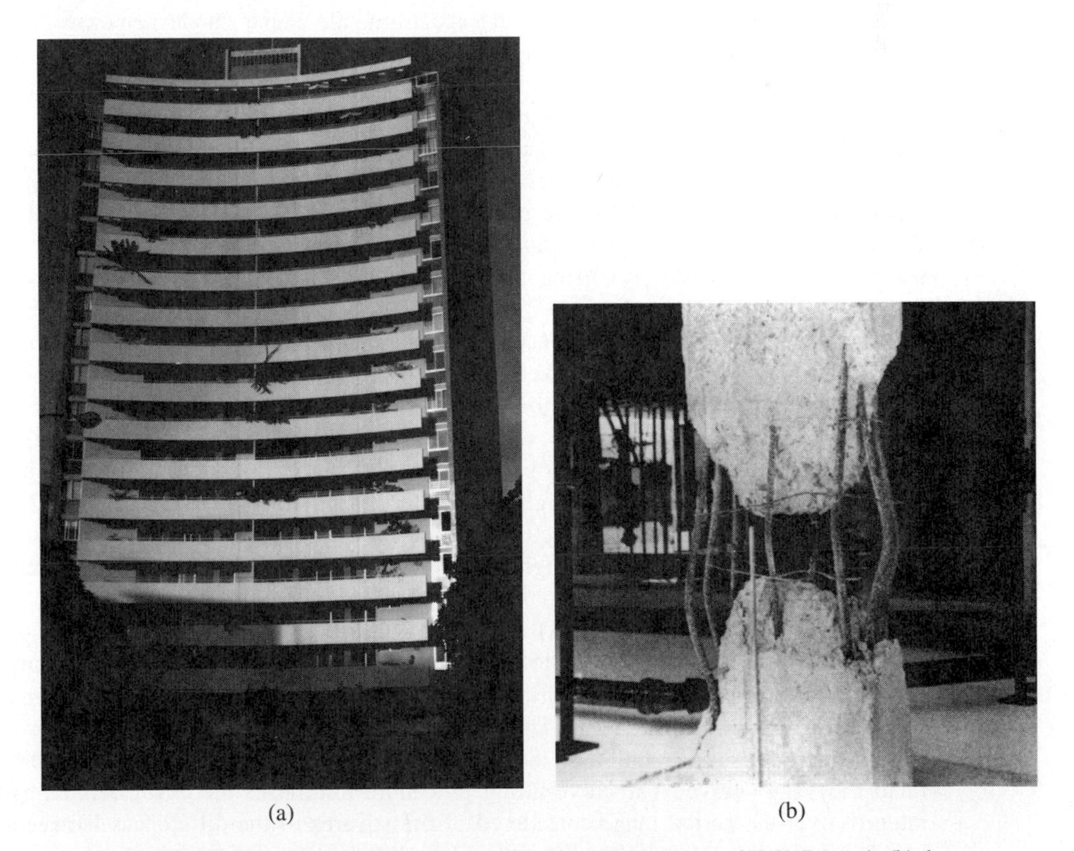

(a) (b)

Figure 21.4.5 (a) Caromay Building, Caracas, Venezuela (courtesy of V. V. Bertero); (b) damage to column B-1 of Caromay Building at garage level (courtesy of M. A. Sozen).

PART B: EVALUATION OF BUILDING CODES

In Part B we evaluate how well the seismic forces specified in building codes agree with the results of dynamic analysis presented in Chapters 18 and 19.

21.5 BASE SHEAR

The significance of the response contributions of the higher vibration modes in the dynamic response of buildings (Chapter 18) plays a central role in evaluating the code forces. For this reason we first recall that the combined responses of the second and higher modes depend mainly on two parameters: fundamental period T_1 and beam-to-column stiffness ratio ρ. With reference to Fig. 18.4.1, we had concluded that the base shear for buildings with T_1 within the acceleration-sensitive region of the spectrum is essentially all due to the first mode. However, for buildings with T_1 in the velocity- or displacement-sensitive regions of the spectrum, the higher-mode responses can be significant, increasing with increasing T_1 and with decreasing ρ, for reasons discussed in Chapter 18.

For buildings with T_1 in the acceleration-sensitive region of the spectrum, these results and Eq. (21.4.3) indicate that the code formula, Eq. (21.1.1), would accurately predict the base shear for elastic buildings if the seismic coefficient C_e were defined as A_1/g and the total weight W were replaced by the first-mode effective weight W_1^*. If W is used instead of W_1^*, as in building codes, the base shear is overestimated. This becomes obvious by renormalizing the base shear data of Fig. 18.4.1 with respect to the total weight, as shown in Fig. 21.5.1; recall that the $W_1^* = 0.679W$, $0.796W$, and $0.880W$ for $\rho = 0$, $\frac{1}{8}$, and ∞, respectively (Table 18.1.1). Therefore, the overestimation varies with ρ, being the least for shear buildings ($\rho = \infty$), largest for flexural buildings ($\rho = 0$), and in between for frame buildings with intermediate values of ρ.

For buildings with T_1 in the velocity- or displacement-sensitive spectral regions, however, the increase in base shear by using the total weight of the building may not be sufficient to compensate for the higher-mode response. This is clear from Fig. 21.5.1, where we observe that V_b/W exceeds A_1/g for longer T_1 and smaller ρ, conditions that produce increasingly significant higher-mode response. For this range of parameters, therefore, the seismic coefficient C_e should be larger than A_1/g.

The dynamic response (RSA) results of Fig. 18.4.1 provide insight into how the A/g spectrum should be modified to obtain the seismic coefficient C_e. For this purpose curves of the form $\alpha T_1^{-\beta}$ are fitted to the base shear versus period curve from dynamic analysis, as shown in Fig. 21.5.2. The parameters α and β for each of the velocity- and displacement-sensitive regions of the spectrum are evaluated by a least-squared-error fit to the V_b–T_1 curve. The curve-fitting procedure minimizes the error, defined as the integral over the period range considered of the squares of the differences between the logarithm of the ordinates of the "exact" and the fitted curve. This curve-fitting procedure is designed to satisfy the following constraints. First, the ordinate of the fitted curve at

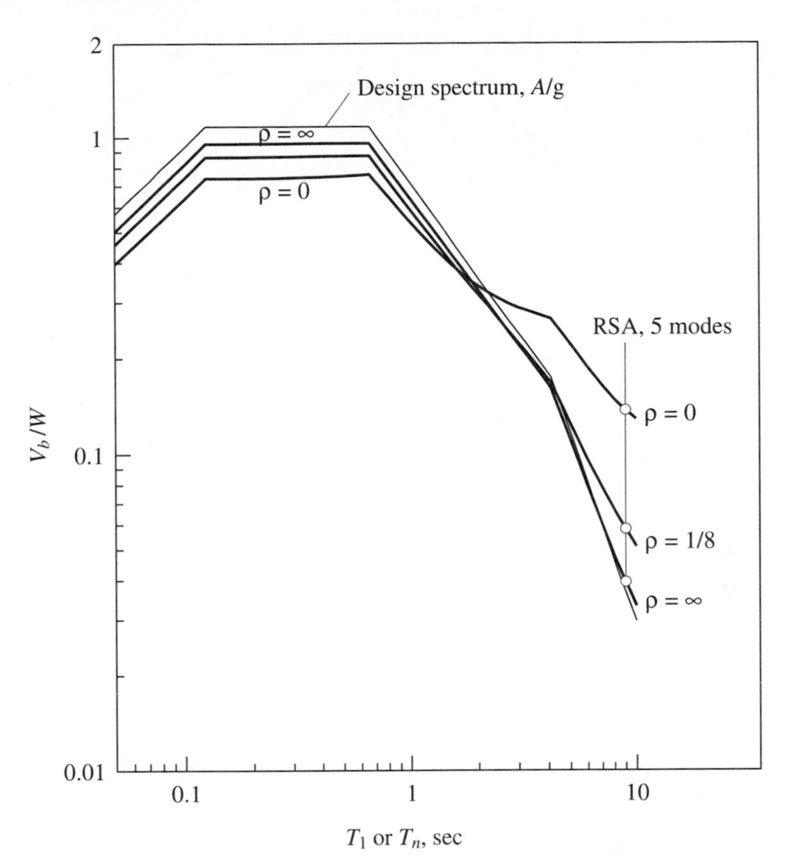

Figure 21.5.1 Base shear V_b (normalized by total weight W) in buildings with $\rho = 0$, $\frac{1}{8}$, or ∞, computed for the design spectrum shown.

$T_1 = T_c$ is equal to the ordinate of the flat portion of the A/g spectrum. Second, the curves fitted to the velocity- and displacement-sensitive regions of the spectrum have the same ordinates at $T_1 = T_d$. Third, the exponent β for the displacement-sensitive region should not be smaller than its value in the velocity-sensitive region. The resulting functions $\alpha T_1^{-\beta}$ are shown in Fig. 21.5.2 together with the V_b–T_1 curves from dynamic analyses.

This comparison suggests that the code seismic coefficient C_e should be defined by decreasing the coefficient β in $\alpha T_1^{-\beta}$ and thus raising the pseudo-acceleration design spectrum to account for the higher-mode response. For this purpose, building codes should define C_e in terms of the design spectrum A/g, which in turn should be specified explicitly. This is the approach followed by *MFDC* but not by most other codes. Once this format is adopted, C_e can be defined by raising the design spectrum in its velocity- and displacement-sensitive regions, based on dynamic response results of the type presented here. The degree to which the spectrum should be raised depends on the beam-to-column stiffness ratio ρ; the spectrum needs to be raised very little for shear

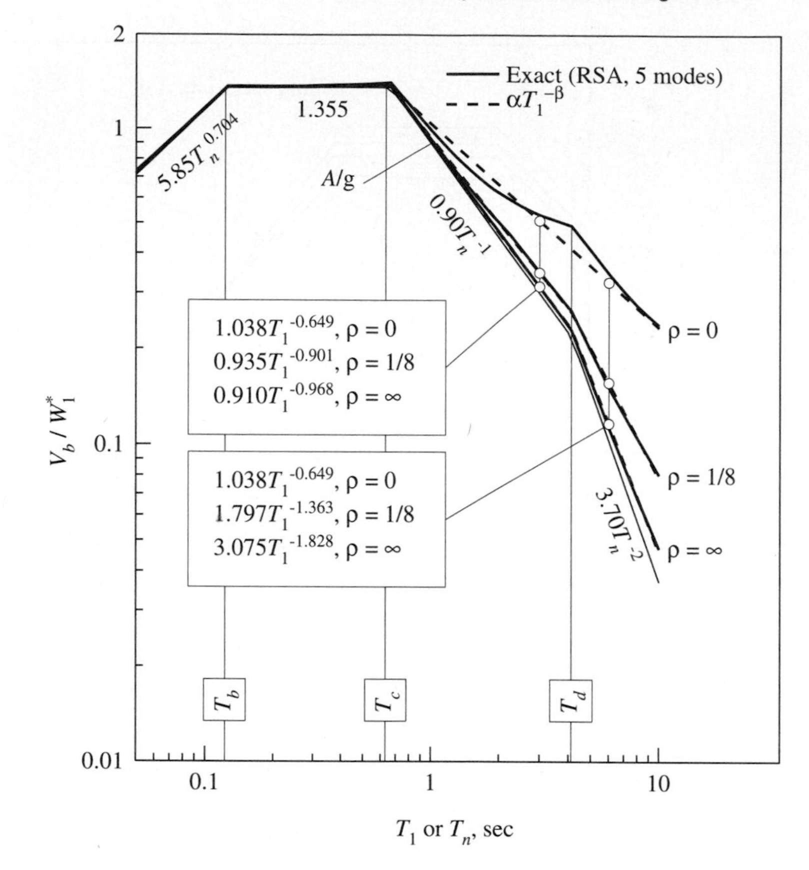

Figure 21.5.2 Comparison of functions $\alpha T_1^{-\beta}$ and "exact" base shear versus T_1 curves.

buildings ($\rho = \infty$) but to an increasing degree with increasing frame action (i.e., decreasing ρ). The spectral modifications presented also depend on the heightwise distribution of mass and stiffness, parameters that have not been varied here.

Having utilized dynamic response results to determine how the design spectrum A/g should be modified for higher-mode response, we now compare these results with building code provisions. For this purpose we return to Fig. 21.4.1, where the $C_e \div A/g$ in three codes was presented as a function of T_1; this ratio is defined explicitly in *MFDC* but only implicitly in *UBC* and *NBCC*. These results are compared in Fig. 21.5.3 with $V_b/W_1^* \div A/g$, the ratio of two values of base shear, the first including responses due to all modes and the other considering only the first mode (Fig. 21.5.2). It is clear that all three codes recognize the dependence of higher mode response on the fundamental period T_1 but not on the stiffness ratio ρ. The *UBC* and *NBCC* appear to overcompensate for the higher mode response over a wide range of T_1 and ρ, whereas *MFDC* seems not to compensate enough in most cases, except for shear buildings.

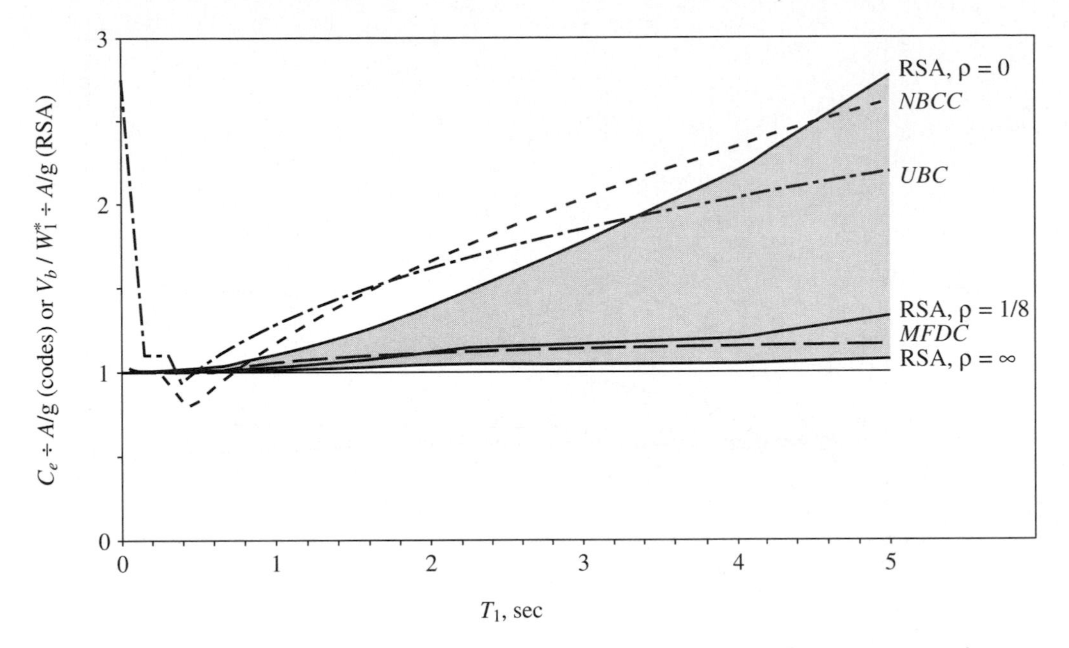

Figure 21.5.3 Comparison of $C_e \div A/g$ in three building codes and $V_b/W_1^* \div A/g$ from RSA for three values of ρ.

Inelastic response behavior in multistory buildings, and its differences relative to SDF systems, is another factor that should be considered in specifying the seismic coefficient in building codes. This important concept is illustrated by returning to Fig. 19.3.7 from nonlinear dynamic analysis, showing the ratio of base shear yield strengths in multistory buildings and SDF systems necessary to limit the ductility demand to the same allowable value, $\mu = 1$, 4, or 8. Superimposed on these results in Fig. 21.5.4 is the ratio $C_e \div A/g$ for three building codes from Fig. 21.4.1; these code ratios also apply to yielding systems irrespective of the allowable reduction factors R, R_w, or Q'. In contrast, dynamic response results indicate that the strength increase required to account for MDF effects depends significantly on the allowable ductility; the strength increase also depends on the failure mechanism, but this is not shown here because the dynamic response results are restricted to shear frames.

21.6 STORY SHEARS AND EQUIVALENT STATIC FORCES

Having compared the code-specified base shear with the predictions of dynamic analysis, we now evaluate the heightwise distribution of story forces. As mentioned in Sections 21.1 to 21.3, the codes specify lateral forces in terms of the base shear and static analysis of the structure for these forces provides the story shears. The story shears and lateral forces determined from three codes are divided by the base shear V_b and presented

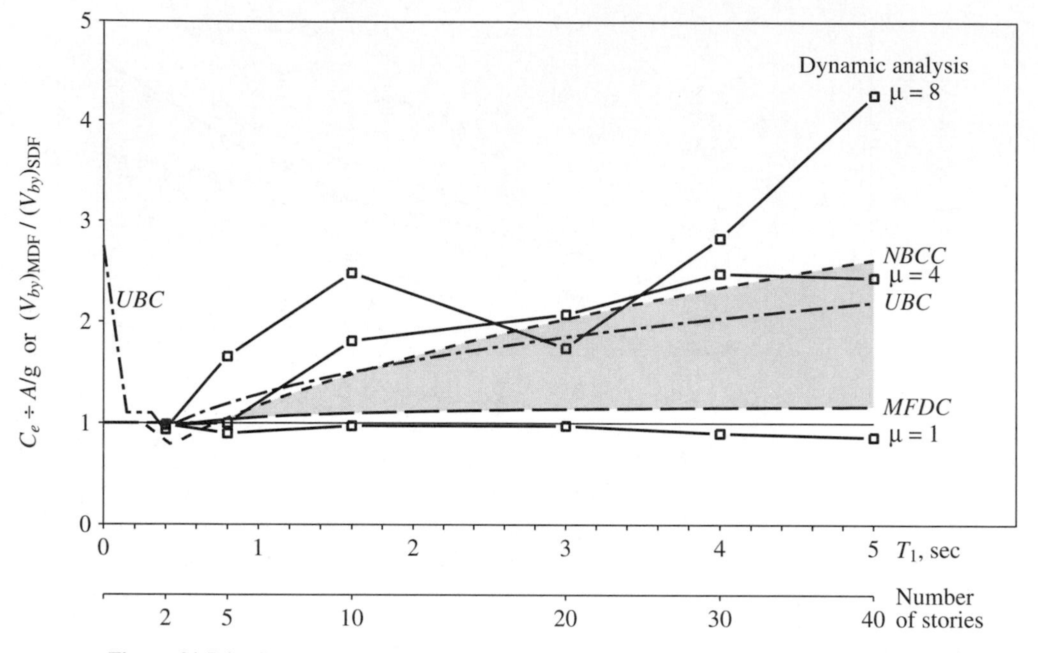

Figure 21.5.4 Comparison of $C_e \div A/g$ in three building codes and $(V_{by})_{MDF}/(V_{by})_{SDF}$ from nonlinear dynamic analysis for three values of allowable ductility μ.

in Figs. 21.6.1 and 21.6.2; included in each case are three values of $T_1 = 0.5$, 3, and 8 sec, chosen to be representative of the three spectral regions. Note that in Fig. 21.6.2, the numerical values of the equivalent static forces that are concentrated at the floor levels have been joined by straight lines between floors for easier visualization. Also included in Fig. 21.6.1 are the story shears of Fig. 18.6.1 from dynamic analysis (RSA), including response contributions of all modes; these have been normalized by the corresponding base shear. Similarly, Fig. 21.6.2 includes the equivalent static forces computed from the story shears of Fig. 18.6.1 as the differences between the shears in consecutive stories (equal to the discontinuity in shears at the floor levels).

Figures 21.6.1 and 21.6.2 permit the following observations: For buildings with T_1 in the acceleration-sensitive region of the spectrum, the heightwise distributions of lateral forces and story shears specified by the three codes are essentially identical to each other and fall between the dynamic response curves for $\rho = 0$ and ∞. With increasing T_1, the code distributions for lateral forces and story shears increasingly differ among the codes, and all three codes increasingly differ from dynamic response. These differences are especially significant for the smaller values of ρ because higher-mode response increases with increasing T_1 and decreasing ρ (Sections 18.4 and 18.5). It is clear that the code formulas do not closely follow the dynamic response results or recognize the effects of the important building parameters on dynamic response. These discrepancies are accentuated when we consider the influence of the heightwise distribution of stiffness and strength on inelastic response of buildings.

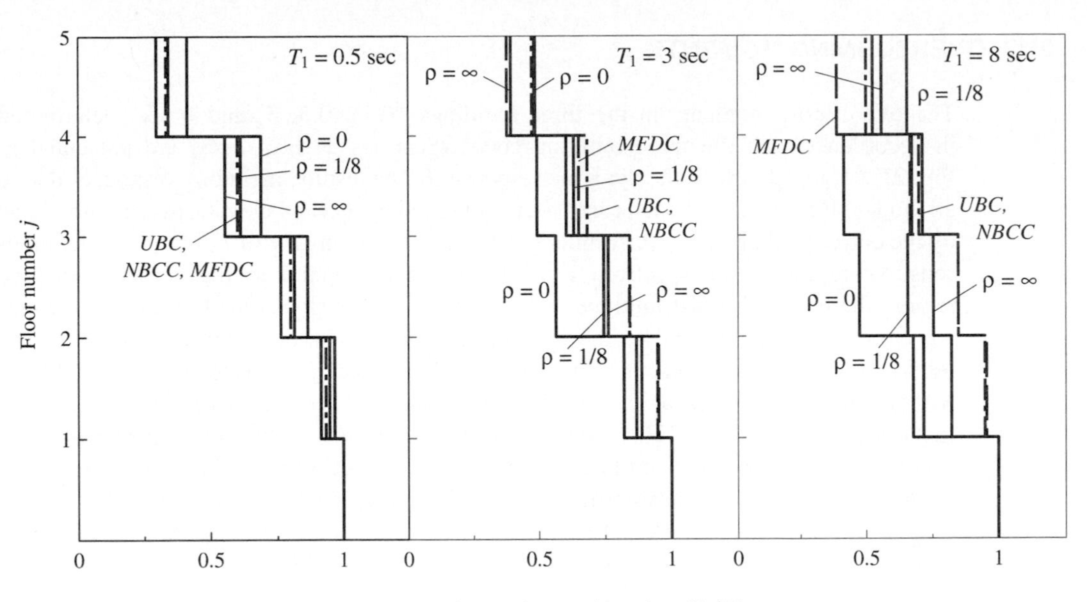

Figure 21.6.1 Comparison of story shear distributions in building codes and from RSA for three values of ρ.

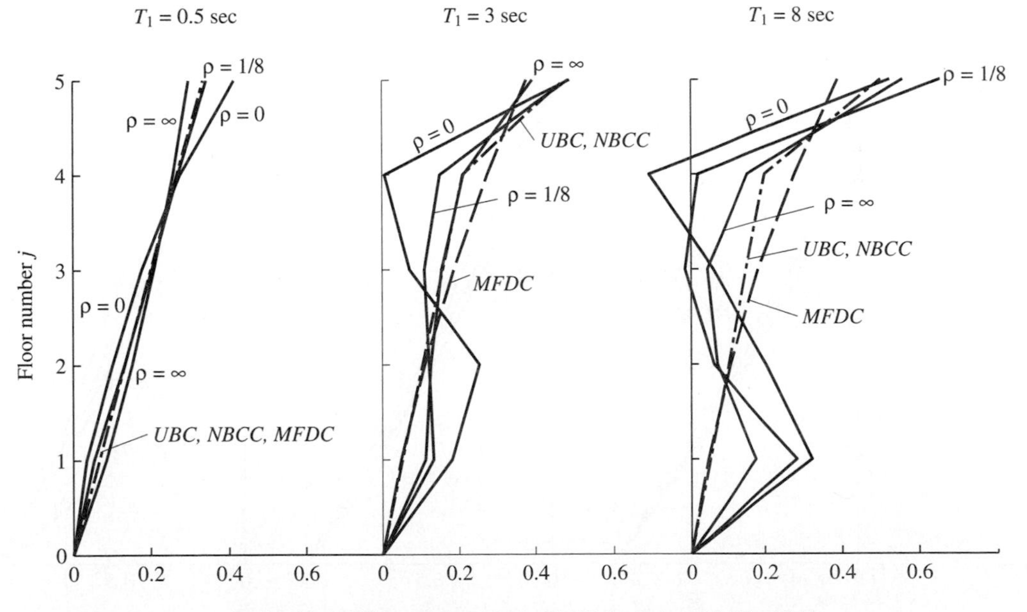

Figure 21.6.2 Comparison of equivalent static force distributions in building codes and from RSA for three values of ρ.

21.7 OVERTURNING MOMENTS

The overturning moments in the three buildings, $T_1 = 0.5$, 3, and 8 sec, determined in accordance with the three building codes (Sections 21.1 to 21.3) are presented in Fig. 21.7.1, together with the dynamic response (RSA) results, including responses due to all modes (Fig. 18.6.2); in each case, overturning moments at all elevations are normalized by the corresponding base overturning moment. For buildings with T_1 in the acceleration-sensitive region of the spectrum and even extending well into the velocity-sensitive region, the heightwise distributions of overturning moments specified by the three codes are close to each other and to the dynamic response. The discrepancy in code values relative to the dynamic response increases with increasing T_1, especially for buildings with smaller values of ρ, as the higher-mode response becomes increasingly significant (Sections 18.4 and 18.5). However, this discrepancy in overturning moments is much smaller than was noted in Section 21.6 for story shears, because the higher-mode response contributions to the overturning moments are less significant (Chapter 18).

We next compare the overturning moments computed by two methods, both based on dynamic analysis: (1) response spectrum analysis (RSA) considering all modes (Fig. 21.7.1), and (2) static analysis of the building subjected to the lateral forces of Fig. 21.6.2 determined, as described in Section 21.6, from RSA predictions of story shears. The latter method will provide the same story shears as the first method because the lateral forces were determined by statics from the story shears predicted by RSA, but not the correct overturning moments. This fact is demonstrated by presenting the ratio

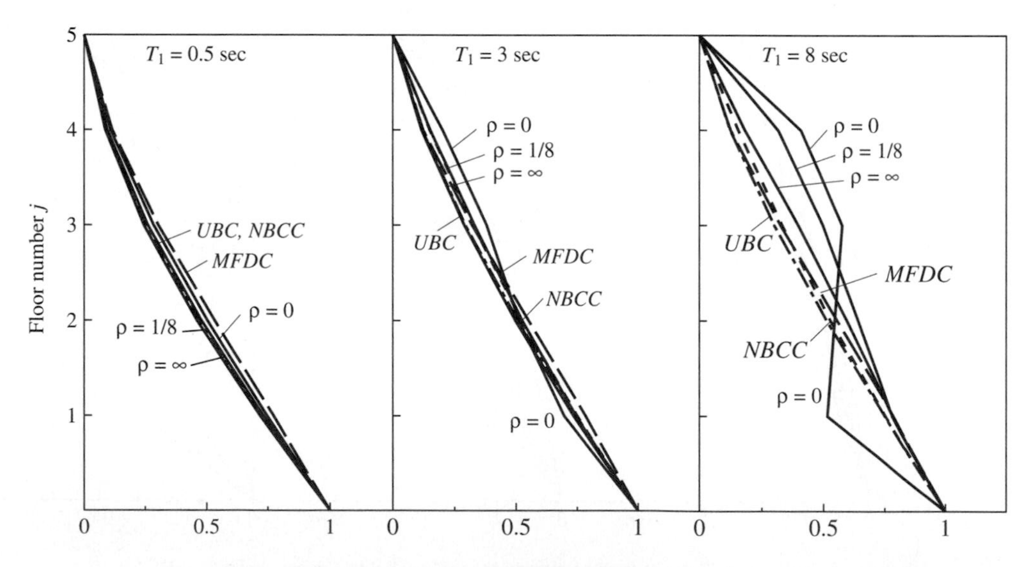

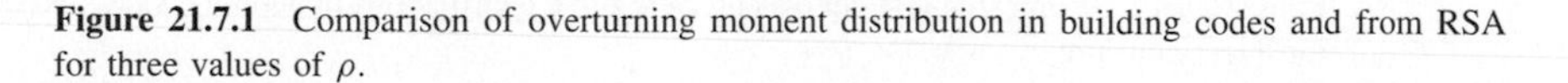
Figure 21.7.1 Comparison of overturning moment distribution in building codes and from RSA for three values of ρ.

of overturning moments computed by the two methods. Akin to the reduction factor J specified in some building codes, this ratio is presented in Fig. 21.7.2 for the base overturning moment as a function of T_1 for three values of ρ; and in Fig. 21.7.3 for overturning moments at all floors of the building for three values of T_1. This reduction factor never exceeds unity, implying that the approximate value of the overturning moment obtained from the lateral forces (second method) always exceeds the "exact" value obtained from RSA (first method). The two values are identical if the response contribution of only the fundamental vibration mode is considered. Thus the reduction factor accounts for the fact that higher vibration modes contribute more to shears than to overturning moments; it decreases (implying greater reduction) with increasing T_1 and decreasing ρ.

Because the lateral forces specified in building codes are intended to provide the dynamically computed story shears, the preceding observations indicate that the overturning moments will be overestimated if they were also computed from the lateral forces by statics. Thus some building codes specify reduction factors by which statically computed overturning moments should be multiplied. These reduction factors, defined earlier for three building codes, are also included in Figs. 21.7.2 and 21.7.3. The reduction factor specified in *MFDC* is independent of T_1, except for the slight variation arising from an equilibrium requirement. As a result, the reduction is excessive in the acceleration-sensitive region of the spectrum, but not enough in parts of the velocity- and displacement-sensitive regions, depending on ρ. However, the *NBCC* specifies a

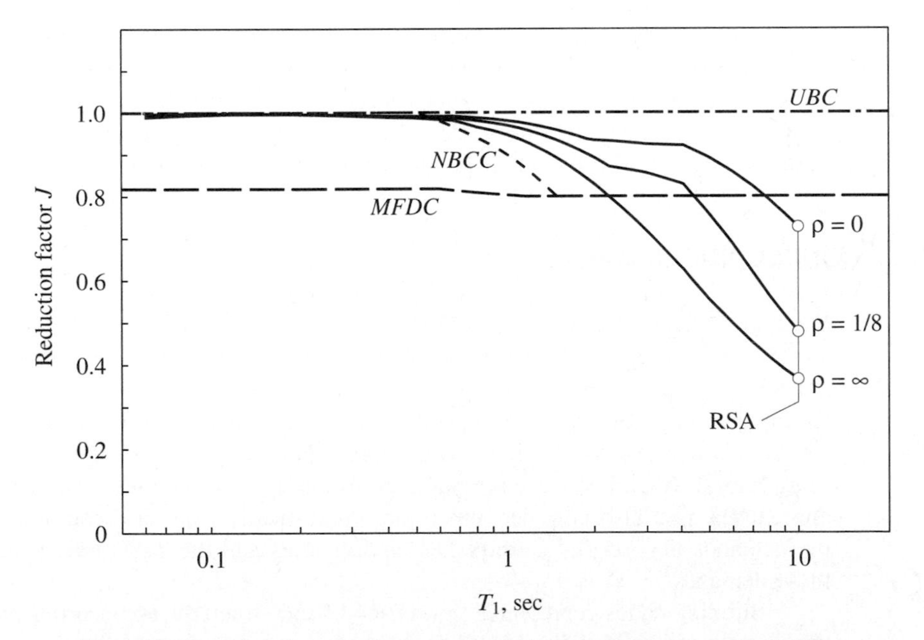

Figure 21.7.2 Comparison of reduction factors for base overturning moment in building codes and from RSA for three values of ρ.

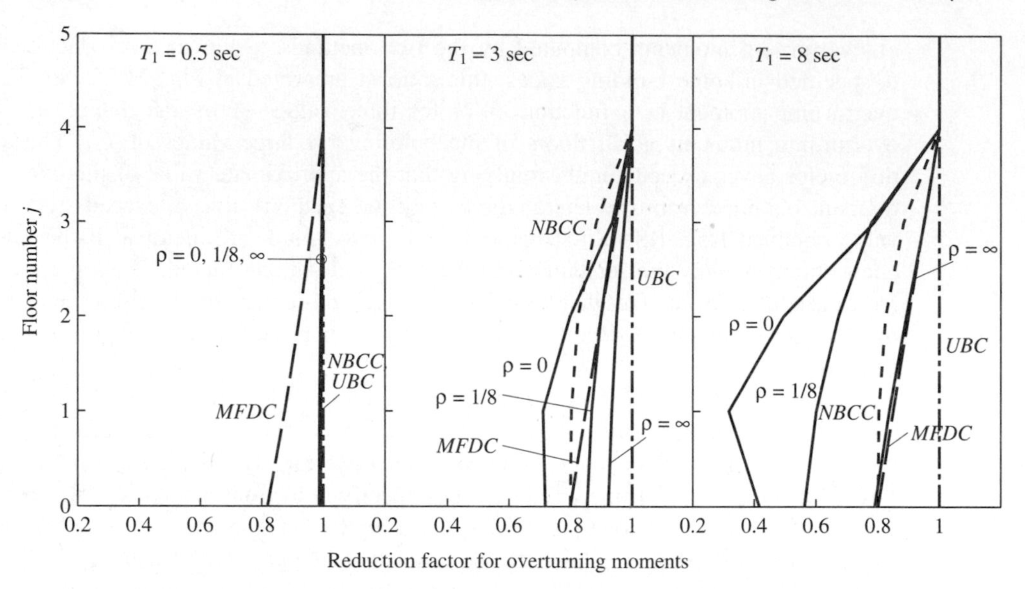

Figure 21.7.3 Comparison of reduction factors for overturning moments in building codes and from RSA for three values of ρ.

reduction factor that varies with T_1 in a manner similar to the dynamic response up to about $T_1 = 1.5$ sec; it does not recognize the dependence on ρ, however, which becomes significant at longer T_1. The *UBC* does not permit any reduction in overturning moments, a provision that is not supported by the results of dynamic analysis. Both the *NBCC* and *MFDC* specify no reduction in the top story and increasing reduction as one moves down to lower levels of the building. Although consistent in this regard with the results of dynamic analysis (Fig. 21.7.3), the code specifications do not fully recognize the dependence of the reduction factor on the building parameters T_1 and ρ.

21.8 CONCLUDING REMARKS

We have demonstrated that many of the concepts developed in this book about earthquake analysis, response, and design of structures are reflected in building codes, but they are not always stated explicitly or applied in accordance with structural dynamics results. Building codes should adopt a different approach, explicitly stating the underlying basis for each provision, so that it can be improved as we develop a better understanding of structural dynamics and earthquake performance of structures. The seismic design approaches must also consider, much more realistically than has been done in the past, the demands imposed by earthquakes on structures and the structural capacity to meet these demands.

Building codes represent a consensus of the structural engineering profession on the seismic design of ordinary buildings where special earthquake considerations are not cost-effective. There can be major design deficiencies if the building code is applied to structures whose dynamic properties differ significantly from those of ordinary buildings. This is suggested by the collapse or irreparable damage of some buildings during major earthquakes. Similarly, building codes should not be applied to special structures, for they require special consideration: because of cost, potential hazard, or the need to maintain function. For critical projects such as high-rise buildings, dams, nuclear power plants, offshore oil-drilling platforms, long-span bridges, major industrial facilities, and so on, special earthquake considerations are necessary—because in the words of Nathan M. Newmark and Emilio Rosenblueth: "Earthquake effects on structures systematically bring out the mistakes made in design and construction, even the minutest mistakes."

FURTHER READING

Associate Committee on National Building Code, *National Building Code of Canada 1995*, National Research Council, Ottawa, 1995.

Chopra, A. K., and Cruz, E. F. "Evaluation of Building Code Formulas for Earthquake Forces," *Journal of Structural Engineering, ASCE*, **112**, 1986, pp. 1881–1899.

Chopra, A. K., and Newmark, N. M., "Analysis," Chapter 2 in *Design of Earthquake Resistant Structures* (ed. E. Rosenblueth), Pentech Press, London, 1980.

Cruz, E. F. and Chopra, A. K., "Improved Code-Type Earthquake Analysis Procedure for Buildings," *Journal of Structural Engineering, ASCE*, **116**, 1990, pp. 679–700.

International Association of Earthquake Engineering, "Mexico," in *Earthquake Resistant Regulations, A World List*, Tokyo, 1992, pp. 24–1 to 24–27.

International Conference of Building Officials, *Uniform Building Code*, ICBO, Whittier, Calif., 1994.

Rosenblueth, E., "Seismic Design Requirements in a Mexican 1976 Code," *Earthquake Engineering and Structural Dynamics*, **7**, 1979, pp. 49–61.

Seismology Committee, *Recommended Lateral Forces and Commentary 1990*, Structural Engineers Association of California, Sacramento, Calif., 1990.

Sozen, M. A., Jennings, P. C., Matthiesen, R. B., Housner, G. W., and Newmark, N. M., *Engineering Report on the Caracas Earthquake of July 29, 1967*, National Academy of Sciences, Washington, D.C., 1968.

Notation

All symbols used in this book are defined where they first appear. For the reader's convenience, this appendix, arranged in three parts to follow the organization of the text, contains the principal meanings of the commonly used notations. The reader is cautioned that some symbols denote more than one quantity, but the meaning should be clear when read in context.

Abbreviations

CM	center of mass
CQC	complete quadratic combination
DOF	degree of freedom
MDF	multi-degree-of-freedom
MFDC	*Mexico's Federal District Code*
NBCC	*National Building Code, Canada*
SDF	single-degree-of-freedom
SRSS	square root of the sum of squares
UBC	*Uniform Building Code*

Accents

$\bar{(\ )}$	modal contribution factor for ( )
boldface	vector or matrix
$\check{(\ )}$	shifted value of ( )
$\dot{(\ )}$	$\frac{d}{dt}(\ )$
$\hat{(\ )}$	Fourier transform of ( )
$\hat{\mathbf{k}}$	condensed **k**
$\tilde{(\ )}$	approximation to ( )
$\tilde{(\ )}$	generalized ( )

Prefixes

δ, $\boldsymbol{\delta}$ increment over extended time step

$\delta(\cdot)$ virtual $(\cdot)$

Δ, $\boldsymbol{\Delta}$ increment over time step

Subscripts

A acceleration

b base; beam; base-isolation system

c column; complementary solution

cr critical

d duration

D damping; damped; displacement

e eccentric; element

eff effective

eq equivalent

f fixed base system

F friction

g ground

i time step number; peak number

i, j floor number; story number; DOF; frame number

I inertia; input

K kinetic

m peak value for inelastic systems; maximum

n natural; mode number

o peak value

p particular solution

sec secant

st static

S spring (elastic or inelastic); strain

T tangent; transmitted

V velocity

x, y, θ directions or components

y yield

Y yielding

Superscripts

s quasi-static

st static

t total

PART I: CHAPTERS 1–8

Roman Symbols

a constant in time-stepping methods

a_j Fourier cosine coefficients

a_y f_y/m

a_0 Fourier coefficient

A integration constant; arbitrary constant; coefficient in Eq. (5.2.5); pseudo-acceleration spectrum ordinate

A' coefficient in Eq. (5.2.5)

$A(t)$ pseudo-acceleration

A_y $\omega_n^2 u_y$

A_1, A_2 arbitrary constants

b constant in time-stepping methods

b_j Fourier sine coefficients

B integration constant; arbitrary constant; coefficient in Eq. (5.2.5)

B'	coefficient in Eq. (5.2.5)
$B_1,\ B_2$	arbitrary constants
c	damping coefficient
$\tilde{c}$	generalized damping
c_{cr}	critical damping coefficient
C	arbitrary constant; coefficient in Eq. (5.2.5)
C'	coefficient in Eq. (5.2.5)
D	arbitrary constant; coefficient in Eq. (5.2.5); deformation spectrum ordinate
D'	coefficient in Eq. (5.2.5)
D_y	yield deformation spectrum ordinate
e	eccentricity of rotating mass
E	modulus of elasticity
E_D	energy dissipated by damping
E_F	energy dissipated by friction
E_I	input energy
E_K	kinetic energy
E_{Ko}	maximum kinetic energy
E_S	strain energy
E_{So}	maximum strain energy
E_Y	energy dissipated by yielding
$EI(x)$	flexural rigidity
f	exciting or forcing frequency (Hz)
f_D	damping force
f_I	inertia force
$f_I(x,t)$	distributed inertia forces
f_{Ij}	inertia force in DOF j
f_{jo}	peak value of force at jth floor
f_n	natural frequency (undamped) (Hz)
$f_o(x)$	peak value of $f_S(x,t)$
f_S	elastic or inelastic resisting force; equivalent static force
$(f_S)_i$	value of f_S at time i
$\tilde{f}_S(u,\ \dot{u})$	defined by Eq. (7.3.3)
$f_{So},\ f_o$	peak value of $f_S(t)$
f_T	transmitted force
f_y	yield strength
$\bar{f}_y$	normalized yield strength
F	friction force
g	acceleration due to gravity
h	height of one-story frame; story height
$H(i\omega)$	complex frequency response
i	time step number
I	second moment of area
$\mathcal{I}$	magnitude of impulse
I_b	I for a beam
I_c	I for a column
I_O	moment of inertia about O
k	stiffness or spring constant
$\mathbf{k}$	stiffness matrix
$\hat{k}$	see Eq. (5.3.5) or (5.4.14)
$\tilde{k}$	generalized stiffness
k_i	$(k_i)_T$
$\hat{k}_i$	defined in Eq. (5.7.6)
$(k_i)_{\text{sec}}$	secant stiffness at time i
$(k_i)_T$	tangent stiffness at time i
k_j	stiffness of jth story
k_T	$(k_i)_T$
$\hat{k}_T$	defined in Eq. (5.7.8)
L	width of frame; length of beam or tower
$\tilde{L}$	see Eqs. (8.3.12) and (8.4.12)
$\tilde{L}^\theta$	defined in Eq. (8.4.18)
$m(x)$	mass per unit length
m	mass
$\mathbf{m}$	mass matrix
$\tilde{m}$	generalized mass
m_e	eccentrically rotating mass
m_j	mass at jth DOF or jth floor
$\mathcal{M}(x,t)$	bending moments in a distributed-mass system
$M_a,\ M_b$	bending moments at nodes a and b
M_b	base overturning moment
M_{bo}	peak value of $M_b(t)$
M_{io}	peak value of ith-floor overturning moment
$\mathcal{M}_o(x)$	peak value of $\mathcal{M}(x,t)$
N	normal force between sliding surfaces; number of DOFs
p	external force
$\hat{p}(i\omega)$	Fourier transform of $p(t)$
$\tilde{p}(t)$	generalized external force
p_{eff}	effective earthquake force
$(p_{\text{eff}})_o$	peak value of $p_{\text{eff}}(t)$
p_i	value of $p(t)$ at time i

$\hat{p}_i$	defined in Eq. (5.3.6)
p_o	amplitude of $p(t)$
$r(t)$	any response quantity
r_o	$\max_t \lvert r(t)\rvert$, the peak response
R_a	acceleration response factor
R_d	deformation (or displacement) response factor
R_v	velocity response factor
R_y	yield reduction factor
s	constant in e^{st}
t	time
t'	time variable
t_d	duration of pulse force
t_i	time at ith peak in free vibration; time at end of ith time step
t_o	time when $u(t)$ is maximum
t_r	rise time
T_a, T_b, T_c, T_d, T_e, T_f	periods that define spectral regions
T_D	natural period (damped)
T_n	natural period (undamped)
T_0	period of periodic excitation
TR	transmissibility
u	displacement; deformation; displacement relative to ground
u^t	total displacement
$\mathbf{u}$	vector of displacements u_j
$\hat{u}(i\omega)$	Fourier transform of $u(t)$
$u(0)$	initial displacement
$\dot{u}(0)$	initial velocity
u_a, u_b	displacements of nodes a and b
u_c	complementary solution
u_F	F/k
u_g	ground (or support) displacement
$\ddot{u}_g$	ground (or support) acceleration
u_{go}	peak ground displacement
$\dot{u}_{go}$	peak ground velocity
$\ddot{u}_{go}$	peak ground acceleration
$u_{g\theta}$	ground rotation about a vertical axis
u_i	displacement at ith peak; displacement at time i
$\dot{u}_i$	velocity at time i
$\ddot{u}_i$	acceleration at ith peak; acceleration at time i
u_j	relative displacement of jth floor
$u_j^c(t)$	response to $p(t) = p_0 \cos j\omega_0 t$
$u_j^s(t)$	response to $p(t) = p_0 \sin j\omega_0 t$
u_j^t	total displacement of jth floor
u_{jo}	peak or maximum value of $u_j(t)$
$\dot{u}_{jo}$	maximum value of $\dot{u}_j(t)$
u_m	$\max_t \lvert u(t)\rvert$ for an inelastic system
u_m^-	$\lvert \min_t[u(t)]\rvert$
u_m^+	$\max_t[u(t)]$
u_o	peak or maximum value of $u(t)$
$\dot{u}_o$	peak value of $\dot{u}(t)$
$u_o(x)$	peak or maximum value of $u(x,t)$
$\dot{u}_o(x)$	maximum value of $\dot{u}_o(x,t)$
u_o^t	peak value of $u^t(t)$
$\ddot{u}_o^t$	peak value of $\ddot{u}^t(t)$
u_p	particular solution; permanent deformation
$u_{st}(t)$	static deformation due to $p(t)$
$(u_{st})_o$	static deformation due to p_o
u_x, u_y	x and y displacements
u_y	yield deformation
$u_0(t)$	response to $p(t) = a_0$
u_θ	rotation about a vertical axis
v	velocity
V	pseudo-velocity spectrum ordinate
$\mathcal{V}(x,t)$	shearing forces in a distributed mass system
V_a, V_b	shearing forces at nodes a and b
V_b	base shear
V_{bo}	peak value of $V_b(t)$
$\mathcal{V}_{bo}$	peak value of $\mathcal{V}_b(t)$
V_j	shear in the jth story

V_{jo}	peak value of $V_j(t)$
$\mathcal{V}_o(x)$	peak value of $\mathcal{V}(x,t)$
V_y	$\omega_n u_y$
w	weight
x, y	Cartesian coordinates
z	generalized displacement
z_o	peak value of $z(t)$
$\mathbf{1}$	vector of ones

Greek Symbols

$\alpha_A, \alpha_D, \alpha_V$	spectral amplification factors
β	parameter in Newmark's method
β_j	$j\omega_0/\omega_n$
γ	parameter in Newmark's method
$\tilde{\Gamma}$	$\tilde{L}/\tilde{m}$
δ	logarithmic decrement
$\delta(\cdot)$	Dirac delta function
δ_{st}	mg/k
$\delta u(x)$	virtual displacements
$\delta\mathbf{u}$	virtual displacement vector
δu_j	virtual displacement u_j
δW_E	external virtual work
δW_I	internal virtual work
δz	virtual displacement
$\delta\kappa(x)$	virtual curvature
Δ_j	story drift in jth story
$\Delta\hat{p}$	$\Delta\hat{p}_i$
Δp_i	increment in $p(t)$ over time step i
$\Delta\hat{p}_i$	see Eqs. (5.3.6) and (5.4.15)
Δt	time step
Δt_i	time step i
Δu	Δu_i
Δu_i	increment in u_i over time step i
$\Delta\dot{u}_i$	increment in $\dot{u}$ over time step i
$\Delta\ddot{u}_i$	increment in $\ddot{u}$ over time step i
ε	duration of an impulsive force
ζ	damping ratio
$\bar{\zeta}$	numerical damping ratio
ζ_{eq}	equivalent viscous damping ratio
η	rate-independent damping coefficient
$\ddot{\theta}$	rotational acceleration
θ_a, θ_b	rotations at nodes a and b
θ_g	ground rotation about a horizontal axis
$\kappa(x)$	curvature
μ	coefficient of friction; ductility factor
ρ	$I_b/4I_c$; coefficient in $\pm\rho e^{-\zeta\omega_n t}$
σ	standard deviation
τ	dummy time variable
ϕ	phase angle
$\psi(x)$	shape function
$\boldsymbol{\psi}$	shape vector
ψ_j	jth element of $\boldsymbol{\psi}$
ω	exciting or forcing frequency (rad/sec)
ω_D	natural frequency (damped) (rad/sec)
ω_n	natural frequency (undamped) (rad/sec)
ω_0	$2\pi/T_0$

PART II: CHAPTERS 9–17

Roman Symbols

$\mathbf{a}_e$	element transformation matrix
a_{in}	defined by Eq. (14.4.11)
a_l	coefficients in Caughey series
$\mathbf{a}_{xi}, \mathbf{a}_{yi}$	transformation matrices

a_0, a_1	Rayleigh damping coefficients
$\mathcal{A}$	direct assembly operator
A_n	pseudo-acceleration spectrum ordinate $A(T_n, \zeta_n)$
$A_n(t)$	pseudo-acceleration of nth-mode SDF system
$A_{nl}(t)$	$A_n(t)$ due to $\ddot{u}_{gl}(t)$
$\mathbf{c}$	damping matrix
$\tilde{\mathbf{c}}$	defined by Eq. (14.3.4)
c_{ij}	damping influence coefficient
c_j	jth-story damping coefficient
$\mathbf{c}_n$	nth-mode damping matrix
$\mathbf{C}$	$\mathbf{\Phi}^T \mathbf{c} \mathbf{\Phi}$, diagonal matrix of C_n
C_n	generalized damping for nth mode
C_{nr}	element of $\mathbf{C}$
$D_n(t)$	deformation of nth-mode SDF system
$D_{nl}(t)$	$D_n(t)$ due to $\ddot{u}_{gl}(t)$
D_{no}	peak value of $D_n(t)$
e_J	error in static response
E	modulus of elasticity
EI	flexural rigidity
$\hat{\mathbf{f}}$	flexibility matrix
$\mathbf{f}_D, \mathbf{f}_D(t)$	damping forces
f_{Dj}	damping force in DOF j
$\hat{f}_{ij}$	flexibility influence coefficient
$\mathbf{f}_I$	inertia forces
f_{Ij}	inertia force in DOF j
$\mathbf{f}_{in}$	peak value of $\mathbf{f}_{in}(t)$
$\mathbf{f}_{in}(t)$	equivalent static forces: frame i, mode n
f_{jn}	jth element of $\mathbf{f}_n$; peak value of $f_{jn}(t)$
$f_{jn}(t)$	equivalent static force: DOF j, mode n
f_{jyn}	jth element of $\mathbf{f}_{yn}$
$f_{j\theta n}$	jth element of $\mathbf{f}_{\theta n}$
f_n	nth natural frequency (undamped) (Hz)
$f_n(x, t)$	equivalent static forces, mode n
$\mathbf{f}_n$	peak value of $\mathbf{f}_n(t)$
$\mathbf{f}_n(t)$	equivalent static forces, mode n
$f_{no}(x)$	peak value of $f_n(x, t)$
$\mathbf{f}_S$	elastic resisting forces
$\mathbf{f}_S(\mathbf{u}, \dot{\mathbf{u}})$	inelastic resisting forces
f_{SA}	lateral force on frame A
$\mathbf{f}_{Sg}, \mathbf{f}_{Sg}(t)$	equivalent static forces in support DOFs
f_{Sj}	elastic or inelastic resisting force in DOF j
$\mathbf{f}_{yn}$	peak value of $\mathbf{f}_{yn}(t)$
$\mathbf{f}_{yn}(t)$	equivalent static lateral forces, mode n
$\mathbf{f}_{\theta n}$	peak value of $\mathbf{f}_{\theta n}(t)$
$\mathbf{f}_{\theta n}(t)$	equivalent static torques, mode n
h	height of one-story frame; story height
h_j	height of jth floor
h_n^*	effective modal height, mode n
I	second moment of area
$\mathbf{I}$	identity matrix
I_b	I for beam
I_c	I for column
$\mathbf{I}_O$	diagonal matrix: $I_{jj} = I_{Oj}$
I_{Oj}	moment of inertia of jth floor about O
J	number of Ritz vectors
$\mathbf{k}$	stiffness matrix
$\check{\mathbf{k}}$	$\mathbf{k} - \mu \mathbf{m}$
$\tilde{\mathbf{k}}$	defined by Eq. (14.3.4); matrix of $\tilde{k}_{ij}$ in Eq. (17.1.4)
$\mathbf{k}_A$	stiffness matrix of frame A in global DOF
$\mathbf{k}_e$	element stiffness matrix in global element DOFs
$\bar{\mathbf{k}}_e$	element stiffness matrix in local element coordinates
$\mathbf{k}_{gg}^l$	lth column of $\mathbf{k}_{gg}$
$\mathbf{k}_i$	stiffness matrix of frame i in global DOF
$\hat{\mathbf{k}}_i$	see Tables 15.3.1 and 15.3.3
k_{ij}	stiffness influence coefficient
k_j	stiffness of jth story
$\hat{\mathbf{k}}_{tt}$	condensed stiffness matrix
$\hat{\mathbf{k}}_T$	$\hat{\mathbf{k}}_i$
k_{xi}, k_{yi}	lateral stiffness of frame i in x and y directions

$\mathbf{k}_{xi}, \mathbf{k}_{yi}$	lateral stiffness matrix of frame i in x and y directions
k_y	lateral stiffness of frame A
$\mathbf{k}_{\theta y}, \mathbf{k}_{y\theta}, \mathbf{k}_{\theta\theta}$	submatrices of $\mathbf{k}$
$\mathbf{K}$	diagonal matrix of K_n
$\hat{\mathbf{K}}$	see Tables 15.2.1 and 15.2.2
K_n	generalized stiffness, mode n
L	length of beam; length of finite element
L_n^h	see Eqs. (13.2.3) or (16.6.2)
L_n^θ	see Eq. (13.2.9b) or (16.6.17)
L_{nl}	defined by Eq. (13.5.3)
m	mass of an SDF system
$\mathbf{m}$	mass matrix
$\tilde{\mathbf{m}}$	defined by Eq. (14.3.4); matrix of $\tilde{m}_{ij}$ in Eq. (17.1.4)
$m(x)$	mass per unit length
$\mathbf{m}_e$	element mass matrix in global element DOFs
$\bar{\mathbf{m}}_e$	element mass matrix in local element coordinates
m_{ij}	mass influence coefficient
m_j	mass at jth DOF or jth floor
$\mathbf{m}_{tt}$	mass matrix for $\mathbf{u}_t$ DOF
$\mathbf{M}$	diagonal matrix of M_j
$\mathcal{M}(x,t)$	bending moment in a distributed-mass system
$\mathcal{M}_b$	bending moment at the base
$\mathcal{M}_{bn}(t)$	$\mathcal{M}_b(t)$ due to mode n
$M_{bn}(t)$	$M_b(t)$ due to mode n
$\mathcal{M}_{bn}^{\text{st}}$	nth modal static response $\mathcal{M}_b$
M_{bn}^{st}	nth modal static response M_b
M_i	ith-floor overturning moment
M_{in}^{st}	nth modal static response M_i
M_n	generalized mass, mode n
M_n^*	effective modal mass, mode n
N	number of DOFs
N_d	number of modes responding dynamically
N_e	number of finite elements
N_g	number of ground (or support) displacements
$\mathbf{O}$	null matrix
p	external force
$\mathbf{p}$	external forces
$\tilde{\mathbf{p}}$	vector of $\tilde{p}_i$ in Eq. (17.2.6)
$p(\lambda)$	polynomial in λ
$\mathbf{p}_e$	element force vector in global element DOFs
$\bar{\mathbf{p}}_e$	element force vector in local element coordinates
p_{eff}	effective earthquake force
$\mathbf{p}_{\text{eff}}$	effective earthquake force vector
$\mathbf{p}_g$	support forces
$\mathbf{p}_g^s(t)$	quasi-static support forces
p_j	external force at jth DOF or jth floor
p_o	maximum value of $p(t)$
$\mathbf{p}_t$	external forces in $\mathbf{u}_t$ DOF
$\mathbf{P}(t)$	$\Phi^T \mathbf{p}(t)$
$\mathbf{P}(t)$	vector of $P_n(t)$
$\hat{\mathbf{P}}_i$	defined in Table 15.2.1
$P_n(t)$	generalized force, mode n
$\mathbf{q}$	modal coordinate vector
$\mathbf{q}_i$	modal coordinates at time i
$q_n(t)$	nth modal coordinate
r	radius of gyration
$r\, I_{On}^*$	defined by Eq. (13.3.10)
$r(t)$	any response quantity
r_n	peak value of $r_n(t)$
$\bar{r}_n$	nth modal contribution factor
$r_n(t)$	$r(t)$ due to mode n
r_{no}	peak value of $r_n(t)$
r_o	peak value of $r(t)$
r^{st}	static response to forces $\mathbf{s}$
r_n^{st}	nth modal static response
R_{dn}	dynamic response factor for nth-mode SDF system
$\mathbf{s}, \mathbf{s}_a, \mathbf{s}_b$	spatial distributions of $\mathbf{p}(t)$
s_{jn}	jth element of $\mathbf{s}_n$
s_{jyn}	jth element of $\mathbf{s}_{yn}$
$s_{j\theta n}$	jth element of $\mathbf{s}_{\theta n}$
$\mathbf{s}_n$	defined by Eq. (12.8.4)
$s_n(x)$	defined by Eq. (16.6.4)
$\mathbf{s}_{yn}, \mathbf{s}_{\theta n}$	subvectors of $\mathbf{s}_n$
t	time variable
t_d	duration of pulse force
T_b	base torque
T_{bn}	peak value of $T_{bn}(t)$
T_{bn}^{st}	nth modal static response T_b
$T_{bn}(t)$	$T_b(t)$ due to mode n
T_i	ith-story torque
$T_{in}(t)$	$T_i(t)$ due to mode n
T_{in}^{st}	nth modal static response T_i

T_n	nth natural period (undamped)
u	displacement or deformation
$\mathbf{u}$	displacement vector
$\mathbf{u}^s$	quasistatic displacements
$\mathbf{u}^t$	total displacements
u_A	displacement at frame A
$\mathbf{u}_e$	element displacements in global element DOFs
u_g	ground (or support) displacement
$\mathbf{u}_g$	ground (or support) displacement vector
$\ddot{u}_g$	ground (or support) acceleration
u_{gl}	lth support displacement
$\ddot{u}_{gx}, \ddot{u}_{gy}, \ddot{u}_{g\theta}$	x, y, and θ components of ground acceleration
u_i	displacement in DOF i
$\mathbf{u}_i$	lateral displacements of frame i; displacements at time i
$\mathbf{u}_{in}$	$\mathbf{u}_i$ due to mode n
u_j	peak value of $u_j(t)$
$u_j(t)$	relative displacement at DOF j or floor j
u_j^s	quasistatic displacement at DOF j
u_j^t	total displacement at DOF j or floor j
u_{jn}	peak value of $u_{jn}(t)$
u_{jn}^{st}	nth modal static response u_j
$u_{jn}(t)$	$u_j(t)$ due to mode n
u_{jx}, u_{jy}	displacements of CM of floor j along x and y axes
u_{jyn}	peak value of $u_{jyn}(t)$
$u_{j\theta}$	rotation of floor j about CM
$u_{j\theta n}$	peak value of $u_{j\theta n}(t)$
$\mathbf{u}_n$	peak value of $\mathbf{u}_n(t)$
$\mathbf{u}_n(t)$	$\mathbf{u}(t)$ due to mode n
$\mathbf{u}_n^{\text{st}}$	nth modal static response $\mathbf{u}$
$u_n(x, t)$	$u(x, t)$ due to mode n
$u_n^{\text{st}}(x)$	nth modal static response $u(x)$
$\mathbf{u}_t$	dynamic DOF
u_x, u_y	x- and y-displacements of CM
$\ddot{u}_x^t, \ddot{u}_y^t, \ddot{u}_\theta^t$	x, y, and θ components of total acceleration
$\mathbf{u}_y$	y-lateral displacements
$\mathbf{u}_{yn}(t)$	$\mathbf{u}_y(t)$ due to mode n
$\mathbf{u}_0$	DOF with zero mass
$\bar{u}_{5n}$	nth modal contribution factor for u_5
u_θ	rotation about CM
$\mathbf{u}_\theta$	floor rotations
$\mathbf{u}_{\theta n}(t)$	$\mathbf{u}_\theta(t)$ due to mode n
$\mathcal{V}(x, t)$	transverse shear force in a distributed-mass system
V_b	peak value of $V_b(t)$
$\mathcal{V}_b$	base shear
$V_b(t)$	base shear
V_b^{st}	V_b due to forces $\mathbf{s}$
V_{bn}	peak value of $V_{bn}(t)$
$\bar{V}_{bn}$	nth modal contribution factor for V_b
$V_{bn}(t)$	$V_b(t)$ due to mode n
$\mathcal{V}_{bn}(t)$	$\mathcal{V}_b(t)$ due to mode n
V_{bn}^{st}	nth modal static response V_b
$\mathcal{V}_{bn}^{\text{st}}$	nth modal static response $\mathcal{V}_b$
V_{bo}	peak value of $V_b(t)$
V_i	peak value of $V_i(t)$
$V_i(t)$	ith-story shear
V_{in}^{st}	nth modal static response V_i
x, y	Cartesian coordinates
x_i, y_i	define location of frame i
$\mathbf{x}_i$	iteration vector
$\mathbf{y}_n$	eigenvector of $\mathbf{A}$; defined by Eq. (14.4.10)
$\mathbf{z}$	generalized coordinate vector
z_j	generalized coordinates
$\mathbf{z}_n$	eigenvector
$\mathbf{0}$	vector of zeros
$\mathbf{1}$	vector of ones

Greek Symbols

β	parameter in Newmark's method
β_{in}	ω_i/ω_n
γ	parameter in Newmark's method
Γ_n	see Eq. (12.8.3) or (13.2.3)

Γ_{nl}	defined by Eq. (13.5.3)
$\delta u(x)$	virtual displacement $u(x)$
$\boldsymbol{\delta}\mathbf{u}_i$	increment in $\mathbf{u}$ over extended time step i
$\boldsymbol{\delta}\ddot{\mathbf{u}}_i$	increment in $\ddot{\mathbf{u}}$ over extended time step i
$\boldsymbol{\delta}\dot{\mathbf{u}}_i$	increment in $\dot{\mathbf{u}}$ over extended time step i
δu_j	virtual displacement u_j
δW_E	external virtual work
δW_I	internal virtual work
$\delta(\cdot)$	Dirac delta function
Δ_j	peak value of $\Delta_j(t)$
$\Delta_j(t)$	jth-story deformation or drift
$\boldsymbol{\Delta}\hat{\mathbf{p}}_i$	see Tables 15.3.1 and 15.3.3
$\boldsymbol{\Delta}\mathbf{q}_i$	increment in $\mathbf{q}$ over time step i
$\boldsymbol{\Delta}\dot{\mathbf{q}}_i$	increment in $\dot{\mathbf{q}}$ over time step i
$\boldsymbol{\Delta}\ddot{\mathbf{q}}_i$	increment in $\ddot{\mathbf{q}}$ over time step i
$\boldsymbol{\Delta}\mathbf{R}^{(j)}$	residual forces
Δt	time step
$\boldsymbol{\Delta}\mathbf{u}$	incremental displacements
$\boldsymbol{\Delta}\mathbf{u}^{(j)}$	$\boldsymbol{\Delta}\mathbf{u}$ in jth iteration
$\boldsymbol{\Delta}\mathbf{u}_i$	increment in $\mathbf{u}$ over time step i
$\boldsymbol{\Delta}\dot{\mathbf{u}}_i$	increment in $\dot{\mathbf{u}}$ over time step i
$\boldsymbol{\Delta}\ddot{\mathbf{u}}_i$	increment in $\ddot{\mathbf{u}}$ over time step i
Δ_j	peak value of $\Delta_j(t)$
$\Delta_{jn}(t)$	$\Delta_j(t)$ due to mode n
Δ_{jn}	peak value of $\Delta_{jn}(t)$
ϵ_{in}	defined by Eq. (13.7.7a)
ζ_n	damping ratio for nth mode
ζ_n'	defined by Eq. (13.7.7b)
θ	parameter in Wilson's method
θ_g	ground rotation about a horizontal axis
$\boldsymbol{\iota}$	influence vector; influence matrix
$\boldsymbol{\iota}_l$	influence vector for u_{gl}
κ	shear stress constant
$\lambda^{(j)}$	estimate of eigenvalue
$\breve{\lambda}$	$\lambda - \mu$
λ_n	nth eigenvalue
μ	absorber mass ratio; shift of eigenvalue spectrum
ν	complex eigenvalue
$\rho(\mathbf{z})$	Rayleigh's quotient
ρ_{in}	cross-correlation coefficient for modes i and n
ρ_n	eigenvalue
ϕ_{jn}	jth element of $\boldsymbol{\phi}_n$
ϕ_{jyn}, $\phi_{j\theta n}$	jth elements of $\boldsymbol{\phi}_{yn}$ and $\boldsymbol{\phi}_{\theta n}$
$\phi_n(x)$	nth natural vibration mode
$\tilde{\phi}_n(x)$	approximation to $\phi_n(x)$
$\tilde{\boldsymbol{\phi}}_n$	approximation to $\boldsymbol{\phi}_n$
$\boldsymbol{\phi}_n$	nth natural vibration mode
$\boldsymbol{\phi}_{yn}$, $\boldsymbol{\phi}_{\theta n}$	subvectors of $\boldsymbol{\phi}_n$
$\boldsymbol{\Phi}$	modal matrix
$\boldsymbol{\psi}$	$\langle \boldsymbol{\psi}_1 \quad \boldsymbol{\psi}_2 \quad \cdots \quad \boldsymbol{\psi}_J \rangle$
$\boldsymbol{\psi}(x)$	$\langle \psi_1(x) \quad \psi_2(x) \quad \cdots \quad \psi_N(x) \rangle$
$\psi_i(x)$	trial, Ritz, or shape function; finite element interpolation function
$\hat{\psi}_i(x)$	beam interpolation function
$\boldsymbol{\psi}_j$	shape vector or Ritz vector
$\hat{\boldsymbol{\psi}}_n$	defined by Eq. (14.4.12)
ω	exciting or forcing frequency
ω_n	nth natural frequency (undamped) (rad/sec)
ω_n'	ω_n of a beam considering rotational inertia and shear effects
$\tilde{\omega}_n$	approximation to ω_n
$\boldsymbol{\Omega}^2$	spectral matrix

PART III: CHAPTERS 18–21

Roman Symbols

A	pseudo-acceleration spectrum ordinate
A_m	maximum A/g, *MFDC*
A_y	A for yielding system
c	damping coefficient of fixed-base system
c_b	damping coefficient of isolation system
$\mathbf{c}_f$	damping matrix of fixed-base system
C	numerical coefficient, *UBC*

C_e elastic seismic coefficient
C_s seismic coefficient
EI_b flexural rigidity of beams
EI_c flexural rigidity of columns
f_{jn} lateral force: floor j, mode n
f_y design yield strength
F foundation factor, *NBCC*
F_j code lateral force at floor j
F_t additional lateral force at the top floor, *UBC*
h story height
h_j height of jth floor
I importance factor, *UBC*; seismic importance factor, *NBCC*
J, J_i reduction factors for overturning moments
k lateral stiffness of fixed-base system
k_b lateral stiffness of isolation system
$\mathbf{k}_f$ stiffness matrix of fixed-base system
k_j stiffness of jth story
L_b length of beams
L_c length of columns
m lumped mass of fixed-based system
m_b mass of base slab
$\mathbf{m}_f$ mass matrix of fixed-base system
Q seismic behavior factor, *MFDC*
Q' seismic reduction factor, *MFDC*
r coefficient in C_e, *MFDC*
R force modification factor, *NBCC*
R_w structural system coefficient, *UBC*
S site coefficient, *UBC*; seismic response factor, *NBCC*
T_b natural period of (1) the isolated building assumed as rigid, or (2) the isolation system with rigid building
T_b, T_c periods defining the constant-A spectral region
T_f natural period of fixed-base system
T_n natural period of SDF system; nth natural period of MDF system
T_{nf} nth natural period of fixed-base system
u_b isolator deformation
u_{bn} u_b due to mode n
u_{bn}^{st} nth modal static response u_b
u_i ith-floor displacement due to forces F_j ($j = 1, 2, \ldots, N$)
u_{jm} $\max_t |u_j(t)|$ for an inelastic system
u_y yield displacement
U coefficient in *NBCC*
v zonal velocity ratio, *NBCC*
V_b base shear
$V_b^{(1)}$, $V_b^{(2)}$ two parts of V_b, *MFDC*
V_{by} yield strength value of V_b
$\bar{V}_{by}$ normalized value of V_{by}
V_j jth-story shear
V_{jy} jth-story yield strength
w weight of SDF system
w_i weight at ith floor
W total weight of building; total dead load and applicable portion of other loads
Z seismic zone factor, *UBC*
Z_a acceleration-related seismic zone, *NBCC*
Z_v velocity-related seismic zone, *NBCC*

Greek Symbols

α, β coefficients in least-square-error fit of V_b–T_1 curve
Δ_j jth-story deformation or drift
Δ_{jm} peak value of $\Delta_j(t)$ for an inelastic system

Δ_{jy} yield deformation for jth story

ζ damping ratio

ζ_b ζ of isolation system with rigid building

ζ_f ζ for fixed-base system

ζ_{nf} ζ for nth mode of fixed-base system

μ ductility factor

ϕ_{nf} nth mode of fixed-base system

ω_f natural frequency of fixed-base system

Answers to Selected Problems

Chapter 1

1.1 $k_e = k_1 + k_2$; $m\ddot{u} + k_e u = p(t)$

1.3 $k_e = \dfrac{(k_1 + k_2)k_3}{k_1 + k_2 + k_3}$; $m\ddot{u} + k_e u = p(t)$

1.5 $\dfrac{mR^2}{2}\ddot{\theta} + \dfrac{\pi d^4 G}{32L}\theta = 0$

1.7 $\dfrac{w}{g}\ddot{u} + \dfrac{3EI}{L^3}u = 0$

1.10 $m\ddot{u} + \left(\dfrac{120}{11}\dfrac{EI_c}{h^3}\right)u = p(t)$

1.11 $m\ddot{u}_x + \left(\sqrt{2}\dfrac{AE}{h}\right)u_x = 0$; $m\ddot{u}_y + \left(\sqrt{2}\dfrac{AE}{h}\right)u_y = 0$

1.13 $\dfrac{mh^2}{6}\ddot{u}_\theta + \dfrac{AEh}{\sqrt{2}}u_\theta = 0$

Chapter 2

2.1 $w = 40$ lb; $k = 16.4$ lb/in.

2.2 $u(t) = 2\cos(9.82t)$

2.4 $u(t) = 0.565\sin(60.63t)$ in.

2.5 $u(t) = \frac{m_2 g}{k}(1 - \cos\omega_n t) + \frac{\sqrt{2gh}}{\omega_n}\frac{m_2}{m_1 + m_2}\sin\omega_n t$

2.7 $EI = 8827$ lb-ft^2

2.10 $j_{10\%} = 0.366/\zeta$

2.12 $T_n = 0.353$ sec; $\zeta = 1.94\%$

2.14 $\omega_n = 21.98$ rad/sec; $\zeta = 0.161$; $\omega_D = 21.69$ rad/sec

2.16 0.636 in.; $8\frac{1}{2}$ cycles

Chapter 3

3.1 $m = 6.43/\text{g}$ lb/g; $k = 10.52$ lb/in.

3.3 $\zeta = 0.0576$

3.5 $u_o = 2.035 \times 10^{-3}$ in.; $\ddot{u}_o = 0.0052\text{g}$

3.9 $\zeta = 9.82\%$

3.10 $\zeta = 1.14\%$

3.12 474.8 lb

3.13 11.6 kips/in.

3.15 Error = 0, 0.9, and 15% at $f = 10$, 20, and 30 Hz, respectively

3.17 $f \leq 20.25$ Hz

3.19 $f \geq 2.575$ Hz

3.23 0.308 in.

3.24 **(a)** $p(t) = \frac{p_o}{2} + \frac{4p_o}{\pi^2}\sum_{j=1,3,5,\ldots}^{\infty}\frac{1}{j^2}\cos j\omega_0 t$

(b) $\frac{u(t)}{(u_{\text{st}})_o} = \frac{1}{2} + \frac{4}{\pi^2}\sum_{j=1,3,5,\ldots}^{\infty}\frac{1}{j^2(1-\beta_j^2)}\cos j\omega_0 t$ where $(u_{\text{st}})_o = p_o/k$, $\beta_j = j\omega_0/\omega_n$, and $\beta_j \neq 1$

(c) Two terms are adequate.

Chapter 4

4.7 $u_o \simeq 12.2$ in.; $u_0 \simeq 6.1$ in.

4.11 $u_o = 0.536$ in.; $\sigma = 18.9$ ksi

4.12 $u_o = 2.105$ in.; $\sigma = 37.2$ ksi

4.15 $u_o = \frac{p_o}{k}\frac{4}{\omega_n t_1}\sin\omega_n t_1 \sin\frac{\omega_n t_1}{2}$

4.16 $\frac{u_o}{(u_{\text{st}})_o} = \frac{4\pi}{3}\frac{t_d}{T_n}$; error = 5.9%

4.18 (a) $V_b = 15.08$ kips, $M_b = 1206$ kip-ft; (b) increase in mass has the effect of reducing the dynamic response.

Chapter 5

5.2 Check numerical results against the theoretical solution in Tables E5.1a, b.

5.4 Check numerical results against the theoretical solution in Table E5.2.

5.8 Check numerical results against the theoretical solution in Table E5.3.

5.9 Check numerical results against the theoretical solution; the numerical results are large in error, but the solution is stable.

5.11 Check numerical results against the theoretical solution in Table E5.4.

Chapter 6

6.2
$$D = \frac{\dot{u}_{go}}{2\pi} T_n \exp\left(-\frac{\zeta}{\sqrt{1-\zeta^2}} \tan^{-1} \frac{\sqrt{1-\zeta^2}}{\zeta}\right)$$

$$V = \frac{2\pi}{T_n} D; \; A = \left(\frac{2\pi}{T_n}\right)^2 D$$

6.4 $u_o = 1.91$ in.; $\sigma = 38.2$ ksi

6.6 (a) $u_o = 0.674$ in., $M = 40.65$ kip-ft; (b) $u_o = 2.7$ in., $M = 81.30$ kip-ft

6.8 $u_o = 3.86$ in.; bending moments (kip-in.) in columns: 6824 at base and 2925 at top; bending moments (kip-in.) in beam: 2925 at both ends

6.9 $A(T_n)/g = 0.5$ for $T_n \le \frac{1}{33}$ sec; $12.28T_n^{0.916}$ for $\frac{1}{33} < T_n \le \frac{1}{8}$ sec; 1.83 for $\frac{1}{8} < T_n \le 0.623$ sec; $1.14T_n^{-1}$ for $0.623 < T_n \le 3.91$ sec; $4.46T_n^{-2}$ for $3.91 < T_n \le 10$ sec; $24.49T_n^{-2.74}$ for $10 < T_n \le 33$ sec; and $1.84T_n^{-2}$ for $T_n > 33$ sec

Chapter 7

7.1 $\mu = 1.44$, 3.13, and 7.36

7.2

	$T_n = 0.02$ sec		$T_n = 0.2$ sec		$T_n = 2$ sec	
μ	f_y/w	u_m (in.)	f_y/w	u_m (in.)	f_y/w	u_m (in.)
1	0.50	1.955×10^{-3}	1.355	0.530	0.448	17.57
2	0.50	3.910×10^{-3}	0.782	0.612	0.224	17.57
4	0.50	5.865×10^{-3}	0.512	0.801	0.112	17.57
8	0.50	11.730×10^{-3}	0.350	1.095	0.056	17.57

7.3 $A_y(T_n)/g = 0.5$ for $T_n \le \frac{1}{33}$ sec; $1.68T_n^{0.348}$ for $\frac{1}{33} < T_n \le \frac{1}{8}$ sec; 0.818 for $\frac{1}{8} < T_n \le 0.465$ sec; $0.380T_n^{-1}$ for $0.465 < T_n \le 3.91$ sec; $1.487T_n^{-2}$ for $3.91 < T_n \le 10$ sec; $8.16T_n^{-2.74}$ for $10 < T_n \le 33$ sec; and $0.614T_n^{-2}$ for $T_n > 33$ sec

Chapter 8

8.2 (a) $\tilde{m}\ddot{\theta} + \tilde{c}\dot{\theta} + \tilde{k}\theta = \tilde{p}(t)$, $\tilde{m} = \dfrac{103mL^2}{64}$, $\tilde{c} = c$, $\tilde{k} = \dfrac{kL^2}{4}$, $\tilde{p}(t) = \dfrac{9L}{8}p(t)$;

(b) $\omega_n\sqrt{16k/103m}$, $\zeta = 8c/\sqrt{103kmL^4}$; (c) $u(x,t) = \dfrac{72x}{103mL\omega_D}e^{-\zeta\omega_n t}\sin\omega_D t$

8.4 (a) $\tilde{m}\ddot{\theta} + \tilde{c}\dot{\theta} + \tilde{k}\theta = \tilde{p}(t)$; $\tilde{m} = \dfrac{mL^3}{12}$; $\tilde{c} = \dfrac{cL^3}{12}\tilde{k} = \dfrac{kL^3}{12}$; $\tilde{p}(t) = \dfrac{L^2}{6}p(t)$

8.5 z = vertical deflection of lower end of spring; $\tilde{m}\ddot{z} + \tilde{c}\dot{z} + \tilde{k}z = \tilde{p}(t)$; $\tilde{m} = \dfrac{m}{3}$; $\tilde{c} = \dfrac{c}{4}$; $\tilde{k} = \dfrac{k}{5}$; $\tilde{p}(t) = -\dfrac{2}{5}p(t)$

8.7 (a) $\mathcal{V}_o(L/2) = 1426$ kips, $\mathcal{M}_o(L/2) = 0.2399 \times 10^6$ kip-ft, $\mathcal{V}_{bo} = 1739$ kips, $\mathcal{M}_{bo} = 0.7368 \times 10^6$ kip-ft; (b) $u_o(L) = 25.1$ in.

8.9 $u_{1o} = 0.519$, $u_{2o} = 0.830$, $u_{3o} = 0.934$ in.; $V_{3o} = 41.2$, $V_{2o} = 114.4$, $V_{bo} = 160.2$ kips; $M_{2o} = 494$, $M_{1o} = 1867$, $M_{bo} = 3789$ kip-ft

8.11 $\omega_n = 0.726\sqrt{EI/mL^3}$

8.12 $\omega_1 = \sqrt{2.536k/m}$, $\psi_1 = \langle 1 \quad 0.366 \rangle^T$; $\omega_2 = \sqrt{9.464k/m}$, $\psi_2 = \langle 1 \quad -1.366 \rangle^T$

8.13 $\omega_n^2 = \dfrac{\pi^4 EI/32L^3}{m\left[1 + (\pi^2/16)(R/L)^2\right]}$

8.14 $\omega_n = \dfrac{1.657}{L^2}\sqrt{\dfrac{EI}{m}}$

Chapter 9

9.2
$$\frac{mL}{3}\begin{bmatrix} 1 & 0 \\ 0 & 1 \end{bmatrix}\begin{Bmatrix} \ddot{u}_1 \\ \ddot{u}_2 \end{Bmatrix} + \frac{162EI}{5L^3}\begin{bmatrix} 8 & -7 \\ -7 & 8 \end{bmatrix}\begin{Bmatrix} u_1 \\ u_2 \end{Bmatrix} = \begin{Bmatrix} p_1(t) \\ p_2(t) \end{Bmatrix}$$

9.4
$$\mathbf{m} = \frac{mL}{6}\begin{bmatrix} 2 & 1 \\ 1 & 2 \end{bmatrix}, \mathbf{k} = \frac{EI}{L^3}\begin{bmatrix} 28 & -10 \\ -10 & 4 \end{bmatrix}$$

9.5 $\mathbf{u} = \langle u_1 \quad u_2 \quad u_3 \rangle$, where u_1 is the horizontal displacement of the masses and u_2 and u_3 are the vertical displacements of the right and left masses, respectively.

$$\mathbf{m}\ddot{\mathbf{u}} + \mathbf{k}\mathbf{u} = \mathbf{p}_{\text{eff}}(t)$$

$$\mathbf{m} = m\begin{bmatrix} 5 & & \\ & 1 & \\ & & 1 \end{bmatrix}, \mathbf{k} = \frac{3EI}{10L^3}\begin{bmatrix} 28 & 6 & -6 \\ 6 & 7 & 3 \\ -6 & 3 & 7 \end{bmatrix}, \mathbf{p}_{\text{eff}}(t) = -\mathbf{m}\iota\ddot{u}_g(t)$$

$$\iota^T = \begin{cases} \langle 1 \quad 0 \quad 0 \rangle & \text{for } \ddot{u}_g(t) = \ddot{u}_{gx}(t) \\ \langle 0 \quad 1 \quad 1 \rangle & \text{for } \ddot{u}_g(t) = \ddot{u}_{gy}(t) \end{cases}$$

9.7 $m\begin{bmatrix}1 & \\ & 0.5\end{bmatrix}\begin{Bmatrix}\ddot{u}_1\\ \ddot{u}_2\end{Bmatrix}+\frac{EI}{h^3}\begin{bmatrix}37.15 & -15.12\\ -15.12 & 10.19\end{bmatrix}\begin{Bmatrix}u_1\\ u_2\end{Bmatrix}=\begin{Bmatrix}p_1(t)\\ p_2(t)\end{Bmatrix}$

9.9 $m\begin{bmatrix}1 & & \\ & 1 & \\ & & 0.5\end{bmatrix}\begin{Bmatrix}\ddot{u}_1\\ \ddot{u}_2\\ \ddot{u}_3\end{Bmatrix}+\frac{EI}{h^3}\begin{bmatrix}40.85 & -23.26 & 5.11\\ -23.26 & 31.09 & -14.25\\ 5.11 & -14.25 & 10.06\end{bmatrix}\begin{Bmatrix}u_1\\ u_2\\ u_3\end{Bmatrix}=\begin{Bmatrix}p_1(t)\\ p_2(t)\\ p_3(t)\end{Bmatrix}$

9.11 $\mathbf{m}\ddot{\mathbf{u}}+\mathbf{k}\mathbf{u}=\mathbf{p}_{\text{eff}}(t)$

$$\mathbf{m}=m\begin{bmatrix}2/3 & -1/6 & 1/2\\ -1/6 & 2/3 & -1/2\\ 1/2 & -1/2 & 1\end{bmatrix},\ \mathbf{k}=k\begin{bmatrix}5 & -2 & 2\\ -2 & 5 & -2\\ 2 & -2 & 6\end{bmatrix},\ \mathbf{p}_{\text{eff}}(t)=-\mathbf{m}\iota\ddot{u}_g(t)$$

$$\iota^T=\begin{cases}\langle 1/2 \quad 1/2 \quad 0\rangle & \text{for ground motion in the } x \text{ direction}\\ \langle 1/2 \quad -1/2 \quad 1\rangle & \text{for ground motion in the } y \text{ direction}\\ \langle 1/\sqrt{2} \quad 0 \quad 1/\sqrt{2}\rangle & \text{for ground motion in the } d\text{–}b \text{ direction}\end{cases}$$

9.13 $m\ddot{u}+\frac{6EI}{L^3}u=-m\,\langle 1/2 \quad 1/2\,\rangle\begin{Bmatrix}\ddot{u}_{g1}(t)\\ \ddot{u}_{g2}(t)\end{Bmatrix}$

9.14 $\mathbf{m}\ddot{\mathbf{u}}+\mathbf{k}\mathbf{u}=-\mathbf{m}\iota\ddot{\mathbf{u}}_g(t)$

$$\mathbf{u}=\begin{Bmatrix}u_1\\ u_2\end{Bmatrix},\ \mathbf{u}_g=\begin{Bmatrix}u_{g1}\\ u_{g2}\end{Bmatrix}$$

$$\mathbf{m}=\begin{bmatrix}3.624 & \\ & 1.812\end{bmatrix},\ \mathbf{k}=10^4\begin{bmatrix}0.9359 & 0.7701\\ 0.7701 & 1.5088\end{bmatrix},\ \iota=\begin{bmatrix}0.6035 & 0.3965\\ -0.2143 & 1.2143\end{bmatrix}$$

Chapter 10

10.3 **(a)** $u_1(t)=0.5\cos\omega_1 t+0.5\cos\omega_2 t;\ u_2(t)=0.5\cos\omega_1 t-0.5\cos\omega_2 t$

10.6 **(a)** $\omega_n=\alpha_n\sqrt{EI/mL^3},\ \alpha_n=0.5259,\ 1.6135,\ 1.7321,$

$$\phi_1=\begin{Bmatrix}1\\ -1.9492\\ 1.9492\end{Bmatrix},\ \phi_2=\begin{Bmatrix}1\\ 1.2826\\ -1.2826\end{Bmatrix},\ \phi_3=\begin{Bmatrix}0\\ 1\\ 1\end{Bmatrix}$$

(b) $\begin{Bmatrix}u_1(t)\\ u_2(t)\\ u_3(t)\end{Bmatrix}=\begin{Bmatrix}0.3969\\ -0.7736\\ 0.7736\end{Bmatrix}\cos\omega_1 t+\begin{Bmatrix}0.6031\\ 0.7736\\ -0.7736\end{Bmatrix}\cos\omega_2 t$

10.8 $\omega_1=1.971\sqrt{EI/mh^3},\ \omega_2=8.609\sqrt{EI/mh^3},\ \phi_1=\langle 0.919 \quad 1\,\rangle^T,$
$\phi_2=\langle -0.544 \quad 1\,\rangle^T$

10.9 **(a)** $u_1(t)=1.207\cos\omega_1 t-0.207\cos\omega_2 t,\ u_2(t)=1.707\cos\omega_1 t+0.293\cos\omega_2 t$

10.11 $\omega_1=2.407\sqrt{EI/mh^3},\ \omega_2=7.193\sqrt{EI/mh^3}$

$$\phi_1=\langle 0.482 \quad 1 \quad -0.490/h \quad -0.490/h \quad -0.304/h \quad -0.304/h\rangle^T$$

$$\phi_2=\langle -1.037 \quad 1 \quad -0.241/h \quad -0.241/h \quad -1.677/h \quad -1.677/h\rangle^T$$

10.14 **(a)** $\begin{Bmatrix} u_1(t) \\ u_2(t) \\ u_3(t) \end{Bmatrix} = \begin{Bmatrix} 1.2440 \\ 2.1547 \\ 2.4880 \end{Bmatrix} \cos \omega_1 t + \begin{Bmatrix} -0.3333 \\ 0 \\ 0.3333 \end{Bmatrix} \cos \omega_2 t + \begin{Bmatrix} 0.0893 \\ -0.1547 \\ 0.1786 \end{Bmatrix} \cos \omega_3 t$

10.17 $\omega_n = 5.96,\ 6.21,\ 10.90$ rad/sec

$\phi_1 = \langle 0 \quad 2.0322 \quad 0.0033 \rangle^T,\ \phi_2 = \langle 2.071 \quad 0 \quad 0 \rangle^T,$

$\phi_3 = \langle 0 \quad -0.3988 \quad 0.0166 \rangle^T$

10.19–10.23 Compare against exact results.

Chapter 11

11.1 $\mathbf{c} = \begin{bmatrix} 0.824 & -0.257 & 0 \\ (\text{sym}) & 0.604 & -0.110 \\ & & 0.229 \end{bmatrix},\ \zeta_2 = 0.0430$

11.2, 11.4 $\mathbf{c} = \begin{bmatrix} 0.848 & -0.234 & -0.023 \\ (\text{sym}) & 0.628 & -0.133 \\ & & 0.252 \end{bmatrix}$

Chapter 12

12.1 $u_1(t) = (p_o/k)(1.207\mathcal{C}_1 - 0.207\mathcal{C}_2)\sin \omega t,$
$u_2(t) = (p_o/k)(1.707\mathcal{C}_1 + 0.293\mathcal{C}_2)\sin \omega t$, where $\mathcal{C}_n = \left[1 - (\omega/\omega_n)^2\right]^{-1}$

12.4 $u_1(t) = \dfrac{p_o}{2m}\left(\dfrac{\sin \omega_1 t}{\omega_1} + \dfrac{\sin \omega_2 t}{\omega_2}\right),\ u_2(t) = \dfrac{0.707 p_o}{m}\left(\dfrac{\sin \omega_1 t}{\omega_1} - \dfrac{\sin \omega_2 t}{\omega_2}\right)$

12.5 **(a)** $u_1(t) = \dfrac{p_o}{k}(1 - 0.853 \cos \omega_1 t - 0.147 \cos \omega_2 t),\ u_2(t) =$
$\dfrac{p_o}{k}(1 - 1.207 \cos \omega_1 t + 0.207 \cos \omega_2 t)$

12.9 $u_{3o} = \omega^2 p_o \sqrt{\left(\dfrac{\mathcal{C}_1}{K_1} + \dfrac{\mathcal{C}_2}{K_2} + \dfrac{\mathcal{C}_3}{K_3}\right)^2 + \left(\dfrac{\mathcal{D}_1}{K_1} + \dfrac{\mathcal{D}_2}{K_2} + \dfrac{\mathcal{D}_3}{K_3}\right)^2},\ \ddot{u}_{3o} = \omega^2 u_{3o},$
where $p_o = 1.242$ lb, $K_1 = 131.16$, $K_2 = 978.97$, and
$K_3 = 1826.80$ kips/in.; and $\mathcal{C}_n$ and $\mathcal{D}_n$ are as defined in Example 12.5.

12.16
$$\begin{Bmatrix} u_1(t) \\ u_2(t) \end{Bmatrix} = \begin{Bmatrix} 1 \\ 1 \end{Bmatrix} 1.736(1 - \cos 10t) + \begin{Bmatrix} 1 \\ -1 \end{Bmatrix}(-0.116)(1 - \cos 38.73t),\ t \le 0.3 \text{ sec}$$
$$\begin{Bmatrix} u_1(t) \\ u_2(t) \end{Bmatrix} = \begin{Bmatrix} 1 \\ 1 \end{Bmatrix}[3.455 \cos 10(t - 0.3) + 0.245 \sin 10(t - 0.3)]$$
$$+ \begin{Bmatrix} 1 \\ -1 \end{Bmatrix}[-0.0483 \cos 38.73(t - 0.3) + 0.09419 \sin 38.73(t - 0.3)],\ t \ge 0.3 \text{ sec}$$

12.18 Combined values of shears (in kips) and bending moments (in kip-in.):

$$V_a = V_b = 52.08, \quad V_c = V_d = -58.16, \quad V_e = V_f = 6.12$$

$$M_a = 0, \quad M_b = M_c = -2604, \quad M_d = M_e = 306, \quad M_f = 0$$

12.19 $\mathbf{u}_o = \begin{Bmatrix} 0.133 \\ 0.529 \end{Bmatrix}$ in., $\ddot{\mathbf{u}}_o = \begin{Bmatrix} 83.125 \\ 330.63 \end{Bmatrix}$ in./sec^2

12.20 $M_{bo} = 948.8$ kip-in., $M_{do} = -1902$ kip-in.

12.23 **(a)** $M(t) = \sum M^{\text{st}}\bar{M}_n\left[\omega_n^2 D_n(t)\right]$, where

$$D_n(t) = \frac{p_o}{\omega_n^2}\frac{1}{1-(T_n/2t_d)^2}\left[\sin\left(\pi\frac{t}{t_d}\right) - \frac{T_n}{2t_d}\sin\left(2\pi\frac{t}{T_n}\right)\right], \quad t \le t_d$$

$$D_n(t) = D_n(t_d)\cos\omega_n(t-t_d) + \frac{\dot{D}_n(t_d)}{\omega_n}\sin\omega_n(t-t_d), \quad t \ge t_d$$

$$M^{\text{st}} = -0.3125L, \quad \bar{M}_1 = 0.3414, \quad \bar{M}_2 = 0.6000, \quad \bar{M}_3 = 0.0586$$

(b) $M(t) = M^{\text{st}}\left\{p(t) + \bar{M}_1\left[\omega_1^2 D_1(t) - p(t)\right]\right\}$

Chapter 13

13.1 **(a)** $\mathbf{s}_1 = m\,\langle 0.849 \quad 0.594\rangle^T$, $\mathbf{s}_2 = m\,\langle -0.849 \quad 0.406\rangle^T$
(b) $u_1(t) = 0.283D_1(t) - 0.283D_2(t)$, $u_2(t) = 0.594D_1(t) + 0.406D_2(t)$
(c) $M_b(t) = 1.443mLA_1(t) - 0.443mLA_2(t)$

13.2 **(a)** $\mathbf{s}_1 = m\,\langle 1.985 \quad -0.774 \quad 0.774\rangle^T$, $\mathbf{s}_2 = m\,\langle 3.015 \quad 0.774 \quad -0.774\rangle^T$, $\mathbf{s}_3 = \langle 0 \quad 0 \quad 0\rangle^T$
(b) $u_1(t) = 0.397D_1(t) + 0.603D_2(t)$, $u_2(t) = -0.774D_1(t) + 0.774D_2(t)$, $u_3(t) = 0.774D_1(t) - 0.774D_2(t)$
(c) $M_b(t) = 3.533mLA_1(t) + 1.467mLA_2(t)$, $M_a(t) = -0.774mLA_1(t) + 0.774mLA_2(t)$

13.4 **(a)** $\mathbf{s}_1 = m\,\langle 0.854 \quad 0.604\rangle^T$, $\mathbf{s}_2 = m\,\langle 0.146 \quad -0.104\rangle^T$
(b) $u_1(t) = 0.854D_1(t) + 0.146D_2(t)$, $u_2(t) = 1.207D_1(t) - 0.207D_2(t)$
(c) $V_1(t) = 1.458mA_1(t) + 0.042mA_2(t)$, $V_2(t) = 0.604mA_1(t) - 0.104mA_2(t)$
(d) $M_b(t) = 2.062mhA_1(t) - 0.062mhA_2(t)$, $M_1(t) = 0.604mhA_1(t) -0.104mhA_2(t)$

13.5 **(a)** Peak values of total responses: $u_1 = 0.679$ in., $u_2 = 0.964$ in., $V_b = 115.11$ kips, $V_2 = 49.56$ kips, $M_b = 1959.25$ kip-ft, $M_2 = 594.65$ kip-ft

13.7 **(a)** Floor displacements: $u_1(t) = 0.647D_1(t) + 0.353D_2(t)$, $u_2 = 1.341D_1(t) - 0.341D_2(t)$
Joint rotations at first and second floors: $u_3(t) = (1/h)[-0.657D_1(t) + 0.082D_2(t)]$, $u_5(t) = (1/h)\,[-0.407D_1(t) + 0.572D_2(t)]$
(b) Bending moments in first-story column:
Top: $M_a = mh\,[0.216A_1(t) + 0.0473A_2(t)]$
Bottom: $M_b = mh\,[0.443A_1(t) + 0.0441A_2(t)]$
Bending moment at ends of second-floor beam:
$M_a = M_b = mh[-0.211A_1(t) + 0.0332A_2(t)]$

13.10 $M_1^* = 2.3213m$, $M_2^* = 0.1667m$, $M_3^* = 0.0121m$
$h_1^* = 2h$, $h_2^* = -h$, $h_3^* = 2h$

13.13 **(c)** Peak responses: $u_{3o} = 44.58$ in., $V_{ao} = 0.879$ kip, $V_{bo} = 159.83$ kips
(d) Seismic coefficients: 4.69 and 0.426 for appendage and for tower, respectively

13.15 **(a)** $\mathbf{s}_1 = \langle 0 \quad 0.1588 \quad 3.816 \rangle^T$, $\mathbf{s}_2 = \langle 0.1649 \quad 0 \quad 0 \rangle^T$,
$\mathbf{s}_3 = \langle 0 \quad 0.0061 \quad -3.816 \rangle^T$
(c) $u_x(t) = 0.7071 D_2(t)$, $u_y(t) = 0.6809 D_1(t) + 0.0262 D_3(t)$, $u_\theta(t) = 0.0011 D_1(t) - 0.0011 D_3(t)$
(d) $V_{bx}(t) = 0.1649 A_2(t)$, $V_{by}(t) = 0.1588 A_1(t) + 0.0061 A_3(t)$, $T_b(t) = 3.816 A_1(t) - 3.816 A_3(t)$

13.16 **(d)** Peak responses: $u_{yo} = 4.182$ in., $(b/2)u_{\theta o} = 1.22$ in., $V_{bo} = 35.6$ kips, $T_{bo} = 1899$ kip-in.

13.18 (i) $u(t) = 0.8889 u_{go} \sin 6.667t$, $u^t(t) = 1.3889 u_{go} \sin 6.667t$,
$M = -(3EI/L^2)u(t)$, (ii) $u(t) = 1.7778 u_{go} \sin 6.667t$,
$u^t(t) = 2.7778 u_{go} \sin 6.667t$, $M = -(3EI/L^2)u(t)$

13.19 **(a)** $u_2^t(t) = -0.2143 u_g(t) + 1.2143 u_g(t - t') - 0.3416 D_1(t) - 0.0150 D_1(t - t') + 0.1273 D_2(t) + 1.2294 D_2(t - t')$
(b) $u_2^t(t) = u_g(t) - 0.3567 D_1(t) + 1.3566 D_2(t)$

13.20 **(a)** $u_1 = 1.43$ in., $u_2 = 2.96$ in.
(b) First-story column: $M_{\text{top}} = 240$, $M_{\text{base}} = 483$ kip-ft
Second-floor beam: $M_{\text{left}} = M_{\text{right}} = 232$ kip-ft

13.21 $u_1 = 0.680$ in., $u_2 = 0.962$ in.
$V_b = 115.22$ kips, $V_2 = 48.24$ kips,
$M_b = 1955.43$ kip-ft, $M_1 = 578.87$ kip-ft

13.24

	u_3 (in.)	V_a (kips)	V_b (kips)
SRSS	165.0	3.158	144.3
CQC	77.27	1.465	182.3

13.28

	u_x (in.)	u_y (in.)	$(b/2)u_\theta$ (in.)	V_x (kips)	V_y (kips)	T (kip-in.):
CQC	6.286	6.309	1.662	56.6	52.4	2331.4
SRSS	6.286	6.306	1.678	56.6	52.3	2357.8

Bending moments (kip-in.):

	M_{ay}	M_{ax}	M_{by}	M_{bx}	M_{cy}	M_{cx}	M_{dy}	M_{dx}
CQC	1108	1055	554.0	845.5	824.2	845.5	1648	1055
SRSS	1405	1050	702.7	847.1	702.7	847.1	1405	1050

Chapter 14

14.3 $\tilde{\omega}_1 = 8.389,\ \tilde{\omega}_2 = 23.59$ rad/sec

$$\tilde{\phi}_1 = \langle 0.2319 \quad 0.4639 \quad 0.6311 \quad 0.7983 \quad 0.9332 \rangle^T$$

$$\tilde{\phi}_2 = \langle -0.4366 \quad -0.8732 \quad -0.3396 \quad 0.1940 \quad 1.2126 \rangle^T$$

14.4 $\tilde{\omega}_1 = 8.263,\ \tilde{\omega}_2 = 23.428$ rad/sec

$$\tilde{\phi}_1 = \langle 0.2319 \quad 0.4449 \quad 0.6753 \quad 0.8249 \quad 0.9038 \rangle^T$$

$$\tilde{\phi}_2 = \langle -0.4366 \quad -0.4200 \quad -0.1042 \quad 0.2095 \quad 0.4108 \rangle^T$$

Chapter 15

15.1–15.4 Check numerical results against the theoretical solution.

15.5 Check numerical results against Table E15.1.

15.6 Check numerical results against the theoretical solution.

15.7 Check numerical results against Table E15.2.

15.8 Check numerical results against the theoretical solution.

15.9 Check numerical results against Table E15.3.

15.10 Check numerical results against the theoretical solution.

Chapter 16

16.2 $\omega_n = \alpha_n \sqrt{EI/mL^4}$; $\alpha_1 = 15.42$, $\alpha_2 = 49.97$, and $\alpha_3 = 104.2$

16.4 $$u\left(\frac{L}{2}, t\right) = -\frac{2WL^3}{\pi^4 EI}\left(\cos\omega_1 t + \frac{\cos\omega_3 t}{81} + \frac{\cos\omega_5 t}{625} + \frac{\cos\omega_7 t}{2401} + \cdots\right)$$

16.6 $$u\left(\frac{L}{4}, t\right) = \frac{8pL^4}{\pi^5 EI}\left(\frac{1-\cos\omega_2 t}{32} - \frac{1-\cos\omega_6 t}{7776} + \frac{1-\cos\omega_{10} t}{100{,}000} - \cdots\right)$$

16.9 $u_o(L) = 7.410$ in., $\mathcal{V}_{bo} = 1365$ kips, $\mathcal{M}_{bo} = 186{,}503$ kip-ft

Chapter 17

17.2 $$\tilde{\omega}_1 = \frac{11.765}{L^2}\sqrt{\frac{EI}{m_o}},\ \tilde{\omega}_2 = \frac{130.467}{L^2}\sqrt{\frac{EI}{m_o}}$$

$$\tilde{\phi}_1(x) = 1.0\sin(\pi x/L) - 0.0036\sin(3\pi x/L)$$

$$\tilde{\phi}_2(x) = 0.2790\sin(\pi x/L) + 0.9603\sin(3\pi x/L)$$

17.3 **(a)** $\omega_n = \alpha_n \sqrt{EI/mL^4}$; $\alpha_1 = 9.9086$, $\alpha_2 = 43.818$, $\alpha_3 = 110.14$, $\alpha_4 = 200.80$

(b) $\omega_1 = 9.798\sqrt{EI/mL^4}$

17.5 **(a)** $\omega_n = \alpha_n \sqrt{EI/mL^4}$; $\alpha_1 = 15.56$, $\alpha_2 = 58.41$, $\alpha_3 = 155.64$

(b) $\omega_1 = 14.81\sqrt{EI/mL^4}$

17.6 **(a)** $\omega_n = \alpha_n\sqrt{EI/mh^4}$, $\alpha_1 = 1.5354, \alpha_2 = 4.0365, \alpha_3 = 10.7471$

$\phi_1 = \langle 0.5440 \quad -0.5933/h \quad -0.5933/h \rangle^T$

$\phi_2 = \langle 0 \quad -1/\sqrt{2}h \quad -1/\sqrt{2}h \rangle^T$

$\phi_3 = \langle -0.0001 \quad 1/\sqrt{2}h \quad 1/\sqrt{2}h \rangle^T$

17.7 $\omega_1 = 1.5354\sqrt{EI/mh^4}$

$\phi = \langle 0.544 \quad -0.593/h \quad -0.593/h \rangle^T$

Index

Absolute sum (ABSSUM) rule, 516, 527, 531
Acceleration resonant frequency, 79
Acceleration response factor, 78
Accidental torsion, 503, 508
Amplitude of motion
  forced harmonic vibration, 72
  free vibration, 37
Average acceleration method, 165, 569, 576

Base-isolated buildings
  multistory, 695
  one-story, 686
  rigid structure approximation for analysis
    multistory buildings, 701
    one-story systems, 691
Base isolation
  applications, 701
  effectiveness of
    dependence on earthquake design spectrum, 692
    dependence on natural period of fixed-base structure, 692
Base isolation, effects of
  multistory buildings, 696
  one-story buildings, 687
Base isolation systems
  bearings, 684
  laminated bearings, 684
  rollers or sliders, 684
Base rotation, 22, 341, 350, 507
Base shear coefficient, 200
Beam-to-column stiffness ratio, 10, 642, 718
Beam, transverse vibration
  effective earthquake forces, 588
  equation of motion
    applied forces, 587
    support excitation, 588
  natural vibration frequencies and modes, 588–595
    cantilever beam, 591
    simply supported beam, 590
  orthogonality of modes, 595
  rotational inertia and shear, 593
    influence of, 594
Bearings
  additional damping in
    hydraulic dampers, 684
    lead plugs, 684
    steel coils, 684
  laminated bearings, 684
  rubber, 87
Braced frames, 16
Bridges, response to traveling load, 289
Building code evaluation, 720–728
  base shear, 720
  equivalent static forces, 723
  higher-mode response, 720, 724, 726
  overturning moment reduction factor, 727

Building code evaluation (*continued*)
  overturning moments, 726
  story shears, 723
Building codes
  Mexico Federal District Code, 711–713
  National Building Code of Canada, 707–710
  Uniform Building Code, 660, 670, 704–707
Building codes, structural dynamics in, 713–719
  design force reduction, 715
  fundamental vibration period, 713
  lateral force distribution, 717
  overturning moment reduction factor, 719
  overturning moments, 718
  seismic coefficient, 704, 707, 711
Buildings, earthquake response of
  influence of beam-to-column stiffness ratio, 646
  influence of fundamental period, 646
Buildings with soft first story (*see* Soft first-story buildings)
Buildings with symmetric plan
  accidental torsion, 503, 508
  effective modal height, 483
  effective modal mass, 482
  equations of motion
    inelastic systems, 355
    linear systems, 340, 350
  five-story shear frame, 485, 523
  four-story frame with an appendage, 489, 530
  modal expansion of earthquake forces, 475
  modal responses, 476
    equivalent static forces, 476
    modal static responses, 476–477
  peak modal responses, 519
    equivalent static forces, 520
    modal static responses, 520
  recorded torsion, 507
  response history analysis, 474–478
  response spectrum analysis, 519–532
Buildings with unsymmetric plan
  arbitrary-plan buildings, 498
  coupled lateral-torsional motion, 342, 347, 349, 381
  effective modal height, 497
  effective modal mass, 497
  equations of motion, 342–350
  modal expansion of earthquake forces, 492
  modal responses
    equivalent static lateral forces, 495
    equivalent static torques, 495
    modal static responses, 496
  peak modal responses
    equivalent static lateral forces, 533
    equivalent static torques, 533
    modal static responses, 532
  Response history analysis, 491
  Response spectrum analysis, 532–540
Buildings with weak first story (*see* Weak first-story buildings)

Caracas, Venezuela earthquake (June 29, 1967), 719
Caughey damping, 421
Caughey series, 421
Central difference method, 161, 568
Characteristic equation (*see* Frequency equation)
Characteristic values (*see* Eigenvalues)
Characteristic vectors (*see* Eigenvectors)
Citicorp Center, New York, 432
Complete quadratic combination (CQC) rule, 516, 527, 531, 534, 536, 540
  correlation coefficient, 517
  correlation coefficient, variation with
    damping, 517
    frequency ratio, 517
Complex frequency-response function, 27
Components of a system
  damping component, 7, 16, 318
  mass component, 7, 16, 318
  stiffness component, 7, 16, 318
Conservation of energy, 52
Conservation of energy, principle of, 299
Convolution integral, 121
Coulomb damping, 53
Coupling terms, 325
  in mass matrix, 327
  in stiffness matrix, 327
Critical damping, 53

D'Alembert's principle, 15, 279, 282, 293, 316, 322
Damped systems
  critically damped, 44
  overdamped, 45
  underdamped, 44

Damping, 7, 321
classical, 386
Coulomb friction, 53
hysteretic (*see* Damping, rate-independent)
nonclassical, 386
numerical (*see* Numerical damping)
rate-independent, 100
solid (*see* Damping, rate-independent)
structural (*see* Damping, rate-independent)
viscous, 13
Damping component of a system, 7, 16, 317
Damping influence coefficient, 321
Damping matrix
Caughey damping, 421
definition, 321
mass-proportional, 418
Rayleigh damping, 418
stiffness-proportional, 418
when it is needed, 416
Damping matrix, computation of
structures with energy-dissipating devices, 426
Damping matrix from modal damping ratios
classical damping, 417
nonclassical damping, 425
Damping ratio, 44
Damping ratios, recommended, 416
Deformation response factor
half-cycle sine pulse, 140
harmonic force, 65, 75
rectangular pulse, 135
Dirac delta function, 120
Discrete fast Fourier transform, 27
Discretization
degrees of freedom, 318
elements, 318
nodal points, 318
nodes, 318
Displacement resonant frequency, 77
Dissipated energy (*see* Energy dissipated)
Distributed-mass systems
difficulty in analyzing practical systems, 607
effective modal height, 603
effective modal mass, 603
Rayleigh's method for, 299
treated as generalized SDF systems, 278
Duhamel's integral, 26, 122, 124–125
Dynamic equilibrium, 15, 316, 322
Dynamic hysteresis, 97
Earthquake analysis of distributed-mass systems
Response history analysis (RHA), 600–604
Response spectrum analysis (RSA), 604–607
Earthquake analysis of linear systems, methods for
Response history analysis (RHA), 468–514
Response spectrum analysis (RSA), 514–540
Earthquake design spectrum
as envelope of two design spectra, 227
distinction relative to response spectrum, 227
Earthquake design spectrum: elastic, 217, 645, 684
amplification factors, 221
comparison with response spectrum, 225
construction of, 222
median, 226
median-plus-one-standard-deviation, 226
Earthquake design spectrum: inelastic, 269
comparison with response spectrum, 274
construction of, 270
normalized strength, 269
relations between peak deformations of elastoplastic and linear systems, 272
relations between yield strengths of elastic and elastoplastic systems, 272
Earthquake excitation, 187
influence matrix, 352
influence vector, 339, 353
Earthquake ground motion
rotational components, 22, 341, 506
translational components, 20, 193, 337, 340, 342, 474, 492
Earthquake response of buildings
influence of beam-to-column stiffness ratio, 646
influence of fundamental period, 646
Earthquake response of elastoplastic systems
response history, 250
response spectrum, 257
Earthquake response of linear SDF systems
deformation response, 195
equivalent static force, 196
pseudo-acceleration response, 196
response history, 195
response spectrum, 197

Earthquake response of generalized SDF systems, 285, 292
Earthquake response spectrum for elastoplastic systems
  construction of, 259
  design yield strength and deformation from, 261
  pseudo-acceleration, 257
  pseudo-velocity, 257
  relative effects of yielding and damping, 265
  yield deformation, 257
Earthquake response spectrum for linear systems
  acceleration, 198
    comparison with pseudo-acceleration, 229
  characteristics at long periods, 213
  characteristics at short periods, 212
  characteristics of, 211
  combined deformation–pseudo-velocity–pseudo-acceleration, 202
  computation of peak structural response, 206
  construction of, 204
  deformation, 198
  effect of damping, 216
  mean, 220
  mean-plus-one-standard-deviation, 220
  median, 221–222
  median-plus-one-standard-deviation, 221–222
  probability distribution, 220
  pseudo-acceleration, 200
  pseudo-velocity, 199
  relative velocity, 198
    comparison with pseudo-velocity, 228
Earthquakes
  Caracas, Venezuela (June 29, 1967), 252, 719
  Guam, U.S. Territory (August 8, 1993), 188
  Imperial Valley, California (May 18, 1940), 192
  Killari, India (September 30, 1993), 188
  Koyna, India (December 11, 1967), 634
  Loma Prieta, California (October 17, 1989), 188, 190, 415
  Long Beach, California (March 10, 1933), 188
  Lytle Creek, California (September 12, 1970), 414, 513
  Mexico City, Mexico (September 19, 1985), 692
  Northridge, California (January 17, 1994), 188, 415
  San Fernando, California (February 9, 1971), 188, 414, 513, 667
  Upland, California (February 28, 1990), 504
Effective earthquake force: SDF systems, 21
Effective earthquake forces
  buildings with unsymmetric plan, 350
  distributed-mass systems, 588
  MDF planar or symmetric-plan systems
    rotational ground motion, 342
    translational ground motion, 338–339
  multiple support excitation, 353
Effective modal height
  buildings with symmetric plan, 483
  buildings with unsymmetric plan, 497
  distributed-mass systems, 603
Effective modal mass
  buildings with symmetric plan, 482
  buildings with unsymmetric plan, 497
  distributed-mass systems, 603
Effective modal weight (first mode), 720
Eigenvalue problem, 369
  complex, 460
  modal matrix, 370
  real, 369
  spectral matrix, 370
  transformation to standard form, 404
Eigenvalue problem, solutions methods for
  determinant search method, 393
  inverse vector iteration method, 394–399
    convergence criterion, 395
    convergence proof, 396
    convergence rate, 398
    evaluation of fundamental mode, 395
    evaluation of higher modes, 398
    tolerance, 396
  inverse vector iteration with shifts, 399
    convergence rate, 400
  Lanczos method, 393
  polynomial iteration techniques, 393
  Rayleigh's quotient iteration, 401
    application to structural dynamics, 403
  subspace iteration method, 393
  transformation methods, 393
  vector iteration methods, 393
Eigenvalues, 369
Eigenvectors, 370

Elastic–perfectly plastic system (*see* Elastoplastic SDF system)
Elastoplastic multistory buildings, 660
allowable ductility, 664, 672
base shear yield strength
modification factor, 677, 679
corresponding SDF system, 672
deformation capacity, 664
ductility capacity, 664
ductility demand
heightwise variation of, 664, 675, 678
variation with fundamental period, 675
permanent displacement after earthquake, 663
Elastoplastic SDF system
allowable ductility, 262, 660
corresponding linear system, 247
ductility demand, 254, 660
ductility factor, 248
effects of yielding on response, 250
influence of yield strength on earthquake response, 253
normalized yield strength, 248, 254
peak deformation, 254
permanent displacement after earthquake, 252
relationship between peak displacements of elastoplastic and linear systems in
acceleration-sensitive region of spectrum, 255–256
displacement-sensitive region of spectrum, 254–255
velocity-sensitive region of spectrum, 255
yield deformation, 247
yield reduction factor, 248
yield strength, 247
yield strength for specified ductility, 257
El Centro ground motion, 192, 226, 232–236
Element forces
computed from displacements, 24, 357, 438
computed from equivalent static forces, 24, 357, 438
Energy
input, 52, 95
kinetic, 52, 96, 299, 300
potential, 52, 95, 298–300
strain, 52, 95, 298–300
Energy conservation, 52
Energy dissipated
in Coulomb friction, 106
in rate-independent damping, 101
in viscous damping, 53, 94
Energy-dissipating devices, 267
Energy-dissipating mechanisms, 13, 417
Energy quantities for elastoplastic systems
earthquake input energy, 264
energy dissipated by viscous damping, 265
energy dissipated by yielding, 265
kinetic energy, 265
strain energy, 265
Equation of motion
buildings with symmetric plan
torsional excitation, 350
translational ground motion, 340
buildings with unsymmetric plan, 342–350
multistory one-way unsymmetric system, 348–350
one-story, one-way unsymmetric system, 346–347
one-story, two-way unsymmetric system, 342–346
coupling terms, 325
distributed-mass systems, 587–588
MDF systems subjected to external forces, 325–334
multiple support excitation, 351–355
one-story symmetric system, 347
planar systems: rotational ground motion, 341
planar systems: translational ground motion, 337–340
SDF systems subjected to earthquake excitation, 21
SDF systems subjected to external force, 15
solution methods, overview of
direct solution, 358
modal analysis, 357
Equivalent static force: SDF systems, 24, 147, 196
Equivalent static forces
generalized SDF systems, 285, 295
MDF systems, 357, 438, 476, 495, 520, 533
Equivalent viscous damping, 13, 98
systems with Coulomb friction, 107
systems with rate-independent damping, 104
Existing buildings
of historical or architectural merit, 701
retrofit of, 701
seismic strengthening, 701

Experimental testing
  forced harmonic vibration, 83, 410
  free vibration, 50, 391
  resonance, 83
Explicit methods, 162, 566

Finite element method, 325, 619–636
  comparison with exact solution, 630
  direct assembly procedure, 622
  element (applied) force vector, 621
    consistent formulation, 627
    simpler formulation, 627
  element degrees of freedom, 622
  element mass matrix, 621
    consistent mass, 625
    lumped mass, 626
  element stiffness matrix, 621, 624
  finite elements, 619
  interpolation functions, 619, 623
  nodal points, 619
  nodes, 619
  three-dimensional finite elements, 632
  trial functions, 619
  two-dimensional finite elements, 632
First Federal Savings, Pomona, California, 504
Floor diaphragms
  flexible, 324
  rigid, 324, 340
Force
  harmonic, 62
  impulsive, 120
  ramp, 125
  step, 123
  step with finite rise time, 126
  varying arbitrarily with time, 121
Force–displacement relation
  elastoplastic, 246
  linear, 9, 320
  nonlinear, 12, 355
Fourier series, 109
Fourier transform, 27
Four-way logarithmic graph paper, 78, 113, 202, 237
Fraction of critical damping (*see* Damping ratio)
Free-body diagram, 15, 18–19, 316
Free vibration equations for MDF system, solution of
  classically damped systems, 390
  nonclassically damped systems, 386
  undamped systems, 383
Free vibration of MDF systems
  classically damped systems, 389–390
  nonclassically damped systems, 387
  undamped systems, 366
Free vibration of SDF systems
  Coulomb-damped, 53
  input energy, 52
  kinetic energy, 52
  potential energy, 52
  strain energy, 52
  undamped, 36
  viscously damped, 44
Free vibration tests, 50, 391
Frequency-domain method, 27, 101
Frequency equation, 369, 393
Frequency-response curve
  analytical solution, 72
  experimental evaluation, 83

Generalized
  damping, 278, 437
  force, 278, 435, 597
  mass, 278, 434, 597
  stiffness, 278, 435, 597
Generalized coordinate (*see* Generalized displacement)
Generalized displacement, 277
Generalized SDF systems, 278
  lumped-mass system: shear building, 292
  rigid-body assemblages, 279
  systems with distributed mass and elasticity, 281
Gram–Schmidt orthogonalization, 398, 558
Guam, U.S. Territory earthquake (August 8, 1993), 188

Half-power bandwidth, 78
Harmonic tests, 80, 411
Harmonic vibration (forced)
  steady state, 62, 68
  systems with Coulomb friction, 104
  systems with rate-independent damping, 100
  transient, 62, 68
  undamped systems, 62
  viscously damped systems, 68

Higher-mode response of buildings
building code evaluation, 718, 720, 724, 726
heightwise variation of, 653
influence of beam-to-column stiffness ratio, 652
influence of fundamental period, 649
number of modes to include
dependence on beam-to-column stiffness ratio, 656
dependence on fundamental period, 656
Hysteresis
dynamic, 97
static, 97, 100
Hysteresis loop, 14, 96

Imperial Valley, California earthquake (May 18, 1940), 192
Implicit methods, 166, 566
Impulse response (*see* Unit impulse response function)
Impulsive force, 120
Inelastic systems, 11, 355
Isolation (*see* Vibration isolation)
Isolator deformation, 689

Killari, India earthquake (September 30, 1993), 188
Kinetic energy, maximum value of, 299–300
Koyna Dam, 634
Koyna, India earthquake (December 11, 1967), 634

Laplace transform, 27
Lateral force coefficient, 200
Lateral stiffness, 9, 24, 41
Linear acceleration method, 165, 569
Logarithmic decrement, 49
Loma Prieta, California earthquake (October 17, 1989), 188, 190, 415
Long Beach, California earthquake (March 10, 1933), 188
Loss factor, 97
Lumped-mass idealization, 323
for multistory buildings
floor diaphragm, flexible, 324
floor diaphragm, rigid, 324
Lytle Creek, California earthquake (September 12, 1970), 414, 513

Mass component of system, 7, 16, 317
Mass influence coefficient, 322
Mass matrix
diagonal, 324
general, 323
Mass–spring–damper system, 18, 316
Matrix eigenvalue problem (*see* Eigenvalue problem)
Mexico City earthquake (September 19, 1985), 692
Mexico Federal District Code, 711–713
design spectrum, 711
lateral forces, 712
overturning moment reduction factor, 712
seismic behavior factor, 712
seismic coefficient, 711
seismic reduction factor, 712
Millikan Library, Pasadena, California, 409, 513
Lytle Creek earthquake, 414
San Fernando earthquake, 411
Millikan Library, vibration properties from motions recorded during
forced harmonic vibration tests, 411
Lytle Creek earthquake, 412
San Fernando earthquake, 412
Millikan Library, vibration properties of
amplitude dependence, 414
damping ratios, 412
natural vibration periods, 412
Modal analysis, 434–439
modal expansion of displacements, 434
modal responses, 438
summary, 439
total response, 438
Modal analysis for $\mathbf{p}(t) = \mathbf{s}p(t)$, 447–451
modal contribution factor, 448–449
modal participation factor, 447
modal response contributions, 449
modal static response, 448–451
number of modes to include, 451
dependence on dynamic response factors, 452
dependence on force distribution, 451
dependence on modal contribution factors, 455
dependence on response quantity, 451
SDF system, nth mode, 447
Modal analysis interpretation, 471

Modal analysis of distributed-mass systems
forced response, 596
modal equations, 597, 601
modal expansion of effective earthquake forces, 600
modal responses, 598, 601
equivalent static forces, 601
modal static response, 602
SDF system, *n*th mode, 601
Modal analysis of earthquake response of lumped-mass systems
modal equations, 469
modal expansion of displacements and forces, 468
modal responses
equivalent static forces, 469
modal static responses, 469
SDF system, *n*th mode, 469
total response, 470
Modal combination rules
absolute sum (ABSSUM), 515
complete quadratic combination (CQC), 516
square-root-of-sum-of-squares (SRSS), 516
Modal contribution factors, 448, 647, 655
base overturning moment, 647
base shear, 647
dependence on force distribution, 451
dependence on response quantity, 451
fifth-story shear, 647
influence of beam-to-column stiffness ratio, 648
top-floor displacement, 647
Modal coordinates, 382
Modal damping ratios, 389, 437
estimation of, 414
Modal equations
damped systems, 436
generalized
damping, 437
force, 435, 597
mass, 434, 597
stiffness, 435, 597
modal coordinates, 435, 437
modal damping ratios, 437
undamped systems, 434, 597
Modal expansion of displacements, 382, 434, 468
Modal expansion of excitation vector, 444, 468
Modal static responses, 448, 450, 469, 476, 496, 603
Mode acceleration superposition method, 458
Mode displacement superposition method (*see* Modal analysis)
Momentum, 121
Multiple support excitation
equations of motion, 351
response analysis, 508–510
dynamic displacements, 508
equivalent static forces, 509
modal equations, 508
quasistatic displacements, 509
quasistatic support forces, 509
SDF system, *n*th mode, 508

National Building Code of Canada, 707–710
design spectrum, 709
force modification factor, 709
foundation factor, 707
lateral forces, 710
overstrength factor, 707
overturning moment reduction factor, 710
seismic coefficient, 707
seismic importance factor, 707
seismic response factor, 707
story forces, 710
zonal velocity ratio, 707
Natural frequencies of MDF system
damped vibration, 391
undamped vibration, 367–382
Natural frequency of SDF system
damped vibration, 45
undamped vibration, 37
Natural period of SDF system
damped vibration, 45
undamped vibration, 37
Natural vibration frequencies and modes, computation of (*see* Eigenvalue problem, solution methods for)
Natural vibration frequency
by Rayleigh's method
distributed-mass systems, 300
lumped-mass systems, 300
from generalized SDF system analysis, 284, 294

Natural vibration modes
fundamental mode, 370
normalization, 372
orthonormal, 373
Natural vibration periods and modes of buildings
dependence on beam-to-column stiffness ratio, 642–644
Natural vibration periods of MDF systems, 367–382
Newmark's method, 164, 174, 569
noniterative formulation, 165
Newton–Raphson iteration, 178, 578–579
convergence criterion, 178, 579
Newton's second law of motion, 15, 18, 314
Nonclassically damped systems, 386, 687, 696
analysis of, 459, 687, 696
Nonstructural elements, 514
Normal coordinates (*see* Modal coordinates)
Normal modes (*see* Eigenvectors)
Normal values (*see* Eigenvalues)
Northridge, California earthquake (January 17, 1994), 188, 415
Nuclear power plant reactor building, 633
Numerical damping, 172, 575
Numerical evaluation of response
linear systems, 157–173
linear systems with nonclassical damping, 567
nonlinear systems, 174–183, 574–582
Numerical time-stepping methods
average acceleration method, 165, 569, 576
based on interpolation of excitation, 157
central difference method, 161, 176, 568
linear acceleration method, 165, 569
Newmark's method, 164, 176, 569
Wilson's method, 172, 579
Numerical time-stepping methods, accuracy of
errors for linear systems, 170
errors for nonlinear systems, 175
Numerical time-stepping methods, requirements for
accuracy, 157, 170, 566
convergence, 157, 566
stability, 157, 170, 566
Numerical time-stepping methods, types of
conditionally stable, 170, 567, 569, 575
explicit methods, 162, 566, 568
implicit methods, 166, 566, 570
unconditionally stable, 170, 569, 575, 579

Olive View Hospital, Sylmar, California, 667
Orthogonality of modes
discretized or lumped-mass systems, 371, 422, 434
distributed-mass systems, 595
interpretation of, 372
Overstrength of buildings, 707, 717

Periodic excitation, 108
steady-state response, 110
Phase angle, 65, 72
Phase lag (*see* Phase angle)
Potential energy, maximum value of, 299–300
Pulse force
approximate analysis for short pulses, 144
effects of pulse shape, 144
effects of viscous damping, 147
half-cycle sine pulse, 137
rectangular pulse, 131
symmetrical triangular pulse, 142
Pulse ground motion, 149

Ramp force, 125
Random vibration theory, 517
Rayleigh damping, 419, 426, 662
Rayleigh–Ritz method for discretized systems, 551–554, 560
generalized coordinates, 552
orthogonality of approximate modes, 554
Ritz transformation, 552
Ritz vectors, 552
force-dependent, 556
mass orthonormal, 558
Rayleigh–Ritz method for distributed-mass systems, 613–619
disadvantages, 618
formulation using conservation of energy, 613
formulation using virtual work, 617
Ritz functions, 614
shape functions, 614
Rayleigh's method, 298, 551, 713
for distributed-mass systems, 299
for lumped-mass systems, 300
Rayleigh's quotient
bounds, 394
for distributed-mass systems, 300
for lumped-mass systems, 301, 553
in Rayleigh–Ritz method, 614

Rayleigh's quotient (*continued*)
properties, 301
Rayleigh's stationarity condition, 394, 553, 614
Reduction of degrees of freedom
kinematic constraints, 550
Rayleigh–Ritz method, 551
Resonance testing, 83
Resonant frequency, 66
acceleration, 77
displacement, 77
velocity, 77
Response factors
acceleration, 76
deformation, 65, 75
velocity, 76
Response spectrum analysis of structures, 514–540, 645, 687
avoidance of pitfall, 529
comparison with response history analysis, 528, 532
interpretation of, 519
modal combination rules, 515–516
absolute sum (ABSSUM) rule, 515
complete quadratic combination (CQC) rule, 516
square-root-of-sum-of-squares (SRSS) rule, 516
modal combination rules, errors in, 518
peak modal responses, 515
peak total response, 515
Response spectrum for step force with finite rise time, 128
Rigid bodies, inertia forces for, 306
Ritz vectors, selection of, 554–560
by visualizing natural modes, 555
force-dependent Ritz vectors, 556

Salt Lake City and County Building, 701
San Fernando, California earthquake (February 9, 1971), 188, 414, 513, 667
Shaking machine (*see* Vibration generator)
Shape function, 278, 282
Shape function selection, 302
displacement boundary conditions, 303
from deflections due to static forces, 302
Shape vector, 292
Shear building, 292, 313
equations of motions for, 314
idealization, 313
Shock spectrum
half-cycle sine pulse, 141
rectangular pulse, 134
symmetrical triangular pulse, 144
Simple harmonic motion, 36, 367
Single-degree-of-freedom system, 7
Soft first-story buildings, 666
concentration of yielding in first story, 667
Soil–structure interaction, 425
Spatially varying ground motion (*see* Multiple support excitation)
Specific damping capacity, 97
Specific damping factor, 97
Spectral regions
acceleration-sensitive, 214, 255, 264, 645, 650
displacement-sensitive, 214, 254, 264, 645, 650
velocity-sensitive, 215, 255, 264, 645, 650
Square-root-of-sum-of-squares (SRSS) rule, 516
Static condensation method, 11, 334, 551
Static correction method, 455, 470, 561
Static hysteresis, 97, 100
Steady-state response (*see* Steady-state vibration)
Steady-state vibration, 62, 102, 104, 110
Step force, 123
with finite rise time, 126
Stiffness
complex, 101
lateral, 9, 24, 41
Stiffness coefficients
uniform flexural element, 11, 30
Stiffness component of a system, 7, 16, 317
Stiffness degradation, 681
Stiffness influence coefficient, 319
Stiffness matrix
computation of
direct stiffness method, 321, 345
direct equilibrium method, 321, 343
condensed, 335
lateral, 341
two-story shear building, 315
Story stiffness, 293, 315
Strain hardening, 681
Strength degradation, 681
Strong-motion accelerograph, 188
Structural Engineers Association of California, 704

Structural idealization, quality of, 513
Structure–fluid system, 28, 425
Structure–soil system, 28, 425
Support excitation (*see* Earthquake excitation)
System identification, 414

Timoshenko beam theory, 593
Transient response (*see* Transient vibration)
Transient vibration, 62, 68
Transmissibility, 86–87
Tributary length, 3
Tuned mass damper (*see* Vibration absorber)
Two-DOF systems, analysis of
  analytical solution for harmonic excitation, 430

Uniform Building Code, 660, 670, 704–707
  design spectrum, 705
  elastic seismic coefficient, 704
  importance factor, 704
  lateral forces, 706
  seismic coefficient, 704
  seismic zone factor, 704
  site coefficient, 704
  story forces, 706
  structural system coefficient, 705
Unit impulse, 120
Unit impulse response function, 121
Upland, California earthquake (February 28, 1990), 504

Velocity resonant frequency, 77
Velocity response factor, 76
Vibration absorber, 432
Vibration generator, 81
Vibration isolation
  applied force excitation, 85
  ground motion excitation, 87
Vibration-measuring instruments, 91
Virtual displacements, principle of, 282, 293, 617
Viscous damping, 13, 321
Viscous damping effects
  in earthquake response, 216, 263
  in free vibration, 45–46
  response to harmonic excitation, 72–75
  response to pulse force, 147

Wasatch fault, Utah, 702
Weak first-story buildings, 665
  concentration of yielding in first story, 666
Wilson's method, 579
Wind-induced vibration of buildings, 432